T0186745

STOCHASTIC HYDRAULICS '96

PROCEEDINGS OF THE SEVENTH IAHR INTERNATIONAL SYMPOSIUM
MACKAY/QUEENSLAND/AUSTRALIA/29-31 JULY 1996

Stochastic Hydraulics '96

Edited by
KEVIN S.TICKLE, IAN C.GOULTER, CHENGCHAO XU,
SALEH A.WASIMI & FRANÇOIS BOUCHART
Central Queensland University, Rockhampton, Queensland, Australia

A.A.BALKEMA/ROTTERDAM/BROOKFIELD/1996

Photo cover: D. Steeley, Central Queensland University
Waterpark Creek, Byfield, Queensland, Australia

The texts of the various papers in this volume were set individually by typists under the supervision of each of the authors concerned.

Published by
A.A. Balkema, P.O. Box 1675, 3000 BR Rotterdam, Netherlands(Fax: +31.10.413.5947)
A.A. Balkema Publishers, Old Post Road, Brookfield, VT 05036, USA (Fax: 802.276.3837)

ISBN 90 5410 817 7
© 1996 A.A. Balkema, Rotterdam
Printed in the Netherlands

Stochastic Hydraulics'96, Tickle, Goulter, Xu, Wasimi & Bouchart (eds) © 1996 Balkema, Rotterdam. ISBN 90 5410 817 7

Table of contents

5 Groundwater hydraulics

6 Dispersion and diffusion

9 Sediment transportation

Stochastic Hydraulics'96, Tickle, Goulter, Xu, Wasimi & Bouchart (eds) © 1996 Balkema, Rotterdam. ISBN 90 5410 817 7

Preface

The International Symposium on Stochastic Hydraulics has become a regular and respected event in the technical conference calendar for engineers and scientists working in all areas of hydraulics. Its importance and reputation was established by the technical successes of the first six symposia, all of which were held in the northern hemisphere. We are delighted to host the 7th International Symposium of Stochastic Hydraulics in Australia recognising that it is the first time that the event has been held in the southern hemisphere. In keeping with the traditions established at the first six symposia, the objectives of this symposium are to provide a forum for exchange of the latest developments in the application of stochastic analysis to river hydraulics, sediment transport, catchment hydraulics, groundwater, waves and coastal processes, hydraulic network and structures, hydrology, risk and reliability in hydraulic design and water resources in general.

This focus of the symposium has particular relevance to a world in which water resource supplies are increasingly nearing full exploitation. Similarly, as the level of development has increased across the globe, the vulnerability of society to the impacts of extreme events has also increased. These two factors give rise to an increasingly critical need to understand and manage the stochastic processes that underly the physical environment in which hydraulic infrastructure is constructed and operated. This symposium is one of the major opportunities for engineers and scientists to meet in order to report on and discuss ways in which hydraulic and stochastic analyses can be integrated in an effective and useful manner in order to meet these challenges.

In this context, it is important to note that the move, in which the first six in this series of symposia have played a pivotal role, over the last twenty years towards more explicit recognition of stochastic processes in the design and operation of hydraulic infrastructure and systems must continue. It is therefore the responsibility of this symposium to continue this role, not just by providing a forum in which new developments and approaches can be presented and debated but also by ensuring, through efforts of the participants once the symposium is over and by wide distribution of the proceedings, that these new developments and approaches are promoted more broadly throughout the professional community.

A large number of papers from around the globe have been accepted for presentation at the symposium. These papers have undergone a careful review process at both the abstract and full paper steps by an International Review Panel and the organising committee is confident and proud of technical content and relevance of the meeting.

The Symposium has been organised by the Centre for Land and Water Development at Central Queensland University and the Institution of Engineers, Australia, Mackay Branch. Financial support for the symposium was also received from the Mackay City Council. We thank these organisations for their willing and generous support of the meeting. The critical role of the members of the International Review Panel in ensuring the technical content of the symposium is gratefully acknowledged.

We would also like to acknowledge the support of Professor Chao-Lin Chui of the University of Pittsburgh for his guidance and enthusiasm during the early stages of planning the conference and for his commitment to the concept and objectives of the meeting. The efficient and enthusiastic administrative assistance of Ms Judy Yewdale is also gratefully acknowledged as is the work of staff of the Mackay Campus of Central Queensland University, in particular, Joe Hallein, Bernadette Howlett and Clive Booth.

Finally, we would like to thank all contributors and participants in the Symposium for their efforts in making this symposium a technical success.

Keynote lectures

Stochastic Hydraulics'96, Tickle, Goulter, Xu, Wasimi & Bouchart (eds)© 1996 Balkema, Rotterdam. ISBN 90 5410 817 7

Synthetic hydrology – Where's the hydrology?

T.A.McMahon & P.B.Pretto
Centre for Environmental Applied Hydrology, The University of Melbourne, Vic., Australia

F.H.S.Chiew
Cooperative Research Centre for Catchment Hydrology, The University of Melbourne, Vic., Australia

ABSTRACT: Techniques for stochastically generating synthetic data sequences have been used extensively in hydrology for over thirty years. Many of the models used have been criticised regarding their inadequacies in relation to modelling hydrologic data. These criticisms and the lack of a true physical basis for most of the stochastic time series models commonly employed have led to a significant level of suspicion regarding the use of stochastic data generation and the interpretation of the results of its use for simulation analyses in hydrology. This suspicion is a good thing, particularly considering the temptations of placing too much faith in the results of stochastic simulation studies which produce reams of apparently precise (although in reality unreliable) estimates of a great range of parameters. However, stochastic models do have a place as very useful tools in the armoury of the practising hydrologist. This review presents some results of a survey of use of and attitudes towards stochastic data generation in hydrology and discusses the role of these models in hydrology, and the important role of hydrology in their application.

1 INTRODUCTION

Fiering (1961) defined synthetic hydrology as the generation of "...a series of simulated streamflows that did not actually occur but that, based on certain statistical considerations, could have been the real record.". Hazen (1914) combined the standardized flow records of a number of streams to create a synthetic record for estimating the probability of occurrence of very dry years in reservoir analysis. However, Sudler (1927) was the first to employ random sampling of historical flows, based on card shuffling and sampling, to produce synthetic flow sequences. Hurst (1965) also performed card sampling techniques for annual flow generation.

Barnes (1954) was the first to generate synthetic flows based on a theoretical distribution of flows and using a random number table, although the simple model used did not incorporate any dependence structure between flows in different periods. Since the early 1960's, stochastic time series models have been widely used for the generation of synthetic data sequences as an input to simulation of hydrologic systems. These models were initially based on a Markov process structure (Fiering, 1961), in which the flow generated depends only upon the flow in the previous period plus a random component (see, for example, McMahon and Mein, 1986). Considerable developments have occurred over the last 30 years, with a wide range of time series structures and models currently available (see Hipel and McLeod, 1994, Salas, 1993 and Salas *et al.*, 1980 for an extensive overview) including multiple site models.

Although the definition by Fiering quoted above includes only streamflow data, stochastic time series models have been used to generate synthetic sequences for a great range of hydrologic data types, particularly rainfall and evaporation (Srikanthan and McMahon, 1985; Chapman, 1994). The stochastic models used are not hydrologic models in the sense of modelling physical hydrologic processes. They are mathematical models which are used to describe the statistical structure of time series and to generate synthetic realisations via numerical techniques. As such, these models are widely used to model many different types of time series totally unrelated to hydrologic applications (Brockwell and Davis, 1991).

Even though stochastic data generation (SDG) can be useful when applied to some hydrologic analyses, it is a technique which is often criticised and viewed with suspicion by practising hydrologists. This relates not just to deficiencies in the models in terms of their abilities to reproduce the statistical structures of the time series modelled, but particularly to the absence of any real connection with hydrologic processes. This paper examines the use of and attitudes towards SDG in hydrology based on a survey of practising hydrologists and addresses the question of whether the application of such techniques can be considered to be included within the sphere of hydrology.

2 SURVEY OF USE AND PERCEPTIONS OF STOCHASTIC DATA GENERATION

Table 1. Geographic location and response of those invited to take part in the survey.

Location	Questionnaires sent	Response
Australia and New Zealand	55	40
South Africa	8	5
Europe	29	16
North America	21	12
Other	5	0
All locations	118	73

Table 2. Fields of interest and response of those invited to take part in the survey.

Field	Questionnaires sent	Response
Research	75	44
Consultant	14	10
Agency	29	19
All fields	118	73

To help gauge attitudes to SDG in hydrology, a survey was carried out involving a sample of engineers and hydrologists from around the world, including those involved primarily in research, in consultancy and in water agency related work. The locations of those invited to take part and their response rates are shown in Table 1. Included within the European classification were those from the UK

and Ireland. Some questionnaires were also sent to India, Japan and Russia and Georgia (former USSR), although no responses were received from these locations. Fields in which those approached work and the associated response rates are shown in Table 2. Overall, 118 questionnaires were sent with 73 responses received, corresponding to a response rate of approximately 60%.

The questions asked related to whether SDG had been used for simulation applications in hydrology over the last 10 years and, if so, what types of data, applications and models had been used. Further, attitudes towards the advantages and limitations of SDG were sought, including a subjective judgement by respondents on their level of satisfaction with the results of their own applications of SDG.

Fifty of the respondents (approximately 70%) indicated that they had used SDG in the last 10 years, with a further twelve (about 15%) having considered its use. This was fairly consistent over all locations and all fields of interest. The aim of the survey was not to determine the level of use of SDG amongst a random sample of hydrologists, but rather to identify patterns of use of different models and users' attitudes. Those invited to take part were chosen based on the authors' perceptions of hydrologists expected to have had experience with using SDG models. The sample of respondents would therefore be expected to be quite biased in this respect. However, as no other criterion was used in inviting participants and as all responses were included for analysis, the results of the survey are useful for indicating the patterns of use of SDG and attitudes towards SDG by hydrologists who use these techniques.

2.1 *Use of stochastic data generation*

Table 3 shows how frequently SDG had been used to generate different data types as a percentage of users of SDG for each data type over the last 10 years. Of the respondents who had used SDG, 76% generated streamflow sequences, 66% generated rainfall sequences and 30% generated other data types. These figures include respondents who generated more than one type of data. Other data types generated included evaporation, water demand, temperature, radiation, storage losses and soil hydraulic properties. The pattern evident is that use of SDG for streamflow and rainfall data was, overall, far more common than for other data types. Users of rainfall data generation generally used SDG more frequently (typically more than once per year) than

4

users of streamflow data generation (typically less than once per year). For the purposes of this paper, the emphasis will be placed upon streamflow data, with only a brief consideration of rainfall.

Table 3. Frequency of use of SDG for different data types; figures are as percentages of users of SDG for each data type shown.

Frequency of use	Streamflow	Rainfall	Other
<1/year	39%	27%	27%
1-2/year	32%	24%	20%
>2/year	29%	49%	53%

The general pattern of SDG use did not reflect the pattern of use found in Europe. The European respondents were significantly greater users of rainfall generation than streamflow generation. Respondents working as consultants and in water agency areas showed a generally lower frequency of use of SDG for all data types than those involved in research.

2.2 *Applications for stochastic data generation*

By far the most common applications for streamflow generation involved reservoir storage-reliability-yield (SRY) analysis. This dominated over all other applications, of which non-specific research and data analysis and analysis of hydrologic extremes were most significant. For rainfall, the most common applications were as input for streamflow generation (rainfall-runoff) models to be used for drainage design and analysis, hydrologic catchment studies and SRY analysis as well as for the assessment of the impacts of potential climate change. A specific application did not dominate for rainfall as was the case for streamflow, indicating a generally greater range of applications for generated rainfall data. This could be related to a greater range of potential for rainfall data as an input to hydrologic models, particularly rainfall-runoff models, compared with streamflow data. It could also give some insight into the observation that where respondents generated rainfall data, they used SDG relatively more frequently than was the case with those who generated streamflow data. A greater range of commonly used applications would be expected to encourage a greater frequency of use.

Table 4 shows the relative use of SDG for modelling streamflow and rainfall data of various durations, expressed as percentages of users of streamflow or rainfall generation. Generation of seasonal data includes time resolutions greater than or equal to a week but less than one year. As would be expected, for streamflow generation annual and seasonal models dominated, whereas for rainfall models of daily or shorter duration were most common. This reflects the dominant applications for the two types of data, rainfall being used typically as an input to models requiring a fine time resolution and generated streamflows typically used to analyse statistical properties of annual and seasonal flows and as input to water balance type models using annual and monthly time steps.

Table 4. Relative use of SDG for modelling streamflow and rainfall data of various durations, expressed as percentages of users of streamflow and rainfall generation respectively. These figures include respondents who modelled data at two or more different time steps.

Data type	Annual	Seasonal	Daily or less
Streamflow	87%	84%	32%
Rainfall	36%	42%	67%

It was apparent from the survey results, however, that stochastic generation of streamflows of daily duration was significantly more common amongst respondents from Europe than amongst respondents from other locations. This may be due to the relatively small sample of respondents from Europe who actually used SDG for streamflow (six in total) being biased toward users of daily flow models. However, it could also be an indication of a greater perception of the importance of water resource system simulation with a daily time step and greater confidence in SDG applied to daily streamflow.

3 MODELS USED

An interesting aspect of the survey results was which models were preferred for generation of different data types, especially when considered in relation to specific problems usually associated in the literature with particular models. Assuming the respondents who used SDG to be a representative sample of SDG users, the results can be expected to give an overall

5

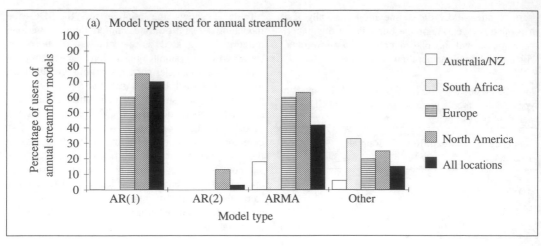

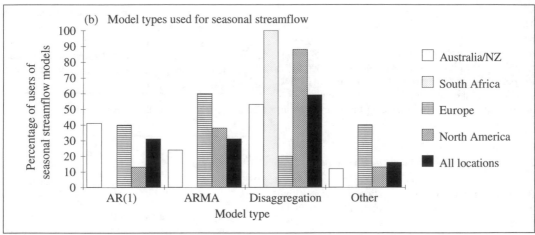

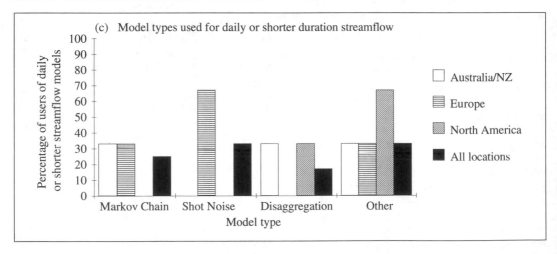

Figure 1. Types of models used for generation of streamflow data of annual (a), seasonal (b) and daily or shorter (c) durations.

insight into model use, including differences between different parts of the world, between different fields of practice and between application to different data types.

3.1 *Streamflow*

Figure 1 (a), (b) and (c) show the preferred models for annual, seasonal and daily or shorter streamflow generation respectively, with comparisons between hydrologists from different locations. For annual streamflow generation, a Markov or first order autoregressive (AR(1)) model was generally used most commonly, although the more general class of autoregressive moving average (ARMA) models was also commonly used. Where autoregressive models were used specifically, only a second order model (AR(2)) was used apart from the AR(1) model, although its use was not common. Other models used included shot noise and multi-lag linear regressive type models.

Seasonal streamflows were modelled most commonly using disaggregation type models, in which annual flows are first generated and then disaggregated to give the desired flows. Interestingly, in North America parametric disaggregation models (for example see Valencia and Schaake, 1973, Mejia and Rousselle, 1976, and Hoshi and Burges, 1979) were used exclusively, whereas respondents from other areas mostly used the empirically based method of fragments (Svanidze, 1980 and Srikanthan and McMahon, 1982). Seasonal, or periodic, AR(1) and ARMA models were also used quite commonly, with the Europeans generally using these models more than the disaggregation types.

Streamflows of daily or shorter duration were generated using a range of techniques, including Markov chain type models and disaggregation models. Shot noise models comprising an event process and a linear response process, as described by Cowpertwait and O'Connell (1992), dominated in Europe but were not used by any respondents from other parts of the world. The bulk of the other models used involved stochastically generating rainfall and using this as an input to rainfall-runoff models to generate streamflow volumes.

The most striking result from this aspect of the survey is the popularity of relatively simple models for annual and seasonal streamflow generation, particularly AR(1) models and disaggregation techniques. None of the respondents used long memory models such as fractional Gaussian noise (Mandelbrot and Wallis, 1968) and broken line

(Mejia *et al.*, 1972) models, despite the criticism that short memory processes such as AR(1) models fail to preserve elements of long term persistence (Mandelbrot and Wallis, 1968, Wallis and Matalas, 1972 and Ashkanasy, 1978).

The predominance of the use of disaggregation techniques to generate seasonal streamflow volumes may reflect the needs of respondents to preserve the statistical structure of annual streamflows as well as seasonal, this being difficult to achieve using seasonal time series models (for example, see Srikanthan and McMahon, 1982). The popularity of the method of fragments as a disaggregation technique would be expected to reflect its significantly reduced parameter estimation requirements compared with the parametric disaggregation techniques, as discussed by Porter and Pink (1991).

3.2 *Rainfall*

The models used by survey respondents for generation of annual, seasonal and daily and shorter duration rainfall are shown in Figure 2 (a), (b) and (c) respectively. In general, simple AR(1) models were preferred for annual rainfall generation, although the European respondents preferred other techniques, including aggregation of stochastically generated daily rainfall, for both annual and seasonal rainfall generation. Apart from the Europeans, seasonal AR(1) and disaggregation models were most popular overall for seasonal rainfall generation.

Markov chain type models incorporating transitional probabilities for the change from one rainfall state to another (for example, see Srikanthan and McMahon, 1985 and Chapman, 1994) were most popular for daily rainfall generation in Australia and New Zealand (NZ). However, in Europe and North America clustered point process models (described by Cowpertwait, 1994) were preferred.

3.3 *Multiple site data generation*

The proportions of users of both streamflow and rainfall generation who also used multiple site generation models are shown in Table 5. The results indicate that multiple site SDG is used quite commonly for both rainfall and streamflow. The survey results showed that this was consistent over all fields of interest as well as over all geographic locations surveyed.

Figure 3 shows the model types preferred by respondents who used multiple site SDG. Multiple

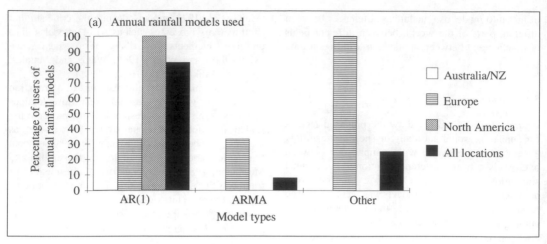

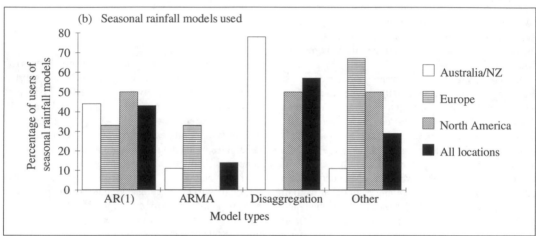

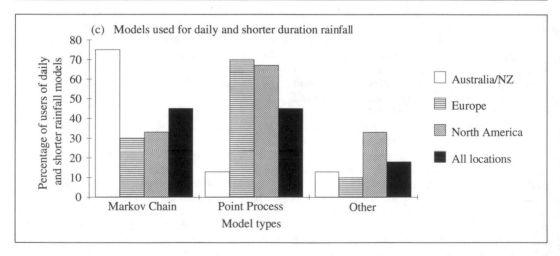

Figure 2. Types of models used for generation of rainfall data of annual (a), seasonal (b) and daily or shorter (c) durations.

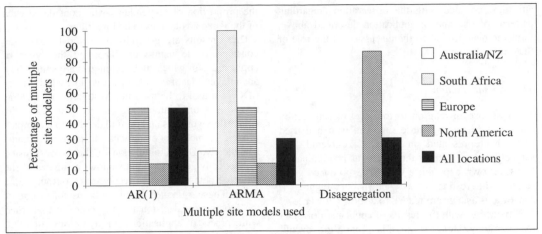

Figure 3. Types of models used for multiple site streamflow generation.

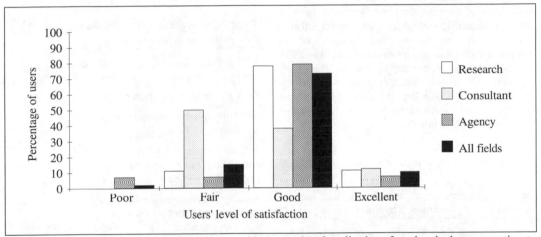

Figure 4. Perceived level of satisfaction of users with the results of application of stochastic data generation to simulation in hydrology.

Table 5. Use of multiple site generation of streamflow and rainfall data; figures shown are as percentages of users of SDG for streamflow and for rainfall respectively.

Data type	Percentage use of multiple site generation
Streamflow	53%
Rainfall	36%

site AR(1) models (Matalas, 1967) were used most commonly in Australia and NZ, multiple site ARMA models (Stedinger *et al.*, 1985) in South Africa, with no clear preference of one over the other apparent from the European respondents. The North American respondents, however, seemed to particularly favour spatial disaggregation (for example, see Stedinger *et al.*, 1985) techniques.

4 ATTITUDES TO USE IN HYDROLOGY

A key component of the survey was to ascertain attitudes to the use of SDG in hydrology. These attitudes, coupled with the observations regarding patterns of use and of application described above, form the background to the discussion of the role of SDG in hydrology.

4.1 *User satisfaction*

The response in relation to the level of satisfaction with their results by those surveyed who had used SDG is represented in Figure 4. Overall, the subjective judgement of the users was predominantly that their own experience with SDG produced good results. Interestingly, as shown in Figure 4, consultants as a group tended to a show a lower level of satisfaction with the results of application of SDG to hydrology. Apart from this group, the overall pattern of user satisfaction was remarkably consistent between groups.

Although these results are of some interest, it is difficult to draw any definite conclusions from such indications of user satisfaction. The lower level of satisfaction among consultants may reflect either a difference in the specific applications used as compared with other groups or a reduced availability of time to investigate the appropriate model for their applications. Another interesting point raised by some respondents was that although they may have felt the results obtained to be good, they found others to be far less satisfied.

4.2 *Advantages of stochastic data generation*

Table 6. Perceived advantages of the use of stochastic data generation in hydrology.

Advantages	% of respondents
Provision of assessment of uncertainty and risk	52%
Facilitates investigation of response of complex systems	52%
Allows simulation of long data sequences	19%
Useful for forecasting models	1%
Useful for assessing potential impacts of climate change	3%

Table 6 shows a classification of the advantages of using SDG in hydrology based on the perceptions of the respondents to the survey. The table also shows

the proportion of respondents who considered each of the listed advantages to be important.

These results are generally as would be expected considering the nature of SDG and the types of applications to which such techniques are suited. The advantages identified reflect a role in hydrology for SDG as a tool to be used in investigation of systems involving some stochastic input (for example, streamflow volumes and rainfall depths) to a complex process which is extremely difficult to model analytically. Direct simulation of the process using synthetically generated inputs allows estimates of the range of potential system response, or output, to be made. These estimates may then be useful as one of the inputs to a decision making process concerned with practical applications in hydrology or with statistical analyses.

4.3 *Limitations of stochastic data generation*

Table 7. Perceived limitations of the use of stochastic data generation in hydrology.

Limitations	% of respondents
Lack of adequate input data	26%
Difficulty in setting up models and lack of software	30%
Potential for inappropriate conclusions to be drawn due to poor understanding of limitations	18%
Assumptions regarding model structure and distributions of variables modelled	39%
Failure to capture long term persistence and poor reproduction of extremes and extreme events	26%
Difficulties implementing models	12%
Lack of a physical basis	8%

Although there is a role for SDG in hydrology, it is of paramount importance to focus attention on the limitations of synthetic hydrology before its various techniques are applied. The reaction of the survey respondents when asked to identify the major limitations of SDG was generally stronger and more varied than their response to the advantages. This would seem to indicate that the group surveyed was, on the whole, aware of many of the limitations which may restrict application of SDG. Such a case is expected given the basis upon which hydrologists were identified for invitation to take part in the survey, as outlined in Section 2 above. A categorisation of the limitations based on their responses is set out in Table 7, along with the proportion of respondents indicating each category to be important.

Once again, the list produced does not hold any great surprises. All of the limitations listed have been recognized in the literature relating to synthetic hydrology as well as in more general criticisms of the applicability of SDG to hydrology. Most of the limitations apply to SDG models in general. Of these, difficulty in setting up models/lack of software and difficulty in model implementation are more of a problem for specific model types, although they are applicable to all models to some extent, depending on the application. However, failure to capture long term persistence and poor reproduction of extremes and extreme events are related limitations which are usually perceived as applying specifically to short memory type models such as the simple AR(1) process. It is interesting to note that although this limitation was identified most often by respondents, an AR(1) model was generally the most commonly used model for annual streamflow generation.

Lack of a physical basis is an important limitation even though it was identified less often than any other category. A number of limitations included within other categories, and which are based on some failing in model performance, are in part due to inadequacies in model structures in terms of their inability to adequately represent the real hydrologic processes being modelled. This relates to the nature of SDG models as purely mathematical or statistical models rather than hydrologic models. For some hydrologists and engineers there is a marked distrust of the results of analyses based on synthetic hydrology because these results have been "fabricated" using "make-believe" data. This distrust is actually founded upon the absence of any hydrologic component in the models commonly used in synthetic hydrology.

In practical terms, one of the most serious and potentially most damaging limitations of synthetic hydrology is related more to the user than to the techniques themselves. This is the application of SDG by users with a poor understanding of the limitations of the techniques or with a poor understanding of hydrology, or both. The problem here is that reams of apparently precise results can still be produced, but their correct interpretation will be jeopardised. As a result, totally inappropriate

conclusions are likely to be drawn which could lead to unfortunate errors in decision making.

5 ASSESSMENT OF LIMITATIONS

5.1 *Inadequacy of input data*

It is well recognized that SDG models are severely hampered by poor data, both from the point of view of adequate identification of an appropriate model structure and errors in the estimates of model parameters (see Klemes *et al.*, 1981). In fact the impact of parameter uncertainty has been shown by Stedinger and Taylor (1982a) to be potentially as important as the choice between different types of models in terms of the results of the analysis carried out. An important point to make, however, is that the unavailability of good data is a limitation to all models and techniques applied to hydrology, including those explicitly incorporating a representation of hydrologic processes. The problem for synthetic hydrology is that some users may believe that SDG compensates for lack of data, even though this is not the case as SDG cannot extract information which is not already contained in the real data. This problem is a question of poor understanding on the part of the user.

5.2 *Model assumptions*

The use of statistical models such as those used in SDG are, of course, limited by the assumptions made to facilitate the fitting of theoretical models and distributions to observations of hydrologic processes. These processes are generally not at all related in a real physical sense to those theoretical models and distributions. The results of the application of SDG in hydrology are therefore limited by the model assumptions. However, once again, this is not peculiar to SDG techniques alone, but to all purely statistically based methods of analysis applied in hydrology (for example, flood frequency analysis and even simple linear regression analysis). The key is that the user understands the limitations of the techniques used and the hydrology of the context in which they are being applied.

5.3 *Failure to capture persistence and to reproduce extremes*

As discussed briefly in Section 3.1, so called short memory models such as the AR(1) process have been criticised for not being capable of modelling long term persistence, this leading to their inability to adequately reproduce extremes and their failure to generate events more extreme than those observed in the historical record (see, for example, Srikanthan and McMahon, 1978a). However, this may not always be the case in practice. For example, Stedinger and Taylor (1982b) found that a simple AR(1) annual flow generator was able to generate drought sequences more severe than the historical worst drought 45% of the time for a 50 year period for the upper Delaware River Basin.

A number of researchers have found that ARMA models (for example see O'Connell, 1971 and Hipel and McLeod, 1994) can be used to preserve long term persistence characteristics, although Srikanthan (1979) found that an ARMA(1,1) model generally failed to perform better in preserving long term persistence than fractional Gaussian noise and broken line models (also see Srikanthan and McMahon, 1978b and 1978c).

Klemes *et al.* (1981) questioned the value of incorporating long term persistence in SDG models, arguing that uncertainty due to usually inadequate hydrologic and socioeconomic data reduces its importance in terms of the results produced. It would appear, based upon the results of our survey, that this is the view adopted in practice by users of SDG. Although failure to reproduce long term persistence was strongly identified as a limitation for SDG models used, analysis of hydrologic extremes was still a common application for these models.

5.4 *Difficulties in setting up and implementing models*

The limitations here mainly related to the need for a relatively high level of expertise in the area of time series analysis and modelling. Many respondents, particularly in the consulting field, felt their own use of SDG models to be limited by their lack of adequate expertise and the resulting difficulties in setting up and running models themselves. This was identified as a problem due to a lack of readily available and accepted software for application of SDG models and may have contributed to the generally lower level of satisfaction with SDG amongst the consultants surveyed.

5.5 *Lack of a physical basis*

Although some aspects of the structure of particular SDG models have been interpreted as being compatible with physical hydrologic processes (Salas *et al.*, 1980, Salas and Smith, 1981, Stedinger *et al.*, 1985, Claps *et al.*, 1993 and Hipel and McLeod, 1994), they are actually based on statistical and mathematical concepts and not on hydrologic principles. As such, they do not provide any insight to physical hydrologic reality. Their use is not necessarily invalid, as long as unjustifiable extrapolations are not made and the existence and crucial nature of the assumptions inherent in their use are fully appreciated. These concepts have been thoroughly discussed and illustrated with a number of interesting examples by Klemes (1986).

Where physical aspects of streamflow generation, for example, are being investigated to determine the actual hydrologic mechanisms at work and their structure, the use of SDG models will have little value. They tell us nothing of such physical realities. However, where the variability and uncertainty in the response of a system with an assumed statistical structure of the input is being investigated, SDG models may be useful. This depends on the assumption that the input to the system is adequately specified by a mathematical model which is based on reproducing the statistical structure of the observations on which it is based. The key here is the understanding that such an analysis models the response from a purely statistical standpoint, its results only being valid in this sense.

6 DISCUSSION - WHERE IS THE HYDROLOGY ?

The cynical view would be that the term "synthetic hydrology" should be taken literally to mean artificial, or even pretend hydrology - that it has nothing to do with hydrologic processes and should have nothing to do with applied hydrology. So, where is the hydrology in synthetic hydrology? Indeed, the hydrology will not be found by looking at the models used for SDG. The hydrology is where it should be - in the understanding and experience of the hydrologist applying the techniques of synthetic hydrology. It is in the interpretation and application of SDG models and in understanding their inherent assumptions and limitations and how these impact upon their usefulness in the context of particular problems in hydrology.

Synthetic hydrology should be seen as a tool which provides an input for decision making, more specifically to explore the potential variability in response of a system to variable (and unknown) inputs. It should not be seen as a means to pinpointing more accurate, or more reliable, estimates of design parameters. Its main strength is in giving an estimate of the degree of potential unreliability of any design estimates based (necessarily) on a single data record.

The mathematics of SDG models is the same, whatever the origin of the time series modelled, having no regard at all for real physical mechanisms of runoff generation. The hydrologist using SDG methods must be capable of applying them only where this fundamental aspect of those methods is not inconsistent with the aim of the analysis being carried out. It is here that the knowledge, experience, skill and intuition of a *good* hydrologist are needed to temper the otherwise almost overpowering seduction of the capability of time series simulations to produce reams of figures which may, in the wrong context, mean absolutely nothing. The hydrology is the essential ingredient required to keep the hydrologist from being tempted into blindly accepting these figures, captivated by the almost magical spell of the elegance and rigour of the mathematical foundations of the methods.

The fact that these methods do not have any real link with the physical aspects of hydrology as a science (the only link is that hydrologic data are often available as time series) makes the importance of the user having a sound understanding of the hydrology underlying the system being examined of even greater significance. This is contrary to the view which may be taken without deeper reflection. It is felt by many practising hydrologists that using SDG is potentially dangerous due to the lack of a need for any understanding of hydrology at all. Generating a set of numbers statistically similar to another set of numbers requires no knowledge of catchment characteristics nor of the hydrologic cycle.

This is of course true. However, we argue that the proper application of SDG techniques in hydrology requires a very good understanding of the hydrologic principles associated with the system analysed. Thus the hydrology is certainly there, not necessarily as part of the time series modelling itself, but as the structure and context in which the modelling is applied. Getting this part (the hydrology) right is at least (and probably more) important than the details of the SDG models. If the basic hydrology is not right, the results of the analysis will have little relevance in practice.

7 CONCLUSION

The main conclusion which needs to be made is that despite the obvious and often considerable limitations involved with the use of synthetic hydrology, SDG techniques do have a place in the statistical armoury of practising hydrologists. They should not be summarily dismissed based on the deficiencies associated with their use (and misuse). In this sense they are no different to other statistically based techniques applied in hydrologic analyses, many of which suffer from similar limitations to those of SDG. The key is the recognition of suitability in application and understanding the limits of the inferences which can be reasonably drawn from the analysis. This is where the hydrology should be found in the application of SDG.

8 REFERENCES

Ashkanasy, N.M. 1978. Generation of synthetic streamflow data for management and planning of water resource systems. *Hydrology Symposium of the Institution of Engineers, Australia, CANBERRA, 5-7 September.* The Institution of Engineers, Australia, National Conference Publication No. 78/9: 75-79.

Barnes, F.B. 1954. Storage required for a city water supply. *J. Inst. Eng., Aust.*, 26: 198-203.

Brockwell, P.J. and Davis, R.A. 1991. *Time Series: Theory and Methods.* Second edition, Springer-Verlag, New York.

Chapman, T.G. 1994. Stochastic models for daily rainfall. *25th Congress of The International Association of Hydrogeologists/International Hydrology and Water Resources Symposium of the Institution of Engineers, Australia. ADELAIDE 21-25 November.* The Institution of Engineers, Australia, NCP No. 94/15: 7-12.

Claps, P., Rossi, F. and Vitale, C. 1993. Conceptual-stochastic modeling of seasonal runoff using autoregressive moving average models at different scales of aggregation. *Water Resour. Res.*, 29(8): 2545-2559.

Cowpertwait, P.S.P. 1994. A generalized point process model for rainfall. *Proc. R. Soc. Lond. A*, 447: 23-37.

Cowpertwait, P.S.P. and O'Connell, P.E. 1992. A Neyman-Scott shot noise model for the generation of daily streamflow time series. Chapter 6 in: *Advances in Theoretical Hydrology - a Tribute to James Dooge*, Elsevier Science Publishers: 75-94.

Fiering, M.B. 1961. Queuing theory and simulation in reservoir design. *Proc. Am. Soc. Civ. Eng., J. Hydraul. Div.*, 87(HY6): 39-69.

Hazen, A. 1914. Storage to be provided in impounding reservoirs for municipal water supply. *Trans. Am. Soc. Civ. Eng.*, 77: 1539-1640.

Hipel, K.W. and McLeod, A.I. 1994. *Time Series Modelling of Water Resources and Environmental Systems*. Elsevier Science B.V., Amsterdam.

Hoshi, K. and Burges, S.J. 1979. Disaggregation of streamflow volumes. *Proc. Am. Soc. Civ. Eng., J. Hydraul. Div.*, 105(HY1): 27-41.

Hurst, H.E., Black, R.P. and Simaika, Y.M. 1965. *Long Term Storage: an Experimental Study*. Constable, London.

Klemes, V. 1986. Dilettantism in hydrology: transition or destiny? *Water Resour. Res.*, 22(9): 177S-188S.

Klemes, V., Srikanthan, R. and McMahon, T.A. 1981. Long-memory flow models in reservoir analysis: what is their practical value? *Water Resour. Res.*, 17(3): 737-751.

Mandelbrot, B.B. and Wallis, J.R. 1968. Noah, Joseph, and operational hydrology. *Water Resour. Res.*, 4(5): 909-918.

Matalas, N.C. 1967. Time series analysis. *Water Resour. Res.*, 3(3): 817-829.

McMahon, T.A. and Mein, R.G. 1986. *River and Reservoir Yield*. Water Resources Publications, U.S.A.

Mejia, J.M., Rodriguez-Iturbe, I. and Dawdy, D.R. 1972. Streamflow simulation: 2. The broken line process as a potential model for hydrologic simulation. *Water Resour. Res.*, 8(4): 931-941.

Mejia, J.M. and Rousselle, J. 1976. Disaggregation models in hydrology revisited. Water Resour. Res., 12(2): 185-186.

O'Connell, P.E. 1971. A simple stochastic modelling of Hurst's law. *Mathematical models in hydrology symposium, Warsaw, July 1971*, IAHS-AISH publication No. 100, 1974 printing, Volume 1: 169-187.

Porter, J.W. and Pink, B.J. 1991. A method of synthetic fragments for disaggregation in stochastic data generation. *International Hydrology and Water Resources Symposium of the Institution of Engineers, Australia. PERTH 2-4 October*. The Institution of Engineers, Australia, NCP No. 91/22, Volume 3: 781-786.

Salas, J.D. 1993. Analysis and modeling of hydrologic time series. Chapter 19 in *Handbook of Hydrology*, David R. Maidment editor in chief, McGraw-Hill, New York: 19.1-19.72.

Salas, J.D., Delleur, J.W., Yevjevich, V. and Lane, W.L. 1980. *Applied Modeling of Hydrologic Time Series*. Water Resources Publications, Colorado, USA.

Salas, J.D. and Smith, R.A. 1981. Physical basis of stochastic models of annual flows. *Water Resour. Res.*, 17(2): 428-430.

Srikanthan, R. 1979. *Stochastic Generation of Annual and Monthly Flow Volumes*. Ph.D. Thesis, Monash University, Clayton, Australia.

Srikanthan, R. and McMahon, T.A. 1978a. A review of lag-one Markov models for generation of annual flows. *J. Hydrol.*, 37: 1-12.

Srikanthan, R. and McMahon, T.A. 1978b. Generation of annual streamflow using a filtered Gaussian noise model. *J. Hydrol.*, 37: 13-21.

Srikanthan, R. and McMahon, T.A. 1978c. Comparison of fast fractional Gaussian noise and broken line models for generating annual flows. *J. Hydrol.*, 38: 81-92.

Srikanthan, R. and McMahon, T.A. 1982. Stochastic generation of monthly streamflows. *Proc. Am. Soc. Civ. Eng., J. Hydraul. Div.*, 108(HY3): 419-441.

Srikanthan, R. and McMahon, T.A. 1985. *Stochastic Generation of Rainfall and Evaporation Data*. Australian Water Resources Council Technical Paper No. 84, Australian Government Publishing Service, Canberra.

Stedinger, J.R. and Taylor, M.R. 1982a. Synthetic streamflow generation, 2: effect of parameter uncertainty. *Water Resour. Res.*, 18(4): 919-924.

Stedinger, J.R. and Taylor, M.R. 1982b. Synthetic streamflow generation, 1: model verification and validation. *Water Resour. Res.*, 18(4): 909-918.

Stedinger, J.R., Lettenmaier, D.P. and Vogel, R.M. 1985. Multisite ARMA(1,1) and disaggregation models for annual streamflow generation. *Water Resour. Res.*, 21(4): 497-509.

Sudler, C.E. 1927. Storage required for the regulation of stream flow. *Trans. Am. Soc. Civ. Eng.*, 91: 622-660.

Svanidze, G.G. 1980. *Mathematical Modelling of Hydrologic series*. Translated from Russian Edition (Gidrometeoizdat, Leningrad, USSR, 1977), Water Resources Publications, Fort Collins, Colorado, USA.

Valencia, D. and Schaake, J.C. 1973. Disaggregation processes in stochastic hydrology. *Water Resour. Res.*, 9(3): 580-585.

Wallis, J.R. and Matalas, N.C. 1972. Sensitivity of reservoir design to the generating mechanism of inflows. *Water Resour. Res.*, 8(3): 634-641.

ACKNOWLEDGEMENTS

The authors wish to thank the Australian Research Council for financial support and the University of Melbourne for providing an APRA scholarship.

Stochastic Hydraulics'96, Tickle, Goulter, Xu, Wasimi & Bouchart (eds)© 1996 Balkema, Rotterdam. ISBN 90 5410 817 7

A natural law of open-channel flows

Chao-Lin Chiu
University of Pittsburgh, Pa., USA

ABSTRACT: This paper presents a natural law of open-channel flows, "The ratio of the mean and maximum velocities of flow in a channel section is constant." The ratio represents the overall characteristics of the system at a channel section. It controls the flow variables including the energy and momentum coefficients, and the distributions of velocity, shear stress and sediment concentration, in interaction with the channel and fluid variables at the section. To establish and maintain the constant ratio, the variables adjust themselves when the discharge fluctuates. The law can be used to gain an insight into unobservable interactions among the variables, and to ease the study, measurements and prediction of open-channel flows.

1 INTRODUCTION

A natural law is "a small class of empirical propositions" or "a domain of regularities" (Hospers 1980). Natural laws form a network in which one law is supported by the other laws. This paper presents a natural law derived from a regularity found in open-channel flows; support for the law; and usefulness of the law in open-channel hydraulics.

The variables used to describe open-channel flow systems are many and can be classified into the following three categories: the channel; the fluid; and the flow. A channel can be described by the geometrical shape, slope, alignment, and bed material and form (roughness). A fluid can be described by the density and viscosity. A flow can be described by the discharge, secondary flow, flow resistance, head loss, and distributions of velocity, shear stress and sediment concentration, et al. These variables interact to produce a very large number of possible combinations. The phenomena observed in open-channel flows are only some small fractions of these combinations. To understand and predict open-channel flows accurately, finding the underlying mechanism or a law of nature responsible for generating the observed phenomena is necessary.

The task of finding such a law requires observation, perception and conception. Perception is to organize the observations according to time and space. Conception is to organize the perception into a theory or a law. Metaphysics as well as science is, therefore, needed to accomplish the task.

In such an effort, the scientific principles of hydrodynamics that have occupied open-channel hydraulics for many years have inherent limitations. This is mainly because hydrodynamics treats all fluid-flow problems as boundary value problems, but the channels often have complex, irregular and variable boundary conditions. The numbers of reliable boundary conditions are insufficient for adequately solving the hydrodynamic equations for the large number of variables.

To derive a natural law, the development of ideas beyond those of rational hydrodynamics is, therefore, needed to synthesize the knowledge gained from the limited observations, hydrodynamic analysis, and any other efforts. Furthermore, it is necessary to think clearly to grasp the essence, or to reach the highest degree of generalization possible, of the fascinating and often chaotic and menacing phenomena in open channels such as rivers and streams.

2 A REGULARITY IN OPEN-CHANNEL FLOWS

Fig. 1(a) includes four straight lines (Chiu and Said 1995), each relating the maximum and mean velocities in one of the four channel sections of wide ranges of discharge, slope, roughness, shape and alignment Each of these lines can be represented by

$$u_{max} = \frac{1}{\phi} \bar{u} \tag{1}$$

in which u_{max} = maximum velocity; $\bar{u}$ = mean velocity; and ϕ is the ratio of the mean and maximum velocities. Each value of $\bar{u}$ in the figure was obtained from a given discharge, and u_{max} from velocity data. The figure shows that each channel section has a constant value of ϕ. Similarly, Fig. 1(b) is plotted with experimental data (Tsujimoto, et al. 1993) from open-channel flows in a circular vinyl conduit, with slope, discharge and flow depth varied in wide ranges. It also shows that ϕ is a constant. Fig. 2 shows four velocity profiles measured at a channel section in a bend of the South Esk River (Bridge and Jarvis 1985) which maintains a constant value of ϕ. It describes how the channel, to maintain the constant value of ϕ, adjusts the velocity distribution by adjusting the maximum velocity u_{max} and its location h below the water surface, when the discharge Q and water depth D fluctuate.

These and many other similar observations show that the constancy of ϕ has the "universality," "invariance" and "regularity" qualities required of a natural law (Hospers 1980). It is universal and invariant since it is applicable to all types of open channels, at any time, in a wide range of discharge. Its regularity is evident from the fact that it occurs repeatedly.

3 EXPLANATION BY VELOCITY DISTRIBUTION IN PROBABILITY DOMAIN

The ratio of the mean and maximum velocities is related to velocity distribution. Chiu (1987, 1988, 1989, 1991) derived a system of velocity distribution equations of the following general form:

$$\frac{\xi - \xi_0}{\xi_{max} - \xi_0} = \int_0^u p(u)du \tag{2}$$

in which u = velocity at ξ; ξ = independent variable with which u develops such that each value of ξ corresponds to a value of u; and ξ_{max} and ξ_0 = maximum and minimum values of ξ, respectively. In the physical space a curve along which ξ has a constant value is an isovel; ξ_{max} occurs at the location of the maximum velocity u_{max}; and ξ_0 occurs along the channel bed where u is zero. Eq. (2) means that if ξ is randomly sampled many times within the range (ξ_0, ξ_{max}) and the corresponding velocity samples are obtained, the probability of velocity falling between u and u+du is p(u)du, and the probability of velocity less than or equal to u is equal to ($\xi - \xi_0$)/($\xi_{max} - \xi_0$). The random sampling of ξ in (ξ_0, ξ_{max}) is equivalent to that of ($\xi - \xi_0$)/($\xi_{max} - \xi_0$) in (0, 1). p(u) in (2) is the probability density function,

$$p(u) = \exp(a_1 + a_2 u) \tag{3}$$

This function is derived by using a statistical method based on the law of uncertainty (Rao 1980), or by maximizing Shannon's information function (Shannon 1948),

$$H = -\int_0^{u_{max}} p(u) \ln p(u) du \tag{4}$$

subject to the following constraints:

$$\int_0^{u_{max}} p(u) \ du = 1 \tag{5}$$

$$\int_0^{u_{max}} up(u)du = \bar{u} = \frac{Q}{A} \tag{6}$$

Eq. (5) is the constraint on p(u) to satisfy the definition of probability density function. Eq. (6) is the constraint on p(u) to make the mean or average velocity in the flow cross section equal to Q/A where Q is the discharge, and A is the cross-sectional area of the flow. The function as defined by (4) was originally derived as a measure of the information content of a message about a variable relative to the prior uncertainty (Shannon 1948). The idea was that information content increases with uncertainty. It was unfortunate that the function was called "entropy," since it caused a great deal of misunderstanding and confusion between the function and the thermodynamic entropy.

To avoid any misunderstanding, it must be clearly

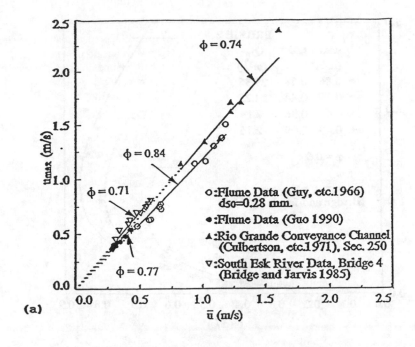

(a)

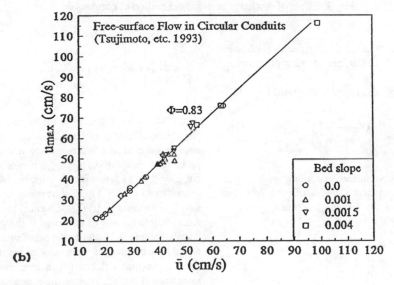

(b)

Fig. 1. Relation between u_{max} and $\bar{u}$: (a) In Open Channels; (b) In Closed Conduits.

understood that H is used only as a measure of uncertainty (among other possible measures) or of uniformity of p(u), and that the method of deriving p(u) by maximizing H is based on the laws of uncertainty and of statistics, not on the laws of thermodynamics. What distinguishes H as a popular and favored measure of uncertainty is essentially that the function is amenable to solution by the calculus of variations to identify p(u). The law of uncertainty and the method based on it are supported by results from their applications in fluid-flow studies (Chiu 1987, 1988, 1989, 1991, Chiu et al. 1993, Chiu and

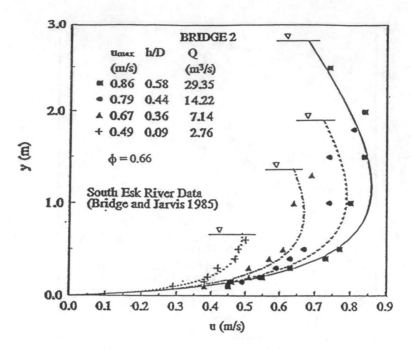

Fig. 2. Effect of discharge at a channel section of constant ϕ

Said 1995) as well as in information theory (Goldman 1953), statistics (Rao 1980) and in many other fields.

With (3) giving p(u), (2) can be integrated to yield

$$\frac{u}{u_{max}}=\frac{1}{M}\ln\left[1+(e^{M}-1)\frac{\xi-\xi_0}{\xi_{max}-\xi_0}\right] \qquad (7)$$

in which M is a parameter that along with u_{max} is related to the parameters a_1 and a_2 of p(u). M is the only parameter of the distribution $p(u/u_{max})$ that is equal to $u_{max}\,p(u)$ or

$$p\left(\frac{u}{u_{max}}\right)=\frac{M}{e^{M}-1}\exp\left(M\frac{u}{u_{max}}\right) \qquad (8)$$

By substituting (3) into (5) and (6), the following equation for the ratio of the mean and maximum velocities can be obtained:

$$\frac{\bar{u}}{u_{max}}=\phi=\frac{e^{M}}{e^{M}-1}-\frac{1}{M} \qquad (9)$$

Since u_{max} is $\bar{u}/\phi$, (7) is equivalent to

$$\frac{u}{\bar{u}}=\frac{1}{M\phi}\ln\left[1+(e^{M}-1)\frac{\xi-\xi_0}{\xi_{max}-\xi_0}\right] \qquad (10)$$

Eqs. (7)-(10) show that M is the common and only parameter of u/u_{max}, $p(u/u_{max})$, $\bar{u}/u_{max}$, and $u/\bar{u}$. A smaller value of M corresponds to: a more uniform probability distribution $p(u/u_{max})$, a smaller value of $\bar{u}/u_{max}$ or ϕ, and a less uniform velocity distribution as shown by Figs. 3-5. Fig. 5 shows that for M greater than 2, $\bar{u}$ occurs at $(\xi-\xi_0)/(\xi_{max}-\xi_0)$ equal to e^{-1} or 0.368. According to the law of uncertainty, the probability distribution $p(u/u_{max})$ attains the maximum uniformity possible and becomes stationary or invariant under constraints. Eqs. (8) and (9) in turn show that the invariance of $p(u/u_{max})$ means the constancy of M and, therefore, ϕ. Conversely, the observed constancy of ϕ supports the invariance of $p(u/u_{max})$ and, therefore, the law of uncertainty. The constant value of ϕ represents the state of an open-channel flow system attained by maximizing uncertainty, at which the system has a propensity to exhibit u/u_{max} according to the probability distribution $p(u/u_{max})$. A propensity is a highly resilient probability, which is a statistical notion of invariance or "necessity" (Skyrms 1980).

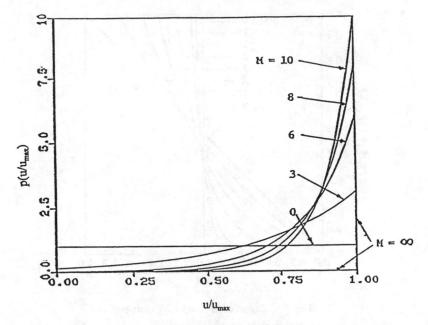

Fig. 3.Relation of M to p (u/u$_{aux}$)

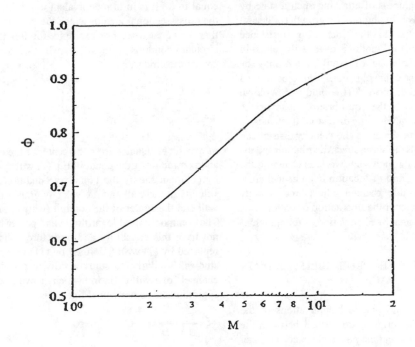

Fig. 4. Relation of $\bar{u}/u_{max}$ or ϕ to M

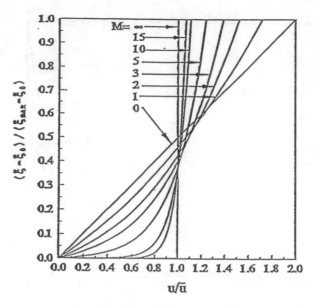

Fig. 5. Relation of M to velocity distribution

Physically, the process of attaining such a state by erodible channels is evolution or a growth (or decay) process that, due to the increasing resistance encountered by any motion, eventually ends in equilibration. The state of equilibration may be represented by the channel pattern said to be formed by the "bankfull discharge" (Leopold and Wolman 1970). To maintain the equilibration state or the constant value of ϕ when the discharge fluctuates, a non-erodible channel adjusts the velocity distribution that in turn adjusts the shear-stress distribution and the secondary flow, through such work as changing the magnitude of u_{max} and its location in a channel cross section. An erodible channel adjusts the velocity distribution to counter the fluctuation of both the discharge and channel-bed elevation or topography.

4 BRIDGES BETWEEN PROBABILITY DOMAIN AND PHYSICAL SPACE

For verifications and practical applications of the preceding results, bridges are needed between the probability domain and the physical space. This can be accomplished through identifying suitable forms of ξ in (2) or (7) as functions of the coordinates in the physical space, for various types of flows (Chiu, et al. 1993). According to (2), the value of $(\xi-\xi_0)/(\xi_{max}-\xi_0)$ is equal to the probability of flow velocity, randomly sampled in a channel cross section, being less than or

equal to u. It is in turn equivalent to the fraction of the cross-sectional area in which the velocity is less than or equal to u. For example, for full flows in circular conduits (pipes) in which isovels are concentric circles, ξ is

$$\xi = \frac{\pi r_0^2 - \pi r^2}{\pi r_0^2} = 1 - \left(\frac{r}{r_0}\right)^2 \qquad (11)$$

in which r = distance from the center of the conduit; and r_0 = radius of the conduit. Eq. (7) with ξ defined by (11) can satisfy the boundary conditions of the velocity at the wall and the velocity gradients at the wall and the center of the conduit (Chiu et al. 1993). The commonly used logarithmic and power laws do not have this capability and, therefore, should be replaced by (7) with ξ defined by (11) in pipe-flow studies. Similarly, to study a flow in a wide open channel of width B in which isovels can be approximated as horizontal lines, ξ is

$$\xi = \frac{By}{BD} = \frac{y}{D} \qquad (12)$$

in which y = vertical distance from the channel bed; and D = water depth. For ξ defined by both (11) and (12), $\xi_0 = 0$ and $\xi_{max} = 1$. For open-channel flows in which the velocity distribution is affected by the two sides of the channel cross section, an equation for ξ

20

derived by Chiu and Chiou (1986) is:

$$\xi = Y(1-Z)^{\beta_i} \exp(\beta_i Z - Y + 1) \qquad (13)$$

in which

$$Y = \frac{y + \delta_y}{D + \delta_y - h} \qquad (14)$$

$$Z = \frac{|z|}{B_i + \delta_i} \qquad (15)$$

Figs. 6(a) and 6(b) show the coordinates chosen, along with other variables and parameters that appear in (13)-(15). The y-axis is selected such that it passes through the point of maximum velocity; D = water depth at the y-axis; B_i for i equal to either 1 or 2 = transverse distance on the water surface between the y-axis and either the left or right side of a channel cross section; z = coordinate in the transverse direction; y = coordinate in the vertical direction; and h = a parameter. If h ≤ 0, ξ increases monotonically from the channel bed to the water surface as shown by Fig. 6(a). However, if h > 0, the magnitude of h is the actual depth of the maximum velocity below the water surface as shown by Fig. 6(b) ; therefore, ξ increases with y only from the channel bed to the point of maximum velocity where $\xi = \xi_{max} = 1$ and, then, decreases toward the water surface. δ_y, δ_i and β_i are parameters that vary with the shape of the zero-velocity isovel (i.e., the channel cross-section). Both δ_y and δ_i are approximately zero for fairly rectangular channels, and increase as the cross-sectional shape deviates from the rectangular shape. The η curves shown in Figs. 6(a) and 6(b) are the orthogonal trajectories of the ξ curves.

The velocity distribution along the y-axis that passes through the point of maximum velocity can be represented by (7) with ξ represented by

$$\xi = \frac{y}{D - h} \exp\left(1 - \frac{y}{D - h}\right) \qquad (16)$$

which can be obtained by making z = 0 and $\delta_y = 0$ in (13). The capability of (7), with ξ represented by (11) and (16), to represent velocity distributions has been verified by experimental data (Chiu 1989, Chiu and Murray 1992, Chiu et al. 1993).

5 EXPLANATION BY HYDRODYNAMIC ANALYSIS

It is known that the Darcy-Weisbach equation is equivalent to

$$\frac{\bar{u}}{u_*} = \sqrt{\frac{8}{f}} \qquad (17)$$

in which f is the friction factor; and u_* is the shear velocity defined as $(\tau_0/\rho)^{1/2}$ where τ_0 is the shear stress at the channel bed and ρ is the fluid density. If the channel is wide,

$$\tau_0 = \rho g D s_f = \rho \epsilon_0 \frac{du}{dy}\Big|_{y=0} \qquad (18)$$

in which g = gravitational acceleration; s_f = energy slope; and ϵ_0 = momentum transfer coefficient at the channel bed, and equal to the kinematic viscosity v of the fluid if the flow is laminar. With the velocity gradient obtained by (7) and ξ = y/D, the boundary condition as represented by (18) can be expressed as

$$F(M) = \frac{f N_R}{8} \left(\frac{\epsilon_0}{v}\right)^{-1} \qquad (19)$$

in which N_R = Reynolds number defined as $\bar{u}D/v$ where water depth D approximates the hydraulic radius; and

$$F(M) = \frac{e^M - 1}{M \phi} \qquad (20)$$

For flows in circular pipes, an equation similar to (19) can be derived (Chiu et at. 1993) in which D in N_R is the diameter of the pipe that is four times the hydraulic radius, and, therefore, the denominator of (19) increases from 8 to 32. Although (19) is applicable only to wide channels, it gives a glimpse of the factors affecting φ. It shows that φ or M represents the combined effect of N_R, f and ϵ_0/v. If φ or M is to remain constant when the discharge and, therefore, N_R fluctuate, the flow adjusts f and ϵ_0/v to keep the right side of (19) constant. If the flow is laminar so that $\epsilon_0/v = 1$, it is known from experiments that f decreases when N_R increases such that the product of f and N_R remains constant (Chow 1959). This also supports the constancy of φ.

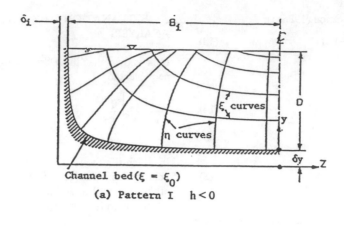

(a) Pattern I $h < 0$

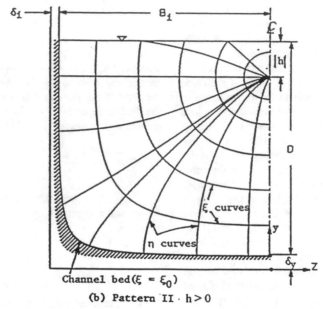

(b) Pattern II · $h > 0$

Fig. 6. Velocity distribution patterns and the network of ξ and η curves

6. UTILITIES OF CONSTANT ϕ

The law of the constant ratio of the mean and maximum velocities is useful in various ways to help explain, predict and measure the variables describing open-channel flows. The following sections illustrate its potential applications.

6.1 Development of an efficient method to measure discharge

Since a channel section has a constant value of ϕ, the cross-sectional mean velocity $\bar{u}$ can be determined from u_{max}, and vice versa. An efficient method to measure discharge in open-channel flows is, therefore, to measure u_{max} that occurs on the y-axis shown in Figs. 6(a) and 6(b). The location of y-axis is stable but may shift without causing serious errors in discharge estimation. The velocity measurement on a single vertical such as y-axis can be accomplished automatically by using a modern acoustic device. Therefore, the method enables measuring the discharge in highly unsteady flows for which reliable discharge data are extremely scarce or nonexistent. Such a system of automatic measurements of velocity

22

and discharge can be coupled with real-time flow forecasting.

6.2 Simplification of determining energy and momentum coefficients

With the probability density function p(u), the cross-sectional mean values of u, u^2 and u^3 can be obtained without integrating them over the cross-sectional area in the physical plane. The cross-sectional mean of u has been expressed by (6); the mean values of u^2 and u^3 can be expressed and obtained by

$$\int_0^{u_{max}} u^2 p(u)du = \overline{u^2} = \beta \overline{u}^2 \qquad (21)$$

$$\int_0^{u_{max}} u^3 p(u)du = \overline{u^3} = \alpha \overline{u}^3 \qquad (22)$$

in which α and β are the energy and momentum coefficients, respectively. These coefficients are the important measures of the rates of transport of energy and momentum through a channel cross section, and can be expressed as the following functions of M (Chiu, 1991):

$$\alpha = \frac{e^M (M^3 - 3M^2 + 6M - 6) + 6}{(e^M - 1)^{-2} [e^M (M-1) + 1]^3} \qquad (23)$$

$$\beta = \frac{(e^M - 1)[e^M (M^2 - 2M + 2) - 2]}{[e^M (M-1) + 1]^2} \qquad (24)$$

Therefore, α and β are also determined by ϕ. The constant ϕ makes α and β to be also constant at a channel section. This is an attractive feature of the probabilistic treatment of velocity distribution, especially when the channel cross section has an irregular and complex geometrical shape such that the channel boundaries cannot be clearly specified. In open-channel hydraulics or river hydraulics, there are no other simple methods to determine them. Due to the difficulty of determining α and β, it is a widespread practice to assume them to be unity. This is equivalent to assuming the velocity distributions to be uniform. Fig. 7 shows that α can be much greater than unity when M is less than 6, or ϕ less than 0.83, which commonly occurs in open-channel flows. Therefore, the assumption of unity should be avoided in hydraulic computations involving the one-dimensional energy equation.

A smaller value of ϕ or M at a channel section means a larger value of α and, therefore, a greater rate of transport or expenditure of kinetic energy at a given value of $\bar{u}$. Figs. 8 (a) and 8(b) are based on the experimental data from a bend of a trapezoidal flume (Yen 1965), and show that when a flow enters a bend from a straight reach, the value of $\bar{u}$ does not change appreciably but u_{max} increases conspicuously to reduce ϕ and, therefore, to increase α. Also, u_{max} dips below

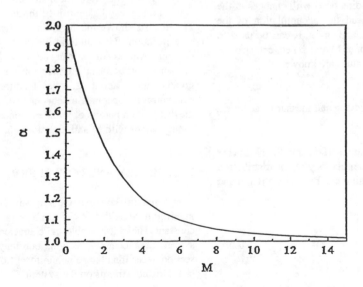

Fig. 7. Relation of M to energy coefficient

the water surface. These changes in the magnitude and the location of u_{max} are accompanied by the corresponding changes in shear stress distribution and the pattern of secondary currents (Chiu and Lin 1983), and explain the nature's way of increasing the rate of energy expenditure to overcome the increased resistance encountered by the flow in the bend. It should be noted that the ϕ value of any section in the bend remains constant. A smaller ϕ value means a smaller value of M and, therefore, a more uniform probability distribution $p(u/u_{max})$ and greater uncertainty that, according to the law of uncertainty, means a channel section is more natural or stable. The ϕ values in meandering parts of a channel are smaller than those in straight reaches; therefore, the sections in bends should be more stable than those in straight reaches of a channel. The variation of ϕ along nonuniform flows can be explained by that of α and, therefore, by the following equation (Chiu and Murray 1992):

$$\alpha = \frac{2gA^2}{Q^2} \int_0^x (s_w - s_f) dx + \alpha_0 \left(\frac{A}{A_0} \right)^2 \qquad (25)$$

Eq. (25) is equivalent to the one-dimensional energy equation, and gives α at a channel section at x along the flow, in which A = cross=sectional area of the channel section at x; s_w and s_f = water-surface slope and energy slope, respectively; and α_0 and A_0 = α and A at x = 0. The equation shows that in nonuniform flows α, M and ϕ may vary along the flow direction (the longitudinal or x direction) with changes in the cross sectional area and the accumulation of the difference between s_w and s_f. It can be used to interpolate the values of α, M and ϕ between channel sections where these values are known.

6.3 Applications in studying and measuring sediment transport

If u_{max} occurs at the water surface and the channel is wide so that ξ may be defined as y/D, the distribution of sediment concentration can be derived (Chiu and Rich 1992) as

$$\frac{C}{C_0} = \left[\frac{1 - \dfrac{y}{D}}{1 + (e^M - 1) \dfrac{y}{D}} \right]^{\lambda'} \qquad (26)$$

in which C_0 is the sediment concentration at y=0, and

$$\lambda' = \frac{v_s \bar{u} (1 - e^{-M})}{\beta u_*^2 M \phi} = \lambda G(M) \qquad (27)$$

in which

$$\lambda = \frac{v_s \bar{u}}{\beta u_*^2} \qquad (28)$$

and

$$G(M) = \frac{1 - e^{-M}}{M \phi} \qquad (29)$$

which is a dimensionless function of M. Under a given set of flow and channel conditions, an increase in λ means an increase in the sediment size. Eq. (26) appears similar to the well-known Rouse equation. However, the sediment concentration given by (26) realistically stays finite at the channel bed, while that by the Rouse equation goes to infinity since it is based on the von Karman's logarithmic velocity distribution equation. Therefore, (26) can give realistic values of mean sediment concentration (Chiu and Rich 1992).

Fig. 9 is based on (26) and shows the distributions of sediment concentration at $\lambda = 1/4$, for M between 1 and 8. It shows that, under a given set of conditions of flow, channel and sediment as represented by λ, the sediment concentration is less uniformly distributed in a channel section of smaller M than that in a section of greater M. Therefore, the distribution of sediment concentration can be controlled through the ϕ value. Fig. 9 also shows the location of mean sediment concentration. This and similar plots obtained for different values of λ can be used to facilitate measurements of mean sediment concentration. For M greater than about 2, the location of mean concentration becomes independent of M, although the distribution pattern of sediment concentration still clearly varies with M and, therefore, ϕ.

7. SUMMARY AND CONCLUSION

It is a natural law that the ratio of the mean and maximum velocities of flow in a channel section is constant. The ratio is stable and characterizes the flow system at a channel section, but may vary from section to section along a nonuniform flow due to the changing constraints on the system.

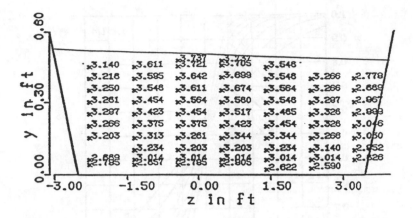

(a) Velocity distribution in a section (facing downstream) upstream from a bend

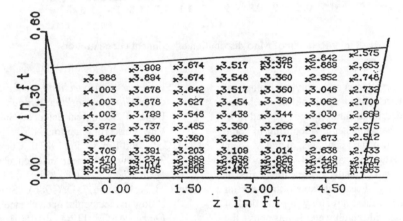

(b) Velocity distribution in a section (facing downstream) in the bend

Fig. 8. Velocity data showing the changes in maximum velocity in a bend.

The ϕ value is connected to the velocity distribution, energy and momentum coefficients, sediment concentration and other related variables such as the shear-stress distribution and secondary currents. A channel section with a smaller value of ϕ or M has a larger value of energy coefficient and, therefore, a greater rate of transport or expenditure of kinetic-energy at a given mean velocity, and less uniform distributions of velocity and sediment concentration. The law of constant ϕ can be used to gain an insight into unobservable interactions among the system variables and to explain the observed phenomena in various channels.

The constant ratio ϕ is equivalent to the constant

values of the energy and momentum coefficients α and β, the parameter M of the probability distribution $p(u/u_{max})$ and the velocity distribution in a channel section. However, among them, only ϕ is directly measurable, and has an easily understandable physical meaning. The ϕ value can be used as an index of the distributions of velocity, shear stress and sediment concentration as well as the secondary currents and any other variables in a channel section. The constant value of ϕ maintained by a channel section can also simplify and increase the efficiency of the measurements of mean velocity (and therefore discharge) and mean sediment concentration.

The experimental data used herein were gathered from

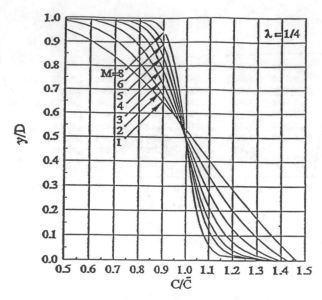

Fig. 9 Relation of M to distribution of sediment concentration

flows in laboratory flumes and alluvial channels of established patterns within the bankfull discharge. Efforts are needed in the future to study φ and M in flows above the bankfull stage. According to some available data from natural channels, φ and M seem to increase in certain orderly ways with the water level, which shows a gradually-decreasing degree of stability or maturity of channels above the bankfull stage. To conduct a thorough study, a great deal of data must be collected at high water levels above the bankfull stage. It will be difficult since the high water levels usually occur during highly-unsteady flow periods. However, acoustic devices for velocity measurements that recently became available should help automate and enable the data collection needed. This type of effort is essential to advancing the frontier of knowledge in open-channel hydraulics.

ACKNOWLEDGMENT

This paper is based on work partially supported by the U. S. National Science Foundation under Grant No. EAR-9415862, and by the Taiwan Provincial Water Conservation Bureau.

REFERENCES

Bridge, J. S., and Jarvis, J. (1985). *Flow and sediment transport data, River South Esk, Glen Glova, Scotland*. Unpublished Rep., Dept. of Geological Sciences, State Univ. of New York, Binghamton, N.Y.

Chiu, C.-L., and Lin, G.-F. (1983). "Computation of 3-D flow and shear in open channels." *J. Hydr. Engrg.*, ASCE, 109(11), 1424-1440.

Chiu, C.-L., and Chiou, J.-D. (1986). "Structure of 3-D flow in rectangular open channels." *J. Hydr. Engrg.*, ASCE, 112(11), 1050-1068.

Chiu, C.-L. (1987). "Entropy and Probability Concepts in Hydraulics." *J. Hydr. Engrg.*, ASCE, 113(5), 583-600.

Chiu, C.-L. (1988). "Entropy and 2-D velocity distribution in open channels." *J. Hydr. Engrg.*, ASCE, 114(7), 738-756.

Chiu, C.-L. (1989). "Velocity distribution in open-channel flow." *J. Hydr. Engrg.*, ASCE, 115 (5), 576-594.

Chiu, C.-L. (1991). "Application of entropy concept in open-channel flow study." *J. Hydr. Engrg.*, ASCE, 117(5), 615-628.

Chiu, C.-L., and Murray, D. W. (1992)."Variation of velocity distribution along nonuniform open channel flow." *J. Hydr. Engrg.*, ASCE, Vol. 118 (7),989-1001.

Chiu, C.-L., and Rich, C. A. (1992). "Entropy-based velocity distribution model in study of distribution of suspended-sediment concentration." *Proc., ASCE National Conf.*

on Hyd. Engrg., Baltimore, Aug., 1992.

Chiu, C.-L., Lin, G.-F., and Lu, J.-M. (1993). "Application of probability and entropy concepts in pipe-flow study." *J. Hydr. Engrg.*, ASCE, 119(6), 742-756.

Chiu, C.-L., and Said, C. A. A. (1995). "Maximum and mean velocities and entropy in open-channel flow." *J. Hydr. Engrg.*, ASCE, 121(1), 26-35.

Chow, V. T. (1959). *Open-channel hydraulics*. McGraw-Hill Book Co., New York, NY.

Culbertson, J. K., Scott, C. H., and Bennett, J. P. (1971)."Summary of alluvial-channel data from Rio Grande Conveyance Channel, New Mexico, 1965-69." Open File Report, U. S. Geological Survey, Water Resources Division, Aug., 1971.

Goldman, S. (1953). *Information theory.* Prentice-Hall, Inc., New York, NY.

Guo, Z.-r. (1990). Personal communication. Southeast China Envir. Sci. Inst., Yuancun, Guangzhou, China.

Guy, H. P., Simons, D. B., and Richardson, E. V. (1966). "Summary of alluvial channel data from flume experiments, 1956-61." Geological Survey Professional Paper 462-I, U.S. Govt. Printing Ofc., Washington, D. C.

Hospers, J. (1980). "What is explanation" and "Law." *Introductory readings in the philosophy of science*, edited by E. D. Klemke et al. Promentheus Books, Bufffalo, NY.

Leopold, L., and Wolman, G. (1970). "River channel patterns."Chapter 7, *Rivers and river terraces*, ed. by G. H. Dury, Praeger Publishers, New York, 197-237.

Rao, C. R. (1965). *Linear statistical inference and its applications*. John Wiley & Sons, Inc., New York, NY.

Shannon, C. E. (1948). "A mathematical theory of communication." *The Bell System Tech. J.*, 27 (Oct.), 623-656.

Skyrms, B. (1980). *Causal necessity.* Yale University Press, New Haven.

Tsujimoto, T., Okada,T., and Motohashi, K.(1993). "Experimental study on velocity distribution of flow with free surface in a circular conduit." *KHL Progress Rept.*, Hydraulic Lab., Kanazawa Univ., Kanazawa, Japan, December, 61-76.

Yen, B. C. (1965). "Characteristics of subcritical flow in a meandering channel." Tech. Rept., Institute of Hydraulic Research, Univ. Of Iowa, Iowa City, Iowa.

Stochastic Hydraulics'96, Tickle, Goulter, Xu, Wasimi & Bouchart (eds) © 1996 Balkema, Rotterdam. ISBN 90 5410 817 7

Uncertainty analysis in water resources engineering

Yeou-Koung Tung
Wyoming Water Resources Center & Statistics Department, University of Wyoming, Laramie, Wyo., USA

ABSTRACT: The existence of various uncertainties in water resources engineering analysis, planning, and design has long been recognized. It is only in the recent decade or so has the uncertainty related issues been addressed on a more formal and rigorous fashion. Our use of probabilistic methods to various water resources engineering problems has already gone far beyond the conventional frequency anlaysis-type approaches. In this paper, an overview of the uncertainties in water resources engineering and the development of their treatments are offered. Furthermore, several obervations and thoughts with regard to uncertainty analysis are presented with the hope to foster new ideas in the future development of uncertainty analysis.

1 INTRODUCTION

Water resources engineering design and analysis deal with the occurrence of water in various parts of a hydrosystem and its effects on environmental, ecological, and socio-economical settings. Due to the extreme complex nature of the physical, chemical, biological, and socio-economical processess involved, tremendous efforts have been devoted by researchers attempting to have a better understanding of the processes. One beneficial product of these research efforts is the development of a model which describes the interrelationships and interactions of the components involved in the processes. Herein, the term 'model' is used in a very loose manner, referring to any structural or nonstructural ways of transforming inputs to produce some forms of outputs. In water resources engineering, most models are structural which take the forms of mathematical equations, tables, graphs, or computer programs. The model is a useful tool for engineers to assess the system performance under various scenarios based on which efficient designs or effective management schemes can be formulated.

1.1 *Uncertainties in water resources engineering*

Despite numerous research efforts made to further our understanding of various processess in hydrosystems,

there is still much more that are beyond our firm grasp. Therefore, uncertainties exist due to our lack of perfect knowledge concerning the phenomena and processes involved in problem definition and resolution.

Yen and Ang (1971) classified uncertainties into two types: objective uncertainties associated with any random process or deducible from statistical samples, and subjective uncertainties for which no quantitative factual information is available. Yevjevich (1972) distinguishes the basic risk due to inherent randomness of the process and uncertainty due to the various other sources. Burges and Lettenmaier (1975) categorize two types of uncertainty associated with mathematical modeling: Type I error results from the use of an inadequate model with correct parameter values; Type II error assumes the use of perfect model with parameters subject to uncertainty.

In general, uncertainty due to inherent randomness of physical processes cannot be eliminated. On the other hand, uncertainties such as those associated with lack of complete knowledge about the process, models, parameters, data, and etc. could be reduced through research, data collection, and careful manufacturing.

In water resources engineering, uncertainties involved can be divided into four basic categories: hydrologic, hydraulic, structural, and economic (Mays & Tung, 1992). More specifically, in water resources engineering analyses and designs uncertainties could arise from the various sources (Yen et al., 1986) including natural uncertainties, model uncertainties, parameter uncertainties, data uncertainties, and operational uncertainties.

Natural uncertainty is associated with the inherent randomness of natural processes such as the occurrence of precipitation and flood events. The occurrence of hydrological events often display variations in time and in space. Their occurrences and intensities could not be predicted precisely in advance.

Types of model used in water resources engineering vary in wide spectrum, ranging from simple empirical equations to sophisticated computer simulation models. It should be recognized that a model is only an abstraction of the reality, which generally involves certain degrees of simplifications and idealizations. Model uncertainty reflects the inability of the model or design technique to represent precisely the system's true physical behavior. Ang and Tang (1984) classify model prediction error into systematic error and random error. Systematic error may arise from factors not accounted for in the model. Hence, model prediction tends to produce biased results which consistently over-predicts or under-predicts the outcomes of the process. Random error is associated with range of possible error primarily due to sampling error. In general, both systematic error and random error co-exist.

Parameter uncertainties resulting from the inability to quantify accurately the model inputs and parameters. All hydrologic and hydraulic equations involve several physical or empirical parameters that cannot be quantified accurately. Parameter uncertainty could also be caused by change in operational conditions of hydraulic structures, inherent variability of inputs and parameters in time and in space, and lack of a sufficient amount of data.

Data uncertainties include (1) measurement errors, (2) inconsistency and non-homogeneity of data, (3) data handling and transcription errors, and (4) inadequate representation of data sample due to time and space limitations.

Operational uncertainties include those associated with construction, manufacture, deterioration, maintenance, and human. Construction and manufacturing tolerances may result in difference between the 'nominal' and actual values. The magnitude of this type of uncertainty is largely dependent on the workmanship and quality control during the construction and manufacturing. Progressive deterioration due to lack of proper maintenance could result in changes in resistance coefficients and structural capacity reduction. This would add additional uncertainty to the design and evaluation of hydraulic structure performance.

1.2 Measures of uncertainty

Several expressions have been used to describe the degree of uncertainty of a parameter, a function, a model, or a system. In general, the uncertainty associated with the latter three is a result of combined effect of the uncertainties of the contributing parameters.

The most complete and ideal description of uncertainty is the probability density function (PDF) of the quantity subject to uncertainty. However, in most practical problems such a probability function cannot be derived or found precisely.

Another measure of the uncertainty of a quantity is to express it in terms of a reliability domain such as the confidence interval. A confidence interval is a numerical interval that would capture the quantity subject to uncertainty with a specified probabilistic confidence. The methods to evaluate the confidence interval of a parameter on the basis of data samples are well-known and can be found in standard statistics and probability text books (e.g., Ang and Tang, 1975). Nevertheless, the use of confidence intervals has a few drawbacks: (1) the parameter population may not be normally distributed as assumed in the conventional procedures and this problem is particularly important when the sample size is small; (2) no means is available to directly combine the confidence intervals of individual contributing random components to give the overall confidence interval of the system.

A useful alternative to quantify the level of uncertainty is to use the statistical moments associated with a quantity subject to uncertainty. In particular, the variance and standard deviation which measure the dispersion of a random variable are commonly used. Sometimes, the coefficient of variation, which is the ratio of standard deviation to the mean, offers a normalized measure useful and convenient for comparison and for combining uncertainties of different variables.

1.3 Implications of uncertainty and purposes of uncertainty analysis

In water resources engineering design and analysis, the decisions on the layout, capacity, and operation of the system largely depend on the system response under some anticipated design conditions. When some of the components in a hydrosystem are subject to uncertainty, the system responses under the design conditions cannot be assessed with certainty. The presence of uncertainties makes the conventional deterministic design practice inappropriate for its inability to account for possible variation of system responses. In fact, the issues involved in the design and analysis of hydrosystems under uncertainty are multi-dimensional. An engineer has to consider various criteria including, but not limited to, cost of the system, probability of failure, and consequence of failure so that a proper design can be made for the system.

The main objective of uncertainty analysis is to assess the statistical properties of system outputs as a function of stochastic parameters. In water resources engineering design and modeling, the design quantity and system output are functions of several system parameters not all of which can be quantified with absolute accuracy. The task of uncertainty analysis is to determine the uncertainty features of the system outputs as a function of uncertainties in the system model itself and in the stochastic parameters involved. It provides a formal and systematic framework to quantify the uncertainty associated with the system output. Furthermore, it offers the designer useful insights regarding the contribution of each stochastic parameter to the overall uncertainty of the system outputs. Such knowledge is essential to identify the 'important' parameters to which more attention should be given to have a better assessment of their values and, accordingly, to reduce the overall uncertainty of the system outputs.

1.4 An overview of uncertainty analysis techniques

Several techniques can be applied to conduct uncertainty analysis of water resources engineering problems. Each technique has different levels of mathematical complexity and data requirements. Broadly speaking, those techniques can be classified into two categories: analytical approaches and approximated approaches. The selection of an appropriate technique to be used depends on the nature of the problem at hand including availability of information, resources constraints, model complexity, and type and accuracy of results desired.

Section 2 describes several useful analytical techniques for uncertainty analysis including derived distribution technique and integral transform techniques. It describes some well-known integral transforms including the Fourier, Laplace, and exponential transforms. Also, a less known transform technique, called the Mellin transform, is described. Although these analytical techniques are rather restrictive in practical applications due to the complexity of most models, they are, however, powerful tools for deriving complete information

about a stochastic process, including its distribution, in some situations.

In Section 3, several approximation techniques are described. These techniques are particularly useful for problems involving complex functions which cannot be analytically dealt with. They are primarily developed to estimate the statistical moments about the underlying random processes. One such method is the first-order variance estimation (FOVE) method (Benjamin & Cornell, 1970; Ang & Tang, 1975). Two other techniques are the probabilistic point estimation methods and variations of Monte Carlo simulation procedures. In describing these uncertainty analysis techniques, the advantages and disadvantages of each technique will also be discussed.

2 ANALYTICAL TECHNIQUES

In this section, several analytical methods are briefly discussed that would allow an analytical derivation of exact PDF and/or statistical moments of a random variable as a function of several stochastic variables. In theory, the techniques described in this chapter are straightforward. However, the success of implementing these procedures largely depends on the functional relation, forms of the PDFs involved, and analyst's mathematical skill.

2.1 Derived distribution technique

This derived distribution method is also known as the transformation of variables technique. Example applications of this technique can be found in modeling the distribution of pollutant decay process (Patry & Kennedy, 1989) and rainfall-runoff modeling (Patry et al., 1989).

Suppose that a random variable W is related to another random variable X as W=g(X). Furthermore, the PDF and/or CDF of X are known. The CDF of W can be obtained as

$$H_W(w) = P[\ W \le w\ ] = F_X[g^{-1}(w)] \qquad (2.1)$$

where $g^{-1}(w)$ represents the inverse function of g. The PDF of W, $h_W(w)$, then can be obtained by taking the first derivative of $H_W(w)$ with respect to w.

In the case that the functional relation between X and W is either strictly increasing or strictly decreasing, the PDF of W can be derived directly from the PDF of X as

$$h_W(w) = f_X(x)\ |dx/dw| \qquad (2.2)$$

in which $|dx/dw|$ is the Jacobian.

For a multivariate case in which N random variables W_1, W_2, ..., W_N are related to N other random variables X_1, X_2, ..., X_N through a system of N equations as

$$W_i = g_i (X_1, X_2, ..., X_N), i = 1, 2, ..., N$$

When the functions $g_1(.)$, $g_2(.)$, ..., $g_N(.)$ satisfy a monotonic relationship, the joint PDF of random variables Ws can be directly obtained from

$$h(w_1, w_2,..., w_N) = f(x_1, x_2,..., x_N) |\mathbf{J}| \qquad (2.3)$$

where $f(x_1, x_2,..., x_N)$ and $h(w_1, w_2,..., w_N)$ are the joint PDFs of Xs and Ws, respectively; and $|\mathbf{J}|$ is the absolute value of determinant of the $N{\times}N$ Jacobian matrix.

2.2 Fourier transform technique

The Fourier transform of a function, $f(x)$, is defined, for all real values of s, as

$$\mathscr{F}_X(s) = \int_{-\infty}^{\infty} e^{isx} f(x) \, dx \qquad (2.4)$$

where $i= \sqrt{-1}$. If the function $f(x)$ is the PDF of a random variable X, the resulting Fourier transform $\mathscr{F}(s)$ is called the characteristic function. Hence, the characteristic function of a random variable X having a PDF $f(x)$ is

$$\mathscr{F}_X(s) = E\left[e^{isX}\right] \qquad (2.5)$$

The characteristic function of a random variable always exists for all values of the argument s. Furthermore, the characteristic function for a random variable under consideration is unique. In other words, two distribution functions are identical if and only if the corresponding characteristic functions are identical (Patel et al., 1976). Therefore, given a characteristic function of a random variable, its PDF can be uniquely determined through the inverse Fourier transform.

Using the characteristic function, the r-th order moment about the origin of the random variable X can be obtained as

$$E\left[X^r\right] = \mu_r' = \frac{1}{i^r}\left[\frac{d^r \mathscr{F}_X(s)}{d s^r}\right]_{s=0} \qquad (2.6)$$

Fourier transform is particularly useful when random variables are independent and linearly related. In such cases, the convolution property of the Fourier transform can be applied to derive the characteristic function of the resulting random variable. For example, consider that $W = X_1 + X_2 + ... + X_N$ and all Xs are independent random variables with known PDF, $f_i(x)$, i=1,2, ..., N. The characteristic function of W then can be obtained as

$$\mathscr{F}_W(s) = \mathscr{F}_1(s) \, \mathscr{F}_2(s) \cdots \mathscr{F}_N(s) \qquad (2.7)$$

which is the product of the characteristic functions of each individual random variable. The resulting characteristic function for W can be used in Eq. (2.6) to obtain the statistical moments of any order for the random variable W. Furthermore, the inverse transform of $\mathscr{F}_W(s)$ can be made to derive the PDF of W, if it is analytically possible.

2.3 Laplace and exponential transform techniques

The Laplace and exponential transforms of a function, $f(x)$, are defined, respectively, as

$$\mathscr{L}_X(s) = \int_{0}^{\infty} e^{sx} f(x) \, dx \qquad (2.8)$$

and

$$\mathscr{E}_X(s) = \int_{-\infty}^{\infty} e^{sx} f(x) \, dx \qquad (2.9)$$

As can be seen, the Laplace and exponential transforms are practically identical except that the former is applicable to functions with a non-negative argument. In case that $f(x)$ is the PDF of a random variable, the Laplace and exponential transforms defined in Eqs.(2.8) and (2.9) can, respectively, be stated as

$$\mathscr{L}_X(s) = E\left[e^{sX}\right], \text{ for } x \geq 0;$$
$$\mathscr{E}_X(s) = E\left[e^{sX}\right], \text{ for } -\infty < x < \infty \qquad (2.10)$$

The transformed PDFs given by Eq. (2.10) are called the moment generating functions. Similar to the characteristic function, statistical moments of a random variable X can be derived from its moment generating function as

$$E[X^r] = \mu_r' = \left[\frac{d^r \mathscr{E}_X(s)}{d\,s^r} \right]_{s=0} \qquad (2.11)$$

There are two deficiencies associated with the moment generating functions: (1) the moment generating function of a random variable may not always exist for all distribution functions and all values of s, and (2) the correspondence between a PDF and moment generating function may not necessarily be unique. Springer (1978, pp.80-81) stated three theorems describing the conditions under which unique correspondence between a PDF and moment generating function exists. However, these conditions are generally satisfied in most situations.

Due to the fact that Fourier and exponential transforms attempt to find the expected values of exponentiation of random variables, they are frequently used in uncertainty analysis of a model that involves exponentiation of stochastic parameters. Examples of their applications can be found in probabilistic cash flow analysis (Tufekci & Young, 1987) and probabilistic modeling of pollutant decay (Patry & Kennedy, 1989).

2.4 *Mellin transform technique*

When the functional relation W=g(X) satisfies the following two conditions, the exact moments for W of any order can be derived analytically by the Mellin transform without extensive simulation or using any approximation methods. The two conditions are:
(1)   The function g(X) has a product form as

$$W = g(X) = a_0 \prod_{i=1}^{N} X_i^{a_i} \qquad (2.12)$$

where a_i are constants;
(2)   The stochastic variables are independent and non-negative.

The Mellin transform is particularly attractive in uncertainty analysis of hydrologic and hydraulic problems because many models and the involved parameters satisfy the above two conditions (Tung, 1989, 1990). In general, the non-negativity condition of Xs is not strictly required by the Mellin transform; but it would require some mathematical manipulations to find the Mellin transform of a function involving random variables that can take negative values (Epstein, 1948; Springer, 1978).

The Mellin transform of a function f(x), where x is positive, is defined as (Giffin, 1975; Springer, 1978)

$$M_X(s) = M[f(x)] = \int_0^\infty x^{s-1} f(x)\, dx , \qquad (2.13)$$

where $M_X(s)$ is the Mellin transform of the function f(x). Like Fourier and Laplace transforms, a one-to-one correspondence between $M_X(s)$ and f(x) exists. When f(x) is a PDF, one can immediately recognize that the relationship between the Mellin transform of a PDF and the statistical moments about the origin as

$$\mu'_{s-1} = E(X^{s-1}) = M_X(s), \text{ for } s=1,2,.... \qquad (2.14)$$

Similar to the convolution property of the Laplace and Fourier transforms, the Mellin transform of h(w) can be obtained as

$$M_W(s)=M[h(w)]=M[f(x)*g(y)]=M_X(s)\times M_Y(s) \quad (2.15)$$

in which '*' is the convolution operator. From Eq.(2.15), the Mellin transform of the convolution of the PDFs associated with two independent random variables in a product form is simply equal to the product of the Mellin transforms of two individual PDFs. Equation (2.15) can be extended to the general case involving more than two independent random variables.

From the convolution property of the Mellin transform and its relationship to statistical moments, one can immediately see the advantage of the Mellin transform as a tool for obtaining the moments of a random variable which is related to other random variables in a multiplicative fashion. In addition to the convolution property, which is of primary importance, the Mellin transform has several useful operational properties (Bateman, 1954; Park, 1987). Futhermore, the Mellin transforms of some commonly used PDFs are obtainable from Epstein (1948) and Park (1987). Applications of the Mellin transfrom can be found in economic benefit-cost analysis (Parks, 1987), and hydrology and hydraulics (Tung, 1989, 1990).

Although the Mellin transform is useful for uncertainty analysis under the conditions stated previously, it possesses one drawback which should be pointed out: under some combinations of distribution and functional form, the resulting transform may not be defined for all values of s. This could occur especially when quotients or variables with negative exponents are involved. For example, if the random variable W is related to the inverse of X, i.e. W=1/X, and X has a uniform distribution in (0,1), then $M_W(s)=M_X(2-s) = 1/(2-s)$. In this case, the expected value of W, E(W), which can be calculated by $M_W(s=2)$ does not exist because $M_W(s=2) = 1/0$ which is not defined. Under such circumstances, other

33

transforms such as the Laplace or Fourier transform could be used.

2.5 *Estimations of probabilities and quantiles using moments*

Although it is generally difficult to analytically derive the PDF from the results of the integral transform techniques described above and the approximation techniques in Section 3, it is, however, rather straightforward to obtain or estimate the statistical moments of the random variable one is interested in. Based on the computed statistical moments, one is able to estimate the distribution and quantile of the random variable. The following three subsections describes such approaches: one is based on the asymptotic expansion about the normal distribution for calculating the values of CDF and quantile, and the other is based on the maximum entropy concept.

2.5.1 Edgeworth asymptotic expansion of CDF - In terms of the statistical moments and standard normal distribution, the general Edgeworth asymptotic expansion for the PDF and CDF of any standardized random variable, $X' = (X - \mu_X)/\sigma_X$, can be found in Abramowitz and Stegun (1970) and Kendall et al. (1987). For practical applications, considering that the first four moments are estimated, the Edgeworth asymptotic expansion of the CDF, $F_{X'}(\xi)$, is

$$F_{X'}(\xi) \approx \Phi(\xi) - \phi(\xi) \left[\left(\frac{\gamma_x}{6} \right) H_2(\xi) \right.$$
$$\left. + \left(\frac{\kappa_x - 3}{24} \right) H_3(\xi) + \left(\frac{\gamma_x^2}{72} \right) H_5(\xi) \right]$$

(2.16)

in which $\Phi(\xi)$ is the standard normal CDF and $H_r(\xi)$ is the r-th order Hermite polynomial which can be found in Abramowitz and Stegun (1970). It should be pointed out that, using finite terms in the Edgeworth series expansion, it is possible to obtain negative values for PDF and CDF in both tail portions of the distribution.

2.5.2 Fisher-Cornish asymptotic expansion of quantile - Inversely, to estimate the quantile ξ_p in which $P(X' \le \xi_p) = p$, the Fisher-Cornish asymptotic expansion (Fisher and Cornish, 1960; Kendall et al., 1987), considering the first four moments, can be expressed as

$$\xi_p \approx z_p + \left(\frac{\gamma_x}{6} \right) H_2(z_p) + \left(\frac{\kappa_x - 3}{24} \right) H_3(z_p)$$
$$- \left(\frac{\gamma_x^2}{36} \right) \left[2 H_3(z_p) + H_1(z_p) \right]$$

(2.17)

in which $z_p = \Phi^{-1}(p)$, the p-th quantile of standard normal variate. The quantile of the original scale can be easily computed as $x_p = \mu_X + \xi_p \sigma_X$. For more complete expansion series, which would require higher order moments, readers are referred to Kendall et al. (1987). As can be seen from Eqs.(2.16) and (2.17), if only the first two moments are available, the two asymptotic expansions reduce to the case of normal distribution.

2.5.3 Maximum entropy distribution - The use of entropy concept for measuring the amount of uncertainty in a statistical experiment was originated by Shannon (1948). It is based on Boltzmann's entropy from statistical physics which has been used as an indicator of disorder in a physical system. Shannon's entropy has been used in a widely variety of areas including information and communication, economics, physics, ecology, reliability, etc. Recently, the entropy concept is applied to model velocity distribution in open channel (Chiu, 1987, 1988, 1989; Chiu & Said, 1995) and in pipe flow (Chiu et al., 1993), hydrology (Armorocho & Espildora, 1973; Singh et al., 1986; Singh & Krstanovic, 1987; Singh & Rajagopal, 1987), and water quality monitoring (Kusmulyono & Goulter, 1994).

The Shannon entropy is defined, for discrete case, as

$$H(X) = - \sum_{i=1}^{n} \ln(p_i)\, p_i$$

(2.18a)

and, for continuous case, as

$$H(X) = - \int_{x_{min}}^{x_{max}} \ln[f(x)]\, f(x)\, dx$$

(2.18b)

in which p_i is the probability mass function (PMF) of the discrete random variable $X = x_i$ and $f(x)$ is the PDF of the continuous random variable X. The degree of uncertainty (or information) associated with the realization of a random variable is measured by $I(x_i) = - \ln(p_i)$, for discrete case and by $I(x) = - \ln[f(x)]$, for continuous case. As can be seen, the entropy is the expected information content associated with a

34

random variable X over its whole range, that is, H(X) = E [I(X)]. It should be pointed out that the value of entropy for discrete random variables is non-negative whereas, for continuous random variables, the entropy value could be negative. For detail discussions on the properties of entropy, readers are referred to Guiasu (1977) and Jumarie (1990).

The maximum entropy principle is proposed by Jaynes (1957) stating that of all the distributions which satisfy the constraints supplied by the known information is the one that has the largest entropy for the random variables. Consider a continuous random variable X for which some of its statistical moments are known a priori. Using the maximum entropy principle, the PDF of the random variable X can be derived by solving the following optimization problem (Cover & Thomas, 1991)

$$\text{Maximize} \quad H(X) = - \int_{x_{min}}^{x_{max}} \ln[f(x)] \, f(x) \, dx \qquad (2.19a)$$

subject to

$$\int x^j \, f(x) \, dx = \mu_j' , \; j=0,1,...,k \qquad (2.19b)$$

in which μ_j' is the j-th moment about the origin.

The above maximization problem can be solved by using the Lagrangian multiplier method. The resulting entropy-based PDF is

$$f(x) = \exp\left[-1-\lambda_0 - \sum_{j=1}^{k} \lambda_j \, x^j \right] \qquad (2.20)$$

where λ's are the Lagrangian multipliers. To obtain the entropy-based distribution as given in Eq.(2.20), the values of Lagrangian multipliers λ's must be solved. A system of (k+1) non-linear equations containing (k+1) unknown λ's can be established by substituting Eq.(2.20) into constraint Eq.(2.19b) and the results are

$$-\lambda_0 + \ln\left[\int x^r \exp\left(-\sum_{j=1}^{k} \lambda_j x^j \right) dx \right] = 1 + \ln(\mu_r') \quad (2.21)$$

for r=0,1,2,..., k. The above system of non-linear equation can be solved by using appropriate iterative numerical techniques.

3 APPROXIMATION TECHNIQUES

In the previous section, methods are described that

allow one to analytically derive the PDF and/or statistical moments of functions of random variables. However, many of those analytical methods are restrictive in practical applications because they require simple functional relationships and independence of stochastic model parameters which are most likely to be violated in real-life problems. Most of the models or design procedures used in water resources engineering are nonlinear and highly complex. This basically prohibits any attempt to derive the probability distribution of model output analytically. As a practical alternative, engineers frequently resort to methods that yield approximations to the statistical properties of uncertain model output. In this section, several methods that are useful for uncertainty analysis in hydraulic design are described. They include first-order variance estimation method and probabilistic point estimation procedures and variations of Monte Carlo simulation techniques.

3.1 First-order variance estimation (FOVE) method

This method, also called the variance propagation method (Berthouex, 1975), estimates uncertainty features associated with a model output based on the statistical properties of model's stochastic parameters. The basic idea of the method is to approximate a model involving stochastic parameters by the Taylor series expansion.

Consider that a hydraulic or hydrologic design quantity W is related to N stochastic parameters X_1, X_2, ..., X_N as

$$W = g(X) = g(X_1, X_2, ..., X_N)$$

where $X= (X_1, X_2, ..., X_N)^t$, an N-dimensional column vector of model parameters in which all Xs are subject to uncertainty, the superscript 't' represents the transpose of a matrix or vector. The first-order Taylor series expansion of the function g(X) with respect to a selected point, $X=x_0$, in the parameter space can be expressed as

$$W \approx g(x_0) + s_0^t \cdot (X-x_0) \qquad (3.1)$$

where s_0 is the column vector of sensitivity coefficients with each element representing $\partial W/\partial X_i$ evaluated at $X=x_0$. The mean and variance of W by the first-order approximation can be expressed, respectively, as

$$E[W] \approx g(x_0) + s_0^t \cdot (\mu-x_0) \qquad (3.2)$$

and

$$Var[W] \approx s_o^t \, C \, s_o \qquad\qquad (3.3)$$

in which μ and C are the vector of means and covariance matrix of stochastic parameters X, respectively.

Commonly, the FOVE method consists of taking the expansion point $x_o = \mu$ and the mean and the variance of W reduce to

$$E[W] \approx g(\mu) \qquad\qquad (3.4)$$

and

$$Var[W] \approx s^t \, C \, s \qquad\qquad (3.5)$$

in which s is an N-dimensional vector of sensitivity coefficients evaluated at $x_o = \mu$. When all stochastic parameters are independent, the variance of model output W can be approximated as

$$Var[W] \approx \sum_{i=1}^{N} s_i^2 \, \sigma_i^2 = s^t \, D \, s \qquad\qquad (3.6)$$

in which σ represents the standard deviation and $D = diag(\sigma_1^2, \sigma_2^2, ..., \sigma_N^2)$, a diagonal matrix of variances of involvedd stochastic parameters. From Eq.(3.6), the ratio, $s_i^2\sigma_i^2/Var[W]$, indicates the proportion of overall uncertainty in the model output contributed by the uncertainty associated with the stochastic parameter X_i.

In general, $E[g(X)] \neq g(\mu)$ unless $g(X)$ is a linear function of X. Improvement of the accuracy can be made by incorporating higher-order terms in the Taylor expansion. However, one immediately realizes that as the higher-order terms are included not only the mathematical complication but also the required information increase rapidly. This is especially true for estimating the variance. The method can be expanded to include the second-order term to improve estimation of the mean to account for the presence of model non-linearity and correlation between stochastic parameters.

In practice, the first two moments are used in uncertainty analysis for practical engineering design. To estimate higher-order moments of W, the method can be implemented straightforwardly only when the stochastic parameters are uncorrelated. The method does not require knowledge of the PDF of stochastic parameters which simplifies the analysis. However, this advantage is also the disadvantage of the method because it is insensitive to the distributions of stochastic parameters on the uncertainty analysis. Recently, Yen et al. (1986) provided a very comprehensive evaluation and description of the

FOVE method in uncertainty and reliability analyses.

To circumvent the disadvantages of the FOVE method while keeping the simplicity of the first-order approximation, quantification of uncertainty features of the model output can be made using the advanced first-order second-moment (AFOSM) reliability method by which the expansion point x_* in the first-order Taylor series is located on the limit-state equation defined by $W(x_*)=0$. The AFOSM method can be repeated applied to evaluate $P[W(X)<c]$ for different values of 'c' from which the CDF of the random model output $W(X)$ is defined. With the CDF known or estimated, the pertinent statistical properties of $W(X)$ can be derived. Note that the distributional properties of stochastic parameters can be considered in the AFOSM method. Applications of the AFOSM method for assessing the uncertainty associated with hydrologic model outputs have been made by Melching et al. (1990) and Melching (1992).

The FOVE method is simple and straightforward. The computational effort associated with the method largely depends on the ways how the sensitivity coefficients are calculated. For simple analytical functions the computation of derivatives are trivial tasks. However, for functions that are complex and/or implicit in the form of computer programs, or charts/ figures, the task of computing the derivatives could become cumbersome or difficult. In such cases probabilistic point estimation techniques can be viable alternatives.

There are numerous applications of the FOVE method in the literature. Some example applications of the method are found in open channel flow (Huang, 1986; Cesare, 1991), groundwater flow (Dettinger & Wilson, 1981), water quality modeling (Chadderton et al., 1982), benefit-cost analysis (Dandy, 1986; Wood & Gulliver, 1991; Tung, 1992), gravel pit migration analysis (Yeh & Tung, 1993), storm sewer design (Tang & Yen, 1972; Yen et al., 1976), culverts (Mays, 1979), and bridges (Tung & Mays, 1982).

3.2 *Probabilistic point estimation (PE) methods*

Unlike the FOVE methods, probabilistic PE methods quantify the model uncertainty by performing model evaluations without computing the model sensitivity. The methods generally is simpler and more flexible especially when a model is either complex or non-analytical in the forms of tables, figure, or computer programs. Several types of PE methods have been developed and applied to uncertainty analysis and each has its advantages and disadvantages. It has been shown by Karmeshu and Lara-Rosano (1987) that the

FOVE method is a special case of the probabilistic PE methods when the uncertainty of stochastic parameters are small.

3.2.1 Rosenblueth's PE method

It was firstly developed for handling stochastic variables that are symmetric (Rosenblueth, 1975) and the method is later extended to treat non-symmetric random variables (Rosenblueth, 1981). The basic idea of Rosenblueth's PE method is to approximate the original PDF or PMF of the random variable X by assuming that the entire probability mass of X is concentrated at two points x_- and x_+. The four unknowns, that is, the locations of x_- and x_+ and the corresponding probability masses p_- and p_+, are determined in such a manner that the first three moments of the original random variable X are preserved. The solutions for x_-, x_+, p_-, and p_+ are

$$x_- = \mu - z_- \sigma \qquad (3.7a)$$

$$x_+ = \mu + z_+ \sigma. \qquad (3.7b)$$

$$p_+ = \frac{z_-}{z_+ + z_-} \qquad (3.7c)$$

$$p_- = 1 - p_+ \qquad (3.7d)$$

where

$$z_+ = \frac{\gamma}{2} + \sqrt{1 + \left(\frac{\gamma}{2}\right)^2} \qquad (3.7e)$$

$$z_- = z_+ - \gamma \qquad (3.7f)$$

with γ being the skew coefficient of the stochastic variable X.

When the distribution of random variable X is symmetric ($\gamma = 0$), $z_- = z_+ = 1$ and $p_- = p_+ = 0.5$. This implies that, with a symmetric random variable, the two points are located at one standard deviation on either side of the mean with equal probability mass assigned at the two locations.

For problems involving N stochastic variables, the two points for each variable are computed, according to Eqs. (3.7a-f), and permutated producing a total of 2^N possible points of evaluation in the parameter space. The r-th moment of $W=g(X)=g(X_1,X_2,..., X_N)$ about the origin can be approximated as

$$E(W^r) \approx \sum p_{(\delta1, \delta2,..., \delta N)} W^r_{(\delta1, \delta2,..., \delta N)} \qquad (3.8)$$

in which subscript δ_i is a sign indicator that can only be + or - for representing the stochastic variable X_i having the value of $x_{i+}=\mu_i+z_{i+}\sigma_i$ or $x_{i-}=\mu_i-z_{i-}\sigma_i$, respectively; $p_{(\delta1, \delta2,..., \delta N)}$ can be determined as

$$P_{(\delta_1, \delta_2,..., \delta_N)} = \prod_{i=1}^{N} p_{i,\delta_i} + \sum_{i=1}^{N-1} \left(\sum_{j=i+1}^{N} \delta_i \delta_j a_{ij} \right) \qquad (3.9)$$

in which

$$a_{ij} = \frac{\rho_{ij} / 2^N}{\sqrt{\prod_{i=1}^{N} \left[1 + \left(\frac{\gamma_i}{2}\right)^2 \right]}} \qquad (3.10)$$

where ρ_{ij} is the correlation coefficient between stochastic variables X_i and X_j. The number of terms in the summation of Eq. (3.8) is 2^N which corresponds to the total number of possible combinations of + and - for all N stochastic variables. Recently, Panchalingam and Harr (1994) proposed a modified procedure to handle correlated and skewed random variables.

For each term of the summation in Eq. (3.8), the model has to be evaluated once at the corresponding point in the parameter space. This indicates a potential drawback of Rosenblueth's PE method as it is applied to practical problems. When N is small, the method is practical for performing uncertainty analysis. However, for moderate or large N, the number of required function evaluations of g(X) could be too numerous to implement practically, even on the computer. To circumvent this shortcoming, Harr (1989) developed an alternative PE method that reduces the 2^N function evaluations required by Rosenblueth's method down to 2N.

Example applications of Rosenblueth's PE method for uncertainty analysis can be found in groundwater flow model (Nguyen & Chowdhury, 1985; Emery, 1990), dissolved oxygen deficit model (Emery, 1990), and bridge pier scouring model (Chang et al., 1994).

3.2.2 Harr's PE method

The method (Harr, 1989) utilizes the first two moments (that is, the mean and variance) of the random variables involved and their correlations. Skew coefficients of the random variables are ignored by the method. Hence, the method is appropriate for treating random variables that are symmetric. For problems involving only a single random variable, Harr's PE method is identical to Rosenblueth's method with zero skew coefficient. The theoretical basis of Harr's PE method is built on the orthogonal transformation using eigenvalue-eigenvector decomposition which maps correlated random variables from their original space to a new domain in which they become uncorrelated. Hence, the analysis is greatly simplified.

Consider N multivariate random variables $X=(X_1, X_2, ..., X_N)^t$ having a mean vector $\mu=(\mu_1, \mu_2, ..., \mu_N)^t$ and covariance matrix $C(X)$. The vector of correlated standardized random variable $X'=D^{-1/2}(X-\mu)$ with $X'_i=(X_i-\mu_i)/\sigma_i$ for i=1,2,..., N would have mean vector of zero, 0, and the covariance matrix equal to the correlation matrix $R(X)$ where D is an N × N diagonal matrix of variances of the stochastic variables.

Orthogonal transformation can be made by several ways (Young & Gregory, 1973; Golub & Van Loan, 1989). One frequently used approach is the eigenvalue-eigenvector factorization by which the correlation matrix $R(X)$ is decomposed as

$$R(X) = C(X') = V \Lambda V^t \qquad (3.11)$$

where V is an N×N eigenvector matrix consisting of N eigenvectors as $V=(v_1, v_2, ..., v_N)$ with v_i being the i-th eigenvector and $\Lambda=diag(\lambda_1, \lambda_2, ..., \lambda_N)$ is a diagonal eigenvalues matrix. Using the eigenvector matrix V, the following transformation can be made

$$U = V^t X' \qquad (3.12)$$

The resulting transformed stochastic variables U have the mean 0 and covariance matrix Λ indicating that U are uncorrelated because their covariance matrix $C(U)$ is a diagonal matrix Λ. Hence, each new stochastic variable U_i has the standard deviation equal to $\sqrt{\lambda_i}$, for all i=1,2,..., N.

For multivariate problems involving N stochastic variables, Harr's PE method selects its points for function evaluation which are located at the intersections of the N eigenvector axes and a hypersphere of radius $\sqrt{N}$ centered at the origin in the transformed U-space. The corresponding points in the original parameter space can be obtained as

$$x_{i\pm} = \mu \pm \sqrt{N} \, D^{1/2} v_i \, , \, i=1, 2, ..., N \qquad (3.13)$$

in which $x_{i\pm}$ is the coordinates of the N stochastic variables in the parameter space on the i-th eigenvector v_i.

Based on the 2N points determined by Eq. (3.13), the function values at each of the 2N points can be computed. From that the r-th moment of the function $W=g(X)$ can be calculated according to

$$E[W^r] = \mu'_r(W) = \frac{\sum_{i=1}^{N} \lambda_i \bar{w}_i^r}{N} \, , \quad r = 1, 2, \cdots \qquad (3.14)$$

where

$$\bar{w}_i^r = \frac{w_{i+}^r + w_{i-}^r}{2} = \frac{g^r(x_{i+}) + g^r(x_{i-})}{2} , \qquad (3.15)$$
$$i = 1, 2, \cdots, N ; \, r = 1, 2, \cdots$$

Harr's PE method has been applied to uncertainty analysis of a gravel pit migration model (Yeh & Tung, 1993), regional equations for unit hydrograph parameters (Yeh et al., 1995), groundwater flow models (Emery, 1990; Guymon, 1994), and parameter estimation of a distributed hydrodynamic model (Zhao, 1994).

3.2.3 Li's PE method - Recently, Li (1992) proposed a computationally practical PE method that allows incorporation of the first four moments of correlated stochastic parameters. For a univariate model, it can be shown that a two-point representation is sufficient for preserving the first three moments of a stochastic model parameter. In fact, Rosenblueth's solutions are a special case of Li's solution when $\kappa=\gamma^2+1$ which is the boundary for all feasible probability distributions.

For a multivariate model of the following form

$$W(X) = w_o + \sum_{i=1}^{N} a_i (X_i - \mu_i) + \sum_{i=1}^{N} b_i (X_i - \mu_i)^2$$
$$+ \sum_{i=1}^{N} c_i (X_i - \mu_i)^3 + \sum_{i=1}^{N} d_i (X_i - \mu_i)^4 \qquad (3.16)$$
$$+ \sum_{i=1}^{N-1} \sum_{j=i+1}^{N} e_{ij} (X_i - \mu_i)(X_j - \mu_j)$$

Li's method requires $(N^2+3N+2)/2$ evaluations of the model $W(X)$. When the polynomial order is four or less as shown in Eq.(3.16), Li's method would yield the exact expected value of $W(X)$. However, for higher order moments, its computation of $E(W^m)$ is no longer exact. It should also be pointed out that Eq.(3.16) is not a complete fourth-order Taylor series expansion. For a general model, moments of model output computed by Li's method are only approximation. The method has been applied to quantify the uncertainty in backwater profile computation (Zoppou & Li, 1993).

3.2.4 Some improvements for probabilistic PE method - Among the three probabilistic PE algorithms described above, Harr's method is the most attractive from the computational viewpoint. However, the method cannot incorporate additional distributional information of the stochastic parameters other than the first two moments. Such distributional information could have important implications on the results of uncertainty analysis. To incorporate the information about the marginal distributions of involved stochastic model parameters, Chang et al. (1995) first modified the original Harr's algorithm by selecting the points for model evaluation from the intersections of a hypersphere with radius $\sqrt{N}$ centered at the origin and the axes in the standardized eigen-space, rather than the U-space. Then, a transformation between non-normal parameter space and a multivariate standard normal space (Der Kiureghian & Liu, 1985; Liu & Der Kiureghian, 1986) is incorporated into the modified Harr's method. The resulting method preserves the computational efficiency of Harr's PE method while extends it capability to handle multivariate non-normal stochastic parameters. Detailed descriptions of the extended Harr's method can be found in Chang (1994).

3.3   *Monte-Carlo simulation*

Simulation is a process of replicating the real world based on a set of assumptions and conceived models of reality (Ang and Tang, 1984). Because the purpose of a simulation model is to duplicate reality, it is a useful tool for evaluating the effect of different designs on system performance. The Monte Carlo procedure is a numerical simulation to reproduce random variables preserving the specified distributional properties.

Algorithms to generate random variates from a specified distribution can be categorized into three types: CDF-inverse method, acceptance-rejection method, and variable transformation method. The CDF-inverse method is based on the fact that the CDF, $F_X(x)$, of a random variable X is a non-decreasing function with respect to its value and $0 \leq F_X(x) \leq 1$. For the great majority of continuous probability distributions applied in water resources engineering analysis, $F_X(x)$ is a strictly increasing function of x. Hence, there exists a unique relationship between $F_X(x)$ and u with u being a standard uniform random variable defined over the unit interval [0,1].

Several books have been written for generating univariate random numbers (Rubinstein, 1981; Dagpunar, 1988; Law & Kelton, 1991). A number of computer programs are available in the public domain. The challenge of Monte Carlo simulation lies in generating multivariate random variates. Compared with univariate random variate generators, algorithms for multivariate random variates is much more restricted to a few joint distributions such as multivariate normal, multivariate lognormal (Parrish 1990), multivariate gamma (Ronning 1977), and several others (Johnson 1987). If the multivariate random variables involved are correlated with a mixture of marginal distributions, the joint PDF is difficult to formulate. Similar to Li & Hammond (1975), Chang et al. (1994) proposed a practical multivariate Monte Carlo simulation procedure for problems involving mixtures of non-normal random variables by utilizing the semi-empirical formulas derived by Der Kiureghian and Liu (1985) and Liu and Der Kiureghian (1986).

In uncertainty analysis, stochastic model parameters are generated according to their probabilistic laws which are used, in turn, to compute the corresponding model outputs. With sufficient number of repetitions, the plausible variation of the model output as affected by the stochastic model parameters can be realized, allowing for assessment of statistical features of model outputs. The implementation of this brutal-force type of simulation is straightforward and can be very computationally intensive. Furthermore, because the Monte Carlo simulation is a sampling procedure, the results obtained inevitably involve sampling errors which decrease as the sample size increases. Increasing sample size, for achieving higher precision, generally means an increase in computer time for generating random variates and data processing. Therefore, the issue lies on using the minimum possible computation to gain the maximum possible accuracy for the quantity under estimation. For this, various variance reduction techniques have been developed which include the importance sampling technique, antithetic-variate technique, correlated sampling technique (Rubinstein, 1981; Ang & Tang, 1984), stratified sampling technique (Cochran, 1966), and Latin-hypercubic sampling technique (KcKay 1988).

Applications of Monte Carlo simulation in water resources engineering are abundant. Sample examples can be found in groundwater (Freeze, 1975; Nguyen & Chowdhury, 1985), benefit-cost analysis (Mercer & Morgan, 1975), water quality model (Brutsaert, 1975), pier-scouring prediction (Johnson, 1992; Chang et al., 1994), and open channel (Huang, 1986; Mizumura & Ouazar, 1992).

3.4   Resampling techniques

Note that Monte Carlo simulations are conducted under the condition that the probability distribution and the associated population parameters are known for the random variables involved in the system. The observed data are not directly utilized in the simulation.   In many statistical estimation problems, the statistics of interest are often expressed as function of random observations, that is,

$$\hat{\Theta} = \Theta(X_1, X_2, ..., X_M)   \qquad (3.17)$$

The statistics $\hat{\Theta}$ could be the estimator of an unknown population parameter of interest.   For example, consider that random observations Xs are annual maximum floods.   The statistics $\hat{\Theta}$ could be the distribution of the floods, its statistical properties such as the mean, standard deviation, skew coefficient, the magnitude of 100-year event, a probability of exceeding the capacity of a hydraulic structure, etc.

Note that the statistics $\hat{\Theta}$ is a function of random variables, itself is also a random variable having a PDF, mean, and standard deviation as any random variable. Unlike the Monte Carlo simulation approach, resampling techniques reproduce random data exclusively on the basis of observed ones. The two resampling techniques that are frequently used are jackknife method and bootstrap method.

The bootstrap technique was first proposed by Efron (1979a,b) to deal with variance estimation of sample statistics based on observations.   The technique intends to be a more general and versatile procedure for sampling distribution problems without having to rely heavily on the normality condition on which classical statistical inferences are based. In fact, it is not uncommon to observe non-normal data in water resources engineering problems. Although the bootstrap technique is computationally intensive - a price to pay to break away from dependence on normality theory - such concerns will be gradually diminished as the calculating power of computers increases (Diaconis & Efron, 1983).

Since the introduction of the bootstrap resampling technique by Efron, it has rapidly caught the attention of statisticians and those who apply statistics in their research works. The bootstrap technique and its variations have been applied to various statistical problems such as bias estimation, regression analysis, time series analysis, and others. An excellent overall review and summary of bootstrap techniques, variations, and other resampling procedures are given by Efron (1982) and Efron and Tibshirani (1993).   The bootstrap resampling

technique has been applied to assess the confidence intervals of the optimal risk-based hydraulic design parameters (Tung, 1993) and the uncertainty features of derived unit hydrographs (Zhao et al., 1993).

4.   SOME OBSERVATIONS AND THOUGHTS

It is not the intention here to list all methods that are used or applicable to uncertainty analysis of water resources engineering problems.   More specifically, the methods described above are limited to those that are either frequently used or potential useful to assess the uncertainty of a model used in solving water resources engineering problems. Even for that, the list is by no means complete which can be easily proven in this Symposium showing there are many other methods used or developed to deal with uncertainty related problems.   Within the scope of this presentation, I would like to share with you a few of my observations and thoughts.

4.1   Uncertainty   analysis   versus   sensitivity analysis

It is a common practice to conduct sensitivity analysis when some of the inputs or parameters are subject to uncertainty. The primary objective of the exercise is to assess how much the model output would be effected due to change in values of inputs or parameters. Information such as this is important in model calibration and design of data collection program.

However, information obtained about the sensitivity effect of an input provides only a partial information with regard to its contribution to the overall uncertainty of the model output. Referring to Eqs.(3.3) or (3.6), note that Var[W] depends not only on the sensitivity of model output to the parameters at the point of evaluation but also on the uncertainty of the parameter.   A stochastic parameter with large sensitivity may not be a major contributor to the overall model uncertainty if its corresponding uncertainty is small. On the other hand, a stochastic parameter with small sensitivity could be an important player if it had large variance.

Sensitivity analysis and uncertainty analysis are closely related but different in concept. The former is used to analyze the internal mathematical responses of model outputs as affected by changes in model inputs/parameters, whereas the latter is used to analyze the stochasticity of the model through these internal   relationships.   Computationally,   the determination of input/parameter sensitivity, in fact, constitutes the major part of uncertainty analysis because the main computation burden lies on the evaluation of sensitivity coefficients. If the sensitivity analysis is performed, what is left to complete an uncertainty analysis is generally trivial.

4.2   *Local versus global analysis*

The frequently used measure of sensitivity is the sensitivity coefficient, s_o, defined in Eq.(3.1) or variations of it. The sensitivity measure based on first-order partial derivative indicates the change in model output due to one unit change in the input at the neighborhood of the selected point. This, in effect, approximate a model by a linear function at the selected point, x_o. Since the model output uncertainty is dependent, in part, on the sensitivity coefficients, consequently, any conclusion made on the important inputs/parameters with respect to sensitivity or uncertainty would depend on the point selected in the parameter space and the corresponding sensitivity coefficients. If the model responses vary drastically at different points in the parameter space, the conclusion drawn on the basis of local model behavior could be of limited usefulness and misleading.

Sensitivity and uncertainty analyses of model behaviors could be performed from a local or global viewpoint. Local sensitivity analysis is conerned with the model output variability due to differential changes in parameters at a selected point in the parameter space. The selected point often is the nomial values of the parameters. The sensitivity vector, s_o, or its variations are measures of local sensitivity. For a model whose sensitivity feature varies from one region of the parameter space to another, the local sensitivity measures at a selected point would not shed much light in understanding the behavior of the model over the entire domain of the parameter space. The validity of this argument is also true if one attempts to use local measures to address the global uncertainty feature of a model output. Although local measures can provide, in principle, a more detailed description of model behaviors, the use of local measures, in practice, is often restricted by the computational effort required for their evaluations. This is especially true if the model is complex and the number of parameters is large.

Analysis from a global viewpoint, on the other hand, focus on the general model behaviors over the defined parameter space. Global sensitivity analysis is concerned with the pattern of change in model output due to change in parameters over the parameter space.

In general, global sensitivity analysis can be accomplished with less computation. The lack of resolution could limit its usefulness, especially when the effect of an input on an output is drastically different in various parts of the parameter space (Yeh & Tung, 1993). However, if such global analysis is performed properly, the results could be much more valuable and useful than those from the local analysis. An example of local and global sensitivity and uncertainty analyses can be found in Chang et al. (1993) in that they used the Latin Hypercubic sampling technique to conduct the global analysis.

4.3   *A unified view of approximation methods*

The approximation methods for uncertainty analysis discussed herein can all be viewed as some form of sampling procedures by which values of stochastic parameters from the parameter space are sampled. The sampled parameter values are then used to compute the corresponding model output values based on which the uncertainty features of model output are assessed. Although the FOVE method or its variations estimate the statistical moments of model output based on a particular point in the parameter space x_o, such as the means, and the parameter sensitivity in the neighborhood of x_o, the computation of the sensitivity coefficients, when they are calculated by numerical differencing procedures, requires evaluations of model responses at various points in the vicinity of x_o. Therefore, the sampling nature of the FOVE method is implicit and the domain of concentration is in the vicinity of x_o.

As for the Monte Carlo simulation and various sampling schemes such as stratified sampling, Latin hypercubic sampling and importance sampling, the domain of interest is generally extended to cover the entire parameter space. Nevertheless, these methods have the flexibility allowing analysts to focus the model performance on a sub-domain of the parameter space.

Compared with the FOVE methods and the various sampling techniques, the domain of parameter space covered by the points for model evaluations from a probabilistic PE method lies somewhere between the two. Therefore, the scope of view in uncertainty analysis by the probabilistic PE methods is certainly larger than the local, but is not yet wide enough to be global.

4.4   Identification of important parameters

In addition to the quantification of uncertainty features of the model output, one important purpose of the uncertainty analysis is to gain some insight for identifying important parameters for their contributions to the overall uncertainty of the model output. For most of the methods described herein, there is no clear way to directly address the issue.

When the stochastic variables are uncorrelated, the assessment on the contributiony by each individual parameter to overall uncertaint can be made rather easily as shown below Eq. (3.6) by the FOVE method. Depending on the domain of the parameter space examined, the issue of whether the parameter is globally or locally important would surface. For a model whose parameter sensitivity varies drastically from one domain to the other, the rating of the relative importance of the parameters could be different by different views.

Relative importance of involved stochastic parameters can also be identified indirectly based on the correlation analysis of model outputs and parameters from the numerical simulation. The common meausres to evaluate the relative importance of the stochastic parameters are partial correlation coefficient, partial rank correlation coefficient, or standardized regression coefficient. Examples of this type can be found elsewhere (McKay, 1988; Chang et al., 1994; Saltelli et al., 1993).

In general, different measures may not always lead to a consistent identification of important parameters. The performance of different measures depends on the characteristics of the model and statistical properties of the stochastic parameters. The situation becomes even more difficult when stochastic parameters are correlated.

4.5   Uncertainty in model selection

The methods for uncertainty analysis describe herein are only applicable to the case when the model for the design and analysis is chosen. In a real-life design and analysis, one may face the problem of selecting an appropriate model to perform the task. This is especially true because many of the processes involved in water resources engineering are not entirely known and there could be many existing theories or hypotheses floating around which are all plausible to a certain degree. Another commonly encountered problem is the how to choose the appropriate model(s) to fit a set of data obtained from the field measurements or laboratory expriments. It is well known that the model that best fits the existing data may not be the most accurate when it comes to predict the future events. This brings up the issue of model validation and the ways how to conduct it.

In engineering practice, the widely used approach is to select the most appropriate model among several candidate models and, then, treat the selected model as if it is the correct one. Draper (1995) recently address this issue and made the following remarks:

"In general, this approach fails to assess and propagate structural uncertainty fully and may lead to miscalibrated uncertainty assessment about y (the model output) given x (the model inputs). When miscalibration occurs it will often result in understantement of inferential or predictive uncertainty about y, leading to inaccurate scientific summaries and overconfident decisions that do not incorporate sufficient hedging against uncertainty."

REFERENCES

Abramowitz, M., & Stegun, I.A. (eds.) 1972. *Handbook of Mathematical Functions With Formulas, Graphs, and Mathematical Tables*, 9th ed., Dover Publications, New York.

Amorocho, J. & Espildora, B. 1973. Entropy in the assessment of uncertainty of hydrologic systems and models, *Water Resources Research*, 9(6): 1515-1522.

Ang, A.H.S. & Tang, W.H. 1975. *Probability Concepts in Engineering Planning and Design, Vol. 1: Basic Principles*, John Wiley and Sons, Inc, New York.

Ang, A.H.S. & Tang, W.H. 1984. *Probability Concepts in Engineering Planning and Design: Decision, Risk and Reliability, Vol. 2: Decision, Risk, and Reliability*, John Wiley & Sons, Inc., New York.

Bateman, H. 1954. *Tables of Integral Transforms*, Vol. I, McGraw-Hill Book Co., New York.

Benjamin, J.R. & Cornell, C.A. 1970. *Probability, Statistics, and Decisions for Civil Engineers*, McGraw-Hill, Inc., New York, NY.

Berthouex, P.M. 1975. Modeling concepts considering process performance, variability, and uncertainty. In T.M. Keinath & M.P. Wanielista (eds.) *Mathematical Modeling for Water Pollution Control Processes*. 405-439, Ann Arbor Science, Ann Arbor, MI.

Brutsaert, W.F. 1975. Water quality modeling by monte carlo simulation. *Water Resources Bulletin*, 11: 115-130.

Burges, S.J. & Lettenmaier, D.P. 1975. Probabilistic methods in stream quality management. *Water Resources Bulletin*, 11: 115-130.

Cesare, M.A. 1991. First-order analysis of open channel flow. *J. of Hydraulic Engineering*, ASCE, 117(2): 242-247.

Chadderton, R.A., Miller, A.C. & McDonnell, A.J. 1982. Uncertainty analysis of dissolved oxygen model. *J. of Environmental Engineering*, ASCE, 108(5): 1003-1012.

Chang, C.H. 1994. Incorporating non-normal marginal distributions in uncertainty analysis of hydrosystems. *Ph.D. Dissertation*, Civil Enginerring Department, National Chiao-Tung University, Hsinchu, Taiwan.

Chang, C.H., Tung, Y.K. & Yang, J.C. 1994. Monte carlo simulation for correlated variables with marginal distributions. *J. of Hydraulic Engr.*, ASCE, 120(2): 313-331.

Chang, C.H., Tung, Y.K. & Yang, J.C. 1995. Evaluation of probabilistic point estimate methods. *Applied Mathematical Modelling*, 19(2): 95-105.

Chang, C.H., Yang, J.C. & Tung, Y.K. 1993. Sensitivity and uncertainty analyses of a sediment transport model: a global approach. *J. of Stochastic Hydrology and Hydraulics*, 7(4): 299-314.

Chiu, C.L. 1987. Entropy and probability concepts in hydraulics. *J. of Hydraulic Engineering*, ASCE, 113(5):583-600.

Chiu, C.L. 1988. Entropy and 2D velocity distribution in open channels. *J. of Hydraulic Engineering*, ASCE, 114(7):738-756.

Chiu, C.L. 1989. Velocity distribution of open channel flow. *Journal of Hydraulic Engineering*, ASCE, 115(5):576-594.

Chiu, C.L., Lin, G.F. & Lu, J.M. 1993. Application of probability and entropy concepts in pipe-flow study. J. of Hydraulic Engineering, ASCE, 119(6): 742-756.

Chiu, C.L. & Said, C.A.A. 1995. Maximum and mean velocities and entropy in open-channel flow. J. of Hydraulic Engineering, ASCE, 121(1): 26-35.

Cochran, W. 1966. *Sampling Techniques*, 2nd ed., Wiley, New York, N.Y.

Cover, T.M. & Thomas, J.A. 1991. *Elements of Information Theory*. John Wiley and Sons, Inc. New York, N.Y.

Dagpunar, J. 1988. *Principles of Random Variates Generation*, Oxford University Press, New York, N.Y.

Dandy, G.C. 1986. An approximate method for the analysis of uncertainty in benefit-cost ratios. *Water Resources Research*, 21(3): 267-271.

Der Kiureghian, A. & Liu, P.L. 1985. Structural reliability under incomplete probability information. *J. of Engineering Mechanics*, ASCE, 112(1): 85-104.

Dettinger, M.D. & Wilson, J.L. 1981. First order analysis of uncertainty in numerical models of groundwater flow. Part 1. Mathematical development. *Water Resources Research*, 17(1): 149-161.

Diaconis, P. & Efron, B. 1983. Computer-intensive methods in statistics. *Scientific America*, 116-131, May.

Draper, D. 1995. Assessment and propagation of model uncertainty. *J. Royal Statistical Society, Series B*, 57(1): 45-70.

Efron, B. 1979a. Bootstrap methods: another look at the jackknife. *The Annals of Statistics*, 3:1189-1242.

Efron, B. 1979b. Computers and theory of statistics: thinking the unthinkable. *SIAM Reviews*, 21:460-480.

Efron, B. 1982. *The Jackknife, the Bootstrap, and Other Resampling Plans*. CBMS 38, SIAM-NSF.

Efron, B. & Tibshirani, R.J. 1993. *An Introduction to the Bootstrap*, Chapmann & Hall, New York, 1993.

Emery, J. 1990. Comparison of reliability approximation techniques. *M.S. Thesis*, Department of Statistics, University of Wyoming, Laramie, Wyoming.

Epstein, B. 1948. Some appliations of the mellin transform in statistics. *Annals of Mathematical Statistics*, 19: 370-379.

Fisher, R.A. and Cornish, E.A. 1960. The percentile points of distributions having known cumulants. *Technometrics*, 2(2): 209-225.

Freeze, R.A. 1975. A stochastic conceptual analysis of one-dimensional groundwater flow in nonuniform homogenous media. *Water Resources Research*, AGU, 11(5): 725-741.

Giffin, W.C. 1975. *Transform Techniques for Probability Modeling*, Academic Press.

Golub, G.H. & Van Loan, C.F. 1989. *Matrix Computations*, 2nd ed., The Jonhs Hopkins University Press, Baltimore, PA.

Guiasu, S. 1977. *Information Theory with Applications*. McGraw-Hill Book Company, New York, N.Y.

Guymon, G.L. 1994. *Unsaturated Zone Hydrology*, Prentice-Hall, Englewood Cliffs, New Jersey.

Harr, M.E. 1989. Probabilistic estimates for multivariate analyses. *Applied Mathematical Modelling*, 13: 313-318.

Huang, K.Z. 1986. Reliability analysis of hydraulic design of open channe. In B.C. Yen (ed.), *Stochastic and Risk Analysis in Hydraulic Engineering*, 59-65, Water Resources Publications, Littleton, CO.

Jaynes, E.T. 1957. Information theory and statistical mechanics. *Physics Review*, 106: 620-630; 108: 171-182.

Johnson, M.E. 1987. *Multivariate Statistical Simulation*. John Wiley & Sons, New York, N.Y.

Johnson, P.A. 1992. Reliability-based pier scour engineering. *J. of Hydraulic Engineering*, ASCE. 118(10): 1344-1358.

Jumarie, G. 1990. *Relative Information: Theories and Applications*. Springer-Verlag, New York, N.Y.

Karmeshu & Lara-Rosano, F. 1987. Modelling data uncertainty in growth forecasts. *Applied Mathematical Modelling*, 11: 62-68.

Kendall, M., Stuart, A. & Ord, J. K. 1987. *Kendall's Advanced Theory of Statistics, Vol. 1: Distribution Theory*, 5th Edition, Oxford University Press, New York, 1987.

Kusmulyono, A. & Goulter, I. 1994. Entropy principles in the prediction of water quality values at discontinued monitoring stations. *J. of Stochastic Hydrology & Hydraulics*, 8(4): 301-317.

Law, A.M. & Kelton, W.D. 1991. *Simulation Modeling and Analysis*, McGraw-Hill Book Company, New York, NY.

Li, K.S. 1992. Point estimate method for calculating statistical moments. *J. of Engineering Mechanics*, ASCE, 118(7):1506-1511.

Li, S.T. & Hammond, J.L. 1975. Generation of psuedorandom numbers with specified univariate distributions and covariance matrix. *IEEE Transcation on Systems, Man, and Cybernatics*, 557-561, Sept.

Liu, P.L. & Der Kiureghian, A. 1986. Multivariate distribution models with prescribed marginals and covariances. *Probabilistic Engineering Mechanics*, 1(2): 105-112.

Mays, L.W. 1979. Optimal design of culverts under uncertainty. *Journal of Engineering Mechanics*, ASCE, 105(5): 443-460.

Mays, L.W. & Tung, Y.K. 1992. *Hydrosystems Engineering and Management*, McGraw-Hill Book Compnay, New York, NY.

McKay, M.D. 1988. Sensitivity and uncertainty analysis using a statistical sample of input values. In Y. Ronen (ed.) *Uncertainty Analysis*, CRC press, Inc., Boca Raton, FL.

Melching, C.S. 1992. An improved first-order reliability approach for assessing uncertainties in hydrologic modeling. *J. of Hydrology*, 132: 157-177.

Melching, C.S., Yen, B.C., & Wenzel, H.G. Jr. 1990. A reliability estimation in modeling watershed runoff with uncertainties. *Water Resources Research*, 26(10): 2275-2286.

Mercer, L.J. & Morgan, W.D. 1975. Evaluation of a probability approach to uncertainty in benefit-cost analysis. *Technical Report*, Contribution No. 149, California Water Resources Center, University of California, Davis.

Mizumura, K. & Ouazar, D. 1992. Stochastic characteristics of open channel flow. In J.T.Guo & G.F. Lin (eds.) *Stochastic Hydraulics '92*, 417-424.

Nguyen, V.U. & Chowdhury, R.N. 1985. Simulation for risk analysis with correlated variables. *Geotechnique*, 35(1): 47-58.

Panchalingam, G. & Harr, M.E. 1994. Modelling of many correlated and skewed random variables. *Applied Mathematical Modelling*, 18(11): 636-640.

Park, C. S. 1987. The mellin transform in probabilistic cash flow modeling. *The Engineering Economist*, 32(2):115-134.

Parrish, R.S. 1990. Generating random deviates from multivariate Pearson distributions. *Computational Statistics & Data Analysis*, 9: 283-295.

Patel, J.K., Kapadia, C.H. & Owen, D.B. 1974. *Handbook of Statistical Distributions*, John Wiley and Sons, Inc., New York, N.Y.

Patry, G.G. & Kennedy, A. 1989. Pollutant washoff under noise-corrupted runoff conditions. *J. of Water Resources Planning and Management*, ASCE, 115(5): 646-657.

Patry, G.G., Kennedy, A. & Potter,S. 1989. Runoff modeling under noise-corrupted rainfall conditions. *Canadian J. of Civil Engineering*, 16(5): 669-677.

Ronning, G. 1977. A simple scheme for generating multivariate gamma distributions with non-negative covariane matrix. *Technometrics*, 19(2): 179-183.

Rosenblueth, E. 1975. Point estimates for probability moments. *Proceedings*, National Academy of Science, 72(10): 3812-3814.

Rosenblueth, E. 1981. Two-point estimates in probabilities. *Applied Mathematical Modelling*, 5: 329-335.

Rubinstein, R.Y. 1981. *Simulation and The Monte Carlo Method*, John Wiley and Sons, New York, N.Y.

Saltelli, A., Andres, T.H. & Homma,T. 1993. Sensitivity analysis of model output: an investigation of new techniques. *Computational Statistics & Data Analysis*, 15: 211-238.

Shannon, C.E. 1948. A mathematical theory of communication. *Bell System Technical Journal*, 27: 379-423; 623-656.

Singh, V.P., Rajagopal, A.K. & Singh, K. 1986.

Derivation of some frequency distributions using the principle of maximum entropy (POME). *Advances in Water Resources*, 9: 91-106.

Singh, V.P. & Krstanovic, P.F. 1987. A stochastic model for sediment yield using the principle of maximum entropy. *Water Resources Research*, 23(5): 781-793.

Singh, V.P. & Rajagopal, A.K. 1987. Some recent advances in the application of the principle of maximum entropy (POME) in hydrology. *Water for Future: Hydrology in Perspective*, IAHS Publication no. 164: 353-364.

Springer, M.D. 1979. *The Algebra of Random Variables*, John Wiley and Sons, Inc., New York.

Tang, W.H. & Yen, B.C. 1972. Hydrologic and hydraulic design under uncertainties. *Proceedings*, International Symposium on Uncertainties in Hydrologic and Water Resources Systems, 2:868-882, 3:1640-1641, Tucson, AZ.

Tufekci, S. & Young, D.B. 1987. Moments of the present worths of general probabilistic cash flows under random timing. *The Engineering Economist*, 32(4): 303-336.

Tung, Y.K. 1987. Uncertainty analysis of National Weather Service rainfall frequency atlas. *J. of Hydraul. Engr.*, ASCE, 113(2): 178-189.

Tung, Y.K. 1989. Uncertainty on Travel Time in Kinematic Wave Channel Routing. *Proceedings* of International Conference on Channel Flow and Catchment Runoff, University of Virginia, 767-781, Charlottesville, VA.

Tung, Y.K. 1990. Mellin transform applied to uncertainty analysis in hydrology/ hydraulics. *J. of Hydraul. Engr.*, ASCE. 116(5): 659-674.

Tung, Y.K. 1992. Investigation of probability distribution of benefit/cost ratio and net benefit. *J. of Water Resour. Plan. and Mgmt.*, ASCE, 118(2): 133-150.

Tung, Y.K. 1993. Confidence intervals of optimal risk-based hydraulic design parameters. In B.C. Yen & Y.K. Tung (eds). *Reliability and Uncertainty Analyses in Hydraulic Design*, 81-96. ASCE.

Tung, Y.K. & Mays, L.W. 1982. Optimal risk-based hydraulic design of bridges. *J. of Water Resources*

Planning and Management, ASCE, 108(2): 191-202.

Woods, J. & Gulliver, J.S. 1991. Economic and Financial Analysis. In J.S. Gulliver & R.E.A. Arndt (eds), *Hydropower Engineering Handbook*, 9.1-9.37, McGraw-Hill Inc., New York, NY.

Yeh, K.C. & Tung, Y.K. 1993. Uncertainty and sensitivity of a pit migration model. *J. of Hydraulic Engr.*, ASCE, 119(2): 262-281.

Yeh, K.C., Yang, J.C. & Tung, Y.K. 1995. Regionalization of unit hydrograph parameters: 2. Uncertainty Analysis. *J. of Stochastic Hydrology and Hydraulics*. (to appear).

Yen, B. C. & Ang, A.H.S. 1971. Risk analysis in design of hydraulic projects. In C.L. Chiu (ed.) *Stochastic Hydraulics*, 694-701, Proceedings of First Internat. Symp. on Stochastic Hydraulics, University of Pittsburgh, Pittsburgh, PA.

Yen, B.C., Wenzel, H.G., Jr., Mays, L.W. & Tang, W.H. 1976. Advanced methodologies for design of storm sewer systems. *Research Report*, No. 112. Water Resources Center, University of Illinois at Urbana-Chanpaign, IL.

Yen, B.C., Cheng, S.T. & Melching, C.S. 1986. First-order reliability analysis. In B. C. Yen (ed.) *Stochastic and Risk Analysis in Hydraulic Engineering*, 1-36, Water Resources Publications, Littleton, CO.

Yevjevich, V. 1972. *Probability and Statistics in Hydrology*, Water Resources Publications, Littleton, CO.

Young, D.M. & Gregory, R.T. 1973. *A Survey of Numerical Mathematics - Vol.2*, Dover Publication, Inc., New York, NY.

Zhao, B. 1994. Stochastic optimal control and parameter estimation for an estuary system. *Ph.D. Thesis*, Department of Civil Engineering, Arizona State University, Tempe, Arizona.

Zhao, B., Tung, Y.K., Yeh, K.C. & Yang, J.C. 1993. Uncertainty analysis of hydrologic model and its implications on reliability of hydraulic structures (II). *Report*, Agricultural Council, Executive Yuan, Taiwan, Republic of China.

Zoppou, C. & Li, K.S. 1993. New point estimate method for water resources modeling. *Journal of Hydraulic Engineering*, ASCE, 119(11): 1300-1307.

Stochastic Hydraulics'96, Tickle, Goulter, Xu, Wasimi & Bouchart (eds) © 1996 Balkema, Rotterdam. ISBN 90 5410 817 7

Determination of peak contaminated bed load profile

Hsieh Wen Shen & Jenq Tzong Shiau
University of California at Berkeley, Calif., USA

ABSTRACT : Contaminated particle movements in a river receive significant concerns. However, it's difficult to measure the peak contaminated profiles in the fields. According to our current knowledge, the step lengths of a particle are a gamma distribution, while the rest periods are exponential distributed. In this study, available laboratory data of step length and rest period are collected and analyzed first. Then a procedure is proposed to determine the distribution parameters based on the results of laboratory data analysis. A few numerical examples are also presented here to demonstrate this procedure. The results show that this procedure is useful to determine the contaminated particles distribution and its corresponding longitudinal distribution of peaks based on these laboratory measurements.

INTRODUCTION

Contaminants can attach to sediment particles and move as bed load in a stream, and the transport and dispersion of contaminants can be affected by the dispersion characteristics of sediment particles. The location of peaks and the longitudinal distribution of the contaminated particles are of great concerns. However, the peaks of contaminated particles are difficult to measure in the fields. A stochastic model of bed load transport can describe the movements of a sediment particle as an alternate sequence of rest and step period. Current studies indicate that the probability distribution of step lengths are gamma distribution, and the rest periods are exponential distribution. The scope of this study is to provide a procedure to determine the parameters of above distributions and the location of peak contaminated bed load profiles based on the hydraulic and sediment properties as determined from the laboratory data.

BACKGROUND

It is assumed that the bed load movements follow a stochastic process. Specifically, sediment particle movements can be characterized by an alternate series of step length and rest period. Experiment results (Grigg, 1970 , Yang et al. 1971) indicate that the probability density function of step length is gamma distribution with shape parameter varying between 1 and 3, and the rest period is exponential distribution.

The probability density function of step lengths and rest periods are:

$$f_X(x) = \frac{k_1^r \, e^{-k_1 x}}{\Gamma(r)} \, x^{r-1} \quad , \quad x \geq 0 \qquad (1)$$

$$f_T(t) = k_2 \, e^{-k_2 t} \quad , \quad t \geq 0 \qquad (2)$$

where X and T : random variable describing the step lengths and rest periods of a single particle; r, k_1 , and k_2 : parameters of distribution.

The mean step length and rest period, according to above distributions, are:

$$\overline{X} = \frac{r}{k_1} \quad , \quad \overline{T} = \frac{1}{k_2} \qquad (3)$$

Based on these two distributions, the function describing the longitudinal distribution of the contaminated particles at time t for a straight alluvial channel with the steady and uniform flow is (Shen et al.,1973 , Yang et al. 1971):

$$f_t(x) = k_1 \, e^{-k_1 x - k_2 t} \sum_{n=1}^{\infty} \frac{(k_1 x)^{nr-1} \, (k_2 t)^n}{\Gamma(nr) \; n!} \qquad (4)$$

The mean and variance of contaminated particles movements of above distribution are:

$$\overline{x} = \frac{rk_2 t}{k_1} \quad , \quad \sigma_x^2 = \frac{k_2 r(r+1)}{k_1^2} \qquad (5)$$

APPROACHES

The laboratory data of mean step lengths and rest periods can be expressed in term of dimensionless effective shear stress, such as

$$r \,/\, k_1 = f(\tau_* - \tau_{*c})$$
$$1 \,/\, k_2 = f(\tau_* - \tau_{*c})$$

Based on these relationships, we still can't determine the distribution and peak of contaminated particles. To determine the distribution and peak of contaminated particles, we must solve the following two equations, the mean rate of movement and spreading of contaminated particles, simultaneously.

$$\frac{d\overline{x}}{dt} = \frac{rk_2}{k_1} \qquad (6)$$

$$\frac{d\sigma_x^2}{dt} = \frac{k_2 r(r+1)}{k_1^2} \qquad (7)$$

According to Shen and Cheong (1973), the location of peaks doesn't depend on the value of r. Therefore, by assuming a value of r and solving Eq.(6) and (7) simultaneously, we can have k_1 and k_2 in term of $(\tau_* - \tau_{*c})$, such as

$$k_1 = f(\tau_* - \tau_{*c})$$

$$k_2 = f(\tau_* - \tau_{*c})$$

Substituting the given r and k_1 , k_2 into Eq.(4), we can determine the distribution and peak of contaminated particles for specific value of r. However, the relationships between r/k_1 , $1/k_2$ and $(\tau_* - \tau_{*c})$ must be established first by the laboratory measurement, which are derived in the following section.

ANALYSIS OF LABORATORY DATA

Three different set of laboratory data of step lengths and rest periods are used in this study to establish the relationship between r/k_1 , $1/k_2$ and $(\tau_* - \tau_{*c})$. These experiment conditions are described briefly below:
(1) Grigg (1970) :
    d = 0.32 ~ 0.45 mm , S = 0.00088 ~ 0.00212 ,
    h = 0.141 ~ 0.181 m
(2) Lee et al. (1994) :
    d = 1.36 ~ 2.47 mm , S = 0.002 ~ 0.023 ,
    h = 0.037 ~ 0.121 m
(3) Nino et al. (1993) :
    d = 15 ~ 31 mm , S = 0.03 ~ 0.07 ,
    h = 0.05 ~ 0.149 m

These laboratory measurements of mean step length and rest period are shown in Figure 1 and 2. Obviously, each set of data are fitted well with its regression equation which are also shown in Figure 1 and 2. The regression equation, step length and rest period versus dimensionless effective shear stress ($\tau_* - \tau_{*c}$), are:
(1) Step Length :
    (i) Grigg : $\overline{X}/d = 4837.8 \, (\tau_* - \tau_{*c})^{1.269}$
    (ii) Lee et al. : $\overline{X}/d = 196.3 \, (\tau_* - \tau_{*c})^{0.788}$
    (iii) Nino et al. : $\overline{X}/d = 12.3 \, (\tau_* - \tau_{*c})^{0.245}$
(2) Rest Period :
    (i) Grigg : $\overline{T} = 0.075 \, (\tau_* - \tau_{*c})^{-1.314}$
        where $\overline{X}$: mean step length expressed in m;
            d = particle size;
            $\overline{T}$ = mean rest period expressed in hour;
            $\tau_* = \tau_o/(\gamma_s - \gamma_w)d$ = dimensionless shear stress;
            $\tau_{*c} = \tau_{oc}/(\gamma_s - \gamma_w)d$ = dimensionless critical shear stress;

Due to the particle size used in these three experiments have significant difference, the derived step length regression equations are also different, which implies that the coefficients of step length

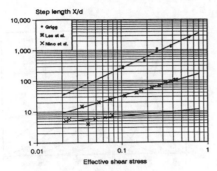

Figure 1. Mean Step Length Data and
Regression Equations.

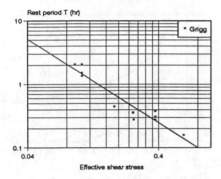

Figure 2. Mean Rest Period Data and
Regression Equations.

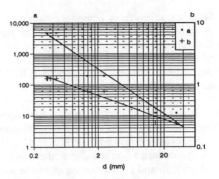

Figure 3. Relationship Between Regression
Coefficients and particle size.

regression equation might be a function of particle size. It is assumed that the step length equation can be expressed as:
$$\overline{X}/d = a \ (\tau_* - \tau_{*_c})^b$$
The regression coefficients derived above and corresponding particle size are shown in Figure 3, which show the step length regression coefficients are a function of particle size. The coefficients a and b are:

$$a = 887.0 \ d^{-1.445}$$
$$b = 0.878 \ d^{-0.391}$$

where d is particle size expressed in mm.

Based on this equation, the above three step length regression equations can be combined into a single equation which can be written as:

$$\frac{\overline{X}}{d} = 887.0 \ d^{-1.445} \ (\tau_* - \tau_{*c})^{0.878 \ d^{-0.391}} \quad (8)$$

DETERMINATION OF k_1 AND k_2

For simplicity, r is assumed as an integer and equal to 1, 2, and 3 in this study. Grigg's data are used here to demonstrate how to derive the distribution parameters k_1 and k_2, because only Grigg's data include step length and rest period measurement. From above analysis, we have

$$r \ / \ k_1 = 4837.8 \ (\tau_* - \tau_{*_c})^{1.269}$$
$$1 \ / \ k_2 = 0.075 \ (\tau_* - \tau_{*_c})^{-1.314}$$

By substituting these two equations into Eq.(6) and (7), we have

$$r \ k_2 \ / \ k_1 = 64504.0 \ d \ (\tau_* - \tau_{*_c})^{2.583}$$
$$k_2 \ r \ (r+1) \ / \ k_1^2 = 624114902.4 \ d^2 \ (\tau_* - \tau_{*_c})^{3.852}$$

Assuming that r=1, 2, 3 and solving above equations simultaneously, we can have k_1 and k_2 expressed in term of $(\tau_* - \tau_{*_c})$:

$$r = 1 \ , \ k_1 = 1 \ / \ [4837.8 \ d \ (\tau_* - \tau_{*_c})^{1.269}]$$
$$k_2 = 13.333 \ (\tau_* - \tau_{*_c})^{1.314}$$
$$r = 2 \ , \ k_1 = 1 \ / \ [3225.2 \ d \ (\tau_* - \tau_{*_c})^{1.269}]$$
$$k_2 = 10.0 \ (\tau_* - \tau_{*_c})^{1.314}$$
$$r = 3 \ , \ k_1 = 1 \ / \ [2418.9 \ d \ (\tau_* - \tau_{*_c})^{1.269}]$$
$$k_2 = 8.889 \ (\tau_* - \tau_{*_c})^{1.314}$$

For given particle size and flow condition, from above equation we can solve distribution parameters r, k_1 and k_2. Then by Eq.(4), the contaminated particle distribution and its corresponding peak can be determined.

PEAKS OF CONTAMINATED PARTICLES DISTRIBUTIONS

Three cases with different particles size and flow

condition are taken from laboratory data of Grigg, Lee et al., and Nino et al. respectively. The particle and flow condition are described below:

Case 1 : d = 0.33 mm , S = 0.00119 , h = 0.156 m
Case 2 : d = 1.36 mm , S = 0.016 , h = 0.0454 m
Case 3 : d = 15 mm , S = 0.07 , h = 0.0495 m

It is assumed that the rest period equation derived from Grigg's data can be also applied to the other data. From Eq.(8) and the rest period equation, with the procedure described above, the distribution parameters r, k_1 and k_2 of above cases can be solved:

Case 1 : r = 1 , k_1 = 6.762 , k_2 = 1.452
r = 2 , k_1 = 10.143 , k_2 = 1.089
r = 3 , k_1 = 13.524 , k_2 = 0.968
Case 2 : r = 1 , k_1 = 4.146 , k_2 = 1.452
r = 2 , k_1 = 6.219 , k_2 = 1.089
r = 3 , k_1 = 8.293 , k_2 = 0.968
Case 3 : r = 1 , k_1 = 8.233 , k_2 = 1.452
r = 2 , k_1 = 12.349 , k_2 = 1.089
r = 3 , k_1 = 16.559 , k_2 = 0.968

The distribution of above cases for t = 4, 8, 12, 16, 20 hours are shown in Figure 4, 5, and 6 (in case 3, t = 5, 10, 15, 20, 25 hours). In each case, no matter the value of r, all peaks are collapsed into a single curve. Shen and Cheong (1973) indicate that the approximate peak envelop is:

$$f_t(x) = k_1 \, e^{-2k_1 x} \, I_1(2k_1 x) \qquad (9)$$

where I_1 = modified Bessel function of order 1.

This theoretic curve is also shown in Figure 4, 5, and 6 for each case. From these figures, this curve is fitted well with the peaks. Therefore, by derivation of k_1 from laboratory measurements, we still can determine the peaks of contaminated particles without knowing the value of r

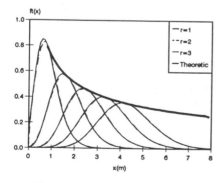

Figure 4. Peak Profile of Case 1.

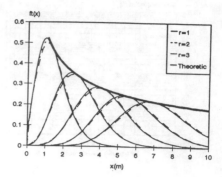

Figure 5. Peak Profile of Case 2.

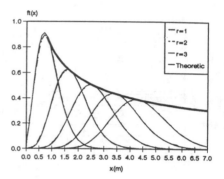

Figure 6. Peak Profile of Case 3.

CONCLUSIONS

1. From different laboratory data, the regression coefficients of step length is a function of particle size, which is described in Eq.(8).
2. The proposed procedure can determine the distribution parameters k_1 and k_2 for specific r by using the laboratory data of step lengths and rest period.
3. The peaks of contaminated particles distribution are determined by the derivation of k_1 from laboratory measurements, without knowing the value of r.

REFERENCES

Grigg, Neil S., 1970, "Motion of Single Particles in Alluvial Channels", *Journal of the Hydraulics Division, ASCE*, Vol.96, No.HY12, pp.2501-2518.
Lee, Hong-Yuan, and In-Song Hsu, 1994, "Investigation of Saltating Particle Motions", *Journal of Hydraulic Engineering, ASCE*, Vol.120, No.7, pp.831-845.

Nino, Yarko, Marcelo Garcia, and Luis Ayala, 1993, "Video Analysis of Gravel Saltation", *Hydraulic Engineering '93*, edited by Hsieh Wen Shen, S. T. Su, and Feng Wen, pp.983-988.

Shen, Hsieh Wen, and Hin Fatt Cheong, 1973, "Dispersion of Contaminated Sediment Bed Load", *Journal of the Hydraulics Division, ASCE*, Vol.99, No.HY11, pp.1947-1965.

Yang, Chih Ted, and William W. Sayre, 1971, "Stochastic Model for Sand Dispersion", *Journal of the Hydraulics Division, ASCE*, Vol.97, No.HY2, pp.265-288.

Stochastic Hydraulics'96, Tickle, Goulter, Xu, Wasimi & Bouchart (eds) © 1996 Balkema, Rotterdam. ISBN 90 5410 817 7

Review of reliability analysis of water distribution systems

Larry W. Mays
Arizona State University, Tempe, Ariz., USA

ABSTRACT: Recently, there has been considerable emphasis on the state of decay in the nation's infrastructure because of its importance to society's needs and industrial growth. Water distribution systems are one of the many kinds of infrastructure systems amenable to higher levels of serviceability. Conventional design methods often fail to recognize, analyze, and account for, systematically, the effects of the various uncertainties that prevail in the design and operation of water distribution systems. Because of the existence of design and operation uncertainties, water distribution systems have an associated risk or probability of failure and an associated reliability or probability of not failing. The definition of failure can take on several meanings for a water distribution system, ranging from not meeting required pressure heads to the mechanical failure of a pipe or component to not meeting water quality standards. For water distribution systems, there are no universally accepted definitions for risk or reliability. The purpose of this paper is to review methods for risk and reliability evaluation of water distribution systems and to comment on future research directions.

1 INTRODUCTION

The urban water distribution system is comprised three major components: pumping stations, distribution storage, and distribution piping. These components may be further divided into subcomponents that can in turn be divided into sub-subcomponents as shown in Figure 1. For example, the pumping station component consists of structural, electrical, piping, and pumping unit subcomponents. The pumping unit can be further divided into sub-subcomponents: pump, driver, controls, power transmission, and piping and valves. The exact definition of components, subcomponents, and sub-subcomponents is somewhat fluid and depends on the level of detail of the required analysis and, to a somewhat greater extent, the level of detail of available data.

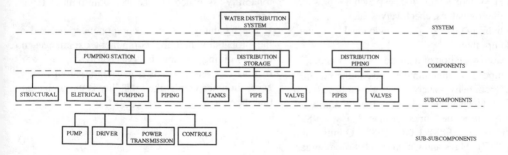

Figure 1. Hierarchical relationship of system, components, subcomponents, and sub-subcomponents for a water distribution system

The reliability of water distribution systems from a quantity view point is concerned with two types of failure: mechanical failure and hydraulic failure. Mechanical failure considers system failure due to pipe breakage, pump failure, power outages, control valve failure, etc. Hydraulic failure considers system failure due to demands and pressure heads being exceeded that could be the result of changes in demand and pressure head, inadequate pipe size, old pipes with varying roughness, insufficient pumping capacity, and insufficient storage capability. Because either the mechanical measure or the hydraulic measure alone is inadequate to measure the system reliability, it seems reasonable to unify this definition by specifying the reliability as the probability that the given demand nodes in the system receive sufficient supply with satisfactory pressure head. In other words, the failure occurs when the demand nodes receive either insufficient flowrate and/or inadequate pressure head. Similarly, a nodal reliability is the probability that a given demand node receives sufficient water flowrate with adequate water pressure head.

Mechanical reliability is the ability of distribution system components to provide continuing and long-term operation without the need for frequent repairs, modifications, or replacement of components or subcomponents. Mechanical reliability is usually defined as the probability that a component or subcomponent performs its mission within specified limits for a given period of time in a specified environment. Hydraulic reliability is a measure of the performance of the water distribution system. The hydraulic performance of the distribution system depends to a great degree on the following factors:

- interaction between the piping system, distribution storage, distribution pumping, and system appurtenances, such as pressure reducing valves, check valves, etc
- reliability of the individual system components
- spatial variation of demands in the system
- temporal variation in demands on the system

Network reliability analysis models based on considering mechanical failure in the reliability of water distribution networks include Mays (1989), Hobbs and Beim (1988); Duan (1988); Quimpo and Shamsi (1987); Mays and Cullinane (1986); Wagner, Shamir, and Marks (1986, 1988a, 1988b); Duan and Mays (1987); and Tung (1985). The reliability of water systems due to the hydraulic failure resulting from mechanical failure was considered by Su et al (1987) and Cullinane, Lansey, and Mays (1992) in an optimization model These papers focus on new methodologies for water distribution system reliability; reliability analysis of pumping systems; and reliability-based optimization models for water distribution systems.

The reader is referred to the following publications for a complete understanding of published methodologies: Bao and Mays (1990); Coals and Goulter (1985); Fujiwara and Tung (1991); Goulter and Bouchart (1987); Goulter and Coals (1986); Hobbs, Beim, and Gleit (1987); Kettler and Goulter (1985); Lansey and Mays (1987, 1989); Lansey et al. (1989); Mays and Cullinane (1986); Mays, Duan, and Su (1986); Mays (1989); Mays et al. (1989); Ormsbee and Kessler (1990); Quimpo and Shamsi (1987); Su et al. (1987); Tung (1985); Tung et al. (1987); Woodburn, Lansey, and Mays (1987); Goulter and Kazemi (1988, 1989); Yen and Tung (1993); Goulter, Dividson, and Jacobs (1993); Wu et al. (1993); and Quimpo (1994).

2 METHODS FOR RELIABILITY ANALYSIS OF WATER DISTRIBUTION SYSTEM COMPONENTS

2.1 Reliability concepts

The analysis of reliability and availability requires an understanding of some basic terms, which are defined in this section. The concepts represented by these terms will be used in later sections to quantify reliability and availability. The common thread in the analysis of reliability and availability is the selection of an appropriate failure density function. Failure density functions are used to model a variety of reliability-associated events, including time to failure and time to repair.

The reliability $R(t)$ of a component is defined as the probability that the component experiences no failures during the time interval $(0,t)$ from time zero to time t, given that it is a new or repaired at time zero. In other words, the reliability is the probability that the time to failure T exceeds t, or

$$R(t) = \int_{t}^{\infty} f(t)\mathrm{d}t \qquad (2.1)$$

where $f(t)$ is the probability density function of the time to failure. Values for $R(t)$ range between 0 and 1. The probability density function $f(t)$ may be developed from equipment failure data, using various statistical methods. In many cases, a simple exponential distribution is found appropriate. The unreliability $F(t)$ of a component is defined as the probability that the component will fail by time t. Unrealiability can be defined mathematically as

$$F(t) = \int_0^t f(t)\mathrm{d}t = 1 - R(t) \qquad (2.2)$$

The failure rate $m(t)$ is the probability that a component experience a failure per unit of time t given that the component was operating at time zero and has survived to time t. Note that the failure rate $m(t)$ is a conditional probability. The relationship of $m(t)$ to $f(t)$ and $F(t)$ is given as $m(t) = f(t)/F(t)$. Sometimes the failure rate is called the hazard function. The quantity $m(t)\mathrm{d}t$ is the probability that a component fails during time $(t, t + \mathrm{d}t)$. Values for $m(t)\mathrm{d}t$ range from 0 to 1. Given the failure rate, the failure density function and the component reliability can be obtained.

2.2 Time-to-failure analysis

Because the time to failure of a component is not certain, it is always desirable to have some idea of the expected life of the component under investigation. Furthermore, for a repairable component, the time required to repair the failed component might also be uncertain. This section briefly describes and defines some of the useful terminology in the field of reliability theory that is relevant in the reliability assessment of water distribution systems.

The mean time to failure (MTTF) is the expected value of the time to failure, stated mathematically as

$$\text{MTTF} = \int_0^\infty tf(t)\mathrm{d}t \qquad (2.3)$$

which is expressed in hours.

Similar to the failure density function, the repair density function, $g(t)$ describes the random characteristics of the time required to repair a failed component when failure occurs at time zero. The probability of repair, $G(t)$ is the probability that the component repair is completed before time t, given that the component failed at time zero. Note that the repair process starts with a failure at time zero and ends at the completion of the repair at time t.

Similar to the failure rate, the repair rate $r(t)$ is the probability that the component is repaired per unit time t given that the component failed at time zero and is still not repaired at time t. The quantity $r(t)\mathrm{d}t$ is the probability that a component is repaired during time $(t, t + \mathrm{d}t)$ given that the components failure occurred at time t. The relation between repair rate, repair density, and repair probability function is $r(t) = g(t)/G(t)$.

The mean time to repair (MTTR) is the expected value of the time to repair a failed component. The MTTR is defined mathematically as

$$\text{MTTR} = \int_0^\infty tg(t)\mathrm{d}t \qquad (2.4)$$

where $g(t)$ is the probability density function for the repair time.

The mean time between failures (MTBF) is the expected value of the time between two consecutive failures. For a repairable component, the MTBF is defined mathematically as

$$\text{MTBF} = \text{MTTF} + \text{MTTR} \qquad (2.5)$$

The mean time between repairs (MTBR) is the expected value of the time between two consecutive repairs and equals the MTBF.

2.3 Availability and unavailability concepts

The reliability of a component is a measure of the probability that the component would be continuously functional without interruption through the entire period $(0, t)$. This measure is appropriate if a component is nonrepairable and has to be discarded when the component fails. However, many of the components in a water distribution system are generally repairable and can be put back in service again. In that situation, a measure that has a broader meaning than that of the reliability is needed.

The availability $A(t)$ of a component is the probability that the component is in operating condition at time t, given that the component was good as new at time zero. The reliability generally differs from the availability because reliability requires the continuation of the operational state over the whole interval $(0, t)$. Subcomponents contribute to the availability $A(t)$ but not to the reliability $R(t)$ if the subcomponent that failed before time t is repaired and is then operational at time t. As a result, the availability $A(t)$ is always large than or equal to the reliability $R(t)$, i.e., $A(t) \geq R(t)$. For a nonrepairable component, it is operational at time t, if and only if, it has been operational to time t, i.e., $A(t) = R(t)$. As shown in Figure 2, the availability of nonrepairable component decreases to zero as t becomes larger, whereas the availability of a repairable component converges to a nonzero positive number.

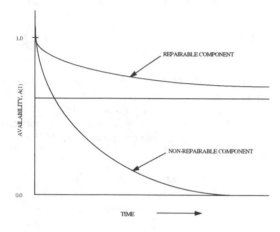

Figure 2. Availability for repairable and nonrepairable components

The unavailability $U(t)$ at time t is the probability that a component is in the failed state at time t, given that it started in the operational state at time zero. In general, the $U(t)$ is less than or equal to the unreliability $F(t)$; for nonrepairable components they are equal. A component is either in the operational state or in the failed state at time t, therefore,

$$A(t) + U(t) = 1 \tag{2.6}$$

Using exponential failure and repair density functions, the resulting failure rate μ and repair rate η, according to the definitions given previously, are constants equal to their respective parameters. For a constant failure rate and a constant repair rate, the analysis of the whole process can be simplified to analytical solutions. Henley and Kumamoto (1981) use LaPlace transforms to derive the unavailability and the availability.

The steady state or stationary unavailability $U(\cdot)$ and the stationary availability $A(\cdot)$ for i approaches ∞ are, respectively

$$U(\infty) = \frac{\mu}{\mu + \eta} = \frac{\text{MTTR}}{\text{MTTF} + \text{MTTR}} \tag{2.7}$$

and

$$A(\infty) = \frac{\eta}{\mu + \eta} = \frac{\text{MTTF}}{\text{MTTF} + \text{MTTR}} \tag{2.8}$$

As time gets large, the steady state (or stationary) unavailability and availability for the pump can be calculated.

3 MODEL FOR WATER DISTRIBUTION SYSTEM RELIABILITY

The reliability of a water distribution system can be defined as the probability that the system will provide demanded flowrate at a required pressure head. Due to the random nature of pipe roughness, water demands, and required pressure heads, the estimation of water distribution system reliability is subject to uncertainty. Bao and Mays (1990) presented a methodology by which to estimate the nodal and system reliabilities of a distribution system accounting for such uncertainty using Monte Carlo simulation.

The hydraulic uncertainty is considered by treating the demand, pressure head, and pipe roughness as random variables. Assuming that the randomness of water demand (Q_d) and the pipe roughness coefficient (C) follows a probability distribution, a random number generator is used to generate the values of Q_d for each node and C for each pipe. For each set of values of Q_d and C generated, a hydraulic network simulator is used to compute the pressure heads at the demand nodes, provided that the demands are satisfied. The required pressure head (H_d) at given nodes can be treated as constant with both lower and upper bounds or as a random variable. The corresponding nodal and system hydraulic reliabilities are then computed.

The framework for the methodology is based on a Monte Carlo simulation consisting of three major components: random number generation, hydraulic simulation, and computation of reliability. The random number generator is the core of the methodology and is used to generate values of the random variables of demand (Q_d), pressure head

(H_d), and the Hazen-Williams coefficient for pipe roughness (C). For each set of values of Q_d, H_d, and/or C generated, the University of Kentucky hydraulic simulation model (KYPIPE) is used to determine pressure heads for the nodes throughout the water distribution system. After a certain number of iterations, the nodal or system reliability is computed.

The nodal reliability (R_n) is the probability that a given node receives sufficient flowrate at the required pressure head. So theoretically the nodal reliability is a joint probability of water flowrate and pressure head being satisfied at the given nodes. However, it is difficult mathematically to derive and compute this joint probability. For instance, the flowrate and pressure head at a node are not independent. The approach used herein is to compute the conditional probability in terms of pressure head provided that the water demand has been satisfied or vice versa. This approach assumes that the water demand is satisfied $(Q_s = Q_d)$. The nodal reliability can be defined as the probability that the supplied pressure head (H_s) at the given node is greater than or equal to the minimum required pressure head (H_d^l),

$$R_n = P\left(H_s > H_d^l \middle| Q_s = Q_d\right)$$
$$= \int_0^\infty f_s(H_s)\left[\int_0^{H_s} f_{dl}(H_d^l)\mathrm{d}H_d^l\right]\mathrm{d}H_s \qquad (3.1)$$

Alternatively, both lower and upper bounds of required pressure head, H_d^l and H_d^u can be considered. In this case the nodal reliability is the probability that the supplied pressure head (H_s) at a given node is greater than or equal to the minimum required pressure head H_d^l and less than or equal to the maximum required pressure head (H_d^u),

$$R_n = P\left(H_d^u \geq H_s > H_d^l \middle| Q_s = Q_d\right)$$
$$= \int_0^\infty f_s(H_s)\left[\left(\int_0^{H_s} f_{dl}(H_d^l)\mathrm{d}H_d^l\right)\left(1 - \int_0^{H_s} f_{du}(H_d^u)\mathrm{d}H_d^u\right)\right]\mathrm{d}H_s \qquad (3.2)$$

where $f_s(\ )$ represents the probability density functions of supply, and $f_{dl}(\ )$, $f_{du}(\ )$ represent the minimum and maximum requirements of pressure head at a given node, respectively. If the required pressure head (H_d) is considered as a constant with

lower bound (H_d^l), the nodal reliability is given, respectively, below

$$R_n = P\left(H_s > H_d^l\right) = \int_{H_d^l}^\infty f_s(H_s)\mathrm{d}H_s \qquad (3.3)$$

or with lower and upper bounds, (H_d^l) and (H_d^u), the nodal reliability is

$$R_n = P\left(H_d^u \geq H_s \geq H_d^l\right) = \int_{H_d^l}^{H_d^u} f_s(H_s)\mathrm{d}H_s \qquad (3.4)$$

Although the nodal reliability depicts a fairly complete reliability measure of the water distribution system, it is also convenient to use a single index such as "system reliability" to represent the composite effect of the nodal reliabilities. Such an index is difficult to define because of the dependence of the computed nodal reliabilities. Three heuristic definitions of the system reliability were considered by Cullinane (1989).

The system reliability (R_{sm}) could be defined as the minimum nodal reliability in the system, $R_{sm} = \min\{R_{ni}\}, i = 1, 2, ..., I$, where R_{ni} is the nodal reliability at node i, and I is the number of demand nodes of interest. Another system reliability could be the arithmetic mean (R_{sa}), the mean of all nodal reliabilities, $R_{sa} = \sum_{i=1}^{I} R_{ni} \Big/ I$.

A third approach could be to define the system reliability as a weighted average (R_{sw}), which is a weighted mean of all nodal reliabilities weighted by the water supply at the node, $R_{sw} = \sum_{i=1}^{I} R_{ni}\overline{Q}_{si} \Big/ \sum_{i=1}^{I} \overline{Q}_{si}$, where $\overline{Q}_{si}$ is the mean of the water supply at node i.

4 ALGORITHM

The procedure to evaluate the nodal and system reliability of a water distribution for hydraulic failure is illustrated in the flowchart in Figure 3. The algorithm includes the following basic steps:
1. Assign distributions to Q_d, H_d, and/or C.
2. Generate Q_d, H_d, ad/or C using Monte Carlo simulation.
3. Compute H_s at all nodes using the hydraulic simulator KYPIPE, assuming Q_d is satisfied $(Q_s = Q_d)$.
4. Compute nodal and system reliabilities.

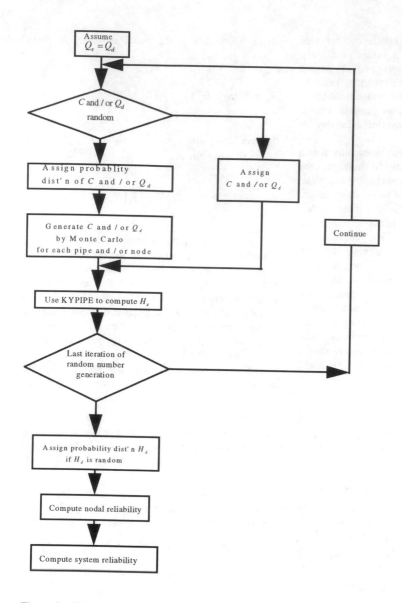

Figure 3.   Flowchart of algorithm to evaluate system reliability

5 RELIABILITY ANALYSIS

Another methodology for the reliability analysis of pumping stations for water supply systems was developed by Duan and Mays (1990) that considers both mechanical failure and hydraulic failure. The reliability methodology models the available capacity of a pumping station as a continuous-time Markov process, using bivariate analysis and conditional probability approaches in a frequency and duration analysis framework.   A supply model, a demand model, and a margin model are developed that are used to compute the expected duration of a failure, expected unserved demand of a failure, expected number of failures in the period of study, expected total duration of failures in the period of study, and expected total unserved demand in the period of study.

The frequency and duration analysis (FD), referred to herein as the FD approach, allows derivation of various reliability indices and is well suited for analyzing the reliability performance of a pumping system. Not only the failure probability, but also the failure frequency and the cycle time between failures can be analyzed by this approach. Hobbs (1985) was one of the first to recognize the importance and applicability of frequency and duration analysis to water supply systems. Some of his work includes the development of a methodology to compute the expected unserved demand and the inclusion of storage in the FD computations.

Duan and Mays (1990) presented a new methodology to analyze the reliability of pumping systems using a modified FD analysis to make the reliability analysis more realistic and complete. Both the mechanical failure and hydraulic failure of pumping systems are analyzed in computing the reliability parameters for the methodology. Hydraulic failure in this context refers to not meeting required demands and required pressure heads. The methodology has been programmed into a computer code called RAPS (Reliability Analysis of Pumping Systems). RAPS is used to determine the following eight reliability parameters for pumping systems: (1) failure probability; (2) failure frequency; (3) cycle time between failures; (4) expected duration of a failure; (5) expected unserved demand of failure; (6) expected number of failures in the period of study; (7) expected total duration of failures in the period of study; and (8) expected total unserved demand in the period of study. RAPS has been tested on example problems ranging from two pumps and five demand states to ten pumps and twenty-five demand states.

6 OPTIMIZATION MODEL FOR RELIABILITY-BASED (AVAILABILITY) DESIGN OF WATERDISTRIBUTION NETWORKS

The overall optimization problem for a general water distribution network design can be mathematically stated as a function of the nodal heads, $\mathbf{H}$, and the design parameters. The pipe flows, $\mathbf{Q}$, are a second set of system variables but are not included since they can be written in terms of $\mathbf{H}$ via the flow equations. Since the nodal pressures are generally considered the restricting constraints in design, the general model can be formulated, with respect to this set, as

Minimize Cost $f(\mathbf{D}, \mathbf{H})$ $\qquad$ (6.1)

subject to

a. Conservation of Flow and Energy Constraints

$$\mathbf{G}(\mathbf{H}, \mathbf{D}) = \mathbf{0} \qquad (6.2)$$

b. Head Bounds

$$\underline{\mathbf{H}} \leq \mathbf{H} \leq \overline{\mathbf{H}} \qquad (6.3)$$

c. Design Constraints

$$\underline{\mathbf{j}(\mathbf{D})} \leq \mathbf{j}(\mathbf{D}) \leq \overline{\mathbf{j}(\mathbf{D})} \qquad (6.4)$$

d. Reliability Constraints

$$\underline{\mathbf{r}(\mathbf{R})} \leq \mathbf{r}(\mathbf{R}) \leq \overline{\mathbf{r}(\mathbf{R})} \qquad (6.5)$$

where $\mathbf{D}$ is the vector of decision variables that are defined for each component in the system and represents the dimension of each component, such as diameter of the pipes, pipe size, valve setting, tank volume, tank elevation, etc. The design constraints are usually simple bounds but are shown as functions for the general case. The vector $\mathbf{R}$ represents the reliability constraints. The vector $\mathbf{H}$ represents the heads at specified locations in the systems, with $\underline{\mathbf{H}}$ and $\overline{\mathbf{H}}$ being the lower and upper bounds, respectively.

The solution approach employs a technique whereby the problem is reduced to a form that is more manageable by large-scale nonlinear programming (NLP) codes. The technique reduces the problem by writing some variables called "state" variables that are dependent in terms of other "control" (independent) variables using equality constraints. This steps results in a smaller, reduced problem with a new objective and a smaller set of constraints, many of which are simple bounds, that can now be efficiently solved by existing NLP codes. In this problem, the pressure heads, $\mathbf{H}$, will be defined as the state or basic variables and written with respect to the design parameters, $\mathbf{D}$, the control or nonbasic variables. This variable reduction technique has been successfully applied to problems in econometric control, oil and gas reservoir management, groundwater management, and large water distribution systems without reliability constraints (Lansey 1987). A water distribution simulation model can be used to solve the network equations for the nodal heads given a set of design parameters. A computer program, WSAVOPT

(Water Supply AVailability OPTimization), has been developed (Cullinane 1989; Cullinane, Lansey, and Mays 1992).

7 RELIABILITY OF WATER QUALITY

The Safe Drinking Water Act of 1974 and its Amendments of 1986 (SDWAA), requires the U. S. Environmental Protection Agency (U. S. EPA) to establish maximum contaminant level (MCL) goals for each contaminant which may have adverse health effect on humans. Each goal must be set at a level at which no known or anticipated adverse effect occurs on human health, allowing for a safety of margin. As a result of the SDWAA there has been increased interest in developing an understanding of the variation of water quality found in municipal water distribution systems. The U. S. EPA is required to regulate chemical contaminants and pathogenic microorganism in drinking water so that emphasis has changed from the treatment source to the points of consumption.

Computer models for simulating the hydraulics of water-distribution systems have been available for many years. It is only more recently, as a result of the SDWAA that these models have been extended to analyze the water quality. The new EPANET model is probably the best example of these developments (Rossman, 1994; Rossman and Boulos, 1996; Rossman et al., 1993, 1994).

There is also little known about the effect of distribution system storage on water quality. Studies have indicated that water quality is degraded as a result of long residence times in storage facilities. Recent effort by Grayman and Clark (1993) and Mau, et at. (1994) have developed computer models to consider the effect of storage on water quality.

To the writer's knowledge there have been no efforts reported in the literature to develop reliability methods for analyzing the water quality of distribution system and no reliability-based optimization methods. There needs first to be developed: (1) methods for analyzing water quality reliability measures for distribution systems; (2) operation models that can be used to optimally operate systems to consider water quality; and (3) reliability based optimization models. The operation problem can be posed as a discrete-time optimal control problem similar to the operation problem posed and solved by Brion and Mays (1991). These approaches should take full advantage of previous methods to interface simulation models with optimizers.

A key point of these new methodologies would be to define reliability as the probability that MCL's defined in terms of concentration $C_{x,t}$ as a function of time t and space x, as

$$\text{Reliability} = P(C_{x,t} < C^*) \qquad (7.1)$$

The overall reliability-optimization model would determine optimal operation pumping strategies to minimizing energy cost that would satisfy the reliability constraint (7.1). This approach would require writing (7.1) as a chance constraint formulation.

REFERENCES

Bao, Y., and L. W. Mays. 1990. Model for water distribution system reliability. *Journal of Hydraulic Engineering*, ASCE 116(9), 199-217.

Brion, L. M. and L. W. Mays, 1991. Methodology for optimal operation of piping stations in water distribution systems, *Journal of Hydraulic Engineering*, ASCE, 117(11).

Coals, A., and I. C. Goulter. 1985. Approaches to the consideration of reliability in water distribution networks. In *Proceeding of the 1985 International Symposium on Urban Hydrology, Hydraulic Infrastructures and water Quality Control*. The University of Kentucky, Lexington, Kentucky, July 23-25.

Cullinane, M. J., Jr. 1989. Methodologies for the evaluation of water distribution system reliability /availability. Ph. D. dissertation, Department of Civil Engineering, University of Texas, Austin, Texas.

Cullinane, M. J., Jr.; K. E. Lansey; and L. W. Mays. 1992. Optimization-availability based design of water distribution networks. *Journal of Hydraulic Engineering*, ASCE 118(3), 420-441.

Duan, N., and L. W. Mays. 1987. Reliability analysis of pumping stations and storage facilities. In *Proceedings of the American Society of Civil Engineers National Conference on Hydraulic Engineering*, edited by R. M. Ragan. New York: American Society of Civil Engineers.

Duan, N., and L. W. Mays. 1990. Reliability analysis of pumping systems. *Journal of Hydraulic Engineering*, ASCE 116 (1), 230-248.

Duan, N. 1988. Optimal reliability-based design and analysis of pumping systems for water distribution systems. Ph. D. dissertation, Department of Civil Engineering, The University of Texas, Austin, Texas.

Fujiwara, O., and H. D. Tung. 1991. Reliability improvement for water distribution networks through increasing pipe size. *Water Resources Research* 27 (July): 1395-1402.

Goulter, I., and F. Bouchart. 1987. Joint consideration of pipe breakage and pipe flow probabilities. In *Proceeding of the American Society of Civil Engineers National Conference on Hydraulic Engineering*, edited by R. M. Ragan. New York: American Society of Civil Engineers.

Goulter, I., and A. Coals. 1986. Quantitative approaches to reliability assessment in pipe networks. *Journal of Transportation Engineering*, ASCE 112(March): 287-301.

Goulter, I., J. Davidson, and D. Jacobs., 1993. Predicting water main breakage rates. *Journal of Water Resources Planning and Management*, ASCE 119 (July/August).

Goulter, I., and A. Kazemi. 1988. Spatial and temporal groupings of water main pipe breakage in Winnipeg, *Can. Journal of Civil Engineering* 15(1):91-97.

Goulter, I., and A. Kazemi. 1989. Analysis of water distribution pipe failure types in Winnipeg, Canada. *Journal of Transportation Engineering*, ASCE 115(2): 95-111.

Henley, E. J., and H. Kumamoto. 1981. *Reliability engineering and risk assessment*. Emglewood, N. J.: Prentice Hall.

Grayman, W. M. and R,. M. Clark. 1993. Using computer models to determine the effect of storage on water quality, *Journal AWWA*, pp. 67-77, July.

Hobbs, B. 1985. Reliability analysis of water system capacity. In *Proceedings of the American Society of Civil Engineers Specialty Conference, Hydraulics and Hydrology in Small Computer Age*, edited by W. Waldrop. New York: ASCE.

Hobbs, B. F., and G. K. Beim. 1988. Analytical simulation of water system capacity reliability 1. Modified frequency-duration analysis. *Water Resources Research,* 24(9), 1431-1444.

Hobbs, B., G. K. Beim, and A. Gleit. 1987. Reliability analysis of power and water supply systems. In *Strategic Planning in Energy and Natural Resources*, edited by B. Lev, et al. Amsterdam: North-Holland.

Kapur, K. C., and L. R. Lamberson. 1977. *Reliability in engineering designs*. New York: Wiley.

Kettler, A. J., and I. C. Goulter. 1985. An analysis of pipe breakage in urban water distribution networks. *Can. J. Civil Eng.* 12:286-293.

Lansey, K. E. 1987. Optimal design of large-scale water distribution systems under multiple loading conditions. Ph. D. dissertation, The University of Texas, Austin, Texas.

Lansey, K. E., and L. W. Mays. 1987. Optimal design of large scale water distribution systems. In *Proceedings of the American Society of Civil Engineers National Conference on Hydraulic Engineering*, edited by R. M. Ragan. New York: America Society of Civil Engineers.

Lansey, K. E., and L. W. Mays. 1989. Optimization model for water distribution system design. *Journal of the Hydraulics Division*, ASCE 115 (10), 1401-1418.

Lansey, K. E., N. Duan, L. W. Mays, and Y. -K. Tung. 1989. Water distribution system design under uncertainties. *Journal of Water Resources Planning and Management*, ASCE 115 (10), 630-695.

Mau, P. E., et al. 1995. Explicit mathematical models of distribution storage water quality, *Journal of Hydraulic Engineering*, ASCE, 121(10), 699-709.

Mays, L. W., and M. J. Cullinane, Jr. 1986. A review and evaluation of reliability concepts for design of water distribution systems. Miscellaneous paper EL-86-1, U. S. Army Corps of Engineers, Environmental Laboratory, Waterways Experiment Station, Vicksburg, Mississippi, January.

Mays, L. W., N. Duan, and Y. C. Su. 1986. Modeling reliability in water distribution network design. In *Proceedings*, Water Forum 1986: World Water Issues in Evolution, Edited by M. Karamouz, et al. New York: American Society of Civil Engineers.

Mays, L. W., ed. 1989. *Reliability analysis of water distribution systems*. Compiled by the Task Committee an Reliability Analysis of Water Distribution Systems. New York: American Society of Civil Engineers.

Mays, L. W., et al. 1989. Methodologies for the assessment of aging water distribution systems. Report No. CRWR 227, Center for Research in Water Resources, The University of Texas at Austin, July.

Ormsbee, L., and A. Kessler. 1990. Optimal upgrading of hydraulic-network reliability. *Journal of Water Resources Planning and Management*, ASCE 116 (November/December): 784-802.

Quimpo, R. G. 1994. Reliability analysis of water distribution systems, Risk-based Decision Making in Water Resources VI, ed. by Haimes, et al., ASCE. pp. 45-55.

Quimpo, R. G., and U. M. Shamsi. 1987. Network analysis for water supply reliability determination. In *Proceedings of the American Society of Civil Engineers National Conferences on Hydraulic Engineering*, edited by R. M. Ragan. New York: American Society of Civil Engineers.

Rossman, L. A. 1994. EPANET-User Manual, EPA-600/R-94/057, U. S. Environmental Protection Agency, Risk Reduction Engrg. Lab., Cincinnati, Ohio.

Rossman, L. A., P. E. Boulos, and T. Altman. 1993. Discrete volume element method for network water quality models, *Journal of Water Resources Planning and Management*, ASCE, 119(5), 505-517.

Rossman, L. A. and P. E. Boulos. 1996. Numerical Methods for modeling water quality in distribution systems: A comparison. *Journal of Water Resources Planning and Management*, ASCE, 122(2), 137-146.

Rossman, L. A., R. M. Clark, and W. M. Grayman. 1994. Modeling chlorine residuals in drinking-water distribution systems. *Journal of Environmental Engineering*, ASCE, 120(4), 803-820.

Su, Y. C., L. W. Mays, N. Duan, K. E. Lansey. 1987. Reliability-based optimization model for water distribution systems. *Journal of Hydraulic Engineering*, ASCE 114 (12), 1539-1556

Tung, Y.-K. 1985. Evaluation of water distribution network reliability. In *Proceedings of the ASCE Hydraulic Division Specialty* Conference, Orlando, Florida.

Tung, Y.-K., K. E. Lansey, N. Duan, and L. W. Mays. 1987. Water distribution systems design by chance-constrained model. In *Proceedings of the 1987 ASCE National Conference on Hydraulic Engineering*.

Wagner, J. M., U. Shamir, and D. H. Marks. 1986. Reliability of water distribution systems. Report No. 312, Ralph M. Parson Laboratory, Massachusetts Institute of Technology, Cambridge, Mass.

Wagner, J. M., U. Shamir, and D. H. Marks. 1988a. Water distribution reliability: Analytical methods. *Journal of Water Resources Planning and Management*, ASCE 114 (3), 253-275.

Wagner, J. M., U. Shamir, and D. H. Marks. 1988b. Water distribution reliability: Simulation methods. *Journal of Water resources Planning and Management*, ASCE 114(3), 276-294.

Woodburn, J., K. E. Lansey, and L. W. Mays. 1987. Model for the optimal rehabilitation and replacement of water distribution system components. In *Proceedings of the American Society of Civil Engineers National Conference on Hydraulic Engineering*, edited by R. M. Ragan. New York: American Society of Civil Engineers.

Wu, S.-J., Y. -H. Yoon, and R. G. Quimpo. 1993. Capacity-weighted water distribution system reliability. *Reliability Engineering & System Safety*, 42(1), 39-46.

Yen, B. C., and Y.-K. Tung, eds. 1993. *Reliability and uncertainty analysis in hydraulic design*. Compiled by ASCE Subcommittee on Uncertainty and Reliability Analysis in Design of Hydraulic Structures of the Technical Committee on Probabilistic Approaches to Hydraulics of the Hydraulics Division of ASCE.

1 Water resources management

Stochastic Hydraulics'96, Tickle, Goulter, Xu, Wasimi & Bouchart (eds) © 1996 Balkema, Rotterdam. ISBN 90 5410 817 7

Stochastic optimization techniques in hydraulic engineering and management

D.A. Savic & G.A. Walters
School of Engineering, University of Exeter, UK

ABSTRACT: Many problems that arise in hydraulics and water resources in general may be formulated as optimization problems and solved using available techniques. However, most of these problems are complex, highly non-linear and have discontinuous objective functions and constraint sets. In the absence of a deterministic optimization technique capable of finding the global optimum, stochastic optimization techniques provide a good alternative. Stochastic optimization refers to a family of optimization techniques in which the solution space is searched by generating candidate solutions with the aid of a pseudo-random number generator. As the run proceeds, the probability distribution by which new candidate solutions are generated may change, based on results of trials earlier in the run. Genetic Algorithms are probably the best known type of stochastic optimization method. This paper reviews published research on applications of stochastic optimization techniques to hydraulic engineering and management. The work done by the Water Systems Group at University of Exeter on genetic algorithm applications in hydraulic pipeline systems serves as a basis for evaluation of the techniques. Examples from other areas such as groundwater management and hydrology are also provided.

1 INTRODUCTION

The analysis and optimal design of water resources systems are extremely complex. The complexity is due to the numerous simultaneous and often conflicting objectives and purposes of these systems, the variable character of their elements, their stochastic nature and the different scales under consideration. A systems analysis approach and operational research provide the philosophical framework and quantitative techniques, respectively, for handling complex physical and socio-economic considerations within optimization processes. Over the decades the development and application of optimization models for design, planning and operation of water resources systems have attracted growing attention among engineers, planners and managers. These techniques range from simple search and simulation to more advanced linear programming, dynamic programming and non-linear programming techniques (Yeh, 1985; Walters, 1988). In addition to these, so-called traditional methods, hydraulic/hydrologic engineers have been looking into the field of Artificial Intelligence (AI) and Expert Systems (ES) to develop computer programs which emulate the logic and reasoning processes the human would use to solve a problem in the water resources field (Delleur, 1988). For many years research in AI followed a symbolic paradigm which required a level of knowledge described in terms of rules. More recently applications based on another AI subdiscipline often referred to as Computational Intelligence (CI), which incorporates Evolutionary Computation (EC), Fuzzy Systems (FS), and Artificial Neural Networks (ANN), started to emerge in the water resources field.

There are many search mechanisms which can be used to arrive at the optimal solution, and in recent years general search methods based on models of natural evolution have become increasingly popular. Evolution Programs (EP) are general artificial-evolution search methods based on natural selection and mechanisms of population genetics (Michalewicz, 1992). They emulate nature's very effective optimisation techniques of evolution which are based on preferential survival and reproduction of the fittest members of the population, the maintenance of a population with diverse members, the inheritance of genetic information from parents, and the occasional mutation of genes. Genetic Algorithms (GAs), which are probably the best known type of EPs, are also referred to as *stochastic optimization* techniques. Stochastic optimization designates a family of optimization techniques in which the solution space is searched by generating candidate solutions with the aid of a pseudo-random number generator. As the run proceeds, the probability distribution by which new candidate solutions are generated may change, based on results of trials earlier in the run. Because of their stochastic nature there is no guarantee that the global optimum will be found using GAs although the number of applications suggests a good rate of success in identifying good solutions. After reported successes in many problem domains (Goldberg, 1989), water resources systems optimisation has started to benefit from the use of EPs.

2. GENETIC ALGORITHMS

The theory behind GAs was proposed by Holland (1975) and further developed by Goldberg (1989) and others in the 1980's. These algorithms rely on the collective learning process within a population of individuals, each of which represents a search point in the space of potential solutions. They draw their power from the theoretical principle of implicit parallelism (Holland, 1975). This principle enables highly fit solution structures (schemata) to receive increased numbers of offspring in successive generations and thus lead to better solutions.

There are many variations of GAs but the following general description encompasses most of the important features. The analogy with nature is established by the creation within a computer of a set of solutions called a *population*. Each individual in a population is represented by a set of parameter values which completely describe a solution. These are encoded into *chromosomes*, which are, in essence, sets of character strings analogous to the chromosomes found in DNA. Standard GAs use a binary alphabet (characters may be 0's or 1's) to form chromosomes. For example a two-parameter solution $x = (x_1, x_2)$ may be represented as an 8-bit binary chromosome: 1001 0011 (i.e., 4 bits per parameter, $x_1 = 1001$, $x_2 = 0011$). At this point it should be noted that not all EPs restrict representation to the binary alphabet which makes them more flexible and applicable to a variety of decision-making problems.

The initial population of solutions, which is usually chosen at random, is allowed to evolve over a number of *generations*. At each generation, a measure (*fitness*) of how good each chromosome is with respect to an objective function is calculated. This is achieved by simply decoding binary strings into parameter values, substituting them into the objective function and computing the value of the objective function for each of the chromosomes. Next, based on their fitness values individuals are selected from the population and recombined, producing offspring which will comprise the next generation. This is the *recombination* operation, which is generally referred to as *crossover* because of the way that genetic material crosses over from one chromosome to another. For example, if two chromosomes are $x = (x_1, x_2) =$

1111 1111 and $y = (y_1, y_2) = 0000\ 0000$, the two offspring may be $z = 1100\ 0000$ and $w = 0011\ 1111$.

The probability that a chromosome from the original population will be selected to produce offspring for the new generation is dependent on its fitness value. Fit individuals will have higher probability of being selected than less fit ones resulting in the new generation having on average a higher fitness then the old population. Mutation also plays a role in the reproduction phase, though it is not the dominant role, as is popularly believed, in the process of evolution (Goldberg, 1989). In GAs mutation randomly alters each bit (also called *gene*) with a small probability. For example, if the original chromosome is $x = (x_1, x_2) = 1111\ 1111$, the same chromosome after mutation may be $x' = 1110\ 1111$. If the probability of mutation is set too high, the search degenerates into a random process. This should not be allowed as a properly-tuned GA is not a random search for a solution to a problem. As a simulation of a genetic process a GA uses stochastic mechanisms, but the result is distinctly better than random.

3 STOCHASTIC OPTIMIZATION AND HYDRAULIC ENGINEERING

After reported successes in many problem domains, hydraulic engineering systems design and management have started to benefit from the use of stochastic, evolution-based optimization techniques. In the last few years many researchers have begun to investigate the use of evolution based computer methods for calibration of various hydraulic/hydrologic models. Wang (1991) investigated the use of GAs combined with fine-tuning by a local search method for calibration of a conceptual rainfall-runoff model. Models were calibrated by minimising the residual variance defined as the sum of square of differences between computed and observed discharges. Duan et al. (1994) introduced the shuffled complex evolution method for a similar problem by hybridising a GA with the Simplex search method. The objective function used was the mean daily square root of the difference between the observed flows and simulated flows. Babovic et al. (1994) used a GA and the hydrodynamic MOUSE package to fit Manning numbers to pipes while Mohan and Loucks (1995) reported on use of GAs for estimating parameter values of some linear and non-linear flow routing and water quality prediction models.

Groundwater simulation models have been used in conjunction with GAs to solve groundwater management problems. McKinney and Lin (1994) report on using the standard GA to solve three problems: maximum pumping from an aquifer, minimum cost water supply development and minimum cost aquifer remediation. Ritzel and Eheart (1994) and Cieniawski, et al. (1995) use multiobjective GA techniques to solve a groundwater pollution containment problem and a groundwater monitoring problem, respectively. They show that GAs have the ability to find multiobjective trade-off curves for complex groundwater problems that have proven to be intractable using traditional optimization techniques.

Goldberg and Kuo (1987) applied the standard (binary) GA to solve a steady-state optimisation problem of serial liquid pipelines consisting of ten pipes and ten pump stations. In recent years there has been a growing interest in developing methods for optimal design of networks, both for the sizing of pipe diameters and for the choice of layouts for a system. Murphy and Simpson (1992) introduce GAs to pipe sizing and rehabilitation of water distribution networks. They used a standard GA to find the optimal solution of a network, where the method chooses the optimal combination among the eight alternative decisions possible for each of the eight decision variables (pipes). Murphy et al. (1994) solve the "Any town" network (1987), introducing, in addition to the pipes, pumps and reservoirs as decision variables. Walters and Lohbeck (1993) and Walters and Smith (1995) developed EPs for optimal layout of tree networks based on directed and undirected base graphs, respectively. Cembrowicz (1992) combined classical optimization techniques, Linear Programming (LP) and Dynamic Programming (DP), to determine the least-cost diameters, with evolution programs to determine the optimal layout of water distribution systems.

4. EVOLUTION PROGRAMS AND HYDRAU-LIC NETWORKS RESEARCH AT EXETER

4.1 *Layout Design of Branched Hydraulic Networks*

One possible aim of optimising the layout of a branched network is to select the minimum cost set of arcs that is necessary to supply all given demands. Walters and Lohbeck (1993) analysed the layout problem of branched networks for water, gas and sewer systems. The optimisation problem was defined as a search for the best tree layout from a directed base graph of possible pipeline connections (Figure 1). The main assumption of the method employed was that flow directions in the base graph are specified in advance. The simplification of using a directed base graph greatly reduces the number of candidate solutions, i.e., tree-like networks. The authors have shown that for a 25-node *grid* network (whose nodes are located in a rectangular pattern with each node served by at most one of two upstream arcs) the assumption of flow directions reduces the search space from approximately 3.3×10^{13} possible layouts to about 3.4×10^7 possibilities. Even for this small example the solution space still remains large with a great number of local minima.

Undirected base-graph networks are not only larger problems to solve but also pose more difficulties than directed networks. These problems are highly constrained since the connectivity of the network has to be preserved. Walters and Smith (1995) used an integer-coded EP for the selection of a branched network from a non-directed base graph. The program developed ensures the generation of feasible solutions. This is achieved through innovative coding, recombination and mutation operators.

By running the program a large number of times on different networks and with different population sizes and mutation rates, the best EP parameters were identified. It was found that a population of $n = 16$-20 and mutation rate averaging about one additional arc added to the genetic pool were the most effective. The program was successfully applied to a two-source example problem with 100 nodes and

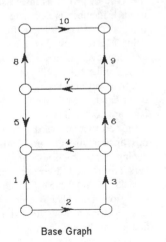

Base Graph

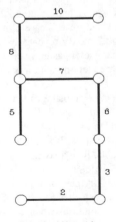

Branched Network

Figure 1. A directed base graph and a possible branched network

232 arcs (3.65×10^{54} possible solutions). To reach the best identified solution each run required on average 32000 evaluations of network cost, 50 orders of magnitude fewer than for complete enumeration of the problem. When solutions obtained from the directed and undirected base graph problems were compared it was found that the new EP had identified less expensive designs.

4.2 *Pressure Regulation in Hydraulic Networks*

As sections of the network may have to be isolated to repair leaks or for other purposes, isolating valves are frequently installed at specific intervals, the spacing being dependent on cost and operational considerations. Closing one or more of these valves without isolating any part of the network changes the configuration and hence the distribution of flows and pressures in the whole system. If nodal requirements (i.e. demand and minimum pressure head) stay unchanged, the problem can be posed as: find the optimal settings of all isolating valves (i.e., open or closed) to attain the best possible pressure distribution without compromising network performance (i.e., required flow is supplied to each node and minimum head requirement is satisfied). The objective function for this optimisation problem is given in the form

$$\min_{CV \subseteq V} J = \sum_{i=1}^{N} \left| H_i - H_i^{min} \right| \qquad (1)$$

where V is a set of all valves in the network, CV is a set of closed valves, H_i is the head at node i, and H_i^{min} is the minimum required head at the same node.

To assess the reduction in excess pressures, the pressure distribution for an example problem is first obtained for the case where there are no closed valves in the network (Savic and Walters, 1995a). The objective function value for this case of no pressure control is $J = 1257.5$ m. By repeatedly running the EP with the same parameters as in the first example, the lowest objective function value was found to be $J = 473.7$ m.

The best solution identifies closure of 9 valves. In contrast, an alternative, near-optimal solution, having an objective function value of $J = 476.2$ m, identifies a very different set of 10 valves. If a smaller number of valves is more desirable, then a solution which requires only 5 valves to be closed can be used. The objective function value for this solution ($J = 494.3$ m) is only 4.3% greater than that for the best solution found. In addition, if solutions which are infeasible with respect to the minimum head requirement are analysed, those with relatively small minimum

head violations may be considered acceptable in some circumstances. For example one such solution is to close only five valves. This gives an objective function value of $J = 371.0$ m, an improvement of 21.7% over the best solution identified but with a minimum head violation of 2.0 m at one node. It is also important to note that the EP was allowed 50,000 evaluations (number of generations × number of members in a population) per run which represents only 0.00036% of the expected number of feasible (connected) networks.

4.3 *Pump Scheduling for Water Supply*

Due to increased levels of urbanisation and consumer demand, most water distribution systems have become increasingly complex and so too has the task of efficient scheduling of pump operation. Classical optimization methods may become inadequate when there are more than two reservoirs in the system or even for one-reservoir systems which have several different pump combinations or complicated system constraints. The amount of necessary calculation increases so rapidly with the number of reservoirs and/or possible pump combinations that the computing requirements become unacceptable. An attempt was made to investigate the use of GAs for pump scheduling. GAs treat discrete values used in pump scheduling models naturally and thus they seem well suited to this kind of optimisation.

For pump scheduling, the use of a standard binary coding is the obvious choice, since each binary value can represent one pump that is either on or off during a particular time interval. The use of other forms of coding were considered to have no advantages for this type of problem. In the standard GA the fitness of each member is calculated as an inverse function of the cost. The relative fitness and probability of being a parent are then derived. The relative fitness can, however, cause difficulties. If the fitnesses are quite similar there will not be a proper selection between good and less good solutions, but if, on the other hand, the relative fitnesses of a very few members are too high compared to the others, this will soon lead to a lack of diversity in the population. For this application, the problem was solved by introducing a ranking selection where population members are ranked in order according to their costs (Mäckle et al., 1995). For simplicity, a fitness equal to the order number is used, the most expensive solution getting a fitness of 1, the next 2 etc. The introduction of the ranking function leads to a remarkable increase of the efficiency of the GA.

To ensure that the system constraints are not violated in the solution, it is necessary to introduce penalty functions for unfulfilled demand, violation of reservoir capacity, etc. The cost function, now including all the different cost components and constraint violation penalties is generally non-linear and non-differentiable, thus difficult to optimize.

The example used for demonstration of the method is a simple system with 4 different pumps (16 pump combinations) delivering water to a single reservoir. The day was divided up into 24 time intervals, each of one hour, with the demand drawn from the reservoir changing for each hour. The difference between the rate at which water is pumped to the reservoir and the demand is accommodated by changes in the volume of water stored in the reservoir, subject to maximum and minimum storage volumes. The complete binary string for the solution space is 24 x 4 = 96. With this binary string all the 2^{96} (8×10^{28}) theoretically possible pump schedules are represented.

The best solution identified by the GA shows, as expected, that as much of the water as possible should be pumped using the cheap night time tariff. As shown in Figure 2, the reservoir is consequently as full as possible at the end of the cheap period and as empty as allowed at the end of the expensive period.

Current work is directed towards the development of a multiobjective GA which will incorporate other objectives, like to minimise the number of pump switches, discourage complex operation decisions, etc. A hybrid model is also being developed which combines a GA and a hill-climbing (local search) technique.

4.4 *Calibration of Hydraulic Networks*

Instead of working with binary coding and applying problem-independent genetic operators, the size of hydraulic network calibration problems dictates direct representation of decision variables. This simply means that binary strings of the standard GA are replaced with real numbers. The change simplifies the algorithm in that no additional mapping is necessary since these numbers represent unknown parameters of the model. However, other alterations to the GA are required. Firstly, random binary initialisation is replaced with random real number initialisation. In addition to the standard crossover operator, an operator suitable for continuous parameter optimization may be used. An operator analogous to binary mutation, but suitable for continuous parameter optimization must be used since simple bit inversion is not possible with the floating-point representation. An obvious way to mutate a real-valued parameter x is to randomly select a number that falls within parameter limits $x_m \in [x_{min}, x_{max}]$. Alternatively, the new parameter may be given by

$$x_m = x + z \qquad (2)$$

where z is a number in the mutation range interval. This range can be a constant value throughout the evolution process or it may be a function of the generation number. By exploiting an analogy with an-

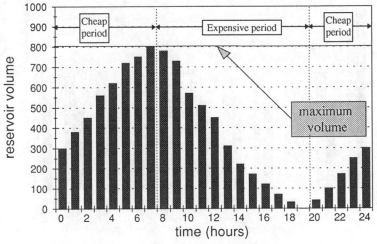

Figure 2. Stored volume vs. time

nealing processes the range should become smaller with the evolution process approaching final stages.

The proposed algorithm is used to provide a calibrated network model for the Danes Castle Zone of Exeter City (Devon, UK). This network was chosen for the study because it provides a complex calibration problem to solve and because the necessary input and output data were readily available (Savic and Walters, 1995b). The skeleton of the network consists of 197 nodes and 242 elements. Of the 197 nodes, one is a fixed-head reservoir (at treatment works) and one is the Danes Castle service reservoir. Of the 242 network elements, two are pumps and three are modelled as throttling valves. A 48 hour field test was undertaken between 13th and 15th August 1991. Flows were monitored into or within the system at 15 locations while pressures were monitored at 23 locations including the water level variation at the reservoir and the pump suction and delivery pressures at the pump station. Three loading conditions were considered in the analysis.

Although the number of parameters to be estimated cannot exceed the number of total observations available for all the loading conditions, the solutions obtained using the GA technique were based on fitting friction factor values to each of the pipes. This assumption was used for several reasons:
(a) to demonstrate the model ability to deal with a large number of variables; (b) to obtain an initial grouping of pipes in the absence of detailed knowledge of the age and the service condition of the pipes; (c) to investigate how different and unrealistic solutions can result from attempting to acquire more

information from collected data than is available.

The objective function used in this work is:

$$f(x) = p_1 \sum_i (H_i^o - H_i^p)^2 + p_2 \sum_j (Q_j^o - Q_j^p)^2 \qquad (3)$$

where, $p_{1,2}$ are normalising coefficients, and H_i^o and H_i^p are the observed and predicted heads at node i, respectively, and Q_j^o and Q_j^p are the observed and predicted flows through the pipe j.

Since GAs are stochastic-search techniques, solutions obtained running the program with different random seed values used to initialise the evolution process may be different. Therefore, several GA runs were necessary to ensure that the solutions identified were of good quality. Figure 3 shows results, in terms of sums of squared errors for both pressure heads and flows in the network, obtained by the original calibration study (Ewan, 1991) and the 10 GA runs.

4.5 Optimal Design and Expansion of Hydraulic Networks

Design of water distribution networks is often viewed as a least-cost optimisation problem with pipe diameters being decision variables. Even that, somewhat restricted, formulation of the optimal network design represents a difficult problem to solve because the optimization problem is non-linear due to energy conservation constraints and because pipes for water supply are manufactured in a set of discrete-sized diameters. In order to solve this NP-hard problem

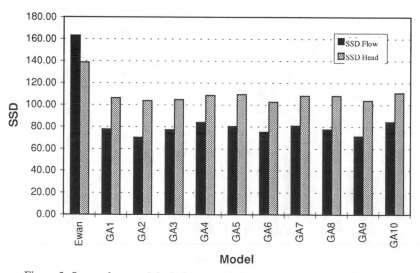

Figure 3. Sums of squared deviations (SSD) for a trial-and-error solution and 10 GA runs

exactly it is suggested that only explicit enumeration or an implicit enumeration technique, such as Dynamic Programming, can guarantee the optimal solution. Murphy et al. (1994) and Savic and Walters (1994) used a binary GA to solve several example problems from the literature. They found that GA solutions compared favourably in terms of cost and minimum head requirements to those obtained by several other techniques. Namely, consistently better discrete-diameter results were obtained, while they were up to 3% more costly than the continuous-diameter or split-pipe solutions from the literature.

Water distribution systems age with time and begin to experience problems of deteriorating infrastructure, unaccounted-for water and service disruption. Their sensitivity toward leaks and breaks increases and accelerates their deterioration resulting not only in a loss of water, but also in contamination through cracked pipes. This situation, combined with expansion of existing systems, gives rise to tremendous maintenance decision-making problems. The rehabilitation, replacement, and/or expansion of existing systems to meet current and future demands becomes a centre of primary interest, when increased amounts of money are necessary to re-establish and/or maintain an adequate level of service, and the funds for this type of work are very limited, allowing at best only partial solutions to the problem.

Halhal et al. (1995) demonstrate the effectiveness of combining Structured Messy Genetic Algorithms (SMGA) with multiobjective optimisation to form a model capable of handling real network improvement problems, in particular the selection of solutions involving small sets of pipe improvements from within a very large water distribution network. This may be achieved by improving its carrying capacity, its pipes' physical integrity, its flexibility and its water quality by means of duplicating, replacing and/or cleaning some of its pipes, and/or installing new pipes to convey water to new areas. The problem is formulated as:

Maximise $f(i) = Benefit(i)$

Minimise $F(i) = Cost(i)$ (4)

Subject to $Cost(i) \leq Fund$

where $Benefit(i)$ is the benefit resulting from adopting solution i, $Cost(i)$ is the cost of the solution i and $Fund$ is the maximum budget available for the improvement of the system.

The multiobjective aspect is tackled here using the concept of Pareto-optimality, a way of ranking a multiobjective function on a rational basis. The benefit and the cost are split into separate terms, and are calculated for each solution. The population of solutions is divided into different groups or sub-populations according to cost. To effect an adequate distribution of solutions throughout the cost and benefit ranges, a technique of niche formation and speciation was introduced through a "fitness sharing" scheme. The final optimal results are all non-dominated solutions, known as Pareto optimal solutions, which are neither inferior nor superior to one another (topmost solutions in Figure 4). The results in Figure 4 are for a typical real life water distribution network of a town of 50000 inhabitants in Morocco. The SMGA developed starts with short string solutions (including only a small number of pipes) and builds up more complex solutions as the process continues, mirroring the evolution of complex life forms from very simple beginnings.

CONCLUSIONS

Various complex problems that arise in hydraulics and water resources in general have been solved using evolution-based programs and their hybrids. The applications of these techniques yield remarkable results with respect to the number of possible solutions an engineer may be faced with when dealing with the design and management of hydraulic systems. It can be anticipated that the number of applications in this area will steadily grow since GAs are not only effective, they are also easily realisable due to the con-

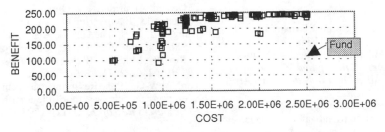

Figure 4. Best solutions of each generation of SMGA

ceptual simplicity of the basic mechanisms. Their potential is even greater when parallel forms of the algorithms can be developed and executed in low-cost multiprocessor computing systems.

ACKNOWLEDGEMENT

This work was supported by the UK Engineering and Physical Sciences Research Council, grant GR/J09796. The authors are also grateful to Ewan Associates and South West Water Services Ltd. for providing the data for the network calibration example.

REFERENCES

Babovic, V. and Larsen, L.C. 1994. Calibrating Hydrodynamic Models by Means of Simulated Evolution, UT Delft, personal communication.

Cembrowicz, R.G. 1992. Water supply systems optimisation for developing countries, *Pipeline Systems*, B. Coulbeck and E. Evans (eds.), Kluwer Academic Publishers, 59-76

Cieniawski, S.E., Eheart, J.W. and Ranjithan, S., 1995. Using Genetic Algorithms to Solve a Multiobjective Groundwater Monitoring Problem, *Water Resources Research*, 31(2): 399-409.

Delleur, J.W. 1988. Expert systems in hydrology. *Proc. 1st International Conference on Computer Methods and Water Resources: Computational Hydraulics*, Ouazar, D. Brebbia C.A. and H. Barthet (eds.): 119-133.

Duan, Q., Sorooshian, S., and Gupta, V.K. 1994. Optimal Use of the SCE-UA Global Optimization Method for Calibrating Watershed Models, *Journal of Hydrology,* 158: 265-284.

Ewan Associates. 1991. Network Analysis Report for the Danes Castle Zone of Exeter City. Report commissioned by South West Water Services Limited.

Goldberg, D.E. and Kuo, C.H.1987, Genetic Algorithms in Pipeline Optimization. *ASCE Journal of Computing in Civil Engineering*, 1(2): 128-141.

Goldberg D.E.1989.*Genetic Algorithms in Search, Optimisation and Machine Learning*, Addison Wesley.

Halhal D, Walters G.A., Ouazar D, and Savic D.A. 1995. Multi-Objective Optimal Expansion of Water Distribution Systems Using a Structured Messy Genetic Algorithm Approach, submitted to *ASCE J. Water Res. Planning and Management*.

Holland, J.H. 1975 *Adaptation in Natural and Artificial Systems*. MIT Press.

McKinney, D.C. and Lin, M-D.1994. Genetic Algorithm Solution of Groundwater Management

Models. *Water Resources Research*, 30(6):1897-1906.

Mäckle, G., Savic D.A. and Walters G.A. 1995. Application of Genetic Algorithms to Pump Scheduling for Water Supply, *Proc. GALESIA '95*, IEE Conf. Publ. No. 414, Sheffield, UK: 400-405.

Michalewicz Z. 1992. *Genetic Algorithms + Data Structures = Evolution Programs*, Springer-Verlag.

Mohan, S. and Loucks, D.P. 1995. Genetic Algorithms for Estimating Model Parameters. *Proc. Integrated Water Resources Planning for the 21st Century*, ASCE, M.F. Domenica (ed.): 460-463.

Murphy, L.J. and Simpson, A.R. 1992. Genetic Algorithms in Pipe Network Optimization, Report No. R93, Department of Civil and Environmental Engineering, University of Adelaide, Australia.

Murphy, L.J., Dandy, G.C. and Simpson, A.R. 1994. Optimum Design and Operation of Pumped Water Distribution System. *Proc. Conf. on Hydraulics in Civil Engineering, Institution of Engineers*, Australia, Brisbane, February, 1994

Ritzel, B.J., Eheart, J.W. and Ranjithan,S.1994. Using Genetic Algorithms to Solve a Multiple Objective Groundwater Pollution Containment Problem, *Water Res. Research*,30(5): 1589-1603.

Savic, D.A. and Walters G.A. 1995a. An Evolution Program for Optimal Pressure Regulation in Water Distribution Networks, *Eng. Opt.* 24(3): 197-219.

Savic, D.A. and Walters G.A. 1995b. Genetic Algorithm Techniques for Calibrating Network Models, *Centre for System and Control*, Report 95/12 University of Exeter, UK.

Walski, T.M, Brill, E.D., Gessler, J., Goulter, I.C., Jeppson, R.M., Lansey, K., Han-Lin Lee, Liebman, J.C., Mays, L., Morgan, D.R., and Ormsbee, L. 1987. Battle of the Network Models: Epilogue. *J. Water Resources Planning and Management*, ASCE, 113(2), 191-203.

Walters, G.A. 1988. Optimal design of pipe networks: a review, *Proc. 1st International Conference on Computer Methods and Water Resources: Computational Hydraulics*, Ouazar, D., C.A. Brebbia and H. Barthet (eds.): 21-32.

Walters, G.A. and Lohbeck, T.K. 1993. *Optimal layout of tree networks using genetic algorithms*, *Engineering Optimization*, 22: 27-48.

Walters, G.A. and Smith, D.K. 1995. Evolutionary Design Algorithm for Optimal Layout of Tree Networks, *Engineering Optimization*,24:261-281.

Wang, Q.J. 1991. The Genetic Algorithm and its Application to Calibrating Conceptual Rainfall-Runoff Models. *Water Resources Research*, 27(9): 2467-2471.

Yeh, W.W-G. 1985. Reservoir management and operations models: a state-of-the-art review, *Water Resources Research*, 21(12): 1797-1818.

Stochastic Hydraulics'96, Tickle, Goulter, Xu, Wasimi & Bouchart (eds) © 1996 Balkema, Rotterdam. ISBN 90 5410 817 7

Application of stochastic dynamic programming for the operation of a two-reservoir system on the Blue Nile in the Sudan

Darko Milutin & Janos J. Bogardi
Wageningen Agricultural University, Netherlands

Mamdouh M.A. Shahin
International Institute for Infrastructural, Hydraulic and Environmental Engineering (IHE), Delft, Netherlands

Hesham E.S.A. Ghany
Permanent Joint Technical Commission for Nile Waters, Giza, Egypt

ABSTRACT: The paper presents the application of two stochastic dynamic programming (SDP) based optimization approaches in deriving a long-term operating strategy of a two-reservoir system. Both models consider stochasticity explicitly by taking into account only the uncertainty of river flows. The conventional SDP-based optimization algorithm derives the joint optimal operating strategy for both reservoirs simultaneously. Thus, it is a direct application of the SDP optimization methodology. To alleviate dimensionality problems, another model relies on decomposition of the system into single-reservoir subsystems with subsequent iterative optimization and simulation of an individual reservoir operation. The methodology models the interaction between reservoirs through consideration of non-utilized releases and supply deficits associated with each of the reservoirs. The case study system consists of two real-world reservoirs forming a cascade on the Blue Nile in the Sudan primarily providing water for local domestic and irrigation purposes. The paper presents a comparison of the two approaches through analyses carried out over a number of alternatives involving various system and model configurations.

1 INTRODUCTION

Life in Egypt and certain parts of the Sudan is almost completely dependent on water supplied by the River Nile system. The explosive increase of population in Egypt and the Sudan coupled with the limited water resources in both countries call for more efficient management of available resources.

The major contributor to the Nile's flow is the Blue Nile which provides approximately four-seventh's of the Nile's total flow. The Sennar and the Roseires dams were built on the Blue Nile in 1925 and 1968 respectively. The reservoirs created by these dams aim at releasing the water stored in a flood season to augment the natural flow in the subsequent low-flow season. The government of the Sudan has a plan to heighten the Roseires reservoir thereby increasing its live capacity from about $2.3 \times 10^9 \text{m}^3$ to about $6.7 \times 10^9 \text{m}^3$. The Sennar reservoir has presently a live storage capacity of about $0.6 \times 10^9 \text{m}^3$ declining at a rate of 0.7% of the present capacity each year (Shahin 1985).

The Blue Nile brings a long-term mean annual flow of about $49 \times 10^9 \text{m}^3$ at Roseires reduced by 1 to $2 \times 10^9 \text{m}^3/\text{yr}$ at Sennar. This loss is compensated for by $4 \times 10^9 \text{m}^3/\text{yr}$ supplied by the Dinder and Rahad rivers as they reach the Blue Nile below Sennar. A schematic drawing of the Roseires-Sennar system

including the related demand centres are shown in Figure 1. Both reservoirs primarily serve for irrigation water supply having a secondary purpose of hydro-power generation. Roseires provides water for the Rahad irrigation scheme (RH), a number of pumping schemes immediately downstream of the reservoir (RP), and the Abu Niam irrigation scheme (AN). Sennar regulates the agreed flow rate of Egypt's share of the Blue Nile discharge (EG) and supplies the Gezira (GZ) and Suki (SK) irrigation schemes, the local (SS) and Geneid (GN) sugar-cane cultivation schemes, and provides water for mainly domestic and irrigation demands in the local area downstream of the reservoir (SP).

Irrigation is the principal water consumer in arid zones. Shortage of water may occur any time, thus affecting agriculture adversely. Accordingly, a plan is needed to distribute any expected shortage in the best possible pattern both temporally and spatially to reduce the adverse impacts on agriculture.

2 OBJECTIVES AND SCOPE OF STUDY

The aim of the present study is a comparison of two stochastic dynamic programming (SDP) based optimization techniques within the derivation of a long-term operating strategy of the presented

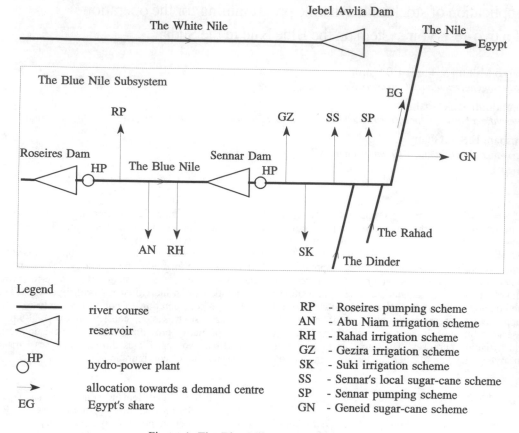

Figure 1. The Blue Nile two-reservoir system.

two-reservoir system. The first method employs decomposition of the system into single-reservoir subsystems wherein the subsequent optimization and simulation of each reservoir's policy are carried out separately. The second approach considers the two-reservoir system as a whole and applies direct SDP-based optimization procedure to derive their joint operating strategy. The analyses were carried out over a number of alternatives which included variations of the system configuration and initial assumptions with respect to the features of the methods themselves. Namely, four issues were considered to be of major importance:

· the use of 'real' and 'hypothetical' demands in deriving the optimal operating policy;
· the impact of increasing the capacity of the upstream reservoir;
· the effects on the operating policy caused by associating all the demands to either of the reservoirs; and

· optimization of the system's operation wherein the system itself is represented by a composite hypothetical reservoir.

3 SYSTEM OPERATION

Long-term operational assessment of a multiple-reservoir water resources system operation is confronted with a series of difficulties of which the dimensionality of the optimization appears to be one of the major problems. This complexity is induced by a number of factors: (1) reservoirs may interact by means of both serial and parallel interconnections; (2) water transfer from one basin to another should also be allowed; (3) possible supply-demand patterns may include a single reservoir supply towards multiple demand targets while any of the demand centres could be associated with more than one reservoir;

(4) uncertainty of the processes involved; etc. Each new element (e.g. reservoir, demand centre, water transfer structure) of a complex reservoir system require additional (state and/or decision) variables and sets of constraints to describe the system thus introducing new dimensions to the problem. This may subsequently result in prohibitive computational requirements imposed by any straightforward optimization application.

Dynamic programming (DP) and its stochastic extension (SDP) have long been recognized as effective tools in water resources management, and especially in reservoir operation. This popularity is due to DP's inherent flexibility to accommodate most important features of water resources systems, such as: their dynamic nature, nonlinear interrelationships, and stochasticity of processes involved. However, the discrete nature of DP causes that the computational load may easily become prohibitive by increasing the number of elements of the system considered, thus increasing the number of state and decision variables needed to describe the system. Namely, DP requires that system state(s) and decisions be mapped over and represented by their respective discrete domains. The subsequent DP-based optimization includes a search for the set of optimum decisions over all possible combinations of state and decision variables. Ultimately, the increase of the number of state and decision variables and the refinement of the discretization of their representative domains may easily cause the dimensionality of the problem to explode beyond any acceptable limits thus reflecting the well known "curse of dimensionality" that follows DP (Bellman 1957).

3.1 Decomposition

One of the means to alleviate the inherent computational burden could be sought in representing the system by a set of simpler subsystems. The proposed decomposition algorithm (Bogardi et al. 1995) reduces the dimensionality of an optimization problem while, at the same time, preserving some important features of a multiple-reservoir system operation. The method is based on a physical decomposition of a system into a series of single-reservoir subsystems. Thus, a multidimensional decision problem is reduced to a sequence of one-dimensional optimization tasks. To derive the operational strategy of the system as a whole a combined optimization/simulation procedure is applied to each reservoir individually. The reservoirs are introduced into the computational process in a sequence which is primarily determined by their physical position in the system. The ordering is based on the mechanism named

"sequential downstream moving decomposition" (further in the paper referred to as *sequential SDP*). Application of this principle to the Blue Nile case study system could be summarized in the following: (1) being the uppermost reservoir in the system the Roseires reservoir's operation is derived and assessed first; (2) optimization of Sennar's operation follows taking into account the relevant results obtained in the previous step.

The algorithm is essentially an iterative procedure. One iteration cycle comprises optimization and simulation of operation of each reservoir within the system. These principle iterative cycles are repeated until a stable system return is reached. Information transfer between two consecutive iteration runs includes sets of individual reservoirs' monthly supply shortage estimates derived in the preceding iteration. These deficits are to be considered as additional demands associated with each reservoir situated directly upstream of the reservoir in question. In other words, Sennar's supply deficit obtained in one iteration is considered as an additional demand associated with Roseires in the subsequent iteration.

The methodology's intrinsic stochastic dynamic programming optimization procedure (Loucks et al. 1981) derives the optimal, expectation oriented, long-term operational strategy for a single reservoir. The operating policy is expressed in distinctive operating tables derived for each month (stage) within an annual cycle. The state variable that describes the system (reservoir) is the volume of water stored at the beginning of the time step. Uncertainty is explicitly incorporated into the optimization procedure: monthly inflow to a reservoir represented by a set of different classes with their respective independent or transitional probabilities is considered as an additional state variable in the SDP-based optimization procedure. Thus, and regardless whether the inflows are considered random or Markovian, the system's state is described by two state variables: (1) reservoir storage at the beginning of the month, and (2) the inflow to the reservoir during the month. With respect to the case study system, the preliminary analyses of the available historical inflow records for the Blue Nile (67-year-long time series of monthly flow volumes for both reservoirs) showed weak correlation between subsequent monthly flows. Therefore, inflow uncertainty within the SDP optimization procedure has been represented by independent probabilities of monthly river flows.

The resulting SDP operating policy is expressed in terms of the optimal decision to be taken as a function of system states. The decision to be taken during each time step is represented by the storage volume of the reservoir at the end of the time interval. Having the initial storage, inflow and final storage defined and assuming that reservoir losses could be derived at each stage, both the consumptive and non-consumptive releases could be estimated from the continuity equation which describes the balance of water in the reservoir during the given stage. The state transformation equations for both reservoirs are:

$$S_{t+1,1} = S_{t,1} + I_{t,1} - R_{t,1} - F_{t,1} - E_{t,1} \qquad (1)$$

and

$$S_{t+1,2} = S_{t,2} + I_{t,2} + F_{t,1} - R_{t,2} - F_{t,2} - E_{t,2} \qquad (2)$$

where (note that index r with its values 1 and 2 depicts Roseires and Sennar reservoirs respectively)
$S_{t,r}$ initial storage volume in period t,
$S_{t+1,r}$ final storage volume in period t,
$I_{t,r}$ inflow to a reservoir during period t,
$R_{t,r}$ consumptive release in period t,
$F_{t,r}$ non-consumptive release during period t,
$E_{t,r}$ evaporation loss during period t.

Equations 1 and 2 have two binding constraints reflecting minimum and maximum storage and consumptive release limits:
· storage constraint

$$S_{t,r}(min) \leq S_{t,r} \leq S_{t,r}(max) \qquad (3)$$

· release constraint

$$0 \leq R_{t,r} \leq R_{t,r}(max) \qquad (4)$$

A multiple-decision problem that arises from the envisaged complex water allocation pattern is reduced to a single-objective optimization by aggregating individual requirements for water from a reservoir into a single composite demand. This simplification is justified by the arrangement of individual demands with respect to a predetermined priority order which is conformed with in the subsequent allocation of available releases from the reservoir. The aggregated demand and its distribution can further be used in two different ways in the optimization process:
· Directly, by optimizing the operational policy of a reservoir towards reaching the 'real demand' as close as possible. This is achieved by minimizing the expectation of the annual aggregate of the objective function which penalizes, in terms of a squared value, both shortage and surplus of a monthly reservoir supply.
· Indirectly, forming only the shape of the demand curve. The final estimates of monthly demand values constitute the so-called 'hypothetical

monthly demands' which are oriented towards the value of the incremental annual median inflow to the reservoir. This alternative arose from the intention to create an overwhelmingly large demand. The 'hypothetical demand' is assumed to constitute a theoretical maximum demand a reservoir of unrestricted size, while having no losses whatsoever, would be able to fulfil without any shortage to occur. It is obvious that these prerequisites are not met by real-world reservoirs. Thus the "hypothetical demand" might be approximated, but never achieved. This transformation provides maximum challenge towards the utilization of the reservoir storage capacity, while the demand distribution remains unchanged. The annual median inflow to the reservoir is redistributed with respect to the real (monthly) demand distribution within an annual cycle, i.e. the "hypothetical demand" values are reflecting the monthly distribution of the components of the supply requirements. The objective pursued in optimization with respect to the "hypothetical demand" is to minimize the expected value of the annual sum of squared monthly shortage of releases towards the corresponding demands for water. Note that, unlike with the "real demand" objective function, possible oversupply is not penalized in this case.

With respect to the case study system it should be emphasized that the "real demand" concept has been tested in optimization of both Roseires and Sennar reservoirs. However, the "hypothetical demand" optimization approach could have only be applied to Roseires reservoir. This is due to a very low incremental inflow to Sennar while it faces an overwhelmingly large demand regarding the Egypt's share of the Blue Nile's water. Thus, a "hypothetical demand" based optimization would have resulted in an unrealistic and ineffective operating strategy with respect to satisfying the Egypt's rights to the Blue Nile's water.
The SDP-based optimization procedure is driven by a well-known backward recursive relationship written for the case of random monthly flows:

$$f_t^{(n)}(k,i) = min_l \left[C_{k,i,l,t} + \sum_{j=1}^{NI_{t+1}} p_j^{(t+1)} \cdot f_{t+1}^{(n-1)}(l,j) \right] \qquad (5)$$

$$\forall k,i; \; l \; feasible$$

where
k, i, l indices depicting representative discrete values of the initial storage and inflow state variables, and the final storage decision variable respectively;

76

$C_{k,i,l,t}$ is the total deterministic cost (penalty) accumulated if the system has endured a transition from state S_k to state S_l at stage t, given the inflow to the system during the stage was I_i;

$p_j^{(t+1)}$ is the independent probability of inflow occurrence of class j during period $t+1$;

NI_{t+1} is number of discrete values describing the inflow process in period $t+1$;

$f_t^{(n)}(k,i)$ accumulated expected penalty over the stages $(1, 2, ..., n)$ given that the system states at stage n are (k,i);

$f_{t+1}^{(n-1)}(l,j)$ accumulated expected penalty over the stages $(1, 2, ..., n, n+1)$ given that the system states at stage $n+1$ are (l,j).

Penalty component $C_{k,i,l,t}$ can assume two forms depending on whether a 'real" or a 'hypothetical' demand representation is used in optimization:
· 'real' demand

$$C_{k,i,l,t}^{(r)} = (D_t^{(r)} - R_{k,i,l,t}^{(r)})^2 \qquad (6)$$

· 'hypothetical' demand

$$C_{k,i,l,t}^{(r)} = \left\{ \begin{array}{ll} (D_t^{(r)} - R_{k,i,l,t}^{(r)})^2 ; & D_t^{(r)} \geq R_{k,i,l,t}^{(r)} \\ 0 & ; \quad otherwise \end{array} \right. \qquad (7)$$

where
r assumes values 1 and 2 representing Roseires and Sennar reservoirs respectively;
$D_t^{(r)}$ total demand imposed upon reservoir r in time period t;
$R_{k,i,l,t}^{(r)}$ total consumptive release from reservoir r during time period t.

Following the optimization, simulation takes place to evaluate the derived policy. As a part of simulation output the updated values of the expected remaining demands, available non-utilized releases over the whole simulation period, and the expected supply shortages of the reservoir are passed through to computational cycles involving reservoirs which operation is directly influenced by these factors. Finally, the release volumes obtained by simulation are allocated to individual users according to the pre-determined priority assigned to each demand associated with the reservoir.

3.2 Joint operation

Due to its relatively simple structure the presented case study system offers a possibility to compare the presented *sequential SDP* algorithm with the direct SDP optimization of the operation of the two-reservoir system as a whole. Direct optimization of the entire system's operation (further to be referred to as a *conventional SDP*) concentrates on the analyses of the joint operation of system elements (reservoirs). This inevitably increases the number of state and decision variables required to describe the system within the optimization procedure. In this particular case, the system's state is represented by four variables, i.e. two reservoirs depicted by their respective storage states and the related stochastic inflow variables. At the same time, the values of two decision variables remain to be optimized: the respective final storage volumes of each of the reservoirs.

The *conventional SDP* optimization model retains the same sets of continuity equations and related constraints as its iterative counterpart (Equations 1 through 4). Similarly, Equation 5 may also represent the backward SDP recursion in the *conventional SDP*. However, the interpretation of the indices within the *conventional SDP* is somewhat more complex:

k depicts a representative discrete value of the initial system state, thus reflecting every possible combination of the initial storage volumes of both reservoirs in period t;

l is a representative discrete value of the final storage decision variable depicting every possible combination of the final storage volumes of both reservoirs in period t;

i, j are representative discrete values of the inflow to the system including every possible combination of inflow to both reservoirs in periods t and $t+1$ respectively;

NI_{t+1} number of all possible combinations of representative discrete inflow values of both reservoirs in period $t+1$.

Within the *conventional SDP* optimization there has been only one alternative as to the selection of the objective function and demand representation. Namely, the joint objective function has been to minimize the expectation of the annual aggregate of squared deviation of a monthly system release from the respective "real demand" imposed upon the two reservoirs jointly. Therefore, the penalty component $C_{k,i,l,t}$ (Equation 5) is assigned a value according to the following relation (the superscripts *(1)* and *(2)* depict Roseires and Sennar reservoirs respectively):

$$C_{k,i,l,t} = (D_t^{(1)} - R_{k,i,l,t}^{(1)})^2 + (D_t^{(2)} - R_{k,i,l,t}^{(2)})^2 \qquad (8)$$

Similarly to a single-reservoir operational analysis, within the *conventional SDP* the optimization itself has to evaluate the operation of the system over all possible combinations of initial states (initial storage and inflow associated with both reservoirs)

77

and decisions (final storage states of both reservoirs). Clearly, the number of state-decision combinations increases dramatically with the number of reservoirs in the system. As a consequence, the choice of storage discretization becomes severely limited. Therefore, the comparison of the two approaches has been confined to the cases (Table 2) wherein storage discretization levels have been kept at 13 storage classes for Roseires and 6 storage classes for Sennar. The *sequential SDP* does allow finer storage discretization (i.e. up to 55 storage classes for each of the reservoirs) and these alternatives have been analyzed (Ghany 1994). However, the results obtained under various combinations of refined storage discretization within the *sequential SDP* have not shown significant improvement over those presented in Table 2 (see alternative *S12*).

Table 1. Initial system characteristics.

Reservoir	Active storage ($10^9 m^3$)	Annual demand ($10^9 m^3$)	Demands (arranged according to the associated supply priorities)
Roseires	2.3	2.06	- Rahad irrigation scheme; - aggregated domestic and irrigation demands supplied by a pumping scheme downstream of Roseires; - Abu Niam irrigation scheme;
Sennar	0.6	43.88	- Egypt's share of the Blue Nile's water; - Gezira irrigation scheme; - Suki irrigation scheme; - Sennar sugar-cane cultivation scheme; - Geneid and Geneid extension of sugar-cane cultivation scheme; - aggregated domestic and irrigation demands supplied by a pumping scheme downstream of Sennar;

Table 2. Alternatives analyzed in the study.

Alternative	Demand type used in optimization		Specific assumptions and modifications regarding the system configuration, in addition to the original system structure and water allocation patterns (Table 1)
	Roseires	Sennar	
S1 - sequential	fictive	real	-
S2 - sequential	fictive	real	Optimization only: A fictitious Egypt's share demand is associated with Roseires while the real Egypt's share demand remains assigned to Sennar.
S3 - sequential	real	real	Optimization only: A fictitious Egypt's share demand is associated with Roseires while the real Egypt's share demand remains assigned to Sennar.
S4 - conventional	fictive	real	-
S5 - conventional	real	real	-
S6 - sequential	fictive	real	As *S2*, with Roseires active storage augmented to 6.7 [$10^9 m^3$].
S7 - conventional	real	real	Roseires active storage augmented to 6.7 [$10^9 m^3$].
S8 - conventional	real	real	All demand targets associated with Roseires.
S9 - conventional	real	real	All demand targets associated with Sennar.
S10 - conventional	composite: real		The system is represented by a composite reservoir with an active storage of 2.9 [$10^9 m^3$].
S11 - conventional	composite: real		The system, assuming the augmented Roseires, is represented by a composite reservoir with an active storage of 7.3 [$10^9 m^3$].
S12 - sequential	fictive	real	As *S2*, with Roseires storage discretization set to 25 classes.

4 ANALYSES AND RESULTS

The presented optimization methodologies have been tested over a number of alternative system configurations. All the alternatives have been created on the basis of the initial system characteristics which are presented in Table 1. The options composed for the presentation are displayed in Table 2.

All the alternatives are generated in a relatively straightforward fashion with the exception of options $S10$ and $S11$ which require some further clarification. Namely, the two-reservoir system is represented by a single hypothetical reservoir with the following characteristics:

· The active storage of the composite reservoir equals the aggregated active volumes of Roseires and Sennar i.e., $2.9 \times 10^9 m^3$ and $7.3 \times 10^9 m^3$ given the original Roseires size and the planned augmented Roseires volume respectively.
· Due to Sennar's relatively small contribution towards the composite reservoirs size the characteristic curves of the hypothetical reservoir as well as the relative evaporation losses are assumed to be those of Roseires reservoir.
· Inflow to the composite reservoir equals the aggregate of the inflows to Roseires and Sennar.
· The composite reservoir's storage representation within the optimization procedure is set at 25 discrete storage values.

Clearly, these aggregation principles have not involved any calibration of the composite reservoir characteristics (i.e., characteristic curves, evaporation losses, inflow time series). However, the principal intention behind this simplified approach has been to provide an additional piece of information for the appraisal of the *sequential SDP* decomposition approach. Similarly, the $S12$ alternative has been included on the list of presented options due to the adopted storage discretization. This alternative may be used to draw a parallel to options $S10$ and $S11$.

Table 3 summarizes the outcomes of the performed optimization and the respective simulation analyses over the 12 selected alternatives. The choice of a type of results to be presented has been confined to only the expected annual system supply deficit to provide a clear-cut distinction between different methods. In addition, the aim of optimization has been to derive long-term operating strategy based on the assumption that all the system characteristics and processes except river flows assume constant, expected values. Therefore, it has been assumed that the expected annual system supply deficit may provide sufficient information to differentiate the major characteristics of the proposed algorithms.

Table 3. The results.

Alternative	Demand type used in optimization: Roseires/Sennar	Expected annual supply deficit $[10^6 m^3]$
$S1$ - sequential	fictive / real	3314.35
$S2$ - sequential	fictive / real	3110.12
$S3$ - sequential	real / real	3141.25
$S4$ - conventional	fictive / real	1472.85
$S5$ - conventional	real / real	1048.72
$S6$ - sequential	fictive / real	2390.07
$S7$ - conventional	real / real	244.53
$S8$ - conventional	real / real	3886.38
$S9$ - conventional	real / real	751.60
$S10$ - conventional	composite: real	2865.30
$S11$ - conventional	composite: real	2146.55
$S12$ - sequential	fictive / real	3070.07

5 CONCLUSIONS AND RECOMMENDATIONS

The selected alternatives have provided a set of results that reflect a number of features that could be attributed to the methods used in the analyses. The principal lines that could be drawn from the presented results are summarized in the following:

· As expected, *conventional SDP* outperforms the *sequential SDP* throughout all the alternatives, even in the case where the original size of the Roseires reservoir is considered within the *conventional SDP* as opposed to the enlarged Roseires volume assumed in the decomposition option. This is clearly due to the fact that the *conventional SDP* optimization utilizes the advantages of the two reservoirs' joint operation up to the maximum possible level provided by the initial set of assumptions.
· The alternative wherein a fictitious Egypt's share demand component is associated with the Roseires reservoir brought about some improvement into the operation of the system derived by *sequential SDP* thereby justifying the assumption that it would "swing" the Roseires' policy towards a more synchronized Roseires-Sennar release decisions with respect to Egypt's share of the Blue Nile's water.

· Both optimization approaches show significant improvement in the system's operation given that the Roseires' active storage is augmented.

· Probably the most interesting conclusion might be drawn from the *S8* and *S9* alternatives which show that considerable operational gains could be achieved if all the demand components are associated with the downstream reservoir. In this case the larger upstream reservoir improves significantly the temporal distribution of available water at the downstream reservoir site. Namely, the operation of Sennar is optimized under the assumption that all the demands imposed upon the two-reservoir system are associated with this reservoir only. The subsequent simulation of its operation results in the distribution of supply shortage that would occur in this case. This shortage is in turn used as a single requirement put upon Roseires which optimal operating strategy would modify the temporal distribution of its huge incoming flow towards the estimated distribution of shortage obtained by simulation of Sennar operation.

The *sequential SDP* has repeatedly resulted in a performance inferior to that of the *conventional SDP*. In addition to those allready mentioned, another reason for this may be in the fact that there is a considerable mismatch between the Blue Nile's flow volumes and the imposed demands on one side and the reservoir sizes on the other. It has, however, been reported (Bogardi *et al.* 1995) that the *sequential SDP* offers flexible means for operational analyses of large, multiple-reservoir systems. This can also, to a certain extent, be argued by the results obtained out of the composite reservoir concept (alternatives *S10* and *S11*). Obvious advantages of the method may clearly be seen within a long-term operational assessment of a water resources system consisting of 3 or more reservoirs, the cases where computational requirements in terms of the dimensionality of a direct optimization by means of the *conventional SDP* become prohibitive.

Further research involving the presented two reservoirs may be directed towards operational analysis of the extended system including also the Jabel Awlia reservoir on the White Nile and Khashm el-Gerba reservoir which is located on the Atbara river approximately 350km upstream of its confluence with the Nile.

Another research direction stems from the fact that both Roseires and Sennar are confronted with serious problems of active storage reduction due to siltation. This raises the need for a study of the effects of siltation on the optimal long-term operation of the two-reservoir system. In addition, the envisaged augmentation of Roseires brings about the inevitable increase in the installed capacity of the adjoining hydro-power installation. Consequently, an optimization study involving a maximum energy generation as an objective, including the analysis of the respective impacts on water supply, inevitably arises as a necessary research option.

REFERENCES

Bellman, R. 1957. *Dynamic Programming*, Princeton University Press, Princeton, N.J.

Bogardi, J.J., B.A.H.V. Brorens, M.D.U.P. Kularathna, D. Milutin, & K.D.W. Nandalal 1995. Long-term assessment of a multi-unit reservoir system operation: The ShellDP program package manual, *Report Series*, Report 59, 272p, Department of Water Resources, Wageningen Agricultural University, The Netherlands.

Ghany, H.E.S.A. 1994. Application of stochastic dynamic programming for the operation of a two-reservoir system on the Blue Nile in the Sudan, *MSc Thesis*, H.H. 186, IHE, Delft, The Netherlands.

Loucks, D.P., J.R. Stedinger & D.A. Haith 1981. *Water Resources Systems Planning and Analysis*, 320-332, Prentice Hall, N.J.

Shahin, M.M.A. 1985. *Hydrology of the Nile Basin*, Developments in Water Science Series, 21, 575p, Elsevier Publ. Co., Amsterdam, New York, Tokyo.

Stochastic Hydraulics'96, Tickle, Goulter, Xu, Wasimi & Bouchart (eds) © 1996 Balkema, Rotterdam. ISBN 90 5410 817 7

Application of probabilistic point estimation methods to optimal water resources management under uncertainty

Bing Zhao
Civil and Environmental Engineering Department, Arizona State University, Tempe, Ariz., USA

Yeou-Koung Tung
Water Resources Center and Department of Statistics, University of Wyoming, Laramie, Wyo., USA

ABSTRACT: Stochastic programming deals with optimization problems under uncertainty. In particular, a distribution problem in stochastic programming attempts to find the probability distribution function or statistical moments of the optimal objective function and decision variables. This paper applies two efficient probabilistic point-estimate (PE) methods to distribution problems. A numerical example of linear programming-based groundwater hydraulic model is used to demonstrate the methodology. Transmissivity and boundary conditions in groundwater hydraulic model are spatial random variables. The numerical results show that there exist significant uncertainties in the solutions. The results computed from using the PE methods are compared with those computed from using Monte-Carlo simulation. It is found that the PE methods are more efficient than the Monte Carlo method while achieving the same mean of optimal objective function.

1 INTRODUCTION

Uncertainties in water resource engineering projects arise from various aspects including, but not limited to, hydrologic, hydraulic, structural, and economic uncertainties (Mays and Tung, 1992). Uncertainties in hydrology and hydraulics are further attributed to the following sources (Yen et al., 1986): (1) natural uncertainties associated with the inherent randomness of natural processes; (2) model uncertainty reflecting the inability of a model or design technique to represent precisely the system's true physical behavior; (3) model parameter uncertainties resulting from inaccurate quantification of the model inputs and parameters; (4) data uncertainties including measurement errors, inconsistency and non-homogeneity of data, and data handling and transcription errors; and (5) operational uncertainties including those associated with construction, manufacture, deterioration, maintenance, and other human factors that are not accounted for in the modeling or design procedure.

Optimization using mathematical programming is useful and has been widely applied to water resources management. Due to the presence of various uncertainties mentioned above, the results of an optimization model for a water resources system are subject to uncertainty. Stochastic programming deals with optimization problems whose parameters are random variables. Problems in the stochastic programming can be classified into distribution problems, chance-constrained problems, recourse problems, and discrepancy cost problems (Dempster, 1980). A distribution problem is aimed at finding the probability distribution functions or statistical moments of the optimal solutions and the objective function value.

The determination of the probability distribution function of the objective function was first studied by Tintner (1955) in the area of agricultural economics. Since then distribution problems have attracted notable attention in literature (Bereanu, 1967; Dempster and Papagaki-Papoulias, 1978; Wets, 1980; Ermoliev and Wets, 1988). Most of the distribution problems are related to stochastic linear programming (LP). In general, the distribution problems arise in practice from four related but distinct sources (Dempster, 1980): error analysis for LP, post-optimality analysis for LP, sensitivity analysis of LP in a random environment, and recourse problem.

Stochastic programming has been applied to water resources managements. Aquado et al. (1977) and Gorelick (1982) applied sensitivity analysis to evaluate the optimal solution for groundwater management problems. Chance-constrained programming has been applied to various problems such as groundwater management problem (Tung, 1986) and groundwater quality problem (Wagner and Gorelick 1987), stream-aquifer management problem (Hantush & Marino, 1989), and estuary management (Tung et al., 1990; Bao and Mays, 1994). Wagner and Gorelick (1989) used the Monte Carlo simulation to generate a large number of realizations of hydraulic conductivity to characterize the probability distribution function of the optimal pumping rates. Chance-constraint, stochastic approximation, and stochastic quasi-gradient methods were applied to a large irrigation system (Dupačová,

1991). The application of recourse problems in groundwater quality management can be found in Wagner et al. (1992). .

This paper applies two efficient methods to solving distribution problems based on probabilistic point-estimate (PE). They are Harr's PE method (Harr, 1989) and a modified Harr's PE method (Chang et al., 1995). Other PE methods can be found in Rosenblueth (1975, 1981), Li (1992), and Zoppou and Li (1993). The basic idea of the PE methods is to approximate the original probability distribution function of a random variable by "concentrated" points in such a manner that certain statistical characteristics of the random variable are preserved. Then, the "concentrated" points are used to evaluate the uncertainty features of the model outputs resulting in significant computation saving.

In this paper, a numerical example of LP-based groundwater optimization model is performed to demonstrate the methodology in which the transmissivity and boundary conditions are random. The results by the Monte Carlo method and the PE methods are compared. The results show that the PE methods are efficient than the Monte Carlo method and yield comparably accurate estimation of the mean and standard deviation of the optimal pumping rates (optimal solution) and the optimal sum of hydraulic heads (objective function).

2 GENERAL FORM FOR STOCHASTIC PROGRAMMING

A general form for the deterministic mathematical programming model can be stated as

$$\min \quad g_0(x)$$

$$\text{s.t.} \quad g_i(x) \le 0, \ i=1,...,m \qquad (1)$$

$$x \in X \subset \Re^n$$

in which x is the vector of decision variables to be determined and $g_0(x)$ and $g_i(x)$ for $I=1,2,...,m$ are the objective function and constraints, respectively. In a stochastic system uncertainties may exist in the parameters of the objective function $g_0(x)$ and constraints, $g_i(x)$.

Suppose ξ is a random vector of dimension $(k \times 1)$. Then, the stochastic programming model for (1) can be expressed as (Dupačová, 1987; Kall, 1993)

$$\min \quad g_0(x,\xi)$$

$$\text{s.t.} \quad g_i(x,\xi) \le 0, \ i=1,...,m$$

$$x \in X \subset \Re^n \qquad (2)$$

$$\xi \in \Xi \subset \Re^k$$

in which ξ is a vector of random parameters and is assumed to have a known joint probability distribution

function P. A more appropriate general form for the stochastic programming can be stated as (Kall, 1993)

$$\min \quad E_\xi[g_0(x,\xi)]$$

$$\text{s.t.} \quad E_\xi[f_i(x,\xi)] \le 0, \ i=1,...,s$$

$$E_\xi[f_i(x,\xi)] = 0, \ i=s+1,...,m \qquad (3)$$

$$x \in X \subset \Re^n$$

$$\xi \in \Xi \subset \Re^k$$

in which f_i may be constructed in various ways yielding different stochastic programming problems. It can be seen that a stochastic programming problem has been converted to a deterministic equivalent.

3 DISTRIBUTION PROBLEMS OF STOCHASTIC PROGRAMMING

As stated previously, the distribution problem of stochastic programming is to find a probability distribution function or statistical moments of the optimal solution and the objective function value. Consider the stochastic programming model (2) in which ξ is a random vector of dimension $(k \times 1)$ with known probability distribution P. By using Monte Carlo method, many realizations of ξ can be generated. For each of the realizations, the model described by (2) is solved by an appropriate deterministic mathematical programming methods. Therefore, many optimal decision vectors and objective function values, each of which corresponds to a realization, can be obtained from which the probability distribution functions and/or the statistical moments for the optimal decision vector and the objective function value can be estimated.

The Monte Carlo analysis solves a distribution problem by repeatedly solving the deterministic optimization model with the realizations of random parameters being the inputs and with optimal solution and objective function value being the outputs. However, this would require tremendous computations. When the dimension of the random parameter vector is large, solving distribution problems by the Monte Carlo method may not be practical.

4 SOLVING DISTRIBUTION PROBLEMS BY PROBABILISTIC PE METHODS

The algorithms for Harr's and the modified Harr's PE methods will be presented and followed by a discussion on how to apply the PE methods to solve a distribution problem. Consider a system with inputs, outputs, and kernel function with parameters. The PE method was originally proposed to assess the uncertainty features of the model outputs as affected by the uncertainties of the stochastic model parameters and inputs. Instead of using the information on the entire probability distribution function of the involved random variables, the PE methods only use the selected points in the parameter space to evaluate model outputs from which uncertainty features of model outputs are estimated..

4.1 Harr's PE Method

Consider a model $W = h(\xi)$ in which W is the output, ξ is a vector k random variables with known mean vector (μ) and covariance matrix (**C**), and $h(\cdot)$ is the kernel function which can be a deterministic simulation model or deterministic optimization model. The procedures of Harr's PE method for assessing the model output uncertainty is as follows:

a.   Apply eigenvalue-eigenvector decomposition to the correlation matrix **R** for the random vector ξ as

$$\mathbf{R} = \mathbf{V} \, \Lambda \, \mathbf{V}^t \qquad (4)$$

in which **V** is an eigenvector matrix consisting of k columns of eigenvectors, $\mathbf{V} = [v_1, ..., v_k]$; Λ is a diagonal matrix with k eigenvalues on the diagonal, $\Lambda = \mathrm{diag}(\lambda_1, ..., \lambda_k)$; and the superscript 't' denotes the transpose operator for a matrix.

b.   Compute 2k points in the parameter space for model evaluations by

$$\xi_{i\pm} = \mu \pm \sqrt{k} \, \mathbf{D} \, v_i, \quad i = 1, ..., k \qquad (5)$$

in which μ is the mean vector for the random vector ξ, **D** is the diagonal matrix containing the standard deviations of the random vector ξ, and v_i is the ith eigenvector of the correlation matrix **R**.

c.   Compute 2k model output values by

$$w_{i\pm} = h(\xi_{i\pm}), \quad i = 1, ..., k \qquad (6)$$

d.   The mth moment about the origin for the output W is computed by

$$E(W^m) = \frac{\sum\limits_{i=1}^{k} \lambda_i \overline{w_i^m}}{k} \qquad (7)$$

in which

$$\overline{w_i^m} = \frac{w_{i+}^m + w_{i-}^m}{2}, \quad i = 1, \cdots, k \qquad (8)$$

e.   The mean and standard deviation for the output W are computed, respectively, by

$$E(W) = \frac{\sum\limits_{i=1}^{k} \lambda_i \overline{w_i}}{k} \qquad (9)$$

and

$$STD(W) = \sqrt{E(W^2) - E^2(W)} \qquad (10)$$

It may be noted that above procedures only involve one output variable W. For multiple model outputs, the above procedure can still be applied from which the correlation between the output variables can be derived.

4.2 Modified Harr's PE Method

Chang et al. (1995) modified Harr's algorithm by selecting points at the intersections of the eigenvectors and the hypersphere in the standardized eigen-space. Their numerical experimental results indicated that the modified Harr's method is comparable with Harr's method and Rosenblueth's method. Most of the procedures in the modified Harr's PE algorithm are the same as those in Harr's. However, there are two differences. First, the 2k points in the parameter space for model evaluations are obtained by

$$\xi_{i\pm} = \mu \pm \sqrt{k} \, \mathbf{D} \, \mathbf{L}^{0.5} \, v_i, \quad i = 1, ..., k \qquad (11)$$

Second, the eigenvalues in Harr's algorithm are replaced by 1.0.

4.3 Solving Distribution Problems by PE Methods

The PE methods can be used to solve the distribution problems by properly defining $W = h(\xi)$. The procedures for solving the optimization model is the kernel function $h(\cdot)$ which transforms the inputs to the outputs. The inputs are the random vector ξ in the optimization model while the outputs are the optimal decision variables x and the corresponding objective function value.

When Harr's point-estimate or modified Harr's point-estimate is used, 2k points in the parameter space are generated by (5) or (11) at each of which the optimization model is solved. Then, 2k sets of optimal solutions and objective function values are computed by (6) from which the means and standard deviations of the optimal solutions and the optimal objective function value can be computed by (9)-(10).

5   NUMERICAL APPLICATION IN GROUNDWATER HYDRAULIC MANAGEMENT

The proposed methodology is applied to a groundwater optimization-based hydraulic

Table 1. Non-Homogeneous, Stationary (same mean and variance at each node for T), T is spatially correlated, T and h_0 are uncorrelated, mean of T = 10,000 ft^2/day, mean of h_0 = 100 feet, C.O.V. of T = 20%, C.O.V. of h_0 = 10%, an exponential model for spatial covariance for T : Cov(d)=sigma^2*exp(-d/L) (Maidment, 1993, p20.6) where d is distance between 2 pts., range (alpha) = 200 ft, L = alpah/3 = 67 ft.

	Harr's Method		Modified Harr's Method		Monte Carlo Simulation	
Computing Time	4.6 minutes (75 runs)		4.6 minutes (75 runs)		10.51 hours (10,000 runs)	
	Mean	STD	Mean	STD	Mean	STD
$h_{1,1}$	95.5558	14.0152	96.8600	13.2111	95.6214	13.7379
$h_{2,1}$	97.1412	11.2792	97.9416	11.1729	97.3189	10.4953
$h_{3,1}$	96.2517	11.4743	97.6881	11.2063	96.8372	10.5159
$h_{4,1}$	91.4186	16.0482	95.9201	13.7278	92.6893	15.1948
$h_{1,2}$	97.3083	11.2701	97.8057	11.1751	97.3258	10.5124
$h_{2,2}$	96.7781	11.0416	97.2166	11.0187	96.9040	10.1982
$h_{3,2}$	96.1327	11.0871	96.7871	11.0277	96.4862	10.1755
$h_{4,2}$	95.6237	11.4704	96.8569	11.1943	96.0922	10.6808
$h_{1,3}$	96.9896	11.3090	97.0533	11.3227	96.8695	10.5332
$h_{2,3}$	96.4688	11.0490	96.3437	11.0824	96.5014	10.1828
$h_{3,3}$	95.7386	11.1291	95.3671	11.1555	96.0328	10.1468
$h_{4,3}$	94.7784	12.0396	94.1280	12.2365	95.4708	10.6959
$h_{1,4}$	93.8855	14.6598	93.6897	14.9020	92.8090	15.1781
$h_{2,4}$	96.4032	11.3381	95.5069	11.9856	96.1148	10.6936
$h_{3,4}$	95.4905	11.5427	93.3539	12.5153	95.4918	10.6806
$h_{4,4}$	89.2818	16.8118	86.1221	16.5481	89.2731	16.0499
$Q_{1,1}$	13.6122	33.5448	8.7837	27.3961	13.1500	33.7963
$Q_{2,1}$	0.0000	0.0000	0.0000	0.0000	0.0200	1.4141
$Q_{3,1}$	0.0000	0.0000	0.0000	0.0000	0.0500	2.2356
$Q_{4,1}$	27.9514	43.0876	11.4864	31.0810	23.8200	42.6003
$Q_{1,2}$	0.0000	0.0000	0.0000	0.0000	0.0000	0.0000
$Q_{2,2}$	0.0000	0.0000	0.0000	0.0000	0.0000	0.0000
$Q_{3,2}$	0.0000	0.0000	0.0000	0.0000	0.0000	0.0000
$Q_{4,2}$	0.4220	4.5743	0.0000	0.0000	0.7000	8.1000
$Q_{1,3}$	0.0000	0.0000	0.0000	0.0000	0.0000	0.0000
$Q_{2,3}$	0.0000	0.0000	0.0000	0.0000	0.0000	0.0000
$Q_{3,3}$	0.0000	0.0000	0.0000	0.0000	0.0000	0.0000
$Q_{4,3}$	2.6063	15.9325	4.0540	19.7223	0.9000	9.5000
$Q_{1,4}$	19.9217	37.9206	19.5945	39.0491	23.4000	42.4000
$Q_{2,4}$	0.4220	4.5743	2.7027	16.2162	0.7000	8.4000
$Q_{3,4}$	0.0000	0.0000	6.7567	25.1002	0.8000	9.1000
$Q_{4,4}$	35.0640	47.1830	46.6216	49.3752	36.3000	48.1000
obj.	1525	176	1529	176	1528	161

The unit for h is feet; the unit for Q is ft/day; the unit for obj. is ft.

management model. The groundwater system is a two-dimensional, steady-state, and non-homogeneous confined aquifer. The optimization technique is the linear programming (LP) technique. The optimization-based hydraulic management model is to determine the optimal locations for pumping wells and the associated pumping rates such that the summation of hydraulic heads for the nodes is maximized while the following constraints are satisfied: (1) the partial differential equations governing the groundwater system (with the boundary conditions); (2) the summation of pumping rates is greater than a prescribed value; and (3) the

pumping rates and heads are greater than zero. Since the boundary conditions and transmissivity are random variables, the solutions from the groundwater optimization-based hydraulic management model are also random. The PE methods can be used to find the uncertainties associated with the optimized objective function and optimal pumping rates.

The transmissivity, T, and boundary condition, h_0, are assumed to be independent lognormal random variables having the means 10,000 ft^2/day and 100 ft, respectively. The coefficients of variation for T and h_0 are Ω_T=0.2 and Ω_{ho}=0.1, respectively. There are 16 interior nodes and each node is a potential pumping location. The Harr's PE and the modified Harr's PE methods are used to compute the uncertainties in the optimal objective function and pumping rates. For each point generated by the PE methods, the LP technique is used to solve the optimization model from which the uncertainties in the optimal objective function and pumping rates are computed. The results from using the PE methods are compared with those from the A number of realizations for the random vector are generated by Monte Carlo analysis. For each of the realizations, the groundwater optimization-based management model is solved. Then, the mean and standard deviations are computed.

Table 1 indicates that there exist great uncertainties in the optimal pumping rates and the heads by three methods (please see the columns of standard deviations or STD). It may imply that the optimal solutions of many deterministic models may not be really optimal in the viewpoint of real-world application. It may also imply that it is very important to do uncertainty analysis for the exist groundwater models. It is also found that the mean of optimal objective function for two PE uncertainty analysis methods and Monte Carlo analysis is approximately the same. The optimal pumping rates and heads from the Harr's PE method are approximately the same as those from Monte Carlo analysis. The differences in the optimal pumping rates and heads between the modified Harr's PE method and other two methods imply that the optimal solutions are not unique. Nevertheless, all these three methods give the same mean optimal objective function, which is the goal of the management model. More importantly, it can be found in Table 1 that the computing time for the PE methods is less than that for Monte Carlo analysis.

6 SUMMARY AND CONCLUSIONS

This paper proposed efficient and general methods to solve distribution problems of stochastic programming based upon PE uncertainty analysis, thus, unifying uncertainty analysis and stochastic programming. Stochastic programming deals with optimization problems under uncertainty. To find probability distribution functions (or moments such as mean and standard deviation) of an optimal solution and an objective function is defined as a distribution problem of stochastic programming. Although the most general method for solving a distribution problem is Monte Carlo analysis, Monte Carlo analysis may not be efficient for solving distribution problems of large-scale stochastic programming. Therefore, efficient and general methods are needed. PE uncertainty analysis methods are such methods. The considered PE uncertainty analysis methods are Harr's PE and the modified Harr's PE.

A numerical example of groundwater linear-programming-based optimization model was performed to demonstrate the methodology. Transmissivity and boundary conditions are considered as random variables. It was found that there exist great uncertainties in the optimal solutions indicating the importance of uncertainty analysis for the real-world groundwater management. It was also found that PE uncertainty analysis methods can yield comparably accurate results as does Monte Carlo analysis with a large number of realizations. The computational time for PE methods is less than that for Monte Carlo analysis. The saving of computational time is particularly significant for distribution problems for large-scale stochastic programming. PE uncertainty analysis methods are promising methods for solving the distribution problems of large-scale stochastic programming.

REFERENCES

Aquado, E., Sitar, N, & I. Remson 1977. Sensitivity analysis in aquifer studies. *Water Resources Research*, 13: 733-737.

Bao, Y.X., & L.W. Mays 1994. Optimization of freshwater inflows to the Lavaca-Tres Palacios Estuary. *J. of Water Resources Planning. and Management.*, ASCE, 120(2): 218-236.

Bereanu, B. 1967. On stochastic linear programming distribution problems, stochastic technology matrix. Z. *Wahrsch. Verw. Gebiete*, 8: 148-152.

Chang, C.-H., Tung, Y.-K., & J-C. Yang 1995. Evaluating performance of point estimates methods. *Applied Math. Modelling*, 19(2).

Dantzig, G.B. & A. Madansky 1961. On the solution of two-state linear programs under uncertainty. In Neyman (ed), *Proc.* 4th Berkeley Symp. Math. Stat. Prob., I.J. Berkeley, 165-176.

Dempster, M.A.H. 1980. Introduction to stochastic programming. In M.A.H. Dempster (ed), *Stochastic Programming*, Academic Press, Inc., New York, New York.

Dempster, M.A.H. & A. Papagaki-Papoulias 1978. Monte Carlo analysis of two small scale test problems for the evaluation of methods for the distribution problem of stochastic linear programming. *Tech. Report* TR7806, Oxford Systems Associates Limited.

Dupačová, J. 1987. Stochastic programming with incomplete information: a survey of results on postoptimization and sensitivity analysis. *Optimization*, 18: 507-532.

Dupačová, J., Gaivoronski, A., Kos, Z., & T. Szántai 1991. Stochastic programming in water management: a case study and a comparison of solution techniques. *European Journal of Operational Research*, 52: 28-44.

Dyson, R.G. & G. Swaithes 1976. An approach to stochastic programming for medium term planning. *Omega*, 4: 479-485.

Ermoliev, Yu. & R.J-B. Wets 1988. *Numerical Techniques for Stochastic Optimization*, Springer-Verlag, New York, New York.

Gorelick, S.M. 1982. A model for managing sources of groundwater pollution. *Water Resources Research*, 18(4): 773-781.

Guymon, G.L. 1994. *Unsaturated Zone Hydrology*, Prentice-Hall, Englewood Cliffs, New Jersey.

Hantush, M.M.S. & M.A. Marino 1989. Chance-constrained model for management of a stream-aquifer system. *Journal of Water Resources Planning and Management*, ASCE, 115(3): 259-277.

Harr, M.E. 1989. Probabilistic estimates for multivariate analysis. *Applied Math. Modelling*, 13(5), 313-318.

Kall, P. 1993. "Solution methods in stochastic programming." In J. Henry and J.-P. Yvon (eds), *System Modelling and Optimization*, Proceedings of the 16th IFIP-Tc7 Conference, Springer-Verlag.

Maidment, D.R. (ed.) 1993. *Handbook of Hydrology*, McGraw-Hill, Inc.

Mays, L. W. & Y.-K. Tung 1992. *Hydrosystems Engineering and Management*, McGraw-Hill, New York.

Rosenblueth, E. 1975. Point estimates for probability moments. *Proc. Nat. Academy of Science*, 72(10): 3812-3814.

Rosenblueth, E. 1981. Two-point estimates in probabilities. *Applied Math. Modelling*, 5(5): 329-335.

Tinter, G. 1955. Stochastic linear programming with applications to agricultural economics. In: Proceedings of 2nd symp. linear programming, H.A. Antosiewicz, ed., 197-209.

Tung, Y.-K. 1986. Groundwater management by a chance-constrained model. Journal of Water Resources Planning and Management, ASCE, 112(1): 1-19.

Tung, Y.-K., Bao, Y.X., Mays, L.W., & W.H. Ward 1990. Optimization of freshwater inflow to estuaries. *Journal of Water Resources Planning and Management*, ASCE, 116(4): 567-584.

Tung, Y.-K. & Yen, B.C. Some recent progress in uncertainty analysis for hydraulic structure designs. In B.C. Yen and Y.K. Tung (eds), *Reliability and Uncertainties in Hydraulic Designs*, 17-34, American Society of Civil Engineers, New York, NY, 1993, p.17-34.

Wagner, B.J. & S.M. Gorelick 1987. Optimal groundwater quality management under parameter uncertainty. *Water Resources Research*, 23(7): 1162-1174.

Wagner, B.J. & S.M. Gorelick 1989. Reliable aquifer remediation in the presence of spatially variable hydraulic conductivity: from data to design. *Water Resources Research*, 25(10): 2211-2225.

Wagner, J.M., Shamir, U., & H.R. Nemati 1992. Groundwater quality management under uncertainty: stochastic programming approaches and the value of information. *Water Resources Research*, 28(5): 1233-1246.

Wets, R.J-B. 1980. The distribution problem and its relation to other problems in stochastic programming. In M.A.H. Dempster (ed), *Stochastic Programming*, Academic Press, Inc., New York, New York.

Wolsey, L.A. 1970. Bounds in stochastic linear programming. *Research Paper*, Manchester Business School.

Yeh, K.-C. & Y.-K. Tung 1993. Uncertainty and sensitivity analyses of pit-migration. *Journal of Hydraulic Engineering*, ASCE, 119(2): 262-281.

Yen, B.C. Cheng, S.T. & C.S. Melching 1986. First order reliability analysis. In B.C. Yen, (ed), *Stochastic and Risk Analysis in Hydraulic Engineering*: 1-36. Water Resources Publications, Littleton, Colo.

Stochastic Hydraulics'96, Tickle, Goulter, Xu, Wasimi & Bouchart (eds) © 1996 Balkema, Rotterdam. ISBN 90 5410 817 7

Acceptable risk: A normative evaluation

J.K.Vrijling
Delft University of Technology, Faculty of Civil Engineering, Netherlands

W.Van Hengel
Rijkswaterstaat, Building Division, Utrecht, Netherlands

R.J.Houben
Simtech, Rotterdam, Netherlands

ABSTRACT: Human civilisations have always tried to protect themselves against natural and manmade hazards. The degree of protection is a matter of political choice. This choice should be expressed in terms of risk and acceptable probability of failure to form the basis of the probabilistic design of the protection. A normative framework for the evaluation of risk is proposed and tested in two cases.

1 INTRODUCTION

Human civilisations are threatened by many natural hazards like floods, earthquakes, etc. that have increasing consequences in developed societies. Others hazards are man-made and result from the technological progress in civil, chemical and nuclear engineering. Human civilisations try to protect themselves against these hazards after their occurrence has shown the consequences or when the risks are felt to be high.

The idea of acceptable risk or safety may change quite suddenly due to a single spectacular accident as the examples of the catastrophe at Chernobyl, the plane crash at Schiphol airport in 1992, and the Dutch river floods of 1993 and 1995 have shown. Public opinion is influenced not only by the accident itself, but also by the attention paid to it by the media.

However in an advanced technological society politicians should not only base their decisions upon above mentioned subjective interpretations of risk. As the notion of acceptable risk forms the basis for the design of many technological wonders, ranging from simple river levees to advanced multi-purpose dams, that contribute to the welfare of the western nations, politicians should have a more or less objective framework for risk evaluation. This paper proposes the outlines of a possible framework, that can serve as a rational basis for technological design.

2 COMMON POINTS

Generally two points of view appear in studies of acceptable risk levels [1,2,3]. The point of view of the individual, who decides to undertake an activity weighing the risks against the direct and indirect personal benefits and the point of view of the society, considering the question whether an activity is acceptable in terms of the risk/benefit trade off for the total population.

An important aspect is the degree of voluntariness with which the decision is taken and the risk is endured. In the personal sphere these decision are freely and quickly made knowing that the choice can be immediately amended if the risks turn out to be higher than expected. In the case of societal decisions involving risk however the individual can still make his appraisal in accordance with his own set of standards, but his influence on the final outcome is democratically limited. This might imply a sense of involuntariness and compel him to adopt a sceptical attitude towards (involuntary) risks imposed by societal decisions. The following characteristics results:

> *the decision to accept risk has a*
> *cost/benefit character and depends*
> *on the degree of voluntariness*

The first point of view leads to the personally

acceptable level of risk or the acceptable individual risk, defined in [8] as "the frequency at which an individual may be expected to sustain a given level of harm from the realisation of specified hazards". The specified level of harm is narrowed down to the loss of life in many practical cases.

Society looks to the *total* damage done by the occurrence of an accident, which may comprise a number of casualties, material and economic damage and the loss of or harm to immaterial values. Commonly the notion of risk in a societal context is reduced to total number of casualties [1,2,3] using a definition as in [11]: "the relation between frequency and the number of people suffering from a specified level of harm in a given population from the realisation of specified hazards". If the specified level of harm is narrowed down to the loss of life, the societal risk may be modelled by the frequency of exceedance curve of the number of deaths, also called the FN-curve due to a specific hazard. The consequence part of a risk may also be limited to the

total material damage expressed in monetary terms [3,8]. It should be noted however, that the reduction of the consequences of an accident to the number of casualties or the economic damage may not adequately model the public's perception of the loss. It is clear that the societal risk is judged at a national level i.e. the total risk in a year (casualties as well as material and immaterial damage related to the frequency) connected to a certain activity. The distinction between local and national risk seems necessary noting that "small unrestrained developments could add up to a noticeable worsening of the overall situation" (HSE [2]).

3 RISK

The probability density function (p.d.f.) of the number of deaths N_{dij} given an accident for activity i at place j can have many forms. Three forms are presented here to facilitate further thinking. The first

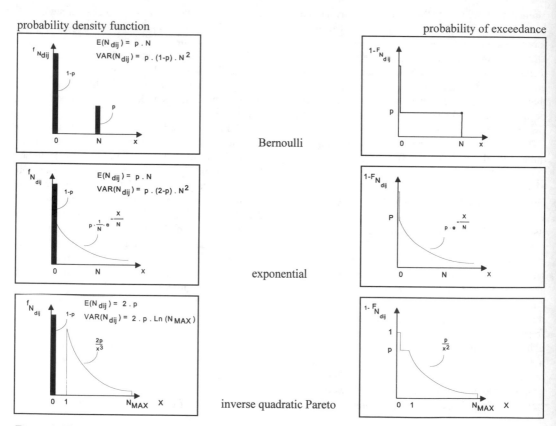

Figure 1: Theoretical p.d.f's and probability of exceedance curves for the number of deaths

is a Bernoulli one, that limits the outcomes to zero or N fatalities. The second, that allows for a greater variation in the outcome, is the exponential distribution. The probability of exceedance curve of the number of fatalities, that can be derived from the exponential form reflects to some extent the FN-curves that result from practical quantitative risk assessment (QRA) studies.

The third is the inverse quadratic Pareto distribution that coincides with the type of norm put forward by the Ministry of VROM [1]. Exactly the same models could be applied for the material damage that results from a disaster, if the horizontal axis is measured in monetary units.

4 ACCEPTABLE RISK; A NORMATIVE EVALUATION

In all concepts the most stringent of the *personally* and the *socially* acceptable level of risk determines the acceptable level of risk. So both criteria have to be satisfied.

A philosophy for acceptable risk comparable to [1], that takes into account the cost/benefit and the voluntariness aspect, was developed by the Technical Advisory Committee for Water Retaining Structures (TAW [3]).The safety standard consists of a flexible evaluation of the individual and the societal acceptable risk but adds to these an economic approach taking the material damage into account. The latter provides the link with the safety philosophy of the Dutch dikes that was developed after the 1953 flood [8,9].

4.1 *Personally acceptable level of risk*

The smallest component of the socially accepted level of risk is the personal assessment of risks by the individual. As an attempt to model this appraisal procedure quantitatively is not feasible, it is proposed to look with the insight gained to the preferences revealed in the accident statistics.

The actual personal risk levels inherent to various activities show statistical stability over the years and are approximately equal for the Western countries, indicating a consistent pattern of preferences. The probability of losing ones life in normal daily activities such as driving a car or working in a factory appears to be one or two orders of magnitude lower than the overall probability of dying. Only a

purely voluntary activity such as mountaineering entails a higher risk (Fig.2). This observation of public tolerance of 1000 times greater risks from voluntary than from involuntary activities with the same benefit was already made by Starr (1969, 1972).

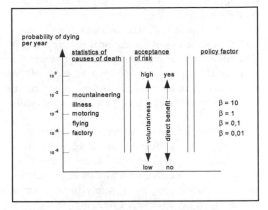

Figure 2: Personal risks in Western countries, deduced from the statistics of causes of death and the number of participants per activity

In view of the consistency and the stability of the death risks presented, apart from a slightly downward trend due to technical progress, it would appear permissible to deduce therefrom a guideline for decisions with regard to the personally acceptable probability of failure:

$$P_{fi} = \frac{\beta_i \cdot 10^{-4}}{P_{d|fi}} \qquad (1)$$

where $P_{d|fi}$ denotes the probability of being killed in the event of an accident. In this expression the policy factor β_i varies with the degree of voluntariness with which an activity is undertaken and with the benefit perceived. It ranges from 100 in the case of complete freedom of choice like mountaineering, to 0.01 in case of an imposed risk without any perceived direct benefit. This last case includes the individual risk criterion proposed by VROM for the siting of a hazardous installation near a housing area without any direct benefit to the inhabitants.

4.2 Socially acceptable level of risk

The basis of the framework with respect to societal risk is an evaluation of risks due to a certain activity on a national level. The risk evaluation on a national level has to be translated to local installations or activities in order to support a systematic appraisal by the local authorities.

If a risk criterion is defined on a local level as in [1] and [2] the height of the national risk criterion is determined by the number of locations, where the activity takes place and the type of p.d.f. of the consequences of an accident. The resulting national norm has to be evaluated, as it was not intentionally formulated.

It seems preferable to start with a risk criterion on a national level and to evaluate the acceptable local risk level in view of the actual number of installations, the cost/benefit aspects of the activity and the general progress in safety in an iterative process with say a ten year cycle (Fig. 3).

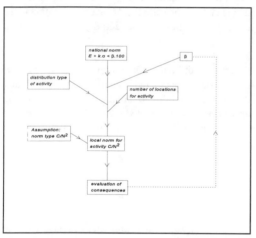

Figure 3: Framework for risk management

4.2.1 Nationally acceptable level of risk

The determination of the socially acceptable level of risk starts from the proposition that the result of a social process of risk appraisal is reflected in the accident statistics. It seeks to derive a standard from these revealed preferences. Using the circle of acquaintances as an instrument of observation, the very low probabilities of a fatal accident, which appear socially acceptable, are perceptible. The recurrence time is within the order of magnitude of a human life span.

In seeking to establish a norm for the acceptable level of risk for engineering structures it is more realistic to base oneself on the probability of a death due to a non-voluntary activity in the factory, on board a ship, at sea, etc. which is approximately equal to : $1.4 \ 10^{-3}$ /year.

If this observation-based frequency is adopted as the norm for assessing the safety of activity i, then after re-arranging the expression, and adopting a rather arbitrary distribution over some 20 categories of activities, each claiming an equal number of lives per year, the following norm is obtained for an activity i with N_{pi} participants in the Netherlands:

$$P_{fi} \cdot N_{pi} \cdot P_{d|fi} \ < \ \beta_i \cdot 100 \qquad (2)$$

This norm should be interpreted in the sense that an activity is permissible as long as it is expected to claim fewer than $\beta_i * 100$ deaths per year (in general: $\beta * 7.10^{-6} *$ national population size). The formula does not account for the standard deviation, which will certainly influence acceptance by a risk averse community.

Risk aversion can be represented mathematically by increasing the mathematical expectation of the total number of deaths, $E(N_{di})$ by the desired multiple k of the standard deviation before the situation is tested against the norm :

$$E(N_{di}) \ + \ k \cdot \sigma(N_{di}) \ < \ \beta_i \cdot 100 \qquad (3)$$

where: k = risk aversion index

To determine the mathematical expectation and the standard deviation of the total number of deaths occurring annually in the context of activity i, it is necessary to take into account the number of independent places N_A where the activity under consideration is carried out.

The model is tested for several activities in [13]. The agreement between the norm derived in this study for reasonable values of N_A and $0.01 < \beta_i < 100$ and the risk accepted in practice in the Netherlands seems to support the model.

4.2.2 Locally acceptable level of risk

The translation of the nationally acceptable level of

risk to a risk criterion for one single installation or location where an activity takes place depends on the distribution type of the number of casualties for accidents of the activity under consideration as shown above. In order to relate the new framework to the present one, it is assumed that on a local level the societal risk criterion is of the type proposed by VROM.:

$$1 - F_{N_{dij}}(x) < \frac{C_i}{x^2} \tag{4}$$
$$for\ all\ x \geq 10$$

Assuming a Bernoulli distribution of the number of casualties, in the national criterion, equation (3), and taking account of N_{Ai} independent locations, gives for the value of C_i:

$$C_i = \left[\frac{-k \cdot \sqrt{N_{A_i}} + \sqrt{k^2 \cdot N_{A_i} + 4 \frac{N_{A_i}}{N} \cdot \beta_i \cdot 100}}{2 \frac{N_{A_i}}{N}} \right]^2 \tag{5}$$

If the expected value of the number of deaths is much smaller than its standard deviation, which is often true for calamities, the previous result reduces to :

$$C_i - \left[\frac{\beta_i \cdot 100}{k \cdot \sqrt{N_{A_i}}} \right]^2 \tag{6}$$

Similar results are obtained for the exponential distribution.

The national societal acceptable risk criterion leads to a local acceptable risk criterion of the VROM-type, which is inversely proportional to the number of independent places N_A and the square of the policy factor β_i:

$$1 - F_{N_{dij}}(x) \leq \frac{C_i}{x^2} \quad for\ all\ x \geq 10$$

$$where \quad C_i = \left[\frac{\beta_i \cdot 100}{k \cdot \sqrt{N_{A_i}}} \right]^2 \tag{7}$$

The VROM-rule [1] is a special case of this general framework for acceptable risk: with $C_i = 10^{-3}$, $N_A = 1000$ (the approximate number of chemical installations) and $k = 3$, it follows that $\beta = 0.03$ which is according to Figure 2 not unreasonable for an involuntarily imposed risk.

4.3 Economically optimal level of risk

The problem of the acceptable level of risk can also be formulated as an economic decision problem. The expenditure I for a safer system is equated with the gain made by the decreasing present value of the risk (Figure 4).

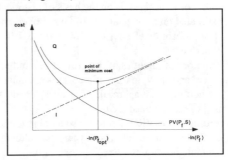

Figure 4: The economically optimal probability of failure of a structure

The optimal level of safety indicated by P_{fopt} corresponds to the point of minimal cost.

$$min(Q) = min(I(P_f) + PV(P_f \cdot S)) \tag{8}$$

where :
Q = total cost
PV = present value operator
S = total damage in case of failure

If despite ethical objections, the value of a human life is rated at s, the amount of damage is increased to:

$$P_{d|fi} \cdot N_{pi} \cdot s + S \tag{9}$$

where:
N_{pi} = number of participants in activity i

This extension makes the optimal failure probability a decreasing function of the expected number of deaths.

The valuation of human life is choosen as the present value of the nett national product per inhabitant. The advantage of taking the possible loss of lives into account in economic terms is that the safety measures are affordable in the context of the national income.

5 PRACTICAL APPLICATIONS

5.1 *Airports*

At Schiphol airport, surrounded by inhabited areas, 90,000 planes leave and arrive every year. So the total number of movements is 180,000 per year. The probability of an accident is on the basis of historical data estimated on average at $5.0 \cdot 10^{-7}$ per movement [12]. The probability of a crash is $180,000 * 5.0 \ 10^{-7}$ = 0.09 per year.

The number of fatalities at the ground (excluding passengers and crew) in case of a crash is estimated at 50, when in a first approximation every crash is assumed to hit inhabited areas.

According to the VROM-rule for societal risk one single flight movement (per year) is already unacceptable because :

$$5.0 \cdot 10^{-7} \ > \ \frac{10^{-3}}{N_{di}^2} \ = \ \frac{10^{-3}}{50^2} \ = \ 4.0 \cdot 10^{-7}$$

(10)

Moreover due to the large number of aircraft movements the expected value and the standard deviation of the total number of fatalities in a year are not very small.

$$E(N_{di}) = N_{A_i} \cdot p \cdot N =$$
$$= 180.000 \cdot 5.0 \cdot 10^{-7} \cdot 50 = 4.5$$
$$\sigma(N_{di}) = \sqrt{(N_{A_i} \cdot p)} \cdot N =$$
$$= \sqrt{180.000 \cdot 5.0 \cdot 10^{-7}} \cdot 50 = 15$$

(11)

A dramatical improvement of aircraft safety would be required, if the total airport operations were to meet the VROM requirement. If the risk of Schiphol is judged on a national level as seems appropriate for a national airport, the result is :

$$E(N_{di}) + k \cdot \sigma(N_{di}) = 49.5 \leq \beta_i \cdot 100 \qquad (12)$$

The societal risk is only acceptable if $\beta \geq 0.5$. This means that the situation depicted here will not be acceptable without discussion.

The refined computer calculations [12] show a more acceptable picture than the crude computations presented above. However the 10^{-5} and the 10^{-6} individual risk contours are respectively just and far outside the perimeter of Schiphol. This may be unacceptable according the VROM-rule for personal risk, but using the framework developed here the situation might be accepted if $\beta = 0.1$.

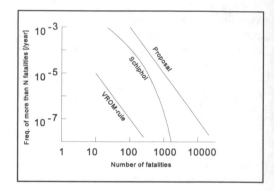

Figure 5: FN-curve for Schiphol, in relation to the VROM-criterion, and the new proposed criterion

The FN-curve calculated in [12] is more favourable than the simple approximation presented above, but unacceptable by several orders of magnitude compared with the VROM-rule for societal risk (Figure 5). If the framework of this paper is applied and C_i is adapted with $N_{Ai}=1$, one national airport, and $\beta_i = 0.1$, in other words if judgement is placed at a national level and the benefits are taken into account, Figure 5 shows that the FN-curve might be acceptable .

The benefits of the airport have to be weighed against the external risk and the possibilities of improvement have to be studied, before a political decision to increase β_i to 0,1 can be taken. Additionally one has to decide that Schiphol will be the only major airport in Holland. Implicitly the Dutch gouvernment has taken both decisions, when it proposed to accept the personal as well as the societal risk connected to Schiphol.

5.2 Polders

The half of Holland that lies below the sea level is divided in $N_A = 40$ more or less independent polders surrounded by dike-rings. If it is assumed that at some future date each polder will house $N_{pij} = 1,000,000$ inhabitants, an estimate of the number of casualties in case of flooding can be made. In 1953 approximately 1% of the inhabitants drowned, giving a value of $p_{d|f} = 0.01$. Little is known of the influence of modern technological development on this number, but the failure of energy and communication networks during the minor floods in Limburg point to a reduced beneficial influence.

The expected value and the standard deviation of the number of deaths in 40 independent polders per year are equal to:

$$E(N_{di}) = N_{A_i} \cdot p_f \cdot p_{d|i} \cdot N_{pi} =$$
$$= 40 \cdot p_f \cdot 0.01 \cdot 10^6$$

$$\sigma^2(N_{di}) = N_{A_i} \cdot p_f \cdot (1 - p_f) \cdot (p_{d|i} \cdot N_{pi})^2 =$$
$$= 40 \cdot p_f \cdot (1 - p_f) \cdot (0.01 \cdot 10^6)^2$$

(13)

If these expressions are substituted in the national norm equ. 3 the solution for $\beta = 1$ becomes $p_f = 3.10^{-7}$ per year. In case the aversion of the inhabitants against flooding is more extreme and $\beta = 0.1$ the acceptable probability of failure of the dikering is

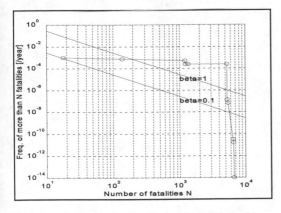

Figure 6: FN-curve for flooding of the Brielse polder

$p_f = 3.10^{-9}$ per year.

For the Brielse dikering near Rotterdam a FN-curve (Fig.6) has been drawn estimating the probability of failure of the existing dikes at 10^{-4} per year. The FN-curve shows that there are five equally likely scenario's with death counts variing from 15 to appr. 5000 people. As these scenario's are assumed to be independent the combinations, that are less likely by an order of magnitude, claim even more casualties. Developing the local criterion for a dikering using the values mentioned above the constant becomes $C_i = 27.8 - 0.278$ for $\beta = 1-0.1$. Thus the present situation based on the philosophy developed by the Deltacommittee [8] in 1960 seems insufficiently safe in the light of modern developments. Following the normative framework developed here the acceptable probability of failure of the dike equals $6.3 \ 10^{-7}$ to $6.3 \ 10^{-9}$ depending on the value of β.

6 CONCLUSIONS

From the personal point of view, the probability of failure (a fatal accident) should meet the following requirement :

$$P_{fi} \leq \frac{\beta_i \cdot 10^{-4}}{P_{d|fi}}$$

(14)

The societal acceptable risk is judged at a national level by placing an upper-bound upon the expected number of fatalities per activity per year. However limiting only the expected number of deaths does not account for risk aversion. Risk aversion can be represented mathematically by adding a confidence requirement to the norm:

$$E(N_{di}) + k \cdot \sigma(N_{di}) < \beta_i \cdot 100$$

(15)

where : k = 3; risk aversion index

The synthesis of this national risk criterion and the VROM-type of local societal risk criterion approach leads to an upper-bound to the FN-curve of the local activity, which is inversely proportional to the number of independent places N_A and the square of the policy factor β_i :

$$1 - F_{N_{dij}}(x) \leq \frac{C_i}{x^2} \quad \text{for all } x \geq 10$$

$$\text{where} \quad C_i = \left[\frac{\beta_i \cdot 100}{k \cdot \sqrt{N_A}} \right]^2 \tag{16}$$

The numerical value of the tolerable frequency can, within certain limits mentioned above, be tuned by the factor β_i. This factor β_i reflects the relative voluntariness and economical benefits of the activity under consideration.

An mathematical-economic approach of the acceptable risk should be included in the philosophy of acceptable risk. It is important to weigh the reduction of risk in monetary terms against the investments needed for additional safety. In this way an economic judgement of the safety level proposed by the two other approaches is added to the information available in the decision making process.

In assessing the required safety of a system the three approaches described above should all be investigated and presented. The most stringent of the three criteria should be adopted.

Finally it should be realised that the philosophy and the techniques set out above are just means to reach a goal. One should not loose sight of the goal *managed safety*, when dealing with the tools, that are provided as instruments to measure an aspect of the entire situation.

REFERENCES

[1]   MINISTRY OF HOUSING, LAND USE PLANNING AND ENVIRONMENT, National Environmental Plan, The Hague, 1988.

[2]   HEALTH AND SAFETY EXECUTIVE, Risk criteria for land-use planning in the vicinity of major industrial hazards, HM Stationery Office 1989

[3]   TECHNICAL ADVISORY COMMITTEE ON WATER RETAINING STRUCTURES, Some considerations on acceptable risk in the Netherlands, Dienst Weg- en Waterbouwkunde, Delft 1984.

[4]   MINISTRY OF HOUSING, LAND USE PLANNING AND ENVIRONMENT, LPG Integral Study (in Dutch), The Hague, 1985.

[5]   CUR, Probabilistic design of flood defences, Gouda, 1988

[6]   VRIJLING, J.K., WESSELS, J.F.M., HENGEL, W. VAN., HOUBEN, R.J., What is acceptable risk. Delft University of Technology and Bouwdienst RWS,Delft 1993.

[7]   VAN DE KREEKE, J., PAAPE, A., On the optimum breakwater design, Proc. 9-th Int. Conf. Coastal Eng.

[8]   DANTZIG, V.D, KRIENS, J. The economic decision problem of safeguarding the Netherlands against floods. Report of Delta Commission, Part 3, Section II.2 (in Dutch), The Hague, 1960.

[9]   DANTZIG V.D, Economic Decision Problems for Flood Prevention, Econometrica 24, pp 276-287, New Haven, 1956.

[10]   MINISTRY OF HOUSING, LAND USE PLANNING AND ENVIRONMENT, Relating to risks (in Dutch), The Hague, 1992.

[11]   INSTITUTE OF CHEMICAL ENGINEERING, Nomenclature for hazard and risk assessment in the process industries, 1985, ISBN 85 295184 1

[12]   NATIONAL AEROSPACE LABORATORY NLR, Analyse van de externe veiligheid rond Schiphol, CR 93485 L,Amsterdam, 1993

[13]   VRIJLING, J.K. ET ALT,Framework for risk evaluation, Hazardous Materials 43 ,1995

[14]   VROUWENVELDER,A.,VRIJLING, J.K., Normstelling acceptabel risiconiveau, PM-95-29 TAW,1995

Stochastic Hydraulics'96, Tickle, Goulter, Xu, Wasimi & Bouchart (eds)© 1996 Balkema, Rotterdam. ISBN 90 5410 817 7

Application of fuzzy set theory and neural networks to the control system of reservoir operation from the viewpoint of water resource

Masahiko Hasebe
Department of Civil Engineering, University of Utsunomiya, Ishii, Japan

Yasutoshi Nagayama
Division of Civil Engineers, Tochigi Prefectural Government, Sano, Japan

ABSTRACT: In this paper, the possibility of the dam control system applied to both neural networks and fuzzy set theory for supporting reservoir operation is investigated. The authors put the hydrological information of precipitation, release discharge, inflow discharge and predicted inflow discharge by filter separation AR method (M.Hino & M.Hasebe 1981 and M.Hasebe etc.1989) etc., the control rules of reservoir operation and the information obtained by inquires to reservoir operator about the operating method of the past, into the dam supporting system.

Neural networks are applied to the decision of the operational line of reservoir operation for the selection of release discharge from reservoir, constant water level of reservoir and storage volume. On the other hand, fuzzy set theory is applied to the decision of the operational volume of reservoir operation for the release discharge from reservoir. As the a result, the dam control system applied to both neural networks and fuzzy set theory is possible and effective for reservoir operation.

1. INTRODUCTION

The automatic operation of dam gate is generally designed to lighten a troublesome workload for the administrator of dam in Japan. The neural networks and fuzzy systems can be applied to the dam control system for supporting reservoir operation. The reasons why two systems are applicable is that the mathematical expressions of dam operating system are difficult and vague in a minor point as for operating dam gate. For the above mentioned reasons, it is considered to be effective to adapt the dam control system combined both the fuzzy set theory and the neural networks for supporting to release discharge from reservoir.

Neural networks and fuzzy set theory have initiated a growing interest not only on the part of mathematicians but also civil engineers. A major portion of this current interests and research is oriented toward the aplications. The hydrological information for dam management is abundant in quality and quantity because the technique of hydrological observation has made great advances recently. Thus, if the hydrological characteristics in the basin are obvious from hydrological information, high level prediction of inflow discharge to flow into reservoir can be performed.

We can better understand the growing popularity of numerical methods that deal with a wide range of real world problems, problems AI has failed to solve if even address. Prominent among these techniques are neural networks and fuzzy set theory. Separately and in combination, neural networks and fuzzy systems have helped to solve a wide variety of problems ranging from process control and signal processing to fault diagnosis and system optimization.

In the control system for reservoir operation, we join together two techniques-neural networks and fuzzy systems-that seem at first quite different but that share the common ability to work well in this natural environment. Although there are other important reasons for interest in them, from an engineering point of view much of the interest in neural networks and fuzzy systems has been for dealing with difficulties arising from uncertainity, imprecision, and noise.

The more a problem resembles those encountered in the real world--and most interesting problems are those--the better the system must cope with these difficulties (B.Kosko 1992).

In this study, the dam control system for supporting reservoir operation, based on the information obtained by inquiries to reservoir operators, is constructed. Consequently, it becomes obvious that the application of both neural networks and fuzzy set theory to the dam control system is possible and effective.

2. FUZZY SYSTEMS AND NEURAL NETWORKS

Neural networks and fuzzy systems estimate input-output functions. Both are trainable dynamical systems. Unlike statiscal estimators, they estimate a function without a mathematical model of how outputs depend on inputs. They *learn from experience* with numerical and, sometimes, linguistic sample data. Neural and fuzzy systems encode sampled information in a parallel distributed framework. Both frameworks are numerical.

2.1 *Fuzzy systems*

In many applied fields, such as engineering, the social sciences and medical diagnostics, the sources of vague or fuzzy data are numerous and diverse both in origin and in magnitude.

Fuzzy set theory has initiated a growing interest not only on the part of mathematicians, but also among engineers. A major portion of this current interests and research is oriented toward the applications. We now explore fuzziness as an alternative to randomness for describing uncertainty. We develop the new sets-as-points geometric view of fuzzy sets. This view identifies a fuzzy set with a point in a unit hypercube, a nonfuzzy set with a vertex of the cube, and a fuzzy system as a mapping between hypercubes.

An element belongs to a multivalued or fuzzy set to some degree in [0,1]. An element belongs to a nonfuzzy set all or none, 1 or 0. More fundamentally, one set contains another set to some degree. Sets fuzzily contain subset as well as elements. Subsethood generalizes elementhood. We shall argue that subsethood generalizes probability as well.

2.2 *Neural networks*

Artificial neural networks consist of numerous, simple processing units or "neurons" that we can globally program for computation. We can program or train neural networks to store, recognize, and associatively retrieve patterns; to solve combinatorial optimization problems; to filter noise from measurement data; to control-illdefined problem; in summary, to estimate sampled functions when we do not know the form of the functions. Artificial neural systems may contain millions of nonlinear neurons and interconnecting synapses.

Many feedback neural networks can learn new patterns and recall old patterns simultaneously, and ceaselessly. Supervised neural networks can learn far more input output pairs, or stimulus-response associations, than the number of neurons and synapses in the network architecture. Since neural networks do not use a mathematical model of how a system's output depends on its input. we can apply the same neural network architecture, and dynamics, to a wide variety of problems (Clifford.G.Y.Lau 1992).

3. THE OUTLINE OF THE DAM CONTROL SYSTEM FOR RESERVOIR OPERATION

The content of the dam control system for supporting reservoir operation is constructed as follows; This control system is composed of two subsystems. One is the decision of the operational line of reservoir operation for the selection of release discharge from reservoir, constant water level of reservoir and storage volume, the other is one of the operational volume of reservoir operation for the release discharge from reservoir.

3.1 *Subsystem for decision of the operational line*

The sub-control system for the selection of release discharge, constant water level of reservoir and storage volume to operate dam gate is applied to the neural networks that imitates the human-like performance in the field of image recognition. This system of the neural networks is composed of three-layer perceptron, that is, input layer (sensory units), second hidden layer (association units) and output layer (response units). Seven neurons are included in the input layer. These neurons are precipitation, river discharge, inflow discharge into reservoir, inflow discharge predicted by filter separation AR method, changing inflow discharge, water level in reservoir and release discharge from reservoir. Three neurons are included in second hidden layer. These are neuron to respond to the hydrologic system (precipitation, river discharge and inflow discharge.), to respond to the discharge (inflow discharge, predictive inflow one ,changing inflow one) and to respond to the state of reservoir (inflow discharge, release discharge and water level in reservoir). One neuron is included in output layer. This is neuron for the selection of release discharge, storage volume and preservation of water level in reservoir. The flow chart of the operational line is shown in Fig.1

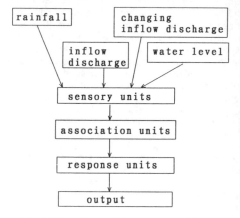

Fig.1 Flow chart of the operational line

3.2 *Subsystem for decision of the operational volume*

This subsystem is the decision of the operational volume of reservoir operation, that is, the release discharge is determined from fuzzy system which is infered from information about inflow discharge into reservoir and predictive inflow one (M.Hasebe &

Y.Nagayama 1993,1994). The flow chart of the operational volume is shown in Fig.2.

Last, with reference to the above-mentioned two subsystem, the outline of the dam control system for supporting reservoir operation is shown in Fig.3.

4. IDENTIFICATION OF PARAMETERS FOR DAM CONTROL SYSTEM

Parameters of the dam control system for reservoir operation are identified on the part of the decision of the operational line which is determined from neural networks system and that of the decision of the operational volume determined from fuzzy system. Neural networks are generally divided into two types. One is the mutually connected net, and another is the hierarchically connected net. In the part of the identification of the operational line for reservoir operation, we use the perceptron which belongs to the hierarchically connect type. Further, parameters are identified through Back Propagation.

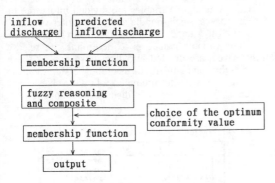

Fig.2 Flow chart of the operational volume

In the part of the operational volume for reservoir operation, the structures of menbership functions of inflow discharge and predicted inflow discharge, and fuzzy inference, that is, fuzzy reasoning and composition, are identified.

4.1 *Sigmoid function*

Reservoir operation including various elements is modeled with neural nets as already shown in Fig.3 Output function combined association units with response unit is sigmoid function as shown in Fig.4.

4.2 *Identification of fuzzy inference*

There are some methods of fuzzy reasoning, fuzzy composition and the consequence value (M.Mizumoto et. al. 1981, M.Mizumoto. 1988). In this paper, the fuzzy systems on the part of the operational volume of reservoir operation are applied to the fuzzy inferences of *Minimum-Maximum-Rule* and *Algebraic product-Maximum-Rule* judging from the fuzzy plain explained subsequently. These figures show in Fig.5 and Fig.6.

The author's call the three dimensional graph, which represents X-axis as predicted inflow discharge, Y-axis as inflow discharge and Z-axis as release discharge, as *fuzzy plain*. This fuzzy plain represents the characteristics of fuzzy inference and the difference in the fuzzy reasoning. The fuzzy plains are shown in Fig.7 and 8. Which fuzzy inference method is better ?. The judgement is difficult because there is no criterion of evaluation to judge rightly.

From the viewpoint that fuzzy control is very consistent with human thinking, it is understood that fuzzy inferences by mim-max-type and algebraic product-max-one is in accord with human thinking because these fuzzy plains are better smooth.

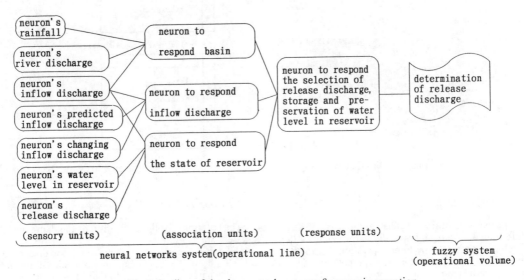

Fig.3 Outline of the dam cotrol system of reservoir operation

97

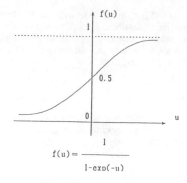

$$f(u) = \frac{1}{1-\exp(-u)}$$

Fig.4 Sigmoid function

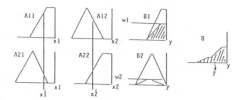

Fig.5 Fuzzy inference of Min-Max-type

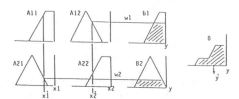

Fig.6 Algebraic product-Max-type

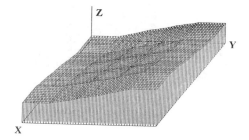

Fig.7 Fuzzy plain of Min-Max-type

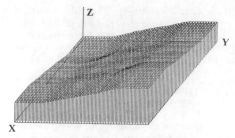

Fig.8 Algebraic product-Max-type

discharge. The relationship between fuzzy labels of input variables and fuzzy labels of output variables is determined from two dimensional fuzzy matrix as shown in Fig.9. And this fuzzy matrix is determined with reference to both the control rules of reservoir operation and the information obtained by inquires to an expert reservoir operator about the operating method of the past. An example of this fuzzy matrix is shown in Fig.10. The intervals of menbership fuctions as input variables are divided into three or four, that is, small, middle and big, further very big. On the other hand, those of menbership functions as output variable ,which is to determine the release discharge, are divided into five or seven. As the a result, the optimum structures of menbership function as inflow discharge, predicted inflow discharge and release discharge are identified as shown in Fig.11.

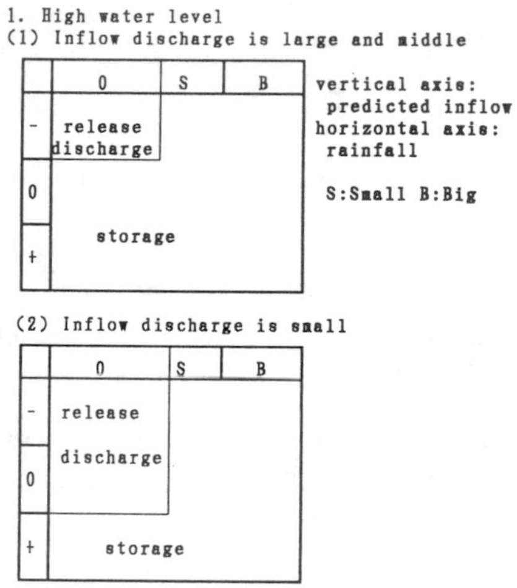

1. High water level
(1) Inflow discharge is large and middle

	0	S	B
-	release discharge		
0			
+	storage		

vertical axis: predicted inflow
horizontal axis: rainfall

S:Small B:Big

(2) Inflow discharge is small

	0	S	B
-	release		
0	discharge		
+	storage		

Fig.9 Two dimensional fuzzy matrix

4.3 *Fuzzy Modelling*

The intervals of fuzzy labels as input variables are determined from inflow discharge and inflow discharge predicted by filter separation AR method. In the same way, those of labels as output variables are determined from storage volume and release

4.4 *Evaluation of the dam control system*

The criterions for the evaluation of the dam control system are as follows;
(1) To decrease the peak volume of release discharge.
(2) To delay the beginning time of peak of release discharge compared with that of peak of inflow discharge.

1 High water level
(1) Inflow discharge is large

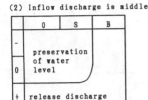

vertical axis: predicted inflow
horizontal axis: rainfall

S:Small B:Big

(2) Inflow discharge is middle

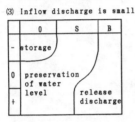

(3) Inflow discharge is small

Fig .10 Fuzzy matrix by inquires of reservoir operator

Inflow Discharge

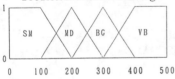

SM: small MD: middle BG: big
VB: very-big

Predicted Inflow Discharge

SM: small MD: middle BG: big
VB: very-big

Release Discharge

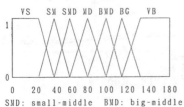

SMD: small-middle BMD: big-middle

Fig.11 Optimum menbership function

(3) To make smooth the release curve of discharge from reservoir.
(4) To secure the storage volume for effective use of the surface of the reservoir as the lake.

5. APPLICATION OF THE REAL DAM BASIN

The above mentioned method has been applied to a A dam basin in Japan with drainage area of 271 km^2. Rainfall patterns using for this analysis are middle scale rainfall type and large scale rainfall one.

5.1 *An example of flood of middle scale rainfall type.*

Comparison the result operated by reservoir operator with the result simulated by the dam control system which is driven by neural networks - fuzzy system are shown in Fig.12 and Fig.13. Fig.12 shows the result operated by the dam control system which is applied to the fuzzy inference of *Minimum-Maximum-Rule* on the part of the operational volume of reservoir operation. In Fig.13, the dam control system is applied to the fuzzy inference of *Algebraic product-Maximum-Rule*.

From these figures, the method operated by neural networks-fuzzy system (hereinafter refered to as simulator) are better than that operated by reservoir operator for the reservoir operation, judging from the criterion of the evaluation of this supporting system. Namely, though the peak value of release discharge by simulator is a little numerous compared to the reservoir operator, the beginning time of the release diccharge by simulator is later than that of the reservoir operator, the release curve is also smooth and the storage volume is many compared to the reservoir operator. On the other hand, in comparison with the fuzzy inference of Min-Max-Rule and that of Alge.pro-Max-Rule, the simulator applied to the fuzzy inference of Min-Max type is better.

5.2 *An example of flood of large scale rainfall type*

Comparison the operating method of the reservoir operator with that of the simulator are shown in Fig.14 and Fig.15. The fuzzy inference of Fig.14 is applied to min-max type and Fig.15 is algebraic product-max-type. From these figures, it becomes obvious that, as the peak value of release discharge by simulator is almost equal, storage volume is many and the release curve is smoth by reservoir operator, the operation of simulator is better than that of reservoir operator.

On the fuzzy inference, min-max-type is better as the same as middle scale flood type.

Last, comparison the total storage volume in reservoir and the peak value of release discharge operated by the simulators applied to the fuzzy inferences of min-max type and algebraic product-max type with those operated by reservoir operator is shown in Table 1.

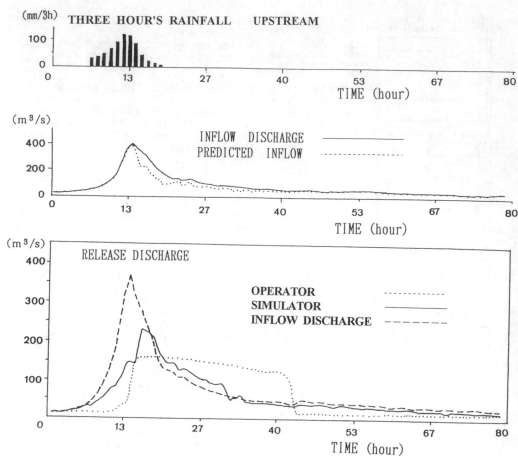

Fig.12 Comparison the result operated by reservoir operator with that by simulator applied to neural networks and fuzzy system (max-min-type)(middle scale flood)

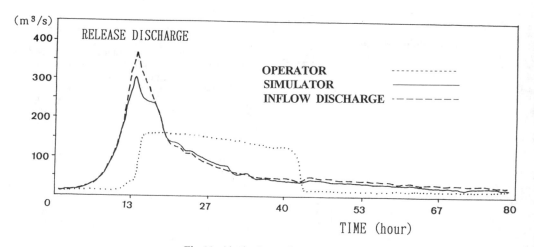

Fig.13 Algebraic product-max -type

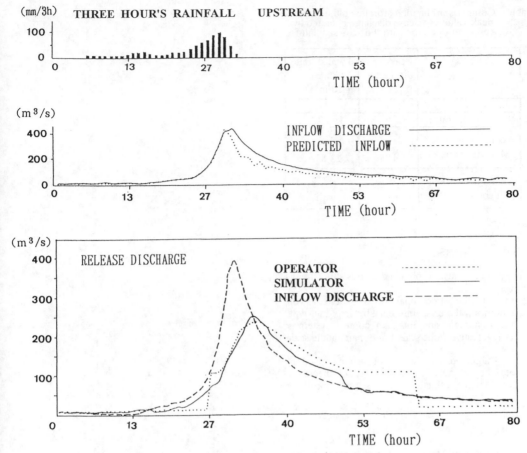

Fig.14 Comparison the result operated by reservoir operator with that by simulator applied to max-min-type (large scale flood)

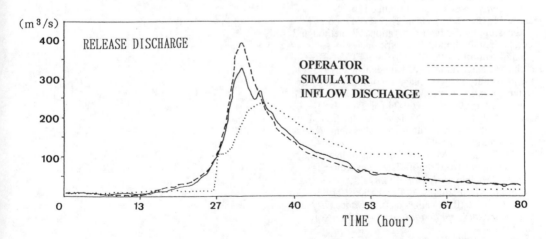

Fig.15 Algebraic product-max-type

Table 1 Comparison the total storage volume and the peak value of release dischage operated by reservoir operator with those by simulators.

	Peak discharge (m^3/s)	Total storage volume ($10^5 m^3$)
Flood data	257	
Resevoir operator	206	19.19
Neural networks (Min-Max-type)	185	23.83
Neural networks (Algebraic product)	219	12.13

6. CONCLUSIONS

The main conclusions obtained by this study are as follows.
(1) As for the operational volume of reservoir operation, the fuzzy inference by max-min-type rule is suitable and this dam control system is better effective large scale flood than middle scale one.
(2) It is suggested that the control system applied to both neural networks and fuzzy set theory is possible and effective for the flood control of reservoir.

REFERENCES

Clifford.L (eds) 1993. *Neural Networks : Theoretical foundations and analysis.* Newyork; IEEE Press.
Kosko.B 1992. *Neural Networks and Fuzzy Systems ; A dynamical systems approach to machine intelligence.* Newjersey : Prentice Hall.
Hasebe.M.,Hino.M. & Hoshi.K. 1981. Flood forecasting by the filter separation AR method and comparison with modelling efficiencies by some rainfall- runoff models. *J. Hydrol.*, 49:287-313.
Hasebe.M., Nagayama.Y. & Kumekawa.T. 1993. On the possibility of the application of fuzzy set theory to the operation system of dam. *Proc. of Hydraulic Engineering, JSCE*, 37: 69-74 (in Japanese).
Hasebe.M. & Nagayama.Y. 1994. Application of fuzzy set theory to the dam control system. *Trends in Hydrology*, 1: 35-47.
Hino.M. & Hasebe.M. 1981. Analysis of hydrologic characteristics from runoff data-A hydrologic inverse problem. *J.Hydrol.*,49:287-313.
Mizumoto. M.,Fukami.S. & Tanaka.K., 1979. In M.M.Gupta & R.R.Yager (eds). *Advances in fuzzy set theory and application*, pp.117-136 : North Holland.
Mizumoto.M 1988. *Fuzzy control and its application*, Tokyo: Science Publishing Company (in Japanese).

Stochastic Hydraulics'96, Tickle, Goulter, Xu, Wasimi & Bouchart (eds)© 1996 Balkema, Rotterdam. ISBN 90 5410 817 7

Stochasticity in environmental releases from reservoirs

M.E.Gubbels
South East Queensland Water Board, Brisbane, Qld, Australia

I.C.Goulter
Central Queensland University, Rockhampton, Qld, Australia

ABSTRACT: A stochastic dynamic programming approach for the explicit consideration of both the 'economic' and 'environmental' water requirements of a reservoir system is proposed. The stochastic nature of the inflows is incorporated into the model through the use of transitional probability matrices relating the probability of a flow in one period given a known flow in the previous period. The approach uses the steady state operating policies derived in response to specified environmental and economic targets for water to derive the probabilistic distribution of the instream flows available for environmental purposes. Each probability distribution can then be compared with the variability associated with the natural flow regime to assess the 'environmental' acceptability of the associated reservoir operating policy.

1 INTRODUCTION

Reservoir operation has traditionally been dominated by the need to 'optimise' the 'economic' returns achievable from the reservoir. With the growing emphasis on environmental considerations, especially in water resource development, reservoir operating policies that enable the best economic performance to be obtained while meeting specified environmental requirements are now being developed.

A stochastic dynamic programming approach to optimising the 'economic' operation of a reservoir while giving consideration to maintenance of the statistical characteristics (variation) of the natural flow regimes in the system has been developed. The approach is run on a scenario basis in which the environmental release strategies are specified as targets for environmental flows which, in line with recent research in Australia, are expressed predominantly in terms of percentiles of historical flows. The stochastic nature of the inflows is incorporated into the model through the use of transitional probability matrices relating the probability of a flow in one period given a known flow in the previous period. The probability distribution of total actual, opposed to decision, releases (for economic and environmental purposes) from the reservoir, as determined from the steady state operating policy, is used to derive the probability distribution of the flow available for environmental purposes. This probability distribution, or more correctly this variation, of flows can then be compared to the variation associated with the natural regime. In

this manner, the trade-off between economic and environmental demands on a reservoir can be assessed on the basis of both total flow released and how closely the pattern of releases of that total flow matches that of the natural regime. The model is demonstrated on the Fairbairn Dam system in Central Queensland, Australia.

2 MATHEMATICAL FORMULATION OF THE MODEL TO DEVELOP RESERVOIR OPERATING POLICIES

2.1 *Justification for choice of model*

Optimisation was chosen to solve this problem since it "looks at (implicitly) all possible decision alternatives, while simulation is limited to a finite number of input decision alternatives" (Yeh, 1985, p. 1810). The optimisation technique used to develop the optimal operating policies which consider the stochastic nature of environmental flows explicitly was a stochastic dynamic programming approach based upon the approach of Loucks *et al.* (1981). Dynamic programming "has the advantage of effectively decomposing highly complex problems with a large number of variables into a series of subproblems which are solved recursively" (Yeh, 1985, p.1802). A stochastic model was developed to ensure the resulting recommended operating policies reflect the uncertainty of future streamflows and the stochasticity of natural flows as they reflect baseline environmental conditions. The system is assumed to be 'stationary', whereby phenomena, such as the greenhouse effect, are not taken into consideration. Hence, the distribution of inflows in any time period is assumed not to vary from year to year.

A 'no forecast' type model was chosen since it reflects the more common practical situation whereby the previous time period's inflows and storage volume at the beginning of this time period are known with certainty and the inflows in the current time period are uncertain. This type of model is conservative in comparison to a 'perfect forecast' type model which assumes that the current time period's inflows are known with certainty and that there is uncertainty with respect to the inflows in the next month.

The stage or time interval used for this model is one month. The formulation requires the use of two state variables, specifically the storage volume at the beginning of the month and the inflow volume that occurred in the previous month, to define "The condition the system may find itself in at any stage" (Goulter, undated, p.89).

The Markov chain method, which allows the storage, inflow and release volumes to be discretised, was used in this study. In recognition of the 'curse of dimensionality' associated with dynamic programming, while still providing a framework in which sufficiently accurate results for practical consideration are obtained, nine discrete volumes were used for the release decision and the storage and inflow state variables. Savarenskiy's discretisation method was used, as recommended by Kelmes (1977).

2.2 Inflows - transitional probability matrices

A stochastic process is defined as "A random variable whose value changes through time according to probabilistic laws" (Loucks et al., 1981, p.116). The stochastic nature of the inflows was incorporated into the model through the use of transitional probability matrices relating the probability of a flow in one period given a known flow in the previous period. The set of inflow data used for the analysis of a particular reservoir system may be purely historical or simulated data may be utilised as appropriate. In the demonstration example used in this study, simulated flows were used to generate the transition matrices.

Each row in the transitional probability matrices used in this study represents a discrete previous month inflow. The elements in a particular row are equal to the probability of getting each possible discrete inflow in the current month, knowing the inflow in the previous month. The sum of the probabilities in each row therefore equals 1.0, since there is a one hundred percent chance that a flow from within the complete range of inflows for that month will occur in the month.

2.3 Targets

Targets, also known as the 'allocations' or 'demands' on the reservoir system, are defined as the annual water volumes required for municipal, industrial, irrigation and environmental purposes. Of these, the environmental target is the most difficult target to define and quantify. Various methods which recommend general guidelines for the specification of environmental flow conditions in terms of quantity aspects have been summarised in Karim et al. (1995).

This model has been developed to consider both fixed and total recommended monthly environmental flow targets. [Fixed monthly demands are defined as the environmental water requirements that must be satisfied with very high priority. Total monthly environmental demands include the fixed environmental demands as well as the additional environmental water requirements which have lower priorities assigned to them]. The fixed monthly environmental targets are considered in conjunction with the other fixed demands (municipal and industrial), which are satisfied with as close to one hundred percent reliability as possible. The remaining recommended monthly environmental targets (total minus fixed) are handled in the same manner as the irrigation targets, which may or may not be completely satisfied depending on the availability of water. The 'base' model requires the recommendation of one combination of fixed and total environmental targets for each month. These targets may be based on any one of the numerous methods available to quantify the instream flow requirements or a method found to be suitable for the river system under consideration. Flushing flows required at certain times of the year may be incorporated within the monthly fixed or total environmental flow targets, depending on their priority and need for explicit timing.

A second version of the model has been developed to consider a 'special' case when the recommended environmental targets depend on the storage in the supply reservoir at the beginning of the month. This model is essentially the same as the base model except that the fixed environmental targets are assumed to be zero and the total recommended environmental targets are in a matrix form; one target volume for each discrete storage state combination, for a total of 108 values over the year.

2.4 Mathematical formulation

The actual objective function for the model is to minimise the 'expected' deficits from each of the fixed, irrigation and environmental targets. The recursive equation for this objective is shown in Equation 1.

$$f^*[k,h,t,y] = \text{Minimise} \sum_{i=1}^{ni} [P[h,i,t] * Return] \quad (1)$$
$$rd \, \varepsilon \, \underline{R[t]}$$

for all k, h, t and y, where t = month index; y = year index; k = index for the discrete storage states at the beginning of month 't'; h = index for the discrete inflow states in the previous month '$t-1$'; i = index for the discrete inflow states in month 't'; ni = number of discrete inflow states 'i' or 'h'; rd = discrete release decision; $\underline{R[t]}$ = set of possible discrete release decisions 'rd' in month 't'; $f^*[k,h,t,y]$ = optimal return in year 'y' and month 't' for storage volume 'k' and previous month inflow 'h'; $P[h,i,t]$ = probability of getting inflow 'i' in month 't', knowing inflow 'h' occurred in the previous month; and $Return$ = the sum of the short term and long term returns for the combination of inflow 'h' in month '$t-1$', storage 'k', release decision 'rd' and inflow 'i' in month 't' and storage 'l' in month '$t+1$', as described below.

$$Return = STR[k,i,rd,t] + f^*[l,i,t+1,y] \quad (2)$$

where l = index for the discrete storage states at the beginning of next month '$t+1$'; $STR[k,i,rd,t]$ = short term return for release decision 'rd' in month 't' for storage volume 'k' and inflow 'i' (see Equation 3); and $f^*[l,i,t+1,y]$ = long term optimal return in year 'y' and next month '$t+1$' for storage volume 'l' and the current month inflow 'i'.

$$STR[k,i,rd,t] = \sum_{d=1}^{nd} [AR[d,rd,i,t] - T[d,t]] \quad (3)$$

where d = index for the various targets (fixed, irrigation and environmental demands); nd = number of targets 'd' (demands); $AR[d,rd,i,t]$ = 'actual' release to target 'd' associated with release decision 'rd' and inflow 'i' in month 't'; $T[d,t]$ = target volume for target 'd' in month 't'; and $[AR[d,rd,i,t] - T[d,t]]$ = deficit from target for each target 'd', release decision 'rd' and inflow 'i' in time 't'.

The physical constraints on the reservoir system include the maximum (capacity) and minimum (dead) storage volumes, as shown in Equation 4.

$$reservoir\ capacity >= storage >= dead\ storage \quad (4)$$

Similarly, the maximum and minimum monthly release volumes for each of the reservoir outlets are physical constraints on the reservoir system, as described below.

$$\text{max. } release_r >= release_r >= \text{min. } release_r \quad (5)$$

for all release outlets 'r', where r = index for release outlets.

An additional constraint, referred to as the continuity or transformation equation, is used to ensure that the storage volume at the beginning of a month minus releases and losses (evaporation, seepage, etc.) plus inflows during the month equals the storage volume at the beginning of the next month, as summarised in Equation 6.

$$storage[t+1] = storage[t] - losses[t] - release[t] + inflow[t] \quad (6)$$

where $storage[t+1]$ = storage volume at the beginning of next month '$t+1$'; $storage[t]$ = storage volume at the beginning of the current month 't'; $losses[t]$ = net losses due to evaporation, seepage, precipitation and overflow at the outlet in month 't'; $release[t]$ = release decision volume from the reservoir at the beginning of month 't'; and $inflow[t]$ = inflow volume in month 't'.

2.5 Steady state solution

The solution generated by the model reaches steady state when two conditions are met.

1. When the optimal return from one year to the next increases by the same amount for every combination of inflow and storage at each stage (see Equation 7). A small tolerance (for example 0.1%) is required to allow for precision and round-off in the model.

$$\begin{aligned} f^*[k,h,t,y] - f^*[k,h,t,y+1] \\ \cong f^*[k,h,t,y+1] - f^*[k,h,t,y+2] \end{aligned} \quad (7)$$

2. When the optimal discrete releases from one year to the next are the same for every combination of inflow and storage at each stage (Loucks et al., 1981), as shown in Equation 8.

$$R^*[k,h,t,y] = R^*[k,h,t,y+1] \quad (8)$$

where $R^*[k,h,t,y]$ = optimal discrete release associated with the optimal return '$f^*[k,h,t,y]$'.

Steady state usually occurs within 15 annual cycles of a stochastic dynamic program.

2.6 'Optimal' operating policy

The release decisions and thus the optimal operating policy developed by the model do not explicitly consider the possibility of uncontrolled spills that occur when the reservoir capacity is exceeded. The 'actual' release may therefore be partially made up of this spill volume, which is larger than the original release decision. Similarly, if the storage state calculated for the next month is less than the dead storage volume, the 'actual' release volume will be less than the release decision. However, 'actual' release volumes associated with a release decision, rather than the release decision itself, are used in calculating the returns. The returns are determined by allocating a portion of the 'actual' release volume to satisfy the fixed demands (municipal, industrial and fixed environmental) first. The remaining water volume (if any) is divided in proportion to the relative magnitudes of the irrigation and environmental (total minus fixed) target demands. Any water allocated for irrigation purposes in excess of the irrigation target is re-allocated for environmental purposes. As a result, there are nine 'actual' fixed, irrigation and environmental releases associated with one release decision.

2.7 'Expected' irrigation and environmental releases

In this study, the nine possible irrigation and environmental releases (derived from the 'actual' releases, as they relate to the nine discrete inflows in each month, rather than the release decision itself) are collapsed into one value representing the 'expected' irrigation and environmental releases for a given release decision. This collapsing of values is achieved by multiplying each 'actual' irrigation release (including the extreme values) by the corresponding probability of occurrence of the inflow which results in the 'actual' irrigation release, and summing these values into a single number, as described below.

$$EIR[k,h,rd,t] = \sum_{i=1}^{ni} \left[P[h,i,t] * AIR[k,h,rd,i,t] \right] \quad (9)$$

where $EIR[k,h,rd,t]$ = 'expected' irrigation release for storage 'k', previous month inflow 'h' and release decision 'rd' in month 't'; $P[h,i,t]$ = probability of getting inflow 'i' in month 't', knowing inflow 'h' occurred in the previous month '$t-1$'; and $AIR[k,h,rd,i,t]$ = 'actual' release to irrigation for storage 'k', previous month inflow 'h', release decision 'rd' and inflow 'i' in month 't'.

The same procedure is used for the 'actual' environmental releases to determine the 'expected' environmental release in each month.

2.8 Probability of releases

The probabilities associated with each optimal discrete release of a steady state solution in a particular month are a function of the storage volume at the beginning of the month, the inflow in the previous month, the storage volume in the next month and the probability of getting a particular discrete inflow in the current month. Since all of these variables are inter-related, a set of simultaneous equations must be solved in order to determine the probability of each combination of

discrete storage and previous month inflow. The necessary equations were derived for the 'no forecast' model based on the theory outlined by Loucks et al. (1981, p.326) for a 'perfect forecast' model. Each combination of previous month inflow and storage volume at the beginning of the current month has a probability of occurring. Therefore the summation of the probabilities for each of these combinations in each month must equal one, i.e.,

$$\sum_{k=1}^{ns} \sum_{h=1}^{ni} P[k,h,t] = 1.0 \quad (10)$$

for all t, where ns = number of discrete storage states 'k' or 'l'; and $P[k,h,t]$ = probability of having an inflow 'h' in month '$t-1$' and storage volume 'k' at the beginning of month 't'.

In addition, the probability of getting a certain combination of inflow 'i' in a particular month and storage volume 'l' at the beginning of the next month, depends on the combination of inflow 'h' in the previous month and storage volume 'k' which existed at the beginning of the current month 't'. The continuity equation (Equation 6 re-arranged in Equation 11 below) relates l, i, k and the steady state optimal release $SSR[k,h,t]$ (with associated 'actual' releases). However, not all combinations of storage 'k' and steady state optimal release '$SSR[k,h,t]$' in a current month are physically capable of attaining each combination of inflow 'i' during month 't' and storage volume 'l' at the beginning of next month. For example, if the storage volume at the beginning of the month is equal to the dead storage volume and the inflow during the month is zero or very small (a common feature in Australian conditions), it is physically impossible (for most reservoir systems) to have a full reservoir at the beginning of the next month, even with the minimum possible release.

Consider the continuity equation of Equation 6 rearranged as follows:

$$storage[t+1] - inflow[t]$$
$$= storage[t] - losses[t] - SSR[k,h,t] \quad (11)$$

in symbols: $\{l,i,t+1\} ==> \{k,h,t\}$

where $SSR[k,h,t]$ = steady state optimal release for storage 'k' in month 't' and previous month inflow 'h'; and $\{-,-,-\}$ = combination of storage at beginning of the month and inflow during the previous month at the specified time interval.

The probability '$P[l,i,t+1]$' of getting the combination of next month storage 'l' and current month inflow 'i' is equal to the summation of the probabilities '$P[k,h,t]$' of the combinations of current storage volume 'k' and previous month inflow 'h', which give, on the basis of the steady state release '$SSR[k,h,t]$' for the combination of 'k' and 'h' and inflow 'i', a storage volume 'l' in the next month, multiplied by the corresponding probability 'P[h,i,t]', which links the previous month inflow 'h' to the current inflow 'i'. Mathematically, this process is represented by:

$$P[l,i,t+1] = \sum_{k=1}^{ns} \sum_{h=1}^{ni} \left[P[k,h,t] * P[h,i,t]\right] \quad (12)$$

(Only if $\{l,i,t+1\} ==> \{k,h,t\}$)

for all l, i, and t, where $P[l,i,t+1]$ = probability of having an inflow 'i' in month 't' and storage volume 'l' at the beginning of month 't+1'.

The probability of each combination of storage and previous month inflow in each month can be determined for the steady state solution obtained from the Stochastic Dynamic Programming model by solving the corresponding set of 984 simultaneous equations (= 12 + (9*9*12), from Equations 10 and 12, respectively). In this study, this complex set of simultaneous equations was solved using the Equality Constrained Least Squares Method (Golub and VanLoan, 1989, p.566).

The probability of each discrete steady state optimal release volume is then determined by summing the probabilities for each combination of storage and previous month inflow that gives rise to that discrete release. The probabilities associated with each of the 'expected' irrigation and 'expected' environmental releases can be determined in the same general fashion as the probabilities of the optimal releases.

On first examination, it may appear to be inappropriate, or even theoretically unsound, to determine the probability of an 'expected' release. However, the process allows an easy comparison between the distributions of 'expected' environmental flows with the monthly historical inflow distributions. In this manner, the environmental flow combination (fixed and total target) that is most similar to the natural flow regime, both temporally and spatially, can be identified for each month.

Similarly, the probabilities associated with the 'expected' irrigation releases may be used to determine the probability of satisfying the entire monthly irrigation targets. However, this probability does not take into consideration the probabilities associated with satisfying portions of the monthly irrigation targets. The simplest method available of addressing this issue is to sum the products of the discrete 'expected' irrigation releases and their respective probabilities of release. This 'expected' total irrigation release is therefore based on a double expectation, which is not mathematically rigorous. However, it does give an indication of the amount of water 'expected' to be available for irrigation purposes in each month. Furthermore, if the reservoir system under consideration is capable of satisfying all of the monthly fixed and irrigation demands without consideration of environmental requirements, then an 'expected' total irrigation release equal to less than the total fixed and irrigation demands will give an indication of the trade-offs that may be occurring between the economic and environmental demands on the system.

3 DEMONSTRATION OF THE MODEL

The model was applied to the Fairbairn Dam, located on the Nogoa River near the town of Emerald in Central Queensland, Australia. All of the required information and data for the Fairbairn Dam system were supplied by the Queensland Water Resources Commission.

Construction of Fairbairn Dam was completed in 1972. The resulting reservoir, Lake Maraboon has a catchment area of 16,320 square kilometres and a capacity of 1,443,000 Megalitres. The average annual rainfall in the catchment is 635 millimetres, seventy percent of which is concentrated in the summer months.

The Fairbairn Dam system supplies water for municipal, industrial and irrigation demands in the area. The primary industry in the area is coal mining and cotton is the predominant crop. The annual irrigation and fixed water demands (targets), as defined by the Queensland Water Resources Commission in August 1993, are 56,706 Megalitres and 36,172 Megalitres respectively. Water is distributed to the farms through a channel system made up of the Weemah Main Channel, the Selma Main Channel and the Nogoa River itself. The three main tributaries that join the river system downstream of Fairbairn Dam are Retreat Creek and the Comet and Isaac Rivers. The Nogoa River becomes the Mackenzie River at the junction of the Nogoa and Comet Rivers.

The critical environmental point used to demonstrate both versions of the model was defined as the location on the river where the environmental flow recommendations are most critical. Water released for 'economic' purposes (municipal, industrial or irrigation) were considered environmental flows until extracted from the river system. As a result, the critical environmental point is assumed to be the point just downstream of where all of the water has been extracted for 'economic' purposes. The model uses flow contributions of tributaries downstream of the reservoir to satisfy some or all of the water demands downstream of the point at which the tributary flows enter the river. Since the Isaac River has a high probability of being more than able to satisfy all of the downstream water demands up to the end of the 'regulated section', as defined by the Queensland Water Resources Commission, the 'critical environmental point' is assumed to be just upstream of the point where the Isaac River joins the Mackenzie River.

Ninety five years of inflow data from 1898 to 1992 were available and used to generate the transitional probability matrices. However, the resulting matrices were very sparse due to the highly variable hydrological conditions of the study area. It was therefore decided to generate data in an attempt to 'fill-in' the matrices, which resulted in the development of a 'less-sparse' set of transitional probability matrices.

The trade-offs between the economic and environmental demands for this system were analysed by considering nine different recommended environmental flow combinations (cases). The municipal, industrial and irrigation targets remained constant for each case considered. Eight of the cases were analysed using the 'base' stochastic dynamic programming model (described previously), where each case represents a specific combination of fixed and total recommended environmental flow target volumes, which, for this study, are various combinations of percentiles of the historical inflow data consistent with the recommendation of the Environmental Study of Barker-Barambah Creek (Arthington et al., 1992). The ninth case considered was analysed using the 'special' version of the stochastic dynamic programming model, wherein the recommended environmental flow strategy specifies percentiles which vary according to the storage volume in the reservoir at the time. Specifically, this case recommends releases up to the fiftieth percentile when the reservoir is greater than eighty percent full. The percentiles then reduce progressively as the volume of the reservoir decreases until, when the reservoir is at or near the dead (minimum) storage volume, it recommends that no water be released for environmental purposes.

Trade-offs between economics and the environment were found to appear for recommended total environmental targets greater than the fiftieth percentile. The sixtieth percentile case satisfies all of the irrigation demands in most months, but falls slightly short in November, January and February.

In order to obtain an indication of the nature and magnitude of 'trade-offs' between the economic and environmental demands that occur, the total annual 'expected' irrigation releases were calculated and compared to the total annual irrigation target. The difference between the two is the total annual 'expected' deficit. All cases up to the sixtieth percentile recommended total environmental target have negligible deficits. The sixtieth percentile has a three percent deficit, which is also considered to be insignificant in comparison to the twenty percent deficit for the eightieth percentile case. A fifty three percent deficit resulted for the one hundredth percentile case, which was considered for comparative theoretical purposes only.

Based on these results, a recommended environmental target equal to the sixtieth percentile appears to be the largest environmental target which could be accommodated without significantly reducing the volume of water available to satisfy the economic (irrigation) demands.

An analysis of the results was then performed to determine which combinations of recommended environmental targets are able to maintain a flow regime variability similar to that of the historical flow regime. The monthly historical histogram data was compared to the distributions of the probabilities of 'expected' environmental releases. Given the existence of the Fairbairn Dam on the Nogoa River and the dominant consumptive uses of the water in the system, it would be very difficult to find a flow regime exactly the same as the natural regime. However, by comparing the probability distributions of the 'expected' environmental releases to the historical probability distributions, observations can be made as to which combinations of recommended environmental demands would be able to maintain monthly flow regimes most similar to the historical regimes. These comparisons are described in further detail in Gubbels (1994).

It was found, not unexpectedly that the differences between the historical and reservoir derived flow patterns vary by season and case. The 'ideal' recommended environmental target appears to be less than or equal to the sixtieth percentile in the winter (April to October). In the summer months, the 'ideal' target appears to be less than or equal to the environmental release strategy which is based on the storage volume. Recall that this strategy specifies an environmental flow of zero for the smallest storage volumes increasing to releases corresponding to the fiftieth percentile for the largest storage volumes. However, the true 'ideal' strategy may be in a different range, say the twentieth to sixtieth percentiles. (The evaluation of these other ranges is not included in this study and is an area for future work).

These 'ideal' recommended environmental targets are based on the findings of the stochastic dynamic program. The findings need to be verified by a detailed simulation before they can be used by the Queensland Water Resources Commission as an operating strategy for the Fairbairn Dam. In addition, percentiles of historical flows were used as reasonable first estimates of environmental flow requirements. Once the ecological and environmental sensitivities of the Nogoa / Mackenzie River system are known, the model should be re-run with the new set of recommended environmental flows that are not necessarily based solely on percentiles of historical flows. The results of these runs would provide a more appropriate indication of any 'trade-offs' required between economics and the environment for the Fairbairn system. Hence the results would assist in defining a more refined operating strategy for the system which is able to recognise, at a finer level of detail, the environmental requirements and the trade-offs between those requirements and the economic demands.

4 CONCLUSIONS

An important goal in the development of reservoir operating policies which take explicit consideration of environmental demands is to develop approaches that can identify reservoir operating policies which result in the best economic performance for the reservoir while maintaining an acceptable environmental flow regime similar in both total volume and variability of flow to that of the natural flow regime. The stochastic dynamic programming reservoir optimisation model together with the methodology, described in this paper, which defines the statistical characteristics of both the flows in the system available for environmental purposes and the flows available for economic (industrial, agricultural and municipal) purposes is a significant step toward achieving this goal.

ACKNOWLEDGMENTS

We would like to thank Stephen Smith, Central Queensland University, Rockhampton, Australia, for developing the simultaneous equation solution model used in the approach. In addition, we would like to thank the Water Resources Commission, Department of Primary Industries, Queensland, Australia, and A.H. Arthington, Centre for Catchment and In-stream Research, Griffith University, Brisbane, Australia, for supplying the necessary data and information used to test the model.

REFERENCES

Arthington, A.H., Conrick, D.L. and Bycroft, B.M. 1992. *Environmental Study Barker-Barambah Creek*, Vol.2 Scientific Report: Water Quality, Ecology and Water Allocation Strategy. Centre for Catchment and In-stream Research, Griffith University/Water Resources Commission, Department of Primary Industries: Brisbane.

Golub, G.H. and VanLoan, C.F. 1989. *Matrix Computations*, 2nd Ed. Johns Hopkins University Press, Baltimore.

Goulter, I.C. undated. *An introduction to water resources systems theory and application - Course notes*. Dept. of Civil Engineering, University of Manitoba, Winnipeg.

Gubbels, M.E. 1994. *'Optimal' reservoir operating policies with explicit consideration of economic and environmental requirements*. Master of Engineering Thesis, Central Queensland University, Rockhampton.

Klemes, V. 1977. Discrete representation of storage for stochastic reservoir optimization. *Water Resources Research*. Vol.13, No.1:149-158.

Karim, K., Gubbels, M.E. and Goulter, I.C. 1995. Review of determination of instream flow requirements with special application to Australia. *Water Resources Bulletin*. To be published in Dec. 1995.

Loucks, D.P., Stedinger, J.R., and Haith, D.A. 1981. *Water Resource Systems Planning and Analysis*. Prentice-Hall Inc., Englewood Cliffs, New Jersey.

Yeh, W.W-G. 1985. Reservoir management and operations models: a state-of-the-art review. *Water Resources Research*. Vol.21, No.12:1797-1818.

Stochastic Hydraulics'96, Tickle, Goulter, Xu, Wasimi & Bouchart (eds) © 1996 Balkema, Rotterdam. ISBN 90 5410 817 7

Effects of fuzzy constraints in reservoir operation strategy

Suharyanto
James Goldston Faculty of Engineering, Central Queensland University, Rockhampton, Qld, Australia

Saleh A. Wasimi
Department of Mathematics and Computing, Central Queensland University, Rockhampton, Qld, Australia

ABSTRACT :
The formulation of reservoir operating rule through the application of stochastic dynamic programming (SDP) requires reservoir storage and release values to be constrained within allowable ranges. In some cases, these constraints are not strictly crisp constraints in the sense that whenever the system's state is approaching the extreme boundary, the operator is alarmed and makes operating decisions with added precaution. The degree of additional precaution is a function of many aspects including operator's judgement, expected damages, political and social consequences, and environmental effects. An application of fuzzy (soft) constraint is, therefore, more realistic in generating reservoir operating rule and the process is demonstrated in this paper. An application of this fuzzy constraint based reservoir operating policy has been made to the Fairbairn reservoir in Emerald, Central Queensland, Australia.

1. INTRODUCTION

The operation of a reservoir is to be bounded by some constraints for technical, environmental, or socio-economic reasons such as minimum and maximum storage volumes, and minimum and maximum releases. Often, these constraints are not truly crisp constraints, i.e., the boundary is not sharp value or an exact number. In case of storage constraints, whenever the reservoir volume is approaching the extreme values, the operator may already be alarmed and may take some additional precautions in operating the reservoir. When the reservoir storage is rising, for example, and reaches close to its maximum volume, the operator may release more water than what is indicated on the crisp operation rule based on his personal judgement or unwritten advice from others. Such an approach is not totally unwarranted or illogical because there is always some uncertainty associated with any reservoir operating policy. The reservoir operator is probably in a best position to evaluate the most recent state of affairs and make some rationing (hedging) on the release depending upon the current status and other precautionary measures.

To illustrate the practicality of using boundaries which are not crisp, let us cite the case of minimum reservoir release. For environmental conservation, most reservoirs should supply a small amount of water to maintain the low-flow requirement of the down-stream river throughout the year, while the absolute minimum release constraint is zero. The minimum release for low-flow augmentation may act as an upper bound of the minimum constraint. It is necessary to always supply the low-flow requirement, however, should the conditions become very extreme, the reservoir release could be made to be less than the low-flow requirement.

For the prevailing weather conditions of Central Queensland where most rainfall tend to occur during the summer season, and during the winter period there is very little amount of rain which combined with the chill temperature do not suit some plants and forces the release of water from the reservoir to be adaptive on factors beyond what can possibly be included in an operation policy guide. In these conditions, the release constraints realistically are fuzzy constraints.

2. STOCHASTIC FUZZY DYNAMIC ROGRAM (SFDP)

The formulation of stochastic dynamic programming as outlined in Loucks et al. (1981) is used as the basic formulation for the fuzzy case. The general recursive equation of the stochastic fuzzy dynamic programming is as follow :

$$f_t^n(k,i) = \min_l \left\{ K\delta\left(R_{k,i,l,t}, T_t\right)^2 + \sum_j^{NI} P_{i,j}^t \cdot f_{t-1}^{n-1}(l,j) \right\}$$

(1)

where $f_t^n(k,i)$ is the optimum long-term return at the beginning of period t with n periods of the optimisation stages remaining if the beginning period storage state is in class k and the current period inflow is in class i, K is a constant, P_{ij}^t is the probability that the inflow during the period $t+1$ will be $Q_{j,t+1}$ given that the inflow during the previous time period t is $Q_{i,t}$ and $\delta\left(R_{r,t}, T_t\right)$ is the deviation of release, $R_{k,i,l,t}$, from its target demand, T_t. The index l is showing the end-period storage state class. The value of $R_{k,i,l,t}$ is to be obtained from the continuity equation below.

$$R_{k,i,l,t} = S_{k,t} + Q_{i,t} - S_{l,t+1}$$

(2)

where $S_{k,t}$ is the beginning-period (t) storage state which is at class k, $S_{l,t+1}$ is the beginning-period storage state in the next time period $(t+1)$ which is in class l, and $Q_{i,t}$ is the inflow during the current-period which is in class i.

Since any variable in the system could be fuzzy number, it is further assumed that the storage state and the target demand could be either fuzzy number or singleton (crisp equivalent) number. The arithmetical operation between two fuzzy numbers or mixed numbers is detailed in Kaufmann and Gupta (1991). In this study, the inflow state is considered as a crisp number. The resulting $R_{k,i,l,t}$ from equation (2) will, therefore, be either fuzzy number or singleton number.

To evaluate the closeness of the deviation of $R_{k,i,l,t}$ from the target demands, the following 'dis-resemblance' index (Kaufmann and Gupta, 1991) is adopted. The index shows the standardised distance between two fuzzy numbers. This index, denoted by $\delta(A,B)$, has a value in the range from 0 to 1, and is defined as :

$$\delta(A,B) = \frac{Aa + Ab + 2As - Ai}{2(\beta_1 - \beta_0)}$$

(3)

where :

Aa, Ab = Area under the membership functions for fuzzy numbers A and B, respectively,

As = Area of separation between the membership functions for the two fuzzy numbers A and B,

Ai = Area of intersection between the membership functions for the two fuzzy numbers A and B,

β_0, β_1 = lower and upper boundaries of the universe respectively, where the two fuzzy numbers A and B would possibly falls in.

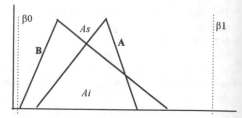

Figure 1. Definition in 'dis-resemblance' Index

3. FUZZY CONSTRAINT

The constraints envisaged in the stochastic dynamic programming formulated above is in the release values only. The storage domain has been implicitly bounded by its discretised values. To account for the possible fuzziness in the release constraints and in the release from the reservoir, the following measures are introduced (Dubois and Prade, 1988).

a) Possibility measure

$$Poss\left(\overline{A} \geq \underline{B}\right) = \sup_x \, \min\left(\mu_A(x), \, \sup_{y \leq x} \mu_B(y) \right)$$

(4)

where A and B are two fuzzy numbers which are bounded by their membership function $\mu_A(x)$ and $\mu_B(y)$, respectively. $\mu_A(x)$ is the membership level of fuzzy number A at x, while $\mu_B(y)$ is the membership level of fuzzy number B at y. $Poss\left(\overline{A} \geq \underline{B}\right)$ is the possibility that x will be at least as large as y, which is interpreted as the possibility that the largest value that x can take are at least as great as the smallest values that y can take (Dubois and Prade, 1988).

b) Necessity measure

$$Nec\left(A \geq \overline{B}\right) = \inf_x \; \max\left(1 - \mu_A(x), \; \inf_{y \geq x} \left[1 - \mu_B(y)\right]\right)$$

(5)

where $Nec\left(A \geq \overline{B}\right)$ is the necessity that x can take only the values which are greater than the values that y can take, and interpreted as the smallest value that x can take are greater than the largest value that y can take (Dubois and Prade, 1988).

In this paper, the necessity measure is used to constraint the minimum release while the possibility measure is used for the maximum release constraint. These constraints are written as :

$$Poss\left(R_{k,i,l,t} \geq R_{max}\right) \leq \alpha_{R\,max}$$

(6)

$$Nec\left(R_{k,i,l,t} \geq R_{min}\right) \geq \alpha_{R\,min}$$

(7)

where $\alpha_{R\,max}$ and $\alpha_{R\,min}$ are the maximum acceptance level of $Poss\left(R_{k,i,l,t} \geq R_{max}\right)$ and the minimum acceptance level of $Nec\left(R_{k,i,l,t} \geq R_{min}\right)$, respectively. It should be noted that the higher value of $\alpha_{R\,max}$ reflects more relaxation on the requirement to satisfy the proposition of $R_{k,i,l,t} \geq R_{max}$, while the higher value of $\alpha_{R\,min}$ reflects more stringency on the requirement to satisfy the proposition $R_{k,i,l,t} \geq R_{min}$.

4. FAIRBAIRN RESERVOIR

The Fairbairn reservoir is located on the Nogoa river and close to Emerald city in Central Queensland, Australia. The reservoir is used to supply water requirement for cotton dominated irrigation, industrial and municipal use, and riparian flow (Gubbels, 1994, Tickle and Goulter, 1992, and Water Resources Commission, 1992). The reservoir has minimum and maximum storage capacity of 123,000.0 megalitres and 1,443,000.0 megalitres respectively. Two main irrigation channels called Weemah channel on the right and Selma channel on the left have capacities of 350 megalitres/day and 777 megalitres/day, respectively. For weed control program and maintenance purposes, these channels are to be shut down for periods of three weeks each in the months of May, June, July, and August (Water

Resources Commission, 1992). Another outlet from the reservoir is for riparian requirement along the reservoir with an estimated delivery of 3,200 megalitres/month. The gated outlet of the reservoir and 5 syphons have an estimated total capacity of about 1500 megalitres/day (Gubbels, 1994).

5. APPLICATION AND DISCUSSION

The above formulation of SFDP is applied to the Fairbairn reservoir to generate yearly operation policy. The fuzzy constraints on the minimum and maximum releases are shown in Figure 2. The upper portion of the minimum release constraint is determined as the capacity of supplying riparian flow requirement. The maximum release constraint is determined as in Table 1.

The results of execution of the SFDP with different values of $\alpha_{R\,min}$ and $\alpha_{R\,max}$ are shown in Table 2, which gives the value of expected annual deviation (EAD) for each run. In this yearly execution of SFDP, the number of storage interval, NS, is 50 and the number of inflow interval, NI, is 20. In this case, the storage interval is considered as triangular fuzzy interval with 50% of each class-width overlapping.

From these results the EAD values on the Table 2 demonstrates that the change in stipulation on the minimum release constraint does not really change the EAD values. It gives some indication that the generated operating policy may not be influenced by different stipulation on the fuzzy minimum release constraint defined above. This indication is made obvious by the following : The underlying reason of this fact may be partly due to the implementation of necessary measure on this minimum release constraint, and in yearly operation, this constraint is not binding, which is obvious by the fact that the minimum yearly inflow is greater than the upper-bound of the minimum constraint. In monthly operation, the minimum constraint may be binding in some months.

The influence of the stipulation level on the maximum constraint seems more pronounced than on the minimum constraint, since the different stipulation level does change the feasible region of the release. The bigger value of $\alpha_{R\,max}$ is widening the feasible release by relaxing the acceptable maximum release constraint. Therefore, it is expected that the bigger value of $\alpha_{R\,max}$ will result in higher release policy.

113

A more obvious results on the influence of different levels of $\alpha_{R\,min}$ and $\alpha_{R\,max}$ are shown in Figures 2, 3, and 4. The figures show the optimum policy, i.e., the end-storage class to go as a function of the beginning-period storage class and the inflow during the current period. Figures 2, 3, and 4 are showing the optimum policy for beginning-period storage class equal to zero, five, and ten respectively. Figures 2.a, 3.a, and 4.a show the influence of the maximum constraint stipulation level which are $\alpha_{R\,max}=0.0$, $\alpha_{R\,max}=0.4$, and $\alpha_{R\,max}=0.8$, respectively. Figures 2.b, 3.b, and 4.b highlight the combined influence of the maximum and minimum release constraints. In these cases, the values of $\alpha_{R\,min}=\alpha_{R\,max}=0.0$, $\alpha_{R\,min}=\alpha_{R\,max}=0.4$, and $\alpha_{R\,min}=\alpha_{R\,max}=0.8$, respectively.

In Figure 2.a, 3.a, and 4.a, it is shown that the increase in the value of $\alpha_{R\,max}$ causes the decrease in the end-period storage class, which also means that greater release volume is delivered from the reservoir. In these figures, the influence is more pronounced in the higher values of inflow. In Figures 2.b, 3.b, and 4.b the combined influence on the increase values of $\alpha_{R\,min}$ and $\alpha_{R\,max}$ are occurring almost for the whole range of inflow values.

6. CONCLUSSION
It has been demonstrated that in some conditions where the release constraint is not truly crisp, a more realistic approach is to consider fuzziness in the constraints.

Since the fuzzy constraint in this formulation is only on the release, different stipulation level on the fuzzy constraint does not really do any hedging on the reservoir operation policy due to the fact that the stipulation level change only changes the feasible range. Therefore, the fuzzy constraint on the release will only shift up or down the operation policy. It may be worthwhile to re-formulate the SFDP formulation so that it can consider the constraint on the reservoir storage explicitly as well.

REFERENCES
Dubois, D. and Prade, H., 1988. Possibility Theory: An Approach to Computerised Processing of Uncertainty. Plenum Press, New York.
Gubbels, M. E., 1994. Optimal Reservoir Operating Policies with Explicit Consideration of Economic and Environmental Requirements (unpublished MEng. Thesis), Central Queensland University, Rockhampton, Australia.
Kaufmann, A. and Gupta, M. M., 1991. Introduction of Fuzzy Arithmetic : Theory and Applications, Van Nostrand Reinhold, New York.
Klemes, V., 1977. Discrete Representation of Storage for Stochastic Reservoir Optimisation. Water Resources Research, 13(1) :149-158, February 1977.
Loucks, D. P., Stedinger, J. R., and Haith, D. A., 1981. Water Resources Systems Planning and Analysis. Prentice-Hall, Inc., New Jersey.
Queensland Water Resources Commission, 1992. Emerald Irrigation Area : Management Guidelines, Water Resources Commission, Department of Primary Industry, Australia.
Tickle, K. And Goulter, I. C., 1992. Assessment of Performance Metrics for a Reservoir Under Stochastic Conditions, Proceedings of the sixth International Symposium on Stochastic Hydraulics (Kuo, J. T. and Lin, G. F. eds.), Taipei, May 18-20, 1992.

Table 1. Determination of Maximum Release Constraint

	(mgl/yr)	Rounded (mgl/yr)
i. The Least Max. Possible Capacity : Two main channels only operated for 331 days	373037.00	350000.00
ii. Lower Median : 50% of time two main channels deliver at FULL capacity 50% of time two main channels deliver at HALF capacity 25% of time Dam Outlets deliver at FULL capacity 75% of time Dam Outlets deliver at QUARTER capacity	510777.75	550000.00
iii. Upper Median and Maximum Possible Capacity : 50% of time two main channels deliver at FULL capacity 50% of time two main channels deliver at HALF capacity 50% of time Dam Outlets deliver at FULL capacity 50% of time Dam Outlets deliver at HALF capacity	675777.75	700000.00

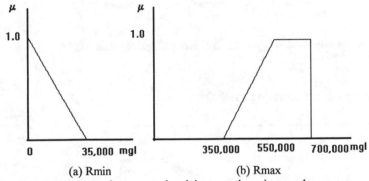

(a) Rmin   (b) Rmax

Figure 2. Fuzzy constraints on yearly minimum and maximum releases.

Table 2. Expected Annual Deviation of the SFDP Outputs

α_{Rmin}	α_{Rmax}					
	0.0	0.2	0.4	0.6	0.8	1.0
0.0	1.254E5	1.129E5	1.041E5	1.016E5	1.009E5	1.003E5
0.2	1.254E5	1.129E5	1.041E5	1.016E5	1.009E5	1.003E5
0.4	1.247E5	1.112E5	0.999E5	0.816E5	0.773E5	0.519E5
0.6	1.235E5	1.101E5	0.868E5	0.474E5	0.321E5	0.268E5
0.8	1.233E5	1.102E5	0.879E5	0.522E5	0.389E5	0.181E5
1.0	1.232E5	1.101E5	0.905E5	0.428E5	0.562E5	0.175E5

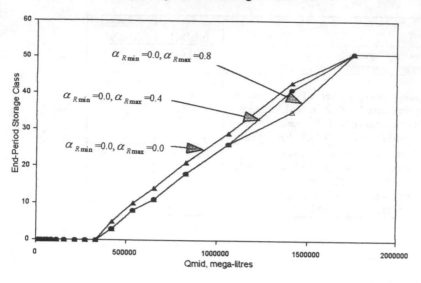

Figure 2.a. Optimum Policy for $\alpha_{Rmin} = 0.0$ and $\alpha_{Rmax} = 0.0,\ 0.4,$ and 0.8, and $S_{k,t}$ class = 0.

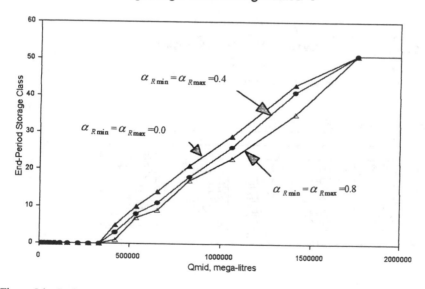

Figure 2.b. Optimum Policy for $\alpha_{Rmin} = \alpha_{Rmax} = 0.0, \alpha_{Rmin} = \alpha_{Rmax} = 0.4,$ and $\alpha_{Rmin} = \alpha_{Rmax} = 0.8$, and $S_{k,t} = 0$.

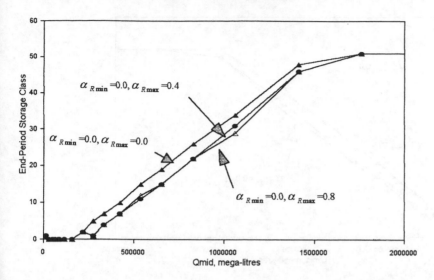

Figure 3.a. Optimum Policy for $\alpha_{R\min} = 0.0$ and $\alpha_{R\max} = 0.0$, 0.4, and 0.8, and $S_{k,t}$ class = 5.

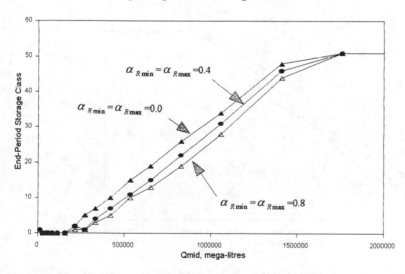

Figure 3.b. Optimum Policy for $\alpha_{R\min} = \alpha_{R\max} = 0.0$, $\alpha_{R\min} = \alpha_{R\max} = 0.4$, and $\alpha_{R\min} = \alpha_{R\max} = 0.8$, and $S_{k,t} = 5$.

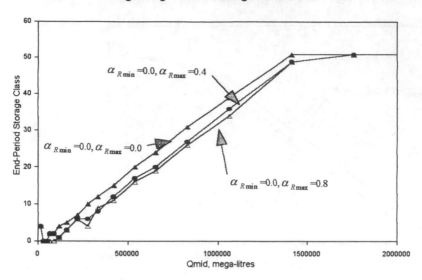

Figure 4.a. Optimum Policy for $\alpha_{Rmin} = 0.0$ and $\alpha_{Rmax} = 0.0,\ 0.4,$ and 0.8, and $S_{k,t}$ class $= 10$.

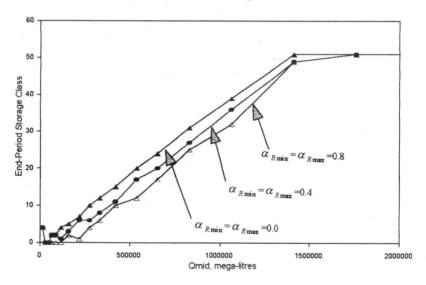

Figure 4.b. Optimum Policy for $\alpha_{Rmin} = \alpha_{Rmax} = 0.0,\ \alpha_{Rmin} = \alpha_{Rmax} = 0.4$, and $\alpha_{Rmin} = \alpha_{Rmax} = 0.8$, and $S_{k,t} = 10$.

Stochastic Hydraulics'96, Tickle, Goulter, Xu, Wasimi & Bouchart (eds) © 1996 Balkema, Rotterdam. ISBN 90 5410 817 7

A more efficient approach to consideration of uncertainty in multiple reservoir operation

Khalid Karim & Ian C. Goulter
Central Queensland University, Rockhampton, Qld, Australia

ABSTRACT: Explicit consideration of uncertainty in the development of optimal operating policies in multiple reservoir systems has traditionally required complex models and approaches. This paper describes a technique wherein Limiting State Probabilities derived from underlying Markovian characteristics of flows, rather than the steady state operating policies, are used to define the deterministic inflow structures. These inflow structures are then used to reduce the complexity of models required for consideration of the uncertainty. This simplification, in turn, leads to the opportunity to use deterministic optimisation approaches in the development of operating policies for both simple one reservoir systems, and the more complex multiple reservoir systems, in a computationally effective and theoretically sound fashion. The paper summarises the theory behind the transformation of the optimisation problem from stochastic to deterministic and describes the application of the methodology, using deterministic Network Linear Programming, to a multiple reservoir system.

1 INTRODUCTION

Uncertainty is a key consideration in planning and operation of water resources systems. Operation of reservoir systems is no exception. Of the various parameters involved in planning reservoir operation, inflows are often both the most important, and the most uncertain variables. While elements affecting the demand may also be predictable to some degree, or at least can be controlled to some extent to conform to the predictions, inflows are generally beyond control. However, they are one of the most important components because, notwithstanding the dependency of inflows among themselves or other meteorological variables, all the variables in an optimisation or simulation formulation are affected, directly or indirectly, of the inflows. As such, since the success or failure of a water resources project to deliver the planned or anticipated outcomes depends a great deal upon how well the uncertainties have been accounted for and evaluated, there is a need to address the uncertainty, or stochasticity of inflows, while planning the long term or medium term

operations of reservoirs.

Various approaches have been used to incorporate, or rather address the impact of, the uncertainty of inflows in a planning process. The actual approach used, of course, depends upon the technique being used in the planning process, i.e., whether simulation or optimisation is being used and, in case of optimisation, which of the various available techniques is being adopted. The uncertainty of inflows in an optimisation framework is generally addressed by use of either the expected values (mean or median), critical period values (the worst case scenario), a large number of equally likely flows, or the assumption of a Markov process. Each approach has its advantages and shortcomings from a statistical perspective with their strengths and weaknesses being imposed on the characteristics of the optimisation algorithm used.

For example, expected value is acceptable only if the variation due to uncertainty lies within reasonable limits. Any optimal solution generated on the basis of expected values should be verified by a sensitivity analysis (Loucks et al. 1981). In reality the possible variation in inflows is rarely sufficiently insignificant to justify the use of this approach. Secondly, the requirement for a sensitivity analysis greatly increases the number of trials needed to obtain sufficient information for the development of final 'optimal' operating policy. Critical period inflows, on the other hand, can result in the prediction of a storage requirement greater than what might actually be needed. Such a situation may result in greater losses through evaporation, while still being susceptible to both spills in periods of higher inflows (Hall and Dracup 1970), and possible damage because of the inability of the reservoir system to absorb floods to any greater extent than a smaller sized reservoir because the larger reservoir may, at the time of extreme floods, be full or close to full with no more capacity to absorb the floods than a smaller reservoir. Again, the operating policy developed by such an approach would also need to be verified by a sensitivity analysis.

Use of equally likely inflows can encompass a greater level of uncertainty but would also require a large number of trials and, even more importantly, would result in extensive output which would again need to be analysed by the decision maker to determine a final operating policy. The Markovian assumption for the relationship between inflows in successive time intervals, i.e., the probability of a particular inflow in one period is dependent to some extent on the inflow that occurred in the previous time period, is a common approach to incorporation of uncertainty in models for optimisation or reservoir operation (Loucks et al. 1981). However, due to the nature of the optimisation involving such processes it's application has generally been limited to only one optimisation algorithm, i.e., dynamic programming, which, to date, has limited its utility primarily to single reservoir systems.

As mentioned earlier, the utility of each type of representation of the stochasticity of inflows also depends upon the actual technique, i.e., simulation or optimisation, and if optimisation then which optimisation approach is used, and as such needs to be evaluated in that context. Simulation is essentially deterministic in nature, not only in the way in which inflows are represented but also by virtue of the fact that each policy and condition has to be pre-set and response of the system obtained for all possible scenarios and conditions, one at a time. Given

the large number of trials generally required of this approach, even without incorporating a large number of equally likely flows, such inclusion greatly increases the computational requirements, particularly for multiple reservoir systems. As stated by Loucks et al. (1981), "Simulation relies on trial and error to identify near optimal solutions. The value of each decision variable is set, and the resulting objective values are evaluated. The difficulty with simulation is that there is often a frustratingly large number of feasible solutions or plans."

Of the optimisation techniques, linear and dynamic programming have been fairly popular for reservoir operation problems. Various approaches have been developed to incorporate stochasticity in LP models. Loucks (1968) developed a stochastic LP for a single reservoir using net inflows and noted the existence of the dimensionality problem. Chance-constrained LP is another form of stochastic LP which imposes the probabilistic consideration by using the probability distributions of the parameters and coefficients in the constraints. However, this process may be quite difficult and reservations about the limitation of this technique have been expressed by Hogan et al. (1981). Another possible option is stochastic programming with recourse. However Yeh (1985) has noted that a complete analysis using this technique would be difficult and computationally expensive. Success in the application of LP based techniques to multiple reservoir systems has been very limited, primarily due to the exponential increase in computational requirements, which occurs as the number of reservoir increases. Although LP models for multiple reservoir systems using spatial and temporal decomposition (Yeh 1985) have been proposed to address the problem of computational requirements, improvements were achieved at the cost of further approximation of the physical representation of the system. Lately, interest in use of LP to plan the operation of reservoirs seems to have waned and only occasional examples of this technique appear in the current literature.

Dynamic programming (DP) has been by far the more popular technique in recent times, in part due to the relatively non-complex manner in which the technique is able to incorporate stochasticity. The most common and effective approach to the consideration of stochasticity in

DP has been to assume a Lag-1 Markov relationship between the inflows in successive time periods. This assumption requires the probability of a particular inflow in a time period to be specified in relation to the inflow in the previous period. However, even without the consideration of stochasticity, DP is subject to the 'curse of dimensionality', a situation which is aggravated when multiple reservoir systems are being considered. As with linear programming, the dimensionality problem may be overcome to some degree by the use of decomposition or aggregation of the system and additional simplifications. These steps results in a greater degree of inaccuracy however and most, if not all, efforts have concentrated on addressing the dimensionality problems at a more fundamental level (Saad and Turgeon 1988, Georgakakos and Yao 1993).

Notwithstanding the computational burden associated with optimisation techniques, in case of multiple reservoir systems there is still a need not just to address the general uncertainty of inflows and their period to period correlations, but also to recognise and consider the temporal and spatial cross correlations between various reservoir sites. A further complication arises from the fact that natural inflows to downstream reservoirs may be modified by the operation of upstream reservoirs. This paper looks at addressing the incorporation of stochasticity for optimisation of multiple reservoir system in a computationally feasible manner.

2. THEORY

2.1 Stochastic dynamic program

As mentioned earlier, dynamic programming is considered the most efficient technique for optimisation of single reservoir systems, when uncertainty or stochasticity is to be considered. Dynamic programs, known as stochastic dynamic programs or SDP, which explicitly incorporate uncertainty do so by assuming that the process of inflows can be described by Markov chains whereby the transition probability matrices are obtained from historic records. These transition probability matrices provide the probability of a discrete inflow in a time period, given that a certain inflow has occurred in the previous period. These matrices

also require an additional assumption of stationarity, i.e., the process does not change over time, which in the case of inflows means that, although the transition probabilities change from month to month, they would not change from year to year. Under these conditions, the set of transition probability matrices is considered sufficient to define the behaviour of the system described by a single Markov chain process (Loucks et al. 1981).

As described by Loucks et al. (1981), the steady state operating policy obtained from an SDP defines the optimal storage volume for each initial storage volume and inflow in each period. Thus the values of PR_{kit}, i.e., the joint probabilities of initial storage volume, the inflow and the final storage volume, in period t are obtained from:

$$PR_{ljt+1} = \sum_{k} \sum_{i} PR_{kit} \, P^t_{ij} \qquad \forall \; l,j,t+1 \qquad (1)$$

$$\sum_{k} \sum_{i} PR_{kit} = 1 \qquad \forall \; t \qquad (2)$$

where PR_{ljt+1} is the joint probability associated with obtaining storage l with inflow j in period t+1, and P^t_{ij} is the transition probability of inflow from state i at the beginning of period t to state j at the end of period t. The right hand side of Equation (1) is a selective summation over only those initial storage and inflow indices k and i which result in a final storage volume having index l.

One equation from Equation set (1) in each time period is redundant. Loucks et al. (1981) show that, once the solution of PR_{kit} is obtained, the corresponding marginal distribution of inflows PQ_{it} can be obtained by:

$$PQ_{it} = \sum_{k} PR_{kit} \qquad \forall \; i, t \qquad (3)$$

In summary, the inflows for SDP are generally assumed to be described by a single first order Markov chain. This assumption, coupled with an assumption of stationarity, enables the transition probability matrices to be used to arrive at an operating strategy based on steady state conditions. These steady state policies are, in turn, used to derive the probabilities of the recommended releases as well as the marginal probabilities of inflows associated with those probabilities. This paper is derived from the hypothesis that, if the marginal inflow probabilities PQ_{it} can be determined a priori, then they can be used directly as input, without recourse to SDP. The following sections examine this hypothesis.

2.2 Limiting state probabilities

A potential approach to the explicit recognition of the stochastic nature of inflows in a computationally tractable fashion is to use the concept of limiting state probabilities. Limiting state probabilities constitute a vector of probability of inflows for each month, with the values of this vector being independent of all previous values in previous months. However, their use in the development of optimal operating policies requires the underlying Markov process to be monodesmic, i.e., to consist of a single Markov chain, which is of course the assumption used in stochastic dynamic programming.

If the transition probability matrices for each period, obtained on the basis of the assumption of a Lag-1 Markov chain in the process, are multiplied successively a sufficient number of times the resulting vector will approach the limiting state probability vector for that process. These limiting state probabilities constitute the probability vector which i) describes the conditions existing at the end of the period and ii) provides the probability of being in each inflow state irrespective of any preceding inflow states, including the starting state.

The mathematical methodology for direct determination of limiting state probabilities is provided in Howard (1971) and is constituted by the simultaneous equations:

$$\Pi_j = \sum_{i=1}^{N} \Pi_i \, p_{ij} \qquad j=1,2,......,N \qquad (4)$$

$$\sum_{i=1}^{N} \Pi_i = 1 \qquad (5)$$

where Π_j is the limiting state probability vector for period j and p_{ij} is the transition probability from state i at the beginning of a period to state j at the end of a period. The set of equations defined by Equation (4) is linearly dependent.

However the addition of Equation (5) enables a unique solution as proved by Howard (1971). Howard (1971) has extended the above analysis to cyclic processes, such as the annual cycle of inflows, where the Markov process is governed by a succession of transition probability matrices (12 in case of monthly inflows) which are repeated over and over again in that order. To solve for the limiting state probabilities in such cyclic systems, Howard (1971) shows that a composite stochastic matrix may be formed by:

$$[^C P] = [^1 P][^2 P]......[^{12} P] \qquad (6)$$

where $^C P$ is the composite stochastic matrix and $^1 P$ is the transition probability matrix for period 12 to period 1, $^2 P$ for period 1 to 2 etc.

The limiting state probability vector $^L \pi$ that holds at the end of transitions can be obtained from:

$$[^L \pi] = [^L \pi][^C P] \qquad (7)$$

and

$$\sum_{i=1}^{n} {}^L \pi_i = 1 \qquad (8)$$

The limiting state probability vector $^k \pi$ for each month k may then be obtained from:

$$^1 \pi = [^L \pi][^1 P]$$
$$^2 \pi = [^1 \pi][^2 P]$$
$$\vdots$$
$$^k \pi = [^{k-1} \pi][^k P]$$

2.3 SDP and limiting state probabilities

In order to examine the relationship between the marginal probabilities of inflow, PQ_{it}, obtained from the steady state solution of an SDP, and the limiting state probabilities obtained directly from the transition probability matrices that are used as input for the SDP, an SDP model for optimisation of the operation of a single reservoir, namely, the Fairbairn Dam which is located upstream of Emerald in Central Queensland, Australia, was formulated.

Historical data from the 95 years of record at the reservoir was used to compute the transition probability matrices. For this purpose, a range of inflow intervals from 4 to 9 was used. Although it is desirable for accuracy purposes that the inflow intervals be as fine as possible, in this case the maximum number of intervals was limited to 9 in order to ensure that no column or row in any of the matrices was composed entirely of zeros. This step is essential in order to meet the basic condition of Markov processes which requires that there be a possibility to move from each state to every other state. The other mathematical condition, of course, is that the elements of every row should sum to 1.

The model was formulated as a standard backwards stochastic dynamic program. The probabilities of releases PR_{kit} together with the marginal probabilities of inflows PQ_{it}, were computed from the steady state solution derived from the model in accordance with Equations (1) - (3). Using the same transition probability matrices that were used for the SDP, the limiting state probability vectors were also computed following Howard's methodology as described previously. The limiting state probability vectors and the marginal probabilities derived from the result of the SDP, shown in Table 1. Comparison of the two probabilities reveals that the two are essentially equal.

It is also evident from Table 1, the maximum difference between the limiting state probabilities and the marginal probabilities of inflows is smaller than 0.001, indicating that the two vectors are, for all intents and purposes, the same.

On the basis of this similarity, the SDP was modified so that limiting state vectors, rather than the transition probability matrices, were used for the optimisation. It is important to re-emphasise that these limiting state vectors are by definition independent probability vectors. Use of the limiting state vectors in this manner therefore implies that the probability of a particular inflow in a given period is independent of the inflow in the previous period. In other words, P^t_{ij} is the same for all i and the recursive equation becomes:

$$f^n_t(k,i) = \underset{r}{minimum}[B_{kilt} + \sum_{j=1}^{m} P_{j,t+1}\, f^{n-1}_{t+1}(l,j)] \quad (9)$$

The purpose of the above formulation is to determine the applicability of PQ_{it}, if it can be

Table 1. Comparison of marginal (PQ_{it}) and limiting (Lim.) state probabilities.

	Inflow Intervals				
	1	2	3	4	5
January					
PQ_{it}	0.053	0.685	0.137	0.063	0.063
Lim.	0.053	0.685	0.136	0.063	0.063
February					
PQ_{it}	0.064	0.695	0.116	0.074	0.053
Lim.	0.064	0.696	0.116	0.074	0.053
March					
PQ_{it}	0.085	0.695	0.095	0.063	0.063
Lim.	0.084	0.696	0.095	0.063	0.064
April					
PQ_{it}	0.158	0.727	0.021	0.021	0.074
Lim.	0.159	0.728	0.021	0.021	0.074
May					
PQ_{it}	0.453	0.453	0.042	0.010	0.042
Lim.	0.454	0.454	0.042	0.010	0.042
June					
PQ_{it}	0.485	0.347	0.021	0.063	0.084
Lim.	0.486	0.348	0.021	0.064	0.084
July					
PQ_{it}	0.474	0.432	0.011	0.032	0.052
Lim.	0.475	0.433	0.011	0.032	0.052
August					
PQ_{it}	0.548	0.389	0.021	0.021	0.021
Lim.	0.549	0.390	0.021	0.021	0.021
September					
PQ_{it}	0.558	0.347	0.021	0.042	0.032
Lim.	0.559	0.348	0.021	0.042	0.032
October					
PQ_{it}	0.453	0.484	0.010	0.032	0.021
Lim.	0.454	0.485	0.010	0.032	0.021
November					
PQ_{it}	0.358	0.610	0.011	0.011	0.011
Lim.	0.359	0.612	0.011	0.011	0.011
December					
PQ_{it}	0.126	0.759	0.042	0.042	0.032
Lim.	0.127	0.760	0.042	0.042	0.032

obtained a priori, by computing the limiting state probabilities as discussed previously.

The steady state solutions for the month of January obtained from the two SDP formulations are shown in Table 2.

123

Table 2. Steady state releases obtained from standard (S) and modified (M) SDP. (Megalitres, ML)

| Storage | Previous Month's Inflow | | | |
| | 62710 | 109623 | 210987 | 682095 |
(10^3)				
123 (S)	0	0	0	2890
(M)	(0)	(0)	(0)	(0)
217 (S)	84637	84637	84637	84637
(M)	(84637)	(84637)	(84637)	(84637)
405 (S)	84637	84637	84637	84637
(M)	(84637)	(84637)	(84637)	(84637)
594 (S)	84637	84637	84637	84637
(M)	(84637)	(84637)	(84637)	(84637)
783 (S)	84637	84637	84637	84637
(M)	(84637)	(84637)	(84637)	(84637)
971 (S)	84637	84637	84637	84637
(M)	(84637)	(84637)	(84637)	(84637)
1160 (S)	84637	84637	84637	84637
(M)	(84637)	(84637)	(84637)	(84637)
1348 (S)	84637	84637	84637	84637
(M)	(84637)	(84637)	(84637)	(84637)
1443 (S)	84637	84637	72959	84637
(M)	(84637)	(84637)	(84637)	(84637)

Note that the figures shown in brackets in Table 2 are obtained from the modified SDP, while the values above them result from the standard SDP. It needs to be noted here that the exercise resulting in Table 2 was carried out with nine inflow intervals. However, due to space limitations this table only contains the steady state solutions pertaining to the last four intervals. In this regard it is sufficient to point out that, in the portion omitted, the two solutions were exactly the same, as is the situation with almost all cases included in the table.

It can be seen in Table 2 that there are only two values that differ in any manner. Firstly, the second last entry relating to the maximum storage of 1443000 ML shows that the standard SDP using transition probability matrices recommended an optimal release of 72959 ML compared to 84637 ML from the modified SDP using limiting state vectors. Since the magnitude of inflow increases from left to right, this result indicates that, for the same storage, the standard SDP recommends a lower release for a higher inflow. Such discrepancies can arise as a result of the effect of transients but clearly the solution returned by the modified SDP is more logical. The second difference is in the last element relating to the minimum storage where the standard SDP recommends a release of 2890 whereas the modified SDP recommends no release. Here it is important to note that the probability associated with both these releases are 0, which diminishes the significance of this difference. Further, the difference between the magnitude is actually quite small. Keeping this in mind and examining the rest of the solution, it can be asserted with some confidence that the solutions provided by both the techniques are, in effect, exactly the same. This is a very important assertion because it translates into the hypothesis that limiting state probability vectors can replace the transition probability matrices as inputs to the SDP. However, the use of these vectors eliminates the need for consideration of period to period correlation while still preserving the stochastic character of the inflows, derived as it is from the Markov assumption. Even more importantly, the independence of these vectors enables the specification of stochastic input in a deterministic structure.

A further exercise was carried out where the magnitude of inflows associated with transition probabilities was reduced arbitrarily, while retaining the same transition probabilities. In this case the optimal solution with regard to releases and their associated probabilities changed, which is quite obvious. However in this case the marginal probabilities of inflows, PQ_{it}, still remained the same. In other words, as long as the transition probabilities are not changed, incorporation of limiting state vectors would always result in an optimal solution. Needless to say that if the transition probabilities are changed, their associated limiting states would also change, but with the same general applicability of results.

3. APPLICATION

3.1 *Network linear program*

Network linear programming (NLP) is a special form of linear programming based on an approach of minimising the cost of conveying a commodity through a network. This general structure can be exploited to optimise the operation of reservoir systems for long term planning. The advantage of this technique is that convergence is not only guaranteed but it is also very fast. Due to these fast solution times, network linear programs are very suitable for multiple reservoir systems.

However, one disadvantage of using NLP in its traditional form is that it is deterministic in structure and, therefore, the inflow inputs must also have a deterministic structure. Given the importance of stochasticity in the development of optimal operating policies for reservoirs, one way around this problem is to use a large number of equally likely flows, based upon the probability distributions of natural inflows. However, this process would not only again significantly increase the computational time and effort but would also result in a large output which would need to be evaluated for final development of policy.

In order to overcome this problem, the concept of limiting state probabilities discussed previously can be applied. It should be recalled at this point that limiting state probability vectors are derived from the transition probability matrices and hence incorporate the elements of stochasticity. On the other hand, such vectors provide a deterministic input structure which is, by definition, independent from one time period to the next and as such does not require any consideration of correlations, because they are embedded in the derivation from Markov chains. Most important of all, as mentioned before, these vectors can be used to develop the optimal operating policy for an individual reservoir. However they are equally applicable for use in an NLP to determine the optimal joint operation of multiple reservoir systems.

3.2 *Case study*

The concept was applied to system of 6 reservoirs with multiple demand areas, most of which are joint demand centres for two or more reservoirs, directly or indirectly (indirectly in the sense that, e.g., an upstream reservoir may not be able to supply a particular demand area directly but can supplement a downstream reservoir which provides water for that demand).

The actual node and arc formulation is not shown here due to space limitations. However it is important to note that there are a total of seventeen demand centres comprising domestic, industrial and irrigation uses. More importantly, additional demands were placed on these reservoirs to provide flows for maintaining the ecological integrity of the streams.

The mathematical formulation of the network linear program is very simple, i.e.,

$$\text{minimise cx} \tag{10}$$

where c is a vector of dimension of costs and penalties for flow in each arc and x is an n dimensional vector representing the volumes of flow along the arc.

The above formulation is subject to the following constraints:

$$\text{nodal inflow-nodal outflow} = 0 \ \forall \text{ nodes} \tag{11}$$

$$x_{max} >= x >= x_{min} \tag{12}$$

The network is, of course, formulated for optimal operation of reservoirs through an objective function in which shortages of water are minimised.

The algorithm that was used for the model was RELAX, which is a public domain code presented by Bertsekas (1991). This algorithm is a dual ascent method and has the fastest solution time, according to the comparison of various codes provided by Bertsekas (1991). It is important to note that this is a general purpose code which can solve any network, irrespective of the actual problem. Although it has to be appropriately modified to incorporate evaporation etc. the problem is actually embedded in the input file thereby providing a significant level of flexibility in the use of this type of algorithms.

The inflows to each reservoir in this formulation were based on the limiting state probability vectors computed from the transition probabilities which were derived, in turn, from the historic flow records. The order of inflows was generated randomly from the limiting state probabilities. It should be noted that this step was performed only to ensure that the inflow order was random and not 'influenced' in any way. Nevertheless, the flows coming from this random generation process are different from those derived on the basis of generation of equally likely flows in that only those outputs which match with the limiting state probability vector are incorporated.

A range of different trials were performed wherein the traditional demands remained unchanged but the instream demands representing environmental requirements were varied as percentiles of historic flows, i.e., 20th, 50th and 90th percentiles. The problem was formulated for a twenty year period with the network containing a total of 17034 arcs and 6961 nodes.

The program was run on DEC 3000 - M600. The solution times on this machine ranged from 25 to 40 seconds, with the corresponding number of iterations required for convergence ranging from 8500 to 18000. It is interesting to note that the greater number of iterations and longer times were associated with those cases where the shortages could perhaps be termed as medium, i.e., in the case of the instream demands being set as 50th percentile of historic flows. Shortages are bound to occur for all percentile levels but not unexpectedly they tend to be minimal with the 20th percentile demands and very high when the instream demand is set at the 90th percentile).

The results of the trials indicated that, for all practical purposes, the traditional demands are invariably met. This is not surprising in view of the fact that reservoirs in the system are actually designed to meet these demands on the basis of available hydrologic records or, conversely, demands are planned according to what reservoirs can deliver. Accordingly, the conflicts in meeting the demands tend to occur only as a result of environmental demands.

The percentages of various instream demands that were satisfied are shown in Figure 1. It is clearly evident that, under the current conditions, the reservoirs can only be expected to satisfy 20th percentile environmental demands, and then only by stretching the system to its maximum. For environmental demands greater than the 20th percentile, there has to be a trade off between the existing traditional demands and the environmental demands.

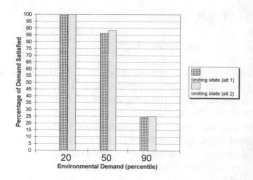

Fig. 1: Percentage of instream demands satisfied

As was noted previously, the modified formulation of SDP, as given by Equation (9), gave essentially the same results as the standard SDP, despite the fact that, in this case, the probabilities of a given inflow were independent of the probabilities in any preceding time period. This result clearly indicates that the random order of inflows does not affect the outcome, as long as they conform to the limiting state vector. However, to be more certain, a second trial was made by incorporating a different order of flows, generated randomly as before, and matching the limiting state vector. As can be seen from Figure 1, there are insignificant differences between the results of the two alternatives.

4 CONCLUSION

It has been shown that by using the concept of limiting state probability, a deterministic structure for the input of inflows which retains the stochastic basis of these inputs can be obtained. This concept, in turn, enables the utilisation of deterministic algorithms like NLP for optimisation of the operation of multiple reservoir systems with implicit consideration of stochasticity in the inflows. These algorithms have very fast solution times and have the capability of both simultaneously optimising multiple reservoir systems while incorporating a better representation of the physical relationships in the system than existing approaches such as SDP.

REFERENCES

Bertsekas, Dimitri P. 1991. *Linear network optimization, algorithms and codes:* 359 p. London: The MIT Press.

Georgakakos, Aris P. & Huaming Yao 1993. New control concepts for uncertain water resources systems. *Water Resources Research.* 29(6): 1505-1516.

Hall, W.A. & J.A. Dracup 1970. *Water resources systems engineering:* 372 p. New York: McGraw-Hill.

Hogan, A.J., J.G. Morris & H.E. Thompson 1981. Decision problems under risk and chance constrained programming: dilemmas in the transition. *Management Science.* 27(6): 698-716.

Howard, R.A. 1971. *Dynamic probabilistic systems, volume I: markov models:* 576 p. New York: John Wiley & Sons, Inc.

Loucks, D.P. 1968. Computer models for reservoir regulations. *Journal of The Sanitary Engineering Division,* ASCE. 94(SA4): 657-669.

Loucks, D.P., J.R. Stedinger & D.A. Haith 1981. *Water resources system planning and analysis,* 559 p. Englewood Cliffs: Prentice Hall.

Saad, Maarouf & Andre Turgeon 1988. Application of principal component analysis to long term reservoir management. *Water Resources Research.* 24(7): 907-912.

Yeh, W. W-G. 1985. Reservoir management and operation models: a state-of-the-art review. *Water Resources Research.* 21(12): 1797-1818.

Stochastic Hydraulics'96, Tickle, Goulter, Xu, Wasimi & Bouchart (eds) © 1996 Balkema, Rotterdam. ISBN 90 5410 817 7

Use of fuzzy set theory for consideration of storage non-specificity in stochastic dynamic programming for reservoir operation

Suharyanto & Ian C.Goulter
Central Queensland University, Rockhampton, Qld, Australia

ABSTRACT : Applications of dynamic programming in reservoir operation and management generally use the reservoir storage as a state variable and in that context require the reservoir storage to be discretised. While the evaluation of the optimisation function is performed at specified grid points, usually the mid-point, within each discretised storage interval, actual operation of the reservoir occurs in an environment in which the storage volume rarely falls at that grid point. In this paper, an approach to recognising, within a stochastic dynamic programming framework, both the implications of the actual storage occurring at any point within the storage class interval and the influence of the storage intervals or classes adjacent to the storage interval in which the actual storage actually falls, is proposed. The approach uses fuzzy set theory and its associated concept of membership functions and is demonstrated by application to the Fairbairn reservoir in Emerald, Central Queensland, Australia.

1 INTRODUCTION

Most applications of stochastic dynamic programming (SDP) in reservoir management require discretisation of the state variables. Evaluation of the objective function within the optimisation is performed at specified grid points, usually the mid-point, of the discretised variables. Another common strategy in the application of SDP in reservoir management is to assume that the storage volume at the beginning of the current period (stage) is known, while the inflow state during the current period is either known (the perfect forecast condition) or known only by its distribution conditioned on the known (realised) inflow in the previous time period. The controlling variable in this situation is the reservoir release. Transformation of the storage volume from a particular state or condition at the beginning of the current period to another state in the next period is controlled by the reservoir release and can be expressed through the continuity equation. The optimum release for each combination of initial storage and inflow in the previous time period is determined on the assumption that the long-range return associated with the storage volume which results at the end of that period is adequately defined

by the value of the long range return associated with the storage volume corresponding to the grid point of the storage volume class within which that resultant storage falls. In other words, in a *crisp* discrete SDP, it is assumed that the grid point of the state variable is an adequate representation of any value in the corresponding class range or interval. Under this assumption, there is implicitly no specific preference for selecting or obtaining any particular value within the class range.

Now consider the use of storage volume as a state variable for dynamic programming models. In crisp, as opposed to fuzzy, SDP applications, no consideration is given to the location of the storage volume in its circumscribing class. It is assumed, at least implicitly, that it does not matter or make any difference whether the actual storage volume is exactly at the grid point, close to, or even far away from the grid point. Even when the storage volume is close to its class boundaries, no consideration is given to recognising the influence of the conditions in the adjacent storage intervals. Use of fuzzy set theory, which is capable of acknowledging non-sharp boundaries, provides the opportunity for a more appropriate approach for explicit consideration of storage non-specificity, i.e., the association of resultant storage volumes with storage values other than the grid point of the storage intervals in which that resultant storage volumes fall.

This paper describes an approach based on fuzzy set theory wherein reservoir operating rules are generated in an SDP with explicit recognition of storage volume non-specificity. The strategy adopted in the approach is to assume that the reservoir release and the resultant storage volume at the end of the period are two interacting variables which should be solved simultaneously. In this framework, the reservoir releases as well as the reservoir inflow state and the reservoir storage state need to be discretised. The strategy is therefore referred to as a 3 state-variable SDP.

2 THEORETICAL BACKGROUND

2.1 Stochastic Crisp Dynamic Programming (SCDP)

In this formulation, the storage volume is divided into NS intervals with sharp or crisp boundaries between each interval with $S_{k,t}$ representing the kth storage class interval at the beginning of the current time period t. The notation $S_{l,t+1}$ is therefore the storage state at the end of period t, i.e., at the beginning of time period $t+1$, which is in class interval l. The monthly inflow ranges are similarly discretised into NI intervals with the term $Q_{j,t}$ representing the inflow class or interval j during time period t. The possible reservoir releases are also discretised into NR classes, with $R_{r,t}$ representing release class r during month t.

Consider now a reservoir operating within an environment in which it has to meet specified target demands, e.g., irrigation demands. Let the demand in time period t be denoted by T_t. The continuity equation for this simple situation is:

$$S_{l,t+1} = S_{k,t} + Q_{j,t} - R_{r,t} - L_t \qquad (1)$$

where L_t is the loss through evaporation, seepage, etc. from the reservoir during time period t. The resulting $S_{l,t+1}$ is constrained to lie within the range of the feasible storage volumes, i.e., :

$$S_{min,t} \leq S_{l,t+1} \leq S_{max,t} \qquad (2)$$

where $S_{min,t}$ and $S_{max,t}$ are the minimum and maximum storage volumes in time period t, respectively.

A typical simple objective function for this type of system is minimisation of the sum of the square of deviations of the actual releases from target releases if the release value is outside the safe range of the target value. In this situation, the short-term return may be defined as :

$$B_{r,t} = \begin{cases} K_1 \left(0.80 - \frac{R_{r,t}}{T_t} \right)^2 & \text{if } R_{r,t} < 0.80T_t \\ K_2 \left(\frac{R_{r,t}}{T_t} - 1.20 \right)^2 & \text{if } R_{r,t} > 1.20T_t \end{cases} \qquad (3)$$

where the safe or acceptable range of releases is defined to lie between $0.8\,T_t$ and $1.2\,T_t$, K_1 and K_2 are weighting constants on the deviation, and $B_{r,t}$ is short-term return as a consequence of releasing $R_{r,t}$ during period t.

Proceeding backward in time, the general recursive equation of the DP is given by :

$$f_t^n(k,i) = \min_{R_{r,t} \in \bar{R}_t} \left\{ B_{r,t} + \sum_{j=1}^{NI} P_{i,j}^{t-1} \, f_{t+1}^{n-1}(l,j) \right\} \qquad (4)$$

where $f_t^n(k,i)$ is the optimum objective function associated with being in storage interval or class k at the beginning of period t with n periods of the optimisation stages remaining and having an observed inflow during the previous time period which falls in interval i, i.e., $Q_{i,t-1}$. $\bar{R}_t$ is the set of possible discretised releases in time period t. The value of the term $f_{t+1}^{n-1}(l,j)$ in this process is the long-term return associated with the grid point for the storage class interval l which encloses the value of actual storage volume S_{t+1}. $P_{i,j}^{t-1}$ is the probability, given that the inflow during the previous time period $t-1$ was $Q_{i,t-1}$, that the inflow during the period t will be in class j, thereby resulting in storage level in class l at the end of the time period t when the release is $R_{r,t}$. For each storage state k at the beginning of the current period, the optimum objective function value is obtained by minimising the right hand side of Equation (4) with respect to the whole discretised release decision values. The above formulation is basically equivalent to that of Butcher (1971).

The following additional chance constraints may also be imposed in this formulation to exercise control on the probability of violation on the minimum and maximum storage volumes.

$$p\left[S_{t+1} \geq S_{min,t} \right] \leq P_{min,t} \qquad (5)$$

$$p[S_{t+1} \leq S_{max,t}] \leq P_{max,t} \qquad (6)$$

where $P_{min,t}$ and $P_{max,t}$ are the maximum acceptable probability of violation of the minimum and maximum storage volumes, respectively, in time period t.

The implementation of this 3-state-variable SDP has a number of advantages over a the 2-state-variable SDP in which release decisions are not discretised. In 2-state-variable approach, the precision of the release decision is essentially prescribed only by the 'fineness' of the storage discretisation level. In the situation of a large reservoir supplying water demands characterised by high variability, a 2-state-variable SDP will require a large number of storage intervals to avoid the 'trapping' effect reported by Goulter and Tai (1985) and the possibility of overlooking water demands which are low in values. Since the release decision value defined in the continuity equation for a 2-state-variable formulation depends on the accuracy of the storage discretisation level, avoiding the 'trapping' effect may, however, still result in a storage discretisation level which is not sufficiently fine to detect low values of target demands. In this situation, the storage discretisation level should ideally be specified with the same level of precision as the low valued target demands. However, the process may result in an unreasonably high number of storage intervals. In the 3-state-variable SDP, the chance of 'overlooking' of water demand of small values is minimised by an appropriate discretisation of the reservoir release, while the 'trapping' effect can be avoided by applying the guidelines presented in Klemes (1977).

2.2 Stochastic Fuzzy Dynamic Programming (SFDP)

As noted earlier, fuzzification of the storage interval in an SFDP formulation provides a mechanism for assessment of non-specificity of storage volume within the discretised storage volume intervals. More specifically, it provides a framework for recognising the fact that the actual storage occurring in respect to a particular storage class interval can lie anywhere within that class range. It also provides the ability to recognise the influence of conditions in adjacent storage intervals and specifically their associated long-range returns in the determination of the optimal release decision. The storage intervals of the SFDP in the approach are defined as fuzzy intervals with some degree of overlap between adjacent

intervals. The actual membership functions used in this study are triangular and trapezoidal as shown in Figure 1.

The general recursive formulation required for the fuzzification process is shown in Equation (7). In this expression, the previous long-term return is determined by either averaging the 'triggered' previous long-term returns, i.e., those long-term returns which the approach suggests should be considered in the calculation of the long-term return because the actual storage falls into the fuzzy storage intervals with which these long-range returns are associated for, weighted by the value of the membership levels of the actual storage in the various fuzzy storage intervals in which the storage falls; or by selecting that storage interval which gives the minimal weighted long-term return.

$$f_t^n(k,i) = \min_{R_{r,j} \in \bar{R}_t} \left\{ B_{r,t} + \sum_{j=1}^{NI} P_{i,j}^{t-1} \, f^* \right\} \qquad (7)$$

where f* can be defined either determined through :
a. averaging the previous long-term returns

$$f^* = \sum_{l=1}^{m} \left(\frac{\mu(l)}{\sum_{i=1}^{m} \mu(i)} \; f_{t+1}^{n-1}(l,j) \right) \qquad (8)$$

where $\mu(l)$ is the membership level of S_{t+1} in storage class l, and m is the number of 'triggered' storage intervals, i.e., fuzzy storage intervals into which the storage falls, or in other words $\mu(l) > 0$.

or

b. minimal consequence

$$f^* = f_{t+1}^{n-1}(l^*, j) \qquad (9)$$

where l^* is the storage interval which results in :

$$\min_l \left(\frac{f_{t+1}^{n-1}(l,j)}{\mu(l)} \right) \; \forall l \, |\mu(l) > 0.0 \qquad (10)$$

It should be noted that the averaging strategy of Equation (8) can be interpreted as an attempt to account for the influence of the adjacent storage intervals circumscribing the actual storage volume

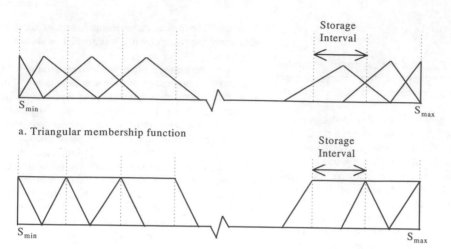

a. Triangular membership function

b. Trapezoidal membership function

Figure 1. Membership Functions of the Storage Intervals.

Table 1. Statistical Characteristics of Monthly Inflows at Fairbairn Dam, Emerald, Australia.

Month	$\bar{x}$	std	ρ	γ	Month	$\bar{x}$	std	ρ	γ
Jan.	92042.49	204784.70	0.091	5.276	Jul.	21496.82	77747.03	0.446	4.940
Feb.	141270.44	335494.19	0.196	5.463	Aug.	4212.06	20163.02	0.138	6.680
Mar.	62617.84	149171.80	0.469	4.307	Sep.	6855.83	27475.23	0.044	6.679
Apr.	41506.49	140425.80	0.294	4.768	Oct.	9169.27	40049.20	0.051	6.798
May	32080.02	136613.25	0.568	5.579	Nov.	28728.02	151277.81	0.013	9.006
Jun.	12124.89	36831.59	0.396	4.303	Dec.	60813.67	194767.33	0.174	6.096

Note : $\bar{x}$ = mean of monthly inflow, in mega-litres
std = standard deviation of monthly inflow, in mega-litres
ρ = coefficient of correlation,
γ = coefficient of skewness.

Table 2. Monthly Target Release of the Fairbairn Dam, Emerald, Australia.

Month	Monthly Demand (megalitres)	Month	Monthly Demand (megaliters)
Jan.	35,101	Jul.	4,493
Feb.	31,667	Aug.	6,858
Mar.	8,733	Sep.	10,576
Apr.	4,002	Oct.	27,308
May	5,185	Nov.	8,613
Jun.	6,166	Dec.	26,817

Table 3. Expected Annual Deviation (EAD) for the SCDP and SFDP Approaches

NR	EAD Results From SCDP Approach	EAD Results From SFDP Approach
5	3.588×10^6	3.570×10^6
10	3.553×10^6	3.542×10^6
15	3.525×10^6	3.514×10^6
20	3.500×10^6	3.491×10^6
25	3.508×10^6	3.498×10^6

Table 4. Comparison of Simulation Results Obtained by Implementing the Operation Policies Generated by the SCDP and SFDP Approaches.

a. First 10 years

		% Empty	% [Rsafe]	%[R<0.8T]	%[R>1.2T]	REL(%)	RES(%)
NR=5	SCDP	43.33	21.67	71.67	6.67	28.33	12.79
	SFDP	46.67	21.67	71.67	6.67	28.33	11.63
NR=10	SCDP	40.83	24.17	69.17	6.67	30.00	13.10
	SFDP	44.17	25.83	68.33	5.83	30.00	15.48
NR=15	SCDP	45.00	24.17	69.17	6.67	30.00	9.52
	SFDP	45.00	28.33	65.83	5.83	32.50	12.35
NR=20	SCDP	44.17	29.17	65.83	5.00	33.33	12.50
	SFDP	48.33	29.17	64.17	6.67	33.33	13.75
NR=25	SCDP	45.00	25.83	67.50	6.67	31.67	10.98
	SFDP	46.67	27.50	65.83	6.67	32.50	13.58

b. First 25 years

		% Empty	% [Rsafe]	%[R<0.8T]	%[R>1.2T]	REL(%)	RES(%)
NR=5	SCDP	18.67	33.67	35.00	31.33	65.00	14.29
	SFDP	20.67	35.00	33.67	31.33	66.00	14.71
NR=10	SCDP	18.00	34.00	33.33	32.67	66.33	17.82
	SFDP	20.00	35.33	32.33	32.33	66.67	18.00
NR=15	SCDP	19.00	35.00	32.67	32.33	67.00	13.13
	SFDP	19.00	36.33	31.00	32.67	68.00	15.63
NR=20	SCDP	19.00	36.67	30.67	32.67	68.67	15.96
	SFDP	21.67	36.33	30.00	33.67	68.67	14.89
NR=25	SCDP	19.00	34.67	31.33	34.00	68.33	13.68
	SFDP	20.00	35.33	30.67	34.00	68.33	15.79

c. First 50 years

		% Empty	% [Rsafe]	%[R<0.8T]	%[R>1.2T]	REL(%)	RES(%)
NR=5	SCDP	12.00	29.67	22.17	48.17	77.83	12.03
	SFDP	13.33	30.67	21.50	47.83	78.17	12.21
NR=10	SCDP	11.67	30.67	21.50	47.83	78.17	15.27
	SFDP	12.83	31.67	20.83	47.50	78.33	15.39
NR=15	SCDP	12.33	30.67	21.17	48.17	78.67	11.72
	SFDP	12.50	32.17	20.00	47.83	79.17	13.60
NR=20	SCDP	12.17	31.67	20.00	48.33	79.50	13.82
	SFDP	13.83	32.00	19.50	48.50	79.50	13.01
NR=25	SCDP	12.33	30.83	20.17	49.00	79.50	11.38
	SFDP	13.00	31.83	19.67	48.50	79.50	13.01

d. First 100 years

		% Empty	% [Rsafe]	%[R<0.8T]	%[R>1.2T]	REL(%)	RES(%)
NR=5	SCDP	6.00	26.08	11.08	62.83	88.92	12.03
	SFDP	6.67	26.92	10.75	62.33	89.08	12.21
NR=10	SCDP	5.83	26.67	10.75	62.58	89.08	15.27
	SFDP	6.42	27.67	10.42	61.92	89.17	15.39
NR=15	SCDP	6.17	26.83	10.58	62.58	89.33	11.72
	SFDP	6.25	28.08	10.00	61.92	89.58	13.60
NR=20	SCDP	6.08	27.17	10.00	62.83	89.75	13.82
	SFDP	6.92	27.83	9.75	62.42	89.75	13.01
NR=25	SCDP	6.17	26.50	10.08	63.42	89.75	11.38
	SFDP	6.50	27.33	9.83	62.83	89.85	13.01

Table 5. Influence of Averaging and Minimal Consequences Long Term Return Strategies on EAD Values with $NS = 100$, $NR = 20$, and $NI = 5$.

Membership Function of Storage Intervals	EAD Values Result From Applying Averaging Option	EAD Values Result From Applying Minimal Option
Triangular	3.653×10^6	3.492×10^6
Trapezoidal	3.670×10^6	3.491×10^6

Table 6. Comparison of the Effects of Averaging and Minimal Consequences Long Term Return Strategies on System Performance.

a. First 10 years

TRIANGULAR membership function

	% Empty	% [Rsafe]	%[R<0.8T]	%[R>1.2T]	REL(%)	RES(%)
minimal	45.00	30.00	64.17	5.83	33.33	11.25
averaging	34.17	10.83	82.50	6.67	16.67	8.00

TRAPEZOIDAL membership function

	% Empty	% [Rsafe]	%[R<0.8T]	%[R>1.2T]	REL(%)	RES(%)
minimal	48.33	29.17	64.17	6.67	33.33	13.75
averaging	35.83	10.00	83.33	6.67	16.67	8.00

b. First 25 years

TRIANGULAR membership function

	% Empty	% [Rsafe]	%[R<0.8T]	%[R>1.2T]	REL(%)	RES(%)
minimal	19.00	37.67	30.00	32.33	69.00	15.05
averaging	13.67	11.00	40.67	48.33	59.00	8.13

TRAPEZOIDAL membership function

	% Empty	% [Rsafe]	%[R<0.8T]	%[R>1.2T]	REL(%)	RES(%)
minimal	21.67	36.33	30.00	33.67	68.67	14.89
averaging	14.33	12.00	41.00	47.00	59.00	8.13

c. First 50 years

TRIANGULAR membership function

	% Empty	% [Rsafe]	%[R<0.8T]	%[R>1.2T]	REL(%)	RES(%)
minimal	12.50	33.00	19.33	47.67	79.83	12.40
averaging	7.83	7.33	26.00	66.67	73.83	7.01

TRAPEZOIDAL membership function

	% Empty	% [Rsafe]	%[R<0.8T]	%[R>1.2T]	REL(%)	RES(%)
minimal	13.83	32.00	19.50	48.50	79.50	13.01
averaging	8.50	7.67	26.33	66.00	73.67	6.96

d. First 100 years

TRIANGULAR membership function

	% Empty	% [Rsafe]	%[R<0.8T]	%[R>1.2T]	REL(%)	RES(%)
minimal	6.25	28.33	9.67	62.00	89.92	12.40
averaging	3.92	3.92	13.00	83.08	86.92	7.01

TRAPEZOIDAL membership function

	% Empty	% [Rsafe]	%[R<0.8T]	%[R>1.2T]	REL(%)	RES(%)
minimal	6.92	27.83	9.75	62.42	89.75	13.01
averaging	4.25	5.25	13.17	81.58	86.83	6.96

S_{t+1}, while the minimum consequence of Equation (10) is directed toward the search of the previous long-term returns which gives the least consequence. Although these strategies do not directly influence the short-term decision, in the long-term this approach does show some changes in the optimum policy and the use of the policy to guide the reservoir operation.

3 APPLICATION

The above formulation was evaluated by application to the Fairbairn reservoir in Emerald, Central Queensland, Australia. This reservoir is mainly used to supply irrigation to an agricultural area dominated by cotton production, for industrial and municipal use, and for riparian flow purposes (Gubbels, 1994; and Queensland Water Resources Commission, 1992). The reservoir itself has a relatively large storage capacity, 1,443,000 megalitres, compared to its mean annual inflow of about 512,000 megalitres. Its minimum capacity is equal to 123,000 megalitres. The statistical characteristics of monthly inflow into the Fairbairn Dam are shown in Table 1, while the monthly demands of the reservoir as used in this study are shown in Table 2. These demands are distributed over the Emerald Irrigation Area which is served through the Selma and Weemah channels systems, over the ponded area around the Fairbairn dam, along the Nogoa and Mackenzie rivers downstream of the dam, and some tributaries along the Nogoa and Mackenzie rivers down to the Springton creek junction (Queensland Water Resources Commission, 1992). A detailed layout of the Fairbairn Dam service areas can be found in Gubbels (1994).

The number of storage intervals, NS, was set at 100 classes, while the number of inflow intervals, NI, was set at 5. The choice of 100 storage intervals was based on the requirement to have an appropriate representation of the storage capacity as envisaged in Klemes' (1977) paper. The historical inflows used to estimate the transitional probability were derived from 90 years of inflow records.

The release policies generated by the SCDP and SFDP techniques were evaluated by simulating the reservoir performances for periods corresponding to 10, 25, 50, and 100 years of operation using the policies generated by the techniques. The inflow sequence used in the simulation was generated by the disaggregation method of McMahon and Mein (1986). Use of the same sequence of inflows for the simulations permitted the investigation of the influence of implementing different operation policies, while keeping the other operational parameters constant. In the SFDP, the storage interval being considered is trapezoidal with the membership function for each storage interval overlapping 50% of the width of the immediately adjacent intervals as shown in Figure 1. The previous long-term return used in generating these results is based on the minimal consequence strategy. The values of K_1 and K_2 used in the analysis were both equal to 1×10^6.

The Expected Annual Deviations (EAD) resulting from the SCDP and SFDP programs for different numbers of release intervals, NR, are shown in Table 3. It can be seen in this Table that the SFDP approach identifies the better solution as indicated by the smaller value of the EAD for that strategy over the complete range of values of NR. The variations in the performance of the reservoir resulting from application of the two policies are shown in Table 4 for NR values ranging from 5 to 25 classes, respectively. The system performances being monitored and presented in this table are :

% Empty   : percentage of time the reservoir is empty,
% [Rsafe]   : percentage of time the release falls in the safe range of target demand (in this study, it is defined as the range between 0.8 to 1.2 of target),
% [R<0.8T]   : percentage of time the release is less than 0.8 of target demand,
% [R>1.2T]   : percentage of time the release is greater than 1.2 of target demand,
REL (%)   : reliability of the reservoir in meeting the target demands, in percent,
RES (%)   : resiliency of the reservoir, in percent.

Table 4 indicates that in most cases the operating policies generated by SFDP result in some improvement in reservoir performance. For example, only rarely does the operating policy generated by the SFDP approach produce a lower system resilience than that derived from the operating policy obtained by the SCDP approach.

The influence on the EAD of implementation of the averaging and the minimal consequences strategy options for the long-term return for the triangular and trapezoidal fuzzy storage membership functions in the SFDP approach is shown in Table 5, while the variation in the performance of the reservoir operated under operating policies generated by the minimal consequences and the averaging strategies for the triangular and trapezoidal fuzzy storage membership function is shown in Table 6. It can be seen that the minimal consequence strategy produces some improvement in reservoir performance in comparison to that obtained by the averaging strategy in terms of both smaller values of EAD and better system performance as defined by the various performance indicators shown in the table. This result may be due in part to the fact that the averaging strategy cannot explicitly determine which storage class the value of S_{t+1} belongs to. Instead, it employs the normalised weighted consequence to average the previous long-term returns and thereby has the potential to minimise the influence of membership level of the 'triggered' intervals. In the minimal consequences strategy, the long-term returns are essentially weighted by the reciprocals of the membership levels of the 'triggered' intervals. The minimal consequences approach is therefore able to identify and react more easily to a pull to a more 'optimum' path.

4 CONCLUDING REMARKS

It has been demonstrated that a more realistic approach to the consideration of storage non-specificity can be achieved by incorporation of fuzzy theory concepts, in particular application of the membership function concept, to the discretised storage intervals in a stochastic dynamic programming approach. In application to the Fairbairn reservoir, a stochastic fuzzy dynamic programming approach incorporating these concepts was shown to result in improved reservoir performance. It was further shown, that the implementation of a minimal consequence strategy for addressing the long-term return in the optimisation process may be a better approach than a normalised weighted strategy. However, further studies on the influences of the type of fuzzy intervals, membership functional form, degree of overlapping between intervals, and different mechanisms for acknowledging non-specificity in storage intervals are needed to provide a better understanding on the methodology.

5. REFERENCES

Butcher, W., S., 1971. Stochastic Dynamic Programming for Optimum Reservoir Operation. *Water Resources Bulletin*, 7(1), 115-123.

Goulter, I. C., and Tai, F. K., 1985. Practical Implications in the Use of Stochastic Dynamic Programming for Reservoir Operation. *Water Resources Bulletin*, 21(1), 65-74.

Gubbels, M. E., 1994. Optimal Reservoir Operating Policies with Explicit Consideration of Economic and Environmental Requirements (unpublished MEng. Thesis), Central Queensland University, Rockhampton, Australia, vi+103p.

Kaufmann, A. and Gupta, M. M., 1991. *Introduction to Fuzzy Arithmetic : Theory and Applications*, Van Nostrand Reinhold, New York, xvii+361p.

Klemes, V., 1977. Discrete Representation of Storage for Stochastic Reservoir Optimisation. *Water Resources Research*, 13(1), 149-158.

Loucks, D. P., Stedinger, J. R., and Haith, D. A., 1981. *Water Resources Systems Planning and Analysis*. Prentice-Hall, Inc., New Jersey, xv+559p.

McMahon, T. A., and Mein, R. G., 1986. *River and Reservoir Yield*. Water Resources Publications, Littleton, Colorado, USA, ix+368p.

Queensland Water Resources Commission, 1992. *Emerald Irrigation Area : Management Guidelines*, Water Resources Commission, Department of Primary Industry, Australia, 8p.

Tickle, K. And Goulter, I. C., 1992. Assessment of Performance Metrics for a Reservoir Under Stochastic Conditions, *Proceedings of the Sixth IAHR International Symposium on Stochastic Hydraulics (Kuo, J. T. and Lin, G. F. eds.)*, Taipei, May 18-20, 583-590.

Stochastic Hydraulics'96, Tickle, Goulter, Xu, Wasimi & Bouchart (eds) © 1996 Balkema, Rotterdam. ISBN 90 5410 817 7

Reliability studies of multireservoir systems using stochastic techniques

Achi M. Ishaq
Department of Civil Engineering, King Fahd University of Petroleum and Minerals, Dhahran, Saudi Arabia

J. M. Euseflebbe
Research Institute, King Fahd University of Petroleum and Minerals, Dhahran, Saudi Arabia

ABSTRACT

Generation of synthetic hydrologic time series is important not only for sizing future reservoirs but also for determining releases for existing reservoir systems. Reliability of these releases becomes an integral and important part of the operating policy of the reservoirs. In the case study reported herein 36 years of stream flow records at a multireservoir system in Sri Lanka were analyzed statistically and seven hundred traces for each reservoir were generated stochastically and maximum releases for each month were determined knowing the reservoir capacities. Fitting an extreme value distribution to these releases, a reliability curve was obtained. These releases were then used to compute the hydroelectric power that can be generated from the reservoir system. When compared with the total power requirement of Sri Lanka, it was seen that 52 percent and 55 percent of the total power requirement can be met, if the releases are considered to be at the probability levels of 99 percent and 42.92 percent respectively.

INTRODUCTION

Existing streamflow records of most streams in the world are not sufficiently extensive enough to provide estimates of the future behavior of those streams. Strict use of historical records alone in operational studies may lead to several serious problems. One answer to this problem is to synthetically generate possible long historical records that may be used to determine several quantities of interest within defined statistical errors. Such streamflows can be generated or synthesized using stochastic techniques, which were first introduced in hydrology to cope with the problems of reservoir design, but could also be used to maximize the release from an existing reservoirs where operational policies have been changed because of the increase in demands. The recently constructed Mahaweli system in Sri Lanka appears to be an ideal candidate for performing the analysis that could lead to the determination of maximum releases from these reservoirs.

Sri Lanka is a tropical island with a land mass of 6.5 million hectares (25,332 square miles), situated in the Indian Ocean, latitude between 6 and 10 degrees north and longitude between 80 and 81 degrees east. The island is subjected to two monsoons; the South-West monsoon prevailing from about April to September and the North-East monsoon from October to March. On the basis of distribution of rainfall, the island has been divided into two distinct areas; the wet and dry zones. The wet zone comprises of the South-Western area, covering about a quarter of the island and has two rainy seasons with an annual average rainfall of 2415 mm (95 inches). The dry zone comprising of over 4.85 million hectares (12 million acres), has only one rainy season, from about October to March, and the average annual precipitation is about 1450mm (57 inches), and well is suited for irrigated agriculture.

Sri Lanka has been divided into 103 natural river basins as shown in Fig. 1. For the purpose of assessing the yield from these basins, stream gauging stations have been established to measure the flow of most of the important rivers of Sri Lanka. Among these river basins, the Mahaweli basin (highlighted in figure 1.) is the largest, and covers nearly one sixth of the area of the island and is 333 km (207 miles) long. It has its sources in the central highlands and drops nearly 2440 m (8000 ft) to flow into the sea, south of Trincomalee (Fig. 1) in the eastern province. At present, the Mahaweli scheme contains five major reservoirs and six hydroelectric power stations. Fig. 2 shows their locations and the schematic diagram in detail. This is the largest source of hydroelectric power in the island, hence it becomes an attractive proposition to find the maximum monthly releases with given reliability of this existing reservoir system in order to maximize hydroelectric power generation.

PREVIOUS WORK

Prior to the introduction of stochastic streamflow models (SSM) for reservoir reliability studies, mass curve techniques introduced by Rippl (1883) or the sequent-peak algorithm introduced by Thomas and Fiering (1963) were used. Thus, until the advent of the field of stochastic hydrology, determination of the reservoir design capacity of the release from a storage reservoir seemed simple: one just determined the minimum storage that would have been required over the N-year historical period to provide the target yield or vice versa with absolutely no water storage. The shortcomings of the strict use of historical records have been provided by Fiering (1967).

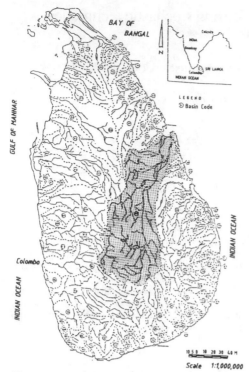

Figure 1. River basin map of SriLanka.

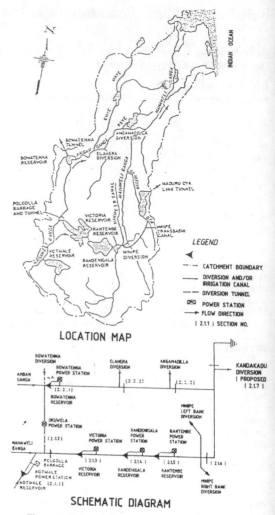

LOCATION MAP

SCHEMATIC DIAGRAM

Figure 2. Reservoir location map and schematic diagram.

Stochastic streamflow models provide answers that can be used to circumvent the shortcomings of the strict use of the historical record alone. It also can provide estimates of reliability in the face of the rising demands for water and can be used to provide estimates of expectation and variance of project net benefits, which is important information in the economic or analysis of storage-water-yield problems. Up until this study there have been no documented case studies of demand calculations from reservoirs of known capacities using stochastic techniques.

Stochastic Streamflow Models

Stochastic monthly streamflow models are often used in simulation studies to evaluate the likely future performance of water resource systems. Stochastic hydrology took several steps before reaching presently available sophisticated models. Lloyd (1963) gave a relatively complete summary of the state of art of stochastic reservoir theory. It was unfortunately long and often the principles involved were not obvious. All the work described depended on Markovion storage relationships and relatively simple releases. The basic Markov model was then extended to the case of seasonal inputs after Lloyd (1967). He next summarized correlated inflows and as a logical extension summarized correlated seasons. Time dependent results and the effects of initial storage conditions were examined in Lloyd's summary. White (1967) surveyed the stochastic aspects of reservoir storage. He advocated use of Markovion methods and pointed to the erroneous results involved in reservoir capacity determination.

Klemes (1967a) developed a technique for computing certainty distribution for specified reservoir life for random input and uniform or random reservoir draft. Later Klemes (1967b) extended his study to examine variability of a storage reservoir having seasonal input. Burges(1970) did a critical analysis on the use of stochastic hydrology to determine storage requirements for reservoirs. In his study, normally distributed Markov model was used to generate monthly and annual flows. It has been shown that most storage cases satisfied the basic extreme value (Type I) distribution when mathematically correlated input parameters were used. But for the cases where the generated coefficients of variation and the demand are small, eye fitting gave a better results.

Later Hoshi and Burges (1978) used annual streamflow volumes, modeled by Autoregressive(AR(1)) and Autoregressive Moving Average (ARMA (1,1)) process to study the impact of seasonal flow characteristics and demand patterns for required reservoir storage. They pointed out that incorporation of the long term persistence into an annually generated flows by using ARMA(1,1) model with Hurst coefficient 0.7 did not significantly influence storage requirements. They concluded that AR(1) is quite suitable for practical application. Procedures for its use for the multi-site case are also given by Hoshi and Burges (1978). Hirsch (1979) used six synthetic models in the study of water supply reliability and pointed out that a new type of auto-regressive-moving-average model was operationally superior to an autoregressive model. Stedinger and Taylor (1982a,1982b) studied the performance of a wide range of monthly streamflow models and compared using data for the Upper Delaware River basin in New York state. It was shown that Thomas-Fierring monthly autoregressive Markov model and the AR(1) annual flow model with disaggregation to obtain monthly flow showed better behavior than other models. Furthermore, among these two models Thomas-Fierring Markov monthly models showed better behavior than AR(1) model. Stedinger, et. al.(1985a) developed a condensed version of Valencia-Schaake disaggregation model which describes the distribution of monthly streamflow sequences, using a set of coupled univariate regression models rather than a multivariate time series formulation. They concluded that condensed disaggregation model is attractive because it has fewer parameters but can still reproduce the mean, variance and the period-to-period correlations of the individual flows while approximately preserving the relationship among the monthly and annual flow values. Howevere, they failed to state how accurate the reproduced mean, variance and correlations were. Vogel and Stadinger (1988) have illustrated in their study, the value of stochastic streamflow models in overyear reservoir design applications that fitting an AR(1) lognormal model leads to more precise estimate of annual storage requirements than if only the historical flows are employed, even in situation when the flows were not generated with an AR(1) lognormal model.

METHODS AND MATERIALS

Streamflows provide valuable information about a stream, traditionally they have been used in the Ripple mass curve analysis or the sequent peak method in reservoir storage studies. Use of such historical records alone however has several drawbacks. Computer generated synthetic data while helping to overcome these drawbacks, increase the scope of the analysis. There are several sophisticated synthetic streamflow models presently in use.

Application of stochastic techniques to a system of river reservoirs needs long historical record of streamflow data, downstream requirements when reservoir capacity is not known, or reservoir capacity when the operational policy is not known, and other related data such as catchment area, precipitation records, evaporation data, key maps of river basins etc.. In the case of Mahaweli scheme in Sri Lanka, all required data including 36 years of historical monthly flow data, key maps of river basins, reservoir configurations, and power data were obtained from the Mahaweli Authority of Sri Lanka.

Historical monthly and yearly flows were plotted against time to determine any possible trend since no visible trends were found in these plots, the data was considered to be random. After a careful check for the disturbance of the data, several tests were made to determine the probability distribution so that the most suitable synthetic scheme may be selected. To perform this exercise the STATGRAF package was used. All historical flow at all locations found to follow the lognormal distribution. Figure (3) shows the lognormal fit for the flows at Kotmale reservoir site. In view of this, the lognormal Markov monthly flow model was selected for generating synthetic flows. This model is given by:

$$q_{i,j} = \mu_j + \rho_j \frac{\sigma_j}{\sigma_{j-1}} (q_{i-1,j-1} - \mu_{j-1}) + n_i \sigma_j \sqrt{1 - \rho_j^2}$$
$$\dots\dots\dots(1)$$

where

i = the sequence index

j = the monthly index

q = flow (log values of depth over the catchment area)

ρ_j = correlation coefficient between months j and j-1

μ = Monthly mean flow

n = unit normal deviate

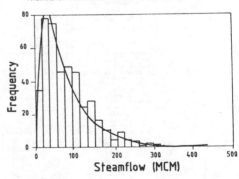

Figure 3. Lognormally fitted frequency histogram of Kotmale reservoir.

To preserve the historical statistics of the flows rather than of their logarithms, mathematically corrected approach was used. In this approach the monthly parameters μ_j, σ_j, ρ_j (they are not population parameters) are defined as

$$\overline{X}_j = \exp(\mu_j + \frac{\sigma_j^2}{2}) \qquad \dots\dots\dots(2)$$

$$\sigma_{xj}^2 = \exp(2\sigma_j^2 + 2\mu_j) - \exp(\sigma_j^2 + 2\mu_j)\dots\dots(3)$$

$$R_{oj} = \frac{\exp(\sigma_{j-1}\sigma_j \rho_j)-1}{[\exp(\sigma_{j-1}^2)-1]^{1/2}\,[\exp(\sigma_j^2)-1]^{1/2}} \quad\text{...........(4)}$$

Where $\overline{X}_j$, σ_{xj}^2, R_{oj} are the observed means, variance, and serial corrilation for month j (R_{oj} is the correlation of month j with j-1). Input parameters to this model were the mean, the standard deviation, and the serial correlation coefficient of historical data. The statistics were computed using SAS regression packages. These statistics were used in equations (2,3,4) to calculate mathematically corrected input to the Markov model explained earlier.

Model Development

A program has been developed using Markov monthly streamflow model as a stochastic generator. The flow chart of the program is given in Figure 4. Input to this program is the mathematically corrected statistics of historical streamflow data and the output is a set of release (demands). First phase of the program reads and writes the input. Second phase of the program calculates the deterministic component, which is denoted by DCOMP(K), of Markov's model. DCOMP(K) is calculated by the equation given below.

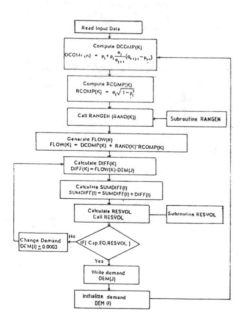

Figure 4. Flow chart of the program developed for the study

$$DCOMP(K) = \mu_J + \rho_J \frac{\sigma_J}{\sigma_{J-1}}(q_{i-1,j-1}-\mu_{j-1}) \quad\text{.............(5)}$$

In the next phase random component, which is denoted by RCOMP(K), of the Markov's model is calculated using the equation given below.

$$RCOMP(K) = \sigma_j\sqrt{1-\rho_j^2} \quad\text{...............................(6)}$$

Random number denoted by RAND(K) is produced by a subroutine with zero mean and unit standard deviation. This subroutine, written by Knuth (1969), is available in the IBM scientific subroutine package. Random numbers from this subroutine is directly used without taking the logarithmic value of them, because the model requires random numbers with zero mean and unit standard deviation. With these three components (DCOMP(K), RCOMP(K), and RAND(K)) the synthetic streamflow , denoted by FLOW(K), is generated using the equation:

$$FLOW\,(K) = DCOMP(K) + RAND(K) \times RCOMP(K)..(7)$$

In this study 600 (50X12) streamflows were generated for each trace and the procedure was repeated for 700 traces. The next phase of program deals with reservoir volume calculations using subroutine RESVOL. This subroutine follows the Sequent Peak Algoritim method to calculate reservoir volume. Assuming the mean of the historical data as the initial demand, the difference between the flow and the demand (DIFF(K)) for each month is calculated. The cumulative values of DIFF(K) known as SUMDIFF(K) is the primary input to the subroutine RESVOL. The reservoir capacity required to meet the given demand is computed by RESVOL. The computed capacity is then compared with the known capacity of the reservoir in question. An iteration is initiated to determine the maximum release using the capacity of the existing reservoir. In this procedure an increment is added or subtracted to the demand computed previously. Then, the new capacity required to meet this demand is obtained using the sequent-peak algorithm. This procedure is continued until the computed capacity is very nearly equal to the capacity of the reservoir in question and the corresponding demand is now chosen as the release from the reservoir. The same procedure is carried out for the other traces as well and the corresponding releases are obtained for each of the traces.

A sample of the mathematically corrected input data is given in Tables (1). for the flows at Kotmale reservoir site. As one can see, the use of the Markov model necessitates the generation of random numbers with zero mean and unit standard deviation.

Table 1: Statistics of historical and synthetic flow data of Kotmale reservoir.

Input Parameters	Oct.	Nov.	Dec.	Jan.	Feb.	Mar.	Apr.	May.	Jun.	Jul.	Aug	Sep
Historical $\overline{x}$	3.037	2.835	2.525	2.077	1.494	1.393	1.797	2.234	2.839	3.104	3.001	2.861
Synthetic $\overline{x}$	3.036	2.833	2.525	2.078	1.495	1.394	1.797	2.238	2.842	3.104	3.002	2.861
Historical σ	0.116	0.204	0.202	0.163	0.188	0.146	0.170	0.483	0.454	0.170	0.212	0.222
Synthetic σ	0.116	0.203	0.201	0.162	0.186	0.145	0.168	0.477	0.450	0.169	0.211	0.221
Historical ρ	0.000	0.479	0.225	0.430	0.765	0.678	0.389	0.218	0.203	0.334	0.460	0.280
Synthetic ρ	0.005	0.468	0.277	0.426	0.759	0.674	0.378	0.220	0.197	0.328	0.458	0.274

$\overline{x}$ = mean σ = Standard Deviation ρ = Correlation Coefficient

After making sure of the zero mean and standard deviation using goodness of fit, IBM's Random number generator package was selected and used in the model to generate the synthetic streamflow.With the random number generator package and statistics of historical data, synthetic data were generated. Next, statistics of synthetic data and the statistics of historical data were compared in order to make sure that the statistics of the historical data were preserved. These comparisons are shown in Table 1. It was concluded, from the results given in Tables 1, that the model generates flows which have the same statistics as the historical flows. Having ascertained that the synthetic flows preserved the statistics of the historical flows, the next stage of operation was the application of the model to flows at Kotmale, the first reservoir in the scheme. Output from the model is a set of releases (demands) which can be met from this reservoir capacity. As explained earlier, Gumbel extreme value distribution was employed to study the reliability of these releases. The values corresponding to the probability level of 42.97 percent as mean releases and the releases at the probability level of 99 percent were chosen from this distribution and recorded. Having obtained these releases, statistics of the next reservoir were updated to correspond with these releases. The new statistics were used as input to the model for the next reservoir. This procedure was continued until all reservoirs were taken into consideration. Finally, mean releases and the releases at the probability level of 99 percent from each reservoir were recorded and an operating policy for the system was prepared.

Reliability Studies

The reliability of a reservoir is defined as the probability that it will deliver the expected demand throughout its life time without incurring a deficiency. In this study, life time was taken as the economic life which is 50 years and 700 traces were generated stochastically. Each trace then may be said to represent one possible example of what might occur during the project lifetime, and all traces are equally likely be representatives of this future period. If the storage required to deliver a specified demand is calculated for each trace, the resulting values of storage can be ranked in order of magnitude and plotted as a frequency curve, or theoretical curve can be calculated from the data. There are several distributions available for this purpose but there is no real proof of their validity. Since the primary objective of this study is to find the maximum demand (extreme value) Gumbel extreme-value distribution appears to be the most appropriate distribution than others. The result is a reliability curve which indicates the probability that the demands during the project life can be met as a function of reservoir capacity.

Alternatively, the same techniques can be adopted for determining the demand that can be met by a given capacity of a reservoir. As explained earlier, demands will be calculated, using sequent-peak algorithm, from each trace. The resulting demands then can be described by an appropriate distribution. In this study, Gumbel extreme-value distribution is used. Gumbel suggested that this distribution of extreme values was appropriate for flood analysis since the annual flood could be assumed to be the largest sample of 365 possible values each year. Based on the argument that the distribution of the floods is limited, i.e. that there is no physical limit to the maximum flood, he proposed that the probability P of the occurrence of a value equal to or greater than any value X be expressed as

$$P = 1 - e^{-e^{-y}} \qquad \qquad \text{...............(8)}$$

Where y is the reduced variate given by

$$y = \frac{1}{.7797\sigma}(x - \bar{x} + 0.45\sigma) \qquad \text{...............(9)}$$

Where

x = flood magnitude with the probability P

$\bar{x}$ = arithmetic mean

$$\sigma = \left[\frac{\sum(x - \bar{x})^2}{n-1} \right]^{1/2} \qquad \qquad \text{...............(10)}$$

Where

n = The number of items in the series.

ANALYSIS OF RESULTS

Since this study deals with a multi reservoir system, the proper way to analyze the results is to consider a single reservoir in the system initially and then to consider the other cases wherever necessary. A more detailed analysis of the results of the operation and reliability studies are presented here. As discussed earlier, historical data was tested for possible trends and appeared to be random. Next, several tests were conducted to determine the probability distribution of the historical flows to select the most suitable synthetic scheme. All historical flows at all reservoir sites best fitted the log normal distribution. Thus, the lognormal Markov monthly flow model was selected as the synthetic flow generator. A key requirement in the synthetic flow generation is that "the statistics of the historical data" must be preserved. Therefore, statistics of the historical flow and the statistics of the synthetic flow were compared to ensure that the statistics of the historical flow data were preserved. From the results given in Table (1), it may be concluded that the model preserves the statistics of the historical flow data. After having drawn this conclusion, the model was applied to the first (Kotmale) reservoir in the system.

Results of Operation Study

The outcome of the application of the model to the Kotmale reservoir is a set of maximum possible releases (demands) that can be met from the known reservoir capacity. These demands (releases) for each month represent the largest event (release) for that month that is possible using the given capacity of the reservoir. Thus, each trace provides 12 values of releases for the life of the reservoir. 700 such traces will provide a matrix of 700 X 12 entries, which is in fact 700 maximum possible values for each of the 12 months.

For certain months, the releases were kept to remain constant. This is because of the following constraints being placed in the iterative process.

1. To economize the power generation the turbine must be run as efficiently as possible. There is a given minimum supply required to do this. Wherever the release from the

141

reservoir falls this minimum it is set at the minimum requirement level.

2. To prevent negative releases during the iterative process.

In operation studies the primary objective is to determine either the capacity of the reservoir that can meet a given demand or to determine the demand (release) that can be met by the reservoir of a given capacity. The study undertaken in this research pertains to the latter case. Having predetermined releases (demands) for certain months does not interfere in the synthetic flow generation.

In view of the demand being fixed for the months of the dry season, reliability curves obtained by the process explained earlier should be a straight line, as shown in Figure 4 for the Kotmale reservoir.

Results of Reliability Studies

As shown earlier there are 700 maximum possible releases for each month for a given reservoir. All these releases have the same likelihood of occurrence. Thus, these values may be ranked in the order of magnitude and plotted as a frequency curve. As pointed out by Burges (1970), the Gumbel extreme value distribution appears to be the appropriate one for this purpose. The result, therefore, should be a reliability curve for each month of the year for the reservoir. In this study, Weibull plotting position formula with the Gumbel probability paper were used to obtain the reliability curve. In order to have this plot, selected probability levels and the corresponding releases (demands) were taken and plotted on Gumbel's extreme value distribution paper. The results of this plot is a set of straight lines. Each line represents the reliability distribution of that particular month. This plot can be used to select releases (demands) with certain reliability according to the necessity of the requirement. For example, as seen from Figure 5, at a probability level of 95 percent, the release for the month of July is 91 MCM. In this study, the mean releases at the probability levels of 42.97 and 99 percent were chosen as the required demands.

study and the releases were tabulated. These are shown in Tables 2 and 3 for all 5 reservoirs at probability levels of 42.92 per-cent and 99 percent respectively. From this

Table 2: Final release at the probability level of 42.92 percent (mean) from all reservoirs after simulation (values are in MCM)

Reservoir	Oct.	Nov.	Dec.	Jan.	Feb.	Mar.	Apr.	May.	Jun.	Jul.	Aug.	Sep.
Kotmale	92.5	71.9	49.3	49.3	49.3	49.3	49.3	49.3	72.2	100.4	58.5	74.3
Victoria	104.8	60.5	60.5	60.5	60.5	60.5	60.5	60.5	36.3	135.1	93.7	80.4
Randenigale	131.8	89.5	89.5	89.5	89.5	89.5	89.5	89.5	208.3	285.9	197.0	173.3
Rantembe	98.6	83.4	122.0	121.0	71.8	44.0	32.7	38.5	16.4	163.8	324.7	140.0
Bowatenne	224.6	161.0	64.3	41.7	43.4	52.4	81.5	163.0	380.3	432.7	299.1	365.9

Table 3: Final release at the probability level of 99 percent from all reservoirs after simulation (values are in MCM)

Reservoir	Oct.	Nov.	Dec.	Jan.	Feb.	Mar.	Apr.	May.	Jun.	Jul.	Aug.	Sep.
Kotmale	77.6	57.0	49.3	49.3	49.3	49.3	49.3	49.3	57.3	85.4	73.6	59.4
Victoria	127.8	57.6	45.3	45.3	45.3	45.3	45.3	45.3	114.5	168.6	118.4	112.6
Randenigale	129.0	89.5	89.5	89.5	89.5	89.5	89.5	89.5	205.4	283.0	194.0	170.0
Rantembe	99.1	57.9	44.1	81.9	81.2	52.1	29.1	29.1	29.1	123.0	280.9	98.4
Bowatenne	217.3	153.6	57.0	41.7	41.7	45.0	74.2	155.6	373.0	425.3	291.7	358.5

Table 4: Total power that can be generated from all hydropower stations in Mahaweli scheme at the probability level of 42.92 percent (GWH)

Reservoir	Oct.	Nov.	Dec.	Jan.	Feb.	Mar.	Apr.	May.	Jun.	Jul.	Aug.	Sep.
Kotmale	45.3	35.2	24.1	24.1	24.1	24.1	24.1	24.1	35.4	49.2	43.3	36.4
Victoria	26.3	13.9	7.5	4.9	5.1	6.1	9.5	19.1	44.5	50.7	35.0	42.8
Randenigale	25.6	17.4	17.4	17.4	17.4	17.4	17.4	17.4	40.5	55.6	38.3	33.7
Rantembe	43.6	25.2	25.2	25.02	25.2	25.2	25.2	25.2	35.9	56.3	59.0	33.5
Bowatenne	7.5	6.3	9.2	9.1	5.4	3.3	2.5	2.9	1.2	12.4	24.6	10.6
Polgolla	4.8	4.8	4.8	4.8	4.8	4.8	4.8	4.8	4.8	4.8	4.8	4.8
Total	153.1	107.8	38.2	35.5	82.0	80.9	83.5	93.5	162.3	229.0	135.0	161.8

Table 5: Total power that can be generated from all hydropower stations in Mahaweli scheme at the probability level of 99 percent (GWH)

Reservoir	Oct.	Nov.	Dec.	Jan.	Feb.	Mar.	Apr.	May.	Jun.	Jul.	Aug.	Sep.
Kotmale	38.0	27.9	24.1	24.1	24.1	24.1	24.1	24.1	23.1	41.8	36.1	29.1
Victoria	53.2	24.0	18.8	18.8	18.8	18.8	18.8	18.8	47.7	70.1	49.3	46.8
Randenigale	25.1	17.4	17.4	17.4	17.4	17.4	17.4	17.4	39.9	55.0	37.7	33.1
Rantembe	7.5	4.4	3.3	6.2	6.2	2.4	2.2	2.2	2.2	9.3	21.2	7.4
Bowatenne	25.4	18.0	6.7	4.9	4.9	5.3	8.7	18.2	43.7	49.8	34.1	42.0
Polgolla	3.4	3.4	3.4	3.4	3.4	3.4	3.4	3.4	3.4	3.4	3.4	3.4
Total	152.6	95.1	73.7	74.8	74.8	71.4	74.6	84.1	165.0	229.4	181.8	161.8

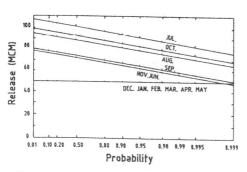

Figure 5. Reliability curve of Kotmale reservoir.

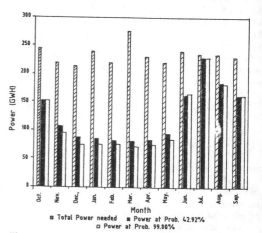

Figure 6. Comparision of the power requirement of SriLanka in 1989 and the power can be generated from the Mahaweli scheme.

The releases thus obtained are maintained for the rest of the life of that reservoir and are used in the computation of the power generation. As explained earlier, next reservoir inflow statistics were updated corresponding to the above release. This procedure was continued for all cases in the information total power that can be generated from the watershed for each month was calculated. These are given in tables 4 and 5. The total average monthly power requirement in Sri-Lanka and the power that can be generated from Mahaweli scheme are compared in Figure 6. From the figure, it is seen that when releases are considered at the probability levels of 42.92 and 99 percent nearly 55 and 52 percent of the total power requirement respectively can be met from Mahaweli scheme alone.

CONCLUSION

Subject to the constraints with regard to the usage of the Markov model and the Sequent-peak analysis the following broad based conclusions may be drawn.

1.   Monthly streamflows at all reservoir locations of Mahaweli scheme of Sri Lanka fit the lognormal distribution.

2.   Stochastic approach is not only useful in determining reservoir capacities but also in determining probable releases given the reservoir capacities.

3.   The lag-one lognormal monthly Markov model appears to be suitable model for performing operation studies in multi reservoir system.

4.   Reliability studies of reservoirs can be successfully performed using stochastic techniques combined with either the Ripple mass curve or the sequent-peak algorithm and the Fischer-Tippet Type 1 (Gumbel) extreme value distribution.

5.   At 99 percent reliability, nearly 52 percent of the total 1989 power requirements of SriLanka may be met from Mahaweli watershed alone, if the suggested operating policy of this study was maintained, however, at reliability level of 42.92 percent (mean monthly flow) the requirement that may be met is approximately 55 percent.

RECOMMENDATIONS

This study has come up with operating regulation pertaining to maximizing power generation only. However, this system of reservoirs has yet another important purpose of providing water for irrigation requirements in the developed lands in the system. Thus, it is recommended that further studies be conducted to come up with a set of operating regulations where benefits from power generation and irrigation are maximized. A possible approach for optimizing the benefits from irrigation and power generation is to use Chance-Constrained programming (Duffuaa (1991)) reach to an overall operating policy. The study in this thesis can be extended for the purpose of forecasting future releases from the reservoirs and time series analysis is recommended for such study.

ACKNOWLEDGMENT

This study was supported by the department of Civil Engineering and The Graduate school of the King Fahd University of Petroleum and Minerals. This generous support is gratefully acknowledged. Comments and suggestions for this study were received from Drs. Munir Ahmed, S.O. Duffuaa, Tahir Husain, and A.N. Shuhaib. Their contribution is also gratefully acknowledged.

REFERENCES

Burgess, S.J.1970. Use of Stochastic Hydrology to Determine storage requirements for reservoirs-a critical analysis. Ph.D Desertation. Stanford University.

Duffuaa, S.A. 1991.. A Chance-constrained model for the operation of the Aswan Dam. Engineering Optimization. 17(1-2): 109-121

Fiering, M. B.1967. Streamflow synthesis. Harvard University Press.

Hirsch, R.M.1979, Synthetic hydrology and water supply reliability. Water Resour. Res., 15(6): 1603-1615.

Hoshi, K., and S.J.Burgess1978. The impact of regional flow characteristics and demand patterns on required reservoir storage. J. Hydrol. 37: .241-260.

Klemas, V.1967a. Reliability of water supply by means of a storage reservoir within a limited period of time. J. Hydrology. 5: 70-92.

Klemas, V.1967b. Reliability estimates for a storage reservoir with seasonal input. Hydrology Symp. Colorado.

Knuth, D. 1969. The Art of Computer Programming. 2: 155-156.

Lloyd, E.H.1963a. A probability theory with serially correlated inputs. Hydrology. l: 99-128.

Lloyd, E.H.1967. Stochastic reservoir theory. Advances in Hydroscience Ven T. Chow, Ed. 4: 281-339. Academic Press. New York.

Rippl, W.1883. The capacity of storage reservoirs for water supply. Proc. Inst. Civil Eng., 71:270.

Stedinger, J.R., and M.R.Telor 1982a. Synthetic streamflow generation, 1. model verification and validation. Water Resour. Res. 18(4): 909-918.

Stedinger, J.R., and M.R.Telor1982b. Synthetic stream flow generation, 2. model verification and validation. Water Resour. Res. 18(4): 919-924.

Stedinger, J.R., D.Pei, and T.A. Cohn 1985. Condensed Disaggregation model for incorporating parameter uncertainty into monthly reservoir simulations. Water Resour. Res. 21(5): 665-675.

Thomas, H.A., Jr. and Myron B. Fiering 1963. The nature of the storage yield function, in Operations Research in Water Quality Management Chap. 2, Harvard University Water Program. Cambridge, Mass.

Vogel, R.M., and J.R.Stedinger 1988. The value of stochastic streamflow models in overyear reservoir design applications. Water Resour. Res. 24(9):1483-1490.

White, J.B.1967. Stochastic Aspects of reservoir storage. International Hydrology Symposium. 354-360, Fort Collins. Colorado.

Stochastic Hydraulics'96, Tickle, Goulter, Xu, Wasimi & Bouchart (eds) © 1996 Balkema, Rotterdam. ISBN 90 5410 817 7

Design and operation of engineering systems using regularized stochastic decomposition

Walid E. Elshorbagy & Kevin E. Lansey
Department of Civil Engineering and Engineering Mechanics, University of Arizona, Tucson, Ariz., USA

Diana S. Yakowitz
Southwest Watershed Research Center, Agricultural Research Service, USDA, Tucson, Ariz., USA

ABSTRACT: Many engineering systems are affected by uncertainty in future demands or inputs. Decisions regarding their design, however, typically must be made in the present. Two-stage stochastic programming can consider this type of problem but, in the past, procedures to fully incorporate the uncertainty have come with a high computational cost. New algorithmic developments, such as Regularized Stochastic Decomposition (RSD), now allow more complex systems to be considered. This paper provides an overview of the RSD method and its extensions and demonstrates the application of two-stage stochastic programming to two water resources problems with different problem structures and types of uncertainty.

1 INTRODUCTION

Design and analysis of engineering systems usually involve many uncertainties. Oftentimes, these uncertainties are neglected because they are not known, their values are low, or the uncertain parameters play an insignificant role in the process. However, when uncertain parameters can significantly affect the design, their uncertainties must be taken into account.

One way to account for these uncertainties in mathematical programming models during design is to use a probabilistic representation instead of the best estimates of the uncertain coefficients. Some versions of this type of model, were introduced in the late 1950's by Charnes and Cooper (1959). One such approach is chance constrained programming that is widely used for design problems.

Many engineering applications, however, involve design and operation decisions, such as irrigation canal layout and water allocation. These problems can be formulated as two-stage programs that consider the future system operations when making design decisions. The uncertainties are accounted for by assigning a probability distribution to the uncertain coefficients.

Although this approach is well known in the area of operations research, it has not been broadly applied to hydraulic and water resources engineering problems. This paper describes and applies one of the more recently developed algorithms to solve two-stage stochastic linear programs with recourse. Higle and Sen (1991) introduced the Stochastic Decomposition (SD) approach that combines the strengths of decomposition based algorithms and stochastic gradient methods.

SD produces a piecewise linear approximation of the objective function, then solves one subproblem and one master program at each algorithm iteration. A major problem with SD is that the master program progressively increases in size of as a result of a new cut (constraint) being added during each iteration. This problem can result in severe computational effort, especially for large problems with many random parameters.

Recently, Yakowitz (1994) introduced a quadratic regularizing term in the master program that limits the movement of the Master problem solutions so that the function estimates remain adequate. In this procedure, the size of the master program can be limited by introducing a cut-dropping scheme similar to that given in Mifflin (1977) and Kiweil (1985).

Regularized Stochastic Decomposition (RSD), has been applied to solve simple engineering applications under limiting conditions. The present work employs this approach and modifies it so that it can be applied to a wider and more practical range of applications. Two applications are presented to demonstrate its capabilites.

2 BACKGROUND

Two-stage stochastic LP with recourse problems have a first-stage set of decisions that must be made at present. The set of second-stage variables are determined in the future based on the actual future conditions while satisfying restrictions resulting from the first-stage decisions.

A general formulation of this type of problem is:

$$Min \quad f(x) = cx + E_{\overline{\omega}}[Q(x,\overline{\omega})] \qquad (1)$$

$$s.t. \quad x \in X \subseteq R^{nl} \qquad (2)$$

$$where \quad Q(x,\overline{\omega}) = Min \ qy \qquad (3)$$

$$s.t. \quad Wy = \overline{\omega} - Tx \qquad (4)$$

The problem consists of: [1] a first-stage objective function, cx, with its associated n1 first-stage decision variables, x, and m1 first-stage constraints, and [2] a second stage objective function $Q(x,\varpi)$, with second-stage solution y of n2 variables, and m2 second-stage constraints based on some observation ϖ. The random vector, ϖ, is defined on a probability space (Ω,A,P) where Ω is a compact set. The distribution probability function, F_ϖ, is associated with ϖ, and $E_\varpi[.]$ is the mathematical expectation with respect to ϖ. The set of feasible first-stage decisions, X, is assumed to be convex and bounded. With these conditions, the total objective function is a piecewise linear convex function of x.

3 SOLUTION ALGORITHM

At each iteration of the RSD algorithm, one Master Program and one second stage Subprogram is solved. The Master program consists of the first stage objective and a piecewise linear approximation of the expected second stage objective. First stage constraints include cutting planes developed in the second stage problem.

The second stage program is solved at the present best Master program solution for the optimum future decisions at a realization of the uncertain coefficients. The new candidate solution is compared to the current solution and accepted or rejected. Iterations continue until the algorithm terminates by satisfying appropriate stopping rules. The steps required in the algorithm are briefly listed below:

Step 0. Initialize with a feasible current first stage decision (e.g., optmal solution using the expected values of random variables) and set the candidate first stage decision equal to the current first stage decision.

Step 1. Randomly generate a single observation of all random variables according to their distributions.

Step 2. Solve the second stage problem that results from Step 1 at the candidate first stage decision and save the solution.

Step 3. Estimate the cutting plane at the current first stage decision to be added to the Master program using the current and past solutions to the second stage problem.

Step 4. Determine if the objective estimate at the candidate solution is significantly lower than the estimate of the objective function at the current first stage decision. If so, the candidate becomes the current solution.

Step 5. Update, re-evaluate, or eliminate past cutting planes and solve the current Master program.

Step 6. Determine if the stopping criteria are met. If so, stop. Otherwise, return to Step 1.

The algorithm was coded in Fortran so that any two-stage stochastic problem can be solved without changing the code. GRG2 (Lasdon. 1985), a nonlinear programming model, was used to solve the master problcm at cach itcration. GRG2 applies the generalized reduced gradient method as a basis for solving the NLP.

4 RSD EXTENSIONS

The RSD approach, to date, has been limited to solving linear two-stage problems with uncertainty only in the RHS of the second-stage constraints. Since many engineering problems behave in a nonlinear manner, the algorithm was modified to handle nonlinearity of the first-stage objective function. In this case, the convexity assumption is violated. Therefore, global optimality of the optimal solution is no longer guaranteed.

However, local optimal solutions are often adequate in engineering practice. Prudent selection of the initial point can potentially improve the solution and bring it closer to the global optimal solution.

The nonlinearity in the first-stage objective is introduced by including the nonlinear function, $g(x_k)$, in the objective of the master program. Cuts are identified in the same manner as the linear first-stage problem, however, they now only approximate the second-stage stochastic function.

The second limitation to practical application of SD and RSD is the requirement of deterministic coefficients in the second-stage objective function. These terms can include future revenues and/or prices that may also be uncertain. The algorithm was successfully modified to handle these uncertainties. However, computational problems during implementation point out the difficulty of its practical application. Details of the RSD and its extensions can be found in Elshorbagy et al (1995). Two applications are presented in the next sections to demonstrate the utility of RSD to water resources and hydraulics problems.

5 REGIONAL WATER SUPPLY PLANNING

Consider a region which has two communities. Each community has demands for both potable water for municipal use, and reused water for irrigation and other purposes. The goal is to size the water supply facilities required to satisfy consumer demands over a 20-yr period. Potable water demands can be met by direct supply from the aquifer and/or treated water from the water treatment plant which is supplied from a surface source (Figure 1). The demands of reused water can be also met from direct supply from the aquifer or from a tertiary treatment plant which is supplied from a secondary wastewater treatment plant. The aquifer is recharged through a infiltration basin system with water from the river or the wastewater treatment plant after secondary treatment.

The planning problem is to determine the design capacities of the recharge basin, water treatment plant, secondary wastewater treatment plant, and tertiary treatment facility. These decisions represent the first-stage decision variables in the two-stage formulation. The second-stage variables represent the water allocations (in million gallons per day, mgd) from

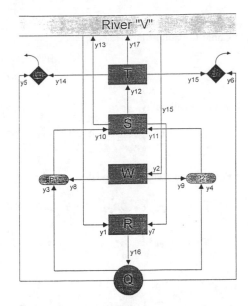

Figure 1: Regional water supply system

the supply facilities to different users during different time periods. The variables y_1 to y_{17}, shown on the system outline of Figure 1, are the second-stage operation variables for the first period. The total number of the second-stage variables for the two periods is 34.

The first-stage objective function represents the present construction cost of the four supply facilities. The second-stage objective represents the expected value of the uncertain future operation costs. These costs result from treating and pumping water during the two 10 year periods. They are assumed to be linear functions of the amounts of treated and delivered water, respectively, and were brought to a present value.

The structure of the water supply planning problem, given in the two-stage formulation, can be written as:

$$
\begin{aligned}
\underset{(x,y)}{MIN} \quad & \sum_{l=1}^{4} c_l * x_l + 6.145 * 365 * \\
& E[\sum_{r=1}^{17} q_r^1 * y_r^1 + 0.386 * \sum_{r=18}^{r=34} q_r^2 * y_r^2] \\
& + q_e * [\sum_{\xi=1}^{2} [eq^{\xi} + \sum_{v=1}^{2} (ep_v^{\xi} + eu_v^{\xi})]]
\end{aligned} \tag{5}
$$

Subject to
First-stage constraints

$$
x_l \geq 0 \qquad l \in \{1,4\} \tag{6}
$$

147

Second-stage constraints (For $\xi=1,2$)
[1] Canal Capacity

$$\sum_{-x_l} y_l^\xi \le x_l \qquad\qquad l\in \{1,4\} \qquad (7)$$

[2] Water Availability

$$\sum_{-V} y_v^\xi \le AV^\xi \qquad\qquad (8)$$

[3] Potable and Reuse Demands

$$\sum_{-P} y_v^k + ep_v^\xi \ge DP_v^\xi \qquad v=1,2 \qquad (9)$$

$$\sum_{-U} y_v^\xi + eu_v^\xi \ge DU_v^\xi \qquad v=1,2 \quad (10)$$

[4] Aquifer Storage

$$QI^\xi + \sum_{-Q} y_v^\xi - \sum_{-Q} y_v^\xi + eq^\xi \ge QS^\xi \qquad (11)$$

[5] Quality of Reuse Demands

$$y_{Q-U_v}^\xi \ge PCU * DU_v^\xi \qquad v=1,2 \qquad (12)$$

[6] Quality of Potable Demands

$$y_{W-P_v}^\xi \ge PCP * DP_v^\xi \qquad v=1,2 \qquad (13)$$

[7] Temporal Continuity

$$QI^\xi + \sum_{-V} y_v^\xi + \sum_v (ep_v^\xi + eu_v^\xi) + eq^\xi \qquad (14)$$
$$\ge (1+loss_{avg}) * \sum_v DP_v^\xi + DU_v^\xi$$

[8] Mass Continuity

$$(1-loss_j) * \sum_j \sum_{-j} y_{-j}^\xi = \sum_j \sum_{-j} y_{-j}^\xi \quad j\in \{R,W,S,T,P1,P2\}$$
$$(15)$$

where x_i is the design capacity of the supply units with x_1, x_2, x_3, and x_4 being capacities of the recharge basin, water, secondary, and tertiary treatment plants, respectively. q_r^1 is an objective function coefficient related to the allocation, y_r^1 (the superscript 1 means the first period), and depends on its treatment and pumping costs. q_e is

a unit price of the penalty water used to maintain feasibilty. The first-stage constraints are only simple bounds to maintain non-negative values of the capacities. The subscript of y on the second-stage constraints identifies the allocated water. For example, $y_{\to U}$, defines all y's entering unit U, and $y_{Q\to U1}$ defines the allocated water from unit Q to unit U1. The second-stage constraints, are divided into eight groups as follows:
[1] Capacity constraints ensure that the total delivered amount of water to any unit during any time period, ξ, will be less than the capacity of the unit.
[2] River Availability constraints ensure that the available water in the river, AV, exceeds the amount diverted to the system during any time period.
[3] Demand constraints guarantee that the potable demands, DP, and the reuse demands, DU, are satisfied for the two communities during any period, ξ. ep, and eu, are external penalty water required to maintain feasibility during random generated constraints which may cause the demand to exceed the supply.
[4] Aquifer Storage Constraints assure that the amount of water stored in the aquifer at the end of each period is greater than a pre-specified reserve amount, QS. The amount of stored water equals the initial storage, QI, plus entering water minus withdrawn water plus external penalty water.
[5] Reuse Quality Constraints maintain a pre-specified ratio of the total reuse demands, PCR, to be direct supply from the aquifer.
[6] Potable Quality Constraints maintain a pre-specified ratio of the total potable demands, PCP, to be delivered from the water treatment plant.
[7] Temporal Continuity Constraints insure that all demands and losses are met using true sources of water.
[8] Mass Balances Constraints preserve the mass balances at different nodes and accounting of their losses. The nodes of concern are the supplying units and the two nodes of potable demands (P).
    The total number of second-stage constraints in this problem is forty two. The stochastic parameters in the right hand side (RHS) of the second-stage constraints are AV, DP, and DU. The number of independent random parameters considered in this case is 10. Stochasticity in the treatment costs of the four supplying units along with the pumping costs, are also considered which represent eight independent random parameters.

148

5.1 *Results and discussion*

5.1.1 *Linear first-stage objective function and stochastic RHS*

The design capacities for this condition were obtained in 15.75 hours using a SPARC-station LX computer system. The four capacities obtained using this design were larger than those of a deterministic design but provided an overall 5% improvement in the total objective function.

5.1.2 *Non-linear first-stage objective function and stochastic RHS*

To introduce nonlinearity, a power function was used as the first-stage objective function. Designs were determined for two values of the power function exponent. Facility components were enlarged when a concave objective was used (power coefficient = 0.80) while they were reduced in the convex case (power coefficient = 1.5). This result is expected due to the economy of scale with a concave objective. In both cases, the optimal solution was improved compared to a deterministic design with the same objective with 11 and 23% gain for the concave and convex objectives, respectively.

5.1.3 *Linear first-stage objective function and stochastic second-stage objective function*

The solution for this case was identical as that obtained from a deterministic model. Thus, the variability in the objective coefficients had no effect on the first-stage decisions for this example which is not expected in all cases.

This solution was found after a large number of algorithm iterations that required significant computation time. The cause of this problem was that the number of vertices was growing with the number of iterations. The vertices are related to the number of independent dual variables that are possible in the subproblem. Stochastic RHS problems have a finite number of these vertices, so convergence is possible. The growth in dual variables, however, was not expected nor reported in the literature when stochastic objective coefficients are considered. Numerous unsuccessful alternative schemes were developed to alleviate this problem.

6 IRRIGATION CANAL SYSTEM DESIGN

Canal capacities of an irrigation system have a great impact on future farm revenues under different operational conditions. Allocation models, developed to plan for the most economical way of distributing water to crops at different growth stages, are all constrained by predefined canal capacities. Using RSD, it is possible to determine the best canal capacities during the initial design stage while considering varying operation conditions.

A demonstration system consisting of 9 canals, 6 fields, and one source of water is considered. The planning horizon was 10 years with two 6-month growing seasons in each year. Figure 2 shows the outline of a symmetrical branch system of canals and the crops grown in each field during each season.

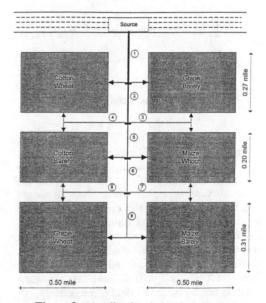

Figure 2: Application irrigation system

The first stage represents the canal construction cost. Linear construction cost functions were applied and reflect their dependence on the canal length and the excavation cost. The second stage is the negative of the net benefit of different crop yields from future time periods. Two sets of constraints corresponding to the two stages are also included. The first-stage constraints are simple bounds to maintain positive canal capacities within a defined range. The second-stage constraints are divided into three types; [1] crop-demand requirements; [2] canal-capacity limitations; and [3] water-availability constraints. This formulation can be stated as:

$$Min \sum_{i=1}^{nc} \alpha_i c_i - E \left\{ \sum_{k=1}^{nk} \sum_{j=1}^{nf} [\overline{Py_{jk}} * A_j * Y_{jk} - \overline{Px_k} * \right.$$

$$\left. \sum_{l=1}^{nx} (NFX_{jl} * A_j * x_{lk})] \right\} \tag{16}$$

subject to:
First-stage Constraints

$$C_l \leq C_i \leq C_u \qquad \forall i \tag{17}$$

Second-stage Constraints

$$\sum_{l=1}^{nx} NXF_{jl} * x_{lk} \leq \overline{D_{jk}} \qquad \forall j,k \tag{18}$$

$$\sum_{l=1}^{nx} [(1.+conv.\ loss\ ratio)^{(LR_{li}-1)} * x_{lk}]$$
$$< c_i \qquad \forall\ i,\ k \tag{19}$$

$$\sum_{l=1}^{nx} [(1.+\ conv.\ loss\ ratio)^{(LR_{li}-1)} * x_{lk}]$$
$$< \overline{AW_k} \qquad \forall\ k \tag{20}$$

where C_i is the i-th canal capacity of nc canals; α_i is a construction cost-capacity coefficient; A_j is the area of field j of nf fields; Py_{jk} is the commodity price per harvested crop unit weight ($/unit weight); Y_{jk} is the crop yield in field j harvested during period k (unit weight/unit area); Px_k is the cost of water during period k ($/unit volume); X_{lk} is the lth allocated water of nx allocations (unit depth); and NFX_{jl} is an identifier flag. NFX equals 1 or 0 defining that allocation l is or is not connected to field j, respectively. C_l and C_u are lower and upper bounds for the canal capacities. D_{jk} is the crop demands of field j during period k and AW_k is the amount of available water during period k. Since D_{jk}, AW_k, Py, and Px are future parameters, they are treated as random variables and denoted with overbars. LR_{li} is the order of the allocation l with respect to canal i. For example, $LR_{li} = 1$ means that the lth allocation has a 1st order rank to the ith canal, or the lth allocation is directly connected to canal i where no conveyance loss is considered. If the allocation is upstream of the canal so that it does not contribute anything to the flow in the canal,

then the whole term corresponding to that allocation is omitted.

The crop yield (Y_{jk}) is determined using the FAO crop response function (Doorenbos and Kassam, (1979)).

$$Y_{jk} = Ym_{jk} [1 - Ky_{jk}[1-\frac{IE_{jk}}{ETm_{jk}} *$$
$$\sum_{l=1}^{nx} NFX_{jl} * x_{lk}]] \tag{21}$$

where Ym_{jk} is the maximum yield of field j harvested during period k (Mg/ha); Ky_{jk} is the FAO yield response factor in fraction for field j; ETm_{jk} is the maximum evapotranspiration of field j during period k (mm); and IE_{jk} is the irrigation efficiency (fraction) for field j during period k.

6.1 Results

The stochastic optimization model has two types of uncertain parameters which appear in the right hand side of the model constraints; crop demands and available water. The twelve objective function coefficients (crops selling prices), were also considered as uncertain. It was assumed that they were described by normal distributions with different coefficients of variation (CV). The effect of each on canal capacities and future revenues was evaluated using the RSD approach for different levels of parameter uncertainty. All runs were made on a SPARC station LX. The results are compared with an optimal deterministic solution which was developed using the mean parameter values.

6.1.1 Available Water

Sufficient amounts of available water during the two periods (condition of zero deficit level) were assumed to follow normal distribution with mean values equal to 3.22 and 2.91 millions of cubic meters/season, respectively. These flows were determined the water demand in a deterministic optimization problem with unlimited available water.

Stochastic problems were solved with different coefficients of variation (CV) for the random available water with the mean values noted above. In all cases, the canal capacities were identical to those obtained in the deterministic design.

A more important problem is balancing the

150

canal capacity under shortage conditions. Thus, a like analysis was repeated for different water deficit levels. In these cases, the mean available water was computed by decreasing the sufficient available water by the deficit level percentage.

Table 1 lists the percentage increase of the stochastic solution in the net revenues compared to the deterministic result for different deficit levels and available water coefficient of variations. Canal capacities were the same for deficit levels of 10% and 25% and were similar to the design obtained from deterministic design with zero deficit level. For the 50% deficit level, however, the design capacities using the RSD approach dramatically changed, as did the percentage increase in revenues. The last observation points out the importance of using the stochastic approach in design when a significant shortage of available water can be expected.

Table 1. Percent increase in net revenues between deterministic and stochastic irrigation system designs

	Coefficient of Variation, CV			
Deficit level	0.25	0.5	0.75	1.0
10%	2.4	2.9	3.3	3.6
25%	4.9	8.0	9.9	11.2
50%	4.7	10.9	17.3	22.4

6.1.2 Crop Demands

The influence of the variability of crop water demand on the canal design capacities was also evaluated. This demand is related to the potential evapotranspiration and the irrigation efficiency. Continuous normal distributions for crop demands were assumed and the deterministic demands were assumed to be the mean values. Canal capacities from the deterministic and stochastic designs (CV=0.50) and showed a slight increase in revenues using capacities obtained from the stochastic design.

Another design was carried out when both the available water (of 25% deficit level) and the demands, were considered stochastic. The percentage increase in net revenues compared to the deterministic design was 9.6%. The percentage increase in net revenues when only stochastic water availability was considered was

8.5%. The minor change between the two cases indicates that the canal capacities are not very sensitive to the variabilities of crop demands for this system.

6.1.3 Influence of the construction cost coefficient

Another set of cost coefficients was chosen to evaluate the influence of these coefficients on the canal capacities, as well as the revenues obtained using different design approaches. The inital cost coefficients were multiplied by a factor of 5 and new stochastic and deterministic designs were computed at the 50% deficit level for different CV's of available water.

The improvement over the deterministic solution was smaller compared to results using the lower cost coefficients shown in Table 1. The percentage improvements were 2.13, 5.36, 9.19, and 13.03 for CVs of 0.25, 0.5, 0.75 and 1.0, respectively. Since the canal sizes are decreased with larger cost, the high available flows and their benefits are not available so the stochastic and deterministic returns become closer to each other.

6.1.4 Stochastic Objective Coefficients

Since RSD had poor convergence in the water supply problem, an alternative method, known as the L-shaped algorithm (Van Slyke and Wets, 1969) was used to solve this problem. However, a large sample of linear prgograms must be solved during each iteration of the L-Shaped algorithm, so that all possible outcomes are covered in the stochastic second stage. Three values were defined for each random objective function coefficient. Each coefficient was randomly generated from a continuous distribution then approximated to the nearest value of the three discretized values. For the discretized distributions, the number of possible outcomes is tremendous ($3^{12}=531441$ outcomes). As an approximation, a finite sample with reasonable size is sufficient in most situations. For the current irrigation system analysis, by examining different sample sizes, it was determined that this condition was reached after 4000 outcomes.

At the zero deficit level and all coefficients of variation, the design capacities of the canals were identical to the deterministic design with the mean coefficient values.

The available water was then lowered to 25%

and 50% deficit levels and the problem was solved for different coefficients of variation (CV=0.25, 0.50, 0.75, and 1.0). The stochastic designs were compared to the determinstic solution as a percentage increase in net revenues (Table 2). In the deterministic design case, the expected return was computed by Monte-Carlo analysis with the same CV's of the objective function coefficients and a sample size of 4000. Although not listed, the design capacities changed considerably from one CV to another. The improvements are significant which demonstrates the sensitivity of the design to net crop selling prices.

Table 2. Percent increase in net revenues for stochastic objective coefficients and different CVs

	Coefficient of Variation, CV			
Deficit level	0.25	0.5	0.75	1
25%	0.1	1.1	3.3	6.3
50%	0.6	4.6	12.5	23.3

6.1.5 *Stochastic objective and RHS coefficients*

The final analysis treated the available water and the objective function coefficients as random. As in the case of only stochastic objective coefficients, the parameter variability at the zero deficit level had no effect on the recommended canal capacities. This conclusion was consistent for all CVs.

At 25% deficit levels, design capacities differed from the deterministic design. However, for CV >= 0.50, the capacities did not change and the system design was identical to zero deficit level case. The gain in net revenues increased with CV until it reached a maximum value of 9.8% at CV=0.50 after that CV the gain decreased. The decrease at higher CV's is attributed to the increase of high level of variability. Similar results and behavior were observed for the 50% deficit level.

Stochasticity in both the objective and RHS may result in a non-convex problem and non-global optimal solutions. Therefore, the L-shaped model was executed using different starting points. All initial points resulted in the same solution which indicates that the solutions are likely globally optimal.

7 CONCLUSIONS

Engineered systems that face an uncertain future are design/operation problems. Two-stage stochastic optimization has been demonstrated to be a useful tool in examining these types of problems. A promising solution algorithm is regularized stochastic decomposition. In this paper, RSD has been shown to be capable of solving general water resources problems. The benefits of the approach are clearly seen in the improved returns from the irrigation canal design and water supply planning studies.

REFERENCES

Charnes, A. and Cooper, W. W. (1959) Chance-Constrained Programming. Management Science, 6(1), 73-79.

Doorenbos, J. and A. H. Kassam. (1979). "Yield response to water." FAO Irrigation and Drainage Paper No. 33. FAO. Rome, Italy.

Elshorbagy, W., Yakowitz, D., and Lansey, K. (1995a). "Design of engineering systems using stochastic decomposition algorithm." Submitted to J. Eng. Opt.

Elshorbagy, W., Lansey, K., Yakowitz, D., and Slack, D. (1995b). "Design of irrigation system of canals using stochastic optimization approach." Submitted to J. of the Irrigation and Drainage Div., ASCE.

Higle, J.L. and Sen, S. (1991) Stochastic Decomposition: an Algorithm for Two Stage Linear Programs with Recourse. Math. of Oper. Res. 16, 650-669.

Kiwiel, K. C. (1985) Methods of Descent for Nondifferentiable Optimization. Lecture Notes in Mathematics no. 1133, Spring-Verlag, Berlin.

Lasdon, L. S. and Waren, A. D. (1983) GRG2 User's Guide. Dept. of General Business, The University of Texas, Austin.

Mifflin, R. (1977) An Algorithm for Constrained Optimization with Semi-Smooth Functions. Math. Oper. Res., 2, 191-207.

Van Slyke, R. and Wets, R. J-B. (1969). "L-Shaped linear programs with application to optimal control and stochastic programming." SIAM J. Appl. Math., 17, 638-663.

Yakowitz, D. S. (1994b), "A regularized stochastic decomposition algorithm for two-stage stochastc linear programs." Comp. Optimization and Applications, 3, 59-81.

Stochastic Hydraulics '96, Tickle, Goulter, Xu, Wasimi & Bouchart (eds) © 1996 Balkema, Rotterdam. ISBN 90 5410 817 7

Fuzzy prediction of reservoir life

Pingyi Wang & Xiaoling Li
Southwest Research Institute of Water Transport Engineering, Chongqing, China

ABSTRACT: The problem of reservoir life is of fuzzy characteristics, it may be studied and predicted through fuzzy analysis. This paper tries to explore this topic and makes an analyses with the data of twelve reservoirs in Taiwan. The results are satisfactory.

1 INTRODUCTION

The prediction of reservoir life is one of the most important problems in the design and management of reservoirs. At present, the prediction of reservoir life are based on qualitative methods (He, 1988), these prediction models, howeaver, are only to consider one or two of many factors, which affect the reservoir life. Consequently results of these models are not very accurate.

The problem of reservoir life is of fuzzy characteristics. It may be studied and predicted through fuzzy analysis. In this paper, the reservoir life is considered as the object of analysis. The parameters of analysis are as follows: watershed area, first storage capacity, annual average sediment deposit volume, annual average drainage erosion thickness. A fuzzy mathematical model is developed. An example including twelve reservoirs in Taiwan is analyzed to demonstrate the proposed method.

2 FUZZY MATHEMATICAL MODEL

2.1 *The management of original data*

The special feature variables affecting the reservoir life are: watershed area, first storage capacity, annual average sediment deposit volume, annual average drainage erosion thickness, etc. Because of the different dimensions and values of these variables, the calculation with original data will exaggerate the variables that their absolute values are bigger, and minimize the variables that their absolute values are smaller. For the purpose of giving every variable the same dimension and value class, we should make uniform management about original data. This can be achieved by the following method (Wang 1992; Zadeh 1979):

It is assumed that there are m reservoirs U_1, U_2, $\cdots$, U_m, every reservoir U_i ($i=1$, 2, $\cdots m$) has n special feature variables, their original data may be expressed as:

$$U_1 = x_{11}, x_{12}, \cdots, x_{1n}$$
$$U_2 = x_{21}, x_{22}, \cdots, x_{2n}$$
$$\cdots$$
$$U_m = x_{m1}, x_{m2}, \cdots, x_{mn}$$

in which x_{ik} ($i=1$, $2\cdots$, m; $k=1$, 2, $\cdots$, n) is the value of k indication of i reservoir.

The maximum value x_{kmax} and the minimum value x_{kmin} can be picked from every rank (k rank), and original data may be converted as the following data:

$$x'_{ik} = \frac{(x_{ik} - x_{kmin})}{(x_{kmax} - x_{kmin})} \qquad (2.1)$$

Obviously, $0 \leqslant x'_{ik} \leqslant 1$.

After original data is converted, the relative relation of their inter-values is not changed, but every variable keeps the same quantity class.

2.2 *The foundation of fuzzy similar relation*

To classify reservoirs, it is necessary that the similar relation among all reservoirs must be found. This relation can be expressed with a similar factor. In this paper, the similar factor between reservoirs U_i and U_j is expressed as:

$$\lambda_{ij} = \frac{\sum_{k=1}^{n} (x'_{ik})(x'_{jk})}{\sqrt{\sum_{k=1}^{n} (x'_{ik})^2 \sum_{k=1}^{n} (x'_{jk})^2}} \qquad (2.2)$$

in which λ_{ij} $(0 \leqslant \lambda_{ij} \leqslant 1)$ is the similar factor between U_i reservoir and U_j reservoir, i, j = 1, 2…, m, (m is the total of reservoirs).

According to the concepts of fuzzy relation, the subordinate degree $U_R(U_i, U_j)$ is equal to λ_{ij}.

2.3 *The foundation of fuzzy equivalent relation*

The classification of fuzzy set depends on fuzzy equivalent relation. Eq. (2.2), the fuzzy similarity can be expressed with the fuzzy matrix R which consists of actual number between $0 \sim 1$. R is of self-reciprocal and symmetry qualities. It can be changed into a fuzzy equivalent relation by means of self-multiply, that is, $R \circ R \circ \cdots \circ R = R^k (k \leqslant m)$. On the basis of this relation, the classification can be made.

2.4 *Compound operation of fuzzy matrix*

Let Q $(= q_{ij})$ be a fuzzy matrix with m $\times$ n dimensions, and R $(= r_{jk})$ be a fuzzy matrix with n $\times$l dimensions. According to the above mentioned, it can be given:

$$S_{ik} = \bigvee_{j=1}^{n} (q_{ij} \wedge r_{jk}), (i = 1, 2, \cdots, m; k = 1, 2, \cdots, l)$$

$$(2.3)$$

From Eq. (2.3), obviously, $0 \leqslant S_{ik} \leqslant 1$. S_{ik} is named as compound matrix, that is, $S = Q \circ R$.

Here, the compound operation of fuzzy matrix is similar to the multiplication of ordinary matrix. The difference between them is that plus symbol $(+)$ is changed into maximum symbol (V), and multy symbol $(\times)$ is changed into minimum symbol (Λ).

By using the method of compound operation on fuzzy matrix, the fuzzy similar relation can be converted into fuzzy equivalent relation. According to this relation, the classification can be well done.

2.5 *The gathering classification and prediction*

According to the fuzzy equivalent relation, if a proper λ is chozen with actual necessity, the classification needed can be obtained. The averaged values of the factors of the reservoirs with the same classification and the parameters of the reservoir predicted are considered as the new sample. By using the above method, the new sample can be classified, and the reservoir life predicted can be determined.

3 EXAMPLE

In this paper, an example is presented and analyzed. Ten reservoirs among twelve reservoirs in Taiwan are classified first and the life of the other reservoirs is predicted. The steps of fuzzy analysis are as follows.

3.1 *The management of the original data*

The original data of ten reservoirs in Taiwan are shown in Table 1 (He 1988). x_1 is watershed area, x_2 is first storage capacity, x_3 is annual average sediment deposit volume, x_4 is annual average drainage erosion thickness. According to Table 1 and Eq. (2.1), x_1, x_2, x_3, and x_4 can be converted into x'_1, x'_2, x'_3, and x'_4, it is shown in Table 2.

3.2 *The calculation of similar coefficient*

According to x'_1, x'_2, x'_3, x'_4, and Eq. (2.2), the

154

similar coefficient (λ) can be calculated, the fuzzy similar relation between ten reservoirs can be obtained, as shown in Table 3.

3.3 The foundation of fuzzy equivalent relation

To apply the method of compound operation of fuzzy matrix, fuzzy similar relation in Table 3 can be made self-multiply. When R^8 is equal to R^4, R^4 is fuzzy equivalent relation, it is shown in Table 4.

3.4 The classification of reservoirs

According to the values in Table 4, λ may be chozen to equal to 0.86, with actual necessity, the following matrix can be obtained:

$$R_{0.86} = \begin{bmatrix} 1 & 0 & 0 & 0 & 0 & 0 & 0 & 0 & 0 & 0 \\ 0 & 1 & 1 & 0 & 0 & 1 & 0 & 1 & 0 & 0 \\ 0 & 1 & 1 & 0 & 0 & 1 & 0 & 1 & 0 & 0 \\ 0 & 0 & 0 & 1 & 1 & 0 & 1 & 0 & 1 & 1 \\ 0 & 0 & 0 & 1 & 1 & 0 & 1 & 0 & 1 & 1 \\ 0 & 1 & 1 & 0 & 0 & 1 & 0 & 1 & 0 & 0 \\ 0 & 0 & 0 & 1 & 1 & 0 & 1 & 0 & 1 & 1 \\ 0 & 1 & 1 & 0 & 0 & 1 & 0 & 1 & 0 & 0 \\ 0 & 0 & 0 & 1 & 1 & 0 & 1 & 0 & 1 & 1 \\ 0 & 0 & 0 & 1 & 1 & 0 & 1 & 0 & 1 & 1 \end{bmatrix}$$

Thus, ten reservoirs can be classifiedinto three kinds of types: Type I includes five reservoirs (4, 5,

7, 9, 10); Type II includes four reservoirs (2, 3, 6, 8); Type III includes one reservoirs (1).

3.5 The prediction of reservoir life

The life of two reservoirs (ordinal number is 11, 12) in Taiwan will be predictedaccording to the above results of the classification. The parameters of two reservoirs is shown in Table 5.

The averaged values of the factors of the reservoirs with the same classification and the parameters of the reservoir 11 are considered as the new sample (Table 6).

According to the above methods of the gathering classification, the matrix of fuzzy equivalent relation can be obtained:

$$R = \begin{bmatrix} 1 & 0.60 & 0 & 0.91 \\ 0.60 & 1 & 0 & 0.60 \\ 0 & 0 & 1 & 0 \\ 0.91 & 0.60 & 0 & 1 \end{bmatrix}$$

When λ=0.91,

$$R_{0.91} = \begin{bmatrix} 1 & 0 & 0 & 1 \\ 0 & 1 & 0 & 0 \\ 0 & 0 & 1 & 0 \\ 1 & 0 & 0 & 1 \end{bmatrix}$$

So, it can be known that sample IV and sample I belong to the same type. That is, the life of the reservoir 11 is about equal to the life of the

Table 1. The table of special feature variables of ten reservoirs

reservoir name	ordinal number	x_1 (km²)	x_2 (10³m³)	x_3 (10³m³)	x_4 (10³m³)	reservoir life (in year)	μ (Y)
Xishu	1	0.48	580	3.31	0.51	125	0.625
Shimen	2	763.40	315960	2175.00	2.85	140	1
Taiping	3	1000.00	900000	251.05	2.51	140	1
Qingzufu	4	30.30	1100	38.64	1.28	30	0
Mingtai	5	61.08	17700	441.41	7.23	40	0
Tiexi	6	592.00	256000	1388.50	2.35	185	1
Longlaoqi	7	7.50	3780	62.43	8.23	60	0
Zhengwen	8	496.00	712700	4422.00	8.92	160	1
Bailu	9	26.55	27417	575.13	21.66	50	0
Jingbei	10	10.60	7000	236.67	22.36	30	0

Table 2.   The results of the management of original data

ordinal number	x'_1	x'_2	x'_3	x'_4
1	0	0	0	0
2	0.763	0.443	0.491	0.107
3	1	0.126	0.056	0.092
4	0.03	0.001	0.008	0.035
5	0.061	0.024	0.099	0.308
6	0.592	0.359	0.313	0.084
7	0.007	0.004	0.013	0.353
8	0.496	1	1	0.385
9	0.026	0.038	0.129	0.968
10	0.01	0.009	0.053	1

Table 3. The fuzzy similar relation

	1	2	3	4	5	6	7	8	9	10
1	1	0	0	0	0	0	0	0	0	0
2	0	1	0.83	0.65	0.41	0.99	0.14	0.86	0.21	0.14
3	0	0.83	1	0.71	0.29	0.86	0.11	0.46	0.13	0.10
4	0	0.65	0.71	1	0.87	0.66	0.77	0.52	0.78	0.76
5	0	0.41	0.29	0.87	1	0.40	0.95	0.53	0.97	0.95
6	0	0.99	0.86	0.66	0.40	1	0.15	0.85	0.20	0.14
7	0	0.14	0.11	0.77	0.95	0.15	1	0.29	0.99	0.99
8	0	0.86	0.46	0.52	0.53	0.85	0.29	1	0.37	0.29
9	0	0.21	0.13	0.78	0.97	0.20	0.99	0.37	1	0.99
10	0	0.14	0.10	0.76	0.95	0.14	0.99	0.29	0.99	1

Table 4. The fuzzy equivalent relation

	1	2	3	4	5	6	7	8	9	10
1	1	0	0	0	0	0	0	0	0	0
2	0	1	0.86	0.71	0.71	0.99	0.71	0.86	0.71	0.71
3	0	0.86	1	0.71	0.71	0.86	0.71	0.86	0.71	0.71
4	0	0.71	0.71	1	0.87	0.71	0.87	0.71	0.87	0.87
5	0	0.71	0.71	0.87	1	0.71	0.97	0.71	0.97	0.97
6	0	0.99	0.86	0.71	0.71	1	0.71	0.86	0.71	0.71
7	0	0.71	0.71	0.87	0.97	0.71	1	0.71	0.99	0.99
8	0	0.86	0.86	0.71	0.71	0.86	0.71	1	0.71	0.71
9	0	0.71	0.71	0.87	0.97	0.71	0.99	0.71	1	0.99
10	0	0.71	0.71	0.87	0.97	0.71	0.99	0.71	0.99	1

Table 5 The data of the life of reservoirs predicted

reservoir name	ordinal number	x_1	x_2	x_3	x_4	Y	μ (Y)
Furoutao	11	60.60	156236	1046.65	17.28	50	0
Ekuntan	12	31.87	18000	507.14	15.91	35	0

Table 6 The data of the new sample

sample	x_1	x_2	x_3	x_4
I	27.21	11399	270.85	12.15
II	712.85	343665	2059.14	4.16
III	0.48	580	3.31	0.51
IV	60.60	156236	1046.65	17.28

reservoirs in Type I.

According to the concept of the subordinate function in fuzzy mathematics, the fuzzy sub-set of the relative long of reservoir life is defined as:

$$\mu(Y) = \begin{bmatrix} 1 & ,(Y \geqslant 140) \\ (Y - 100)/40 & ,(140 > Y > 100) \\ 0 & ,(Y \leqslant 100) \end{bmatrix} \quad (2,4)$$

From Table 1 and Eq. (2.4), for the reservoirs in Type I, μ is 0. The life (Y) of the reservoir 11 is 50, μ is also 0. Therefore, the reservoir 11 belongs to Type I. The result of the fuzzy prediction is better.

156

By using the same methods, the reservoir 12 belongs to Type I. From Table 1, the result of the fuzzy prediction is also better.

4 CONCLUSION

According to the results of the above analyses and calculation, it can be seen that Fuzzy Mathematical Method has the following advantages: (1) it considers many factors which affect the reservoir life, and improves the accuracy of the prediction of the reservoir life; (2) the more reservoirs are, the better effect of the classification and prediction; (3) the calculation can be achieved with middle or small computers, simple and convenient; (4) it is a new method for the study of reservoir life. So, the new method can be widely applied in the field of the study of reservoir life.

REFERENCES

He, J. S. 1988. Reservoir sedimentation in Taiwan. *J. Sed. Res.* 2: 83-84.

Wang, P. Y. 1992. A fuzzy mathematical method on the classification of river patterns. *Proceedings of 5th International Symposium on River Sedimentation*, 2:205−210. Karlsruhe: P. R. Germany.

Zedeh, L. A. 1979. Fuzzy sets and information granularity advances in fuzzy sets theory and applications. 3-25. North-Holland.

Stochastic Hydraulics'96, Tickle, Goulter, Xu, Wasimi & Bouchart (eds) © 1996 Balkema, Rotterdam. ISBN 90 5410 817 7

Incorporating marginal distributions in point estimate methods for uncertainty analysis

Che-Hao Chang
Research, Energy & Resources Laboratory, Industrial Technology Research Institute, Hsinchu, Taiwan, China

Jinn-Chuang Yang
Department of Civil Engineering, National Chiao-Tung University, Hsinchu, Taiwan, China

ABSTRACT: The model performance of a engineering system is affected by many variables subject to uncertainty. Point estimate (PE) methods are practical tools to assess uncertainty features of a model involving multivariate stochastic parameters. Two PE methods have been developed for engineering applications. One is Rosenblueth's PE method which preserves the first three moments of random variables and the other is Harr's PE method which reduces the computations of Rosenblueth's method but only appropriate to be applied to random variables with normal distributions. In this study, two algorithms are proposed to encompass the advantages of the two PE methods: computational practicality and the handling of mixture distributions. The two proposed methods were also examined on estimating statistical moments of a pier scouring model output to demonstrate their performance in an engineering application.

1. INTRODUCTION

In engineering design and analysis, one frequently employs models involving parameters that are subject to uncertainty. Therefore, model outputs on which engineering design and analysis are based are also subject to uncertainty. To perform uncertainty analysis for a model involving many stochastic parameters, point estimate (PE) methods are practical tools. The PE methods evaluate uncertainty of a model by computing the model responses at specified points in the parameter space. Proper points for model evaluation should be selected to preserve probabilistic information of the random variables.

Rosenblueth (1975) proposed a PE method for handling random variables with symmetric distributions which was later extended to handle random variables with non-symmetric distributions (Rosenblueth 1981). To estimate the statistical moments of the model output, 2^n model evaluations are required for a model involving n stochastic parameters. As the number of stochastic parameters increases, Rosenblueth's algorithm becomes computationally less practical. An alternative PE method is proposed by Harr (1989) to circumvent the computationally explosive nature of Rosenblueth's algorithm. However, only the first two moments of random variables are considered in Harr's PE method. Chang et al. (1995) recently showed that the estimated uncertainty feature of model output could be inaccurate if the skewness of a random variable is not accounted for. Nevertheless, the contribution of Harr's PE method to practical uncertainty analysis of engineering problems is valuable.

By incorporating a set of semi-empirical formulas developed by Der Kiureghian and Liu (1985), this study extended Harr's PE algorithm to allow handling random variables with mixture of known marginal distributions. Based on the given information about the marginal distributions of random variables, these formulas transform the original non-normal random variables into equivalent ones in the multivariate standard normal space. Therefore, in the equivalent multivariate standard normal space, the proposed methods, which adopt the fundamental concepts of Harr' algorithm, can properly operate. The selected points in the multivariate standard normal space are transformed back to the original parameter space for evaluating statistical moments of model outputs. Accordingly, the applicability of Harr's PE algorithm for uncertainty analysis is expanded to handle problems involving multivariate non-normal random variables.

Following the idea described above, two

algorithms which consider different expansion points are proposed. They are (1) median-expansion algorithm and (2) mean-expansion algorithm. In this paper, an application is made to a pier scouring model. Specifically, the proposed PE algorithms were compared with Rosenblueth's on the accuracy of uncertainty analysis under different model types and number of stochastic parameters. Furthermore, the overall performances of the three PE methods were evaluated by fitting Johnson distribution (Johnson and Kotz, 1970) curves based on the computed moments.

2. PROPOSED METHOD (1): MEDIAN-EXPANSION ALGORITHM

By incorporating transformation between non-normal and a multivariate standard normal space, the algorithm can be extended to handle random variables with known non-normal marginal distributions.

2.1 Multivariate normal space

In the median-expansion algorithm, the vector of stochastic parameters X having a multivariate normal distribution are standardized as

$$Y = D^{-0.5} (X - \mu) \qquad (2.1)$$

in which Y is the vector of the multivariate standard normal random variables; D is a diagonal matrix containing the variances of the stochastic parameters; and μ is the vector of the mean values of X.

Through an orthogonal transformation, the correlated standard normal variables, Y, are decomposed into independent standard normal variables, Z, as

$$Z = L^{-0.5} V^t Y = L^{-0.5} U \qquad (2.2)$$

in which U is the vector of uncorrelated random variables in the eigen space having the mean 0 and covariance matrix L with L and V, respectively, being the eigenvalue and eigenvector matrices associated with the correlation matrix of the stochastic parameters, R_X. The eigenvector and eigenvalue matrices satisfies

$$R_X = V L V^t \qquad (2.3)$$

where $V = [v_1, v_2, ..., v_n]$, with $v_1, v_2, ..., v_n$ being the column vectors of the eigenvectors; $L = \text{diag}(\lambda_1, \lambda_2, ..., \lambda_n)$ with $\lambda_1, \lambda_2, ..., \lambda_n$ being the

corresponding eigenvalues. The transformations provided by Eqs.(2.1) and (2.2) are linear. Therefore, if all the original stochastic parameters were normally distributed, the transformed parameter spaces for U were also normal.

In the median-expansion algorithm, a hypersphere with radius $\sqrt{n}$ centered at the origin in the n-dimensional standardized eigen space is constructed. The points at which model output to be evaluated are located at the intersections of the hypersphere and the eigenvectors of the correlation matrix of the stochastic parameters. For problems involving n stochastic parameters, there are a total of 2n intersection points at which model evaluations are performed.

Due to the normal distribution and the same scale on each component in the standard space for Z, the 2n proposed points for model evaluation are located on a hypersurface with equal probability density function (PDF) value. Figure 1 schematically shows the point selections by Rosenblueth's algorithm and the proposed method for a bivariate case in the standard normal space. Noted that the selected points by the proposed method are located on the ellipse which is a circle in the standardized eigen space.

By Eqs.(2.1) and (2.2), the points for model evaluation in the original multivariate normal variable space can be obtained as

$$x_{k\pm} = \mu \pm \sqrt{n} D^{0.5} L^{0.5} v_k, \quad k=1,\cdots,n \qquad (2.4)$$

where $x_{k\pm} = (x_{k1\pm}, x_{k2\pm}, ..., x_{kn\pm})^t$ is a column vector containing the coordinates of the two intersection points on the k-th eigenvector in the original normal variable space. At each selected point, the corresponding model output value $w_{k\pm} = g(x_{k\pm})$, for $k=1$ to n, can be computed.

The m-th order moment about the origin of the model output can be calculated as

$$E(W^m) = \frac{1}{n} \sum_{k=1}^{n} \overline{w_k^m} \qquad (2.5)$$

in which

$$\overline{w_k^m} = \frac{w_{k+}^m + w_{k-}^m}{2} \qquad (2.6)$$

The m-th order central moment of the model output W, $\mu_{W.m}$, can be obtained by

$$\mu_m = \sum_{i=0}^{m} (-1)^i \ C_i^m \ \mu^i \ \mu'_{m-i} \qquad (2.7)$$

where $C_i^m = m!/[i! \ (m-i)!]$, a binomial coefficient. Eq. (2.7) can then be employed to obtain the central moments of the model output W.

2.2 Incorporating marginal distributions of random variables

In many practical engineering problems, one often has to deal with random variables having different types of distribution. Such distributional information could have important implication on the results of engineering uncertainty and reliability analyses. The incorporation of marginal distributions information of random variables further enhanced the capability of Harr's PE algorithm which presently accounts for the first two moments (including correlation) of the involved random variables.

For a mixture of correlated random variables (not necessarily all normal), the proposed methods incorporate the available marginal distribution information by using the set of semi-empirical formulas derived by Der Kiureghian and Liu (1985). These formulas transform the correlation coefficient of a pair of non-normal random variables to the equivalent one in the standard normal space. Through this transformation, the above algorithm for multivariate normal parameters described previously can be performed appropriately.

The distribution types for the correlated random variables pair to which the formulas are applicable include: (1) normal, (2) uniform, (3) shifted exponential, (4) shifted Rayleigh, (5) type-I largest value (Gumbel Max), (6) type-I smallest value (Gumbel Min), (7) lognormal, (8) gamma, (9) type-II largest value, and (10) type-III smallest value (Weibull).

The median-expansion algorithm consists of the following three steps:

Step 1. <u>Transformation of correlation in non-normal space to the equivalent normal space</u> - The formulas transform the correlation from the original space to the standard normal space having the equivalent probability content as in the original space by

$$\rho^*_{ij} = T_{ij} \cdot \rho_{ij} \qquad (2.8)$$

in which ρ^*_{ij} is the correlation between two standard normal random variables, Y_i and Y_j, whereas ρ_{ij} is the correlation between the non-normal stochastic

parameters X_i and X_j in the original space; and T_{ij} is a transformation factor which is a function of the marginal distributions and correlation of the two stochastic parameters considered. For each combination of two distributions mentioned above, one corresponding formula exists to compute T_{ij} (see Figure 2). Given the marginal distributions and correlations for the stochastic parameters, the formulas of Der Kiureghian and Liu (1985) compute the corresponding transformation factor to obtain the equivalent correlation ρ^*_{ij}. After all pairs of stochastic parameters are treated, the correlation matrix in the multivariate standard normal space, R_Y, can be obtained.

Step 2. <u>Determine points for model evaluation in the standard normal space</u> - Through the transformation by Eq. (2.8), the operation domain is switched to the space in which the transformed random variables are treated as if they were multivariate standard normal random variables with the correlation matrix R_Y. Note that the transformed space is already standard normal. Therefore, the standardization by Eq. (2.1) in the median-expansion algorithm is not needed. Accordingly, Eq. (2.4) can be used to determine the points for model evaluation in the correlated standard normal space as

$$y_{k\pm} = \pm \sqrt{n} \ L^{0.5} \ v_k, \qquad k = 1, \cdots, n \qquad (2.9)$$

Step 3. <u>Inverse transformation</u> - To generate appropriate points for model evaluations in the original space, the points selected in the standard normal space are transformed back to the original space for evaluating the corresponding model output values. The inverse transformation from the standard normal space to the original space can be established by preserving probability content. From the given marginal distribution of the i-th parameter X_i, the cumulative distribution function (CDF) is known as F_i. The k-th pair of selected points in the standard normal space, $y_{k\pm} = (y_{k1\pm}, y_{k2\pm}, \cdots, y_{kn\pm})^t$, can be transformed back to the original space as

$$x_{ki\pm} = F_i^{-1}[\Phi(y_{ki\pm})] \ , \ i = 1, 2, \cdots, n \qquad (2.10)$$

in which $\Phi(\ )$ is the standard normal CDF. Thus, the k-th pair of the selected points in the original parameter space $x_{k\pm}$ are obtained for model evaluation. The moments of model output can be estimated by Eqs. (2.5) to (2.7).

Note that the selected points by the median-expansion algorithm for model evaluation are, in essence, the expansion with respect to the mean in the standard normal space which also is the median. Through the inverse transformation by Eq.(2.10), it preserves the median of each stochastic parameter in the original space as shown in Figure 3. However, when the distribution of the stochastic parameter in the original space was not symmetric, the mean and median are different in the original space, .

3. PROPOSED METHOD (2): MEAN-EXPANSION ALGORITHM

The mean-expansion algorithm adjusted the expansion point from the median to the mean of the original stochastic parameters. In doing so, the property of equal PDF for each selected points would no longer be held. Therefore, Eqs.(2.5) and (2.6) must be modified.

Let $\mu = (\mu_1, \mu_2, ..., \mu_n)^t$ represent the means of the stochastic parameters in the original space. In the standard normal space, $y^* = (y_1^*, y_2^*, ..., y_n^*)^t$, the equivalent point for the mean of the original distributions can be determined by inverting Eq.(2.10) as

$$y_i^* = \Phi_i^{-1} [F(\mu_i)] , \quad i = 1 \text{ to } n \quad (3.1)$$

Then, the mean-expansion algorithm selects points for model evaluation around the point y^*. Figure 4 shows the idea of the mean-expansion algorithm for a univariate case. More specifically, the selected points which encompass the origin in the standard normal space now is shifted with respect to y^*. Consequently, the equal PDF is no longer valid at the shifted points. Therefore, for the two k-th shifted points in the standard multivariate normal space, $y_{k\pm}^* = (y_{k1\pm}^*, y_{k2\pm}^*, ..., y_{kn\pm}^*)^t$, the associated PDF values are used as the weighing factors, $\alpha_{k\pm}$,

$$\alpha_{k\pm} = \exp \left[-\frac{1}{2} z_{k\perp}^{*t} R_Z^{-1} z_{k\perp}^* \right] \quad (3.2)$$

which is the exponential part of the multivariate standard normal PDF since the remaining part is a constant. By Eq.(2.10), the shifted points in the original space can be obtained. The model output values are computed at these points and are weighed by Eq.(3.2) to estimate the moments. That is, Eq. (2.5) for computing moments of the model output is modified as

$$E(W^m) = \frac{1}{n} \sum_{k=1}^{n} \frac{1}{\alpha_{k+} + \alpha_{k-}} (\alpha_{k+} w_{k+}^m + \alpha_{k-} w_{k-}^m) . \quad (3.3)$$

4. APPLICATION

Bed scouring is a phenomenon in river caused by the interaction of flow and river bed. Hydraulic structures such as bridge piers is susceptible to failure under long term and continuous bed scouring. As required for engineering design as well as for the precaution of undesirable consequences, the knowledge of bed scouring around bridge piers is essential. Many models have been developed to predict the potential scour depth around bridge piers. Using such a computer model to aid the design of pier depth is common in modern hydraulic engineering. However, the existence of various uncertainties involved in bed scouring models results in uncertainty in scour depth prediction required for design.

For purpose of illustration, a simple pier scour model developed by Johnson (1992) is used herein for uncertainty analysis. Focus is placed on the relative performances of the various PE methods in the uncertainty analysis as compared with the Monte Carlo simulation (Chang et al., 1994).

4.1 Pier scouring model

Johnson (1992) proposed an empirical pier scouring model based on experimental data from various sources

$$D_s = 2.02 \lambda \; y \left(\frac{b}{y} \right)^{0.98} F_r^{0.21} \; \sigma^{-0.24} \quad (4.1)$$

in which D_s is the predicted scour depth; λ is the model correction factor; y is the flow depth; b is pier width; F_r is the Froude number; and σ is the sediment gradation. Because the model is empirical by nature, uncertainties exist in both model itself and the inputs/parameters involved (Yeh and Tung, 1993). Consequently, the scour depth computed from Eq. (4.1) is subject to uncertainty and it is likely that a specified design pier depth could be exceeded resulting in potential threat to bridge safety.

4.2 Uncertainty analysis of pier scouring model

The stochastic parameters considered in Eq. (4.1) are λ, y, F_r, and σ. The stochasticity of model correction factor, λ, represents the model uncertainty associated with the pier scouring model whereas the randomness of y, F_r, and σ are resulted from model input uncertainties. Their means and coefficients of variation are listed in Table 1. According to Johnson (1992), all stochastic parameters, except the model correction factor λ, are correlated random variables with the correlation matrix given in Table 2. The model correction factor λ is treated herein as an independent random variable.

The three PE methods are used herein for the uncertainty analysis of the pier scouring model including: Rosenblueth's, proposed median-expansion, and mean-expansion PE methods. In the uncertainty analysis, mixture distributions were adopted to explore the applicability of each method. The distributions used for the stochastic parameters in the pier scouring model were: gamma distribution for λ, lognormal distribution for σ and y, and Weibull distribution for F_r.

To compare the relative performance in uncertainty analysis among the different methods, results from the Monte Carlo simulation can be used as the true values serving as the basis for comparison. Based on the given marginal distributions and correlations for the stochastic parameters, 100,000 samples were generated from which the statistical moments of scour depth from the pier scouring model were computed.

Table 3 lists the estimated moments of the scour depth from the three PE methods and their error percentages (in italic). Under the consideration of the mixture distributions for the correlated stochastic parameters, all the methods are capable of estimating the first two moments accurately. For higher moments, Table 3 indicates that Rosenblueth's method fails to yield a good estimate for the kurtosis whereas the median-expansion PE method's estimation for the skewness is undesirable. The mean-expansion PE method, however, yields closer estimations for both skew coefficient and kurtosis. From the aspect of computation efficiency, 8 model evaluations are needed by using the proposed PE methods since 4 stochastic parameters were involved. However, using Rosenblueth's PE method requires double the amount of computation.

5. SUMMARY AND CONCLUSION

In this study, two PE methods are proposed to incorporate the marginal distributions of correlated random variables. The proposed methods integrate Harr's PE procedure along with the formulas which transform the original correlation to the equivalent one in the standard normal space. The performance of the proposed PE methods was evaluated against Rosenblueth's method. The input requirements are marginal distributions of involved random variables and their correlations.

In the application, the uncertainty analysis of a

pier souring model was performed to demonstrate the relative performance of each PE method for a practical engineering problem. Under the mixture distributions and correlated stochastic parameters considered in this particular application, the three PE methods showed that the estimations of the first two moments of the predicted scour depth are as accurate as those obtained from the Monte Carlo simulation with 100,000 model evaluations. However, only the mean-expansion method can yield closer estimates for the higher moments. It is shown in this application that uncertainty feature estimated by the mean-expansion PE method can achieve comparable accuracy with the one from the Monte Carlo simulation with significantly less computations. The latter point is especially important for those uncertainty analyses of models requiring great amount of computation in themselves.

REFERENCES

Chang, C. H., Tung, Y. K., and Yang, J. C. 1995. Evaluation of probability point estimate methods. *Appl. Math. Modelling*, **19**(2):95-105.

Chang, C. H., Tung, Y. K., and Yang, J. C. 1994. Monte Carlo simulation for correlated variables with marginal distributions. *Journal of Hydraulic Engineering*, ASCE, **120**(3):313-331.

Der Kiureghian, A. and Liu, P. L. 1985. Structural reliability under incomplete probability information. *Journal of Engineering Mechanics*, ASCE, **112**(1):85-104.

Harr, M. E. 1989. Probabilistic estimates for multivariate analyses. *Appl. Math. Modelling*, **13**(5):313-318.

Hill, I. D., Hill, R., and Holder, R. L. 1976. Algorithm AS 99: Fitting Johnson curves by moments. *Appl. Statist.*, **25**:180-189.

Johnson, N.L. and Kotz, S. 1970. *Continuous Univariate Distributions-1*, John Wiley and Sons, New York, NY.

Johnson, P. A., (1992). "Reliability-based pier scour engineering." *Journal of Hydraulic Engineering*, ASCE., **118**(10):1344-1358.

Rosenblueth, E. 1975. Point estimates for probability moments. *Proc. Nat. Acad. Sci. USA*, **72**(10):3812.

Rosenblueth, E. 1981. Two-point estimates in probabilities. *Appl. Math. Modelling*, **5**(10):329-335. Yeh, K. C., and Tung, Y. K., 1993. Uncertainty and sensitivity of a pit migration model. *Journal of Hydraulic Engineering*, ASCE, **119**(2):262-281.

Table 1. Means and coefficients of variation (CV) of stochastic parameters used in the pier scouring model (from Johnson, 1992).

Variables	Mean	CV
λ	1.000	0.18
y	4.250	0.20
Fr	0.537	0.38
σ	4.000	0.20

Table 2 Correlation among stochastic parameters used in the pier scouring model (from Johnson, 1992).

Variables	λ	y	Fr	σ
λ	1.00	0.00	0.00	0.00
y	0.00	1.00	-0.33	-0.79
Fr	0.00	-0.33	1.00	0.29
σ	0.00	-0.79	0.29	1.00

Table 3 Comparison of the first four moments for the random scour depth by various methods.

Moments	Methods			
	ROSEN	Md-PE	Mn-PE	SIMUL
μ	2.5843 (-0.03%)	2.5886 (0.14%)	2.5831 (-0.08%)	2.5851
σ	0.4864 (-0.47%)	.4915 (0.57%)	0.4853 (-0.70%)	0.4887
γ	0.3899 (-3.08%)	.5204 (29.36%)	0.3914 (-2.71%)	0.4023
κ	1.4925 (-53.96%)	3.5097 (8.27%)	3.5365 (9.10%)	3.2416

Note: ROSEN = Rosenblueth's method; Md-PE = median-expansion PE method; Mn-PE = mean-expansion PE method; SIMUL = Monte Carlo simulation.

164

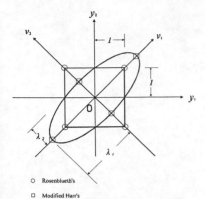

○ Rosenblueth's

□ Modified Harr's

Figure 1 Selections of points for model evaluation by different PE methods

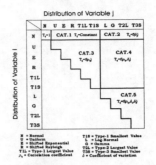

Figure 2 Schematic description of the categories of the transformation factor T_{ij}. (Chang et al., 1994)

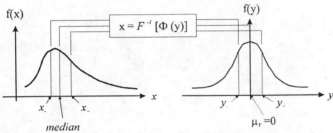

Figure 3 Schematic diagram of the transformation by median-expansion algorithm between non-normal and normal spaces.

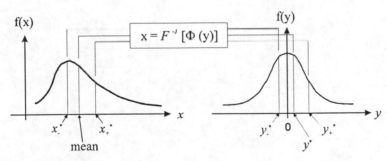

Figure 4 Schematic diagram of the transformation by mean-expansion algorithm between non-normal and normal spaces.

2 Open channel and river hydraulics

Stochastic Hydraulics '96, Tickle, Goulter, Xu, Wasimi & Bouchart (eds) © 1996 Balkema, Rotterdam. ISBN 90 5410 817 7

Shannon's entropy-based bank profile equation of threshold channels

Shuyou Cao
State Key Hydraulics Laboratory, Sichuan Union University, Chengdu, China

Donald W. Knight
School of Civil Engineering, The University of Birmingham, UK

ABSTRACT: The concept of Shannon's entropy based on probability instead of thermodynamics, is applied to develop a new bank profile equation for threshold fluvial channels. A general formula for the lateral distribution of transverse slopes, and hence the cross-sectional bank profile, is derived by the entropy-maximization principle and the calculus of variations. For the case of a threshold channel with non-cohesive boundary material, the Lagrange multiplier λ is equal to zero. L'Hospital's rule of limit theory is used to find the bank profile as a new simple parabolic curve. The results of numerical experiments, based on the boundary shear stress distribution approach, support the conclusion that the entropy-based channel bank profile is at threshold. Bank profile equations are coupled with an appropriate frictional relationship to obtain a new design method. Channel dimensions and bank profiles predicted by this method are compared with those given by other design methods. The predicted channel dimensions are in reasonable agreement with laboratory experimental data.

1 INTRODUCTION

The concept of 'entropy' was first proposed as a thermodynamic quantity by Clausius in 1865 (Simpson and Weiner, 1989). In Clausius' sense, the entropy of a system is the measure of the unavailability of its thermal energy for conversion into mechanical work. The concept of entropy based on probability theory was later introduced into statistical mechanics and information theory in 1928 and 1948 respectively (Jaynes, 1957). It was Shannon (1948) who introduced a measure for information or entropy of general finite complete probability distributions and originally gave a characterisation theorem of the entropy (Aczel and Daroczy, 1975). As a result, the probability-based entropy is also known as information entropy or directly as Shannon's entropy (Jaynes, 1957; Chiu and Jin, 1995). Consequently, this concept is widely used in statistical mechanics and in information theory (Goldman, 1953; Jaynes, 1957; Reza, 1961; Shore and Johnson, 1980). From the probability point of view, entropy is the degree of disorder of a system, measured in terms of the natural logarithm of the probability of occurrence of its particular arrangement of particles (Brown,

1993). Shannon's entropy has only been employed in hydraulics during the last decade for investigating the stochastic nature of certain hydraulic and river mechanics phenomena (Chiu, 1987, 1991; Chiu and Jin, 1995; Chiu and Said, 1995; Cao and Chang, 1988, Cao and Knight, 1995a, b and Cao, 1995).

Shannon (1948) first introduced the definition of entropy to be the measure of uncertainty associated with the sample space of a complete finite scheme for some discrete variable x as

$$H(x) = -\sum_{i=1}^{n} p(x_i) \ln p(x_i) \qquad (1.1)$$

in which $p(x_i)$ = the probability of a system being in state x_i, which is a member of $\{x_i, i = 1,2,...\}$. The entropy, $H(x)$, for a continuous state variable, x, is expressed quantitatively in terms of probability (Shannon 1948) as

$$H(x) = -\int p(x) \ln p(x) dx \qquad (1.2)$$

in which $p(x)$ is the probability density function, $p(x)dx$ is the probability of a state variable being between x and $x + dx$. The entropy defined by Eq.

1.2 is a measure of randomness or uncertainty in the state variable x. It is zero for a deterministic case because the probability density function is equal to unity when that state occurs. Maximising the entropy of a system will make the probability distribution as uniform as possible while satisfying the constraints. The probability law that governs a system, and the corresponding magnitude of entropy, will depend upon the prevailing constraints. The physical meaning of the entropy is dependent on the specific physical variable concerned.

Maximization of entropy of a continuous random variable can be obtained by using mathematical maximisation techniques from the calculus of variations, such as the method of Lagrange multipliers. There are three types of general constraints in common problems of maximisation entropy (Shannon 1948; Goldman, 1953; Reza, 1961; Chiu, 1987; Cao and Chang, 1988 and Cao, 1995).

Type 1. Subject to the constraint:

$$\int_a^b p(x)dx = 1 \qquad (1.3)$$

when maximum entropy is associated with a random variable with a uniform probability density distribution between a and b.

Type 2. Subject to the constraints:

$$\int_a^b p(x)dx = 1 \qquad (1.3)$$
$$\int_a^b xp(x)dx = \bar{x} \qquad (1.4)$$

when maximum entropy corresponds to an exponential probability density distribution. In Eq. 1.4 $\bar{x}$ = the expected value of the variable, x.

Type 3. Subject to the constraints:

$$\int_a^b p(x)dx = 1 \qquad (1.3)$$
$$\int_a^b x^2 p(x)dx = \sigma^2 \qquad (1.5)$$

when the random variable x has a standard deviation σ and zero mean.

The method of the calculus of variations may be applied to determine $p(x)$ in Eq. 1.2. The detail of this method can be found in Shannon (1948), Goldman (1953), Reza (1961), Chiu (1987), Cao and Chang (1988) and Cao (1995).

The concept of entropy, based on probability theory instead of thermodynamic principles, has been applied in this paper to study the bank profile of a threshold channel cross section. A formula for the lateral distribution of transverse slopes, and hence the cross-sectional bank profile, is derived by the entropy-maximisation principle and the calculus of variations. Two boundary conditions are used to give some physical meaning to this entropy approach. One is that the dimensions and shape are dependent on the discharge and boundary sediment size (or the angle of respose of the particles). Another is that the shape curve must satisfy the continuity condition at the junction point of the two bank profile curves.

2 ENTROPY-BASED TRANSVERSE SLOPE EQUATION

The state variable, x, in Eq. 1.2 studied in this paper is specified as the lateral slope of the bank profile, S_t. From the point of view of physical meaning, maximising the entropy of the bank profile will make the probability distribution of the bank profile as uniform as possible while satisfying the constraints. Every point on the bank profile is therefore at a state at which the sediment is at threshold, since the entropy of the bank profile is a maximum. The probability law that governs the bank profile and the corresponding magnitude of entropy will depend on the prevailing constraints.

The right half of a symmetric cross section of an alluvial channel with a channel centre boundary elevation z = 0, a semi-width of L, a transverse slope of $\theta = \tan^{-1}\mu$ at the water margin, is studied. For an ideal threshold cross section, the transverse slopes, $S_t = \tan\theta$, increase monotonically in the lateral direction, from zero at the centreline of the channel (y = 0) to a maximum value equal to the submerged static coefficient of Coulomb friction, μ, at the water margin (y = L). Let S_t be the transverse slope at the lateral distance y from channel centreline, then at any lateral distance less than y, the transverse slope is less than S_t. It is obvious that all values of y between zero and L along the y axis possess the same probability or are equally likely. It can therefore be stated that the probability of the transverse slope being equal to or less than S_t is y/L. The cumulative distribution function of S_t is then

$$F(S_t) = y/L \qquad (2.1)$$

and the probability density function of S_t is

$$p(S_t) = \frac{dF(S_t)}{dy}\frac{dy}{dS_t} = \left(L\frac{dS_t}{dy}\right)^{-1} \qquad (2.2)$$

The Type 2 constraints are selected for this study. The first constraint (Eq. 1.3) is the definition of general probability, for transverse slope as

$$\int_0^\mu p(S_t)dS_t = 1 \qquad (2.3)$$

where the lower limit of integration, 0, and the upper limit of integration, μ, are the transverse slopes at the centre and at right water margin of a channel respectively. The second constraint (Eq. 1.4) is related to the lateral distributions of transverse slope to the expected value of S_t, which is described as

$$\int_0^\mu S_t\, p(S_t)dS_t = \overline{S_t} \qquad (2.4)$$

where $\overline{S_t}$ is the average transverse slope of the whole section boundary. The entropy of transverse slope is expressed according to Eq. 1.1 as

$$H(S_t) = -\int_0^\mu p(S_t)\ln p(S_t)dS_t \qquad (2.5)$$

In order to maximise $H(S_t)$, defined by Eq. 2.5 and subject to the two constraints given by Eqs. 2.3 and 2.4 the technique of calculus of variations is used to get the distribution equation of transverse slope of a channel cross section as

$$p(S_t) = Exp(\lambda_1 + \lambda\, S_t - 1) \qquad (2.6)$$

where λ_1 and λ are the Lagrange multipliers. Equating the right side of Eqs.2.2 and 2.6 gives

$$\frac{dS_t}{dy} = \frac{1}{L}Exp(1-\lambda_1-\lambda S_t) \qquad (2.7)$$

The integration of Eq. 2.7 finally gives the distribution equation of transverse slope as

$$S_t = \frac{1}{\lambda}\ln(1+\lambda e^{1-\lambda_1}y) = \frac{1}{\lambda}\ln(1+(e^{\lambda\mu}-1)\frac{y}{L}) \qquad (2.8)$$

The Lagrange multiplier, λ, should be connected with some suitable parameters, either hydraulic or linked to the boundary condition of channel. In

order to test the effect of λ, a simple channel with a semi-width of 1 m is considered. Some preliminary relationships between the lateral distribution of transverse slope and λ are shown in Fig. 1. Figure 1 shows that the smaller the Lagrange multiplier λ is (i.e. $\lambda < 1.0$), the closer to the diagonal the lateral distribution of transverse slope is. The curves of $\lambda < 1$ are almost identical and lie on the diagonal. At the other extreme, Fig. 1 shows that the lateral distributions of lateral slopes are close to a horizontal line of $S_t = \mu$ for λ values larger than 50.

3 BANK PROFILE EQUATIONS

The transverse slope of the bank profile of a threshold channel, S_t, is given as

$$S_t = \frac{dz}{dy} \qquad (3.1)$$

where z is the boundary elevation above the section centreline. Integration of Eq. 3.1 gives the shape equation, $z = f(y)$. The boundary condition at the centreline is used to determine the constant of integration, $c = 0$ when $z = 0$ and $y = 0$. The bank profile equation is then

$$z = \frac{1}{\lambda}\left(\frac{1+\beta y/L}{\beta/L}\ln(1+\beta y/L)-y\right) \qquad (3.2)$$

in which $\beta = Exp(\lambda\mu)-1$. Equation 3.2 gives the elevation of the right water margin when $y = L$, which is equal to the depth at the channel centre, h_c:

$$h_c = \frac{1}{\lambda}\left(\frac{\lambda\mu Le^{\lambda\mu}}{\beta}-L\right) \qquad (3.3)$$

The effects of Lagrange multiplier, λ, on the bank profiles of threshold channels are numerically tested and shown in Fig 2. In this test λ values are varied from $1 \sim 50$, and $\mu = 0.5$. This approach shows that the Lagrange multiplier, λ, is one of the key parameters in the entropy approach. There are two extreme cases to consider, one in which λ approaches zero, and the other in which λ approaches infinity. L'Hospital's rule of limit theory is used to find the results. When λ approaches zero, Eq. 2.8 is reduced to

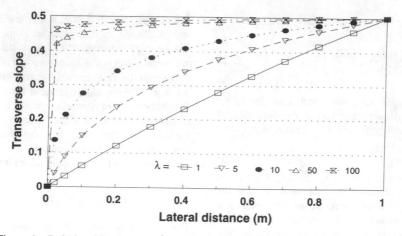

Figure 1• Relationships between λ and lateral distribution of transverse slope (μ = 0.5)

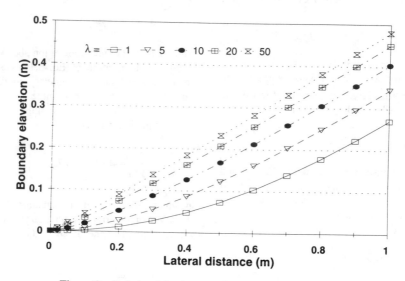

Figure 2• Relationships between λ and bank profiles (μ = 0.5)

$$S_t = \frac{\mu}{L} y \qquad (3.4)$$

When λ approaches infinity Eq.2.8 is reduced to

$$S_t = \mu \qquad (3.5)$$

The boundary elevations above the bed at the centre of these two cases are then obtained by integration of Eqs. 3.4 and 3.5 in the form

$$z = \frac{\mu}{2L} y^2 \qquad \text{(when } \lambda \to 0) \qquad (3.6)$$

$$z = \mu y \qquad \text{(when } \lambda \to \infty) \qquad (3.7)$$

Substituting y = L into Eqs. 3.6 and 3.7, the centreline channel depths are given as

$$h_c = \mu L / 2 \qquad \text{(when } \lambda \to 0) \qquad (3.8)$$

$$h_c = \mu L \qquad \text{(when } \lambda \to \infty) \qquad (3.9)$$

and the aspect ratios are given as

172

$B/h_c = 4/\mu$ (when $\lambda \to 0$) (3.10)

$B/h_c = 2/\mu$ (when $\lambda \to \infty$) (3.11)

where $B = 2L$ = surface width.

Eq. 3.6 gives a very simple parabolic curve bank profile as λ approaches zero, while a trapezoidal bank profile can be obtained from Eq. 3.7 as λ approaches infinity. Eqs. 3.10 and 3.11 show that when λ approaches zero the aspect ratio is twice that when λ approaches infinity. The channel shapes are therefore very sensitively dependent upon λ values, which should be connected with flow and channel boundary conditions. For this reason λ is logically defined as

$\lambda = \zeta w^*$ (3.12)

where ζ is an empirical coefficient and w^* is the dimensionless effective Shields parameter:

$w^* = \dfrac{\Theta - \Theta_c}{\Theta_c}$ (3.13)

where Θ and Θ_c are the actual and critical Shields parameters respectively. It is very interesting to note that a flat bed upper regime flow will be generated for w^* greater than 25 (Chang, 1988). From an engineering point of view a trapezoidal channel will be generated for $\lambda > 50$, making $\zeta = 2$ for an upper regime flat bed flow. Otherwise, when the actual Shields parameter, Θ, is equal to the critical Shields parameter, Θ_c, such that $w^* = 0$ means that the channel is at threshold. For this situation, Eq. 3.13 gives $\lambda = 0$. On the assumption that Eqs. 3.12 and 3.13 are correct, then the shape

of a threshold channel is given by Eq. 3.6. Further work is however clearly needed on the formulation and physical basis of Eq. 3.12. The coefficient ζ also requires further study.

Eq. 3.10 gives the depth of the channel at the centreline and subtracting it from Eq. 3.8 gives the lateral distribution of depths, $h(y)$, as

$$h(y) = \frac{\mu L}{2}\left[1 - \left(\frac{y}{L}\right)^2\right]$$ (3.14)

4 APPLICATION

From an engineering point of view, the most important task of all research into channel shape equations is to provide design criteria for threshold channels.. Bank profile equations are coupled with an appropriate frictional relationship to obtain a new design method. A FORTRAN program was developed to compute the procedure.

Various cross-sectional shape equations have been proposed to describe the bank profiles of straight threshold channels. Among these are those developed by the US Bureau of Reclamation (USBR, see Henderson, 1966) and Diplas and Vigilar (1992). These two design methods are now compared with the new design approach of the case as λ approaching zero in Fig. 3. This figure shows that the shapes given by the entropy approach lie between those of Diplas et al and USBR approaches, but are closer to Diplas values for the given condition (Q = 0.05 m³s⁻¹, d = 0.8 mm, and μ = 0.5).

Stebbings's (1963) laboratory experiments can be used to test the top width and cross section area of the proposed entropy approach. A total of 34 sets

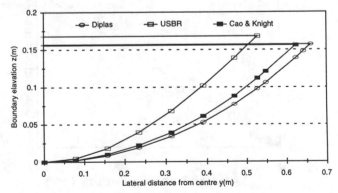

Figure 3• Design bank profiles by three approaches (Q = 0.05 m³s⁻¹)

of data in two series of experiments are used here. The diameter of the channel boundary material was $d_{50} = 0.88$ mm with $\mu = 0.51$. The average value of the aspect ratio in all 34 data sets from Stebbings's experiments was 7.31, compared with 7.8 given by the entropy approach. The aspect ratio given by Diplas is 8.3, and by USBR is 6.16. The test results are shown in Figs. 4 and 5, which show that the predicted values for the top width of the channel and the cross-sectional area are in satisfactory agreement with the laboratory data.

5 CONCLUSIONS

The concept of entropy, based on probability theory instead of thermodynamic principles, has been applied in this paper to study the hydraulic geometry of a threshold channel. A general formula for the lateral distribution of transverse slopes has been derived for threshold channels using the entropy-maximisation principle and the calculus of variations. When the Lagrange multiplier, λ, approaches zero for a threshold channel with non-cohesive material on the boundary, a formula for the lateral distribution of transverse slopes is derived for a threshold channel. The predicted values of top width and area of the cross section agree well with existing laboratory data. The predicted dimensions and shapes are compared with other existing approaches and are shown to be in between those predicted by Diplas et al (1992) and USBR (Henderson, 1966).

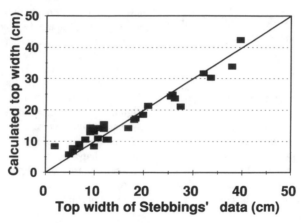

Figure 4• Comparison of top width ($\mu = 0.51$)

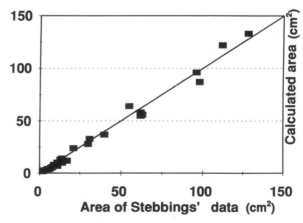

Figure 5• Comparison of cross section area ($\mu = 0.51$)

REFERENCES

Aczel, J. & Daroczy, 1975, *On measures of information and their characterizations*, Academic Press, New York, 1-6.

Brown, L., 1993, *The new shorter Oxford English dictionary*, Vol. 1, Oxford Press, 831.

Cao, S. and Chang, H.H., 1988, Entropy as a probability concept in energy-gradient distribution, *Proc. Nat. Con. Hyd. Engr.*, Colorado Springs, CO, USA, Aug. 8-12, ASCE, New York, 1013-1018.

Cao, S. and Knight, D.W., 1995a, Design of threshold channels, *Hydra 2000, Proc. 26th IAHR Congress*, London, UK, September, Vol. 1, 516-521.

Cao, S. and Knight, D.W., 1995b, New concept of hydraulic geometry of threshold channels, *Proc. 2nd Symposium on the Basic Theory of Sedimentation*, Beijing, China, October (in Chinese).

Cao, S., 1995, Regime theory and a geometric model for stable alluvial channels, *PhD thesis*, School of Civil Engineering, The University of Birmingham, England, UK. 1-350.

Chang, H., 1988, *Fluvial processes in river engineering*, John Wiley & Sons, Inc., 1-432.

Chiu, C. L., 1987, Entropy and probability concepts in hydraulics, *J. Hydr. Engrg.*, ASCE, 110(1), 583-600.

Chiu, C.-L., 1991, Application of entropy concept in open-channel flow study, *J. Hydr. Engrg.*, ASCE, 117(5), 615-628.

Chiu, C.-L., and Said, A., 1995a, Maximum and mean velocities and entropy in open-channel flow, *J. Hydr. Engrg.*, ASCE, 121(1), 26-35.

Chiu, C.-L.and Jin, Weixia, Entropy and open-channel flow properties, *Hydra 2000, Proc. 26th IAHR Congress*, London, UK, September, Vol. 1, 585-590.

Diplas, P., and Vigilar, G. 1992, Hydraulic geometry of threshold channels. *J. Hydr. Engrg.*, ASCE, 118(4), 597-614.

Goldman, S, 1953, *Information theory*, Prentice-Hall, Inc., New York, New York.

Henderson, F.M., 1966, *Open channel flow*, Macmillan, New York, 450-455.

Jaynes, E.T., 1957, Information theory and statistical mechanics I, *Physics Review*, 106, 620-630.

Reza, F.M., 1961, *An introduction to information theory*, McGraw-Hill Book Company, Inc., New York, 278-280.

Shannon, C.E., 1948, A mathematical theory of communication, *The Bell System Technical Journal*, 27, 623-656.

Shore, J.E., and Johnson, R.W., 1980, Axiomatic derivation of the principle of maximum entropy and the principle of minimum cross-entropy, *Transactions on Information Theory*, IEEE, II-26, No. 1.

Simpson, J.A., and Weiner, E.S.C., 1989, The Oxford English dictionary, Vol. V, Oxford Press, 2nd ed., 308.

Stebbings, J., 1963, The shape of self-formed model alluvial channels, *Proc. Inst. Civ. Engrs.*, London, England, 25, 485-510.

Stochastic Hydraulics'96, Tickle, Goulter, Xu, Wasimi & Bouchart (eds) © 1996 Balkema, Rotterdam. ISBN 90 5410 817 7

Analysis of the secondary currents of the outer bank erosion in channel bends

Wang Pingyi
Southwest Research Institute of Water Transport Engineering, Chongqing, China

ABSTRACT: In bend flow near the apex, there is usually an outer bank cell. In this paper, the influence of the channel geometric parameters on the existence of the outer bank cell is discussed first. Next, the existence of the outer bank cell is shown to be related to the shift of the maximum longitudinal velocity below the water surface. According to the measured data of the Fall River in American, the calculations and analysis of the vorticity components are given as a proof the existence of this outer bank cell. Finally, according to the above results and the boundary shear stress distribution along outer bank, the general mechanism of the outer bank erosion is given.

1 INTRODUCTION

In channel bends, the secondary currents is one of the most important characteristics. However, near the apex, there is usually an outer bank cell, which is a small cell of reverse circulation occupying the channel to a distance one average water depth away from the outer bank. This is illustrated with the Lei Section 19 in Xiajian River and the section in Fall River in Fig. 1 (Zhang 1964, Thorne 1985). Although this is a small portion of the cross-section in natural rivers, the outer bank cell is still important because it strongly affects outer bank erosion and river bends migration processes.

Einstein & Harder (1954), Rozovskii (1957), Zhang Zhitang et al (1964), Hey & Thorne (1975), De Vriend & Gelodof (1983), and Thorne & Rais, et al (1983), found the existence of an outer bank vortex in the outer bank region of the flow, and explained that the outer bank cell is a wall effect, it is the result of the interaction of the outward flow with the outer bank. But, they did not comment further and study on its existence and effects on the outer bank erosion.

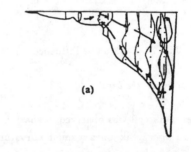

(a)

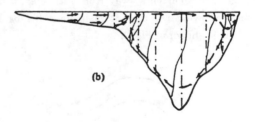

(b)

(a) Lei Section 19 in Xiajing River, China
(b) Section 6 in Fall River, American
Fig. 1   Structure of secondary flows
     in meandering channel

2 INFLUENCE OF THE CHANNEL GEOMETRIC PARMETERS ON THE EXISTENCE OF OUTER BANK CELL

The occurrence or existence of the outer bank cell depends on many parameters of the bend: the bank slope, the curvature of the bend and the width-depth ratio.

2.1 *Bank slope*

Bathurst et al. (1979) noted that if the bank is shelving there is no outer cell. According to Thorne & Rais (1985) studies, one explanation of this is that if the bank slope is small (shelving bank), this ensures that the vertical component V_z is so small that its dynamic effect is negligible.

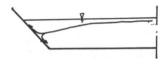

Fig. 2   Bank slope influence

2.2 *Curvature of the bend*

The outer bank cell was observed in sharp bends but not in the weak ones (small curvature). Because, in the latter, the transverse circulation moves the currents away from the outer bank, not allowing the current to impinge the bank. This is shown in Fig. 3.

Fig. 3   Influence of curvature of the bend

2.3 *Width-depth ratio*

In 1957, Rozovskii observed that the wall effect is negligible in the bends with large width-to-depth ratios. Thorne et al. (1985) recognized that, if the bend is wide, the curvature of the bend will direct the currents downstream before they can reach the outer bank. Consequently, there will be no stagnation point to give rise to the two cell configuration.

Stage is also a factor, when the flow is low, there is no outer bank cell effect. One of the reasons for that is that for low stages the water is still in the shelving part of the bank so there is less chance for the occurrence of the outer bank cell. For higher stages, when the flow encounters the steep upper bank, the outer cell does develop. It is shown in Fig. 4.

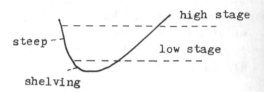

Fig. 4   Stage influence

3 RELATION BETWEEN THE OUTER BANK CELL AND LONGITUDINAL VELOCITY DISTRIBUTION

It is generally observed that the maximum longitudinal velocity is located below the surface near the outer bank. But, the outer bank cell is also produced in this region. It is shown that the depression of the maximum velocity below the water surface could be related to the existence of the outer bank cell. If secondary currents consisted of a single cell, with the velocity directed toward the inner bank near the bottom, and outward near the water surface, the core of the maximum velocity should be at the water surface. If a second cell with an opposite circulation is formed near the surface in the outer bank, this cell tends to shift the velocity maximum downwards to the boundary between the two cells.

The main secondary cell causes an outward

transfer of momentum, and thus an outward shifting of the high velocity core. Momentum associated with the faster surface flow is transmitted below the water surface in the zone of converging surface flow, between main and outer bank cells. This depresses the maximum velocity filament below the water surface, carrying the low velocity fluid upwards to the water surface very close to the outer bank. It should be added that the growth of secondary currents along the bend is first produced by the interaction of the curvature and the vertical gradient of the main velocity. Therefore, the depression of the maximum main velocity makes its vertical gradient changing sign near the water surface so that the secondary currents cells would also change sign there, giving rise to the outer bank cell.

4 ANALYSIS OF THE VORTICITY COMPONENTS

According to the expressions of the three-dimensional flow velocities deduced by the author (1992), the sign of the different components of vorticity derived from the expressions are examined. The expressions of the components of the vorticity vector, in the system of coordinates $(\vec{e_\theta}, \vec{e_r}, \vec{e_z})$ are in general equl to:

$$\tilde{\Omega} = \begin{cases} \tilde{\Omega}_\theta = \dfrac{\partial \tilde{V}_z}{\partial r} - \dfrac{\partial \tilde{V}_r}{\partial z} \\[2mm] \tilde{\Omega}_r = \dfrac{\partial \tilde{V}_\theta}{\partial z} - \dfrac{1}{r}\dfrac{\partial \tilde{V}_z}{\partial \theta} \\[2mm] \tilde{\Omega}_z = \dfrac{1}{r}\dfrac{\partial \tilde{V}_r}{\partial \theta} - \dfrac{\partial (\tilde{r}\tilde{V}_\theta)}{\partial r} \end{cases} \qquad (4.1)$$

for the flow full development, $\partial/\partial\theta=0$, these become:

$$\tilde{\Omega} = \begin{cases} \tilde{\Omega}_\theta = \dfrac{\partial \tilde{V}_z}{\partial r} - \dfrac{\partial \tilde{V}_r}{\partial z} \\[2mm] \tilde{\Omega}_r = \dfrac{\partial \tilde{V}_\theta}{\partial z} \\[2mm] \tilde{\Omega}_z = \dfrac{1}{r}\dfrac{\partial (\tilde{r}\tilde{V}_\theta)}{\partial r} \end{cases} \qquad (4.2)$$

Next, consider these components of vorticity separately.

4.1 Analysis of the radial vorticity component

Consider $\tilde{\Omega}_r = \partial \tilde{V}_\theta / \partial z$. For the profile of V_θ, where the maximum velocity is at the water surface, $\tilde{\Omega}_r$ is positive for all values of z.

Having the maximum velocity of V_θ shifted below the water surface, the situation is shown in Fig. 5, with a negative Ω_r in the upper part of the profile and a positive Ω_r below the point corresponding to the maximum of V_θ (i.e, point at elevation z_m).

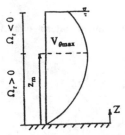

Fig. 5 Sign of Ω_r

4.2 Analysis of the vertical vorticity component

Consider:

$$\tilde{\Omega}_z = \frac{1}{r}\frac{\partial (\tilde{r}\tilde{V}_\theta)}{\partial r}$$

Having expressed V_θ empirically (in dimensionless form) as: $V_\theta = k(z) r^{-m}$, the dimensionless form of $\tilde{\Omega}_z$ can bede duced:

$$\tilde{\Omega}_z = \frac{1}{rR_c}\frac{1}{R_c}\frac{\partial (rR_cV_\theta \overline{V})}{\partial r} = \frac{\overline{V}}{R_c}\frac{1}{r}\frac{\partial (rV_\theta)}{\partial r} \qquad (4.3)$$

Taking $\Omega_0 = \overline{V}/R_c$ as a reference vorticity, the dimensionless vertical vorticity becomes:

$$\Omega_z = \frac{1}{r}\frac{\partial (rV_\theta)}{\partial r} \qquad (4.4)$$

Consequently, Ω_z is deduced from V_θ as:

$$\Omega_z = (-m+1)K(z)r^{-(m+1)} \qquad (4.5)$$

This equation shows that, for a fixed bend, Ω_z

179

(the component of vorticity about the vertical axis) keeps the same sign (despite the shift of the maximum velocity). This is because the signs of the terms in the expression of Ω_z are not affected by the position of the maximum velocity.

4.3 Analysis of the longitudinal vorticity component

Consider:

$$\tilde{\Omega}_\theta = \frac{\partial \tilde{V}_r}{\partial \tilde{z}} - \frac{\partial \tilde{V}_z}{\partial \tilde{r}}$$

Introducing the dimensionless quantities, this becomes:

$$\tilde{\Omega}_\theta = \frac{1}{d}\frac{\partial(\overline{V}\cdot\varepsilon\cdot V_r)}{\partial z} - \frac{1}{R_c}\frac{\partial(\overline{V}\cdot\varepsilon\cdot V_r)}{\partial r}$$

$$\tilde{\Omega}_\theta = \frac{\overline{V}}{d}\cdot\varepsilon\left(\frac{\partial V_r}{\partial z} - q\frac{\partial V_z}{\partial r}\right) \qquad (4.6)$$

Taking again Ω_0 as a reference vorticity the dimensionless longitudinal vorticity is then:

$$\Omega_\theta = \frac{\varepsilon}{q}\left(\frac{\partial V_r}{\partial z} - q\frac{\partial V_z}{\partial r}\right) \qquad (4.7)$$

Substituting the expressions of V_z and V_r as given by the author (Wang 1992) into this, Ω_θ becomes:

$$\tilde{\Omega}_\theta = \frac{3\varepsilon}{(m+1)\ q^2}\cdot$$
$$\{2\beta_3 r(z-z_m)[z_m^2-(z-z_m)^2]^{-2}-Br^{-m}\} \qquad (4.8)$$
$$+\varepsilon(m+1)B'\ r^{-(m+2)}[z_m^2-(z-z_m)^2]$$

In Fig. 6 are represented some graphs of Ω_θ versus z for two values of z_m ($z_m=0.7$ and $z_m=0.9$, with $m=10$, $\varepsilon=q\approx0.1$, and $r=r_{02}=1.30$). These graphs show that Ω_θ takes negative values for elevations below a point slightly smaller than z_m and positive values above that

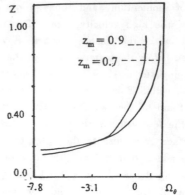

Fig. 6 Variation of Ω_θ along the depth z

point, near the water surface.

From Fig. 6, it is seen that the signs of the longitudinal component of the vorticity are consistent with the sense of rotation of the main and outer bank cells. That is, a positive sense of rotation of the outer bank cell near the water surface and a negative sense for the main cell in the lower part of the cross section as sketched in Fig. 7. It is important to emphasize that the stream channel erodes and deposits material in response to the longitudinal component of vorticity which explains the erosion of the outer bank due to the distribution of Ω_θ along the outer bank. That is the most strong position of outer bank erosion is in the point of intersection (called the stagnation point) between the tangent line of the split point of two cells and the outer bank, and nearby it. In this position, the transverse migration velocity in channel bends is maximum.

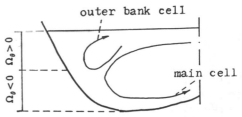

Fig. 7   The relation between the signs of Ω_θ and the sense of rotation of two cells

5 THE MECHANISM OF THE OUTER BANK EROSION

It is discussed just above that Ω_θ takes negation values for elevations below a point slightly smaller than z_m and positive values above that point, near the water surface. Near the stagnation point, the outer bank erosion is most strongly. In fact, boundary shear stress distribution along outer bank proved also the fact.

The distribution given by Chiu and Hsiung (1981) is shown in Fig. 8, which is for a trapezoidal channel of bank slope (1.5 : 1). From Fig. 8, it is seen that there is a maximum of the longitudinal bank shear stress at about mid-height on the bank. This can be explained by the fact that the maximum longitudinal shear stress on the bank is in the zone z_m of the depressed maximum V_θ, and the longitudinal shear stress increases in the zone of downwelling (below z_m) and decreases where there is upwelling. The relation between the distribution of the outer bank erosion and the shear stress distribution is that the outer bank erosion is most strongly in the position of the maximum value of the distribution of the longitudinal shear stress. Therefore, the most strong position of the outer bank erosion is in the stagnation point and nearby it.

According to the above analysis, it can be known that the sense of rotation of the outer bank cell near bank is upward, its motive force affect will give rise to the wash erosion above the stagnation point (called the upper wash). On the position, the sense of rotation of the main cell is downward, its motive force affect will give rise to the wash erosion below the stagnation point (called the down wash). In addition, the position of the maximum value of the longitudinal velocity distribution near bank is corresponding to the stagnation point. Their common affects cause the serious erosion in the stagnation point and nearby it, the bank shape after erosion is parabolic one. This is the general mechanism of the outer bank erosion as sketched in Fig. 9.

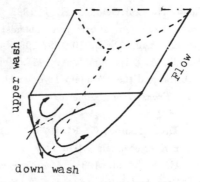

Fig. 9   The mechanism of the outer bank erosion

LIST OF SYMBOLS

V_θ Longitudinal velocity

V_r Radial velocity

V_z Vertical velocity

r   Radial distance from O_c to the point considered in the cross-section

z   The height of a point above bed

m   Power in the expression of V_θ

k   Function of z in V_θ expression

z_m The height of the maximum V_θ

R_c Radius of curvature

q   Ratio of depth of flow to radius of curvature

$\overline{V}$ Mean longitudinal velocity

r_{02} Radial distance of the outer bank

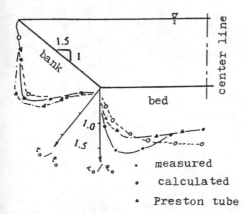

Fig. 8   Distribution of boundary shear in trapezoidal channel

$\vec{\Omega}$ Vorticity vector

Ω_θ Longitudinal component of vorticity

Ω_r Radial component of vorticity

Ω_z Vertical component of vorticity

$\sim$ Characterizes a dimensional variable

β_3 Constants: $\beta_3 = 1/Re \cdot \varepsilon$

ε $\varepsilon = d/R_c$

REFERENCES

Bathurst, J. C. , Thorne, C. R. & Hey, R. D. 1979. Secondary flow and shear stress in river bends. *J. Hydraul. Div.* ASCE. 105 : 997-1007.

Chiu, C. L. & Hsiung, D. E. 1981. Secondary flow, shear stress and sediment transport. *J. Hydraul. Div.* ASCE. 107:1111-1211.

De Vriend, H. J. & Gelodof, H. J. 1983. Main flow velocity in short river bends. *J. Hydraul. Eng.* ASCE. 109:297-299.

Einstein, H. A. & Harder, J. A. 1954. Velocity distribution and the boundary layer at channel bends. *Transactions American Geophysical Union.* 35:1.

Hey, R. D. & Thorne, C. R. 1975. Secondary flows in river channels. *Area* 7:3.

Thorne, C. R. et al. 1985. *Measurements of bend flow hydraulics on the Fall River at bankfull stage.* Colorado State University, Colorado.

Wang, P. Y. 1992. *Flow movement and sediment transport.* Ph. D. Dissertation. Chengdu University of Science and Technology. Chengdu. PR China. (in Chinese)

Zhang Zhitang, et al. 1964. Analysis and study on river bends flow in Xiajing River. *J. Yangtze River.* 2:7-12. (in Chinese)

Stochastic Hydraulics'96, Tickle, Goulter, Xu, Wasimi & Bouchart (eds) © 1996 Balkema, Rotterdam. ISBN 90 5410 817 7

Entropy-based approaches to assessment of monitoring networks

Nilgun B. Harmancioglu, Ahmet Alkan & Necdet Alpaslan
Dokuz Eylul University, Turkey

Vijay P. Singh
Louisiana State University, La., USA

ABSTRACT: With respect to design of water quality monitoring networks, the entropy principle can be effectively used to develop design criteria on the basis of quantitatively expressed information expectations and information availability. The study presented focuses on the applicability of the entropy method in network assessment procedures in the case of a water quality network in a Turkish river basin. The results are discussed with respect to three basic design criteria: sampling points, sampling frequencies and sampling duration.

1 INTRODUCTION

In recent years, the adequacy of collected water quality data and the performance of existing monitoring networks have been seriously evaluated for two basic reasons. First, an efficient information system is required to satisfy the needs of water quality management plans. Second, this system has to be realized under the constraints of limited financial resources, sampling and analysis facilities, and manpower. Problems observed in available data and shortcomings of current networks have led researchers to focus more critically on the design procedures used. Recently, it is observed that several countries have already started to redesign their monitoring systems. The redesign process requires first a sound assessment of existing networks. At this point, the major problem is the selection of an appropriate method for network evaluation purposes.

The methodology proposed herein for development of efficient and cost-effective design of water quality monitoring networks is based on the entropy concept defined in information theory. Fundamental to accomplishment of efficient and cost-effective design is the development of a quantitative definition of "information" and "value of data". Entropy is used to measure the information content of available data and assess the goodness of information transfer between temporal or spatial data points. Maximization of entropy permits the solution of the design questions of what, where, when, and how long to observe. Because the efficiency of a network is a function of the information it yields

for various objectives, the entropy concept serves as the basis for measures of the information altering the efficiency of the network. In a similar vein, the entropy concept is employed as a basis for maximizing the value of data while minimizing the accruing costs of obtaining these data. The design procedure then matches the objectives of monitoring and data needs of each objective under special constraints.

The study presented focuses on the applicability of the entropy method in network assessment procedures within the above described context. The methodology is applied to a water quality network in a Turkish river basin, and the results are discussed with respect to three basic design criteria: sampling points, sampling frequencies and sampling duration.

2 ENTROPY THEORY AS APPLIED TO MONITORING NETWORK DESIGN

Entropy is a measure of the degree of uncertainty of random hydrologic processes. Since the reduction of uncertainty by means of making observations is equal to the amount of information gained, the entropy criterion indirectly measures the information content of a given series of data (Harmancioglu, 1981). According to the entropy concept of communication theory, the term "information content" refers to the capability of signals to create communication. The basic problem is the generation of correct communication by sending a sufficient amount of signals, leading neither to any loss nor to repetition of information (Shannon and Weaver, 1949).

In hydrometric data networks, each sample collected represents a signal from the natural system which has to be deciphered so that the uncertainty about the real system is reduced. Application of engineering principles to this problem calls for a minimum number of signals to be received to obtain the maximum amount of information. Redundant information does not help reduce the uncertainty further; it only increases the costs of obtaining the data. These considerations represent the essence of the field of communications and hold equally true for hydrologic data sampling. On the basis of this analogy, a methodology based on the entropy concept of information theory has been proposed for the design of hydrologic data networks. The basic characteristic of entropy as used in this context is that it is able to represent quantitative measures of "information". As a data collection network is basically an information system, this characteristic is the essential feature required in a monitoring network (Alpaslan et al., 1992; Harmancioglu et al., 1992b).

The entropy measures of information were applied by Krstanovic and Singh (1993) to rainfall network design, by Husain (1989) to design of hydrologic networks, by Harmancioglu and Alpaslan (1992) to design of water quality monitoring networks, and by Goulter and Kusmulyono (1993) to prediction of water quality at discontinued water quality monitoring stations in Australia. In these studies, the entropy concept has been shown to hold significant potential as an objective criterion which can be used in both spatial and temporal design of networks.

With respect to water quality in particular, the entropy principle can be used to evaluate five basic features of a monitoring network: temporal frequency, spatial orientation, combined temporal/spatial frequencies, variables sampled, and sampling duration. The third feature represents an optimum solution with respect to both the time and the space dimensions, considering that an increase in efforts in one dimension may lead to a decrease in those in the other dimension. To determine variables to be sampled, the method can be employed, not to select from a large list of variables but to reduce their number by investigating information transfer between the variables (Harmancioglu et al., 1992a and b). Assessment of sampling duration may be approached in a number of ways. If station discontinuance is the matter of concern, decisions may be made in an approach similar to that applied in spatial orientation. The problem is much simpler when a sampling site is evaluated for the redundancy of information it produces in

the time domain. If no new information is obtained by continuous measurements, sampling may be stopped permanently or temporarily (Harmancioglu, 1994).

The studies carried out so far show that the entropy method works quite well for the assessment of an existing network. It appears as a potential technique when applied to cases where a decision must be made to remove existing observation sites, and/or reduce frequency of observations, and/or terminate sampling program (Harmancioglu and Alpaslan, 1992).

Essentially, the design process is an iterative procedure initiated by the selection of preliminary sampling sites and frequencies. After a certain amount of data is collected, initial decisions are evaluated and revised by statistical methods. It is throughout this iterative process of modifying decisions that the entropy principle works well (Harmancioglu et al., 1994).

3 APPLIED METHODOLOGY

3.1 Definition of entropy for multivariables

The definitions of entropy given in information theory (Shannon and Weaver, 1949) to describe the uncertainty of a single variable can be extended to the case of multiple variables (Harmancioglu, 1981; Harmancioglu and Alpaslan, 1992). The total entropy of M stochastically independent variables X_m (m=1,...,M) is :

$$H(X_1,X_2,...,X_M) = \sum_{m=1}^{M} H(X_m) \qquad (3.1)$$

where $H(X_m)$ represents the marginal entropy of each variable X_m in the form of:

$$H(X_m) = K \sum_{n=1}^{N} p(x_n) \log [1/p(x_n)] \qquad (3.2)$$

with K=1 if $H(X_m)$ is expressed in napiers for logarithms to the base e. Eq.(3.2) defines the entropy of a discrete random variable X_m with N elementary events of probability $P_n = p(x_n)$ (n=1,...,N) (Shannon and Weaver, 1949). For continuous density functions, $p(x_n)$ is approximated as $[f(x_n).\triangle x]$ for small $\triangle x$, where $f(x_n)$ is the relative class frequency and $\triangle x$, the length of class intervals. Then the marginal entropy for an assumed density function $f(x_n)$ is:

184

$$H(X_m; \Delta x) = \int_{-\infty}^{+\infty} f(x) \log [1/f(x)] \, dx + \log [1/\Delta x] \qquad (3.3)$$

If significant stochastic dependence occurs between M variables, the total entropy has to be expressed in terms of conditional entropies $H(X_m/X_1...,X_m)$ added to the marginal entropy of one of the variables:

$$H(X_1,X_2,...,X_M) = H(X_1) + \sum_{m=2}^{M} H(X_m | X_1,...,X_{m-1}) \qquad (3.4)$$

Since entropy is a function of the probability distribution of a process, the multivariate joint and conditional probability distribution functions of M variables need to be determined to compute the above entropies (Harmancioglu, 1981):

$$H(X_1,X_2,...,X_M) = -\int_{-\infty}^{+\infty} ... \int_{-\infty}^{+\infty} f(x_1,...,x_M) \log f(x_1,...,x_M)$$
$$dx_1 \, dx_2...dx_M \qquad (3.5)$$

$$H(X_m | X_1,...,X_{m-1}) = -\int_{-\infty}^{+\infty} ... \int_{-\infty}^{+\infty} f(x_1,...,x_m) \log f(x_m | x_1,...,x_{m-1})$$
$$dx_1 \, dx_2...dx_m \qquad (3.6)$$

The common information between M variables, or the so-called transinformation $T(X_1,...,X_M)$, can be computed as the difference between the total entropy of Eq.(3.1) and the joint entropy of Eq.(3.5). It may also be expressed as the difference between the marginal entropy $H(X_m)$ and the conditional entropy of Eq.(3.6). It follows from above that the stochastic dependence between multi-variables causes their marginal entropies and the joint entropy to be decreased. This feature of the entropy concept can be used in the spatial design of monitoring stations to select appropriate numbers and locations so as to avoid redundant information (Harmancioglu and Alpaslan, 1992).

On the other hand, the marginal entropy of a single process that is serially correlated is less than the uncertainty it would contain if it were independent. If the values that a variable assumes at a certain time t can be estimated by those at times t-1, t-2,..., the process is not completely uncertain because some information can be

gained due to the serial dependence present in the series. In this case, stochastic dependence again acts to reduce entropy and causes a gain in information (Harmancioglu, 1981). This feature is suitable for use in the temporal design of sampling stations. Sampling intervals can be selected so as to reduce the redundant information between successive measurements.

For a single process, the marginal entropy as defined in Eq.(3.3) represents the total uncertainty of the variable without having removed the effect of any serial dependence. However, if the ith value of variable X, or x_i is significantly correlated to values x_{i-k}, k being the time lag, knowledge of these previous values x_{i-k} will make it possible to predict the value of x_i, thereby reducing the marginal entropy of X. To analyze the effect of serial correlation upon marginal entropy, the variable X can be considered to be made up of series X_i, X_{i-1},..., X_{i-k}, each of which represents the sample series for time lags k=0,1,...,K and which obey the same probability distribution function. Then conditional entropies such as $H(X_i | X_{i-1})$, $H(X_i | X_{i-1}, X_{i-2}$,..., $X_{i-k})$ can be calculated. If X_{i-k} (k=1,...,K) are considered as different variables, the problem turns out to be one of the analysis of K+1 dependent multi-variables; thus, formulas similar to Eq.(3.6) can be used to compute the necessary conditional entropies (Harmancioglu, 1981):

$$H(X_i | X_{i-1},...,X_{i-K}) = -\int_{-\infty}^{+\infty} ... \int_{-\infty}^{+\infty} f(x_i,...,x_{i-K}) \log f(x_i | x_{i-1},...,x_{i-K}).$$
$$dx_i...dx_{i-K} \qquad (3.7)$$

For a serially correlated variable, the relation:

$$H(X_i) \geq H(X_i | X_{i-1}) \geq ... \geq H(X_i | X_{i-1},...,X_{i-K}) \qquad (3.8)$$

exists between the variables X_{i-k} (k=0,...,K). Thus, as the degree of serial dependence increases, the marginal entropy of the process will decrease until the condition:

$$H(X_i | X_{i-1},...,X_{i-k}) - H(X_i | X_{i-1},...,X_{i-(k-1)}) \leq \epsilon \qquad (3.9)$$

is met for an infinitesimally small value of ϵ. It is expected that the lag k where the above condition occurs indicates the degree of serial dependence within the analyzed process (Harmancioglu, 1981).

An important step in the computation of any kind of entropy is to determine the type of probability distribution function which best fits the analyzed processes. If a multivariate normal distribution is assumed, the joint entropy of X

185

(the vector of M variables) is obtained as (Harmancioglu, 1981):

$$H(X) = (M/2)\ln 2 + (1/2)\ln |C| + M/2 - M \ln(\triangle x) \quad (3.10)$$

where M is the number of variables and $|C|$ is the determinant of the covariance matrix C. Eq. (3.10) gives a single value for the entropy of M variables. If logarithms of observed values are evaluated by the above formula, the same equation can be used for lognormally distributed variables. In the above formula, the covariance matrix C involves the cross covariances of M different variables. For a single variable, it includes the autocovariances as a measure of the serial dependence within the process and has the dimensions $(K+1)x(K+1)$ (Harmancioglu, 1981).

The calculation of conditional entropies in the multi-variate case can also be realized by Eq.(3.10) as the difference between two joint entropies. For example, the conditional entropy of variable X with respect to two other variables Y and Z can be determined as:

$$H(X|Y,Z) = H(X,Y,Z) - H(Y,Z) \quad (3.11)$$

3.2 Assessment of spatial frequencies

Investigation of spatial frequencies requires the assessment of reduction in the joint entropy of two or more variables due to the presence of stochastic dependence between them. Such reduction is equivalent to the redundant information (transinformation) in the series of the same variable observed at different sites. Thus, the objective in spatial orientation is to minimize the transinformation by an appropriate choice of the number and locations of monitoring stations.

The following procedure is applied to select the best combination of stations with minimum redundancy. First, the marginal entropy of the variable at all stations is computed by Eq.(3.2). The station with the highest entropy is selected as the first priority station to represent the location of the highest uncertainty about the variable. Next, this station is coupled with every other station to select the pair that leads to the least transinformation. The station that fulfills this condition is marked as the second priority location. The same procedure is continued by successively considering combinations of 3, 4, 5,... stations and selecting the combination that produces the least transinformation by satisfying the condition:

$$\min \{H(X_1,...,X_{j-1}) - H(X_1,...,X_{j-1}|X_j)\} =$$
$$\min \{T((X_1,...,X_{j-1}),X_j)\} \quad (3-12)$$

where X_1 is the 1st priority station and X_j is the station with the jth priority. Increasing the number of stations by the above procedure ensures the selection of those locations where the uncertainty about the variable is the highest and redundant information is the lowest.

3.3 Assessment of temporal frequencies

The analysis of temporal frequencies by the entropy method is based on the assessment of reduction in the marginal entropy of a process due to presence of serial dependence. Such a reduction, if any, is equivalent to the redundant information of successive measurements. This analysis has to be carried out separately at each sampling site for the particular variable analyzed. The marginal entropy of the variable is computed by Eq. (3.2). Next, successive reductions in uncertainty at time lags k=1,2,3,... are determined by Eq. (3.7) and transinformations by:

$$T(X_i,...,X_{i-k}) = H(X_i) - H(X_i|X_{i-1},...,X_{i-k}) \quad (3.13)$$

In the above, time lags k=1,2,3,... refer respectively to sampling intervals of $\triangle t=2,3,4,..$ time units. Accordingly, if the redundant information between successive observations, $T(X_i,...,X_{i-k})$, or the ratio $T(X_i,...,X_{i-k}) / H(X_i,...,X_{i-k-1})$ is found to be high at any lag k, the frequency of sampling may be decreased to reduce such redundancy.

3.4 Assessment of sampling duration

The definition given in Eq.(3.2) for the marginal entropy of X involves the term "log $(1/\triangle x)$" which essentially describes a reference level of uncertainty according to which the entropy of the process is evaluated. In this case, the entropy function assumes values relative to the reference level described by "log $(1/\triangle x)$" (Harmancioglu, 1994). On the other hand, some researchers propose the use of a function m(x) such that the marginal entropy of a continuous variable is expressed as:

$$H(X) = - \int_{-\infty}^{+\infty} f(x) \ln f(x)/m(x)] dx \quad (3.14)$$

where m(x) is often considered to be a priori probability distribution function (Harmancioglu et al., 1992b). If a uniform distribution is assumed to be the prior distribution used to describe maximum uncertainty about the process prior to making any observations, then the marginal entropy of Eq.(3.2) becomes:

$$H(X) = \int_{-\infty}^{+\infty} f(x) \log [1/f(x)] \, dx - \log[N] \qquad (3.15)$$

where "log N" becomes the reference level of uncertainty. Another property of this term is that, for a record of N observations, "log N" represents the upper limit of the entropy function (Harmancioglu, 1994). Using this property and rearranging Eq.(3.16), one may write:

$$H'(X) = \log[N] - \int_{-\infty}^{+\infty} f(x) \log [1/f(x)] \, dx \qquad (3.16)$$

where H'(X) now describes the change in information conveyed by data compared to the reference level of "log N". Here, the absolute value of the H'(X) function has to be considered as it defines the difference between the upper limit "log N" and the information provided by data of N records. Since "log N" represents the maximum amount of uncertainty, H'(X) measures the amount of information gained by making observations. This amount, or the rate of change of information conveyed, changes as N is increased. The particular point where H'(X) stabilizes or remains constant with respect to N indicates that no new information is gained by additional observations. If this point is reached within the available record of observations, then one may decide to discontinue sampling, as further observations will not bring further information (Harmancioglu, 1994).

4 APPLICATION

The above methodology is applied to the case of Porsuk river basin in Turkey, where the water quality monitoring network comprises 18 stations. The total available record at 10 of these sampling locations cover a period of 10 years between 1979-1989 with monthly observed values of 40 variables. The remaining stations have shorter records and highly sporadic observations. Essentially, almost all data series include gaps of about 4-5 months each year so that all evaluations based on entropy are carried out on such "messy" data. The results of the application are demonstrated below for those stations and variables which have relatively complete observations within the period of record.

The analysis of sampling locations is applied to available DO, EC, Cl⁻, SS, BOD5 and NH_3-N data from 18 sampling stations. All variables are assumed to be lognormally distributed except for DO where the normal distribution gives a better fit. Joint entropies are computed by Eq.(3.10) for M=2,...,18, which can be used to determine the conditional entropy by Eq.(3.11). The number of stations is increased by fulfilling the condition in Eq.(3.12). Next, transinformations are computed for M=2,...,18. For each variable, the joint entropy of simultaneous observations at all stations represents the total amount of uncertainty that may be reduced by observations at each station. Increasing the number of stations contributes to this reduction so that the total uncertainty is decreased.

Results of such computations are shown in Figs. 1 and 2. For DO of Fig.1, only 10 stations are sufficient to reduce the total uncertainty about the variable, which is represented by their joint entropy. The remaining 8 stations appear to produce redundant information. With 10 stations, the percentage of redundant information (or the ratio of total transinformation to joint uncertainty of 10 stations) remains as low as 7%. Figure 1 also indicates the case when lognormal distribution is assumed for DO. It is evident here that the choice of an appropriate distribution function is significant for reliable evaluations since different entropy levels are obtained for different distributions.

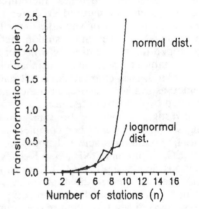

Figure 1. Increases in redundant information for DO with respect to increases in the number of stations in Porsuk network.

Figure 2 shows the results for the other variables, where it is observed that 10 stations are sufficient to describe the total uncertainty about Cl⁻, EC, and NH_3-N. The percentages of redundant information for these variables are 6%, 6%, and 1%, respectively. For SS, 11 sampling locations produce 5% of redundant information; whereas the same number of stations for BOD5 reflect 3% of repetitive information which increases to 12% by the addition of the 12th station.

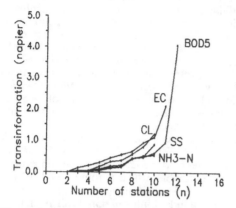

Figure 2. Increases in redundant information for other variables with respect to increases in the number of stations in Porsuk network.

In the above computations, it is observed that the priority list of sampling locations may change from variable to variable. Another observation is that the stations which appear to produce redundant information and which, therefore, are rejected by the entropy method are often those that have large amounts of sporadic data. Thus, it is questionable whether these stations actually produce repetitive information or whether it is the "messy" nature of their data series that lead to this result. Thus, such computations will have to be repeated as available records at these stations are extended by further observations.

To investigate temporal design features, sampling stations with a rather complete set of water quality data are selected for DO, Cl^-, BOD5, EC, SS, and NH_3-N. Results of computations defined by Eqs.(3.7) through (3.9) have indicated that even the first order serial dependence within the analyzed processes are pretty low. Accordingly, reduction of the sampling frequency from monthly to bimonthly observations causes a loss of information in the order of at least 90%. This loss increases further at larger sampling intervals. Figure 3 shows this result for three variables at selected stations, where the highest percentage of redundant information is in the order of 10% for DO at $\triangle t=2$ months. This percentage is reduced further as the sampling frequency is reduced. Thus, it is concluded here that the current practice of monthly observations should be continued.

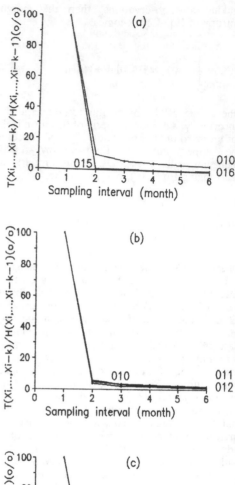

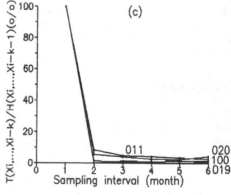

Figure 3. Effects of temporal frequencies upon information gain about: (a) DO; (b) Cl^-; (c) SS (The numbers adjacent to curves indicate station codes).

For the assessment of sampling duration, H'(X) values are computed by Eq.(3.16) with a

10-year data record for different variables at a station which has the most complete data series. As shown in Fig.4, the change of H'(X) with respect to sampling duration of N years has indicated that the rate of information gain is high within the first couple of years of sampling. Although such a trend relatively slows down as N increases, no point of stabilization has yet been reached for any of the variables. This result indicates the need for continuation of sampling beyond 1989 which marks the end of records used in the presented study.

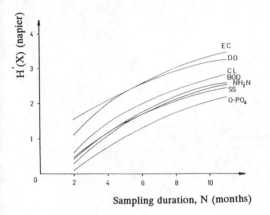

Figure 4. The rate of change of information gain about Porsuk variables with respect to sampling duration.

5 CONCLUSION

The work presented shows an application of the entropy principle to cases where a decision must be made to remove existing observation sites, and/or reduce frequency of observations, and/or terminate sampling program. Application of the entropy principle in assessing the technical (temporal, spatial, and combined temporal/spatial) features of an existing network provides promising results because the benefits of alternative monitoring practices can be evaluated quantitatively as a function of information.

On the other hand, some limitations of the method must be mentioned. With respect to the time dimension, all evaluations are based on the temporal frequencies of available data. Thus, the method inevitably appears to be data dependent. However, the problem of decreasing the sampling intervals may also be investigated by the entropy concept provided that the available monthly data are reliably disaggregated into short interval series. This aspect of entropy applications has to be investigated in future research.

Furthermore, some numerical difficulties are encountered in application of the method. These are essentially related to the properties of the covariance matrix. When the determinant of the matrix is too small, entropy measures cannot be determined reliably since the matrix becomes ill-conditioned. This often occurs when the available sample sizes are very small.

Another point is that the method also requires the assumption of a valid distribution-type. It works quite well with multivariate normal and lognormal distributions. The mathematical definition of entropy is easily developed for other skewed distributions in bivariate cases. However, the computational procedure becomes much more difficult when their multivariate distributions are considered.

One problem that has to be considered in future research is the mathematical definition of entropy concepts for continuous variables. Shannon's entropy definition is developed for discrete random variables, and the extension of this definition to the continuous case entails the problem of selecting the class intervals $\triangle x$ as in Eq.(3.3). Marginal and joint entropies vary with $\triangle x$ such that each selected $\triangle x$ constitutes a different base level or scale for measuring uncertainty. This problem does not apply to determination of transinformation since it is computed as differences of marginal, joint and conditional entropies.

Despite the above problems, the work presented shows that the entropy principle is a promising method in hydrologic network design problems since it can be used to develop design criteria on the basis of quantitatively expressed information availability. Some mathematical difficulties pointed out in the paper need to be solved as part of future research efforts.

REFERENCES

Alpaslan, N.; Harmancioglu, N.B.; Singh, V.P. 1992. "The role of the entropy concept in design and evaluation of water quality monitoring networks", in: V.P. Singh & M. Fiorentino (eds.), *Entropy and Energy Dissipation in Water Resources*, Dordecht, Kluwer Academic Publishers, Water Science and Technology Library, pp.261-282.

Goulter, I. and Kusmulyono, A. 1993. "Entropy theory to identify water quality violators in environmental management", in: R.Chowdhury and M. Sivakumar (eds.),*Geo-Water and Engineering Aspects*, Balkema Press, Rotterdam, pp.149-154.

Harmancioglu, N. 1981. "Measuring the information content of hydrological processes by the entropy concept", Centennial of Ataturk's Birth, *Journal of Civil Engineering*, Ege University, Faculty of Engineering, pp.13-38.

Harmancioglu, N. 1994. "An Entropy-based approach to station discontinuance", in: *Stochastic and Statistical Methods in Hydrology and Environmental Engineering, vol.3: Time Series Analysis and Forecasting* (eds: K.W.Hipel & I. McLeod, Kluwer Academic Publishers, Dordrecht, the Netherlands, pp.163-176.

Harmancioglu, N.B., Alpaslan, N. 1992. "Water quality monitoring network design: a problem of multi-objective decision making", AWRA, *Water Resources Bulletin*, Special Issue on "*Multiple-Objective Decision Making in Water Resources*", vol.28, no.1, pp.1-14.

Harmancioglu, N.B.; Singh, V.P.; Alpaslan, N. 1992a. "Versatile uses of the entropy concept in water resources", in: V.P. Singh & M. Fiorentino (eds.), *Entropy and Energy Dissipation in Water Resources*, Dordecht, Kluwer Academic Publishers, Water Science and Technology Library, pp.91-117.

Harmancioglu, N.B.; Singh, V.P.; Alpaslan, N. 1992b. "Design of Water Quality Monitoring Networks", in: R.N. Chowdhury (ed.), *Geomechanics and Water Engineering in Environmental Management*, Rotterdam, Balkema Publishers, ch.8.

Harmancioglu, N.B.; Alpaslan, N.; Singh, V.P. 1994. "Assessment of the entropy principle as applied to water quality monitoring network design", in: *Stochastic and Statistical Methods in Hydrology and Environmental Engineering, vol.3: Time Series Analysis and Forecasting* (eds: K.W.Hipel & I. McLeod, Kluwer Academic Publishers, Dordrecht, the Netherlands, pp.135-148.

Husain, T. 1989. "Hydrologic uncertainty measure and network design", *Water Resources Bulletin*, 25(3), 527-534.

Krstanovic, P.F. and Singh, V.P. 1993a. "Evaluation of rainfall networks using entropy:I.Theoretical development, II.Application", *Water Resources Management*, v.6,pp.279-314.

Shannon, C.E. and Weaver, W. 1949. *The Mathematical Theory of Communication*, The University of Illinois Press, Urbana, Illinois.

Stochastic Hydraulics'96, Tickle, Goulter, Xu, Wasimi & Bouchart (eds) © 1996 Balkema, Rotterdam. ISBN 90 5410 817 7

The influence of the turbulent shear stress on the initiation of sediment motion in an open channel flow

Alireza Keshavarzy & James E. Ball
Department of Water Engineering, School of Civil Engineering, The University of New South Wales, Sydney, N.S.W., Australia

ABSTRACT: The initiation of sediment motion is an important component in the study of sediment transport. Most existing critical shear stress models used in sediment transport models are based on an time-averaged channel shear stress, which is defined in term of depth, density of the flow and energy gradient, whereas the particles on the bed sustain instantaneous shear stresses which differ from the temporal average shear stress. These differences are derived from the turbulent nature of the flow. Since particle entrainment into the flow occurs as a result of the applied shear stress at the bed, it is necessary to consider the temporal variation in the shear stress in the development of a sediment entrainment model. Presented herein are the results obtained from a study on the effect of flow turbulence on the entrainment of sediment particles from the bed of an open channel. The turbulence characteristics of the flow were measured in a flume with a rough bed under differing flow conditions. The instantaneous turbulent shear stresses were applied in a simple force balance model to define an entrainment function for the prediction of incipient sediment motion. It was found that particle entrainment but not general motion occurred at a lower critical shear stress than that predicted using an average channel shear stress model.

1. INTRODUCTION

The bed load sediment in a river comprises those particles which travel downstream more slowly than the flow whereas suspended sediment are those particles which travel downstream with the flow. Bed load sediment in a river channel moves in a downstream direction by the rolling, sliding and saltation of the sediment particles. The continuing motion of these particles requires the application of forces which exceed the critical force for that particle which is defined as the force required to induce motion of the particle. An estimation of the critical force is obtained the resistance force or the immersed weight of the particle in the initial condition of motion. When the applied force exceeds the resistance force, the particle commences movement along the bed; this movement may be by rolling, sliding or saltating depending on the magnitude of the available shear stress. For situations near the threshold of movement, *i.e.* the initiation of motion, rolling is the dominant mode of bed load transport for sand and gravel particles.

Raudkivi (1990) observed that, particularly where the shear stress was very small or close to zero, the particles were entrained by the influence of turbulent agitation. In a similar manner, Graf (1971) pointed out that incipient motion of similar sized particles under a given flow condition was statistical in nature due to the turbulence of the flow. The studies of Cheng and Clyde (1971), Christensen (1972), Blinco and Simon (1974), and Grass (1970, 1982) have found that fluctuations of the instantaneous shear stress about the temporal mean are the result of the flow turbulence. This turbulence of the flow is also the main mechanism resulting in the entrainment of particles from the bed. Furthermore, Grass (1982) and Thorne *et al.* (1989) noted that the mode and rate of sediment transport changes as a function of turbulence.

Due to the temporally variable nature of turbulence, there is a difficulty in defining initiation of motion for a particle resting on the channel bed. This arises from the need to consider the instantaneous turbulent shear stress which is sometimes lower and sometimes higher than the critical shear stress for that particle. Schober (1989) indicated that even for a plane bed there are difficulties in defining incipient motion because of the random interference of particle stability and flow behaviour. Similar conclusions have been noted by, for example, Chiew and Parker (1994) who pointed out the difficulty of sediment entrainment in open channel flow, and Ball and Keshavarzy (1995) who discussed the difficulty of defining incipient motion of sediment particles. The mechanism of the bursting process which is a process or a mechanism by which momentum is transferred into the boundary layer and its influence on the entrainment and movement of particles has been a focus of many previous studies. The bursting process was introduced by Kline et al. (1967) as a process which consists of four categories of events; these categories are the sweep $(u'>0, v'<0)$, ejection $(u'<0, v'>0)$, outward interaction $(u'>0, v'>0)$ and inward interaction $(u'<0, v'<0)$ with each event having a different phase of action. Where u' and v' are velocity fluctuations about temporal average in horizontal and vertical directions, respectively. Bridge and Bennet (1992) noted that these four alternative types of bursting events have different effects on the mode and rate of sediment transport. Studies by Thorne et al. (1989), William (1990), Nelson et al. (1995), Bennet and Best (1995) and Drake et al. (1988) indicate that close to the bed most of the sediment entrainment occurs during the sweep event; this is particularly the case for coarse sands and gravel sized particles. Shown in Figure 1 is a schematic representation of how the sweep event applies force towards the bed and induces motion of sediment particles.

In this study, the instantaneous shear stress of the flow was applied in a force balance model in order to develop a stochastic entrainment function that included the influence of turbulence.

2. INITIATION OF SEDIMENT MOTION

Flow in most naturally occurring channels varies temporally. Consequently, the sediment particles on the bed of a channel will be subjected to periods where they will be in motion and periods where they will be stationary. Long-term analysis of sediment motion, therefore, requires investigation of the interface between the periods of sediment motion and the periods where the sediment is stationary. This requires the investigation of the initiation of sediment motion.

Yalin (1972) states that in steady uniform flow of water and sediment particles seven basic parameters are needed to define the flow conditions; these parameters are:

> density of water (ρ),
> density of sediment (ρ_s),
> dynamic viscosity (v),
> particle size (D),
> flow depth (H),
> channel slope (S), and
> acceleration of gravity (g).

These seven basic parameters can be reduced to a set of three dimensionless parameters which are:

> a mobility parameter $u*^2/(s-1)gD$,
> a particle Reynold's number Re=$u*D/v$ and
> a specific density parameter ρ_s/ρ.

The forces applied to a sediment particle on the bed of a horizontal plane are the tractive shear stress, a resistance force and a horizontal drag force due to a pressure differential between the upstream and downstream sides of the particle. This drag force induces an agitation force for coarse sand and gravel

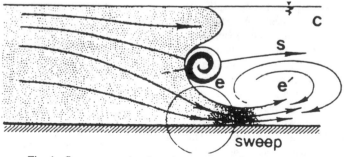

Fig. 1: Sweep event in a bursting process (after Yalin, 1992)

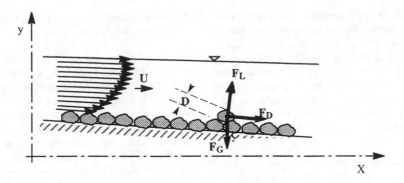

Fig. 2. Applied Forces on Sediment Particles at the bed

bed sediments which causes particles to move by rolling and sliding.

From theoretical considerations, the forces influencing entrainment into the flow of sand and gravel from the bed of an open channel are the drag force and the resistance force. These applied forces are shown in Figure 2. The magnitude of the agitating force in the form of a drag force is given by

$$F_D = \frac{1}{2} C_D \rho A v_r^2 \qquad (1)$$

while the resistance force, or the gravity force, is given by

$$F_r = F_g \times \beta \qquad (2)$$

where;

$$F_g = \frac{1}{6} \pi D^3 (\rho_s - \rho) g \qquad (3)$$

where; F_D is the drag force, F_r is the resistance force, F_g is the gravitational force or weight of the sediment particle, β is friction coefficient C_D is the drag coefficient for the sediment particle, A_p is the cross sectional area of the grain exposed to drag which is given by $1/4\pi D^2$, ρ is the density of water ρ_s is the density of sediment particle, D is the particle diameter, g is the acceleration due to gravity and Vr is relative velocity which is given by

$$V_r = (u* - u_p) \qquad (4)$$

In which $u*$ is the shear velocity and u_p is the particle velocity. Figure 2 shows the particles and applied forces on the bed particles.

Engelund and Fredsoe (1976) introduced a model for sediment transport model. They used a force balance model for estimation of bed load and particle velocity at the bed of open channel. This model is given as follow;

$$\frac{1}{2} C_D \rho A_p (\alpha u_* - u_p)^2 = \frac{\pi}{6} d^3 (\rho_s - \rho) g \cdot \beta$$
$$(5)$$

where $u*$ is average shear velocity, u_p is the average particle velocity, $\beta = \tan\phi$ ($\phi = 27°$) is a dynamic friction coefficient, α is a coefficient which is equal to 6-10 and C_D is drag coefficient and equal to 0.6

3. INFLUENCE OF TURBULENCE

The initiation of motion for sediment particle resting at the bed is controlled by the applied instantaneous shear stress at the bed. These forces are generated by bursting processes and impinge on the boundary layer. The most important event for sediment motion in the bursting process is the sweep event which has a dominant role in entrainment of sediment particles at the bed. The sweep event applies shear in the direction of the flow and provides additional forces to the viscous shear stress. Keshavarzy and Ball (1995) reported that the magnitude of instantaneous shear stress in sweep event is much larger than temporal mean shear stress.

During a sweep event, the impact of the flow velocity at the bed causes an initiation of particle motion due to the applied instantaneous force on the particle. The applied shear stress resulting from a sweep event depends on the magnitude of the turbulent shear stress and its probability distribution.

This magnitude of the shear stress in a sweep event is particularly important close to the bed. In Figure 3 it is shown that how the magnitude of normalised instantaneous shear stress in a sweep event varies in different point in depth (d) which is normalised by total depth of flow (H). The magnitude of the instantaneous shear stress in a sweep event was found to be approximately 1.4 times of the temporal mean shear stress for a region close to the bed.

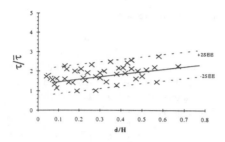

Fig. 3. Instantaneous shear stress in sweep event

If the instantaneous shear stress in the sweep event is τ and the time averaged shear stress at each point of flow is $\bar{\tau}$, the ratio of shear stress at a point of flow for sweep event is:

$$\frac{\tau}{\bar{\tau}} = a, \quad \text{or} \quad \tau = a.\bar{\tau} \tag{6}$$

and

$$\tau = \bar{\tau} + \tau' \tag{7}$$

where τ' is turbulent shear stress and is given by:

$$\tau' = \rho.u_*'^2 \tag{8}.$$

Therefore, the value of turbulent shear stress is calculated as follow;

$$\tau' = \tau - \bar{\tau} \tag{9}$$

If equation 6 is substituted in equation 9, the turbulent shear stress is calculated by

$$\tau' = \bar{\tau}(a-1) \tag{10}.$$

At the bed the time-averaged shear velocity is given as

$$u_* = (\tau_o/\rho)^{0.5} = (gRS_f)^{0.5} \tag{11}$$

where u_* is the shear velocity, τ_o is the shear stress at the bed, ρ is the flow density, g is the

gravitational coefficient, R is hydraulic radius and S_f is energy gradient of the flow.

The magnitude of the shear stress also depends on the impinging angle towards the bed. If the flow velocity in a sweep event is more tangential with the bed, then the applied force on particle movement is more effective in entrainment. If the velocity in the sweep event tends to be normal to the bed, the applied force for sediment entrainment has a minimum effect on entrainment. Thus the angle of velocity in sweep event has an important influence on sediment entrainment. It is shown in Figure 4 how the angle of the sweep event influences the applied force on a sediment particle.

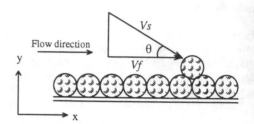

Fig. 4. Inclination angle of the sweep event

The angle of sweep events (θ) was calculated from the experimental data. It was found to be about 22° from the horizontal for a region very close to the bed. The variation of this angle with flow depth is shown in Figure 5.

Another important parameter in the entrainment of sediment particles is the probability of a sweep event (p). The probability of occurrence of sweep event was calculated from the experimental data and is shown in Figure 6. It was found that the frequency

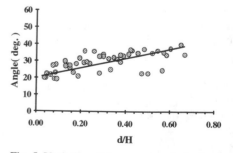

Fig. 5. Variation of the sweep angle in depth

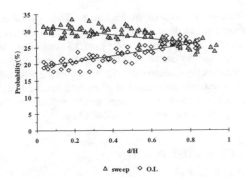

Fig. 6. frequency of Sweep event in bursting process

of sweep events was 30 percent of the total time for bursting process.

The effect of instantaneous turbulent shear stress is investigated as those parameters are applied in a force balance model in order to define the initiation of sediment particles with the influence of turbulence. This model is based on the following relationships which include the effects of the instantaneous shear stress.

$$\frac{1}{2}C_D \rho A_p (\alpha \hat{u}_* - \hat{u}_p)^2 = \frac{\pi}{6}d^3\left(\rho_S - \rho\right)g.\beta \quad (12)$$

$$k_1\left(\alpha \hat{u}_* - \hat{u}_p\right)^2 = k_2 \quad (13)$$

$$\left(\alpha \hat{u}_* - \hat{u}_p\right)^2 = k_2 / k_1 \quad (14)$$

$$\hat{u}_* = u_* + u_*'' \quad (15)$$

$$u_*'' = (\sqrt{a-1}.\cos\theta. p)u_* \quad (16)$$

where; $\hat{u}_*$ is the total instantaneous shear velocity, $\hat{u}_p$ is the particle velocity along the bed, u_*'' is the instantaneous sweep shear velocity, θ is the sweep angle and p is the probability of occurring sweep event in a bursting process.

The values of k_1 and k_2 in equation 13 and 14 are defined as;

$$k_1 = \frac{1}{2}C_D \rho A_p \quad (17)$$

$$k_2 = \frac{\pi}{6}d^3\left(\rho_S - \rho\right)g.\beta$$

$$k_2 / k_1 = \frac{4(S-1)gd\beta}{3C_D} \quad (18)$$

$$\text{or } k_2 / k_1 = \frac{4u^2_* \beta}{3\lambda.C_D} \quad (19)$$

in which, $\lambda = $ shield's parameter $= \dfrac{U_*^2}{(S-1)gd}$

4. EXPERIMENTAL DETAIL

Experimental tests for the analysis described above were carried out in a sediment transport flume of 30 m length, 0.60 m width and 0.60 m height at the Water Research Laboratory, UNSW. The velocity components in two directions of the flow were measured using a very small electromagnetic velocity meter. The bed of the flume was covered by concrete sheet and a sediment particles with $D_{50}=2$ mm size diameter. More detail of the experimental equipment are described by Keshavarzy and Ball (1995). The velocity components were recorded for subsequent analysis to calculate time averaged velocity in two directions and the instantaneous shear stress for bursting events. The turbulence intensity of the flow determined and compared with previous data.

5. RESULTS AND DISCUSSIONS

From the experimental study of the turbulence it was found that the magnitude of the shear stress in a sweep event is much higher than that of time-averaged shear stress in a point of the flow. The magnitude, probability, and angle of instantaneous shear velocity in the sweep event are reported by Keshavarzy and Ball (1995, 1996a). These bursting parameters influence the particle entrainment in the bed of an open channel flow. The magnitude and

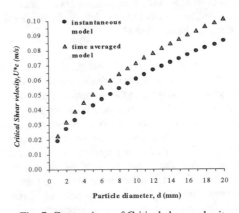

Fig. 7. Comparison of Critical shear velocity

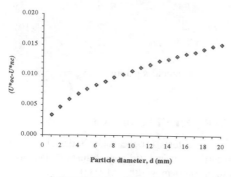

Fig. 8. Differences in critical shear velocity

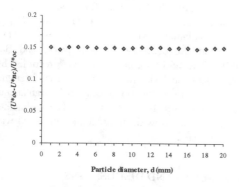

Fig.9. normalised differences for two models

probability of the instantaneous shear stress is the most important factor for particle entrainment in a mobile bed. Those parameters were derived by experimental data in a flume over a rough bed. The instantaneous applied forces on the particles influence on the estimation of sediment rate in open channel. Those turbulence parameters are applied in a force balance model to calculate the actual particle velocity in the mobile bed.

After solving equations 12 and 5, the velocity of particles for different shear velocity for two different cases (with and without the influence of turbulence) were calculated. The magnitude of critical shear velocity for instantaneous shear stress and time-averaged shear stress were compared as shown in Figure 7. For a particular particle diameter, it could be seen that the shear velocity for initiation of particle motion in proposed model is lower than time averaged model. For different particle diameter (1mm-20mm) these difference between critical shear

velocity are shown in Figure 8. The normalised difference in shear stress is shown in Figure 9. As a result, the initiation of particle influence by instantaneous shear stress could start earlier than time average shear stress. This agitation which occurred into the initiation of motion caused is due to the turbulent shear stress which imposed to the particles on the bed by turbulence.

6. CONCLUSION

The influence of turbulence on the entrainment of sediment particle from the bed was investigated through an analysis of the bursting process and in particular the sweep event as applied in a force balance model. The effect of sweep in the entrainment of sands and particles has been reported by many researchers, however no attempt was made to predict the actual sediment rate with the influence of turbulence. In this study the influence of sweep event on initiation of motion resulted an indication of sediment motion with the effect turbulence. It was found that due to instantaneous turbulent shear stress in sweep event particles start to move earlier than predicted by time averaged shear stress models.

7. REFERENCES

Ball, J.E. and Keshavarzy, A., 1995. Discussion on Incipient Sediment motion on non-horizontal slopes by Chiew and Parker. *J. of Hydraulics Research*, 33(5):723-724.

Bennett, S.J. and Best, J.L., 1995. Mean flow and turbulence structure over fixed, two dimensional dunes: Implications for sediment transport and bed form stability. *Sedimentology*, 42:491-513.

Blinco, P.H. and Partheniades, E., 1971. Turbulence Characteristics in Free Surface Flows Over Smooth and Rough Boundaries. *J. of Hydraulics Research*, 9(1):43-71.

Blinco, P.H. and Simon, D.B., 1974. Characteristics of Turbulent Boundary Shear Stress, ASCE, *J. of Eng. Mech. Div.*, 100:203-220.

Bridge, J.S., 1981. Hydraulics interpretation of grain-size distributions using a physical model for bed load transport, *J. of Sediment Petrology*, 51(4):1109-1124.

Bridge, J.S. and Bennet, S.J., 1992. A Model for the Entrainment and Transport of Sediment Grains of Mixed Sizes, Shapes and Densities, *Water Resources Research*, 28(2):337-363.

Cheng, E.D.H. and Clyde C.G., 1971. Instantaneous

hydrodynamic lift and drag forces on large roughness elements in turbulent open channel flow, in *Sedimentation* edited by Shen E.W., pp. [3-1]-[3-20], Fort Colin, Colo.

Drake, T.G., Shreve, R.L., Dietrich, W.E., Whiting, P.J. and Leopold, L.B., 1988. Bed load transport of fine gravel observed by motion-picture photography, *J. of Fluid Mechanics* 192:193-217.

Engelund, F. and Fredsøe, J., 1976. A Sediment Transport Model for Straight Alluvial Channels, *Nordic Hydrology*, 7:293-306.

Graf, W.H., 1971. *Hydraulics of Sediment Transport,* McGraw-Hill, New York.

Grass, A.J., 1971. Structural features of turbulent flow over smooth and rough boundaries, *J. of Fluid Mechanics*, 50(2):233-255.

Grass, A.J., 1982. The influence of boundary layer turbulence on the mechanics of sediment transport, Euromech 156, *Mechanics of Sediment Transport*, Balkema, Rotterdam, .

Keshavarzy, A. and Ball, J.E., 1995. Instantaneous shear stress on the bed in a turbulent open channel flow, *Proc. of XXVI IAHR Congress*, London.

Keshavarzy, A. and Ball J.E., 1996a. Characteristics of turbulent shear stress in open channel, accepted, *7th IAHR International Symposium on Stochasic Hydraulics*, Mackay, QLD, Australia.

Nelsen, J.M., Shreve, R.L., Mclean, S.R. and Drake, T.G., 1995. Role of near-bed turbulence structure in bed load transport and bed form mechanics, *Water Resources Research*. 31(8):2071-2086.

Raudkivi, A.J., 1990. *Loose Boundary Hydraulics*, 3rd Ed., Pergamon Press, Oxford.

Thorne, P.D., Williams, J.J. and Heathershaw, A.D. 1989. In situ acoustic measurements of marine gravel threshold and transport, *Sedimentology*, 36:61-74.

Williams, J.J., 1990. Video observations of marine gravel transport, *Geo. Mar. Lett.*, 10:157-164.

Yalin, M.S., 1972. *Mechanics of Sediment Transport*, Pergamon Press, Oxford.

Yalin, M.S., 1992. *River Mechanics*. Pergamon Press, Oxford.

Stochastic Hydraulics'96, Tickle, Goulter, Xu, Wasimi & Bouchart (eds)© 1996 Balkema, Rotterdam. ISBN 90 5410 817 7

A study of particle motion in turbulent layer boundary flows

Qingchuan Zeng
Department of Civil Engineering, University of Queensland, Brisbane, Qld, Australia

Xiaobing Liu
Department of Power Engineering, Sichuan Institute of Technology, Chengdu, China

Li Guo
Department of Electrical Engineering, Huazhong University of Science & Technology, Wuhan, Hubei, China

ABSTRACT: An analytical solution of the Lagrangian equation of motion for a particle in turbulent boundary layer flows has been presented. The equation include the Saffman lift force and additional drag force created by the wall presence. With the spectral method, analytical expressions relating the components of particle response statistics to those of flow field are developed. The particle's Lagrangian power spectra, autocorrelation function, turbulent Reynolds stresses, turbulent diffusivities are evaluated.

1 INTRODUCTION

Surface erosion of material by solid-particle impact is an important problem in multiphase flow industrial devices and the characteristics of the particle motion in a turbulent boundary layer flow is the base of the study of the material surface erosion. Many computational models have been presented in order to study the particle motion in a turbulent flow. The first study of solid particle motion near a wall was carried out by Soo and Tian (1960). Rouhiainen and Stachiewicz (1970) discussed the importance of the Saffman (1965, 1968) lift force in the viscous sublayer in their study of the deposition of small particle from turbulent streams. Their equation of motion of the particles included this lift force in the direction normal to the wall and neglected the wall effect on the drag forces of the particle. Rizk and Elghobshi (1985) analyzed motions of particle suspended in a turbulent channel flow by including the wall effects. Maclaughlin (1989) studied motions of small spherical particles in a numerically simulated wall bounded turbulent shear flow. The Lagrangian equations of motion of small solid particle suspended in two-dimensional turbulent boundary layer flows is modified in this study. The effects of lift force and additional drag force caused by the wall presence is considered. Then we apply a Fourier transform to the modified equations to obtain a solution for the particle motion. Using the spectral method, analytical expressions relating the Lagrangian power spectra of the particle velocity to that of the fluid are developed and the results are used to study various response statistics. In this paper, our results clearly show that the diffusivity of the particle may be larger than that of the fluid for period of long-time.

2 EQUATIONS OF PARTICLE MOTION

2.1 *Unbounded free flow*

Tchen (1947) extended the equation for the slow motion of a spherical particle in a stagnant fluid derived by Basset (1888), Bousineseq (1903) and Oseen (1927) to the case of a particle suspended in an unsteady velocity field. Here, the modified Tchen's equation of motion for a solid particle in an arbitrary flow field by the writers (Liu and Zeng,1994) in the i direction is given as

$$\frac{dV_{pi}^*}{dt^*} = \frac{1}{K_m + S} \left[\frac{3C_D}{4d^*} \left| \vec{V}_f^* - \vec{V}_p^* \right| (V_{fi}^* - V_{pi}^*) \right]$$
$$+ \frac{3K_B}{2d^*\sqrt{\pi Re_L}} \int_{-\infty}^{t^*} \left(\frac{dV_{fi}^*}{d\tau^*} - \frac{dV_{pi}^*}{d\tau^*} \right) \frac{d\tau^*}{\sqrt{t^* - \tau^*}}$$
$$+ \frac{3}{4} C_M \Omega_i^* \left(V_{fi}^* - V_{pi}^* \right)$$
$$+ \frac{6K_s}{\pi d^*\sqrt{Re_L}} (V_{fj}^* - V_{pj}^*) \left| \frac{\partial V_{fj}^*}{\partial x_i^*} \right|^{1/2} Sign \left(\frac{\partial V_{fj}^*}{\partial x_i^*} \right)$$
$$K_m \frac{dV_{fi}^*}{dt^*} + \frac{DV_{fi}^*}{Dt^*} - \frac{1}{Re_L} \nabla^2 V_{fi}^* - (1 - S)g_i^* \quad (1)$$

Where, V_i is the velocity tensor; $\Omega_i = \omega_{pi} - 0.5 \nabla \times V_i$ (ω_p is the angular speed of particle freely rotating);d is the particle diameter; t,τ denote the time; S is the particle-fluid mass density ratio ρ_p/ρ_f; g_i is the gravitational acceleration tensor; x_i are the components of coordinate; C_D, K_m, K_B, K_s, and C_M are coefficients of the drag, the virtual mass, the Basset, the Saffman lift and Magus (Rubinow,1961) lift forces, respectively; Re_L is the characteristic Reynolds number. *Sign* is the signum function.

The derivative $d()/dt$ denotes differentiation with respect to time following the moving sphere, so that $d()/dt = \partial()/\partial t + V_{pj}\partial()/\partial x_j$. In contrast, $D()/Dt$ denotes time derivative following a fluid element, so that $D()/Dt = \partial()/\partial t + V_{fj}\partial()/\partial x_j$.

The subscripts i and j are the coordinate tensors. f and p refer to the fluid and the particle, respectively.

Here, all quantities have been nondimensionalized with the aid of a turbulent length macroscale l_0, the characteristic velocity V_0 and the fluid kinematic viscosity coefficient ν_f. These dimensionless quantities are defined as, respectively

$$
\begin{aligned}
d^* &= \frac{d}{l_0}, \quad V_{pi}^* = \frac{V_{pi}}{V_0}, \quad V_{fi}^* = \frac{V_{fi}}{V_0}, \\
x_i^* &= \frac{x_i}{l_0}, \quad g_i^* = g_i\frac{l_0}{V_0}, \quad t^* = t\frac{V_0}{l_0}, \\
\tau^* &= \tau\frac{V_0}{l_0}, \quad \Omega_i^* = \Omega_i\frac{l_0}{V_0}, \quad Re_L = \frac{V_0 l_0}{\nu_f},
\end{aligned} \quad (2)
$$

The superscript $*$ refer to the dimensionless quantity.

2.2 Boundary layer flow

In this section we consider the case of particle suspended in a turbulent boundary layer flow field and the particle-wall interaction, the velocity gradient of the fluid parallel to the wall ($\partial V_{fy}/\partial x$), the gravitational as well as the Magus lift force are neglected in this paper.

We may approximate the coefficients C_D (the Stokes' drag coefficient),K_m, K_B, and K_s as $Re_p/24$ (Re_p is the particle Reynolds number), 0.5, 6.0, and 1.615, and the derivatives $d()/dt \approx D()/Dt$ (Liu 1995).

The presence of the wall increases the Stokes' drag coefficient of the sphere as compared to that in the case of unbounded free flow. C_x and C_y are correction factors which account for the effects of

the wall on the Stokes' drag. We adopt Faxen's (1923) expression for C_x and Brenner's (1961) and Maude's (1961) expression for C_y:

$$
\begin{aligned}
C_x &= (1 - 9\delta/10 + \delta^3/8 - 45\delta^5/256 - \delta^5/16)^{-}1, \\
C_y &= 1 + 9\delta/8 + 81\delta^2/64
\end{aligned} \quad (3)
$$

Where $\delta = d/(2y)$, y denotes the coordinate normal to the wall.

The main difficulty in solving Eq.(1) is the nonlinear nature of the lift force. Usually, numerical integration is needed to overcome this difficulty. And, the lift forces will be linearized in order to obtain an analytical solution. Hinze (1975) discussed the behavior of a small particle suspended in a turbulent flow. He showed, a particle of diameter equal to or smaller than the Kolmogorff microlength scale η [$d/\eta \ll 1, \eta = (\nu_f/\epsilon)^{1/4}, \epsilon$ is the dissipation rate of the turbulence kinetic energy] (as in the present study), the velocity gradient of the fluid normal to the wall ($\partial V_{fx}/\partial y$) could be modelled as

$$
\frac{\partial V_{fx}}{\partial y} = \frac{\nu_f}{\eta^2}, \quad (4)
$$

$$
\frac{\partial V_{fx}}{\partial y^*} = \frac{1}{Re_L \eta^{*2}} \quad (5)
$$

In addition, it is assumed that the $\nabla^2 V_{fi}^*$ term may be approximated as

$$
\nabla^2 V_{fi}^* \approx -V_{fi}^*/l^{*2} \quad (6)
$$

Where, the length scale l_1^* may be estimated by minimizing the mean- square error. This leads to

$$
\begin{aligned}
l_i^{*-2} &= \frac{<V_{fi}^* \nabla V_{fi}^*>}{<V_{fi}^* V_{fi}^*>} \\
&\left(\langle\frac{\partial V_{fi}^*}{\partial x_j^*}\frac{\partial V_{fi}^*}{\partial x_j^*}\rangle - \frac{1}{2}\frac{\partial^2 <V_{fi}^* V_{fi}^*>}{\partial x_j^{*2}}\right) / <V_{fi}^* V_{fi}^* >
\end{aligned} \quad (7)
$$

Where $<$ $>$ denotes the expected value. Now, combining Eqs.(1)-(6), the equations of particle motion in a turbulent boundary layer flow in the x and y directions can be modeled as,z respectively (asterisk dropped)

$$
\frac{dV_{px}}{dt} = \alpha\beta C_x(V_{fx} - V_{px})
$$

$$+ \ \beta\sqrt{\frac{3\alpha}{\pi}}\int_{-\infty}^{t}(\frac{dV_{fi}}{d\tau} - \frac{dV_{pi}}{d\tau})\frac{d\tau}{\sqrt{t-\tau}}$$

$$+ \ \beta\frac{dV_{fx}}{dt} + \frac{1}{18}\alpha\beta\varphi_x V_{fx}$$

$$\frac{dV_{py}}{dt} = \alpha\beta C_y(V_{fy} - V_{py})$$

$$+ \ \beta\sqrt{\frac{3\alpha}{\pi}}\int_{-\infty}^{t}(\frac{dV_{fi}}{d\tau} - \frac{dV_{pi}}{d\tau})\frac{d\tau}{\sqrt{t-\tau}}$$

$$+ \ \beta\frac{dV_{fy}}{dt} + \frac{1}{18}\alpha\beta\varphi_y V_{fy}$$

$$+ \ \alpha\beta e(V_{fx} - V_{px}) \qquad (8)$$

Where,

$$\alpha = \frac{12}{d^2 Re_L}, \quad \beta = \frac{3}{1+2S}, \quad e = \frac{K_s}{3\pi}\theta,$$

$$\theta = \frac{d}{\eta}, \quad \varphi_i = \frac{d}{l_i} \qquad (9)$$

Eq.(8) is linear differential equation. Hence, this equation may represent either the particle mean motion equation or the particle fluctuation motion equation.

3 SPECIAL ANALYSIS

3.1 Fourier Transform

Define the Fourier transform $\check{V}_f(\omega)$ and $\check{V}_p(\omega)$ of the fluid and the particle fluctuation velocities $V_f(t)$ and $V_p(t)$ as, respectively

$$\check{V}_f(\omega) = \frac{1}{\pi}\int_{-\infty}^{+\infty}V_f(t)exp(-i\omega t)dt$$

$$\check{V}_p(\omega) = \frac{1}{\pi}\int_{-\infty}^{+\infty}V_p(t)exp(-i\omega t)dt \qquad (10)$$

Where ω is the fluctuation frequency, $i = \sqrt{-1}$. Define the Fourier transform of the particle relative fluctuation velocity between the particle and the fluid as

$$\check{W}(\omega) = \check{V}_p(\omega) - \check{V}_f(\omega) \qquad (11)$$

Using Eq.(10), taking the Fourier transform of Eq.(8) lead to

$$A_x\check{V}_{px}(\omega) = B_x\check{V}_{fx}(\omega)$$
$$A_y\check{V}_{py}(\omega) = B_y\check{V}_{fy}(\omega) + e[\check{V}_{fx}(\omega) - \check{V}_{px}(\omega)] \qquad (12)$$

Where,

$$A_x = C_x + M + i[\frac{\omega}{(\alpha\beta)} + MSign(\omega)],$$

$$B_x = C_x + M + \frac{\varphi_x^2}{18} + i[\frac{\omega}{\alpha} + MSign(\omega)],$$

$$A_y = C_y + M + i[\frac{\omega}{(\alpha\beta)} + MSign(\omega)],$$

$$B_y = C_y + M + \frac{\varphi_y^2}{18} + i[\frac{\omega}{\alpha} + MSign(\omega)] \qquad (13)$$

Here $M = \sqrt{3|\omega|}/(2\alpha)$. Solving Eqs.(12), the particle velocity components in Fourier domain are given by

$$\check{V}_{px}(\omega) = H_{11}(\omega)\check{V}_{fx}(\omega) + H_{12}(\omega)\check{V}_{fy}(\omega)$$
$$\check{V}_{py}(\omega) = H_{21}(\omega)\check{V}_{fx}(\omega) + H_{22}(\omega)\check{V}_{fy}(\omega) \qquad (14)$$

Where the system function $H(\omega)$ are defined as

$$H_{11} = B_x/A_x, H_{12} = 0, H_{21} = e(A_x - B_x)/(A_x A_y),$$
$$H_{22} = B_y/A_y \qquad (15)$$

Combining Eqs.(11) and (14), we get

$$\check{W}_x(\omega) = [H_{11}(\omega) - 1]\check{V}_{fx}(\omega) \qquad (16)$$

$$\check{W}_y(\omega) = H_{21}\check{V}_{fx}(\omega) + [H_{22}(\omega) - 1]\check{V}_{fy}(\omega) \qquad (17)$$

3.2 Lagrangian Power Spectrum and Autocorrelation Function

Define the superscript ' as the complex conjugates. The Lagrangian velocity power spectrum tensor is then given by

$$E_{ij}(\omega) = \check{V}_i(\omega)\check{V}_j'(\omega) \qquad (18)$$

Thus, according to Eq. (14) the spectral components are given as

$$E_{pxx}(\omega) = G_{11}E_{fxx}(\omega)$$
$$E_{pxy}(\omega) = G_{21}E_{fxx}(\omega) + G_{22}E_{fxy}(\omega)$$
$$E_{pyy}(\omega) = G_{31}E_{fxx}(\omega) + G_{32}E_{fxy}(\omega)$$
$$+ \ G_{33}E_{fyy}(\omega) \qquad (19)$$

Where the G function matrix can be written as

$$[G_{ij}] =$$

$$\begin{bmatrix} |H_{11}|^2 & 0 & 0 \\ 0.5(H'_{11}H_{21} + H_{11}H'_{21}) & 0.5(H'_{11}H_{22} + H_{11}H'_{22}) & 0 \\ |H_{21}|^2 & H'_{21}H_{22} + H_{21}H'_{22} & |H_{22}|^2 \end{bmatrix} \quad (20)$$

The Lagrangian autocorrelation function for the particle fluctuation velocity is defined as

$$R_{ij}(\tau) = \overline{V_i(t)V_j(t+\tau)} \quad (21)$$

Where, overbar denotes time-averaging. The power spectrum and the autocorrelation function are Fourier pair. That is

$$R_{ij}(\tau) = \int_0^{+\infty} E_{ij}(\omega)\cos(\omega\tau)d\omega \quad (22)$$

$$E_{ij}(\omega) = \frac{2}{\pi}\int_0^{+\infty} R_{ij}(\tau)\cos(\omega\tau)d\tau \quad (23)$$

Now, we introduce the fluid power spectrum density function $f_{ij}(\omega)$ such that

$$E_{fij}(\omega) = \overline{V_{fi}(t)V_{fj}(t)}f_{ij}(\omega) = R_{fij}(0)f_{ij}(\omega) \quad (24)$$

Where $f_{ij}(\omega)$ is defined such that

$$\int_0^{+\infty} f_{ij}(\omega)d\omega = 1 \quad (25)$$

In fact, with the present knowledge of turbulence it is difficult to theoretically specify these functions in a real turbulent flow (including anisotropy effects). According to all the published theoretical forms of the power spectrum density functions are based on the assumption of local isotropy. Frenkicl(1948) obtained the relation between the longitudinal and transverse spectra of turbulence as

$$f_{yy}(\omega) = \frac{1}{2}f_{xx}(\omega) - \frac{1}{2}\omega\frac{df_{xx}(\omega)}{dt} \quad (26)$$

The power spectrum density function in the x direction, $f_{xx}(\omega)$ is given by Hinze(1975) as

$$f_{xx}(\omega) = (2/\pi)T_L/(1 + \omega^2 T_L^2) \quad (27)$$

From Eqs. (26) and (27), we have

$$\begin{aligned} f_{yy}(\omega) &= 0.5f_{xx}(\omega)[3 - 2/(1 + \omega^2 T_L^2)] \\ &= (T_L/\pi)(1 + 3\omega^2 T_L^2)/(1 + \omega^2 T_L^2)^2 \quad (28) \end{aligned}$$

Unfortunately, at present, the function $f_{xy}(\omega)$ is not available in any literature. Therefore $f_{xy}(\omega)$ is approximated as (Liu, 1995),

$$f_{xy}(\omega) = af_{xx}(\omega) + bf_{yy}(\omega) \quad (29)$$

Where both a and b are constants, and $a + b = 1$. The dimensionless Lagrangian integral time scale T_L is calculated from

$$T_L = C_{TL}K(y)/\varepsilon(y) \quad (30)$$

Where, $C_{LT} \approx 0.2 \sim 0.48$; K is the turbulence kinetic energy of the fluid.

The ratios of autocorrelation functions of the particle to that of the fluid can be expressed as

$$\frac{R_{pxx}(\tau)}{R_{fxx}(\tau)} = \int_0^{+\infty} G_{11}f_{xx}\cos(\omega\ tau)d\omega / \int_0^{+\infty} f_{xx}\cos(\omega\tau)d\omega$$

$$\frac{R_{pxx}(\tau)}{R_{fxy}(\tau)} = \frac{\int_0^{+\infty}[\frac{R_{fxx}(0)}{R_{fxy}(0)}G_{21}f_{xx} + G_{22}f_{xy}]\cos(\omega\tau)d\omega}{\int_0^{+\infty} f_{xy}\cos(\omega\tau)d\omega}$$

$$\frac{R_{pyy}(\tau)}{R_{fyy}(\tau)} = \int_0^{+\infty}[\frac{R_{fxx}(0)}{R_{fyy}(0)}G_{31}f_{xx} + \frac{R_{fxy}(0)}{R_{fyy}(0)}G_{32}f_{xy} + G_{33}f_{yy}]$$

$$\frac{\cos(\omega\tau)d\omega}{\int_0^{+\infty} f_{yy}\cos(\omega\tau)d\omega} \quad (31)$$

4 TURBULENT MASS DIFFUSIVITY

The random continuous motion of a single particle is defined by a time-dependent displacement tensor $\overline{X_iX_j(t)}$:

$$\begin{aligned} \overline{X_iX_j}(t) &= 2\int_0^t (t - \tau)R_{pij}(\tau)d\tau \\ &= 2\int_0^{+\infty} E_{pij}(\omega)\frac{1 - \cos(\omega t)}{\omega^2}d\omega \quad (32) \end{aligned}$$

We can define a particle turbulent diffusivity tensor $D_{pij}(t)$ as

$$D_{pij}(t) = \frac{1}{2}\frac{d}{dt}\overline{X_iX_j}(t)$$

$$= \int_0^{+\infty} E_{pij}(\omega)\frac{\sin(\omega t)}{\omega}d\omega$$

$$= \int_0^t R_{pij}(\tau)d\tau \qquad (33)$$

and similarly,

$$D_{fij}(t) = \int_0^{+\infty} E_{fij}(\omega)\frac{\sin(\omega t)}{\omega}d\omega$$

$$= \int_0^t R_{fij}(\tau)d\tau \qquad (34)$$

Now, combining Eqs. (19), (33) and (34), the ratios of diffusivities of the particle to that of the fluid can be written as

$$\frac{D_{pxx}(t)}{D_{fxx}(t)} = \int_0^{+\infty} G_{11}f_{xx}\frac{\sin(\omega\tau)}{\omega}d\omega \bigg/ \int_0^{+\infty} f_{xx}\frac{\sin(\omega\tau)}{\omega}d\omega$$

$$\frac{D_{pxx}(t)}{D_{fxy}(t)} = \frac{\int_0^{+\infty}\left[\frac{R_{fxx}(0)}{R_{fxy}(0)}G_{21}f_{xx} + G_{22}f_{xy}\right]\frac{\sin(\omega\tau)}{\omega}d\omega}{\int_0^{+\infty} f_{xy}\sin(\omega\tau)d\omega \bigg/ \omega}$$

$$\frac{D_{pyy}(t)}{D_{fyy}(t)} = \int_0^{+\infty}\left[\frac{R_{fxx}(0)}{R_{fyy}(0)}G_{31}f_{xx} + \frac{R_{fxy}(0)}{R_{fyy}(0)}G_{32}f_{xy}\right.$$

$$\left. + G_{33}f_{yy}\right]\frac{\frac{\sin(\omega t)}{\omega}}{\int_0^{+\infty} f_{yy}\frac{\sin(\omega t)}{\omega}d\omega}d\omega \qquad (35)$$

Substituting $t \to +\infty$ into Eqs. (33) and (34), we have

$$D_{ij}(+\infty) = \int_0^{+\infty} R_{ij}(\tau)d\tau = \frac{\pi}{2}E_{ij}(0) \qquad (36)$$

Thus, the components of the long-time diffusivity tensor $D_{pij}(\infty)$ can be written as

$$\begin{aligned}D_{pxx}(\infty) &= G_{11}D_{fxx}(\infty)\\ D_{pxy}(\infty) &= G_{21}D_{fxx}(\infty) + G_{22}D_{fxy}(\infty)\\ D_{pyy}(\infty) &= G_{31}D_{fxx}(\infty) + G_{32}D_{fxy}(\infty)\\ &\quad + G_{33}D_{fyy}(\infty)\end{aligned} \qquad (37)$$

The numerical results show the coefficients $G_{11}\doteq G_{22}\doteq G_{33}, G_{21}\doteq 0, G_{31}\doteq 0, G_{32}\doteq 0$. The figure

displays the variations of G_{11} (or G_{22}, or G_{33}) with different values of the coefficient α, the density ratio S as well as the frequency ω for $y = 100, \theta = 0.5, \varphi = 0.5$. Thus, Eq.(37) shows the long-time diffusivity of the particle can be larger than that of fluid, while the coefficient α is very large(or the particle size is very small). The numerical results also show that the variation of the long-time diffusivity of the particle with the coordinate normal to wall y is very small.

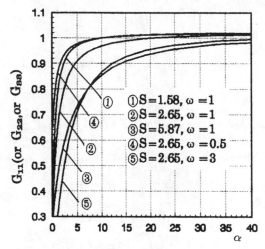

Fig. The variation of G_{11} (or G_{22}, or G_{33}) with different values of the coefficients α, the density ratio S as well as the frequency ω for $y = 100$, $\theta = 0.5$, $\varphi = 0.5$.

5 TURBULENT REYNOLDS STRESS

Substituting $\tau = 0$ into Eqs.(21) and (22), we have

$$\overline{V_i(t)V_j(t)} = R_{ij}(0) = \int_0^{+\infty} E_{ij}(\omega)d\omega \qquad (38)$$

Combining Eqs.(19) and (38), the components of $\overline{V_{pi}V_{pj}}$ may be written as

$$\overline{V_{px}^2} = \overline{V_{fx}^2}\int_0^{+\infty} G_{11}f_{xx}d\omega$$

$$\overline{V_{px}V_{py}} = \overline{V_{fx}^2}\int_0^{+\infty} G_{21}f_{xx}d\omega$$

$$+ \overline{V_{fx}V_{fy}}\int_0^{+\infty} G_{22}f_{xy}d\omega$$

$$\overline{V_{py}^2} = \overline{V_{fx}^2}\int_0^{+\infty} G_{31}f_{xx}d\omega$$

$$+ \ \overline{V_{fx}V_{fy}}\int_0^{+\infty}G_{32}f_{xy}d\omega$$

$$+ \ \overline{V_{fy}^2}\int_0^{+\infty}G_{33}f_{yy}d\omega \qquad (39)$$

and combining Eqs. (16), (17) and (19), we have

$$\overline{W_x^2} \ = \ \overline{V_{fx}^2}\int_0^{+\infty}|H_{11}-1|^2f_{xx}d\omega$$

$$\overline{W_y^2} \ = \ \overline{V_{fx}^2}\int_0^{+\infty}G_{31}f_{xx}d\omega + \overline{V_{fx}V_{fy}}\int_0^{+\infty}(G_{32}-G_4)f_{xy}d\omega$$

$$+ \ \overline{V_{fy}^2}\int_0^{+\infty}|H_{22}-1|^2f_{yy}d\omega \qquad (40)$$

Where, $G_4 = H_{21} + H_{21}'$. $\overline{W_i^2}$ can be written as follows

$$\overline{W_i^2} = \overline{(V_{pi}-V_{fi})^2} = \overline{V_{pi}^2} - 2\overline{V_{pi}V_{fi}} + \overline{V_{fi}^2} \qquad (41)$$

Where,

$$\overline{V_{px}V_{fx}} \ = \ \frac{1}{2}\overline{V_{fx}^2}\int_0^{+\infty}(H_{11}+H_{11}')f_{xx}d\omega$$

$$\overline{V_{py}V_{fy}} \ = \ \frac{1}{2}[\overline{V_{fx}V_{fy}}\int_0^{+\infty}G_4f_{xy}d\omega \qquad (42)$$

$$+ \ \overline{V_{fy}^2}\int_0^{+\infty}(H_{22}+H_{22}')f_{yy}d\omega]$$

6 CONCLUSIONS

The model in this paper can be used to evaluate various statistics of the particle turbulence, by including the turbulent diffusivity and the turbulent Reynolds stress. The model has very interesting implications. The model offers an explanation for a question which has occupied researchers for many years, namely the possibility of the particle having a larger diffusivity than the fluid. This result contradicts the classical theories for particle dispersion, which state that the particle diffusivity can never exceed the fluid diffusivity for any significant period of time.

REFERENCES

Basset, A. B.,1888. Treatise on Hydrodynamics,Vol.2, Deighton, London. (new Dover edition, Dover Publication,Inc., New York, 1961)

Boussinesq, J.,1903. Theory Analytique de la Chaleur, Paris.

Brenner, H.,1961. The slow motion of a sphere through a fluid towards a plane surface. Chem. Engng Sci., 16, 242-251.

Faxen, H.,1923. Ark Mat Astr Fys. 17, 1-6.

Frenkiel, F. N.,1948. Aeronaut, J. Sci. 15, 57-61.

Hinze, J. O.,1975. Turbulence. 2nd ED. McGraw-Hill, New York.

Liu, Xiaobing,1995. Ph. D. thesis, Huazhong University of Sci. and Tech., Wuhan, China.

Liu, Xiaobing and Zeng, Qingchuan,1994. On the motion of particle in a turbulent flow field by Lagrange's method, J. Huazhong Univ. of Sci. and Tech., 22(10), 35-43.

Maclaughlin,J. B.,1989. Aersol particle deposition in numerically simulated channel flow, Phys. Fluids, 7, 1211-1224.

Maude, A. D.,1961. End effects in a falling-sphere viscometer, Br. J. Appl. Phys, 12, 293-295.

Oseen, C. M.,1927. Hydrodynamik, Leipzig.

Rizk, M. A. and Elghobashi, S. E.,1985. The motion of a spherical particle suspentded in a turbulent flow near a plane wall. Phys. Fluids, 28(3), 806-811.

Rouhiainen, P. O. and Stachiewiz, J. W.,1970. On the deposition of small particles from turbulent stream, J. Heat Transfer, 92, 169-177.

Rubinow, S. I. and Keller, J. B.,1961. The transverse force on a spinning sphere moving in a viscons fluid, J. Fluid Mech, 11, 447-459.

Saffman, P. G.,1965. The lift on a small sphere in a slow shear flow. J. Fluid Mech, 22, 385-398.

Saffman, P. G.,1968. Gorrigendum to 'The lift on a small sphere in a slow shear flow'. J. Fluid Mech, 31, 624.

Soo, L. and Tien, C. L.,1960. Effect of the wall on two-phase turbulent motion, J. Appl. Mech. 27, 5-15.

Tchen, C. M.,1947. Mean value and correlation problems connected with motion of small particles suspended in a turbulent fluid, Ph.D. thesis, University of Delft, Martinus, Nijhoff, Hague.

Stochastic Hydraulics'96, Tickle, Goulter, Xu, Wasimi & Bouchart (eds) © 1996 Balkema, Rotterdam. ISBN 90 5410 817 7

The field test and stochastic evaluation of turbulent diffusivity off the southwestern coast of Taiwan, China

Elliot H. H.Chen
Industrial Technology Research Institute, Hsinchu, Taiwan, China

ABSTRACT: A series of field tests to find the intermediate-scale turbulent diffusion coefficient as well as shear flow dispersion were conducted off the southwestern coast of Taiwan. The methods of dye releasing and drogue tracing were used in finding turbulent diffusivity. The tests on vertical concentration and velocity distribution were introduced to find shear flow dispersion. For limited duration of data sampling of releasing dye and tracing drogue, the relatively long-term current data collected from Eulerian RCM current meter is used to compensate the shortage of relatively short-term Lagrangian data. By comparing with the turbulent diffusivity calculated by Eulerian current data, the turbulent diffusivity calculated from the analysis of ensemble mean by tracing drogue is considered to be successful. However, the tests of finding shear flow dispersion are not successful due to some unreasonable data collected in the field. Nevertheless, the dynamic data of the vertical velocity distribution collected by S4 Eulerian current meter still can tell us the diffusivity difference on different water depth because of the influence of bottom shear stress.

1 INTRODUCTION

Using water quality model to simulate the pollution status of the water bodies has been widely applied. Different reactive as well as hydrodynamic kinetics have to be considered for different kinds of water body while using water quality models. Because of the complicated hydrodynamic status and the influences of wind stress and boundary effect in the coastal zone, the factor of diffusion or dispersion is not that simple as in the river and estuary. In another word, the turbulent diffusion as well as shear flow dispersion are very important factors when simulating water quality in the coastal zone in hydrodynamic point of view.

The southwest part of Taiwan is a wide flat land. The residential and industrial zones are mixed together in this area. The untreated waste water comes out from both residential and industrial area discharges into the adjacent river directly. The final receiving water body is then the coastal water in the southwestern Taiwan Strait. In order to simulate the pollutant transport in this coastal area, the complicated hydrodynamic conditions as well as the turbulent diffusion and shear flow dispersion have to be taken into account.

In the modeler's point of view, it is not uncommon only to set the diffusion coefficient as an adjustable parameter while calibrating and validating water quality model. Here, off the southwestern coastal of Taiwan, a special study was conducted due to the special flow condition in this area. A series of drogue and dye tests were deployed in the field to find the turbulent diffusivity and shear flow dispersion.

2 THEORETICAL BACKGROUND

2.1 Type of diffusion

In still water or laminar flow, molecular diffusion occurs when there exist spatial concentration gradients.

In turbulent and non-uniform flow, velocity gradients act to increase concentration gradients. By smoothing out time fluctuation of the turbulent flow velocity, the additional concentration gradient will be generated. This is called turbulent diffusion.

By considering the vertical velocity gradient of the flow field (along vertical direction), additional diffusion effect is added on the concentration gradient by smoothing out spatial velocity and concentration gradients caused by bottom friction and the surface wind stress. This additional diffusion effect only occurs in a vertical-averaged 1-D and 2-D models, and this is called shear flow dispersion. In the coastal zone, this additional diffusion effect could be relatively large comparing with the turbulent diffusion.

In nearshore waters with tidal influences, the flow is turbulent and the dominant eddy has the size of "intermediate scale" instead of "small scale". Thus, molecular diffusion can be neglected. Besides, the magnitudes of turbulent diffusion and shear flow dispersion need to be justified when running 2-D model in the nearshore area.

2.2 Turbulent diffusion

Basically, vertical diffusion (not shear flow dispersion) is less important than horizontal diffusion in the coastal waters. When considering horizontal turbulent diffusion, the horizontal diffusivity is proportional to the scale of eddy to the power of 4/3 in the open ocean far from land boundaries, i.e.,

$$K \sim \varepsilon^{1/3} l^{4/3} \tag{1}$$

where ε is the rate of dissipation of turbulent energy, and l is the length scale of the eddy.

The 4/3 law of turbulent diffusion (Richardson, 1926) is derived from the energy density of intermediate scale eddy in the inertial subrange (shown as Fig.1) which can be expressed as:

$$E(k) \sim \varepsilon^{2/3} k^{-5/3} \tag{2}$$

where k is the $1/l$.

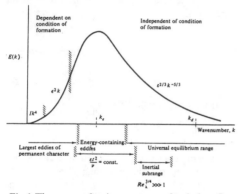

Fig.1 The energy density spectrum of turbulent flow

The intermediate scale eddy in this inertial subrange is considered to be dominant in the big lake area as well as in the coastal zone (Murthy, 1970, 1976; Fischer et al., 1976; Blumberg, 1986).

Actually, in the nearshore waters, the length scale of the eddy available for mixing is restricted by the boundary effects of the land. Thus, the rate of dissipation is curtailed and the power exponent of l is also reduced. Bowden (1983) suggested that the power exponent of l in the nearshore water is approximately 0.5 to 1.0 instead of 4/3.

Despite of the influence of boundary effect on 4/3 law, some other field tests can be introduced to find the turbulent diffusion in the coastal zone, i.e., dye tracer, drogue, and current meter tests. By setting m-axis as the direction of the mean current velocity (v) and r-axis as the lateral direction, the lateral variance of the concentration distribution in dye tracer test can be defined as:

$$S_r^2 = \frac{\int_{-\infty}^{\infty} (r - r_o)^2 \, c \, dr}{\int_{-\infty}^{\infty} c \, dr} \tag{3}$$

where c is concentration and r_o is the center of mass of the concentration distribution.

The lateral diffusivity (K_r) can be calculated by assuming that the concentration distribution in r direction is Gaussian:

$$K_r = \frac{1}{2} \frac{dS_r^2}{dt} = \frac{v}{2} \frac{dS_r^2}{dm} \tag{4}$$

When using single drogue to find turbulent diffusivity, the concepts of ensemble mean and Lagrangian time scale should be mentioned. The Lagrangian time scale (T_L) is defined as the average time needed for a water particle to forget its initial velocity, and is shown as Fig.2.

$$T_L = \int_0^{\infty} R_x(s) \cdot ds \tag{5}$$

$$R_x(\tau_2 - \tau_2) = \frac{\langle U(\tau_2) \cdot U(\tau_1) \rangle}{\langle U^2 \rangle} \tag{6}$$

where R_x is the Lagrangian autocorrelation function and $\langle U^2 \rangle$ is the ensemble mean of $U(0) \cdot U(0)$.

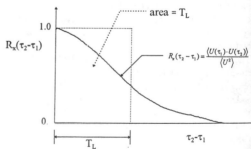

Fig.2 The relationship between Lagrangian autocorrelation function and Lagrangian time scale

The position of a single water particle in the flow field X(t) can be expressed as:

$$X(t) = \int_0^t U(t) dt \tag{7}$$

where U(t) is the velocity of a water particle in the flow field.

The ensemble mean of $X^2(t)$ can be calculated as:

$$\langle X^2(t) \rangle = \int_0^t \int_0^t \langle U(\tau_1) \cdot U(\tau_2) \rangle d\tau_1 d\tau_2 \qquad (8)$$

When tracing time $t \gg T_L$, the ensemble mean of $X^2(t)$ can be expressed as:

$$\langle X^2(t) \rangle = \langle U \rangle^2 \cdot T_L \cdot t + const \qquad (9)$$

Under the assumptions of homogeneous, stationary, and zero mean velocity in the turbulent flow field, the motion of spreading of cloud can be represented by the ensemble mean of the motion of a single drogue. When the conditions of homogeneous, stationary, zero mean velocity, and long tracing time (tracing time $t \gg T_L$) in the turbulent flow field are satisfied, Taylor diffusion theorem can then be applied:

$$K = \frac{1}{2} \frac{d\langle X^2(t) \rangle}{dt} = \langle U^2 \rangle \cdot T_L \qquad (10)$$

Thus, the turbulent diffusivity can be obtained. However, Eq.(10) has to be modified when the tracing time can not reach the standard of $t \gg T_L$.

$$K = \langle U^2 \rangle \int_0^t R_x(s) \cdot ds \qquad (11)$$

In fact, the long-term data of Eulerian velocity sampled by current meter can be applied to find turbulent diffusivity as well. By using the Eulerian velocity, similar to the test of single drogue, the Eulerian turbulent diffusivity can be derived by:

$$K_E = \frac{1}{2} \frac{dX^2(t)}{dt} \qquad (12)$$

The relationship between K and K_E can be expressed as (Wada and Mayaike, 1975):

$$K = \beta K_E \qquad (13)$$

where β is the conversion ratio between Eulerian and Lagrangian turbulent diffusivity.

3 IN-SITU TESTS FOR MEASURING DIFFUSION AND DISPERSION COEFFICIENT

According to the theoretical analysis above, basically, there are three ways to find turbulent diffusivity. One is to use tracer (dye or cluster of drogues), the other is to use single drogue, while the third is to use the data collected by Eulerian current meter. Tracer tests can be conducted either by the instantaneous release of a slug of dye, or by the continuous injection of the dye over a period of time. The test using single drogue can refer to the works of Davis (1991) while the tracer test can refer to the works of Blumberg (1986).

Six in-situ tests using tracer, single drogue, and Eulerian current meter were conducted during the period of September 1994 to June 1995 in the southwestern coast of Taiwan. The brief descriptions of each test are as follow:

1. September 1994, a Eulerian RCM (recording current meter) was deployed to collect the current data for 15 days. Eq.(11) was introduced to calculate the turbulent diffusivity of long-term (33 tidal cycles) current data. The result of calculation is listed in Item-1 of Table 1.

2. November 1994, a Lagrangian single drogue test was conducted and traced for 12 hours. Eq.(11) was introduced to calculate the turbulent diffusivity. The excursion path of the drogue is shown on Fig.3, and the result of calculation is listed in Item-2 of Table 1.

3. November 1994, a dye test of continuous injection was conducted. The spreading of the dye was measured by video camera. Eq.(10) was introduced to calculate the turbulent diffusivity. The result of calculation is listed in Item-3 of Table 1.

4. November 1994, a continuous dye injection test was conducted along a submerged 6m-long vertical mounted diffuser on a mother boat to measure the vertical stratification phenomenon in the coastal zone in order to find the shear flow dispersion coefficient. Two other sampling boats were waiting downstream for 25m and 50m apart to collect the water samples while the vertical velocity profiles were measured by S4 current meters on all three boats. After analyzing the collected data, the velocity can be used while the vertical distribution of dye concentration is considered not reasonable. Thus, only the turbulent diffusivity is calculated. The result of calculation is listed in Item-4 of Table 1.

5. May 1995, a series of six instantaneous releases of dye in 12 hour period was conducted. The duration of six releases covered a whole tidal cycle. The result of calculation is listed in Item-5 of Table 1.

6. May 1995, a Lagrangian single drogue test was conducted and traced for 36 hours with the recording time step of 15 minutes. The result of calculation is listed in Item-6 of Table 1.

4 ANALYSIS OF CALCULATED TURBULENT DIFFUSIVITY

Basically, there are three goals in these field tests. First is to find shear flow dispersion coefficient, second is to find intermediate scale turbulent diffusivity, and third is to find small scale turbulent diffusivity. Through a series of tests stated above, only turbulent diffusivity can be obtained while dispersion coefficient is not.

By assuming that x is the direction of latitude and y is the direction of longitude, the results of calculated turbulent diffusivity can be discussed as follow:

1. From Item-1 of Table 1, the Eulerian turbulent diffusivity is 127.8 m^2/s in y direction and 83.04m^2/s in x direction. The number of diffusivity is similar in both x and y directions.

2. By comparing Item-1 with Item-2 and Item-6 in Table 1, K_x is comparably large for Item-1. When calculating long-term turbulent diffusivity, the result obtained from Eulerian current data is considered to have error and need to be modified.

3. When looking at Item-3, 4, and 5, the turbulent diffusivities are about the same, and they are much smaller than Item-1, 2, and 6. In fact, these results are considered turbulent diffusivity for small scale eddy no matter what kind of testing methods are using when the testing time $t \ll T_L$.

4. In Item-4, the turbulent diffusivity is smaller in the bottom layer (6.3m deep) which is considered to be influenced by the bottom shear stress.

Table 1 The turbulent diffusivity off the southwestern coast of Taiwan

Measurement Method	ITEM-1. Eulerian RCM current meter test (15days)	ITEM-2. Lagrangian single drogue tracing (12hrs)	ITEM-3. Continuous dye test for small scale eddy	ITEM-4. Eulerian S4 current meter test	ITEM-5. Instantaneous dye test along tidal excursion	ITEM-6. Lagrangian single drogue tracing (36hrs)
Turbulent diffusivity (m^2/sec)	$K_x = 83.04$ $K_y = 127.8$	$K_x = 8.47$ $K_y = 96.5$	$K = 0.52$	$K_{x(3.5m\ deep)}$ $= 0.35–4.52$ $K_{x(6.3m\ deep)}$ $= 0.38–0.47$ $K_{y(3.5m\ deep)}$ $= 0.91–4.25$ $K_{y(6.3m\ deep)}$ $= 0.78–1.33$	$K = 0.08–0.52$	$K_x = 11.28$ $K_y = 108.5$

5 DISCUSSION AND CONCLUSION

1. The shear flow dispersion is normally used when simulating water quality in the river and estuary, while the turbulent diffusion is introduced in the open ocean. In shallow water of the coastal zone, the influences of wind stress, bottom shear stress, and tidal current all mix together to generate a somewhat complicated mechanism of diffusion. Thus, both shear flow dispersion and turbulent diffusion need to be investigated.

2. The intermediate-scale eddy is carried by inertial force and acted as a mediator to deliver the energy of large-scale eddy onto the small-scale eddy. The energy of small-scale eddy will then be dissipated by viscosity. In big lake or the coastal zone of the sea, the intermediate-scale eddy is considered dominating the turbulent diffusion. Thus, long-term tracer or dye test is needed.

3. Using single-drogue tracing test to find turbulent diffusivity is only suitable under the assumptions of homogeneous, stationary, and zero mean velocity in the turbulent flow field as well as the condition of tracing time $t \gg T_L$. When using dye test, the tracing time of at least one tidal cycle is recommended.

4. In order to find the shear flow dispersion coefficient in the coastal zone, the vertical concentration and velocity profiles must be obtained dynamically. In this study, the attempts of finding these vertical distributions are not successful. The main reasons are (a) wind and current have different directions so that the sampling boats follow the wind direction while the dye goes with the current heading to another direction, (b) the data collected is unreasonable, i.e., the concentration collected from the third boat is bigger than the second boat. Thus, the field test of finding the shear flow dispersion need to be modified and more efforts must be exerted in the near future.

5. In this study, the tests of finding turbulent diffusivity are successful. Through the single drogue tracing tests, the turbulent diffusivity of about 100m^2/s in the N-S direction and about 10m^2/s in the E-W direction are obtained in the southwestern coast of Taiwan.

REFERENCES

Blumberg, A. F. (1986). "Turbulent mixing processes in lakes, reservoirs, and impoundments", in Physics-Based Modeling of Lakes, Reservoirs, and Impoundments, Published by ASCE, pp79-104.

Bowden, K. F. (1983). "Physical oceanography of coastal waters", Ellis Horward Ltd., Chichester, U.K.

Davis, R. E. (1991). "Lagrangian ocean studies", Annual Review of Fluid Mechanics, Vol.23, pp43-64.

Fischer, H. B., Imberger, J., List, E. J., Koh, R. C. Y., and Brooks, N. H., (1979). "Mixing in inland and coastal waters", Academic press, 483pp.

Murthy, C. R. (1970), "An experimental study of horizontal diffusion in Lake Ontario", Proceeding of 13th Conference of Great Lake Research, International Association of Great Lake Research, pp477-489.

Richardson, L. F. (1926). "Atmospheric diffusion shown on a distance neighbor graph", Proceedings of Royal Society of London, Series A, Vol.110, pp709-737.

Taylor, G. I. (1921). "Diffusion by continuous movements", Proceedings of London Mathematics Society, Series A20, pp196-212.

Wada, A., and Mayaike, Y. (1975). "Study on adaptability of prediction method of simulation analysis for diffusion of discharged warm water in the bay", Central Research Institute of Electric Power Industry, Technical Report No. C374004.

Stochastic Hydraulics'96, Tickle, Goulter, Xu, Wasimi & Bouchart (eds) © 1996 Balkema, Rotterdam. ISBN 90 5410 817 7

Turbulent flow structures from velocity time series

G.M.Smart

National Institute of Water & Atmospheric Research, Christchurch, New Zealand

ABSTRACT: The sizes of eddies are investigated from time series of downstream flow velocities. The data were measured in Canadian and New Zealand gravel-bed rivers with an electromagnetic flow meter and an electronic Pitot tube. Eddy chord lengths are delineated by extrema and zero-crossings in the turbulence velocity signal. The zero-crossing technique indicates that the frequency of eddies of a given size is related to the inverse-square of the eddy chord length.

1 CONCEPTUAL MODEL

The velocity time series measured at a point in a turbulent flow reflects the shape and size of structures passing that point within the flow. As eddies within the flow are advected past the measurement point, the velocity signal may be used to delineate the transected eddies. This process is related to grain-size distribution estimation by measuring the chord lengths of stones transected by a taught string lying across a sample of gravel.

Consider the hypothetical free vortices shown in cross-section in Figure 1.

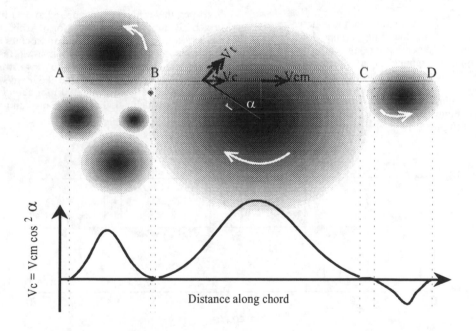

Fig. 1. Velocity component of free vortices sampled along line A - D.

For each free vortex:
$$V_t \, r = \text{const.} \qquad (1.1)$$
where V_t is the tangential velocity at any point distance r from the centre of the vortex. For the velocity component V_c in the direction of line A-D :

$$V_c = V_t \cos(\alpha) \qquad (1.2)$$
but from Eq. (1.1) :
$$V_t = V_{cm} \cos(\alpha) \qquad (1.3)$$
where V_{cm} is the maximum velocity on the chord B-C cut through the vortex.
Therefore:
$$V_c = V_{cm} \cos^2(\alpha) \qquad (1.4)$$
i.e. a turbulent flow comprising isolated, stable, free vortices being advected at constant horizontal velocity past a velocity probe pointing into the flow, would display a series of $\cos^2$ curves superposed on the mean velocity as shown in figure 1. The chord lengths, A-B, B-C, C-D, of eddy vortices could be estimated from successive maxima or minima in the velocity graph or, assuming equivalent frequency and sizes of clockwise and counterclockwise vortices, by measuring the distances over which the velocity graph remains deflected from the mean velocity.

2 DATA AND METHODS

In a steady flow, subtracting the mean downstream flow velocity from the time varying downstream component of velocity leaves the turbulent signal u′. A typical example of u′ measured with a Marsh-McBirney electromagnetic current meter in a gravel bed river (Kirkbride & McLelland, 1994), is shown in Figure 2. Note the similarity between the graph in Figure 1 and the first two-and-a-half seconds of Figure 2.

The time series may be interpreted in two ways:
(a) a sequence of larger structures which cause the flow to remain for some time above (+) or below (-) the mean value, Fig.2 (a) and
(b) a sequence of smaller structures, delineated by maxima and minima in the signal, Fig 2 (b).

Features of type (a) may be sized by measuring the time between successive zero crossings in the u′ time series. Using Taylor's frozen turbulence hypothesis (Monin & Yaglom, 1975), this time may be converted to an eddy chord length by multiplying by the mean flow velocity.

For features of type (b) the time between velocity extrema indicate the passage of a half chord of the smaller structures (successive maxima are separated by a minimum). The half chord length is the inter-extrema time multiplied by the local mean advection velocity.

Note that in both cases a chord length is being measured and not the maximum diameter of the flow structure or eddy. Note also that, by definition, individual eddy sizes are mutually exclusive and space filling, i.e. neither technique by itself can measure eddies within eddies and as long as the flow does not remain constant at the mean value the sum of all eddy lengths will equal the length of the time series. Technique (b) may be achieved by applying technique (a) to the first derivative of the time series.

In this paper, the techniques are applied to data from two rivers; Beauty Creek and the North Ashburton. Beauty Creek is located in the Canadian Rockies and data were recorded at 10Hz with a Marsh-McBirney electromagnetic current meter. The downstream component of veloctiy (mean value 0.638 m/s) is used. Measurements were taken 300 mm above the channel bed (flow depth 450 mm). The median grain size was 29 mm.

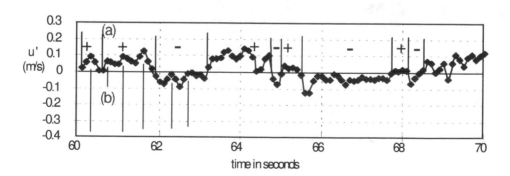

Fig. 2 Turbulent velocity component, Beauty Creek, Canadian Rockies.

The North Ashburton River is located on the Canterbury plain of the South Island of New Zealand. Downstream velocities (mean value 1.54 m/s) were recorded at 28Hz with an electronic Pitot tube (Smart 1994). Measurements were taken 207 mm above the bed (flow depth 352 mm). The median grain size was 39 mm on the channel bed.

3 RESULTS AND DISCUSSION

Figure 3 shows a sequence of type (a) chord lengths derived from the Beauty Creek data, part of which is shown in Figure 2. A similar series was produced for the North Ashburton River. Both series showed insignificant correlation between successive eddy chord lengths, i.e. the sizes of successive type (a) structures are serially independent. Data from other rivers and flumes which we have analysed also show this property.

The frequency distribution of chord lengths from Figure 3 is shown in Figure 4 along with the corresponding distribution for the North Ashburton River data.

Here the chord length is shown relative to flow depth and frequency is shown as a percentage of the total number of eddy chords detected in the time series. The centre points of the histogram intervals are connected by dotted lines on the figure and both distributions are clearly well described by a power law relation.

Application of technique (b) requires knowledge of local mean velocity to convert inter-extrema time to a length scale. From Figure 2 it can be seen that eddies measured by method (b) are superposed on larger flow fluctuations and the problem is to select an averaging period which reflects local mean advection velocity.

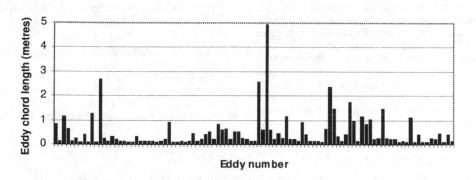

Fig. 3 Eddy chord lengths indicated by u′ zero crossings, Beauty Creek data.

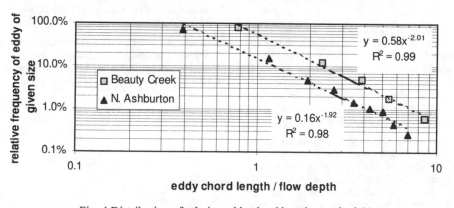

Fig. 4 Distribution of relative eddy chord lengths, method (a).

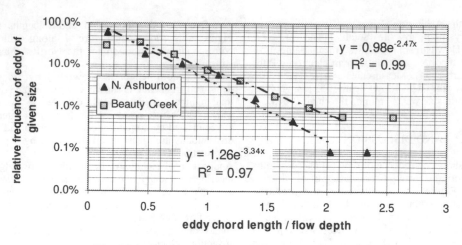

Fig. 5 Distribution of relative eddy chord lengths, method (b).

For both sets of data, local velocity fluctuations are small compared to the mean flow velocity (velocity standard deviation is about 13% of mean velocity in both cases) and it is believed that no great error is introduced by using the series mean velocity in place of a local velocity. Each length between extrema (half chord length) is doubled to be comparable with type (a) lengths. Lag correlation analysis shows successive type (b) chord lengths to be serially independent, however, the frequency of eddy sizes now appears to follow an exponential distribution as shown in Figure 5.

4 SUMMARY & CONCLUSIONS

Work remains to be done on interpreting chord length in terms of a representative eddy diameter, however, for all turbulent flows studied to date, the techniques described above identify serially independent, mutually exclusive space filling eddy sizes.

With the zero crossing technique applied to the turbulent component of the flow, the frequency distribution of eddy chord lengths followed an inverse square power law. This exponent could indicate that the eddies in these gravel bed rivers are two dimensional (note also that the free vortices described in the introduction must theoretically exist as a line terminating on a boundary, or a ring).

The inter-extrema technique for eddy identification detects smaller eddies than the zero crossing technique. However, it has the problem of how to define a local mean eddy advection velocity. In flows with low relative turbulence it may be possible to use mean flow velocity as the advection

velocity without introducing large errors. For the two rivers investigated here the inter-extrema technique indicated an exponential distribution of eddy frequency with eddy size.

5 ACKNOWLEDGEMENTS

The North Ashburton field data were collected with the assistance of C. Beffa. This research was carried out under contract CO1512 from the Foundation for Research, Science and Technology (New Zealand).

REFERENCES

Kirkbride, A.D. & S.J. McLelland (1994), Visualization of the turbulent flow structure in a gravel-bed river, *Earth Surface Processes & Landforms*, Vol. 19, 819-825.

Monin, A.S. & A.M. Yaglom (1975), Statistical fluid mechanics, Vol. 2, MIT Press, ISBN 0-262-13098-X. p.11.

Smart, G.M. (1994), River turbulence appraisal. *Refined flow modelling and turbulence measurements*. Presses Ponts et Chaussees, Paris, ISBN 2-85978-202-8. pp 621-628

Stochastic Hydraulics'96, Tickle, Goulter, Xu, Wasimi & Bouchart (eds) © 1996 Balkema, Rotterdam. ISBN 90 5410 817 7

3D numerical modelling of local scour by stochastic approach

Yuanya Li & Sam S.Y.Wang
Center for Computational Hydroscience and Engineering, The University of Mississippi, Miss., USA

ABSTRACT: Local scour processes are simulated numerically by including the stochastic theory in modeling eddy viscosity of the flow, and diffusivity, fall velocity, bed shear, entrainment, lift and other relevant properties of the suspended sediment particles. Numerical simulations are in reasonable agreement with experimental data.

1 INTRODUCTION

Local scours in stream channels are often induced by engineering structures, such as bridge abutments, groynes and dikes, which alter the flow, intensify turbulence, generate vortices and increase shear stress on the bed. Consequently, the local sediment transport capacity is significantly increased resulting in serious localized bed erosion. Due to the fact that severe local scour can progressively undermine the foundation of the hydraulic structures and cause their failures, the safe and economical design of hydraulic structure requires accurate prediction of effects of maximum scour hole depth and its location, shape and size around the structure under various conditions. Over the last decade, intensive investigations have been conducted primarily using physical experiments, but due to cost-effectiveness, some numerical simulations research have begun.

Many investigations using physical models in laboratories have been reported. Some typical examples are: Laursen (1962), Rajaratnam and Nwachukwu (1983), Chiew and Melville (1989), Vittal, Kothyari, and Morteza (1993). Almost all physical models resulted in empirical relationships between the maximum or equilibrium scour hole depth and flow/sediment conditions of their experiments. However, applications of these relationships are limited by scaling problems, simplified geometry, sediment and flow conditions. The validity of these empirical formulas in application to a site-specific prediction of realistic scour hole depth in a highly complicated nature stream is questionable, especially when the formula is applied to a site where the flow/sediment properties are different from those of the physical modeling study. From their recent report, using the hydrogen bubble technique to measure the flow and turbulence around a bridge abutment, Kwan and Melville (1994), found that the local scour formation around a structure is caused by a primary vortex and its associated downflow and that the vortex is confined predominantly within the scour hole. They also reported that the bed shear stresses in the close vicinity of the structure were amplified up to 5 times of the bed shear stress upstream of the structure.

With advances of computer technology, numerical modeling of 3D turbulent flows, sediment transport, morphological changes, and local scour development have become feasible. Van Rijn (1986) applied a two-dimensional (vertical) model to simulate bed elevation change in a trench normal to flow direction. In his studies the velocity field was considered as steady and the pressure hydrostatic. Additionally, the free surface was approximated by a free slip, rigid-lid boundary and a 2D convection-diffusion equation was used for computing the concentration field. Van Rijn (1987) introduced a stochastical bed-load formula and Hoffmans and Booij (1993) used the stochastical approach of sediment transport in their 2D computational simulation. Both were significant contributions.

A 3D numerical model for solving unsteady, incompressible, free surface, and turbulent flow has been developed at the Center for Computational Hydroscience and Engineering of The University of Mississippi, which had subsequently verified by analytic solutions and physical model results, (Wang and Hu, 1990, Jia and Wang, 1993, Toro and Wang, 1993). More recently, the capability of solving hydrodynamic pressure was added which has enhanced the CCHE3D model's capability to obtain more realistic flow conditions around a scour hole (Li, 1995). In this study, a general stochastical approach to the estimate of the bed-load, suspended sediment transport, and eddy viscosity has been implemented in simulating three dimensional turbulent flow, sedimentation processes and scour hole development.

2 GOVERNING EQUATIONS OF 3D TURBU-
LENT FLOWS

Three dimensional Navier-Stokes equations and continuity equation for incompressible flow are given below.

$$\frac{\partial u_j}{\partial x_j} = 0 \tag{1}$$

$$\frac{\partial u_i}{\partial t} + \frac{\partial}{\partial x_j}(u_i u_j) = \frac{\partial}{\partial x_j}\left[\frac{\mu + \mu_t}{\rho}\left(\frac{\partial u_i}{\partial x_j} + \frac{\partial u_j}{\partial x_i}\right)\right] \\ - \frac{1}{\rho}\frac{\partial p}{\partial x_i} + g_i \tag{2}$$

The free surface elevation is governed by the kinematic equation

$$\frac{D\eta}{Dt} = \frac{\partial \eta}{\partial t} + \vec{u}_s \cdot \nabla \eta = 0 \tag{3}$$

The Prandtl's mixing-length turbulence closure scheme is

$$\mu_t = \iota_m^2 \left\{ \frac{\partial U_i}{\partial x_j}\left[\frac{\partial U_i}{\partial x_j} + \frac{\partial U_j}{\partial x_i}\right]\right\}^{\frac{1}{2}} \tag{4}$$

with ι_m being determined by

$$\frac{\iota_m}{h} = 0.14 - 0.08\left[1 - \frac{y}{h}\right]^2 - 0.06\left[1 - \frac{y}{h}\right]^4 \tag{5}$$

where the subscript j can take the values 1,2,3, denoting the three Cartesian coordinates x_j, u_j = velocity component, t = time, μ_t = turbulent eddy viscosity, μ = fluid dynamic viscosity, ρ = fluid density, η = water surface elevation. The summation convection is adopted.

3 A STOCHASTIC APPROACH TO MODEL-
ING SEDIMENTATION PROCESSES

3.1 General Diffusion Equation

Wang and Ni (1991) succeeded in applying the probability density distribution of particle's fluctuating velocity which they found experimentally to their dilute solid-liquid two-phase flow study. Assuming distribution functions of velocity components being independent random variables, Li and Hwang (1989) proposed

$$p(u_1', u_2', u_3') = \left(\frac{1}{\sqrt{2\pi}}\right)^3 \sum_{i=1}^{3}\frac{1}{\sigma_i}exp\left(-\frac{u'^2}{2\sigma_i^2}\right) \tag{6}$$

Due to the turbulent diffusion, the net suspended sediment diffusion flux ϕ over a unit cross-section area can be written as

$$\phi_i = \int_0^\infty s(x_i - T \cdot u_i')\int_{u_i'}^{+\infty}\int_{-\infty}^{+\infty}\int_{-\infty}^{\infty} p(u_1', u_2', u_3')du_1'du_2'du_3'du_i'$$

$$- \int_0^\infty s(x_i + T \cdot u_i')\int_{-\infty}^{-u_i'}\int_{-\infty}^{+\infty}\int_{-\infty}^{+\infty} p(u_1', u_2', u_3')du_1'du_2'du_3'du_i' \tag{7}$$

where: $i = 1, 2, 3, T$ = the average life time of eddies and σ_i = the standard deviation of fluctuating velocity in i direction.

The general diffusion equation of suspended sediment can be written as:

$$\frac{\partial s}{\partial t} + div(\vec{u}s) = div\left(f_{i1}\frac{\partial s}{\partial x_i} + f_{i3}\frac{\partial^3 s}{\partial x_i^3} + \cdots\cdots\right) \tag{8}$$

$$f_{i,2m+1} = -\frac{T_i^{2m+1}\sigma_i^{2m+2}2^{2m+1}}{(2m+2)!}\prod_{I=0}^{m}\left(9m + \frac{1}{2} - I\right) \tag{9}$$

$$i = 1, 2, 3, \quad and \quad m = 0, 1, 2, 3, \ldots\ldots$$

The coefficients in equation (9) are,

$$f_{i1} = -T\sigma_i^2/2 \tag{10}$$

$$f_{i2} = 0 \tag{11}$$

$$f_{i3} = -\frac{3}{4}T^3\sigma_i^4 \tag{12}$$

High order derivatives may not be neglected when their magnitudes are comparable with the first order term in the diffusion equation. Following the studies on the Brownian movement of tiny particles due to heat, Pais (1982) obtained the diffusion coefficient

$$\nu_h = \frac{\overline{\Delta y^2(t)}}{2T'} \tag{13}$$

where: $\overline{\Delta y^2(t)}$ = standard deviation of particle's displacement and T' = the period of the displacement.

In numerical modeling of a local scour, A coefficient, α_1, is designated to represent the effect of all high order derivative terms instead of solving all of them directly

$$\alpha_1 = \left(1.0 + \frac{f_{3i}\frac{\partial^3 s}{\partial x_i^3}}{f_{1i}\frac{\partial s}{\partial x_i}} + \frac{f_{5i}\frac{\partial^5 s}{\partial x_i^5}}{f_{1i}\frac{\partial s}{\partial x_i}} + \cdots\cdots\right) \tag{14}$$

Experimental studies found that the suspended sediment concentration profiles along vertical lines can not be predicted realistically by numerical models based on the assumption that the diffusivity of sediment is the same as turbulent eddy viscosity. This problem was resolved by the introduction of a coefficient β, Schmidt number (van Rijn 1984; Celik and Rodi 1988) in two dimensional suspended sediment transport simulations, the relation between eddy diffusivity, ν_s and eddy viscosity, ν_t is

$$\nu_s = \nu_t/\beta \tag{15}$$

214

$$\beta = \frac{1}{1 + 2\left(w_0/u_*\right)^2}, \quad \left(0.1 < \frac{w_0}{u_*} < 1\right) \tag{16}$$

For 3D problems, this formula overestimates the diffusion in horizontal directions. In general, the eddy diffusivity of suspended sediment, ν_t, is smaller than the eddy viscosity of turbulent momentum (Brush, 1962). Lately, by using the laser doppler anemometry to measure dilute solid-liquid two phase flow, Wang (1989) showed that the root-square of oscillation velocity of particle's is smaller than that of water's over the entire flow depth and that the relative velocities of suspended particles follow an approximate normal probablistic distribution. See Figure 1.

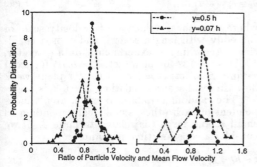

Figure 1. Probabilistic distributions of suspended particle velocity (Wang, 1989)

Boundary conditions for solving the diffusion equation are prescribed below. On the free surface of the flow,

$$\nu_s \frac{\partial s}{\partial x_3} + w_t s = 0 \tag{17}$$

On the interface between bed-load layer and the sediment laden flow, the source term for suspended sediment diffusion equation is defined as (see Figure 2).

$$s_{source} = \nu_s \frac{\partial s}{\partial x_3} + w_t s \tag{18}$$

$$S_{source} = E - D \tag{19}$$

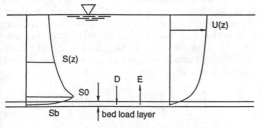

Figure 2. Sediment-laden flow over loose bed

$$D = w_t s_0 \tag{20}$$

If the local flow condition does not satisfy the sediment entrainment criterion, the erosion rate

$$E = 0 \tag{21}$$

otherwise, E is calculated by the modified scheme of Celik and Rodi(1988): For loose bed,

$$E = w_s s_{*b} \tag{22a}$$

For fixed bed with an upstream sand source

$$E = min(w_t s_0, w_t s_{*b}), \tag{22b}$$

where: s_0 = suspended sediment concentration adjacent to the bed load layer, and s_{*b} = the equilibrium suspended sediment concentration at the same location as s_0 (van Rijn, 1986).

$$s_{*b} = s_c \cdot e^{-w_t(y-a)/\varepsilon} \tag{23}$$

$$s_c = 0.18 \cdot s_m \frac{p_1(n = 1.0) + p_2(n = 1.0)}{D_*} \tag{24}$$

with maximum possible volume concentration $s_m = 0.65$, and $D_* = d_{50} \cdot (\Delta \cdot g)/\nu^2)^{\frac{1}{3}}$ and $\Delta = (\rho_s - \rho_w)/\rho_w$.

3.2 Fall Velocity of Sediment Particle in Turbulent Flows

It has been observed that in sediment laden flows the fall velocity of sediment particles is usually reduced by the presence of turbulence. Hwang (1985) reported that nonlinear drag due to turbulence is a major cause of this reduction. He expressed the fall velocity in turbulent flow as

$$\frac{w_t}{w_0} = F\left(\frac{w_0 d}{\nu}, \frac{u_{f,max}}{w_0}\right) \tag{25}$$

in which, w_0 = particle's fall velocity in still water, w_t = particle's fall velocity in turbulent flow, d = particle diameter, $u_{f,max}$ = maximum fluid velocity.

And, the empirical values reported are:

$$\left(\frac{w_t}{w_0}\right) = 0.5, \quad if \quad \frac{u_{f,max}}{w_0} \approx 10 \tag{26}$$

$$w_t/w_o = 0.72 \tag{27}$$

the latter was confirmed by Murray (1970). But, Ludwick and Domurat (1982) found that the fall velocity of fine sand was not reduced significantly.

It can be concluded from the above observations that the ratio of w_t/w_0 is less than unity in turbulent flows but, approach to unity for small particles. In the present local scour simulation, Hwang's findings are used.

3.3 Bed Load Modeling

By considering the probability distribution of shear stresses, van Rijn's (1984) bed load formula can be written as

$$s_b = 0.053 d_{50} \sqrt{\Delta g d_{50}} \, \frac{P_1(n = 2.1) - P_2(n = 2.1)}{D_*^{0.3}} \quad (28)$$

Because the bed load movement is affected by shear stress and gravity, the direction of bed load flux may be different from the flow direction. By considering these forces and bed slope, the angle between directions of the bed load flux and the flow is determined by

$$\beta_1 = arctan[\frac{K_b cos\delta - G_x}{K_b sin\delta - G_y}] \quad (29)$$

$$\delta = arctan \left[\frac{\tau_x}{\tau_y} \right] \quad (30)$$

$$K_b = \frac{\pi}{4} d^2 \tau \quad (31)$$

$$G_x = \frac{1}{6} \pi d^3 (\gamma_s - \gamma) \cdot \frac{\frac{\partial z_0}{\partial x}}{\sqrt{1 + \frac{\partial z_0}{\partial x}}} \quad (32)$$

$$G_y = \frac{1}{6} \pi d^3 (\gamma_s - \gamma) \cdot \frac{\frac{\partial z_0}{\partial y}}{\sqrt{1 + \frac{\partial z_0}{\partial y}}} \quad (33)$$

where τ_x and τ_y are the components of the shear stress τ in x and y directions, respectively.

In fluvial morphology studies, bed elevation change can be determined by applying a two-dimensional sediment continuity equation to relate the bed elevation change with sediment transport fluxes in the flow. A source term to account for the suspension and deposition of suspended sediment near the bed, and the bed load fluxes are considered in the sediment continuity equation. The addition of the source term as shown in Equation (34) has improved the accuracy in predicting the bed elevation change, when the effect of suspended sediment transport cannot be omitted. Computation time required by this approach is less than that needed by the depth integration approach of van Rijn (1991).

$$(1 - \lambda)\frac{\partial z_0}{\partial t} + \frac{\partial s_{bx}}{\partial x} + \frac{\partial s_{by}}{\partial y} = \omega_t(s_0 - s_{*b}) \quad (34)$$

in which, λ is porosity, s_{bx} and s_{by} are the components of bed load transport rate in x and y directions, s_0 is the source term evaluated from the suspended concentration near the bed surface.

3.4 Suspended Load Modeling

The eddy viscosity of turbulence is related to a multitude of eddies of varying sizes (Yalin and Da Silva 1995). The probabilistic density distribution of eddy sizes, and eddy viscosity variation with eddy size are given in Figure 3. The eddy size, ι, will vary in space and time even if the turbulent flow is of steady state. Eddies may be combined with one another and with the mean flow. The smallest possible size of eddies, the microturbulence, is assumed to have $\iota_{min} \approx \nu/\upsilon_*$ (Kolmogrorov scale), and the largest possible scale, the macroturbulence, is approximately equal to the water depth $\iota_{max} \approx h$, so

$$\frac{\nu}{\upsilon_*} \leq \iota \leq h \quad (35)$$

This means that the larger the eddy, the larger the eddy viscosity. The eddy viscosity should, then, be expressed as

$$\nu_t \equiv c_\nu \cdot (\upsilon \cdot \iota) \quad (36)$$

where υ_* is shear velocity, υ is velocity scale of the eddy with length scale ι, and c_ν is a coefficient. According to experimental data (Streeter, 1961), $c_\nu = 0.37$ for plane jet; $c_\nu = 0.025$ for axial symmetrical jet; $c_\nu = 0.09 \sim 0.094$ for plane weak flow. In local scour simulation, $c_\nu = 0.09$ is used.

The random variables, υ and ι, aren't independent of each other, small ι corresponding to small υ, and large ι is related to large υ. If we consider a constant time period of

$$\bar{t} \sim h/u_{max} \quad (37)$$

where u_{max} is the maximum mean velocity of turbulent flow, and assuming that the relationship between υ and ι is

$$\upsilon \sim u_{max} \left(\frac{\iota}{h} \right)^2 \quad (38)$$

then, (36) will becomes

$$\nu_t = c_\nu h \cdot u_{max} \cdot \left(\frac{\iota}{h} \right)^3 \quad (39)$$

The probabilistic distribution of eddy sizes and the functional relation between eddy viscosity and eddy size as proposed for the present study are shown in Figure 3. For open channel flow with logarithmic velocity distribution near the wall is:

$$u_{max} = \left(\frac{1}{\kappa} ln \frac{h}{k_s} + B_s \right) \cdot u_* \quad (40)$$

Due to the similarity of the energy spectrum of turbulence in open channel flow and the Beta distribution (Davis, 1972), the latter was chosen to describe the distribution of eddy occurring frequency. The upper and lower limits of eddy sizes in this situation are h and ν/u_*, respectively. Let $\iota_1 = \iota/h, 0 < \gamma$, and $0 < \eta$

$$f(\iota_1, \gamma_1, \eta) \begin{cases} = \frac{\Gamma(\gamma_1 + \eta)}{\Gamma(\gamma_1)\Gamma(\eta)} \iota_1^{\gamma_1 - 1}(1 - \iota_1)^{\eta - 1}, \\ 0 < \iota_1 < 1, \\ = 0 \qquad otherwise \end{cases} \quad (41)$$

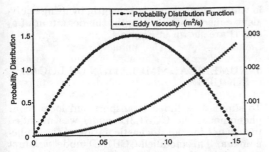

Figure 3. Probabilistic distribution of eddy sizes and thier corresponding values of eddy viscosity

In the vicinity of local scour region (due to the strong vortex movement) the constant value along vertical direction can be used to estimate the mean value of viscosity, which is

$$\overline{\nu_t} = \kappa u_* h / 6 \tag{42}$$

Let $\iota_0 = \nu / u_* h$, so

$$\int_{\iota_0}^{1} f(\iota_1, \gamma_1, \eta) \cdot \nu_t d\iota_1 = \frac{\kappa u_* h}{6 c_\nu} \tag{43}$$

$$\int_{\iota_0}^{1} f(\iota_1, \gamma_1, \eta) \cdot h \cdot u_{max} \cdot \left(\frac{\iota}{h}\right)^3 d\iota_1 = \frac{\kappa u_* h}{6} \tag{44}$$

From (44), $\gamma_1 = 2, \eta = 2$ are resulted.

It is well known that with the increase of the flow velocity and shear stresses acting on the bed, the sediment particles are likely to rotate or saltate along the bed surface. When the lifting force in the turbulent flow becomes comparable to the component of the weight of submerged sediment particles normal to the bed surface, sediment particles could be suspended. Bagnold (1966) stated that a particle only remains in suspension when vertical velocity component of the turbulent eddies exceeds the particle's fall velocity. Following this approach, the criterion for suspension is

$$\theta_{crs} = \frac{u_*^2}{(G-1)gd_{50}} \tag{45}$$

Based on the criterion of van Rijn's (1991) saltation length, sediment particle suspension would occur when the saltation length is equal to or larger than 100 d_{50}:

$$\theta_{crs} = \frac{16}{(D_*)^2} \frac{u_*^2}{(G-1)g \cdot d_{50}}, \quad 1 \le D_* \le 10$$

$$\theta_{crs} = 0.16 \frac{u_*^2}{(G-1)g \cdot d_{50}}, \quad D_* \ge 10 \tag{46}$$

And, on the inclined bed or bank, the sediment entrainment and suspension criterion is modified by using the bed or bank slope angle β.

$$\theta_{crs} = \frac{16}{(D_*)^2} \frac{v^2}{(G-1)g \cdot \cos\beta d_{50}}, \quad 1 \le D_* \le 10$$

$$\theta_{crs} = 0.16 \frac{v^2}{(G-1)g \cdot \cos\beta d_{50}}, \quad D_* \ge 10 \tag{47}$$

The sediment suspension occurs only if $\theta_{crs} \ge 1.0$. It is important to note that u_* is a deterministic variable in van Rijn's formula, while in the present study, lift velocity v is introduced as a random variable. Based on the assumption that the sediment suspension is a random process its occurrence may be estimated by:

$$\Upsilon = P(\theta_{crs} \ge 1.0) \tag{48}$$

In practice, the sediment suspension is neglected, if $\Upsilon \le 5\%$.

4 PROBABLISTIC DISTRIBUTION OF EFFECTIVE BED SHEAR-STRESSES

As mentioned previously, local scours are induced by the presence of obstructures and/or structures in nature rivers. Because they intensify turbulence and generates vortices around obstructions. The empirical approaches to estimate bed erosion based solely on bed shear stress should be modified. Additional factors mentioned above, which affect local scours significantly should be included.

The probability distribution function of shear stresses used for local scour can be written as:

$$p(\tau) = [2\pi\sigma^2]^{-\frac{1}{2}} exp\left\{-\frac{1}{2}\left[\left(\frac{\tau-\tau_0}{\sigma}\right)^2\right]\right\} \tag{49}$$

$$\sigma = \sigma_\tau + \sigma_v + \sigma_t \tag{50}$$
$$\sigma_\tau = \rho u_\tau^2 \tag{51}$$
$$\sigma_t = \rho u_t^2 \tag{52}$$
$$\sigma_v = \rho u_v^2 \tag{53}$$

σ_τ, σ_t and σ_v are standard variation of shear stresses due to ensemble averaged turbulent velocity, turbulent kinetic energy and the strength of vorticity of the flow, respectively.

Assuming that the experimentally obtained logarithmic law of the wall describing of turbulent velocity near the wall is valid approximately (Ikeda 1989 and van Rijn 1991):

$$\frac{u}{u_\tau} = \frac{1}{\kappa} ln\left(\frac{z}{z_0}\right) \tag{54}$$

where: z_0 = zero velocity level($u=0$, $z=z_0$), with

Hydraulic smooth regime: $z_0 = 0.11\frac{\nu}{u_\tau}$ for $\frac{u_\tau k_s}{\nu} \le 5$

Hydraulic rough regime: $z_0 = 0.033 k_s$ for $\frac{u_\tau k_s}{\nu} \ge 70$

217

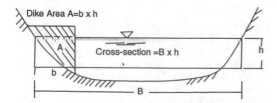

Figure 4. Approximation of channel cross-section

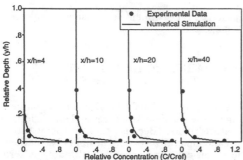

Figure 5. Numerical model verification based on experimental data of nonequilibrium suspended sediment transport (van Rijn, 1981)

Hydraulic transition regime: $z_0 = 0.033k_s + 0.11\frac{\nu}{u_\tau}$ for $5 \le u_\tau k_s/\nu \le 70$

From equation (54), the u_τ can be determined. Additionally, one can evaluate:

$$u_v = \begin{cases} c_1(v_b - v_0)l, & v_b > v_0 \\ 0, & v_b < v_0 \end{cases} \qquad (55)$$

$$u_t = \begin{cases} c_2(\sqrt{k_t} - \sqrt{k_0}), & \sqrt{k_t} > \sqrt{k_0} \\ 0, & \sqrt{k_t} < \sqrt{k_0} \end{cases} \qquad (56)$$

with k_0, v_0 = kinetic energy of turbulence and vorticity of uniform flow, which can be evaluated using the data calculated from the mean flow upstream of the local scour; k_t, v_b = kinetic energy of turbulence and vorticity at each point, and C_1 and C_2 are empirical coefficients.

$$\tau_0 = 1.2\sigma_\tau \qquad (57)$$

Equations (55) and (56) are filter functions, which filter out the fluctuation components generated by uniform flow, so that only excessive vorticity and turbulence energy induced by the structure are taken into account to estimate the effective shear stresses for simulating the local scour development. For the uniform flow, $\sigma_v = 0$ and $\sigma_t = 0$, the equation return its original form.

$$l = b_d/B \cdot b_d \qquad (58)$$

in which, l is characteristic length, by using equivalent concept of dike length, the definition of b_d and B are shown in Figure 4.

5 NUMERICAL SIMULATION OF LOCAL SCOURS

To simulate sediment transport and local scour phenomena, the CCHE3D model was modified to account for the stochastic nature of these phenomena. This modified CCHE3D model was first verified by the general sediment entrainment/suspension experiment of van Rijn (1981). See Figure 5. The flow and sediment properties used in his experiments are: h=0.25m, u_*= 0.03m, k_s/h=0.03, $\frac{\rho_s}{\rho}$=2.65, d_m=0.23mm. The predicted sediment concentration using the general diffusion equation agrees well with the measured nonequilibrium transport process over a flat loose bed. This verification enhanced the confidence in the newly modified model, before it is applied to simulate the development of a local scour hole near an in-stream structure.

Two cases of the experiment of clear water local scour near a groyne-like structure, reported by Rajaranam and Nwachukwu(1983), are selected to test the validity of the newly generalized CCHE3D model in local scour simulation. The details of the two experiments are: (i) u=0.67ft/s, h=0.505ft, $d_m = 1.45$mm, $\frac{\rho_s}{\rho} = 2.65$; and (ii) u = 0.68ft/s, h=0.385ft, d_m=1.45mm, $\frac{\rho_s}{\rho}$=2.65. The coefficients c_1 and c_2 in (56) and (57) are calibrated using data from the initial period of the Case (i). The calibrated coefficients are kept constant and used for longer time simulations. Good agreement of the scour hole depth development has been obtained for both cases, one of them is illustrated in the Figure 6. The size and shape of the scour hole and deposition mount are shown in Figure 7 are also reasonable.

To check the difference between stochastic and deterministic approach, the deterministic formula of shear stress ($\tau_0 = 1.2\sigma$), bed load, suspended load, diffusivity of suspended sediment, and entrainment criterion corresponding to (28), (24), (42) and (46) are used. Results of the comparison of maximum scour depths using these two approaches for an initial period of 700 seconds are given in Figure 8. It is seen that the two approaches predicted the same trend of the growth of the scour hole depth. As expected, the scour hole depth simulated by the stochastic approach grows faster and has a higher value of maximum value than those predicted by deterministic model. The simulated scour hole depth developments exhibit mild fluctuations rather than increase monotonically. These results may be caused by the fact that there are fluctuations in the flow caused by water waves reflected from, and vortices generated around, obstructions. The main difference between these two approaches is that in the deterministic approach, no bed particles can be entrained and suspended, unless the entrainment criterion is satisfied under the particular flow condition. While in the stochastic approach, some bed materials are entrained, suspended, and transported downstream following probability theory even if the deterministic sediment entrainment criterion is not satisfied.

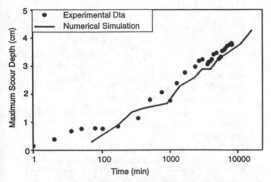

Figure 6. Development of maximum scour hole depth in time. Comparison between numerical simulation and experimental data (Rajaratnam, 1983)

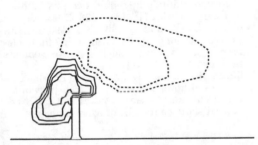

Figure 7. Simulated local scour and deposition (after 237.9 min)

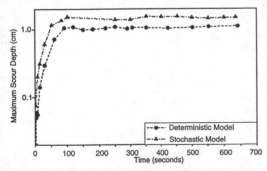

Figure 8. Simulated maximum scour hole depth developments in time. Comparison between deterministic and stochastic models.

6 CONCLUSIONS

By introducing the stochastic theory into the numerical model to account for the probabilistic distribtuions of suspended particle velocity, particle's diffusivity, flow's eddy sizes and energy spectrum, and effective bed shear stresses, etc., the generalized CCHE3D model has been proven effective in simulating both the erosion on a flat loose bed and the development of a local scour around the tip of a spur-dike. Due to the fact that the free surface flows and sedimentation processes in nature are stochastic rather than deterministic, they should be modeled by stochastic approach.

More specifically the general diffusion equation including effects of high order concentration gradients is capable of obtaining more realistic prediction of diffusion phenomena of suspended sediment in turbulent flow. Additionally, the consideration of probabilistic distribution of eddy sizes and their corresponding eddy viscosity and the empirical functional relation between the particles' eddy diffusivity and flow's eddy viscosity has also contributed significantly to the accuracy of the model. Numerical model results agree well to the experimental data for the non-equilibrium sediment transport and bed erosion reported by van Rijn (1981) and local scour development experiments of Rajaratnam and Nwachukwu (1983).

The newly developed scheme to calculate components of the bed load sediment movement on an inclined plane and the modified criterion for sediment entrainment by including the bed slope effect due to gravitation and the stochastic approach to estimating the lifting force have all contributed to the capability of the simulation model in obtaining realistic bed load transport.

The probabilistic distribution of the effective instantaneous bed shear-stress including the effect of both the turbulence kinetic energy and vorticity strength caused by obstructions in the flow is also an important factor to the enhancement of model's accuracy.

In short, comparisons between experimental data and numerical results of the generalized simulation model have proven that the stochastic approach is capable of predicting more realistic local scour phenomena. But, the findings reported in this paper should be considered as preliminary. More research on the applications of stochastic theory to model the local scour and more importantly the fluvial hydraulic processes are underway and shall be reported in the future.

ACKNOWLEDGEMENT

This work is a result of research sponsored in part by the USDA Agricultural Research Service under Specific Cooperation Agreement No. 58-4431-9-075 (monitored by the USDA-ARS National Sedimentation Laboratory) and in part by

The University of Mississippi. Professional contributions and assistance of Dr. Yafei Jia, Ms. Janice Mills and Mr. Keqiang Zheng are gratefully acknowledged.

REFERENCES

Bagnold, R. A. (1966). "An approach to the sediment transport problem from general physics." Prof. paper 422-1, USGS, Washington.

Brush, L. M. Jr. (1962). "Exploratory study of sediment diffusion." J. Geophys. Res., 67(4).

Celik, I. and Rodi, W. (1988). "Modelling suspended sediment transport in Nonequilibrium situations." J. Hydr. Engrg., 114(10), 1157-1191.

Chiew, Y. M. and Melville, B. W. (1989). "Local scour at bridge piers with non-uniform sediments." Proc. Instn Civ. Engrs, Part 2, Paper 9409, 215-224.

Davis, T. T. (1972). "Turbulence phenomena." Academic press, 51-52.

Hoffmans, G. J. C. M. and Booij, R. (1993). "Two-dimensional mathematical modelling of local-scour holes." J. Hydr. Res., 31(5), 615-635.

Hwang, P. A. (1985). "Fall velocity of particles in oscillating flow." J. Hydr. Engrg., 111(3), 485-502.

Ikeda, S. (1989). "Sediment transport and sorting at bends." J. Hydra. Div., ASCE, 108(11), 1369-1373.

Jia, Y. F. and Wang, S. S. Y. (1993). "3D numerical simulation of flow near a spur dike." Proc. First Int'l. Conf. on Hydro-Science and -Engineering. Washington, D.C., 2150-2156.

Kwan, R. T. F. and Melville, B. W. (1994), "Local scour and flow measurements at bridge abutments." J. Hydr. Res., 32(5), 661-673.

Laursen, E. M(1962). "Scour at bridge crossings." Trans. ASCE, Vol. 127.

Li, Y. Y. and Huang, H. (1989). "A turbulent diffusion model for suspended load." Proc. Fourth Int. Symp. on River Sedimentation. Bejing, China, 563-570.

Li, Y. Y. (1995). "Unsteady solution of incompressible Navier-Stokes equation." Report, CCHE, The U. of Mississippi.

Ludwick, J. C. and Domurat, G. W. (1982). "A deterministic model of sediment motion in a turbulent fluid." Marine Geology, Vol. 45, 1-15.

Murray, S. P. (1970). "Settling velocities and vertical diffusion of particles in turbulent water." J. Geo. Res., 75(9), 1647-1654.

Pais, A. (1982). "Subtle is the lord - the science and the life of Albert Einstein." Clarendon Press. 94-98.

Rajaratnam, N and Nwachukwu, B. A. (1983). "Erosion near groyne-like structures." J. Hydr Res., 21(4), 277-287.

Streeter, L. Victor., (1961), "Handbook of fluid dynamics." McGraw-Hill book company, Inc.

Toro, B. F. M. and Wang, S. S. Y. (1993). "Validation of a 3-D open channel flow model ." Proc. First Int. Conf. on Hydro-Science and -Eng'g. Washington, D.C., 2109-2114.

van Rijn, L. C. (1981). "Entrainment of fine sediment particles; development of concentration profiles in a steady, uniform flow without initial sediment load." Report No. M1531, Part 2, Delft Hydr. Lab.

van Rijn, L. C. (1984). "Sediment transport, Part 1: Bed load transport." J. Hydr. Eng'g., 110(10), 1431-1457.

van Rijn, L. C. (1986). "Mathematical modelling of suspended sediment in non-uniform flows." J. Hydr. Engrg., No. 112, 433-455.

van Rijn, L. C. (1987). "Mathematical modelling of morphological processes in the case of suspended sediment transport." Thesis, Dept. of Fluid Mech., Delft Univ. of Technology, Technical Report, Delft, The Netherlands.

van Rijn, L. C. (1991). "Handbook Sediment Transport by Currents and Waves." Delft Hydraulics.

Vittal, N., Kothyari, U. C., & Morteza H. (1993). "Clear-water scour around bridge pier group." J. Hydr Engrg., ASCE, 120(11), 1309-1319.

Wang, G. Q. (1989), "The kinetic theory for solid/liquid two-phase flow." Ph. D. Thesis, Tsinghua Univ., Bejing.

Wang, G. Q. and Ni, J. R. (1991). "The kinetic theory for dilute solid/liquid two-phase flow." J. Engrg. Mech., 116(12), 2738-2748.

Wang, S. S. -Y. and Hu, K. K., (1990), "Improved methodology for formulating finite element hydrodynamic models." Finite elements in fluids, Vol. 8, 457-477.

Yalin, M. S. and Da Silva, A. M. F. (1995). "Mechanics of flows and sediment transport in alluvial channels." lecture notes of CCHE advanced short course, U. of Mississippi.

Stochastic Hydraulics'96, Tickle, Goulter, Xu, Wasimi & Bouchart (eds) © 1996 Balkema, Rotterdam. ISBN 90 5410 817 7

Method for performance evaluation of hydrologic and hydraulic models

J.C.Yang & K.C.Yeh
Department of Civil Engineering, National Chiao-Tung University, Hsinchu, Taiwan, China

Y.K.Tung
Wyoming Water Resources Center & Statistics Department, University of Wyoming, Laramie, Wyo., USA

ABSTRACT: This paper presents a practical methodological framework that allows a systematic evaluation of the performance of a hydrologic or hydraulic model. The method uses the Latin hypercubic sampling technique for generating synthetic system parameters based on which the model outputs are produced and its performance examined. The method will be illustrated through an application to a selected sediment transport model.

1 INTRODUCTION

For most hydrologic and/or hydraulic engineering applications, models are applied to help engineers analyze the problem at hand so that proper design and decision can be made. Most models used in today's engineering works are computerized numerical models that are often very complex. The success of model application is often dictated by the model capability, user's knowledge of the physical process involved, and his/her skill of using the model. Knowledge about a model's capabilities is essential to avoid its misuse. Other than some descriptions in the user's manual, the general behaviour of the model under a variety of conditions is often unknown. Therefore, a method that allows systematic model evaluation is useful. Such a method should provide insight to both model developers and users with regard to model's performance. This is especially true if the model is to be applied to a system in which model inputs and parameters cannot be quantified with certainty.

A model, simple or complex, is an abstract form of the reality. Without exception, all hydrologic and hydraulic models, to a certain extent, involve assumptions, limitations, idealizations, and uncertainties. The models under consideration herein are complicated computer simulation models to predict the responses of hydrologic and hydraulic systems. The exercise of developing a model is a mixture of artistic and scientific endeavor. Due to the subjectivity involved, the performance of a model, when applied to predict the response of a hydrosystem, is inevitably susceptible to some surprises, good or bad. The primary objective in a model evaluation is to explore the model's behavior and its computational performance under a variety of conditions defined by input data.

2 MODEL PERFORMANCE EVALUATION PROCEDURES

Evaluation of model performance can be done by using: (1) measured data including field and experimental data; (2) analytical solution; and (3) synthesized data. Comparison of model results with the measured data, field data in particular, for model evaluation is intuitively appealing. However, the approach suffers from one major disadvantage. Model evaluation cannot be performed for a variety of different situations due to the rigid restrictions of physical condition imposed on the data. It is generally impractical to acquire large amounts of field/experimental data for all conceivable hydrologic and hydraulic conditions because of time, manpower, and fiscal constraints. The second main concern using measured data, especially field data, is the model calibration for empirical coefficients. Calibration often introduces subjectivity of a model user. In addition, the user's ability to grasp the model behavior dictates the success or failure of calibration and model application.

Although analytical solutions generally require significant idealization of real-life systems, it is a practical approach for model evaluation because exact model solution can be obtained relatively inexpensively. The idea behind using the analytical solution for model evaluation is that, if a model would perform satisfactorily in a realistic and complex setting, it should be able to model the behavior of a simplified and idealized system. Application of an analytic model to a simplified hydrosystem requires a minimum amount of calibration and subjectivity in the model evaluation. When the exact solution to the idealized hydrosystem is available, the issue of accuracy of model performance can be addressed.

Using synthesized data for model evaluation lies between using measured data and analytical solutions and is the main focus of this paper. Although the fine details and irregularities of the real-life hydrosystem cannot be duplicated, the main characteristics of the system can be captured through proper synthesis. To achieve a successful synthesis of an hydrosystem, it is essential to capture the important parameter characteristics defining systems observed in the field. Therefore, the use of synthesized data for model evaluation allows one to examine model behavior under a wide spectrum of system conditions. Note that in the use of synthesized data for model evaluation, the true solution to each generated synthetic data set is not observed. Still, model evaluation can be made in a less quantitative manner by judging the reasonableness of computed results based on one's experience and/or by comparing the relative performance of different models.

In remainder of this paper, a framework is presented to systematically evaluate the performance of a hydrologic or hydraulic model based on the use of synthesized data. Specifically, the procedure is applied to the well-known sediment transport model, HEC-6, developed by the U.S. Corps of Engineers.

3 GENERATION OF SYNTHETIC DATA

As stated above, it is important that the synthesized data used in model evaluation captures the essential features of the real-life system that the model is intended to represent. In this particular application the synthetic data set to be generated are limited to the following conditoins:

1. Prismatic channels with rectangular cross-section having subcritical flow condition;
2. Flow rate is at bankfull discharge or less. Therefore, no floodplain cross-section is needed;
3. No lateral flow is considered;
4. Channels are straight with constant initial channel slope;
5. River banks are assumed rigid without bank erosion;

6. No sediment enters from the upstream boundary;
7. Channel bed sediment is homogeneous having gradation described by a lognormal distribution;
8. Physical properties of water such as temperature, viscosity, and those of sediment including specific weight, and porosity are assumed to be constant;
9. The stage-discharge relationship is described by $Q=aD^b$ with Q being the discharge, D being the water depth, and 'a' and 'b' are constants.

Parameters that are essential in defining synthetic fluvial channel conditions are channel cross-section parameters and profile, flow conditions, sediment size distribution, and flow rates. Many of these channel conditions can be described by the following dimensionless parameters (1) Froude Number (F_r); (2) Reynolds Number (R_h); (3) width-depth ratio (ω); (4) initial channel bed slope (S_o); (5) median sediment diameter (d_{50}); (6) sediment geometric standard deviation (σ_g).

The above six basic parameters are treated as random variables and are used to generate channel cross-section, hydraulic characteristics, and sediment composition for a synthetic channel. Some of the above parameters are not statistically independent in physical environments. Such correlative relationships should be considered in the process of generating synthetic data set, if possible. Although the true relationships among the parameters are not entirely known, field data are available in the literature that provide parameter ranges and their correlations. Table 1 summarizes data obtained from the literature that were used to determine the parameter statistical properties and their correlative relations. Information such as this is important for generating synthetic data that are representative of the real-life fluvial system.

4 ALGORITHM FOR GENERATING SYNTHETIC FLUVIAL CHANNEL DATA

The two nondimensional flow condition parameters *Fr* and *Rn*, and width-depth ratio (ω) can be used to generate channel cross-section geometry, dimensions, and the corresponding bankfull discharge. Blench (1969) and Ackers and Charlton (1970) consider the bankfull discharge as the dominant event controlling the channel form. From the Froude number and Reynolds number, the corresponding hydraulic radius (R) can be computed by

$$R = \left(\frac{R_n \upsilon}{F_r} \right)^{2/3} \frac{1}{g^{1/3}} \qquad (4.1)$$

where υ = kinematic viscosity of water and g = gravitational acceleration. The bankfull discharge per unit width (q_b), from the Reynolds number, can be computed as

$$q_b = R_n \upsilon \qquad (4.2)$$

Based on the generated width-depth ratio (ω) the channel width (W), depth (D), and total bankfull discharge (Q_b) can be determined, respectively, as

$$D = \left(\frac{2 + \omega}{\omega} \right) R \qquad (4.3)$$

$$W = \omega D \qquad (4.4)$$

$$Q_b = q_b W \qquad (4.5)$$

In the case of a wide channel ($\omega > 10$), the channel depth can simply be approximated by

$$D = \left(\frac{R_n \upsilon}{F_r} \right)^{2/3} \frac{1}{g^{1/3}} \qquad (4.6)$$

Using a lognormal distribution, the median sediment diameter (d_{50}) and sediment gradation (σ_g) completely define the sediment size distribution. The median sediment size and sediment gradation are considered as independent random variables. Based on the generated d_{50} and σ_g, the mean and standard deviation of log-transformed sediment diameter can be determined by

$$\mu_{\ln(d)} = \ln(d_{50}) \qquad (4.7)$$

$$\sigma_{\ln(d)} = \ln(\sigma_g) \qquad (4.8)$$

The range of sediment size distribution can be divided into several size intervals serving as inputs to the sediment transport model.

The sediment transport model under investigation, namely, HEC-6, requires specifying Manning's roughness coefficient. This surface friction coefficient is not considered separately as an independent stochastic input to the model in synthetic data generation. Instead, Manning's coefficients are generated for each synthetic case indirectly according to the empirical relationships (Camacho and Yen, 1992) which are functions of Froude and Reynolds numbers, sediment properties, viscosity, and others.

5 METHOD FOR GENERATING SYNTHETIC DATA SET

To generate synthetic data for the channel and flow condition, the latin hypercubic sampling (LHS) technique is applied. The LHS technique is a special stratified sampling method that randomly samples each variable over its range in a stratified manner. By the LHS technique, the plausible range of each random variable is divided into several equal-probability intervals from which a random sample for the random variable is taken. An LHS algorithm to generate statistically independent random variables is described by Mckay (1988).

When the basic channel parameters are statistically independent, the generation of synthetic data by the LHS technique can be applied to each variable separately. However, this procedure is not appropriate in the current problem context for generating synthetic channel geometry because the flow and sediment conditions are correlated. When the flow and sediment parameters are correlated normal or lognormal random variables, orthogonal decompositions can be applied to convert the correlated variables into uncorrelated ones. Then, in the transformed space the LHS procedure described above for generating independent variables can be applied. The generated random variables in the uncorrelated space are transformed back to their original space.

6 MODEL RELIABILITY ANALYSIS

The term 'model reliability' is herein referred to as the ability of a model to successfully execute the required simulation task under a variety of conditions. Successful execution requires that no premature termination of the program and the simulated system responses are physically reasonable. For illustration, 50 sets of synthetic fluvial channels were generated in this study by the LHS technique based on the probability distributions and the statistical properties of the basic fluvial channel parameters as shown in

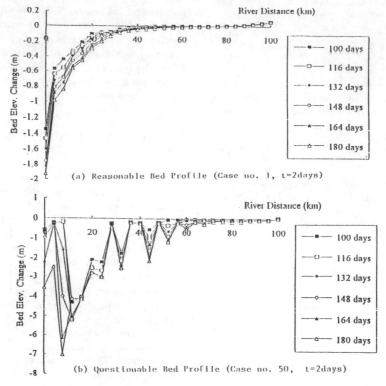

Figure 1. Example Bed Profiles Generated by HEC-6

Tables 2-3. For each of the 50 cases generated, the sediment transport model, HEC-6, was applied to simulate the temporal and spatial variations of water depth, sediment discharge, channel bed elevation, and sediment size distribution under a constant flow condition. In the simulation runs, Yang's sediment discharge equations was used to compute the sediment transport rate. The total simulation period was 180 days and the channel length was 100 km. The computation increments of $\Delta x = 4$ km and $\Delta t = 2$ days or 0.1 days were used.

For all 50 synthesized cases, HEC-6 could successfully complete its numerical computations. However, some of the channel bed profiles produced by HEC-6 appear to be questionable. Since there is no upstream sediment inflow, the expected channel bed profile should appear as shown in Fig.1a with progressive degradation over time and diminishing erosion from upstream to downstream. Unrealistic physical model outputs show significant oscillations in the bed profile as illustrated in Fig.1b.

To examine how the occurrence of oscillation were affected by the hydraulic and sediment characteristics of the fluvial channels, the degree of oscillation between two adjacent points in the simulated

channel bed profile were arbitarily classified into two categories: (1) oscillation less than or equal to 2 feet and (2) oscillation greater than 2 feet. Using this categorization, series of binary scatter plots such as Fig.2a and 2b were constructed. Figure 2a clearly indicates that, when Δt is 2 days, the excessive oscillation occurs predominantly when the initial channel slope is about 0.0014 or steeper. On the other hand, excessive oscillation can occur over the entire range of width-depth ratio, Froude number and Reynolds number as shown in Fig. 2b and other parameters.

Model outputs without significant bed elevation oscillation do not necessarily mean that the model outputs are reasonable. For some cases, the bed profiles simulated by HEC-6 differ significantly when different time increments were used. Using a smaller time increment resulted in much deeper scouring at the upstream end than using a larger time increment. An example of this type of inconsistency are shown in Figs. 3a and 3b. Similar to the study of bed profile oscillation, the degree of inconsistency is categorized into four classes and some representative scatter plots are shown in Figs. 4a and 4b. The plots indicate that the degree of inconsistency in HEC-6 compuation

Table 1. Summary of Fluvial Channel Parameters From Different Sources

	Mean	Stdev	Minimum	Maximum	Source
Q (cfs)	*	*	353	17657	Nixon (1959)
	*	*	106	19423	Charlton et al (1978)
	*	*	5.7	5156	Schumm (1960)
	5288	16859	200	3000000	Blench (1964)
	*	*	35	84755	Bogardi (1974)
	419	1980	13	554617	Leopold/Maddock (1953)
	129	490	4.2	206238	Leopold/Wolman (1957)
	170	423	35	100011	Williams (1978)
W (ft)	87.3	95.5	8	550	Burns (1971)
	*	*	25	800	Schumm (1960)
	640	1138	21	5000	Blench (1964)
	270	431	16	2610	Leopold/Maddock (1953)
	240	253	13	1578	Leopold/Wolman (1957)
	112	116	10	750	Williams (1978)
D (ft)	*	*	0.4	7.10	Hack (1957)
	1.9	1.1	0.5	5.41	Burns (1971)
	4.9	7.5	0.5	51.00	Leopold/Maddock (1953)
	3.6	3.9	0.5	25.92	Leopold/Wolman (1957)
	5.6	5.6	1.1	27.13	Williams (1978)
W/D	*	*	2.0	125.0	Hack (1957)
	*	*	2.5	138.8	Schumm (1960)
	48.1	39.2	8.0	200.0	Blench (1964)
	59.6	44.8	11.7	331.8	Leopold/Maddock (1953)
	81.5	97.3	11.7	837.5	Leopold/Wolman (1957)
	21.3	11.0	6.8	62.0	Williams (1978)
S_o (m/m)	*	*	0.0008	0.17	Hack (1957)
	0.003	0.0057	0.00002	0.035	Leopold/Wolman (1957)
	0.01	0.0135	0.0005	0.118	Ward (1981)
	0.001	0.001	0.00007	0.004	Blench (1964)
	*	*	0.00004	0.002	Bogardi (1974)
	0.0054	0.0087	0.00018	0.0416	Williams (1978)
d50 (mm)	*	*	5	630	Hack (1957)
	77.6	74.4	0.049	268	Leopold/Wolman (1957)
	*	*	6	70	Penning/Townsend (1978)
	*	*	4	174	Limerinos (1969)
	*	*	0.02	8	Schumm (1960)
	*	*	0.14	0.95	Dawdy (1961)
	*	*	0.011	35	Bogardi (1974)
	45.6	43.8	0.5	190	Williams (1978)
	*	*	0.09	0.33	Culbertson et al. (1972)
σ_g	*	*	1.4	9.4	Dawdy (1961)
	*	*	1.22	2.08	Culbertson et al. (1972)

Table 2. Statistical Properties of Fluvial Channel Parameters Adopted in the Application

	Mean	Stdev	Minimum	Maximum	Distribution
Fr	0.3	0.03	*	*	Lognormal
Rn	6,000,000	600,000	*	*	Lognormal
W/D	120	12	*	*	Lognormal
So	0.0015	0.00015	*	*	Lognormal
d50 (mm)	1	0.1	*	*	Lognormal
σ_g	*	*	1.0	3.8	Uniform

225

Table 3. Correlation Among Fluvial Channel Parameters Adopted in the Application

	Fr	Rn	W/D	So	d50	σ_g
Fr	1	0	-0.2	0.5	0	0
Rn	0	1	-0.5	0.1	0	0
W/D	-0.2	-0.5	1	-0.1	-0.4	0
So	0.5	0.1	-0.1	1	0.6	0
d50 (mm)	0	0	-0.4	0.6	1	0
σ_g	0	0	0	0	0	1

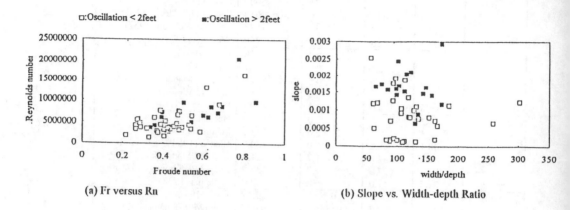

(a) Fr versus Rn

(b) Slope vs. Width-depth Ratio

Figure 2. Scatter Plots for Bed Profile Oscillation by HEC-6 with Δt= 2 days

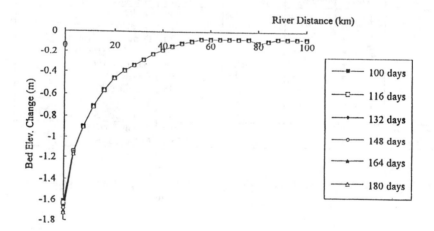

(a) Time Increment, Δt=2 days

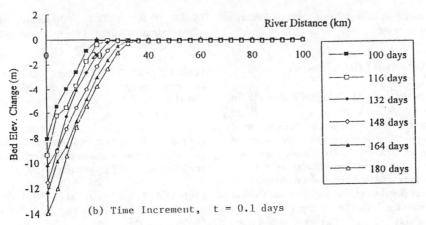

(b) Time Increment, t = 0.1 days

Figure 3. Simulated Bed Profile by HEC-6 Significantly Affected
by Computational Time Increment (Case no. 11)

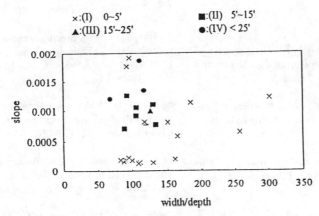

(a) Slope vs. Width-depth Ratio

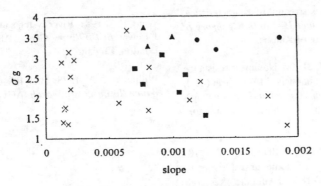

(b) Slope vs. Sediment Gradation

Figure 4. Scatter Plots Showing the Degrees of Inconsistency
in Simulated Bed Profiles by HEC-6

227

increses as initial channel slope and non-uniformity of bed material increase.

7 SUMMARY AND CONCLUSIONS

This paper presents a methodological framework for evaluating the performance of a hydrologic or hydraulic model. In particular, the methodology is illustrated through the use of the well-known sediment transport model, HEC-6. This type analysis provides useful information about the behavior of a model which can be beneficial for both model developers and users. Model developers, on one hand, can utilize such information for further improving and enhancing model's capability. On the other hand, model users can base such information to specify model inputs to avoid possible model failure.

REFERENCES

Ackers, P. & Charlton, F.G. 1970. The geometry of small meandering channels. *Proceedings* of Institute of Civil Engineers, 12: 289-317.

Blench,T. & Qureshi,M.A. 1964. Practical regime analysis of river slopes. *Journal of Hydraulic Div.*, ASCE, 90(HY2): 81-98.

Blench,T. 1969. *Mobile-Bed Fluviology*, The University of Alberta Press. Edmonton, Alberta, Canada.

Borgardi, J. 1974. *Sediment Transport in Alluvial Streams*, Akademiai Kiado, Budapest, Hungary.

Burns, C.V. 1971. Kansas streamflow characteristics, Part-8. *In-Stream Channel Hydraulic Geometry of Streams in Kansas.* State of Kansas.

Camacho, R. & Yen, B.C. 1992. Nonlinear resistance relationship for alluvial channels. In B.C. Yen (eds), *Flow Channel Resistance: Centennial of Manning's Formula*, Water Resources Publications, Littleton, CO.

Charlton, F.G., Brown, P.M., & Benson, R.W. 1978. The hydraulic geometry of some gravel rivers in Britain. *Report*, INT-180, Hydraulic Research Station, Wallingford, England.

Culvertson, J.K., Scott, C.H., & Bennett, J.P. 1972. Summary of alluvial channel data from Rio Grande conveyance channel, New Mexico. USGS *Professional Paper*, 562-J.

Dawdy, D.R. 1961. Depth-discharge relations of alluvial streams - discontinuous rating curves. USGS *Water Supply Paper*, 1498C: 1-16.

Hack, J.T. 1957. Studies of longitudinal stream profiles in Virginia and Maryland. USGS *Professional Paper*, 294-B:45-97.

Leopold, L.B. & Maddock, T. 1953. The hydraulic geometry of stream channels and some physiographic implications. USGS *Profession Paper*, 252.

Leopold, L.B. & Wolman, M.G. 1957. River channel patterns: braded, meandering and straight. USGS *Professional Paper*, 282-B.

Limerinos, J.T. 1969. Relation of Manning's coefficient to measured bed roughness in stable natural channels. USGS *Professional Paper*, 650D: 215-221.

McKay, M.D. 1988. Sensitivity and uncertainty analysis using a statistical sample of input values. In Y. Ronen (eds), *Uncertainty Analysis*: 145-186. CRC Press Inc., Boca Raton, FL.

Nixon,M. 1959. A study of bankfull discharge of the rivers of England and Wales. *Proceedings*, Institute of Civil Engineers, 12: 157-174.

Penning-Rowsell, E.C. & Townsend, J.R.G. 1978. The influence of scale on the factors affecting stream channel slope. *Trans. Inst. Br. Geography*, 3:395-415.

Schumm, S.A. 1960. The shape of alluvial channels in relation to sediment type. USGS *Professional Paper*, 352-B.

Ward, R.C. 1981. River systems and river regimes. In J. Lewin, *Britsh Rivers*, George Allen and Unwin Ltd., London, England.

Williams, G.P. 1978. Bankfull discharge of rivers. *Water Resources Research*, AGU, 14: 1141-1154.

Stochastic Hydraulics'96, Tickle, Goulter, Xu, Wasimi & Bouchart (eds) © 1996 Balkema, Rotterdam. ISBN 90 5410 817 7

Monte Carlo method with spherical process for Dirichlet problem in hydraulics and fluid mechanics

Ke-Zhong Huang
Department of Geography, Zhongshan University, Guangzhou, China

ABSTRACT: The Dirichlet problems in hydraulics and fluid mechanics can be solved by the Monte Carlo method. In this paper we introduce briefly the solution of the Dirichlet problem by the Monte Carlo method with spherical process, this approach is more effective than the Monte Carlo method with traditional lattice-point process. As a demonstration example the problem of the flow around a cylinder between two plates is solved by the spherical process method. The advantages and characteristics of the method are also described.

1 INTRODUCTION

It is well know that the Monte Carlo method can be used to solve probabilistic problems and deterministic problems. The Dirichlet problem is a deterministic problem which can be estimated by the method with lattice-point process(Shreider 1964). In hydraulics and fluid mechanics there are many Dirichlet problems such as the Laplace's differential equation in terms of the velocity potential or stream function in the potential flow. The Monte Carlo simulation is rarely applied to these problems because it requires more computing time than other deterministic numerical methods.

In order to utilize the advantage of the Monte Carlo meethod, we introduce briefly the theory of the method with spherical process presented by M. E. Muller(1956). On the basis of this theory, the solving means of the 2 & 3-dimensional Dirichlet problem are given, and apply it to a solution of the problem of the flow around a cylinder between two plates.

2 FUNDAMENTAL THEORY AND SOLVING PROCEDURE

Throughout this paper, D will denote a bounded finitely connected n-dimensional domain, and $\Gamma(D)$ will denote the boundary of D. A point belonging to D $+\Gamma(D)$ is denoted by x, where x has coordinates$(x_1, x_2, \ldots, x_n)$. Given the domain D and a continuous function f(x) defined on the boundary $\Gamma(D)$, the n-dimensional Dirichlet problem consists of finding a function u(x) that is continuous in D and satisfies

$$\Delta u(x) = \sum_{i=1}^{n} \frac{\partial^2 u(x)}{\partial x_i^2} = 0, \quad x \in D \tag{2.1}$$

$$u(x) = f(x), \quad x \in \Gamma(D) \tag{2.2}$$

The Monte Carlo method for the Dirichlet problem was traditionally together with the lattice-point process. The finite difference equation that corresponds to Eq.(2.1) for a 2 or 3-dimension is

$$u(P) = \frac{1}{2n} \sum_{i=1}^{2n} u(P_i) \tag{2.3}$$

where P is a inner node, and P_i are nearest nodes to point P. It had been shown (Shreider 1964) that the mathematical expectation of the values on the boundary f(x) formed by the random walk starting from a inner node P to a boundary node Q is an approximate solution of Eq.(2.3), i.e.,

$$u(P) = \frac{1}{N} \sum_{i=1}^{N} f(Q_i) \tag{2.4}$$

where N is the simulating times of the random walk starting from the node P to node Q.

In order to use the Monte Carlo method with spherical process, we introduce briefly its some preliminary definitions and theorems given by M. E. Muller (1956) as follow:

Definition 1. Maximum n-sphere: K(x) is the maximum n-dimensional sphere with centre x belonging to D+Γ(D) and radius R if R=inf $\| x'-x \|$, $x' \in \Gamma(D)$; $K_S(x)$ denotes the surface of K(x), where $K_S(x)$ is empty if $x \in \Gamma(D)$.

Definition 2. The spherical process: The n-dimensional spherical process originating from x is $\Phi(x)$, where

(A) $\Phi(x) = \{S(x, \zeta), 0 \quad \zeta \quad 1\}$; i.e., $\Phi(x)$ is the totality of all sequences of points $S(x, \zeta)$.

(B) Each value of ζ specifies a sequence of points $S(x, \zeta) = \{ P_{i+1}(x, \zeta), \ i=0,1,\ldots \}$ generated according to the following stipulations:

(1) About the point $P_0(x, \zeta) = x$, determine the maximum n-dimensional sphere $K(P_0)$;

(2) Select the point $P_1(x, \zeta)$ uniformly at random on $K_s(P_0)$;

(3) The point $P_{i+1}(x, \zeta)$ is determined recursively from $P_i(x, \zeta)$ and $K(P_i)$ in the same manner as $P_1(x, \zeta)$ was determined from $P_0(x, \zeta)$.

Definition 3. The elementary set: A subset E of the boundary Γ (D) will be called an elementary set if it consists of a finite number of mutually disjoint nonabutting simple surfaces on the boundary $\Gamma(D)$, where a simple surface is a homeomorphism of the surface of a n-dimensional shpere.

Theorem 1. The spherical process originating from x belonging to a domain D converges to a point on the boundary $\Gamma(D)$ with probability 1.

Theorem 2. The probability, $Pr(S(x, \zeta), E, D)$, that the spherical process originating from a point x of the domain D will converge to the set E, is a harmonic function of x in D and

$$\lim_{\substack{x \to x_0 \\ x \in D}} Pr(S(x, \zeta), E, D) = \begin{cases} 1, & x_0 \in E \\ 0, & x_0 \in (\Gamma(D) - E) \end{cases}$$

Theorem 3. For any point x_0 belonging to D, the value $u(x_0)$ of the solution $u(x)$ of the Dirichlet problem for the domain D and the boundary value function $f(x)$ is obtained by taking the integral of a Poisson type of $f(x)$ with respect to the kernel $Pr(S(x_0, \zeta), E, D)$ on $\Gamma(D)$ or by taking the mathematical expectation of the composed function $f(P(x_0, \Gamma(D), \omega))$:

$$u(x_0) = \int_{\Gamma(D)} Pr(S(x_0, \zeta), dx, D) f(x)$$
$$= \int_{\Omega} f(P(x_0, \Gamma(D), \omega)) d\omega \qquad (2.5)$$

where ω denotes a sample of the Brownian motion process, $\Omega = \{ \omega \}$ is a set of the samples ω.

On the basis of the preceding theorems of n-dimensional spherical process, we obtain some results about the 3-dimensional spherical process as follows:

The probability that the spherical process originating from a point x belonging to D will converge to the spherical surface S belonging to $\Gamma(D)$ is the solution of the Dirichlet problem, and the value of the solution is 1 on the surface S belonging to $\Gamma(D)$, else 0 on the $\Gamma(D) - S$. Therefore, the probability that the 3-dimensional spherical process originating from an inner point of the sphere will converge to the spherical surface between angles $\theta_1 \leqslant \theta \leqslant \theta_2$ and $\phi_1 \leqslant \phi \leqslant \phi_2$ can be expressed by Poisson's formula of the spherical domain:

$$Pr(r, \theta, \phi) = u(r, \theta, \phi)$$

$$= \frac{R}{4\pi} \int_{\Phi_1}^{\Phi_2} \int_{\Theta_1}^{\Theta_2} \frac{R^2 - r^2}{(R^2 - 2rR\cos\beta + r^2)^{3/2}} \sin\theta' d\theta' d\phi' \qquad (2.6)$$

where

$$\cos\beta = \cos\theta \cos\theta' + \sin\theta \sin\theta' \cos(\phi' - \phi)$$

In respect to the centre, $r=0$, we have

$$Pr = u = \frac{1}{4\pi} \int_{\Phi_1}^{\Phi_2} \int_{\Theta_1}^{\Theta_2} \sin\theta d\theta d\phi$$

$$= \frac{(\phi_2 - \phi_1)(\cos\theta_1 - \cos\theta_2)}{4\pi} \qquad (2.7)$$

i. e., the probability Pr is proportional to the area of the spherical surface restricted by the angles, and the probability distribution that the spherical process originating from the centre will converge to a point on the spherical surface is uniform.

For the same reasons as above, the probability that the 2-dimensional spherical process (i.e., circular process) from an inner point of the circle will converge to the circular arc between angles ϕ_1 and ϕ_2 can be expressed by Poisson's formula of circular domain:

$$Pr(r, \phi) = u(r, \phi) = \frac{1}{2\pi} \int_{\Phi_1}^{\Phi_2} \frac{R^2 - r^2}{R^2 - 2rR\cos(\phi - \phi') + r^2} d\phi' \qquad (2.8)$$

In respect to the centre, $r=0$, we have

$$Pr = u = \frac{\phi_2 - \phi_1}{2\pi} \qquad (2.9)$$

i. e., the probability Pr is proportional to the length of the circular arc restricted by the angles, and the probability distribution that the circular process originating from the centre will converge to a point on the circumference is uniform.

It may be seen that the solving procedure of the solution u (P_0) of 2-dimensional Dirichlet problem solved by the Monte Carlo method with circular process can be proposed as follows:

1. Take the maximum circle with the centre P_0 belonging to the domain D and radius R having the shortest distance between the centre P_0 and the boundary $\Gamma(D)$(Fig. 1).

2. Divide the circumference into m arces with the same length (m is an appropriate large number, e.g., 360). Randomly select an arc with probability 1/m from the m arces, and take the middle point P_1 on the arc selected.

3. The point P_{i+1} is determined recursively from

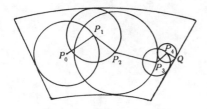

Fig.1

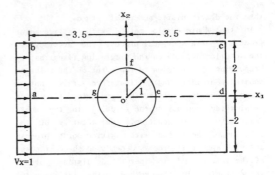

Fig.2

P_i in the same manner as P_1 was determined from P_0.

4. If the distance between the point P_i and the nearest boundary point Q belonging to $\Gamma'(D)$ is less than the permitted distance error δ, we obtain a boundary value $f(Q)$, and the point P_i is referred to as the terminal point P_t.

5. Let N is the simulating times of the preceding random process from the point P_0 to the poin P_t. By doing N times the random process, an approximate estimate for $u(P_0)$ can be obtained, namely

$$u(P_0) = \frac{1}{N} \sum_{i=1}^{N} f(Q_i) \qquad (2.10)$$

where N is a large number.

It is obvious that the solution procedure of 3-dimensional Dirichlet problem by the Monte Carlo method with spherical process is similar to the procedure above.

3 EXAMPLE

Flow around a circular cylinder between two parallel plates is shown in Fig.2. Only the upper half of the flow field should be considered, since the flow is symmetric.

The Problem has the Laplace equation about the stream function ψ (x) and boundary condition as Eq. (2.1)and(2.2), in which the boundary $\Gamma'(D)$ is $\overline{abcdefg}$. Obviously it is a 2-dimensional Dirichlet problem.

Let all the quantities such as velocity, stream function and length having dimensionless. We assume that the velocity Vx=1, and the value of ψ on the boundary $\overline{agfed}$ is zero, then the value of ψ the boundary $\overline{bc}$ is 2, and the value of ψ along the boundaries $\overline{ab}$ and $\overline{dc}$ are variable linearly from 0 to 2, respectively.

At any point(x_1, x_2), the radius of the maximum circle is

$$R(x_1,x_2) = \text{Min} \{ \ |x_1-3.5|, |x_1+3.5|, |x_2-2|, |x_2|,$$

$$\sqrt{x_1^2+x_2^2} - 1 \}, \qquad P(x_1,x_2) \in D \qquad (3.1)$$

Let $\delta = 0.1$, N=10000. On the basis of the solution procedure, we obtain the results of the stream function on three points$(0,1.25),(0,1.50)$and$(0,1.75)$, respectively,as follows:

$$\psi(0,1.25)=0.584; \quad \psi(0,1.50)=1.093; \quad \psi(0,1.75)=1.561$$

The results of the same problem solved by the Monte Carlo method with lattice-point process,in which let the spacing of the square lattice equal to 0.25, are

$$\psi(0,1.25)=0.535; \quad \psi(0,1.50)=1.051; \quad \psi(0,1.75)=1.531$$

The results of the same problem solved by finite element discretizatioin(73 nodes, 111 elements) are

$$\psi(0,1.25)=0.571; \quad \psi(0,1.50)=1.076; \quad \psi(0,1.75)=1.546$$

The results of the analytical solution to the same problem(Chung 1978)are

$$\psi(0,1.25)=0.608; \quad \psi(0,1.50)=1.102; \quad \psi(0,1.75)=1.559$$

4 DISCUSSION AND CONCLUSIONS

It is sometimes not necessary to evaluate $u(x)$ at all inner points, but only at some important points. Consider,for instance,the problem of local scour at bridge abutments. We may only evaluate the stream velocities near by the bridge abutments. In such situation to use the Monte Carlo method for solving Dirichlet problems is appropriate, while the deterministic numerical methods such as the method of finite element or finite difference have to solve the simultaneous equations for $u(x)$ at all nodes as a whole. Such being the case, the necessary storage capacity and the computing time in a computer for the Monte Carlo method may be less than with the deterministic numerical methods. On the contrary, if $u(x)$ at all inner points need to be evaluated, the Monte Carlo method will require more computing time than the deterministic numerical methods.

In the Monte Carlo method with lattice-point process or with spherical process the accuracy of the solution depends on the simulating times that is proportion by inversion to the error of the solution squared (Shreider 1964) . Therefore it is not a effective way to increase the simulating times for raising the accuracy of the solution in the Monte Carlo method. In addition, the accuracy of the solution of the method with lattice-point process and with spherical process depends on the coarseness of the lattice and the permitted distance error, respectively. In the method with lattice-point process if we increase the coarseness of the lattice for raising the accuracy, it will require more computing time for the solution. However, the spherical process method can fit more boundary shape than the lattice-point process method, and its computing time will has a little increase as the permitted distance decreases. Therefore the spherical process method is more effective than the lattice-point process method.

On the basis of the theory of the Monte Carlo method with spherical process, we have given the solving means of the 2 & 3-dimensional Dirichlet problems. It may be seen that the spherical process method provides a special approach to the evaluation of Dirichlet problems in hydraulics, fluid mechanics and other technical fields.

REFERENCES

Chung, T. J. 1978. Finite element analysis in fluid dynamics. USA: McGraw-Hill.
Muller, M.E. 1956. Some continuous Monte Carlo methods for the Dirichlet problem. Ann. of Math. Statistics 27, No.3, 569-589.
Shreider, Yu. A. 1964. Method of statistical testing, Monte Carlo method. Amsterdam: Elsevier Pub. Co.

Stochastic Hydraulics'96, Tickle, Goulter, Xu, Wasimi & Bouchart (eds) © 1996 Balkema, Rotterdam. ISBN 90 5410 817 7

Movement of sediment using a stochastic model with parameters varying as a function of discharge

Hsieh Wen Shen & Guillermo Q. Tabios III
University of California, Berkeley, Calif., USA

ABSTRACT: This paper presents a stochastic model in unsteady flow conditions for determining the movement of the centroid of bedload tracer particles in a mobile bed stream. Field data collected by the U.S. Geological Survey from the East Fork River, Wyoming, U.S.A were used in this study.

INTRODUCTION

The dispersion characteristics of sediment particles moving downstream can be described by an alternate sequence of step lengths and rest periods; where both step lengths and rest periods are conceived as random variables. Stochastic models with this framework have been proposed by Einstein (1937), Hubell and Sayre (1964), Shen and Cheong (1973), among others. Most of these models have been implemented under fairly steady flow conditions. This paper presents an implementation of a stochastic model with parameters varying with flow discharge for determining the movement of the centroid of bedload tracer particles. The bed material tracer particle data as collected by Emmett and Myrick (1985) and Emmett, Myrick and Meade (1980) from the East Fork River, Wyoming, U.S.A. were used here to illustrate the use of this stochastic model.

STOCHASTIC MODELING

It is assumed that the sediment particles movement follows a stochastic process. Specifically, based on the work of Shen and Cheong (1973), the dispersion characteristics of sediment particles moving downstream can be described by an alternate sequence of step lengths and rest periods where both step lengths and rest periods are random variables. Following Shen and Cheong (1973), appropriate probability distribution functions of the

step length X and rest period T are given by:

$$f_X(x) = \frac{k_1}{\Gamma(r)}x^{r-1}e^{-k_1 x}, \quad x \geq 0; \ f_X(x)=0, \ otherwise$$

(1)

and

(2)
$$f_T(t) = k_2 e^{-k_2 t}, \quad t \geq 0; \ f_T(t) =0, \ otherwise$$

The distribution $f_X(x)$ is the two-parameter gamma distribution with scale parameter k_1 and shape parameter r while $f_T(t)$ is the exponential distribution with scale parameter k_2. Applications of these models to sediment transport require two major assumptions: 1) the time it takes for a particle to jump from a point to another is relatively small compared to the rest period; and 2) the sediment movement is one layer so that the transport rate is equal to the ratio of average step length and average rest period.

According to Shen and Cheong (1973), the relationships of rate of movement (i.e., the centroid denoted by $\bar{x}$) and the sediment transport rates (denoted by q_s) as well as the variations of the parameters k_1, r and k_2 for different flow conditions are not well understood. Einstein (1937) attempted to relate $d\bar{x}/dt$ with sediment transport rates and found the $d\bar{x}/dt$ is approximately proportional to $q_s^{2/3}$ for uniform sediment sizes while $d\bar{x}/dt$ is directly proportional to q_s for nonuniform sizes. Sayre and Hubell (1965) also suggested that the transport rate is proportional to the centroid, bed porosity and the depth of the zone of particle movement in which quantifying this depth requires knowledge of bedforms which is not fully known (Shen and Cheong, 1973).

FLOW-VARYING PARAMETERS

In order to account unsteady flow conditions, it is proposed that the scale parameters k_1 and k_2 be functions of the logarithm of discharge (denoted by Q) as follows:

$$k_1^{-1} = a_1 + b_1 \log(Q) \qquad (3)$$

and

$$k_2^{-1} = a_2 + b_2 \log(Q) \qquad (4)$$

where a_1, b_1, a_2 and b_2 are parameters to be estimated and the $\log(.)$ is the logarithm operator. In the model implementation below, the shape parameter r of the gamma distribution in Eq. (1) is set equal to 1.0 or 2.0. Also, the scale parameters as simple linear functions of discharges have also been tried but it is not good compared to being a function of logarithm of discharge.

A nonlinear optimization model proposed by Box (1965) is used here to estimate the parameters a_1, b_1, a_2 and b_2. The objective function is to minimize the sum of squares of *true* centroid gradients (difference of centroid distance today and yesterday) for a given day and computed centroid gradients. Basically, the algorithm is implemented as follows. For given initial values of the parameters a_1, b_1, a_2 and b_2, the distribution functions in Eqs. (1) and (2) are used to generate and track the movement of, say, 1000 particles, with parameters k_1 and k_2 varying as a function of daily flows. With these set of parameters, the centroid gradients can be computed. Then, given the computed centroid gradients and the true centroid gradients, a new objective function value is computed which is subsequently used as a basis for improving the parameter estimates.

BED MATERIAL AND RIVER HYDRAULICS DATA USED

The data used here were collected by Emmett and Myrick (1985) with accompanying data on river hydraulics by Emmett, Myrick and Meade (1980). The data were collected around May through July of 1979 in a 3.3 km reach (between sections 0000 and 3295) of the East Fork River, Wyoming as shown in Fig. 1. The cross section numbers indicated in Fig. 1 correspond to the distance in meters between the cross section and the bedload trap, located downstream of the reach.

Prior to the 1979 data collection, about 6000 kilograms of bed material was excavated from the streambed near the right bank of section 0570. The bed material was mostly coarse sand and fine gravel ranging in size from 0.25 to 8.0 mm. Then the sample bed material were divided into three subsamples of about 2000 kilograms each which were then dyed with pink, blue and orange sufficiently thin fluorescent coating. The dyed material were subsequently reinjected into the streambed as line sources of trace particles. Only the pink particles are analyzed in this study which were injected at section 3037 on May 18, 1979. Then, the locations of the particles were measured daily by bed material sampling at the downstream sections (from the point of injection) for a period of about two months. The pink particles have been monitored at 36 sections. The measurements were expressed in terms of total number of particles per 100 grams of bed material for size ranges of 0.25 to 0.5 mm, 0.5 to 1.0 mm, 1.0 to 2.0 mm, 2.0 to 4.0 mm and 4.0 to 8.0 mm.

The hydraulic data available include: bihourly flow discharges at sections 0000 and 3295; daily water surface elevations at 39 sections; and, bed elevations measured daily at 1-meter intervals at the 39 sections.

DATA ANALYSIS

For the hydraulic data, the daily discharges as shown in Fig. 2 were computed by taking the average of all bihourly readings for that particular day. The discharges at the two end sections are practically the same so that for our analyses, the daily discharges were considered to be uniform throughout the study reach. However, there is a significant daily

234

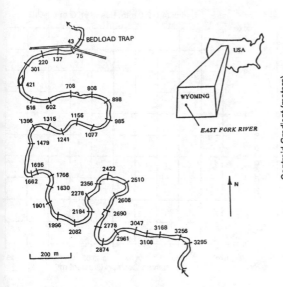

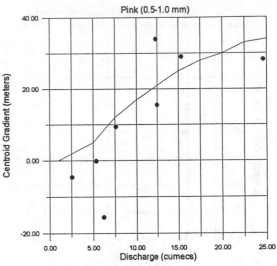

Figure 1.   Location of cross sections along the 3.3-km study reach, East Fork River, Wyoming, U.S.A.

Figure 3.   Raw centroid gradient (solid dots) and fitted centroid gradient (solid line) plotted against discharge for pink particles of sizes 0.5 to 1.0 mm.

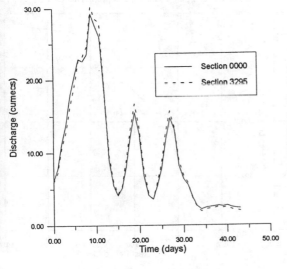

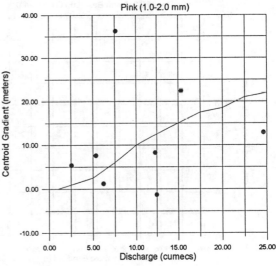

Figure 2.   Time series of daily discharge (day 0 is May 18, 1979).

Figure 4.   Raw centroid gradient (solid dots) and fitted centroid gradient (solid line) plotted against discharge for pink particles of sizes 1.0 to 2.0 mm.

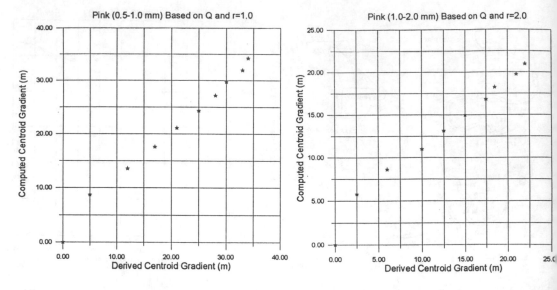

Figure 5. Plot of computed centroid gradient against observed centroid gradient for pink particles of sizes 0.5 to 1.0 mm with shape parameter equal to 1.0.

Figure 7. Plot of computed centroid gradient against observed centroid gradient for pink particles of sizes 1.0 to 2.0 mm with shape parameter equal to 2.0.

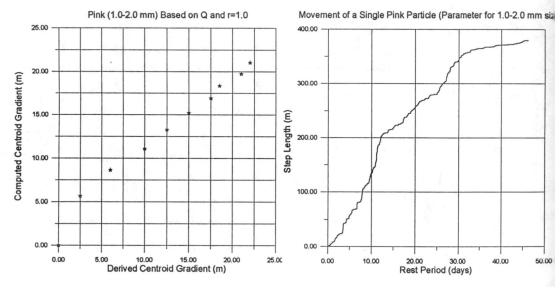

Figure 6. Plot of computed centroid gradient against observed centroid gradient for pink particles of sizes 1.0 to 2.0 mm with shape parameter equal to 1.0.

Figure 8. Movement of a single pink particle with estimated parameters for particles sizes of 1.0 to 2.0 and shape parameter equal to 1.0.

236

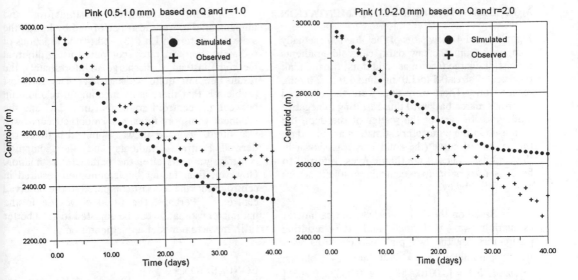

Figure 9.　Time series plots of simulated and observed centroids for pink particles of sizes 0.5 to 1.0 mm with shape parameter equal to 1.0.

Figure 11.　Time series plots of simulated and observed centroids for pink particles of sizes 1.0 to 2.0 mm with shape parameter equal to 2.0.

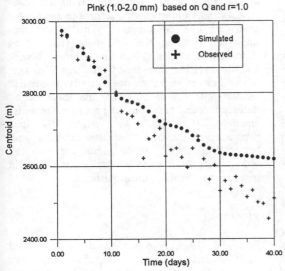

Figure 10.　Time series plots of simulated and observed centroids for pink particles of sizes 1.0 to 2.0 mm with shape parameter equal to 1.0.

variations of the discharges in the two month period which must be considered in the analysis.

The velocities at the different cross sections were also computed. For a given day, the velocities vary only at certain subreaches in the study reach. For the reach between sections 2300 and 3100 in particular (where the pink particles were observed during the entire study period), the velocities did not vary so much for a given day. The assumption that the velocities are about the same in the reach for a given day is desirable especially that we only relate the parameters in our stochastic model as a function of discharge instead of velocity.

For the bed material data, plots of the number of particles as a function of distance for selected days during the sampling period have also been examined. It was noted that generally, the pink particles between 0.5 and 2.0 mm gave the most reasonable and consistent pattern in terms of movement of the bulk of particles as indicated by the peaks and shapes of the particle distribution in time and distance. For this study, only the pink particle sizes of 0.5 to 1.0 mm and 1.0 to 2.0 mm are analyzed.

MODEL ESTIMATION AND IMPLEMENTATION

Figures 3 and 4 show plots of the *observed* (actually computed from the raw data) centroid gradients against discharge (shown as solid dots) for the pink particles of sizes 0.5 to 1.0 mm and 1.0 to 2.0 mm, respectively. There were days were the observed centroids march backward thus resulting in negative gradients. In view of the scatter of the data, the data used in the parameter estimation are based on the smoothed curve (the solid lines) as shown in these plots. These smoothed curves, referred to here as *true* or *derived* centroid gradient where simply fitted by eye.

Based on the derived curves, the parameter estimation was performed and the resulting parameters for the two pink particle sizes with shape parameter r = 1.0 are as follows: $[a_1, b_1, a_2, b_2]$ = [0.00100, 12.2169, 0.00028, 0.2656] for sizes between 0.5 to 1.0 mm; and, [0.00572, 4.7693, 0.00155, 0.8850] for sizes between 1.0 to 2.0 mm. Plots of the derived centroid gradients against the computed centroid gradients for the two pink particle sizes are given in Figs. 5 and 6. A reasonable fit is shown in these plots. For comparison, parameter estimation of pink particles of sizes 1.0 to 2.0 with shape parameter r = 2.0 was performed with resulting parameters $[a_1, b_1, a_2, b_2]$ = [0.00139, 5.9208, 0.000029, 0.62976]. Plot of the derived against computed centroid gradients for this case is given in Fig. 7 which also shows a good fit.

With the estimated parameters, the model was implemented by using the actual time series data of discharges by generating and tracking 1000 particles. Then, the centroids for these generated data were calculated. For illustration purposes, Fig. 8 shows graphically the movement of single pink particle using the parameters for particle sizes of 1.0 to 2.0 mm with shape parameter r = 1.0. For this particular case, this particle move about 380 meter in 45 days. Plots of the simulated together with the actual (observed) centroids are given in Figs. 9, 10 and 11 for the pink particles of sizes 0.5 to 1.0 mm and 1.0 to 2.0 mm with r = 1.0 as well as with the pink particles of size 1.0 to 2.0 with r = 2.0. These plots shows that the model is quite satisfactory. This is especially true for the pink particles of sizes 1.0 to 2.0 mm and also, it may be noted also that there is not so much difference in the simulated centroid whether the shape parameter r is equal to 1.0 or 2.0.

Model parameter estimation and implementation were also performed by letting the scale parameters in Eqs. (3) and (4) be functions of sediment transport rates (computed using different methods) instead of discharge. Unfortunately, the results are not quite satisfactory. The foremost problem is that there is no meaningful relationship between the centroid gradients (for a day) and the sediment transport rates. A smoothed curve was also fitted. A good fit was obtained between the derived centroid gradients and the computed centroid gradients after the parameter estimation. However, final model implementation resulted in centroids moving too fast compared to the observed centroids. Perhaps, the fitted curve use in the parameter estimation can be adjusted to get a better fit in the actual model implementation.

CONCLUSIONS

This paper demonstrates that the movement of centroids of bedload tracer particles can be modeled using a stochastic model with the dispersion of sediment particles as described by an alternate sequence of step lengths and rest periods conceived as random variables. The scale parameters of the model vary as a function of discharge to account for unsteady flow conditions. This model was implemented using the field data collected by Emmett and Myrick (1985) and Emmett, Myrick and Meade (1980) from East Fork River, Wyoming, USA. For the migration of centroids of sediment particles with size of 1.0 to 2.0 mm, there was no difference found between using a shape parameter equal to 1.0 or 2.0. Also, relating the scale parameters of the model with flow discharge were better than relating these parameters with the sediment transport rates for tracking the centroid movement of bedload tracer particles.

REFERENCES

Box, M.J., 1965, A new method of constrained optimization and a comparison with other methods, Computer Journal, Vol. 8, p. 45-52.

Emmett, W.W. and R.M. Myrick, 1985, Field data describing the movement and storage of sediment in the East Fork River, Wyoming, Part V. Bed-material tracers, 1979 and 1980, U.S.G.S. Open-File Report 85-169, Denver, Colorado.

Emmett, W.W., R.M. Myrick and R.H. Meade, 1980, Field data describing the movement and storage of sediment in the East Fork River, Wyoming, Part I. River hydraulics and sediment transport, 1979, U.S.G.S. Open-File Report 85-169, Denver, Colorado.

Hubell, D.W. and W.W. Sayre, 1964, Sand transport studies with radioactive tracers, Jour. of Hydraulics Div., ASCE, Vol. 90, No. HY3, May, pp. 39-68.

Einstein, H.A., 1937, Bed load transport as a probability problem, D.S. Thesis, Federal Inst. of Tech., Zurich, Switzerland.

Sayre, W.W. and D.W. Hubell, 1965, Transport and dispersion of labelled bed material, North Loup River, Nebraska, U.S. Geological Survey Prof. Paper 433-C.

Shen, H.W. and H.F. Cheong, 1973, Dispersion of contaminated sediment bed load, Jour. of Hydraulics Div., ASCE, Vol. 99, No. HY11, Nov., pp. 1947-1965.

Acknowledgment: We particularly thank Dr. William W. Emmett of the U.S. Geological Survey of Denver, Colorado for suggesting to us this study.

3 Waves and coastal processes

Stochastic Hydraulics'96, Tickle, Goulter, Xu, Wasimi & Bouchart (eds) © 1996 Balkema, Rotterdam. ISBN 90 5410 817 7

A new statistical model for extreme water levels along the Dutch coast

Pieter H.A.J.M.van Gelder
Delft University of Technology, Faculty of Civil Engineering, Netherlands

ABSTRACT: The Netherlands is a low-lying country which has to protect itself against flooding from the sea and its rivers. Reliable flood defenses are essential for the safety of the country. The sea dikes are designed to withstand floods with a height of once every 10,000 years. This height is used to be calculated using statistics on sea levels measured along the Dutch coast since 1880. In this paper new calculations will be presented taking account of known sea floods in the period before 1880. A Bayesian framework has been developed for this.

1 INTRODUCTION

Approximately 40% of the Netherlands is below mean sea level; much of it has to be protected against the sea and rivers by dikes. In figure 1, a topographical map of the Netherlands with the North Sea, the river Rhine, its tributaries Waal and IJssel, and the Meuse is shown. The sea - and river dikes are clearly visible. The total length of the dikes is about 3000 km. Hardly any systematic statistical research has been done in connection with the height of the dikes before 1953. On February 1 1953, during a severe storm surge, in several parts of the Netherlands the sea dikes broke, approximately 150,000 ha polderland inundated through 90 breaches and 1800 people and thousands of cattle lost their lives. After this flood the Dutch government appointed a committee to recommend on an appropriate height for the dikes. The statistical work for this committee was under direction of professor D. Van Dantzig (Deltacommissie 1960). The recommendation of the committee was adopted by the government and most of the Dutch sea dikes have since been adapted to meet the new standard. Although the statistical analysis was the decisive argument for determining the new standard, a serious effort has been made to study the problem of determining an appropriate height for the sea dikes from an economic and also from a physical point of view. The economic analysis compared the cost of building dikes of a certain height with the expected cost of a possible inundation given that specific height. In the mathematical-physical analysis the effects of a wind

Figure 1: Topographical map of the Netherlands

field on a rectangular water bassin was studied. Now, more than 30 years later, both the number of observations of sea levels and the statistical methodology have grown considerably. This has led to a new investigation (Dillingh et. al. 1993, De Haan 1990). They investigated the problem solely from a statistical point of view of determining a level of the sea dikes such that the probability that there is a flood in a given year, equals p, where the number p is to be determined by the politics ranging between 10^{-3} and 10^{-4} (depending on the importance of the area under subject). For the data they used the complete set of high tide water levels recorded along the Dutch coast since the end of the 19th century. Although this set consists of lots of observations (over 100,000 per location), the problem of determing the dike level is a problem of far extrapolation: estimating a water level that can occur once every 10,000 years out of a data set of a small 100 years.

After the 1953 flood historical data about floods in previous centuries have been studied systematically (Gottschalk 1977, Jonkers 1989). Their reports give a good overview about the severeness of the floods that took place between 1500 and 1850. Up till now, this information hasn't been used in the calculation of the dike levels. In this paper Bayesian statistics will be shown to be useful to take account of historical data in the extrapolation problem. It will appear that in this Bayesian framework it is possible to include the global warming effect on earth and its result on the sea level rise as well.

The paper is organised as follows: the flood historical data and the assumptions on the sea level rise will be discussed in Sec.2. The measurements of sea level data since the end of 1800 will be discussed in Sec.3. We make a distinction there in POT and AM data (peaks over threshold and annual maxima). In Sec.4, we will look at a Bayesian model for the AM data, followed by a Bayesian model for the POT data in Sec.5. In Sec.7 a decision problem for the dikes. Finally in Sec.7, we will draw the conclusions.

2 FLOOD HISTORICAL DATA

Recently historical research has been undertaken to retrieve information about floods in the period 1500-1850 (Gottschalk 1977, Jonkers 1989). In diaries, signs of flood levels on old churches etc. a collection of old data could be gathered. The old data on flood levels however have to be taken as an indication, not as accurate data points. Different sources on the same sea flood give contradictive reports about the actual height which occured. That's why in this paper the floods will be classified into 4 classes: Class A for very severe floods; B for heavy floods; C for less heavy floods and D for light floods. Every class will be connected with a water level and uncertainty level, based on the historical research data and based on the sea level rise

data. Class A floods are modelled to be realisations from a normally distributed stochast of flood levels with a mean of 390cm and a standard deviation of 10cm, denoted as N(390,10). B - floods are realisations from N(360,10); C - floods are in N(330,10) and D - floods in N(300,10). The following floods in the period 1500-1850 are mentioned including its classification:

1570 - A	1775 - D
1672 - D	1776 - C
1686 - D	1806 - B
1715 - D	1808 - B
1717 - D	1825 - B

It is argued that the sea level rise is, in contrary of popular belief, not a new phenomenon but a known fact from geological observations (Vrijling 1994). About 8,000 BC, the northern edge of the North Sea was dry and connected the British Isles with the continent. As the present depth of this area is equal on average to MSL-35m, a sea-level rise of approximately 35m must have occured in the last 10,000 years. This means a rise of 35cm per 100 year. Since 1888 the sea levels at Hook of Holland are recorded intensively. If we perform a linear regression analysis on the year maxima data of Hook of Holland, we obtain as the estimated regression line:

$$y = \alpha + \beta x = 226.7 + (year-1888)*0.2017 \text{ cm}.$$

This line is dotted in figure 2. We observe a sea level rise of 20 cm in 100 years. From the 90% confidence limits, given in the figure by the 2 solid lines, we notice that the hypothesis $\beta = 0$ would also be accepted. However, in the Delta committee, the sea level rise for the Netherlands during the last centuries is determined at 20cm per 100 year, including the correction for the sinking of the soft soil of the polders (Deltacommissie 1960). We will apply a sea-level rise correction on both the flood historical data as well as the sea level data since 1888.

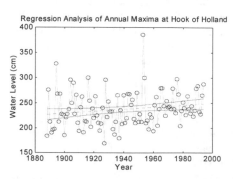

Figure 2: A regression analysis

3 SEA LEVEL DATA SINCE 1888

Since 1888 accurate recordings of the sea levels at Hook of Holland are availabe. For our purpose: determining the level of the sea dikes such that the probability that there is a flood in a given year, equals $p=1/10,000$ yr^{-1} (denoted as $x_{10,000}$ and called the 10^{-4} quantile), we filter the recordings to 2 data sets. The AM-data set, which consists of all the annual maxima of sea levels and the POT-data set, which consists of all the peaks of water levels over a certain threshold.

The AM-data set has the data of the period 1888 - 1995; in total n=108 points. The number of data points of the POT-data set depends on the level of the threshold (see Sec.5). Both the AM and POT data sets are corrected for the sea-level rise influence (of 20 cm per century).

4 MODEL AM DATA

4.1 Extreme value model

Assume that the maximum sea level in a period of 3 days is modelled by a stochastic variable X with (unknown) probability distribution F_X. Divide 1 year in 122 sections of 3 days. Let X_j be the maximum sea level in section j (j=1..122). For Hook of Holland it appears that the X_j's are independent; i.e. the period of 3 days is large enough to make storm 1 independent of storm 2. Let us now look at $X:=\max_{j=1..122} X_j$. From extreme value theory, it follows that X must tend towards an extreme value distribution (Castillo 1988). So X is a Frechet, Weibull or Gumbel distribution, dependent of the parent distribution of X_j. From a graphical analysis (plotting the AM-data on Gumbel paper and examining convexity, concavity or linearity), we suggest for a Gumbel distribution of the AM sea levels at Hook of Holland (see figure 3).

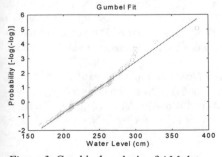

Figure 3: Graphical analysis of AM-data

4.2 *Frequentistic analysis*

With classical methods the calculation of the 10^{-4} quantile and its uncertainty will be presented. From Sec.4.1, we assume the AM data to be fitted by the Gumbel distribution with cumulative distribution function $F(x|\lambda,\delta)=\exp(-(x-\lambda)/\delta)$. The parameters λ and δ are estimated with a maximum likelihood method. This means that the estimators for λ and δ will be approximately (for large n), normally distributed. Their means are asymptotically equal to the true parameters (i.e. are asymptotically unbiased). Their variances are asymptotically equal to:

$$Var(\lambda)=-1/nE(\partial^2 logf(x|\lambda,\delta)/\partial\lambda^2)$$
$$Var(\delta)=-1/nE(\partial^2 logf(x|\lambda,\delta)/\partial\delta^2)$$

in which $f(x|\lambda,\delta)=\exp(-\exp(-(x-\lambda)/\delta))\exp(-(x-\lambda)/\delta)/\delta$ is the Gumbel probability density function. Under the assumption of a Gumbel likelihood model we have:

$$E(x_{10,000})=E(\lambda)-E(\delta)*log(-log(1-p))=E(\lambda)+qE(\delta)$$
$$Var(x_{10,000})=var(\lambda+q\delta)=var(\lambda)+q^2 var(\delta)+$$
$$+2q\rho(\lambda,\delta)\sigma_\lambda\sigma_\delta$$

Here is $p=10^{-4}$ the probability that there is a flood in a given year, $q=log(-log(1-p))=9.2103$ and $\rho(\lambda,\delta)$ the correlation coefficient between λ and δ and related with the covariance which is asymptotically given by $Cov(\lambda,\delta)=-1/nE(\partial^2 logf(x|\lambda,\delta)/\partial\lambda\partial\delta)$.

With the AM-data we get $E(x_{10,000})=483.6$ cm and $var(x_{10,000})=393.7$ or $\sigma(x_{10,000})=19.8$ cm.

The calculation of this 10^{-4} quantile and its uncertainty is based solely on the sea level data from the period since 1888. The data that we have from the period of 1500-1850 is impossible to use in this frequentistic analysis. That is why we like to perform a Bayesian analysis.

4.3 *Bayesian analysis*

The principle of a Bayesian analysis is as follows: First a prior distribution has to be determined in which we put our knowledge before any data is available. This prior distribution is therefore decided on subjective grounds. As soon as data becomes available, the prior distribution can be updated to the so called posterior distribution. Bayes Theorem plays the central role in this analysis:

$$p(\lambda|x) = \frac{f(x|\lambda)p(\lambda)}{\int_0^\infty f(x|\lambda)p(\lambda)d\lambda}$$

In this formula, the prior distribution is denoted by $p(\lambda)$, the likelihood model by $f(x|\lambda)$ and the posterior distribution by $p(\lambda|x)$. The integral assures that the posterior is a distribution function. For a good overview of Bayesian statistics, we refer to Bernardo and Smith 1994.

We continue the sea level analysis with a Bayesian approach in which the unknown parameters λ and δ are treated as random variabels. As argued in 4.1 we assume the likelihoodmodel to be given by Gumbel:

$$f(x_i|\lambda,\delta)=(1/\delta)^n\exp(-(\Sigma x_i-\lambda)/\delta)\exp(-\Sigma\exp(-(x_i-\lambda)/\delta))$$

in which x_j (j=1..108) are the AM data.

We start the Bayesian analysis with vague (or uninformed) prior distributions:

$$p(\lambda,\delta)=\lambda/(\lambda_{max}-\lambda_{min}) \times \delta/(\delta_{max}-\delta_{min})$$

i.e. 2 uniform distributed variabels between the wide bounderies $\lambda_{min}=100$, $\lambda_{max}=300$, $\delta_{min}=10$, $\delta_{max}=40$. The posterior distribution is then given with Bayes by:

$$p(\lambda,\delta|x)=Cp(\lambda,\delta) (1/\delta)^n \exp(-(\Sigma x_i-\lambda)/\delta) \times \exp(-\Sigma\exp(-(x_i-\lambda)/\delta))$$

in which C is the normalisation constant such that $p(\lambda,\delta|x)$ integrates to 1.
The posterior predictive can be calculated from:

$$P(x<x_B)=\int\int F(x_B|\lambda,\delta)p(\lambda,\delta|x)d\lambda d\delta$$

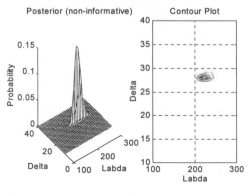

Figure 4: Posterior analysis

From the posterior predictive, $x_{10,000}$ follows from $P(x<x_{10,000})=1-10^{-4}$.

Figure 4 gives us the results of this non-informative analysis. We obtain almost the same estimators for the parameters λ and δ as in the frequentisitc method. $E(x_{10,000})=490.1$cm and $\sigma(x_{10,000})=21.7$cm.

We continue the Bayesian analysis with informed prior distributions. Take in mind that we have historical flood data to our access. We will use this information to get prior information on the δ and λ parameters.

Let X_1, X_2, ..., X_{350} represent the annual maxima in year 1501, 1502, ... 1850. The r-th order statistic of this serie is given by the r-th member in the new sequence $X_{1:350}$, $X_{2:350}$, ..., $X_{350:350}$, which is a rearranging of X_1, X_2, ..., X_{350} in increasing order.

If we assume that the annual maxima of the water level at Hook of Holland is modelled by a Gumbel distribution then the distribution of the r-th order statistic is given by:

$$f_{Xr:350}(x)=350!/(r-1)!/(350-r)!F^{r-1}(x)[1-F(x)]^{350-r}f(x)$$

in which $f(x)$ and $F(x)$ are the Gumbel p.d.f. and c.d.f. respectively. We are interested in the joint probability distribution of $X_{341:350}$, $X_{342:350}$, ..., $X_{350:350}$, given by:

$$f_{X341:350, X342:350,..., X350:350}(x_1,x_2,...,x_{10})=$$
$$350!\Pi_{i=1}{}^{10}f(x_i)\Pi_{j=1}{}^{11} [F(x_j)-F(x_{j-1})]^{r(j)-r(j-1)-1}/(r(j)-r(j-1)-1)!$$

in which $r(0)=0$, $r(1)=341$, $r(2)=342$, ... $r(10)=350$, $r(11)=351$, $x_0=-\infty$, $x_1 = x_{341:350}$,... and $x_{11}=\infty$.

On basis of the historical data we estimate with maximum likelihood the parameters λ and δ. We proceed until both parameters have reached its steady state. After 1000 simulations of $x_1, x_2,..., x_{10}$ this is the case. If we make a plot of λ against δ (figure 5), we observe a high correlation between the two parameters (correlation coefficient = -0.9939).

We have $E(\lambda)=220.3$, $\sigma(\lambda)=28.4$, $E(\delta)=36.18$, $\sigma(\delta)=6.23$ and $E(x_{10,000})=553.6$cm; based on 10 extreme floods in a period of 350 years. We model both parameters by the bivariate normal distribution:

$$f_{\Lambda,\Delta}(\lambda,\delta) = (2\pi \sigma_\Lambda\sigma_\Delta(1-\rho^2)^{-\frac{1}{2}})^{-1} \exp\{-2(1-\rho^2)^{-1}$$
$$[((\lambda-\mu(\Lambda))/\sigma(\Lambda))^2 - 2\rho (\lambda-\mu(\Lambda))$$
$$(\delta- \mu(\Delta))/\sigma(\Lambda)\sigma(\Delta) + ((\delta-\mu(\Delta))/\sigma(\Delta))^2]\}$$
$$\text{for } -\infty<\lambda<\infty, -\infty<\delta<\infty$$

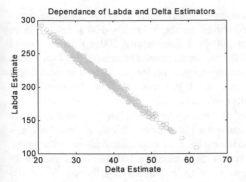

Figure 5: Prior distribution of (δ, λ)

With the informative prior as bivariate normal, we obtain the following results for the posterior: $E(\lambda)=$ 225.5, $var(\lambda)=$ 8.7, $E(\delta)=$ 34.70, $var(\delta)=$ 0.76, $\rho(\lambda,\delta)=-0.6385$ and $E(x_{10,000})=545.1$cm.

We notice that the role of the amount of sea level data since 1888 on the 10^{-4} quantile is of less effect because the quantile decreased from 553.6 to 545.1cm only.

5 MODEL POT DATA

5.1 *The threshold*

POT data is obtained by selecting peaks of sea levels above a certain level. There are no rules known in literature in order to determine a certain threshold. In this paper some sort of stability criterion is suggested to determine the threshold. First of all the POT data is modelled by an exponential distribution $e^{-\lambda x}$. Let x_j (j=1..n) denote the peak sea-levels above threshold T. Then we can calculate the 10^{-4} quantile as a function of T under the assumption of the exponential model.

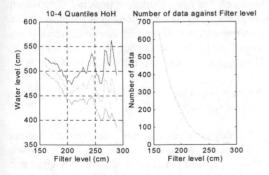

Figure 6: Threshold analysis

Figure 6 give the results (including its standard deviation interval). Where the 10^{-4} quantile remains more or less stable under variation of threshold, we choose the T. For the Hook of Holland data, our choice is T=200cm. The number of peaks over T for this choice is 195.

5.2 *Frequentistic analysis*

In contrary with the Gumbel model for the AM data, a lot of analytic solutions can be found for the exponential model for the POT data. The ML-estimator for λ is given by $\lambda=n/(\Sigma_{i=1..n}x_i)$.
The ML-estimator for the 10^{-4} quantile $x_{10,000,ML}$ follows from:

$P(x>x_{10,000,ML})=exp(-\lambda x_{10,000,ML})=1/10.000$ or
$x_{10,000,ML}=log(10.000)*\Sigma_{i=1..n}x_i/n$

The distribution of $x_{10,000,ML}$ can also be calculated analytically: From $x_{10,000,ML}=log(10,000)*\Sigma_{i=1..n}x_i/n$, and the fact that a summation of n exponential distributed stochasts, each with parameter λ is gamma distributed with parameters (n, λ), and that a constant k times a gamma distributed stochast with parameters (α,β) is also gamma distributed with parameters $(\alpha_1, \beta_1)=(\alpha,\beta/k)$, we have that $x_{10,000,ML}$ is gamma distributed with parameters $(n,n\lambda/log10000)$. So:

$E(x_{10,000,ML})=(log10000)/\lambda$
$Var(x_{10,000,ML})=(log^2 10000)/(\lambda^2 n)$.

5.3 *Bayesian analysis*

In contrary to the Gumbel model for the AM-data where analytical solutions are very difficult to derive, we can derive the Bayesian formulaes of the exponential model analytically very easily. The likelihood model is assumed to be given by an exponential model:

$f(x|\lambda)=\lambda^n exp(-\lambda\Sigma x_i)$

in which x_j (j=1..195) are the POT data. With a non-informative prior $(p(\lambda)=1/\lambda)$, we obtain as posterior:

$p(\lambda|x)=\lambda^{n-1}exp(-\lambda\Sigma x_i)/\int_{\lambda=0..\infty} \lambda^{n-1}exp(-\lambda\Sigma x_i)d\lambda$

This is nothing else than a gamma distribution with parameters $(n,\Sigma_{i=1..n}x_i)$.

The distribution of the 10^{-4} quantile from a Bayesian point of view $x_{10,000,B}$ can be derived from:

$P(x>x_{10,000,B})=\int_{\lambda=0...\infty}(1-F(x_{10,000,B}|\lambda))p(\lambda|x)d\lambda=$
$=(\Sigma_{i=1..n}x_i / (x_{10,000,B} + \Sigma_{i=1..n}x_i))^n=0.0001$ so that:

$x_{10,000,B}=(10.000^{1/n}-1)*\Sigma_{i=1..n}x_i$

Note that for $n\to\infty$, $x_{10,000,B}\to x_{10,000,ML}$; because $n(10.000^{1/n}-1)\to\log10.000\ (n\to\infty)$.

We have $\lambda\sim Ga(n,\Sigma_{i=1..n}x_i)$, so $1/\lambda\sim Ig(n,\Sigma_{i=1..n}x_i)$ in which Ig is the inverted gamma distribution. If $X\sim Ig(\alpha,\beta)$ then $kX\sim Ig(\alpha,k\beta)$. With $x_{10,000,B}=n(10.000^{1/n}-1)(\Sigma_{i=1..n}x_i/n) = (1/\lambda)n(10.000^{1/n}-1)$ it follows that:

$x_{10,000,B}\sim Ig(n,n(10.000^{1/n}-1)\Sigma_{i=1..n}x_i)$

We conclude:

$E(x_{10,000,B})=n(10.000^{1/n}-1)(\Sigma_{i=1..n}x_i)/(n-1)$
$var(x_{10,000,B})=(n/(n-1))^2 (10.000^{1/n}-1)(\Sigma_{i=1..n}x_i)^2 /(n-2)$

Figure 7 gives us the results of the ML-analysis and the non-informative Bayesian analysis. From the 195 data points, it is estimated that $\lambda=0.0375$. The 10^{-4} quantile is calculated in the frequentistic way to be $E(x_{10,000,ML})=461.6$ cm and $\sigma(x_{10,000,ML})=18.7$ cm. The non-informative Bayesian analysis almost gave the same results $E(x_{10,000,B})=462.9$ and $\sigma(x_{10,000,B})=19.1$cm.

We continue the Bayesian analysis with informed prior distributions. In approximately 100 years, we have about 200 POT data (above 200cm). In the period 1500-1850, we therefore assume 700 POT data (in a sensitivity analysis performed later, it appeared that this amount of 700 was not very sensitive for the results). We have 10 POT data available during this period. In fact they are the 10 largest POT data, so we can use order statistics to estimate the most likely λ value if we assume an exponential parent distribution. We repeat this procedure 100 times and use the λ-estimates to fit a gamma distribution (which is conjugate with an exponential likelihood model). In figure 8 (left) the fit is shown. The prior distribution becomes $p(\lambda)=Ga(\lambda|\alpha,\beta)$, with $\alpha=183$ and $\beta=6392$. The conjugate posterior is also gamma distributed: $p(\lambda|x)=Ga(\lambda|\alpha+n,\beta+\Sigma_{i=1..n}x_i)$, where $n=195$ and $\Sigma_{i=1..n}x_i=5204$.

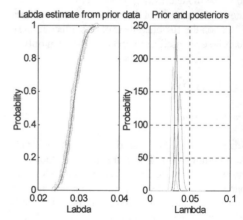

Figure 8: Posterior analysis

In figure 8 (right), the prior distribution (right function), the non-informative posterior distribution $Ga(\lambda|n,\Sigma_{i=1..n}x_i)$ (left function), and the informative posterior distribution (middle function), which is the normalized product of both other functions, are shown. The conjugate posterior predictive is given by:

$P(x<x_B)=\int_{\lambda=0..\infty}F(x_B|\lambda)p(\lambda|x)d\lambda=$
$=Gg(x_B|\alpha+n,\beta+\Sigma_{i=1..n}x_i,1)$

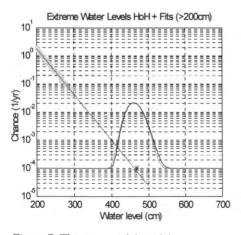

Figure 7: The exponential model

in which Gg stands for the Gamma-gamma distribution. We determine $x_{10,000}$ for which $P(x>x_{10,000})=1-10^{-4}$. We obtain 482.4cm. So notice that the role of the amount of sea level data since 1888 on the 10^{-4} quantile is of larger effect than in the Gumbel

248

model for the AM data, because this time the quantile decreases from 521.4 to 482.4cm. See figure 9 for the progress of updating the prior beliefs with new data points.

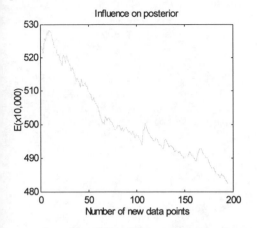

Figure 9: The updating process

6 A DECISION PROBLEM FOR DIKES

Van Dantzig (1956) posed the following economical decision problem for determining the optimal dike height. Taking account of the cost of dike building, of the material losses when a dike break occurs, and of the frequency distribution of sea levels, determine the optimal height of the dikes. If we take in correspondence with the result in section 5, the exponential model as the distribution for the sea levels: $p(h)=ce^{-\alpha h}$. We denote with H_0 the current dike height, X the optimal dike heightening, V the material loss after an inundation, I_0 the initial costs for dike heightening, k the subsequent costs of heightening per meter and δ the rate of interest, then the solution of the decision problem is given by:

$$X = 1/\alpha \ln (100ce^{-\alpha H0}V\alpha/\delta k).$$

In a follow-up research (Van Gelder 1995), a Bayesian analysis on this decision problem, in which the parameters are treated as random variables and the uncertainty in the frequency distribution of sea levels is taken into account, is being performed.

7 CONCLUSIONS

In this paper, we have examined the problem of determining the sea level with a return period of 10,000 years. The known existing statistical models for this problem use the accurate sea level measurements of the period from 1888 up till now. However, flood historical data has been recently published about the period 1500 - 1850. In order to take account of this information into the current statistical model a Bayesian approach has been used in this paper. A classical frequentistic approach wouldn't have been possible in this case. An analysis has been made with a Gumbel model for the Annual Maxima data. The 10^{-4} quantile of the sea level increases significantly if the historical flood data is taken into account; from 490 to 545cm.

A second analysis has been made with an exponential model for the Peaks Over Threshold data. In this case the 10^{-4} quantile increases from 462 to 482cm.

The particular choice for an AM model or a POT model and the choice for the distribution type (Gumbel, exponential, etc.) remains unanswered in this paper. The choice for a Bayesian framework, however, shows to be very useful when taking account of historical data in the analysis of extreme water levels.

REFERENCES

Deltacommissie 1960. *Delta rapport.* Den Haag.
Dillingh, D., De Haan, L., Helmers, R., Können, G.P., Van Malde, J. 1993. *De basispeilen langs de Nederlandse kust.* Den Haag.
De Haan, L. 1990. *Fighting the arch-enemy with mathematics.* Reprint Series No. 602. Erasmus University Rotterdam.
Gottschalk, M.K.E. 1977. *Storm surges and river floods in the Netherlands.* Van Gorcum. Amsterdam.
Jonkers, A.R.T. 1989. *Over den schrikkelijken watervloed.* Den Haag.
Vrijling, J.K. 1994. *Sea-level rise: a potential threat?* Statistics for the environment 2: Water related issues. John Wiley & Sons.
Castillo, E. 1988. *Extreme value theory in engineering.* Academic Press, Inc.
Bernardo, J.M., Smith, A.F.M. 1994. *Bayesian theory.* Wiley Series in Probability.
Van Dantzig, D. 1956. *Economic decision problems for flood prevention.* Econometrica 24, p.276-287.
Van Gelder, P.H.A.J.M. 1995 *Determination of statistical distribution functions for reliability analysis of civil engineering structures.* Technical report. Delft University of Technology.

Stochastic Hydraulics'96, Tickle, Goulter, Xu, Wasimi & Bouchart (eds) © 1996 Balkema, Rotterdam. ISBN 90 5410 817 7

Approximate joint probability analysis of extreme water levels in coastal catchments

Martin Lambert
Department of Civil and Environmental Engineering, University of Newcastle, N.S.W., Australia

George Kuczera
Department of Civil Engineering and Surveying, University of Newcastle, N.S.W., Australia

ABSTRACT: In coastal catchments, water levels in the floodplain and its outlet channel may be jointly affected by storm runoff and ocean water levels. This study considers approximate joint probability methods for estimating annual maximum water level distributions at a particular location subjected to periodic ocean inputs. These methods are based on widely used design storm approaches and, therefore, are much simpler to apply than full joint probability methods. A continuous simulation model is used to assess these approximate methods. It was found that the binned distribution method, which conditions water level distributions on the time within the tidal period when storm runoff commences, was the most reliable. Complementing this is a sensitivity measure to identify locations at which joint probability analysis may be required.

1 INTRODUCTION

Joint probability effects arising from the complex interaction of different inputs can complicate the frequency analysis of extreme water levels. One such problem is the flooding in rivers, streams, coastal lakes and urban trunk drainage systems jointly affected by storm runoff and either high ocean water levels (due to astronomic tides, storm surge and wave setup) or tailwater levels (stream confluences and control structures). The flooding of such interacting systems is of considerable economic and social importance because many urbanised areas in Australia are located on or near such water bodies.

There is little guidance on appropriate joint probability procedures. Indeed, in the Australian context, Spry and Thiering (1989) observe that Australian Rainfall and Runoff (1987), the defacto code of practice, "offers little practical guidance to engineers...". They add that a "conservative approach may lead to high design flood level predictions" which may overestimate water levels by up to 1m over a reach length of several kilometres. With reference to zoning constraints on possible developments, this may have severe economic implications. The significance of joint probability effects is further illustrated by Heideman et al. (1989) who demonstrated, using extensive simultaneous current and wave data, that the conservative combination of the N-year extreme current profile with the N-year extreme wave produces a design load on an offshore structure approximately one-third greater than the load estimated by a joint probability analysis.

Rigorous joint probability analysis is extremely difficult to implement because of the need for massive continuous simulation of complex systems and the problems of identifying the necessary multivariate probability distributions from limited historical data. There appears at the present time to be no simple approach for dealing with the joint probability problem.

This study is part of a long-term project seeking to develop relatively simple methods for joint probability estimation which will be applicable to a wide range of, but not necessarily all, coastal water bodies and which will avoid the need for rigorous joint probability analysis. In this study we:

i) Describe the development of FloodUtopia, a continuous simulation model of coastal catchments interacting with the ocean;

ii) Outline a classification system based on regression of dimensionless catchment parameters to FloodUtopia results to estimate the sensitivity of flood water levels to ocean input factors; and

iii) Evaluate, with the aid of FloodUtopia, three approximate methods for estimating annual maximum water level distributions at locations where water levels are jointly affected by storm runoff and periodic ocean inputs.

2 A CLASSIFICATION SYSTEM

We start by noting the need for a classification system which can identify locations within a coastal system for which joint probability analysis is required. The objective is to apply joint probability analysis only to those input factors which significantly affect peak water levels. Lambert et al. (1994) present a qualitative classification of the significance of input factors affecting peak water levels. It is based on evidence from an examination of over 30 coastal rivers on the eastern Australian coast. It suggests that oceanic inputs may not significantly affect peak water levels for coastal systems having considerable inertia and hence long response times. Such inertia may arise from catchment size, floodplain size or ocean-floodplain hydraulic controls.

Figure 1 illustrates such inertia for a large coastal lake connected to the ocean by a constricted outlet. It presents the results of a dynamic hydraulic analysis during a 36-hour 100-year storm event with astronomic tides being respectively included and excluded from the analysis. It can be seen that the constricted outlet to the ocean and large surface area significantly dampen the ocean water level variation and that during the 100-year storm on the surrounding catchment the peak lake flood levels are minimally affected by the astronomical tides. With a classification system in place, such inertia would be identified obviating the need for a joint probability analysis involving astronomic tides.

3 FLOODUTOPIA: A JOINT PROBABILITY SIMULATION MODEL

Our review of coastal river systems along the east coast of Australia highlighted the paucity of long-term data. To realise our objectives we developed a computer simulation model called FloodUtopia which simulates the dynamic behaviour of a catchment and its coastal floodplain to provide "records" thousands of years long. This enables us to develop simple equations for assessing the sensitivity of peak water levels to oceanic inputs and to develop and test methods for water level frequency analysis based on widely used design storm approaches such as presented in Australian Rainfall and Runoff (1987).

Figure 2 presents a schematic illustrating the FloodUtopia model. FloodUtopia uses the stochastic rainfall model presented by Lambert and Kuczera (1996) to generate rainfall events. Inter-storm duration is modelled using a gamma distribution. Average intensity for a particular storm is sampled using a two-step procedure: First, storm duration is sampled from a modified exponential distribution and, second, the average rainfall intensity is sampled from an exponential distribution with mean dependent on event duration through a power law. This scheme preserves the correlation between average intensity and event duration. The temporal nature of the event rainfall is simulated by randomly jumping through a dimensionless depth (0–1) and time (0–1) space. The dimensionless time space is divided into a finite number of steps. At each step the size of the jump to the next is sampled from a truncated log-normal distribution.

Rainfall is routed through the catchment to the floodplain using the Cordery-Webb unit hydrograph model (Cordery and Webb, 1974). The model is implemented using Chapman's (1983) Laplace transform technique. The unit hydrograph parameters C (hr) and K (hr) were related to catchment area A (km^2) using

$$C = \frac{0.76}{\xi} A^{0.38} \tag{1a}$$

$$K = \frac{0.80}{\xi} A^{0.34} \tag{1b}$$

where ξ is a scaling factor, normally set to 1, used to vary the catchment response to rainfall. Rainfall excess is computed using an initial loss-continuing loss model. The initial loss is modelled as a soil moisture deficit. A water balance is performed daily with evapotranspiration of 2.5 mm/day depleting the soil store and infiltration contributing to the store. The maximum deficit was set to 60mm.

The floodplain is idealised as a level pool because, for the purpose of this study, the dynamic effects on the floodplain are considered to be secondary to storage effects. However, flow in the prismatic channel connecting the floodplain to the ocean is modelled using the full St.Venant equations. The downstream boundary condition is specified as a time series of water levels based on astronomic tides and storm surge. The upstream boundary condition is defined by the water level in the floodplain. The interaction between the floodplain and channel defines the storage-discharge relationship for level-pool routing through the floodplain.

In a typical run FloodUtopia simulates rainfall, catchment, floodplain and ocean dynamics using 15 minutes time steps for periods over several thousand years. It saves various data including annual maximum water levels within the floodplain and along the connecting channel. It needs to be stressed

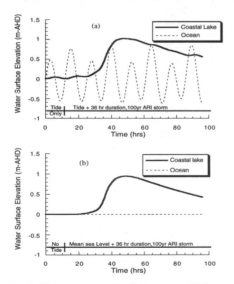

Figure 1. The response of a large coastal lake to: a) Astronomic tide and storm runoff; and, b) Storm runoff ignoring the astronomical tide.

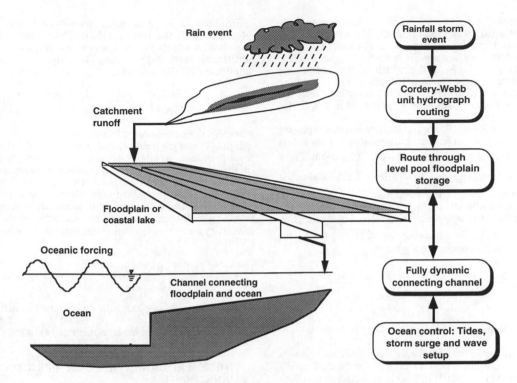

Figure 2. Schematic of the FloodUtopia simulation model during a rainfall event.

that FloodUtopia is not simulating an actual coastal catchment but rather the dominant dynamics in such a system. This is consistent with the study objective, namely to develop and test procedures for joint probability risk assessment.

4 SENSITIVITY OF PEAK WATER LEVEL TO OCEANIC INPUT

Before considering approaches for estimating water level frequency distributions we briefly review and update the sensitivity measure developed by Lambert et al. (1994). This can be used to identify locations in coastal systems where there is sufficient inertia to render peak water levels insensitive to oceanic inputs. The sensitivity measure S_n is defined as

$$S_n = \frac{A-B}{A} \qquad (2)$$

where the peak water levels A and B are defined with reference to the mean water level in Figure 3.

Provided $A > B$, S_n is bounded between 0 and 1. Although $A < B$ is possible, it is of no relevance to the study of annual maximum water levels.

4.1 Dimensional Analysis

A dimensional analysis was undertaken to determine the relevant dimensionless parameters which could be used in a simple statistical description of the sensitivity. The dominant parameters affecting S_n are assembled in the following equation

$$F[(A-B), A, \Delta, Q_p, V_f, A_{fp}, R, T_o, W, y_o, L, n, S_o]$$

$$= 0 \qquad (3)$$

where Δ is the height of rise of the water level in the floodplain, Q_p is the peak inflow, V_f is the flood

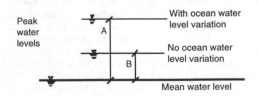

Figure 3. Definition of the peak water levels in the sensitivity measure.

253

volume and is equal to the catchment area A_c multiplied by the total rainfall excess d, which was derived from a 36-hour design storm, A_{fp} is the floodplain surface area, R is the tidal range, T_o is the tidal period, W is the width of the connecting channel, y_o is the initial depth in the connecting channel, L is the length of the connecting channel, n is the Manning roughness coefficient and S_o is the slope.

While R and T_o are particularly obvious repeating variables R produces dimensionless parameters which tend to infinity as $R \to 0$. As a result a new variable $\Delta_o = \dfrac{V_f}{A_{fp}} = \dfrac{A_c d}{A_{fp}}$ which can be defined as the height of rise in the floodplain if all the flood is accommodated in A_{fp}, has been used as the significant length term. Dimensional analysis using Δ_o and T_o as repeating variables produces the following dimensionless groups:

$$\frac{\Delta}{\Delta_o}, \frac{Q_p T_o}{\Delta_o{}^3}, \frac{A_{fp}}{\Delta_o{}^2}, \frac{R}{\Delta_o}, \frac{W}{\Delta_o}, \frac{y_o}{\Delta_o}, \frac{L}{\Delta_o}, S_o, \frac{n\Delta_o{}^{1/3}}{T_o}$$

4.2 Regression Analysis

FloodUtopia was run for thousands of combinations of the dimensionless variables. The rainfall excess depth d corresponds to a 36-hour design storm depth (Australian Rainfall and Runoff, 1987). However, the temporal pattern was derived from a Chicago storm which produces a symmetric, single-peaked temporal pattern constructed so that short-duration design storms are embedded within longer-duration storms. Regression analysis was then used to identify simple predictors of S_n.

For a coastal system subjected to astronomical tides it was found that the most dominant factors were the tidal range (R), the floodplain area (A_{fp}), the width of the connecting channel (W) and the connecting channel length (L). A simple multiplicative relationship of the following form provided an R^2 of 88%.

$$S_n = 458 \left(\frac{R}{\Delta_o}\right)^{1.57} \left(\frac{A_{fp}}{\Delta_o{}^2}\right)^{-0.53} \left(\frac{W}{\Delta_o}\right)^{0.74} \left(\frac{L}{\Delta_o}\right)^{-0.37} \quad (4)$$

where R varied between 0 to 8m, A_{fp} varied between 3.5 and 3000km^2, W varied between 50 and 500m and L varied between 50 and 40000m. It can be observed from eqn. (4) that the sensitivity will increase (as expected) for an increasing tidal range and connecting channel width. On the other hand, it decreases with increasing floodplain surface area and connecting channel length. The sensitivity S_n does have a physical interpretation. For example, if it is less than 0.1 then by neglecting the tidal variation, errors of less than 10% are to be expected in the peak water level determination.

Similarly if a storm surge whose peak level coincides with the peak rainfall of the storm is examined, a similar equation results but includes an additional term representing the storm surge amplitude (SS) ($R^2 = 87\%$).

$$S_n = 458 \left(\frac{R}{\Delta_o}\right)^{1.3} \left(\frac{A_{fp}}{\Delta_o{}^2}\right)^{-0.54} \left(\frac{W}{\Delta_o}\right)^{0.74} \left(\frac{L}{\Delta_o}\right)^{-0.33} \left(\frac{SS}{\Delta_o}\right)^{0.24} \quad (5)$$

where SS varied between 1 to 3m.

Eqns (4) and (5) present sensitivity values for the level pool area. Regions seaward of the level pool (in the connecting channel) are much more likely to be influenced by ocean water level variations. As a result the joint probability problem should be thought of as being one of a continuum rather than single valued. Examination of the variation of the sensitivity along the connecting channel yields ($R^2 = 85\%$)

$$S_c = 1.22 \left(\frac{x}{L}\right)^{0.75} \left(\frac{R}{\Delta_o}\right)^{0.6} (S_n)^{0.35}, \quad S_n \leq S_c \leq 1 \quad (6)$$

where x is the distance along the connecting channel from the floodplain level pool and S_n is the floodplain sensitivity given by either eqn (4) or (5).

5 JOINT PROBABILITIES FOR PERIODIC OCEANIC INPUT

We consider the simplest joint interaction between ocean and catchment, namely when the ocean input is periodic. Some important probability relationships dealing with periodic ocean inputs are developed.

We start by subdividing the periodic ocean input into N equi-spaced intervals denoted as bins. Consider all the storm events during the year which produce runoff into the floodplain commencing during the ith bin - denote these storm events by the set Ω_i. Figure 4 illustrates the bin concept showing a runoff event commencing during a particular ocean input bin.

Let the random variable W_i be the largest water level at some location P produced by the storms Ω_i and let W be the annual maximum water level at location P. Given that

$$W = \max\{W_1,..,W_N\} \quad (7)$$

it follows that

$$P(W \leq w) = P(W_1 \leq w,...,W_N \leq w) = \prod_{i=1}^{N} P(W_i \leq w) \quad (8)$$

given the assumption that storm events are statistically independent. Two cases need to be considered when evaluating the probability $P(W_i \leq w)$:

1) One or more storms with runoff commencing during the ith bin occur in a particular year. Let

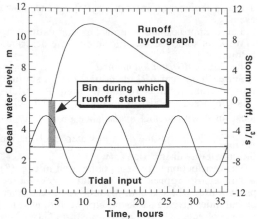

Figure 4. Illustration of storm runoff commencing during a particular ocean input bin.

$F_i(w)$ be the probability that $W_i \leq w$ given that the storm set Ω_i is not empty.

2) No storms occur during the year with runoff commencing during the ith bin. The set Ω_i is empty.

Let p_b be the probability that there is at least one storm with runoff commencing during a bin; that is, p_b is the probability that th set Ω_i is not empty. Application of the total probability theorem yields

$$P(W_i \leq w) = P(W_i \leq w \mid \geq 1 \text{ storms in bin i}) \, p_b +$$

$$P(W_i \leq w \mid 0 \text{ storms in bin i}) \, (1 - p_b)$$

$$= F_i(w) p_b + 1 - p_b \qquad (9)$$

The implicit assumption is made that the water level w exceeds the maximum tidal level at location P. This ensures that $P(W_i \leq w \mid 0 \text{ storms in bin i})$ equals 1. An approximate expression for eqn. (8) can be derived by re-expressing $F_i(w)$ as

$$F_i(w) = \bar{F}(w) + \delta F_i(w) \qquad (10)$$

where $\bar{F}(w)$ is the average probability (averaged over N bins) and $\delta F_i(w)$ is the deviation of $F_i(w)$ about the average. Substituting eqns (9) and (10) in eqn. (8) gives

$$P(W \leq w)$$

$$= [1 - p_b + p_b \bar{F}(w)]^N \prod_{i=1}^{N} \left(1 + \frac{p_b \delta F_i(w)}{1 - p_b + p_b \bar{F}(w)} \right)$$

$$\approx [1 - p_b + p_b \bar{F}(w)]^N \left(1 + \frac{p_b}{1 - p_b + \bar{F}(w)} \sum_{i=1}^{N} \delta F_i(w) \right)$$

$$= [1 - p_b + p_b \bar{F}(w)]^N \xrightarrow[N \to \infty]{} e^{-p_b N (1 - \bar{F}(w))} \qquad (11)$$

by virtue of the fact that $\sum_{i=1}^{N} \delta F_i(w) = 0$ and assuming that the deviations $\delta F_i(w)$ are small.

Eqn. (11) provides a linkage between the annual maximum water level distribution, $P(W \leq w)$, and the maximum water level distributions conditioned on a bin, $F_i(w)$. Note that 1000-year simulations using FloodUtopia have verified eqns (8), (9) and (11).

6 FLOODUTOPIA SIMULATION USING PERIODIC OCEANIC INPUT

Table 1 presents the characteristics of four coastal catchments simulated for 1000 years using FloodUtopia. Catchment 2 is identical to catchment 1 except that the unitgraph scaling ξ was increased to 6 to make the catchment respond more quickly to rainfall. The stochastic rainfall model was calibrated to 80 years of Sydney pluviometer data. For each catchment astronomic tides were represented as a sine wave with a 12-hour period and a range of either 0 or 2m. The sensitivity measure S_n reported in Table 1 was computed for a 36-hour 100-year storm depth of 290 mm. On the basis of S_n one would expect the $P(W \leq w)$ for catchments 1 and 2 to be sensitive to tidal inputs. The parameter T_{ave} is the average time from commencement of runoff to annual maximum water level.

Figures 5a and 5b present annual maximum floodplain water level distributions for catchments 1 and 3 respectively for the no-tide and 2m-tide cases. Catchment 3 has a floodplain which is unresponsive to tidal inputs. As a result the water level distributions for the no-tide and 2m-tide cases are similar with catchment runoff dominating the distribution. Catchment 4 behaves in a similar manner to 3. On the other hand, catchment 1 has a floodplain very responsive to tidal inputs resulting in major differences in water level distributions.

Table 1. Characteristics of coastal catchments used in FloodUtopia simulation.

Parameter	Catchment			
	1	2	3	4
A, km^2	10	10	500	5000
A_{fp}, %	5	5	10	10
ξ	1	6	1	1
L, km	1	1	10	10
W, m	10	10	500	100
n	.030	.030	.035	.035
S_o, %	0.1	0.1	0.01	0.01
S_n	0.118	>0.118	0.027	0.013
p_b	0.48	0.67	0.38	0.32
T_{ave}, hr	10.7	3.2	60.2	150.4

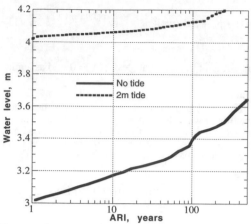

Figure 5a. Annual maximum water level distributions for catchment 1 with 2m tidal range and no tide.

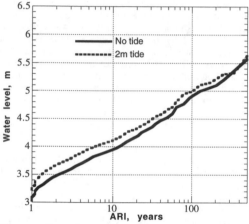

Figure 5b. Annual maximum water level distributions for catchment 3 with 2m tidal range and no tide.

Therefore, the joint interaction between tidal and rainfall inputs must be considered for catchment 1. Catchment 2 is even more responsive than catchment 1.

7 APPROXIMATE JOINT PROBABILITY METHODS USING DESIGN STORMS

We consider several approaches for estimating annual maximum water level distributions which avoid the need for the massive continuous simulation performed by FloodUtopia. The approaches rely on the existence of design methods similar to those in Australian Rainfall and Runoff (1987) which convert annual maximum design storms into hydrographs with peaks having the same

average recurrence interval (ARI) as the storm event.

Three approximate approaches for estimating $P(W \leq w)$ at some particular location P are described below:

1. *Binned distribution method*
For each bin and ARI the critical design storm is found. This storm is routed through a rainfall-runoff/hydraulic model to obtain the water level at location P with annual exceedance probability $1 - \frac{1}{\text{ARI}}$. For the ith bin the derived water levels define the probability distribution $G_i(w)$.

The distribution $G_i(w)$ cannot be used in eqn. (8) because it represents annual exceedance probabilities whereas eqn. (8) requires exceedance probabilities given a particular bin. Accordingly, the inverse of eqn. (11) is used to transform $G_i(w)$ into the distribution of water levels given that runoff started during the ith bin; that is,

$$F_i(w) = \frac{G_i(w)^{\frac{1}{N}} - 1 + p_b}{p_b} \tag{12}$$

Once the $G_i(w)$, $i = 1,..,N$, have been transformed using eqn. (12), eqns (8) and (9) can be used to estimate $P(W \leq w)$. Note that an independent estimate of p_b must be made.

2. *Average water level method*
For each bin and ARI the critical design storm is found. This produces the annual maximum water level distribution given the ith bin $G_i(w)$. For a given ARI the expected or average water level is

$$w(\text{ARI}) = \frac{1}{N} \sum_{i=1}^{N} G_i^{-1} \left(1 - \frac{1}{\text{ARI}} \right) \tag{13}$$

where $G_i^{-1}(u)$ is the water level for the ith bin with an annual exceedance probability of $1 - u$.

3. *Superposition method*
Estimate the annual maximum water level distribution assuming no tidal input $G_0(w)$. Using the total probability theorem superimpose the tidal input to give

$$P(W \leq w) = \frac{1}{N} \sum_{i=1}^{N} G_0 \left(w - twl_i + mwl \right) \tag{14}$$

where mwl is mean water level at location P and twl_i is the tidal water level at the same location during the ith bin.

The three approximate methods were used to estimate the annual maximum water level distribution in the floodplain level pool using the design storm approach presented in Australian Rainfall and Runoff (1987). IFD data were estimated for Sydney using the Australian Rainfall and Runoff regionalised procedure. Following the method of Walsh et al. (1991) the design initial loss was calibrated for each design ARI using the distribution of annual maximum discharge into the floodplain derived from FloodUtopia "observations". Figures 6a to 6d compare the three approximate water level distributions against the FloodUtopia (or exact) distributions for each of the four catchments described in Table 1. The number of bins N used was 20.

Several observations can be made:

1. The binned distribution method is the most robust of the three methods considered. Although the average water level method performed as well, if not better, for the slower responding catchments (1, 3 and 4) it performed notably worse for catchment 2 which responds to storm events in about 3 hours.

This is a reassuring result because the binned distribution method has a theoretical basis whereas the other two methods are better described as heuristics. However, because the binned distribution method requires an estimate of p_b the average water level method may be more suitable in practice. However, its range of applicability needs to be better defined.

2. The discrepancies between the binned and FloodUtopia distributions are attributed to two factors. First, the calibration of the design initial loss in the design storm approach ensured that a Y-year design storm produced a Y-year peak discharge into the floodplain whereas the floodplain dynamics are driven more by runoff volume than peak inflow. Second, the use of the transformation given by eqn. (12) assumes that the probability variations between

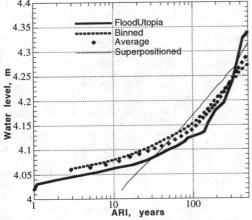

Figure 6a. Comparison of annual maximum water level distributions for catchment 1 with 2m tidal range.

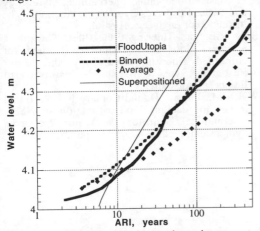

Figure 6b. Comparison of annual maximum water level distributions for catchment 2 with 2m tidal range.

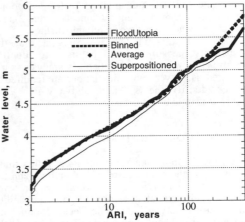

Figure 6c. Comparison of annual maximum water level distributions for catchment 3 with 2m tidal range.

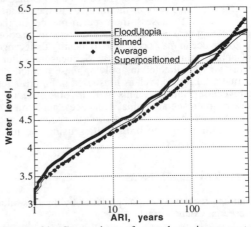

Figure 6d. Comparison of annual maximum water level distributions for catchment 4 with 2m tidal range.

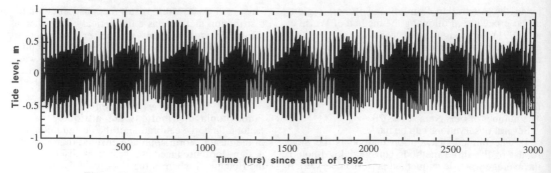

Figure 7. Tidal time series for Newcastle Harbour for first 3000 hours in 1992.

bins $\delta F_i(w)$ are small. This assumption deteriorates as system inertia is reduced.

3. The superposition method performed poorly for the quicker responding catchments (1 and 2) and, therefore, is not considered a reliable method.

4. The results in Table 1 suggest that the sensitivity measure S_n can be used to identify locations for which joint probability analysis is not required. However, more simulations are required to better identify the S_n value above which joint analysis should be undertaken.

8 CONCLUDING REMARKS

The FloodUtopia simulations reported in this study suggest that the binned distribution method (and for less sensitive locations, the average water level method) can adequately account for joint probability interactions between periodic ocean inputs and storm runoff. These methods are based on widely used design storm approaches and, therefore, are much simpler to apply than full joint probability methods. More simulations need to be performed to assess the sensitivity of the methods to the number of bins N, to assess their performance in the connecting channel where much stronger interaction with the ocean may occur, and to identify their range of applicability in terms of the S_n measure. Future work will be directed to applying these methods to actual tidal inputs such as illustrated in Figure 7 for Newcastle Harbour. This will involve selecting several representative tidal sequences, applying the binned distribution method to each sequence, and then combining the probabilities using the total probability theorem.

ACKNOWLEDGMENTS

Funding for this work was provided by the Australia Research Council grant.

REFERENCES

Australian Rainfall and Runoff 1987. A guide to flood estimation, D. Pilgrim (ed.), Institution of Engineers, Australia.

Chapman T.G. 1983. Exact runoff routing algorithms for the Clark and Cordery-Webb models, Civil Eng. Trans., Institution of Engineers, Australia, p. 177-181.

Cordery, I. and Webb S.N. 1974. Flood estimation in Eastern New South Wales - A design method, Civil Eng. Trans., Institution of Engineers, Australia, Vol. CE16, p. 89-93.

Heideman J.C., Hagen, O., Cooper C. and Dahl F. 1989. Joint probability of extreme waves on Norwegian Shelf, ASCE, J. Waterway, Port, Coastal, and Ocean Engineering, Vol. 115 No. 4, p. 534-546.

Lambert M. and Kuczera, G. 1996. A Statistical Model of Rainfall and Temporal Patterns, this issue.

Lambert M., Williams B.J., Field W.G. and Kuczera G. 1994. A taxonomy of joint probability effects in coastal floodplains, International Hydrology and Water Resources Symposium Adelaide, The Institution of Engineers, Australia

Spry R.B. and Thiering C.G. 1989. Towards a practical flood level estimation method for combined flooding, Hydrology and Water Resources Sym.. Christchurch, Institution of Engineers, Australia.

Walsh M., Pilgrim D. and Cordery I. 1991 Initial losses for design flood estimation in NSW, International Hydrology & Water Resources Symposium, Perth, 2-4 October, pp. 283-288.

Stochastic Hydraulics'96, Tickle, Goulter, Xu, Wasimi & Bouchart (eds) © 1996 Balkema, Rotterdam. ISBN 90 5410 817 7

Observations of bispectra of surface waves in deep and shallow waters

S.R. Massel
Australian Institute of Marine Science, Townsville, Qld, Australia

J. Piórewicz
Central Queensland University, Rockhampton, Qld, Australia

ABSTRACT: The paper presents examples of application of the bispectral analysis to study the nonlinear properties of wind induced waves in deep ocean and in shallow water zone. The studies showed the interactions within the spectral peak frequency band (ω_p) are dominant. The additional peaks of the bispectrum close to the main peak are responsible for the formation of the forward face positive energy transfer lobe and the observed shifting of the main peak towards the lower frequencies. Integrating the real part of the bispectrum yields an estimate of skewness. The value and sign of the imaginary part of the bispectrum illustrates the asymmetry of the wave profile against the vertical plane.

1 INTRODUCTION

To a first approximation, a random sea surface may be regarded as a linear superposition of statistically independent free waves and the two-dimensional power spectrum provides a complete description of the ocean wave field. However, for many purposes such a description cannot be adequate. The JONSWAP experiment (Hasselmann et al., 1973) and later theoretical and experimental studies demonstrate clearly that the nonlinear interactions between spectral components play the most important role in the formation of the wave spectrum and its evolution in time.

Particularly in a shallow water zone, the departure of the observed probability density from a Gaussian distribution cannot be neglected. Surface elevation, wave-induced pressure, and orbital velocity time series show increasingly sharper peaks and shallower troughs, caused by higher harmonic generation and their interactions as waves approach the surf zone.

The stochastic properties of such processes can no longer be expressed by the first two statistical moments. To investigate these nonlinear phenomena, second and higher order spectra must be analysed. The lowest order nonlinear spectrum used in the present paper is the so-called bispectrum, first carried out by Hasselmann et al. (1963) Since then, the analysis has been applied in many diverse fields, such as fluid turbulence, wave generation mechanism, internal wave interactions, plasma studies, ship rolling motion, economic time series, etc.

The objective of this paper is to investigate the nonlinear features of the geometry of surface waves and their velocity field in deep and shallow waters, using the bispectral technique. In particular, the temporal evolution of triad interactions in the storm waves is described. Also, the relationships between the real and imaginary parts of the bispectrum and wave profile skewness and asymmetry are identified and estimated.

The paper is organized as follows. In the next section, the basic information on the bispectrum is given. In section three, tha experimantal data used in the analysis are described. Section four contains the basic results of the analysis and in section five, the main conclusions are listed.

2 THE BISPECTRUM

Let us define the two-dimensional autocorrelation function $K_{\zeta\zeta}^{(2)}(\tau_1, \tau_2)$ in a similar way to the one-dimensional function:

$$K_{\zeta\zeta}^{(2)}(\tau_1, \tau_2) = E[\zeta(t)\,\zeta(t+\tau_1)\,\zeta(t+\tau_2)],$$

in which E is the average operator.

Thus, the two-dimensional autocorrelation function is evaluated by shifting the time history of $\zeta(t)$ by the time τ_1 and τ_2, and then averaging the product $\zeta(t)\zeta(t+\tau_1)\zeta(t+\tau_2)$. Applying the Wiener-Khinchine theorem to the two-dimensional autocorrelation function $R_{\zeta\zeta}^{(2)}(\tau_1, \tau_2)$ we obtain the bispectrum in the form:

$$B(\omega_1, \omega_2) = \left(\frac{1}{2\pi}\right)^2 \int_{-\infty}^{\infty}\int_{-\infty}^{\infty} R_{\zeta\zeta}^{(2)}(\tau_1, \tau_2) \cdot$$

$$\cdot \exp\left[-i(\omega_1\tau_1 + \omega_2\tau_2)\right] d\tau_1 d\tau_2 \qquad (1)$$

There are six two-dimensional autocorrelation functions for a given τ_1 and τ_2. Therefore, there also are six bispectra for a given ω_1 and ω_2, having the same value, i.e.:

$$B(\omega_1, \omega_2) = B(\omega_2, \omega_1) = B(\omega_1, -\omega_1 - \omega_2) =$$

$$= B(-\omega_1 - \omega_2, \omega_1) =$$

$$= B(\omega_2, -\omega_1 - \omega_2) = B(-\omega_1 - \omega_2, \omega_1), \qquad (2)$$

and

$$B(\omega_1, \omega_2) = B^*(-\omega_1, -\omega_2), \qquad (3)$$

where the asterisk denotes the complex conjugate. Because of these symmetries, it is sufficient to evaluate the bispectrum only in the domain defined by $0 \le \omega_2 < \omega_1 \le |\omega_1 + \omega_2|$.

For discretely sampled data, the digital bispectrum is more appropriate, i.e. (Kim and Powers, 1979):

$$B(\omega_i, \omega_j) = E\left[A_{\omega_i} A_{\omega_j} A_{\omega_i + \omega_j}^*\right]. \qquad (4)$$

in which A are the Fourier transforms of the surface displacements and an asterisk indicates complex conjugation.

To calculate the Fourier transforms we assume that the wave record $\zeta(t)$ is given for a finite time interval $(0, t)$ and is sampled at N equally spaced points a distance Δt apart. The Fourier representation of the $\zeta(t)$ record becomes:

$$A(\omega, t) = \Delta t \sum_{n=0}^{N-1} \zeta_n \exp\left[-i\omega n \Delta t\right]. \qquad (5)$$

The usual selection of discrete frequency values for the computation of $A(\omega, t)$ is:

$$\omega_k = \frac{2\pi k}{t} = \frac{2\pi k}{N\Delta t}, \quad k = 0, 1, 2, \ldots, N-1. \qquad (6)$$

As the total number of points, N, is usually large, to get a better accuracy spectrum estimation, the record $\zeta(t)$ is divided into K segments, overlapping by one half their length L. Thus, the number of overlapping segments becomes:

$$K_1 = \frac{2N}{L} - 1 = 2K - 1, \qquad (7)$$

in which K is the number of non-overlapped segments.

The Fourier Transform of $\zeta(t)$ for each segment can be viewed as the Fourier Transform of an unlimited time history record multiplied by a rectangular data window:

$$\tilde{\zeta}(t) = \zeta(t)\, v(t), \qquad (8)$$

where:

$$v(t) = \begin{cases} 1 & \text{for } 0 \le t \le L\Delta t, \\ 0 & \text{otherwise.} \end{cases} \qquad (9)$$

Fourier analysis of finite length records results in inherent side lobes in the spectral domain. The large side lobes allow leakage of energy at frequencies well separated from the main lobe. To reduce the leakage problem, a time window that tapers the time-history data was introduced, to eliminate the discontinuities at the beginning and end of the records to be analysed. There are numerous such windows in current use. For our purpose, the Hanning data window has been applied (Otnes and Enochson, 1972):

$$v(t) = \begin{cases} \frac{1}{2}\left[1 - \cos\left(\frac{2\pi t}{L\Delta t}\right)\right], & 0 \le t \le L\Delta t, \\ 0 & \text{otherwise.} \end{cases} \qquad (10)$$

The final estimate of the one-sided spectral density $S(\omega)$ takes the form:

$$\tilde{S}(\omega_k) = \frac{2}{K_1(L\Delta t)^2} \sum_{p=1}^{K_1} |A_p(\omega_k)|^2, \qquad (11)$$

in which $\omega_k = \frac{2\pi k}{L\Delta t}$.

Similarly, the estimate of the bispectrum becomes:

$$\tilde{B}(\omega_1, \omega_2) = \frac{(2\pi)^2}{K_1 (L\Delta t)^3} \cdot$$

$$\sum_{p=1}^{K_1} A_p(\omega_1) A_p(\omega_2) A_p^*(\omega_1 + \omega_2). \qquad (12)$$

Note that $\tilde{S}(\omega_k)$ represents the power spectrum, corresponding to the frequency ω_k.

The quantity $\tilde{S}(\omega_k)$ should be distinguished from the spectral density function $\tilde{G}(\omega_k)$, which is simple given by:

$$G(\tilde{\omega}_k) = \frac{\tilde{S}(\omega_k)}{\Delta \omega} = \left(\frac{2\pi}{L\Delta t}\right)^{-1} \tilde{S}(\omega_k) \qquad (13)$$

Similarly, the $\tilde{B}(\omega_1, \omega_2)$ function is an estimate of total bispectrum, corresponding to frequency square $\Delta \omega \cdot \Delta \omega$, while the function $B(\omega_1, \omega_2)$ in Eq. (1) is a bispectral density function.

Using the representation of the sea surface in terms of the Fourier-Stjeltjes integral we can write the following relationship:

$$\overline{dA(\omega_1)\,dA(\omega_2)\,dA(\omega_3)} =$$

$$= \begin{cases} B(\omega_1, \omega_2)\,d\omega_1\,d\omega_2 \\ 0 \end{cases} \qquad (14)$$

The non-zero value of the product of spectral amplitudes dA is valid for the case when $\omega_1 + \omega_2 + \omega_3 = 0$. Otherwise, the product is equal zero. Therefore, the bispectrum represents the contributions to the mean cube $\overline{\zeta^3}$ from the product of the three Fourier components whose resultant frequency is zero. It can be shown that integrating the real part of the bispectrum yields an estimate of skewness which is proportional to the third moment $E[\zeta^3]$ (Elgar and Guza, 1985; Massel, 1995). The skewness is obtained by normalizing third moment by $E[\zeta^2]^{\frac{3}{2}}$.

Prior to the work of Masuda and Kuo (1981) no physical interpretation had been attached to the imaginary part of the bispectrum. However, they showed that the imaginary part of the bispectrum is related to the vertical asymmetry with respect to the vertical axis of the wave profile. Positive (negative) part of the bispectrum is associated with waves tilted backward (forward). Moreover, Elgar and Guza (1985) related the imaginary part of the bispectrum to a measure of the skewness of the temporal derivative of a time series.

3 DATA BASE

The data for this investigation were collected from three different environments: Macquarie Island in the Southern Ocean, Wheeler Reef in the central Great Barrier Reef on the east coast of Australia and at the Farnborough Beach (Yeppoon), also on the east coast of Australia.

3.1 Macquarie Island data

Wave measurements were made over the period of November 1988 - October 1989 for the 1989-90 ANARE Program. The Datawell Waverider buoy was located in a depth of 130 meters approximately 10 km Northwest of Macquarie Island (54°25.1' S, 158°48.1' E). Data was sampled at $\Delta t = 1sec$ over 1024 samples (about 17 minutes record) every 3 hours (Steedman, 1989).

The 12 hour period on the 13th of May 1989 was chosen as a maximum wave height of 17.7 meters was recorded at 2400 May 13, the significant wave height for this burst was found to be 11 meters. Although the waves are essentially classified as deep water waves it is thought that some insight could be gained from the study of an extreme event.

3.2 Wheeler Reef data

Wheeler Reef is situated in the central section of the Great Barrier Reef, approximately 50 km from Townsville, Australia. Water depth outside the reef is almost constant and equal to approximately 50m. At the reef front the ocean floor rises very rapidly, from approximately 50m to about 2m at the mean tide level. The reef top is horizontal, and on the lee side, water depth gradually increases to 50m over a distance of 800m. Four types of instruments were deployed in a transect across the reef. Seaward of the reef a Datawell Wave Rider Buoy was deployed. At the reef front, in 8m of water, and 5m below sea level an Interocean S4 current meter with pressure sensor was located. Other S4 current meters as well as the Woods Hole current meters with pressure sensors have been deployed on the reef top. Additionally tide gauges were located on the reef top.

Wave records were taken every 1 hour, with S4 records consisting of $N = 2400$ points at a sampling interval of $\Delta t = 0.5sec$, WRB records consisting of $N = 3072$ samples at $\Delta t = 0.3906sec$ and Woods Hole current meters records consisting of $N = 4096$ samples at $\Delta t = 0.25sec$. Three series of observations were conducted; during the first experiment, the instruments were deployed between November 3, 1993, and November 27, 1993. The second experiment lasted from February 27, 1994 till April 8, 1994 and the third experiment was held between October 17 and October 23, 1995. The configurations of the instruments during particular experiments were slightly different.

3.3 Farnborough Beach data

The measurement profile was selected on a reasonably long and straight Farnborough Beach near Yeppoon with clear parallel bottom contour lines. The two S4 current meters had been deployed between 7 and 11 of February, 1993 at water depths from about 5m to 2m, depending on tide phase. During the full period of experiment, the wind was approximately 10m/s ESE, which can be contributed to cyclone 'Oliver'. The cyclone was travelling parallel to the beach at about 500km to 700km distance offshore (Massel, et al., 1993).

4 OBSERVATIONS OF BISPECTRA

For the further analysis the Wave Rider Buoy records taken close to Macquarie Islands and at the front of Wheeler Reef were used. At the Farnborough Beach only the wave induced pressure, recorded by S4 pressure sensor, has been analyzed. Pressure records were not converted to sea-surface elevation due to unrealistically high amplification of small noise level for high-frequency band. However, owing the fact that the depth-correction coefficients do not vary significantly within the low frequency bands, the pressure record can be used as some estimate of the sea-surface elevation.

The selected data were processed by breaking the entire record into consecutive overlapping sections $L = 256$ each, resulting in a frequency resolution for the raw data $\Delta \omega = \dfrac{2\pi}{L \Delta t}$ of 0.0245 rad/sec for Macquarie Island, 0.0628 rad/sec for Wheeler Reef, and 0.0491 rad/sec for Farnborough Beach. Statistical stability of spectral estimates is gained by averaging spectral value over 3 frequencies. Similarly, the stability of bispectral estimates is obtained by averaging bispectral values over 3×3 squares.

As was mentioned above, for bispectrum only $\frac{1}{8}$ of the (ω_1, ω_2) frequency plane is unique. However, the results of a bispectral analysis were plotted below as $3D$ plots for entire (or part) 1st quadrant $(\omega_1, \omega_2 > 0)$ to allow the backface of some of the peaks to be seen.

Figs. 1a-1c show the bispectrum amplitude contours for the three locations (contours levels are relative and they are adopted arbitrary to produce the best distinguishing between observed numerical values). The Figures show clearly, that interactions between fundamental and first harmonic frequencies dominate. Therefore, this interaction

is denoted by (ω_p, ω_p), where ω_p is a peak frequency. The strength of the interaction is dependent on the wave energy at each location. This energy is different. For example, the variance of the surface elevations, which is proportional to the wave energy, is equal $7.42m^2$, $0.2m^2$ and $0.03m^2$ at Macquarie Island, Wheeler Reef and Farnborough Beach, respectively. The corresponding significant wave heights are 10.89m, 1.77m and 0.71m.

The concentration of bispectrum values around the bispectrum peak $B(\omega_p, \omega_p)$ is the highest for Macquarie Island. This means that even at the storm peak, wave steepness is very low and the large waves are not breaking. The nonlinear interaction in the wave field is concentrated around the peak frequency, producing only the Stokes' type harmonic. The departure from the Gaussian model is small and the Gaussian model of the wave field can provide a reasonable approximation of observed reality.

However, the bispectrum amplitude surface for the Wheeler Reef is broader due to interaction of the spectral frequencies, close to the peak frequency. This is confirmed by the spectral density function (not shown here) which has a relatively wide peak. Such spectral form may be caused by two wave systems with the similar peak frequencies approaching to the reef.

Surprisingly weak interaction is shown in Fig. 1c (Farnborough Beach). Probably it is due to low energy of waves (variance $\approx 0.03m^2$).

In Fig. 1a, the peak at (0.343 rad/sec, 0.122 rad/sec) is observed. Physically, this peak suggests that neighboring frequencies within the fundamental peak interact to form a long wave.

In Fig. 2a-c, the imaginary part of the bispectrum is presented. At all three locations, the negative values of the imaginary part are concentrated around the main peak. It means that the wave profiles corresponding to the frequencies close to the peak frequency are tilted forward. The wave profiles corresponding to the other frequencies are symmetric or slightly tilted backward.

The analysis of the real part of the bispectrum showed that the observed small skewness of the wave profile is essentially due to (ω_p, ω_p) interaction. At that frequencies, the real part of the bispectrum is usually positive. The negative values of the real part of the bispectrum are associated with the frequencies out of the spectral peak. The integral over the real part of the bispectrum, normalised with the variance of the surface elevation, represents the skewness of the probability density of surface elevation. Positive skewness corresponds to sharp crests and flat troughs.

5 SUMMARY

The studies showed that the interactions within the spectral peak frequency band (ω_p) are dominant. The additional peaks of the bispectrum, close to the main peak, are responsible for formation of the forward face positive energy transfer lobe and the observed shifting of the main peak towards the lower frequencies. The highest waves are not necessary the most nonlinear ones. Even for very stormy seas (for example, the Macquarie Island records), the nonlinear interaction is weak and concentrated around the peak frequency.

The value and sign of the imaginary part of the bispectrum illustrates the vertical asymmetry of the wave profile. The waves corresponding to the peak frequency are non-symmetric and tilted forward.

Analysis of evolution of bispectra during various storm phases and the relationship between statistical characteristics of wave field and the bispectrum is a subject of further studies.

Acknowledgment
This research is, in part, supported by the Australian Research Council (Stanisław Massel) through grant number A89331580.

6 REFERENCES

Elgar, S. and R.T. Guza 1985. Observations of bispectra of shoaling surface gravity waves. *Jour. Fluid Mech.* 161. 425-448.

Hasselmann, K., Munk, W., and G. MacDonald 1963. Bispectra of ocean waves. In M. Rosenblatt (editor) *Time Series Analysis*: 125-139. New York:Wiley.

Hasselmann, K., Barnett, T.P., Bouws, E., Carlson, H., Cartwright, D.E., Enke, K., Ewing, J.A., Gienapp, H., Hasselmann, D.E., Kruseman, P., Meerburg, A., Müller, P., Olbers, D.J., Richter, K.,Sell, W., H. Walden 1973. Measurements of wind–wave growth and swell decay during the Joint North Sea Wave Project (JONSWAP), *Deutsches Hydr. Zeit.*, A12: 1–95.

Kim, Y.C. and E.J. Powers 1979. Digital bispectral analysis and its applications to nonlinear wave interactions. *IEEE Trans. Plasma Science* PS-7 (2): 120-131.

Massel, S.R. 1995. Ocean Waves; their Physics and Prediction. World Scientific Publ., Singapore, 491pp (in press).

Massel, S.R., Piorewicz, J., Steinberg, C., and P. Boswood 1993. Wave and currents for macrotidal beaches (Field data analysis). *Proc. 11th Australasian Conf. on Coastal and Ocean Eng.*: 235-240. Townsville.

Masuda, A. and Y.Y. Kuo 1981. A note on the imaginary part of bispectra. *Deep-Sea Research.* 28A: 213-222.

Otnes, R.K. and L. Enochson 1972. *Digital Time Series Analysis.* New York: Wiley, 388pp.

Steedman, R.K. 1989. Data Report on Southern Ocean Surface Wave Climate Adjacent Macquarie Island, November 1988 to October 1989. ANARE Program, Project No. MQ/03/89, Antarctic Science Advisory Committee, Australian Antarctic Division.

Stochastic Hydraulics'96, Tickle, Goulter, Xu, Wasimi & Bouchart (eds) © 1996 Balkema, Rotterdam. ISBN 90 5410 817 7

Criteria of bar and berm formation and seadike design

Ching-Ruey Luo
Department of Engineering, Asian Institute of Technology, Bangkok, Thailand

ABSTRACT When wave energy builds up, beach berm and even dunes are quickly eroded with this material being removed offshore to form a bar parallel to the beach. It proceeds seawards during the course of the storm to reach some ultimate location prior to being dismantled and moved back to the beach by subsequent swell. There is difficulty in defining the steepness associated with storm, although it has commonly been accepted that steeper waves often produce a bar or a barred beach bar position and its crest height for a given storm condition. Therefore, criteria for the formation of bar and berm profiles respectively, need to be clarified.
In this paper, the available experimental data of large wave tank (LWT) from U. S. Army Crops of Engineering and Japanese Central Research Institute of Electric Power Industry are used to construct a new relationship. Finally, criteria for bar and berm formation and seadike design are found, and new criteria form are quite simple and widely applicable.

1. INTRODUCTION

When storm waves reach the swell buit profile, they are steep and arrive almost every second. Much water is thrown onto the beach face, which quickly becomes saturated, implying that the groundwater level is almost coincident with the beach face. Less water can percolate and, hence, the downwash almost equals the uprush, causing beach face erosion. Sand-laden water then proceeds seawards and approaches deeper water, where its velocity is reduced and so causing its sedimentary load to be deposited. This material accumulates in the form of a bar parallel to the beach, whick continuous to build up in the course of storm until the depth over it are sufficiently small for the incoming storm wave to be broken over it. At this stage, beach erosion essentially ceases. In the process of building up this bar and its subsequent seawards movement, distribution and sorting of sediment size across the bar mass has been reported, for example, as in Fig. 1. There is difficuly in defining the range of wave steepness (H/L) associated with storms, although it has commonly been accepted that steeper waves often produce a barred beach profile. As a coastal engineer, criteria for the formation of bar and berm profiles and

the guideline of seadike design need to be clarified.
Japanese researchers (Kajima et al. 1982, Takeda 1984, Sunamura and Horikawa 1984, Sunamua and Maruyma 1987) have plotted H_0/L_0 versus $(\tan \beta)^{-0.27}(D/L_0)^{0.67}$ and H_b/gT^2 versus D/H_b for the demarcation of bar migration directions, where D is the mean grain size in millimeters. Whereas workers in the USA have employcd different parameters extensively in using H_0/L_0 versus H_0/wT and H_0/L_0 against $(H_0/wT) \tan \beta$, where "w" being the sediment fall velocity (Dean 1973, Kraus and Larson 1988, Larson 1988, Larson and Kraus 1989). Other large wave tank tests can be found in Vellinga (1982). The results by Takeda (1984) and by Larson (1988) are plotted in Figs. 2 and 3, where H_0 and L_0 are wave height and wave length in deep sea; ß, the beach slope; H_b, the wave height of break wave; g, the gravity acceleration and T, the wave period.
A search for suitable new dimensionless parameters to represent the beach profile change is a worthwhile task. To facilitate this work, the LWT (Large Wave Tank) data of US Army Corps of Engineers (abbreviated as "CE" being for Coastal Engineering Research Center) and Japanese Central research Institute of Electric Power Industry (abbreviated as "CRIEPI"

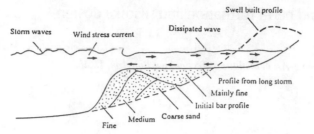

Fig.1 Definition of Beach Profile

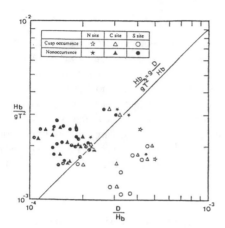

Fig.2 Bar and Berm Grouping
by Takeda

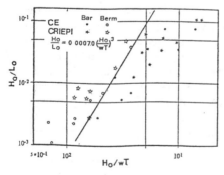

Fig.3 Bar and Berm Grouping
by Larson

data) are summarised and reanalysed. Another data of Tsai et al. (1995) are used to construct new set of seadike design guideline.

2. RESULTS OF LARGE WAVE TANK TESTS

The two independent data sets obtained in using LWT, referred to as CE and CRIEPI experiments respectively, were employed in this study. The 32 data entries were separated into two categories, being 18 data for bar profiles and the remaining 14 entries for berm predominated profiles. Some of the pertinent physical quantities are collected from various reports and others from calculations by the authors. A summary of these basic data is collectively presented in Tables 1 and 2. From reanalysing these data, an alternative definition on storm wave condition can be clearly given, based on the dimensionless sediment fall velocity H_b/wT, evaluated at breaking condition.

For a mean grain size ds (mm) with a fall velocity based on the assumption that the grain is a sphere relating the Reynolds number (wds/υ) to the buoyancy index, B.

$$[(r_s/r)-1] \cdot gds^3/\upsilon = B \qquad (1)$$

where
rs=the specific weight of the solid grain
r=the specific weight of the fluid
g=the gravitational acceleration
υ=the fluid kinematic viscosity
Generally, the fall velocity, w in water of common quartz is defined as:

$$w=[(r_s/r)-1]^{0.7}d_{50}^{1.1}/6\upsilon^{0.4}; (39<B<10^4) \qquad (2)$$

The relationship between Reynolds number and the index B is presented in Fig. 4.
The other dimensionless parameter denoting wave characteristics, such as $\tan \beta/(H_b/L_b)^{1/2}$, is used to define the criterion of bar and berm.

Table 1. Large Wave Tank Data of U.S. Army of Engineers (CE)

Case No.	d_s (mm)	m	H (m)	T (sec)	h (m)	H_o (m)	H_b (m)	H_o/L_o (×10³)	H_o/wT	H_o/ds (×10⁻³)	Profile (E)	Profile (c)
100	0.22	1/15	1.28	11.33	4.57	1.081	1.68	5.40	3.08	4.91	bar	bar
200	0.22	1/15	0.55	11.33	4.57	0.461	1.07	2.30	1.13	2.10	berm	berm
300	0.22	1/15	1.68	11.33	4.27	1.402	2.00	7.00	3.75	6.37	bar	bar
400	0.22	1/15	1.62	5.60	4.42	1.717	2.30	3.51	9.89	7.80	bar	bar
500	0.22	1/15	1.52	3.75	4.57	1.645	1.90	7.50	14.15	7.48	bar	bar
600	0.22	1/15	0.61	16.00	4.57	0.439	1.15	1.10	0.74	2.00	berm	beam
700	0.22	1/15	1.62	16.00	4.11 (3.81)	1.118	2.10	2.80	1.89	5.08	bar	*berm
101	0.40	1/15	1.28	11.33	4.57	1.081	1.80	5.40	1.62	2.70	berm	berm
201	0.40	1/15	0.55	11.33	4.57	0.461	1.90	2.30	0.69	1.15	berm	berm
301	0.40	1/15	1.68	11.33	4.27	1.402	2.40	7.00	2.17	3.51	berm	berm
401	0.40	1/15	1.62	5.60	4.42	1.717	2.40	3.51	5.57	4.29	bar	bar
501	0.40	1/15	1.52	3.75	4.57	1.645	1.60	7.50	7.98	4.11	bar	bar
701	0.40	1/15	1.62	16.00	3.81	1.118	1.95	2.80	1.18	2.80	berm	berm
801	0.40	1/15	0.76	3.75	4.57	0.827	0.76	3.77	3.74	2.07	berm	berm
901	0.40	1/15	1.34	7.87	3.96	1.246	2.00	1.29	2.68	3.12	bar	bar

(1) Profile (E) the experimental data; (c) the aralytical results.
(2) The "h" of Case No. 700 by 3.81 after 10 hours.

Table 2. Large Wave Tank Data of CRIEPI in Japan

Case No.	d_s (mm)	m	H (m)	T (sec)	h (m)	H_o (m)	H_b (m)	H_o/L_o (×10³)	H_o/wT	Hl/ds (×10⁻³)	Profile (E)	Profile (c)
1-1	0.47	1/20	0.44	6.0	4.5	0.461	0.95	8.20	1.202	98.10	berm	berm
1-3	0.47	1/20	1.05	9.0	4.5	0.948	1.40	7.50	1.648	2.02	berm	berm
1-8	0.47	1/20	0.81	3.0	4.5	0.852	0.85	6.07	4.444	1.81	bar	bar
2-1	0.47	3/100	1.80	6.0	3.5	1.758	1.94	3.13	4.585	3.74	bar	bar
2-2	0.47	3/100	0.86	9.0	3.5	0.733	1.54	5.80	1.275	1.56	berm	borm
2-3	0.47	3/100	0.66	3.1	3.5	0.709	08.0	4.73	3.579	1.51	berm	berm
3-1	0.27	1/20	1.07	9.1	4.5	1.040	0.88	7.40	3.294	3.85	bar	bar
3-2	0.27	1/20	1.05	6.0	4.5	2.101	1.58	1.96	5.288	4.08	bar	bar
3-3	0.27	1/20	0.81	12.0	4.5	0.651	1.47	2.90	1.563	2.41	berm	berm
3-4	0.27	1/20	1.54	3.1	4.5	1.619	1.50	1.08	15.050	6.00	bar	bar
4-1	0.27	3/100	0.31	3.5	3.5	0.340	0.50	1.78	2.080	1.26	berm	berm
4-2	0.27	3/100	0.97	4.5	4.0	1.058	1.27	3.35	6.776	3.92	bar	bar
4-3	0.27	3/100	1.51	3.1	4.0	1.604	1.52	1.07	14.910	5.94	bar	bar
5-1	0.27	1/50	0.29	5.8	3.5	0.299	0.63	5.70	1.486	1.11	berm	berm
5-2	0.27	1/50	0.74	3.1	3.5	0.799	0.89	5.33	3.970	2.96	bar	bar
6-1	0.27	1/10	1.66	5.0	4.0	0.778	1.91	4.56	10.250	6.59	bar	bar
6-2	0.27	1/10	1.12	7.5	4.5	1.097	1.42	1.25	4.215	4.06	bar	bar

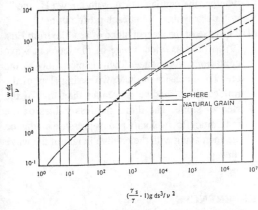

Fig.4 Relationship of Reynolds Number and Buoyancy Index

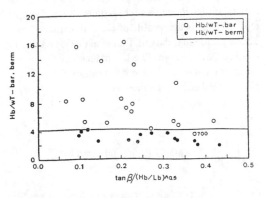

Fig.5 Bar and Berm Grouping by Luo

As shown in Fig. 5 which plots H_b/wT versus $\tan ß/(H_b/L_b)^{1/2}$, a horizontal line at $H_b/wT=4.10$ can be clearly drawn for demarcating the bar/berm profiles, except for the data point representing case 700 in the CE experiment in which its water depth was decreased from 4.11m to 3.81m after 10 hrs of running.

3. DISCUSSION AND CONCLUSIONS

In Tables 1 and 2, only ds, m(=tan ß), wave height H, wave period T, water depth h, deep water height H_0, breaker height H_b, temperature (for getting υ), and the experimentally resulting beach profile are measured, the fall velocity, w, can be obtained from Fig. 3 or Eq. (2), while h_b, the breaker water depth, obtained by

$$\frac{h_b}{H_b} = \frac{1}{b-(aH_b/gT^2)} = f(m, T) \tag{3}$$

$$a=43.75(1-e^{-19m}) \tag{4}$$

$$b= \frac{1.56}{(1+e^{-19.5m})} \tag{5}$$

$$m=\tan ß=\text{initial beach slope} \tag{6}$$
$$C_b=\sqrt{gh_b}=\text{celerity at the breaking wave} \tag{7}$$
$$L_b=C_b \cdot T=\text{Wavelength at the breaking wave position} \tag{8}$$

The above empirical relationships are derived in the Shoreline Protection of U.S.A. Manual by U. S. Army Crops of Engineers.(1984).
By comparing the results in Figures 2, 3 and 5, the new relationship in this study is quite easy and simple to clarify the beach profile of bar and berm.
The experimental data of Tsai et al (1995) on down-rushing flow on a seadike toe induced by waves are summaried and reanalysed by statistics, the new sets of seadike design guideline are:

$$(\frac{H_b}{gT^2})^{3/2} \geq 0.097 \, (h_b/gT^2)^{1.257} \, (\tan ß)^{0.571} \tag{9}$$

for the wave condition of down-rushing on seadike toe; and

$$U+max/g^{0.25}H_b^{0.75}T^{-0.50}=0.018\left[R/(\frac{H_b^{3/2}}{\sqrt{gT^2}}) \right]^{0.600}(\frac{h_b}{gT^2})^{-0.075}$$
$$(\frac{H_b}{gT^2})^{-0.386}(\tan ß)^{-0.400} \tag{10}$$

$$U-max/g^{0.25}H_b^{0.75}T^{-0.50}=0.0165\left[R/(\frac{H_b^{3/2}}{\sqrt{gT^2}}) \right]^{-0.076}(\frac{h_b}{gT^2})^{-0.200}$$
$$(\frac{H_b}{gT^2})^{-0.300}(\tan ß)^{-0.273} \tag{11}$$

$$\overline{P}+max/\rho g^{0.25}H_b^{1.5}T^{-1}=0.941\left[R/(\frac{H_b^{3/2}}{\sqrt{gT^2}}) \right]^{0.832}(\frac{h_b}{gT^2})^{0.240}$$
$$(\frac{H_b}{gT^2})^{-0.018}(\tan ß)^{0.213} \tag{12}$$

$$\overline{P}-max/\rho g^{0.25}H_b^{1.5}T^{-1}=0.192\left[R/(\frac{H_b^{3/2}}{\sqrt{gT^2}}) \right]^{0.298}(\frac{h_b}{gT^2})^{-0.134}$$
$$(\frac{H_b}{gT^2})^{-0.426}(\tan ß)^{-0.918} \tag{13}$$

for the positively and negatively maximum velocity and pressure on the seadike toe with given wave conditions, and R is the height of the wave run-up.

The following conclusions are made:
(1) The new relationship for distingishing bar and berm is convenient and applicable.
(2) For a given beach slope, m; water depth, h; wave height, H_0; the variables of H_b, h_b, are obtained; furthermore, the given beach profile can be told whether a bar (erodible profile) or a berm (deposite profile).
(3) Similarily, for a given beach slope, m; water depth, h; with the storm frequency analysis for wave height, H_0 and period, T, we can adjust whether the coming storm, with known H_0 and T, will destroy the given beach profilc or not.
(4) The case of the steeper slope of seadike from Eqs. (9) to (13)has wider range of the incident wave conditions to produce down-rushing flow to the toe. The intensity of the down-rusing flow of the milder slope seadike is weaker than that of the steeper slope. The dimensionless values of the maximum flow velocity and its pressure on the toe are increasing with the relative height of the wave run-up.

REFERENCES
1. DEAN R.G. 1937. "Heuristic models of sand transport in toe surf zone", Proc. 1st Aust. Conf. Coastal and Ocean Engg., Sydney, 208-214.

2. DALLY, W.R. 1987. "Longshore bar formation surf beat or undertow?", Proc. Coastal Seds, 1987, ASCE, Vol. I, 71-86.

3. KAJRMA, R., SHIMIZU, T., MARUYAMA, K. and SAITO, S. 1982. "Experiments on beach profile change with a large wave flume", Proc. 18th Internl. Conf. Coastal Engg., ASCE, Vol. II, 1385-1404.

4. KRAUS, N. C. and LARSON, M. 1988. "Beach profile change measured in the tank for large waves", 1956-1957 and 1962. Waterways Expt. Station, US Army Corps of Engrs., Tech. Report CERC-88-6.

5. LARSON, M. 1988. "Quantification of beach profile change", Dept. Water Resources Engrg., University of Lund, Sweden, Report 1008.

6. LARSON, M. and KRAUS, N. C. 1989. "SBEACH: Numerical model for simulating storm induced beach change", Report 1-Empirical foundation and model development. Waterways Expt. Station, US Army Corps of Engrs., Tech. Report CERC-89-9.

7. SUNAMURA, T. and HORIKAWA, K. 1984. "Two dimensional beach transformation due to waves", Proc. 14th Internl. Conf. Coastal Engg. ASCE, Vol II, 920-938.

8. SUNAMURA, T. and MARUYAMA, K. 1987. "Wave-induced geomorphic response of eroding beaches-with special reference to seaward migrating bars", Proc. Coastal Seds, 1987, ASCE, Vol. I. 788-801.

9. TAKEDA, I. 1984. "Beach changes by waves", Science Reports, Inst. Geoscience, University of Tsukuba, Section A, Vol. 5, 29-63.

10. VELLINGA, P. 1982. "Beach and dune erosion during storm surges", Coastal Engg., Vol. 6, 361-387.

11. TSAI, C.P., WANG, J.S., LIN, C. 1995. "Downrushing flow on a seadike induced by waves", 17th Ocean Engineering Conforence Tainan Taiwan.

12 U.S. ARMY CROPS. OF ENGINEERING, 1984 "Manual of shoreline Protection".

Stochastic Hydraulics'96, Tickle, Goulter, Xu, Wasimi & Bouchart (eds) © 1996 Balkema, Rotterdam. ISBN 90 5410 817 7

Dynamic characteristics of harbor oscillations in Helmholz mode

Yong Jun Cho
Department of Civil Engineering, Seoul City University, Korea

ABSTRACT: Dynamic characteristics of harbor oscillations were investigated to develope some economic auxiliary measures for improving harbour tranquility inside the port already in service which was frequently malfunctioned due to unexplained randomness at the design stage by extending the wisdom of perforated breakwater. It was shown that harbour with rather narrower entrance could be a effective shelter against a wave packet where the important part of the spectrum is narrow due to enhanced energy dissipation around the entrance and shortened natural frequency of harbour system. Furthermore, it turns out that harbour oscillation excited by long pulse close to the natural frequency can also be controlled by narrowing the harbour entrance.

Introduction

Even though the nature of Harbor resonance like how it can be excited, upto what extent it can grow and the natural frequency of closed basin was well understood [Mei, 1989], it was frequently reported that working condition inside the harbor in service is severely deteriorated during the considerable extension of operating period due to basin oscillation. Even in the harbor where at the design stage the energy containing part of response spectrum inside the harbor was well separated away from the incident design wave spectrum by adjusting the basin surface area and entrance width, there still exists some possibility of oscillation due to the randomness which is the intrinsic property of waves, and the effects of which on harbor oscillation was not known yet. Hence, it is inevitable to introduce additional damping structure into the harbor system for rather complete suppression of basin oscillation unless the possibility of excitation of water mass in the resonance mode is totally removed, which is physically implausible. Among many measures to enhance the energy dissipation, the most promising candidate for this purpose will be artificial enforcing of energy dissipation using the vortex shedding due to sudden expansion. This wisdom was successfully realized in the perforated breakwaters. Due to its own nature, it is very vital in the estimation of amount of additionally introduced energy dissipation and locating the tranquil area within the harbor to accurately model turbulent flow. In 1975, Ünlüata and Mei pointed out that the entrance loss should be considered to remedy the harbor paradox and sequentially derived the boundary condition which should be imposed at the constriction in the so called matched asymptotic expansions technique. With this condition exclusive of the apparent inertia based on the facts that with very likely separation at a constriction, added hydrodynamic length is subject to be reduced so that the apparent inertia is relatively small

especially for larger amplitudes and longer waves, they investigated the effect of entrance loss on harbor oscillation for the rectangular harbor with a centered entrance. Under the situation that the tsunami of sufficiently long duration causes costly delays due to persistent oscillations within the harbor, dymamic characteristics of oscillations should be better known for the optimal operation of the harbor system. Harbor response to transient incident waves was also tackled by Mei [1989] by Fourier representation of the simple harmonic response without apparent inertia term. Furthermore, there should be a practical limit in narrowing harbor entrance to sustain the role of navigational channel. In this case, the apparent inertia term should be considered albeit small. With Ünlüata and Mei's condition inclusive of apparent inertia term, it is the intent of this study to explore the remedation measures of oscillations inside the harbor in service by minimizing the resident time of wave energy within.

Review of Harbor Oscillations

The peak amplitude at resonance can be limited by radiation damping associated with energy escaped seaward from the harbor entrance and frictional loss near the harbor entrance. With matched asymptotic expansions technique based on the facts that various parts of the physical domain are governed by vastly different scales and a friction loss formula which is quadratic in the local velocity, Mei [1989] shows that first representing the net momentum flux through the harbor entrance in terms of loss coefficient f and then utilizing the equivalent linearization technique, the reduction in the amplitude of response, R, when frictional loss at a constriction is taken for account is given by

$$R = \left(\frac{2}{1 + (1 + 16\gamma/(ka)^2)^{1/2}} \right)_{k=\tilde{k}_{mn}} \quad (1)$$

where A, h, r_c and $2a$ is the amplitude of incident waves, water depth at a constriction, the ratio of gross area of the channel to one at Vena contracta, entrance width, respectively, and

$$\gamma = \frac{2fA}{3\pi h}$$

$$f = (r_c - 1)^2$$

and the natural wave numbers of closed basin $\tilde{k}_{mn}$ in terms of basin length L and width W are

$$\tilde{k}_{mn} = \left[(\frac{m\pi}{L})^2 + (\frac{n\pi}{W})^2 \right]^{1/2}$$

. Considering that for usual breakwater dimensions and wave periods, the Keulegan-Carpenter number [or equivalently the Strouhal number] VT/D where V is the velocity amplitude, D the body dimension and T wave period, can be rather large and quadratic friction loss formula is valid only for high values of Keulegan-Carpenter number, relatively accurate information can be deduced from (1). It can be clearly seen from (1) that the reduction of resonant peaks by entrance loss is more pronounced for larger f, larger amplitude, longer waves or lower resonant modes and narrower entrance. With regard to the parameter $16\gamma/(\tilde{k}a)^2$, it should be pointed out that the loss coefficient f may depend on the Strouhal , Reynolds number and the geometry of the breakwater tips at the entrance and eq. 1 is valid only for $16\gamma/(\tilde{k}a)^2 \leq O(1)$. Ito [1974] recommended $f = 1.5$ for the Ofunato tsunami breakwater. For a harbor with comparable dimensions in both horizontal dimensions, the lowest resonance mode [Helmholtz Mode] constitute larger part of response spectrum and in this case $16\gamma/(\tilde{k}a)^2 > O(1)$ violating the underlying assumption of (1).

Equation of motion

The response of water basin in the Helmholz mode to the incident wave can be described by [Mei, 1989]

$$\ddot{x} + 2\beta\omega_n\dot{x} + \omega_n^2 x = \omega_n^2 \zeta \quad (2)$$

where x, ζ, S, $2a$, U_0 and L_0 are the water surface displacement at landward, offshore side of a constriction, surface area of a basin, the width of harbor entrance, the amplitude of velocity at a constriction and added hydrodynamic length, respectively, and and the natural frequency of a basin ω_n is given by

$$\omega_n^2 = 2gha/SL_0$$

$$2\beta\omega_n = c\frac{f}{2L_0}|U_0|$$

and

$$c = 8/3\pi$$

In the derivation of (2), the principles of mass and momentum conservation, quadratic friction loss formula and equivalent linearization technique are again invoked. Following

the Mei's proposal that for a case that $16\gamma/(\tilde{k}a)^2 > O(1)$, U_0 can be approximated by Torricelli's law

$$|U_0| \cong (\frac{?_g A}{4f/3\pi})^{1/2}$$

, the damping coefficient in (2) is reduced to

$$2\beta\omega_n = (cgfA)^{1/2}/2L_0$$

Response spectrum

Following the standard technique in stochastic analysis to explain the randomness intrinsic in incident wave, the impulse response function $h(t)$ is given by

$$h(t) = \frac{\omega_n^2}{\sqrt{1 - \beta^2}\omega_n} \sin(\sqrt{1 - \beta^2}\omega_n t)e^{-\beta\omega_n t} \quad (3)$$

whereas the frequency response function $H(\omega)$ and the response spectral density function $S_{xx}(\omega)$ [Ochi, 1992] is given by, respectively,

$$x(t) = \int_{-\infty}^{\infty} \zeta(\tau)h(t - \tau)d\tau$$

$$H(\omega) = \int_{-\infty}^{\infty} h(\tau)e^{-i\omega\tau}d\tau$$

$$S_{xx}(\omega) = |H(\omega)|^2 S_{\zeta\zeta}(\omega) \quad (4)$$

where $S_{\zeta\zeta}(\omega)$ is the incident wave spectral density function. From eq. (2), $H(\omega)$ is given by

$$H(\omega) = \frac{\omega_n^2}{\omega_n^2 - \omega^2 + i2\beta\omega_n\omega} \quad (5)$$

Numerical Results

To quantify the above results, we must specify added hydrodynamic length, the wave spectrum from which the quantity β may be calculated. In this study, we shall use a transient wave packet with a carrier frequency ω_0 and a slowly varying Gaussian envelope so that ζ in (1) is

$$\zeta = 4B\exp[-\Omega^2 t^2]\cos\omega_0 t$$

and amplitude spectrum takes the form

$$S_{\zeta\zeta}(\omega) = \frac{B}{2\Omega\sqrt{\pi}} \left[\exp[-(\frac{\omega - \omega_0}{2\Omega})^2] + \exp[-(\frac{\omega + \omega_0}{2\Omega})^2] \right]$$

where

$$\frac{\omega_0}{\Omega} \gg 1$$

which implies that the important part of the spectrum is narrow. For added hydrodynamic length, by the analogy of long waves of small amplitude to sound waves , analytical results known for several acoustic orifices may be applied. In 1968, Morse and Ingard showed that for small gaps

$$\frac{L_0}{W} \cong \frac{1}{\pi} \ln \frac{1}{\pi} \frac{2W}{\pi a}$$

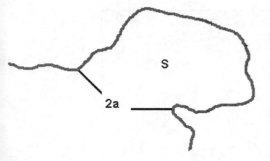

Fig. 1. Definition sketch

, and for large gaps

$$\frac{L_0}{W} \simeq \frac{8}{\pi}(1 - \frac{2a}{W})^2$$

. The impulse response function in (3) and frequency response function in (4) is plotted in Fig. 1 and Fig. 2 , respectively, for $2a = 300$, 210, 160 and 110 m, respectively,. Here, it is obvious that as the harbor entrance is getting narrower, the amplitude of impulse response function is getting diminished and the area under frequency response function also shrinks with the peak frequency where the maxima in frequency function is occurring shifted toward the lower frequency range. It is due to the facts that with narrower entrance, harbor system is getting slender in spite of the increase in added hydrodynamic length for both small and large gaps and the energy loss at the entrance is enhanced. Hence, the possibility of oscillation in resonance mode is very low so that the harbor with a nar-

rower entrance can be a effective shelter against the wind waves whose typical energy containing frequency range is much higher than the peak frequency. In Fig. 4 and Fig. 5, the spectrum of a transient wave packet with $\frac{\omega_0}{\Omega} = 4$ and $\frac{\omega_0}{\Omega} = 2$, respectively, is plotted and corresponding response spectra for $2a = 300$, 210, 160 and 110 m are inculded for the comparison. It is shown that with relatively narrower entrance, the frequency spectrum rather rapidly shrinks in the former case which can be regarded as narrow banded whereas for the latter case, the response spectra cannot be negligible even in the narrowest case. These facts are consistent with our intuition because harbour with narrower entrance is very vulnerable to a relatively long pulse due to its high transmissivity.

Conclusion

Although much progress has been made quite recently in the numerical analysis of harbour tranquility, our understanding of this complicated problem falls short of the complete yet. Hence, some harbors have frequently failed in providing a safe shelter for vessels during storm conditions and the tsunami attack. Furthermore, the complete suppression of harbor oscillation for safe anchorage for vessels is very difficult and expensive goal to achieve since harbor is usually exposed to the waves the spectra of which reside over quite broad frequency range and ship motions are quite sensitive to wave frequency. In this study, dynamic characteristics of harbor oscillations were investigated to develope some economic auxiliary measures for improving harbour tranquility inside the port already in service which was frequently malfunctioned due to unexplained randomness at the design stage by extending the wisdom of perforated breakwater. It was shown that harbour with rather narrower entrance could be a effective shelter against a wave

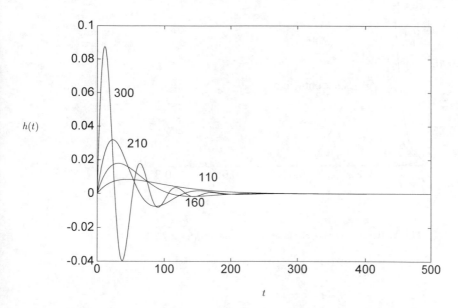

t

Fig. 2. Impulse response function for varying entrance width

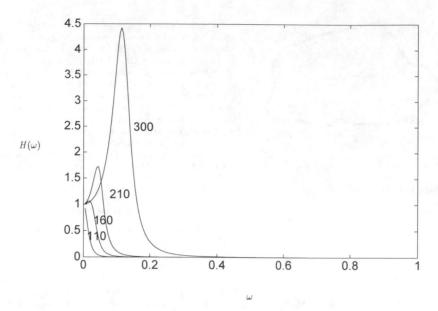

Fig. 3. Frequency response function for varying entrance width

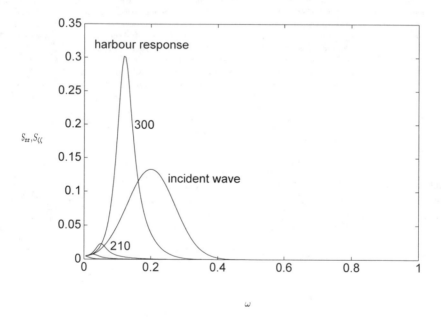

Fig. 4. Harbor response and incident wave spectra (for $\omega_0/\Omega = 4.0$)

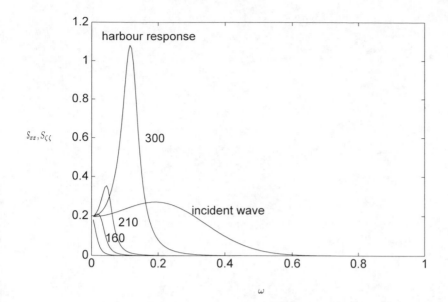

Fig. 5. Harbor response and incident wave spectra (for $\omega_0/\Omega = 2.0$)

packet where the important part of the spectrum is narrow due to enhanced energy dissipation around the entrance and shortened natural frequency of harbour system. Furthermore, it turns out that harbor oscillation excited by long pulse close to the natural frequency can also be controlled by narrowing the harbor entrance.

References

1. Abramowitz, M. and I. A. Stegun 1968. Handbook of mathematical functions. Dover, Mineola, NY. Vol. 237, 212-232.

2. Dean, R. G. and R. A. Dalrymple 1984. Water wave mechanics for engineers and scientists. Prentice-Hall, Inc., Englewood Cliffs, NJ.

3. Huang, N. E., S. R. Long, C. C. Tung, Y. Yuan and L. F. Bliven 1981. A unified two-parameter wave spectral model for a general sea state. J. Fluid Mech., 112, 203-224

4. Huang, N. E., S. R. Long, C. C. Tung, Y. Yuan and L. F. Bliven 1983. A non-Gaussian statistical model for surface elevation of nonlinear random wave fields. J. Geophys. Res., 88, 7597-7606.

5. Lin, Y. K. 1967. Probabilistic theory of structural dynamics. McGraw-Hill book Co., New York, NY.

6. Longuet-Higgins, M. S. 1963. The effect of nonlinearities on statistical distributions in the theory of sea waves. J. Fluid Mech., 17, 459-480.

7. Mei, C. C. 1989. The applied dynamics of ocean surface waves. World scientific publishing Co. Pte. Ltd.

8. Morse, P. M. and K. U. Ingard 1968. Theoretical Acoustics. McGraw-Hill, New York.

9. Ochi, M. K. 1992. Applied probability and Stochastic processes in engineering and physical sciences. John Wiley and Sons, Inc.

10. Papoulis, A. 1965. Probability, random variables and stochastic processes. McGraw-Hill book Co., New York, NY. of breaking limited surface elevation. J. Geophys. Res., 94, No. C1, 967-972.

11. Ünlüata, Ü. and C. C. Mei 1975. Effects of entrance loss on harbor oscillations. J. Waterways, Harbors and Coastal Eng. Div. ASCE 101, 161-180

Stochastic Hydraulics'96, Tickle, Goulter, Xu, Wasimi & Bouchart (eds) © 1996 Balkema, Rotterdam. ISBN 90 5410 817 7

Wave climate off southern Taiwan, China and extreme wave prediction

Ching Dann Juang
North District Division, East-West Expressway Construction, THB, Chungli, Taiwan, China

Chia Chuen Kao
Department of Hydraulic & Ocean Engineering, National Cheng Kung University, Tainan, Taiwan, China

ABSTRACT : Wave climates off the southern Taiwan were measured and analyzed. Surface elevations obtained at the fixed locations by the ultrasonic type wave gauge during 1992-1994. Wave height were fitted to the Weibull distribution function. It is found that the shape factor is different from Rayleigh distribution for both storm and swell waves which provided as an important role of the wave climate study. Wave height proposed as the mean, root mean square and significant wave heights of each record were used as the threshold value for the extreme prediction. Shape factors of the Weibull distribution of each threshold were computed. The distribution is narrower for the higher value of threshold which indicated that wave heights were uniformly distributed for the larger wave height of swell and storm waves. Multiplication factors proposed as the ratio of the maximum wave height predicted by the overall wave distribution to that of the threshold defined above were compared. It was found that the factor varied from 1.1 to 1.4. It can be considered the over prediction which should be noted for the engineering design.

1.INTRODUCTION

An important topic in ocean engineering is the wave climate at a given location, for example, a given pipeline route. For such studies the existence of a sufficient amount of wave measurements becomes crucial. With respect to the Taiwan area most wave data series have a duration of order of a few years, but such duration is considerably shorter than the time period usually adopted as the return period for the design conditions. Accordingly, an evaluation of the long-term representativity of the measuring period is a very important part of the climate consideration. In this note , we addressed our attention to the maximun wave height prediction of the short term measurements. The distributions for the larger wave height considered different threshold were investigated which should be noted for the engineering applications.

Useful as the spectral concept in hindcasting, it is rarely used by the design engineer. The designer is usually most interested in the expected value of maximum wave heights, although the details of the wave shape and the directional spreading of wave energy can also influence forces. The long and complicated path from historical meteorological data to long-term wave statistics has been studied by researchers. One crucial step in this hindcasting process is the determination of the distribution of wave height during the short time interval from the measurements.

Taiwan island have various wave climates along their coasts owing to their topography. As seen in Fig. 1, one side of the island faces the Taiwan Strait, the other side faces the Pacific ocean. Waves in the Taiwan Strait are fetch limited wind waves dominated by the winter monsoon from mid October to March next year. While waves from the Pacific during the typhoon season from mid April to October, are typhoon generated seas and swell.

So far as sea state characteristics such as the significant wave height and the corresponding period are considered, the actual time series will most probably be sufficiently accurate. Their adequacy becomes more questionable as characteristics of the individual waves are to be investigated, especially if the waves are nonlinear. Detail of the accuracy associated with respected to the time series is out of scope of this investigation.

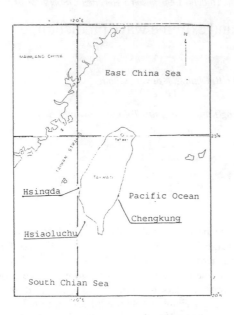

Fig.1   Locations of measurement

Wave height distribution has been discussed concerning the importance of the spectral bandwidth i.e., Longuet-Higgins [1980],Larsen [1981] and Tayfun [1981]. According to the studies of Forristall [1984], if the details of spectral shape are not well known, the Weibull distribution offers some advantage in ease of manipulation. In this note, the Weibull distribution is adopted.

$$p(x) = 1 - \exp(-\alpha x^\beta) \qquad (1)$$

where" x "is the normalized wave height. In this case x=H/Hrms where Hrms is the root mean square value from the individual wave height of the time series. However, the normalized factor can be mean wave height or the significant wave height. " α " is the scale parameter which denoted the peak of the distribution and " β " is the shape factor which represented the narrowness of the distribution. Eq (1) corresponds to Rayleigh distribution when α =1 and β =2. The relation between " α " and " β " can be expressed in term of Gamma function:

$$\alpha = \left[\Gamma\left(1 + \tfrac{2}{\beta}\right)\right]^{\beta/2} \qquad (2)$$

where

$$\Gamma(x) = \int_0^\infty e^{-t} t^{x-1} dt \qquad (3)$$

Statistic characteristic of the distribution function can be derived as

$$H_{1/p}/H_{rms} = (\ln p/\alpha)^{1/\beta} + P\left\{ \tfrac{1}{\beta}\left(\tfrac{1}{\alpha}\right)^{1/\beta} \Gamma(1/\beta) \right.$$
$$\left. - \int_0^{x_0} \exp(-\alpha x^\beta) dx \right\}$$

$$x_0 = \left(\frac{\ln p}{\alpha}\right)^{1/\beta} \qquad (4)$$

$$H_{max}/H_{rms} = (\ln N/\alpha)^{1/\beta} \cdot \left(1 + \frac{\gamma}{\beta \ln N}\right) \qquad (5)$$

γ : Euler constant , γ =0.577275

N : Wave number

2.WAVE DATA MATERIAL

Wave records were obtained by the ultrasonic type wave gauge located at a fixed position as shown in the Fig.1., Recording practice is a 1 Hz, 20 minute record taken every 2 hours during the measuring days. Operation of the wave gauges as well as the routine data analysis is carried out by the Department of Hydraulics & Ocean Engineering National ChengKung University. The following data is used for the present investigation:

Chenkung July, 1994
Hsingta July, 1993 - Sept, 1995
Hsiaoluchu July, 1993 - Sept, 1995

Tab.1. Meaurements of the sea state
Hsingda(swell)   July 1992

Date	Hs	Ts	S
23	62	5.1	0.015
24	89	6.2	0.014
25	174	6.6	0.025
26	371	7.6	0.041
27	328	7.9	0.033
28	245	7.8	0.025
29	132	7.0	0.017
30	88	4.9	0.023

ChengKung (storm)   July 1994

Date	Hs	Ts	S
8	96	3.6	0.047
9	114	3.8	0.050
10	123	4.1	0.047
11	189	3.6	0.095
12	667	8.6	0.058
13	423	5.2	0.098
14	118	4.3	0.039

278

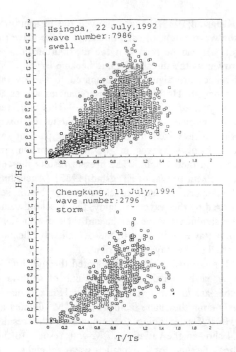

Fig.2　Plot of wave height vs period

Tab.2. Shape factor of wave height and period distribution

Hsingda (swell)		July 1992
Date	H	T
22	1.86	1.64
23	1.48	1.87
24	1.61	1.77
25	1.61	1.88
26	1.67	1.94
27	1.62	1.96
28	1.55	1.84
29	1.60	1.75
30	1.68	1.69

Hsiaoluchu (swell)		Aug. 1993
Date	H	T
25	1.65	2.02
26	1.64	1.97
27	1.64	1.98
28	1.36	1.69
29	1.63	2.14
30	1.61	2.07
31	1.55	1.98

The analyses were done for consecutive 20-min. segments of data. The mean sea level for each segment was calculated and defined as the zero level. A wave was then defined as the part of the record between two consecutive passages of the trace up cross the zero level. This definition is sometimes referred to as the up crossing method. The 20-minute records were thus divided into a number of consecutive sections,each of which constitutes a wave. The period of each wave is then the time between up crossings, and the height of the wave is the difference between the highest and lowest elevations during the wave, that is, the distance between the crest and the trough.

In order to understand the statistic characteristic of the waves, the analyses were carried for the measurements for 24 hours to see the variation of each typhoon or swell. Wave statistics are directly calculated and applied to the Weibull function.

3.REPRESENTATIVITY OF WAVE MEASUREMENTS

3.1 Wave characteristic

Sea states expressed in terms of significant wave height (Hs) and period (Ts) were shown in Table 1. Wave records during 1992-1993 were swell propagated northward due to the storm propagated through the South China sea. Records of 1994 were the storm propagated westward from the Pacific. The transformation of Hs of each day denoted the risen and decay range of the storm or swell. It was found that the growth of Ts was coupled with the height. Physically, the larger wave height might be owing to the long duration of the wind. Totally wave steepness "S" defined as the ratio of height Hs the length corresponding to the Ts ,which can be an indicator of the serverity of sea state. It is important for the wave forces on the offshore structures and the stability for the ship moving.

Fig.2 showed the plot of wave period and height of the records which indicated the composition of the wave period and height. One of the interest topic is the heightwise ranked wave period in the time domain analysis of the wave time series in comparison to the decomposition of the spectral in frequency domain. The analyses of the period ranked wave height distribution for the risen and decay range of the storm and swell is under preparation.

The relation between period and height can be investigated through the H-T plot. A functional relation can be obtained and provided as the prediction of the design wave periods.

$$T = 2.26H^{0.669} \quad \text{(swell)}$$
$$T = 3.51H^{0.697} \quad \text{(storm)}$$

Table 2. showed the distribution of the wave height and period of the swell and storm. There is no significant trend for the risen and decay of the swell and storm since the mechanism is not well known for the wave height and period during the storm or swell. Lower value of the shape factor means that the distribution of wave height is broad and the wave height is mixed. It differed from the Rayleigh distribution pointed out by several reports. Therefore, the prediction of the extreme value of the upper tail will be lower than that of Rayleigh distribution. It will be discussed in the next paragraph. The information of the joint distribution of height and period of storm waves is important for the marine structure dynamic consideration. Some studies derived a closed form to describe the joint distribution with the narrow band assumption i.e., the lower value of height and period correlation. However, it had been pointed out by Juang et al [1992] and Goda [1978] that the coefficient of height and period correlation of storm or swell waves was more than 0.7. According to to the analyses of Goda [1978] using the wave data of Japan, the coefficient of upper portion with high waves was lower so that the theoretical form could be applied. Analysis of the joint distribution of the measurement is under preparation.

3.2 Extreme wave prediction

Since the distribution of the wave height for the swell and storm were broad, the upper tail prediction should be lower than that of Rayleigh. The most important thing is the definition of the threshold value of the upper tail. In this note, the extreme wave analysis considering the portion of the upper tail was stressed. Although the estimate significant wave height from the records itself has some inherent statistical fluctuation. Statistically, it is not easy to find the threshold value as to obtain a reasonable extreme value. In this study, the mean, root mean square, and significant height of each record were choose as the threshold so as to compare to the study

of Krogstad[1985]. Fig.3 showed the distributions of the wave height of each threshold of the wave records. It can be seen that the distribution became narrower as the threshold value became larger. The differences are significant compared to the overall wave height. However, the discrepancies are small among that of different thresholds. This indicated that the number of small wave height in the lower portion with mixed heights the major part of the measurement. Waves in the higher portion are uniform in their height. The present results showed an agreement to that of Krogstad [1985]. According to his report, the surface elevations of North sea measured by the waverider buoy, the threshold value is the significant wave height and their distributions are not Rayleigh in spite of the overall wave heights are fitted to the Rayleigh distribution.

The normalized Hmax computed through Eq.(5) can be easily calculated if the value of and wave number are known. It is clear that Hmax depends mostly upon the shape factor. A ratio "M" defined as M=Hmax/Hs, is going to compare to the results Fig.4 showed the variation of M with respect to the shape factor of overall wave height of

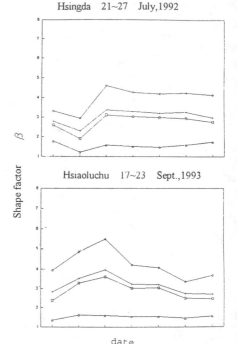

Fig.3 Wave height distributions for the threshold values
□ :H>Hm , + :H>Hrms, ◊ :H>Hs, △:H total

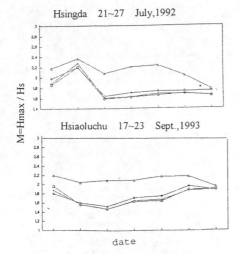

Hsingda 21~27 July,1992

Hsiaoluchu 17~23 Sept.,1993

date

Fig.4 Predicted maximum wave height
to Hs for the threshold values
□ :H>Hm , ✝,H>Hrms, ◊:H>Hs, ▲:H total

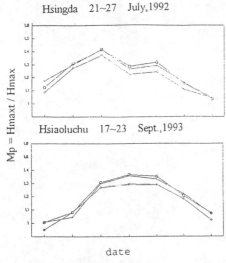

Hsingda 21~27 July,1992

Hsiaoluchu 17~23 Sept.,1993

date

Fig.5 Comparison of factors of predicted
maximum wave height for the threshold
values
□ :H>Hm , ✝,H>Hrms, ◊:H>Hs

Krogstad[1985]. Value of M varied from 1.8 to 2.0
for the threshold values of wave heights where the
wave number varied from 2400 to 3000 respectively.
According to the results of Krogstad, the ratio varied
from 1.6 to 1.8 as wave number N varied from 400
to 30000. However, the shape factor is not known.
From the other hand, we proposed a multiplication
factor defined as the ratio of Hmax of overall wave
to the Hmax of each threshold to see the range we

should be considered. Thus, Mp=Hmaxt/Hmax is
used. As can see in the Fig.5, the value of Mp varied
from 1.1 to 1.4. which should be considered in the
engineering application.

4.CONCLUSION

Wave height distributions of storm and swell waves
are investigated. It is found that the distributions for
the overall and the upper portion wave heights fails
to the Rayleigh distribution.   Extreme height
prediction is modified by the distribution of the upper
portion with high waves. The range of multiplication
is 1.1 to 1.4 corresponding to the threshold value of
mean, root mean square and significant height of the
record. The threshold value of wave height is also
needed in the joint distribution analysis which is
suggestive for the further study.

REFERENCES

Forristall, G. Z., On the statistical distribution of
wave heights in a storm, J. Geophy. Res.,Vol.83,
No.C5, 2353-2358, 1978

Forristall, G.Z.,The distribution of measured and
simulated wave heights as function of spectral
shape, J. Geophy. Res., Vol. 89, No.6, 10547-
10554,1984

Goda,Y., The observed joint distribution of period
and height of sea waves, Chapter 11, 227-240,
ICCE,1978

Haver, S., Wave climate off northern Norway,
Applied Ocean Res.,Vol.7, No.2, 85-92, 1985

Juang,C.D., Kao,C.C., Nearshore wave
characteristics and their evolution on the shoaling
water,Proc.6th IAHR International Symp.on Sto
chastic Hydraulic, 275-282,1992

Krogstad, H.E., Height and period distribution of
extreme waves, Applied Ocean Res., Vol.7,No.3,
158-165, 1985

Larsen,L.H.,The influence of bandwidth on the
distribution of heights of sea waves, J.Geophy.
Res.,86, 4299-4301,1981

Longuet-Higgins, M.S., On the distribution of the
heights of sea waves: Some effect of nonlinearity
and finite bandwidth, J. Geophy. Res., 85,1519-
1523 1980

Tayfun,M.A., Distribution of crest-to trough wave
heights, J. Waterways Harbor,ASCE,107,149-
158,1981

Stochastic Hydraulics'96, Tickle, Goulter, Xu, Wasimi & Bouchart (eds) © 1996 Balkema, Rotterdam. ISBN 90 5410 817 7

Statistical properties of wave groups in nonlinear random waves of finite bandwidth

Yong Jun Cho
Department of Civil Engineering, Seoul City University, Korea

ABSTRACT: Overdue task of investigating nonlinear effects on the statistical properties of wave groups in terms of the average number of waves in a group and the mean number of waves in a high run was resumed in this study utilizing the complex envelope and total phase function, random variable transformation technique and perturbation method. It turns out that phase distribution is modified significantly by nonlinearities, and shows a systematic excess of values near the mean phase and corresponding symmetrical deficiency on both sides away from the mean. For the case of threshold crossing rate, it is noted that threshold crossing rate reaches its maxima at larger reference level as nonlinearity is getting profound. Furthermore, the mean waves in a high run associated with nonlinear wave is shown to has a tendency to be increased than the linear counterpart. Similar trend can also be found in the average number of waves in a group.

Introduction

The grouping of high waves is an important parameter in many engineering problems associated with port development, which may influence long period oscillation of moored vessels and other floating structures and surf beat. After the Gaussian model was first developed in well known papers by Rice [1958], particular attention to properties of wave groups in Gaussian noise was paid by Longuet-Higgins [1957]. Recent interests in the subject [Goda, 1983] has been stimulated by the suggestion that exceptional damage to ships, coastal defences or offshore structures may be caused by the occurrence of runs of successive high waves. Further reason for interest is the relation of wave groups with the formation of wave breaking. In 1984, Longuet-Higgins has obtained the expressions for wave group length and length of a high run using two apparently distinct approaches in the context of Gaussian waves; first, by a wave envelope function and, later, by treating the sequence of wave heights as a Markov chain. It was shown that two approaches are roughly equivalent and spectral bandwidth has a significant influence on the wave group length and length of a high run. On the other hand the probable effect of the nonlinearity on wave groups in random waves of finite bandwidth remains uncertain due to the complicated form of nonlinear random waves which was first developed by Longuet-Higgins in 1963. In a case when the underlying frequency spectrum is narrow, the stochastic representation of a nonlinear sea surface is reduced to a familiar form in which each realization is an amplitude modulated second order Stokes wave [Tayfun, 1986]. In contrast with the intricate complexity of the expression of nonlinear waves of finite bandwidth, such an approximation constitutes a simpler formulation to study numerically or analytically the nonlinear effects on the statistical description of wave properties. But considering the side band instability of Stokes wave, the narrow band assumption at the site away from the generating area is no longer valid. The search for a way simpler than that of Longuet-Higgins [1963] to describe nonlinear waves of finite bandwidth was recently carried out by Tung et al. [1989]. Based on the studies of Tayfun [1980, 1986], Tung et al. [1989] proposed a simple but accurate expression for second order nonlinear wave elevation for waves of moderate bandwidth. This wave model was more elaborated by Cho [1992] to analyze the extreme distributions of wave elevation. It turns out that as the nonlinearity is getting profound, these extreme distributions deviate from the linear counterpart in an increasing manner. The general character of this deviation is in the form of a spreading of the density mass toward the larger and smaller crests. The objective here is to gain some theoretical insight into the nature of nonlinear modifications on the statistical properties of wave groups in terms of wave group length and length of a high run, quantities that are of great importance in the design of ships, coastal defences or offshore structures. In this report, our attention is centered on deep water waves only.

Review of wave groups theory

It is known that for stationary random process $\zeta(t)$ of arbitrary bandwidth, the average number of waves in a group G and the mean number of waves in a high run H are

$$G = N_\zeta(\zeta_o)/N_A(A_0) \tag{1}$$

$$H = N_\zeta(\zeta_o)Q(A_0)/N_A(A_0) \tag{2}$$

where

$$N_\zeta(\zeta_o) = \int_0^\infty \dot{\zeta} f_{\zeta\dot{\zeta}}(\zeta_0, \dot{\zeta})\, d\dot{\zeta} \tag{3}$$

$$N_A(A_0) = \int_0^\infty \dot{A} f_{A\dot{A}}(A_0, \dot{A}) \, d\dot{A} \qquad (4)$$

and

$$Q(A_0) = \int_{A_0}^\infty f_A(A) \, dA \qquad (5)$$

In (3), (4) and (5), $f_{\zeta\dot{\zeta}}(\cdot, \cdot)$ and $f_{A\dot{A}}(\cdot, \cdot)$ are the joint probability density function of ζ and $\dot{\zeta}$ (overdot denotes time derivative) and the joint probability density function of A and $\dot{A}$, respectively, and $N_\zeta(\zeta_0)$ represents the number of up-crossings by ζ of a given level ζ_0 per unit time and $N_A(A_0)$ is the number of up-crossings of a given level A_0 per unit time by the wave envelope A and $Q(A)$ is the exceedance probability of a given level A_0 by the wave envelope. To apply (1) and (2) to nonlinear random waves, it is necessary to have $f_{\zeta\dot{\zeta}}(\zeta_0, \zeta)$ and $f_{A\dot{A}}(A, \dot{A})$ which in turn require a nonlinear wave model of finite bandwidth.

Envelope and phase process of nonlinear random waves

Consider infinitely long-crested waves of arbitrary bandwidth in deep water. The surface displacement is given by [Longuet-Higgins, 1963]

$$\zeta = \sum_{i=1}^\infty a_i \cos \chi_i + \frac{1}{2g} \sum_{i=1}^\infty \sum_{j=1}^\infty a_i a_j \omega_i^2 \cos(\chi_i + \chi_j)$$

$$- \frac{1}{2g} \sum_{i=1}^\infty \sum_{j>i}^\infty a_i a_j (\omega_j^2 - \omega_i^2) \cos(\chi_j - \chi_i)$$

in which $\chi_i = k_i \chi - \omega_i t + \varepsilon_i$, k_i is the wave number, $\omega_i = (gk_i)^{1/2}$ is wave frequency, ε_i is random phase uniformly distributed over the interval $(0, 2\pi)$ and a_i is the amplitude of the component wave. Upon introducing the following random processes

$$\eta_1 = \frac{1}{(M_0)^{1/2}} \sum_{i=1}^\infty a_i \cos \chi_i$$

$$\eta_2 = \frac{1}{(M_2)^{1/2}} \sum_{i=1}^\infty a_i \omega_i \sin \chi_i$$

$$\eta_3 = -\frac{1}{(M_4)^{1/2}} \sum_{i=1}^\infty a_i \omega_i^2 \cos \chi_i$$

$$\eta_4 = \frac{1}{(M_0)^{1/2}} \sum_{i=1}^\infty a_i \sin \chi_i$$

$$\eta_5 = -\frac{1}{(M_2)^{1/2}} \sum_{i=1}^\infty a_i \omega_i \sin \chi_i$$

$$\eta_6 = -\frac{1}{(M_4)^{1/2}} \sum_{i=1}^\infty a_i \omega_i^2 \sin \chi_i$$

, it was shown that the nondimensional nonlinear wave elevation, ζ_1, can be written as [Tung et al., 1989]

$$\begin{aligned}
\zeta_1 &= \zeta/(m_0)^{1/2} \\
&\cong \left(\frac{M_0}{m_0}\right)^{1/2}(\eta_1 - \frac{1}{2}\epsilon \eta_1 \eta_3 + \frac{1}{2}\epsilon \eta_4 \eta_6)
\end{aligned} \qquad (6)$$

where M_i and m_i are the ith spectral moments of the linear and nonlinear wave elevation, respectively and $\epsilon =$

$(M_4)^{1/2}/g$. For a monochromatic wave of amplitude a and frequency ω, $M_4 = a^2 \omega^4 / 2$ so that $\epsilon = ak/2$ is a small quantity. For the problem under consideration, ϵ will be used as a perturbation parameter. The Hilbert transform of ζ_1, $\hat{\zeta}_1$, can be represented by

$$\begin{aligned}
\hat{\zeta}_1 &= \hat{\zeta}/(m_0)^{1/2} \\
&\cong \left(\frac{M_0}{m_0}\right)^{1/2}(\eta_4 - \epsilon \eta_4 \eta_3)
\end{aligned} \qquad (7)$$

On this basis, one defines the complex process

$$W = \zeta_1 + i\hat{\zeta}_1 = A e^{i\varphi}$$

. Hence, the nondimensional envelope and phase process are defined by

$$A = (\zeta_1^2 + \hat{\zeta}_1^2)^{1/2}$$

$$\varphi = \tan^{-1} \hat{\zeta}_1 / \zeta_1$$

Joint distribution of wave envelope and its first derivative

Our task is to obtain the joint distribution of wave envelope, its first derivative to be used in (1) and (2). To this end, we carry out the differentiation of nonlinear wave elevation and its Hilbert transform with respect to time. We first note that

$$\eta_2 = \dot{\eta}_1 \left(\frac{M_0}{M_2}\right)^{1/2}$$

$$\eta_3 = \ddot{\eta}_1 \left(\frac{M_0}{M_4}\right)^{1/2}$$

$$\eta_5 = \dot{\eta}_4 \left(\frac{M_0}{M_2}\right)^{1/2}$$

$$\eta_6 = \ddot{\eta}_4 \left(\frac{M_0}{M_4}\right)^{1/2}$$

and, to the first order of ϵ, $m_0 = M_0$, $m_2 = M_2$ and $m_4 = M_4$. Based on this facts, nondimensional wave elevation, ζ_1, can be rewritten, to the order of ν,

$$\begin{aligned}
A \cos \varphi &= \zeta/(m_0)^{1/2} \\
&\cong \eta_1 - \frac{1}{2}\epsilon \eta_1 \eta_3 + \frac{1}{2}\epsilon \eta_4 \eta_5
\end{aligned} \qquad (8)$$

where

$$\nu = ((M_0 M_2 / M_1^2) - 1)^{1/2} < 1$$

is a measure of the bandwidth of the frequency spectrum which, for all practical purposes, is a small quantity. Then, it follows that, to the order of ν,

$$\begin{aligned}
\dot{A} \cos \varphi - A\dot{\varphi} \sin \varphi &= \dot{\zeta}/(m_2)^{1/2} \\
&\cong \eta_2 - 2\epsilon \eta_2 \eta_3
\end{aligned} \qquad (9)$$

$$\begin{aligned}
A \sin \varphi &= \hat{\zeta}/(m_0)^{1/2} \\
&\cong \eta_4 - \epsilon \eta_4 \eta_3
\end{aligned} \qquad (10)$$

$$\begin{aligned}
\dot{A} \sin \varphi + A\dot{\varphi} \cos \varphi &= \dot{\hat{\zeta}}/(m_2)^{1/2} \\
&\cong \eta_5 - \epsilon \eta_3 \eta_5 - \epsilon \eta_2 \eta_6
\end{aligned} \qquad (11)$$

. The random variables, A, $\dot{A}$, φ and $\dot{\varphi}$ are seen to be functions of η_1, η_2, η_3, η_4, η_5 and η_6 which are random variables having zero mean and unit standard deviation.

Furthermore, the pairs (η_1, η_5, η_3) and (η_4, η_2, η_6) are statistically independent, each of which is jointly Gaussian. Therefore the joint distribution of η_1, η_2, η_3, η_4, η_5 and η_6 is given by

$$f_{\eta_1\eta_2\eta_3\eta_4\eta_5\eta_6}(\cdot,\cdot,\cdot,\cdot,\cdot,\cdot) = f_{\eta_1\eta_5\eta_3}(\cdot,\cdot,\cdot)f_{\eta_4\eta_2\eta_6}(\cdot,\cdot,\cdot)$$

where

$$f_{\eta_1\eta_5\eta_3}(\cdot,\cdot,\cdot) = \frac{1}{(2\pi)^{3/2}\,|\,S_1\,|^{1/2}}\exp[$$
$$-\frac{1}{2\,|\,S_1\,|}\sum_{j=1}^{3}\sum_{k=1}^{3}|\,S_1\,|_{jk}\,\eta_j\eta_k]\quad(12)$$

and

$$f_{\eta_4\eta_2\eta_6}(\cdot,\cdot,\cdot) = \frac{1}{(2\pi)^{3/2}\,|\,S_2\,|^{1/2}}\exp[$$
$$-\frac{1}{2\,|\,S_2\,|}\sum_{j=1}^{3}\sum_{k=1}^{3}|\,S_2\,|_{jk}\,\eta_j\eta_k]\quad(13)$$

. In (12) and (13), $|\,S_i\,|_{jk}$, the cofactor of the element in the jth row and kth column of the matrix of covariances S_i, is given by

$$S_1 = \begin{vmatrix} E[\eta_1^2] & E[\eta_1\eta_5] & E[\eta_1\eta_3] \\ E[\eta_5\eta_1] & E[\eta_5^2] & E[\eta_5\eta_3] \\ E[\eta_3\eta_1] & E[\eta_3\eta_5] & E[\eta_3^2] \end{vmatrix}$$

and

$$S_2 = \begin{vmatrix} E[\eta_4^2] & E[\eta_4\eta_2] & E[\eta_4\eta_6] \\ E[\eta_2\eta_4] & E[\eta_2^2] & E[\eta_2\eta_6] \\ E[\eta_6\eta_4] & E[\eta_6\eta_2] & E[\eta_6^2] \end{vmatrix}$$

, respectively,. By introducing the auxiliary random variables

$$\alpha = \eta_3 \quad (14)$$

$$\beta = \eta_6 \quad (15)$$

, the joint distribution of A, $\dot{A}$, φ and $\dot{\varphi}$, $f_{A\dot{A}\varphi\dot{\varphi}}(\cdot)$, can be obtained by the standard method of transformation of random variables [Papoulis, 1965]. That is,

$$f_{A\dot{A}\varphi\dot{\varphi}\alpha\beta}(\cdot) = f_{\eta_1\eta_5\eta_3}(\eta_1, \eta_5, \eta_3)f_{\eta_4\eta_2\eta_6}(\eta_4, \eta_2, \eta_6)$$
$$\cdot\left|J\left(\frac{\eta_1\eta_2\eta_3\eta_4\eta_5\eta_6}{A\varphi\dot{A}\dot{\varphi}\alpha\beta}\right)\right|$$

where J is the Jacobian of the variable transformation. From (9), (10), (11), (12) and (13), and following the perturbation technique used by Huang et al. [1983], it may be shown that, to the order of ϵ,

$$|\,J\,| = A^2 + \frac{9}{2}\epsilon\alpha A^2$$

Performing the integration with respect to α and β,

$$f_{A\dot{A}\varphi\dot{\varphi}}(\cdot,\cdot,\cdot,\cdot) = \int\int f_{A\dot{A}\varphi\dot{\varphi}\alpha\beta}(\cdot,\cdot,\cdot,\cdot,\cdot,\cdot)d\alpha d\beta \quad (16)$$

the joint distribution of wave envelope and phase process and their first derivative can be obtained. From (9) and (10), and following the similar procedure in the transformation of random variables, the joint distribution of nondimensional wave elevation and its first derivative is given by

$$f_{\zeta\dot{\zeta}}(\zeta,\dot{\zeta}) = \frac{1}{2\pi}(1 + \frac{\epsilon}{2}(7\rho_1 - \rho_2\rho_3)\zeta + \frac{\epsilon}{2}(\rho_2\rho_3 - 4\rho_1)$$
$$\cdot\zeta\dot{\zeta}^2 - \frac{\epsilon}{2}\rho_1\zeta^3)\exp\left[-\frac{1}{2}(\zeta^2 + \dot{\zeta}^2)\right] \quad (17)$$

where the correlation coefficients ρ_i are

$$\rho_1 = E[\eta_1\eta_3] = E[\eta_4\eta_6]$$

$$\rho_2 = E[\eta_2\eta_6] = -E[\eta_3\eta_5]$$

$$\rho_3 = E[\eta_2\eta_4] = -E[\eta_1\eta_5]$$

Here and hereafter the notation $E[\cdot]$ is used to denote the expected value of quantity enclosed in the brackets. As $\epsilon = 0$, the joint distribution in (17) is reduced to jointly Gaussian as expected.

Average number of waves in a group, mean number of waves in a high run

Substituting (17) into (3), we can obtain the number of upcrossings by nondimensional wave elevation of a given level per unit time $N_\zeta(\zeta_0)$. That is, $N_\zeta(\zeta_0)$ is given by

$$N_\zeta(\zeta_0) = \frac{1}{2\pi}\left\{1 + \frac{\epsilon}{2}(\rho_2\rho_3 - \rho_1)\zeta_0 - \frac{\epsilon}{2}\rho_1\zeta_0^3\right\}\cdot\exp(-\frac{1}{2}\zeta_0^2) \quad (18)$$

. Integrating (16) with respect to A, $\dot{A}$ and $\dot{\varphi}$ yields the phase distribution in the form

$$f_\varphi(\varphi) = \frac{1}{2\pi}\left\{1 + \frac{\epsilon}{4}\sqrt{2\pi}\rho_1\cos\varphi\right\} \quad (19)$$

. From the marginal distribution of wave envelope and its first derivative obtainable from (16), (18), (1) and (2), average number of waves in a group G, mean number of waves in a high run H are given by, respectively,

$$G = \frac{1}{\sqrt{2\pi}\sqrt{1 - \rho_3^2}A_0}\left\{1 + \frac{\epsilon}{2}(\rho_2\rho_3 - \rho_1)\zeta_0 - \frac{\epsilon}{2}\rho_1\zeta_0^3\right\}$$
$$\cdot\exp\left[-\frac{1}{2}(\zeta_0^2 - A_0^2)\right] \quad (20)$$

and

$$H = \frac{1}{\sqrt{2\pi}\sqrt{1 - \rho_3^2}A_0}\left\{1 + \frac{\epsilon}{2}(\rho_2\rho_3 - \rho_1)\zeta_0 - \frac{\epsilon}{2}\rho_1\zeta_0^3\right\}$$
$$\cdot\exp\left[-\frac{1}{2}\zeta_0^2\right] \quad (21)$$

Numerical results

To quantify the above results, we must specify the wave spectrum from which the quantities ρ_1, ρ_2, ρ_3 and ϵ may be calculated. In this study, we shall use the Wallops spectrum [Huang et al., 1981] which takes the form

$$\Phi(\omega) = \frac{\alpha g^2}{\omega^m \omega_0^{5-m}}\exp[-\frac{m}{4}(\frac{\omega_0}{\omega})^4] \quad (22)$$

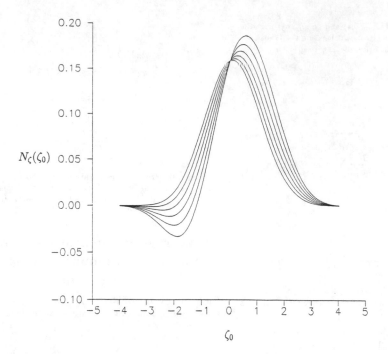

Fig. 1. Threshold crossing rate for varying ξ

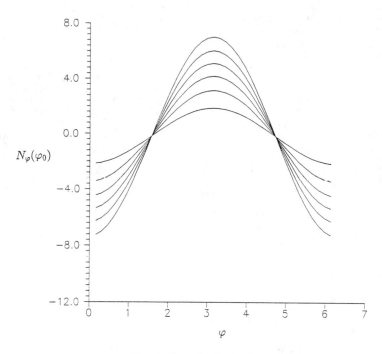

Fig. 2. Phase distribution for varying ξ

286

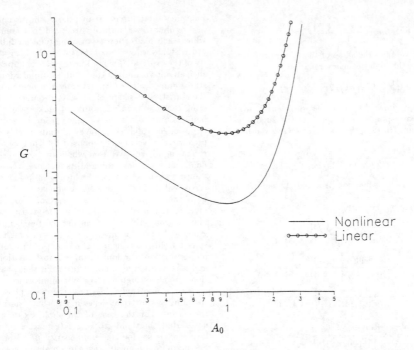

Fig. 3. Mean number of waves in a group with a linear counterpart for $\xi = 0.03$

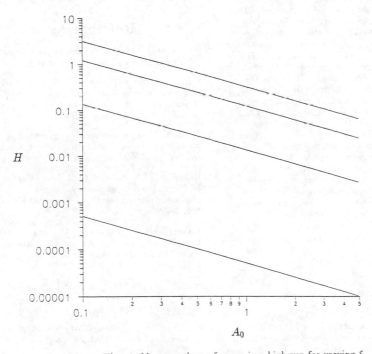

Fig. 4. Mean numbers of waves in a high run for varying ξ

287

where

$$m = \left| \frac{\log(2\pi^2 \xi^2)}{\log 2} \right|$$

is the absolute value of the slope of the spectrum (on the log-log scale) in the high frequency range and

$$\xi = M_0^{1/2}/L_0 = \sigma k/(2\pi) = \varepsilon/(2\pi)$$

is the significant slope, L_0 being the wave length whose frequency ω_0 corresponds to the peak of the single peak Wallops spectrum. In (22), the coefficient α is given by

$$\alpha = \frac{(2\pi\xi)^2 m^{(m-1)/4}}{4^{(m-5)/4}} \frac{1}{\Gamma((m-1)/4)} \qquad (23)$$

where $\Gamma(\cdot)$ is the gamma function [Abramowitz and Stegun, 1968].
From (22), it may be shown that

$$\rho_1 = -\frac{\Gamma[(m-3)/4]}{\Gamma^{1/2}[(m-1)/4]\Gamma^{1/2}[(m-5)/4]}$$

$$\rho_2 = -\frac{\Gamma[(m-4)/4]}{\Gamma^{1/2}[(m-3)/4]\Gamma^{1/2}[(m-5)/4]}$$

$$\rho_3 = \frac{\Gamma[(m-2)/4]}{\Gamma^{1/2}[(m-1)/4]\Gamma^{1/2}[(m-3)/4]}$$

$$\epsilon = 2\pi\xi \left[\frac{m\Gamma[(m-5)/4]}{4\Gamma[(m-1)/4]} \right]^{1/2}$$

so that ϵ, ρ_1, ρ_2 and ρ_3 are solely dependent on the value of ξ which was shown [Huang et al., 1980] to rarely exceed 0.02 in the ocean. In Fig. 1, the number of up-crossings $N_\zeta(\zeta_0)$ is plotted for varying ξ. It is noted that threshold crossing rate reaches its maxima at larger reference level as nonlinearity is getting profound. The phase distribution is plotted in Fig. 2 for $\xi = 0.005$, $\xi = 0.01$, $\xi = 0.015$, $\xi = 0.02$, $\xi = 0.025$ and $\xi = 0.03$. It turns out that the distribution of the phase is modified significantly by nonlinearities, and shows a systematic excess of values near the mean phase and corresponding symmetrical deficiency on both sides away from the mean. In Fig. 3 and 4, the average number of waves in a group G and the mean number of waves in a high run H are plotted for $\xi = 0.03$, respectively, and the linear counterpart obtainable by setting $\epsilon = 0$ is also included for the comparison. It is noted that for random process of moderate bandwidth, the peaks can be negative as well as positive and the peak distribution associated with nonlinear waves differs from the linear counterpart. The general character of this difference is in the form of a spreading of the density mass toward the higher and lower crests and a negative skewness of peak distribution of nonlinear waves, whereas a positive skewness is detected in the peak distribution of a linear wave. In Fig. 3, the peak distributions of nonlinear waves of narrow bandwidth and moderate are plotted together for $\xi = 0.015$.

Conclusion

After the Gaussian model was first developed by Rice [1958], there was a great deal of progress on the theory of nonlinear waves. But these progresses was not extended to the nonlinear statistical properties of wave groups due to the complicated form of nonlinear random waves which deserve much attention in the context of probable damage to coastal defences or offshore structures. In a case when the underlying frequency spectrum is narrow, the stochastic representation of a nonlinear sea surface is reduced to a familiar form in which each realization is an amplitude modulated second order Stokes wave. In contrast with the intricate complexity of the expression of nonlinear waves of finite bandwidth, such an approximation constitutes a simpler formulation to study numerically or analytically the nonlinear effects on the statistical description of wave properties. But considering the side band instability of Stokes wave, the narrow band assumption at the site away from the generating area is no longer valid. For waves of finite bandwidth, an approximate wave model proposed by Tung et al. [1989] is promising alternative from which the joint distribution of nonlinear wave elevation and its first derivative can be obtained and the structure of which is simple enough so that statistical properties of such nonlinear random waves can be obtained. Based on this wave model, overdue task of investigating nonlinear effects on the statistical properties of wave groups in terms of the average number of waves in a group and the mean number of waves in a high run was resumed in this study utilizing the complex envelope and total phase function, random variable transformation technique and perturbation method. It turns out that phase distribution is modified significantly by nonlinearities, and shows a systematic excess of values near the mean phase and corresponding symmetrical deficiency on both sides away from the mean. For the case of threshold crossing rate, it is noted that threshold crossing rate reaches its maxima at larger reference level as nonlinearity is getting profound. Furthermore, the mean waves in a high run associated with nonlinear wave is shown to has a tendency to be increased than the linear counterpart. Similar trend can also be found in the average number of waves in a group.

References

1. Abramowitz, M. and I. A. Stegun 1968. Handbook of mathematical functions. Dover, Mineola, NY.

2. Cho, Y. J. and T. H. Yoon. 1992. Peak distribution of nonlinear random waves of finite bandwidth. Proceedings of the sixth IAHR international symposium on stochastic hydraulics. Teipei

3. Goda, Y. 1983. Analysis od wave grouping and spectra of long travelled swell. Rept. Port and harbour research inst. Vol. 22., No. 1.

4. Huang, N. E., S. R. Long, C. C. Tung, Y. Yuan and L. F. Bliven 1981. A unified two-parameter wave spectral model for a general sea state. J. Fluid Mech., 112, 203-224

5. Huang, N. E., S. R. Long, C. C. Tung, Y. Yuan and L. F. Bliven 1983. A non-Gaussian statistical model for surface elevation of nonlinear random wave fields. J. Geophys. Res., 88, 7597-7606.

6. Rice, S. O., 1954. Mathematical analysis of random noise. Selected papers on noise and stochastic processes. Dover, New York

7. Lin, Y. K. 1967. Probabilistic theory of structural dynamics. McGraw-Hill book Co., New York, NY.

8. Longuet-Higgins, M. S. 1957. Statistical analysis of a random, moving surface. Phil. Trans. R. Soc. Lond. A 966, Vol. 249, 321-387.

9. Longuet-Higgins, M. S. 1963. The effect of nonlinearities on statistical distributions in the theory of sea waves. J. Fluid Mech., 17, 459-480.

10. Longuet-Higgins, M. S. 1984. Statistical properties of wave groups in a random sea state. Phil. Trans. R. Soc. Lond. A 1521, Vol. 312, 219-250.

11. Papoulis, A. 1965. Probability, random variables and stochastic processes. McGraw-Hill book Co., New York, NY.

12. Tayfun, M. A. 1980. Narrow-band nonlinear sea waves. J. Geophys. Res., 85, 1548-1552.

13. Tayfun, M. A. 1986. On narrow-band representation of ocean waves 1, theory. J. Geophys. Res., 91, 7743-7752.

14. Tung, C. C., N. E. Huang, Y. Yuan and S. R. Long 1989. Probability function of breaking limited surface elevation. J. Geophys. Res., 94, No. C1, 967-972.

Stochastic Hydraulics'96, Tickle, Goulter, Xu, Wasimi & Bouchart (eds) © 1996 Balkema, Rotterdam. ISBN 90 5410 817 7

Comparison between a parametric model and measurements of seabed shear stresses under random waves with a current superimposed

Dag Myrhaug
Department of Marine Hydrodynamics, Norwegian Institute of Technology, Trondheim, Norway

Olav H. Slaattelid
Norwegian Marine Technology Research Institute, Valentinlyst, Trondheim, Norway

ABSTRACT: Results of comparison between a parametric model and measurements of seabed shear stresses under random waves with a current superimposed are presented. The model is based on the approach in Myrhaug (1995) where the effect of random waves on the bottom friction is studied by assuming that the wave motion is a stationary Gaussian narrow-band random process, and by using simple explicit friction coefficient formulas for sinusoidal waves. The data used for comparison are from Simons et al. (1994) direct measurements of the bottom shear stresses under the action of combined random waves and orthogonal currents.

1 INTRODUCTION

In shallow and intermediate water depths dominated by waves and currents, the fluid motion in the combined wave and current boundary layer near the seabed controls and affects many phenomena in offshore engineering and oceanography. The boundary layer flow enters in fluid force calculations for seabed pipelines and other near-sea-bottom structures. The shear stress associated with the boundary layer flow represents the dominant mechanism governing sediment transport and erosion phenomena, and contributes also to the dissipation of surface water waves. Boundary layer flow also represents an important component in physical models for predicting coastal and ocean flow circulations.

The combined wave and current boundary layer on the seabed has been investigated by many. Reviews focusing on the interaction between sinusoidal waves and currents are given in Grant and Madsen (1986) and Soulsby et al. (1993b). Results from theoretical models as well as laboratory and field experiments show that the presence of waves increases significantly the bottom roughness parameter for the current boundary layer, i.e., the roughness parameter depends strongly on the seastate. More recent studies have addressed random waves interacting with currents. Zhao and Anastasiou (1993) and Madsen (1994) presented theoretical studies on bottom friction effects for random waves

plus currents. Ockenden and Soulsby (1994) presented a method of predicting sediment transport for the case of currents plus irregular waves. Simons et al. (1994) (hereafter denoted as S94) presented results from experiments of direct measurements of the bottom shear stresses under the action of combined random waves and orthogonal currents.

The purpose of this paper is to compare the approach in Myrhaug (1995) (hereafter denoted as M95) with the S94 data. M95 describes the waves as a stationary Gaussian narrow-band random process, and applies simple explicit friction coefficient formulas for sinusoidal waves. The probability distributions of the maximum bottom shear stress together with some characteristic statistical values of the maximum bottom shear stress are presented. An acceptable agreement is found between measurements and predictions.

2 THEORETICAL BACKGROUND

The maximum bottom shear stress for sinusoidal waves is given as

$$\frac{\tau_m}{\rho} = \frac{1}{2}f_w U^2 \tag{1}$$

where U is the orbital velocity amplitude at the seabed, f_w is the wave friction coefficient, and ρ is the density of the fluid.

For rough turbulent flow Soulsby et al. (1993a) proposed the following friction coefficient formula, obtained as best fit to data

$$f_w = c(\frac{A}{z_0})^{-d} \qquad (2)$$

where $c=1.39$, $d=0.52$, A is the orbital displacement amplitude at the seabed, and z_0 is the seabed roughness parameter.

The basis for the present approach is that the maximum bottom shear stress for sinusoidal waves given in Eq. (1) combined with Eq. (2) is valid for random waves as well. Consequently it is assumed that each wave can be treated individually. The accuracy of this assumption should be validated by using a full boundary layer model to calculate the shear stress under random waves. However, preliminary results by Davies (1994) suggest that this assumption can be used to predict integrated effects such as the bedload transport rate with reasonable accuracy. Thus this assumption is considered to be adequate as a first approximation. Further it is assumed that the free surface elevation $\zeta(t)$ is a stationary Gaussian narrow-band random process with zero expectation and the single-sided spectral density $S_{\zeta\zeta}(\omega)$, where ω is the cyclic wave frequency. Thus the present approach should be applicable for the description of the bottom friction beneath irregular waves occurring in wave groups in intermediate water depths. However, when the water depth decreases the waves will begin to shoal, and the waves become nonlinear. Consequently both the Gaussian and the narrow-band assumption will no longer be valid.

Based on the present assumptions the bed orbital displacement $a(t)$ as well as the bed orbital velocity $u(t)$ will be stationary Gaussian narrow-band random processes with zero expectations and with the single-sided spectral densities

$$S_{aa}(\omega) = \frac{S_{\zeta\zeta}(\omega)}{\sinh^2 kh} \qquad (3)$$

and

$$S_{uu}(\omega) = \omega^2 S_{aa}(\omega) = \frac{\omega^2 S_{\zeta\zeta}(\omega)}{\sinh^2 kh} \qquad (4)$$

respectively, where k is the wave number determined from the dispersion relationship $\omega^2 = gk\tanh kh$, h is the water depth, and g is the acceleration of gravity.

For a narrow-band process the waves are specified as a "harmonic" wave with cyclic frequency ω and with slowly varying amplitude and phase. Then the bed orbital displacement is given as (see e.g. Sveshnikov, 1966) $a(t)=A(\varepsilon t)\cos[\omega t+\Phi(\varepsilon t)]$ where $\varepsilon <<1$ is introduced to indicate that the bed orbital displacement amplitude A and the phase Φ are slowly varying with t. Then the bed orbital velocity is given as

$$u(t)=\frac{da(t)}{dt}=\omega A(\varepsilon t)\sin[\omega t+\Phi(\varepsilon t)-\frac{\pi}{2}]+O(\varepsilon) \qquad (5)$$

where the term $O(\varepsilon)$ represents terms of order ε. As a first approximation, which is consistent with the narrow-band assumption, the terms of $O(\varepsilon)$ are neglected, and accordingly the bed orbital velocity amplitude is related to the displacement amplitude by $U=\omega A$, where U is slowly varying with t as well.

The accuracy of the approximate relation obtained by neglecting the terms of $O(\varepsilon)$ in Eq. (5) is discussed in Sveshnikov (1966). A test of the accuracy is the error in the variance of the derivative of the random function $a(t)$. It appears that this error is small in the case of narrow-band spectrum. Overall some of the main features are covered by using the narrow-band approximation.

Now A and U will both be Rayleigh-distributed with the probability distribution functions

$$P(\hat{A}) = 1 - \exp(-\hat{A}^2) \; ; \; \hat{A} = A/A_{rms} \geq 0 \qquad (6)$$

and

$$P(\hat{U}) = 1 - \exp(-\hat{U}^2) \; ; \; \hat{U} = U/U_{rms} \geq 0 \qquad (7)$$

respectively. A_{rms} and U_{rms} are the root-mean-square (rms) values of A and U, respectively, and are related to the zeroth moments m_{0aa} and m_{0uu} of the amplitude and velocity spectral densities, respectively, or corresponding to the variances of the amplitude (σ_{aa}^2) and the velocity (σ_{uu}^2), given by

$$A_{rms}^2 = 2m_{0aa} = 2\sigma_{aa}^2 = 2\int_0^\infty S_{aa}(\omega)d\omega \qquad (8)$$

and

$$U_{rms}^2 = 2m_{0uu} = 2\sigma_{uu}^2 = 2\int_0^\infty S_{uu}(\omega)d\omega \qquad (9)$$

From Eqs. (9) and (4) it also appears that $m_{0uu} = m_{2aa}$, where m_{2aa} is the second moment of the amplitude spectral density.

A reasonable choice for ω is the mean zero-crossing wave frequency, which is obtained from the spectral moments of $a(t)$ as

$$\omega = \omega_z = (\frac{m_{2aa}}{m_{0aa}})^{1/2} = (\frac{m_{0uu}}{m_{0aa}})^{1/2} = \frac{U_{rms}}{A_{rms}} \qquad (10)$$

where Eqs. (8) and (9) have been used.

By transformation of random variables the probability distribution function of the normalized maximum bottom shear stress for rough turbulent flow is found as (M95)

$$P(\hat{\tau}) = 1 - \exp(-\hat{\tau}^\beta) \; ; \; \hat{\tau} = \frac{\tau_{\hat{m}}}{\rho U_*^2} \geq 0 \qquad (11)$$

where

$$\beta = \frac{2}{2 - d} \qquad (12)$$

$$U_*^2 = \frac{1}{2} c (\frac{A_{rms}}{z_0})^{-d} U_{rms}^2 \qquad (13)$$

Thus it appears that the maximum bottom shear stress is Weibull-distributed for rough turbulent flow.

When the probability distribution is known, the characteristic statistical values of the maximum bottom shear stress can be obtained. The expected (mean) value and the variance of $\hat{\tau}$ are given by (see e.g. Bury, 1975)

$$E[\hat{\tau}] = \Gamma(1 + \frac{1}{\beta}) \qquad (14)$$

$$Var[\hat{\tau}] = \Gamma(1 + \frac{2}{\beta}) - \Gamma^2(1 + \frac{1}{\beta}) \qquad (15)$$

where Γ is the gamma function.

Other statistical quantities of interest are the value of $\hat{\tau}$ which is exceeded by the probability $1/n$, $\hat{\tau}_{1/n}$, and the expected value of the $1/n$ largest values of $\hat{\tau}$, $E[\hat{\tau}_{1/n}]$, given by, respectively (M95)

$$\hat{\tau}_{1/n} = (\ln n)^{1/\beta} \qquad (16)$$

and

$$E[\hat{\tau}_{1/n}] = n \; \Gamma(1 + \frac{2}{\beta}) \; Q[\chi^2 = 2\ln n \,|\, \nu = 2(1 + \frac{2}{\beta})] \qquad (17)$$

$$= n \; \Gamma(1 + \frac{2}{\beta} \,, \, \ln n)$$

where $Q(\chi^2 \,|\, \nu)$ is the χ^2 probability exceedence function with ν degrees of freedom, and $\Gamma(\nu/2, \chi^2/2)$ is the incomplete gamma function (Ch. 26.4, Abramowitz and Stegun, 1972).

Further, according to the Type I Extreme Value asymptote the expected largest value among N values is given by (see e.g. Bury, 1975)

$$E[\hat{\tau}_{max}] \approx (\ln N)^{1/\beta} (1 + \frac{0.5772}{\beta \ln N}) \qquad (18)$$

Here it is assumed that all the values are independent and identically Weibull-distributed, and Eq. (18) is an asymptotic expression for large N. The first term in Eq. (18) can be interpreted as the "characteristic largest value", $\hat{\tau}_N$, which has, on the average, only one exceedence in a sample of size N, i.e. $1 - P(\hat{\tau}_N) = 1/N$, giving

$$\hat{\tau}_N = (\ln N)^{1/\beta} \qquad (19)$$

3 COMPARISON WITH MEASUREMENTS

The present theory will now be compared with measurements done by S94, which give data from laboratory measurements for random waves without and with a current superimposed. The bottom shear stresses were measured directly by using a shear plate device together with simultaneous measurements of three velocity components. The measurements include three sequences of random waves propagating over a fixed rough bed in still water and with two orthogonal currents superimposed. The data which are used here are the friction coefficients calculated from the half-cycle amplitude of shear stress (between consecutive maxima and minima) and the corresponding amplitude of wave-induced velocity. The test conditions are given in Table 1 together with some analysis results which will be discussed subsequently. Here $T_z = 2\pi/\omega_z$ is the mean zero-crossing wave period, and U_{50} is the current velocity at the 50 cm elevation. Re_l and Re_u are the lower and upper values of the roughness Reynolds number $Re = k_N u_* / \nu$, where k_N is the Nikuradse sand roughness, ν is the kinematic viscosity of the fluid, and

Table 1.   Main flow variables and results for Simons et al. (1994) data.

Record	N	T_z (s)	U_{50} (cm/s)	σ_{uu} (cm/s)	$\dfrac{U_{50}}{U_{rms}}$	$\dfrac{A_{rms}}{z_0}$	Re_l	Re_u	c	d	β
WR1P	155	1.28	-0.4	7.2	~0	415	13	44	57.6	1.06	2.13
WR1CC	143	1.28	10.5	7.4	1.00	426	19	47	202.0	1.26	2.70
WR1C	98	1.29	19.5	7.8	1.77	453	20	47	331.1	1.33	2.99
WR2P	202	1.49	0.7	8.9	~0	597	9	49	43.1	1.00	2.00
WR2CC	176	1.48	10.8	9.2	0.83	613	16	48	107.5	1.14	2.33
WR2C	155	1.50	19.7	9.6	1.45	648	19	50	211.4	1.24	2.63
WR3P	171	1.29	-0.8	8.9	~0	517	13	46	63.9	1.07	2.15
WR3CC	153	1.29	10.6	9.3	0.81	540	16	46	154.0	1.22	2.56
WR3C	125	1.31	18.9	9.5	1.41	560	19	50	169.8	1.21	2.53

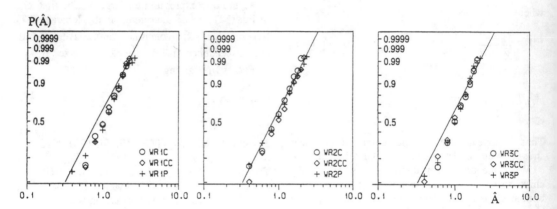

Fig. 1   Probability distribution of normalized bed orbital displacement amplitude in Weibull scale: ——— Rayleigh distribution, Eq. (6); other symbols represent S94 data. See also Table 1.

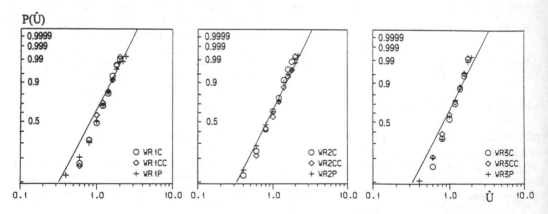

Fig. 2   Probability distribution of normalized bed orbital velocity amplitude in Weibull scale: ——— Rayleigh distribution, Eq. (7); other symbols represent S94 data. See also Table 1.

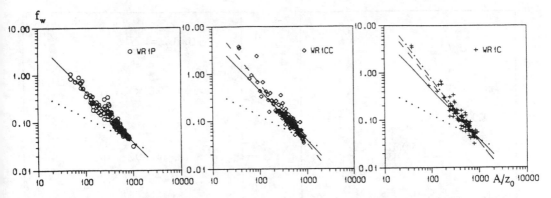

Fig. 3 Friction coefficient versus amplitude to roughness ratio for the S94 half-cycle data:
——— best fit of Eq. (2) to WR1P data; − − − best fit of Eq. (2) to WR1CC data;
− · − · best fit of Eq. (2) to WR1C data; · · · · · Soulsby et al. (1993a);
other symbols represent S94 data. See also Table 1.

$u_* = (\tau_m/\rho)^{1/2}$ is the friction velocity. For these tests $k_N = 0.15$ cm and $z_0 = k_N/30 = 0.005$ cm. From Table 1 it appears that the Re values are in the range 9 to 50, showing that the data represent intermediate turbulent flow conditions (i.e., $5 < Re < 70$, see e.g. Schlichting, 1979). Thus a direct comparison with the Soulsby et al. (1993a) friction coefficient formula for rough turbulent flow in Eq. (2) is not appropriate. Therefore the present approach is extended to cover the intermediate flow regime by making use of the S94 data.

A consequence of the narrow-band assumption is that A and U both are Rayleigh-distributed with the probability distribution functions given in Eqs. (6) and (7), respectively. Figs. 1 and 2 show $P(\hat{A})$ and $P(\hat{U})$ according to Eqs. (6) and (7), respectively, together with the S94 data in Weibull scale. Overall it appears that the Rayleigh distribution gives an adequate representation of the data.

The present approach for comparison with the S94 data is as follows. Firstly, Eq. (2) is fitted to the half-cycle data for random waves as well as random waves with a current superimposed. Thus c and d are determined for each data set. Secondly, the probability distribution function of the normalized bottom shear stress is obtained by using Eqs. (11) to (13). Finally, the characteristic statistical values given in Eqs. (16) to (19) are calculated.

Fig. 3 shows the results of fitting Eq. (2) to the half-cycle data for the WR1P, WR1CC and WR1C records. The Soulsby et al. (1993a) formula representing fully rough turbulent flow conditions are included for comparison. It is noticed that Eq. (2) represents a straight line in log-log scale. Thus c

gives the intersection between the line and the f_w-axis, and d the inclination of the line. Thus an increase in c and d suggests an increase in f_w for lower A/z_0 values. Despite of significant spread in the data, the results in Fig. 3 suggest a slight increase in the friction coefficient for the lower A/z_0 values when the orthogonal current is superimposed, and the effect increases with increasing current velocity. The c and d values are given in Table 1 for all the records, and are plotted versus A_{rms}/z_0 and U_{50}/U_{rms} in Fig. 4 as well. It appears that the WR2P, WR2CC, WR2C as well as the WR3P, WR3CC, WR3C data show the same qualitative behaviour as the data shown in Fig. 3, except for the d values for WR3CC and WR3C which are almost equal. Although modest, Eqs. (1) and (2) together with the results in Fig. 4 suggest a dependence of the maximum bottom shear stress on the amplitude to roughness ratio (A_{rms}/z_0) and the current to wave ratio (U_{50}/U_{rms}).

Figs. 5 to 7 show the probability distribution function of the normalized maximum bottom shear stress in Weibull scale. Both the model, i.e. Eqs. (11) to (13), and the S94 data are shown, and overall it appears that the model gives an adequate representation of the data for larger values of $\hat{t}$, i.e. $\hat{t} \gtrsim 1$. However, the agreement between model and data is best for the WR1 records. The values of the Weibull parameter β are given in Table 1, and are plotted versus A_{rms}/z_0 and U_{50}/U_{rms} in Fig. 8. According to Eq. (12), these results show the same qualitative behaviour as for the d values in Fig. 4.

Fig. 9 shows the measured versus the predicted values of the normalized maximum bottom shear

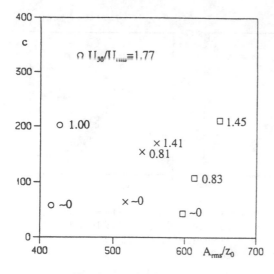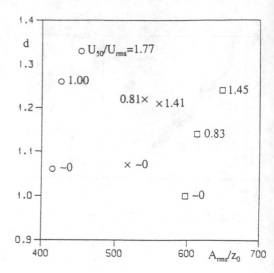

Fig. 4   c and d in Eq. (2) versus A_{rms}/z_0 and U_{50}/U_{rms}: o WR1P, WR1CC, WR1C;
□ WR2P, WR2CC, WR2C; × WR3P, WR3CC, WR3C. See also Table 1.

stress which is exceeded by the probability $1/n$, $\hat{\tau}_{1/n}$, as well as the expected value of the $1/n$ largest values of $\hat{\tau}$, $E[\hat{\tau}_{1/n}]$, for $n=3$ and $n=10$. The predictions are given by Eqs. (16) and (17), respectively. Overall it appears that $\hat{\tau}_{1/3}$ is under-predicted, while $\hat{\tau}_{1/10}$ is fairly well predicted. The predicted to measured ratios of $\hat{\tau}_{1/3}$ and $\hat{\tau}_{1/10}$ are in the ranges 0.87 to 1.01, and 0.94 to 1.14, respectively. However, $E[\hat{\tau}_{1/3}]$, i.e. the significant value, is well predicted, while $E[\hat{\tau}_{1/10}]$ is slightly overpredicted. The predicted to measured ratios of $E[\hat{\tau}_{1/3}]$ and $E[\hat{\tau}_{1/10}]$ are in the ranges 0.93 to 1.11, and 0.96 to 1.20, respectively.

Fig. 10 shows the measured versus the predicted values of the largest normalized maximum bottom shear stress. The N-values are given in Table 1, ranging from 98 to 202. Overall it appears that both predictions, i.e. $E[\hat{\tau}_{max}]$ and $\hat{\tau}_N$ in Eqs. (18) and (19), respectively, give larger values than the measured. The predicted to measured ratios of $E[\hat{\tau}_{max}]$ and $\hat{\tau}_N$ are in the ranges 0.95 to 1.48, and 0.91 to 1.41, respectively.

The most appropriate statistical value to use in practical applications will depend on the problem dealt with. The mean value of the maximum bottom shear stress might be a relevant quantity to use to represent the dissipation of irregular surface water waves in e.g. physical models for predicting coastal and ocean flow circulations. In other applications, such as in suspended sediment calculations beneath irregular waves with a current superimposed, the maximum bottom shear stress which is exceeded by a certain percentage $1/n$, or the expected value of the $1/n$ largest values, might be more appropriate values to use. However, the largest value seems more unreasonable to use in such calculations. Thus the results in Fig. 9 are encouraging from a practical point of view.

In most real situations the bottom boundary layer flow conditions will be in the rough turbulent regime. Since such data are not available at present, no conclusion can be drawn on the ability of the present approach to describe measured data representing such flow conditions. However, the results presented here suggest that the approach can be used as a first approximation to represent the seabed shear stress under random waves with a current superimposed.

4   CONCLUSIONS

Overall the parametric model in Myrhaug (1995) gives an adequate representation of the Simons et al. (1994) data, showing that (1) the probability distribution function of the maximum seabed shear stress is reasonably well represented by the Weibull distribution for larger values; and (2) an acceptable agreement is found between the measured and predicted characteristic statistical values of the maximum seabed shear stress which are reasonable to use in e.g. suspended sediment calculations.

P($\hat{\tau}$)

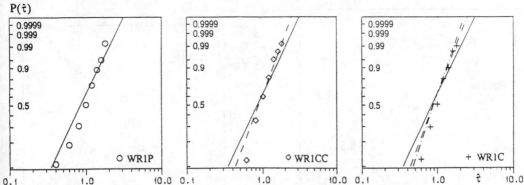

Fig. 5 Probability distribution function of normalized maximum bottom shear stress in Weibull scale:
———— Eqs. (11) to (13) with c, d and β values corresponding to WR1P;
– – – Eqs. (11) to (13) with c, d and β values corresponding to WR1CC;
– · – · Eqs. (11) to (13) with c, d and β values corresponding to WR1C;
other symbols represent S94 data. See also Table 1.

P($\hat{\tau}$)

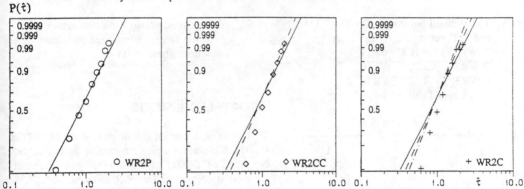

Fig. 6 Probability distribution function of normalized maximum bottom shear stress in Weibull scale:
———— Eqs. (11) to (13) with c, d and β values corresponding to WR2P;
– – – Eqs. (11) to (13) with c, d and β values corresponding to WR2CC;
– · – · Eqs. (11) to (13) with c, d and β values corresponding to WR2C;
other symbols represent S94 data. See also Table 1.

P($\hat{\tau}$)

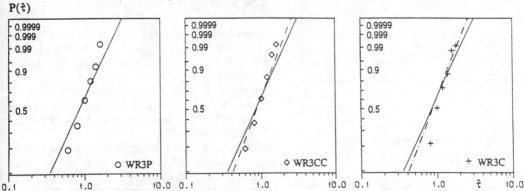

Fig. 7 Probability distribution function of normalized maximum bottom shear stress in Weibull scale:
———— Eqs. (11) to (13) with c, d and β values corresponding to WR3P;
– – – Eqs. (11) to (13) with c, d and β values corresponding to WR3CC;
– · – · Eqs. (11) to (13) with c, d and β values corresponding to WR3C;
other symbols represent S94 data. See also Table 1.

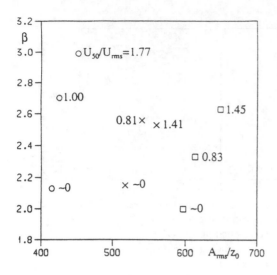

Fig. 8 The Weibull parameter β in Eqs. (11) and (12) versus A_{rms}/z_0 and U_{50}/U_{rms}:
o WR1P, WR1CC, WR1C;
□ WR2P, WR2CC, WR2C;
× WR3P, WR3CC, WR3C.
See also Table 1.

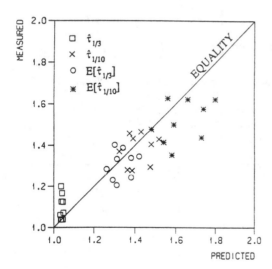

Fig. 9 Measured versus predicted values of $\hat{\tau}_{1/n}$ and $E[\hat{\tau}_{1/n}]$ for n= 3, 10.

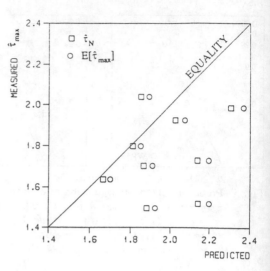

Fig. 10 Measured versus predicted values of the largest normalized maximum bottom shear stress.

ACKNOWLEDGEMENTS

This work was carried out as part of the MASTG8 Coastal Morphodynamics programme under contract No. MAS2-CT-0027. It was funded by the Research Council of Norway. Dr. R.R. Simons and W.M. Saleh at University College London are acknowledged for preparing the data files used in this study.

REFERENCES

Abramowitz, M. & Stegun, I.A., 1972. *Handbook of Mathematical Functions*. Dover, New York.

Bury, K.V., 1975. *Statistical Models in Applied Science*. Wiley, New York.

Davies, A.G., 1994. Personal communication.

Grant, W.D. & Madsen, O.S. 1986. The continental-shelf bottom boundary layer. *Annu. Rev. Fluid Mech.*, 18:265-305.

Madsen, O.S., 1994. Spectral wave-current bottom boundary layer flows. *Proc. 24th Conf. on Coastal Eng.*, ASCE, Kobe, Japan, Vol. 1, pp. 384-398.

Myrhaug, D., 1995. Bottom friction beneath random waves. *Coastal Eng.*, 24: 259 - 273.

Ockenden, M.C. & Soulsby, R.L. 1994. Sediment transport by currents plus irregular waves. Report SR 376, HR Wallingford, Wallingford, UK.

Schlichting, H., 1979. *Boundary-Layer Theory*, 7th Ed. McGraw-Hill, New York, NY.

Simons, R.R.,Grass, T.J., Saleh, W.M. & Tehrani, M.M., 1994. Bottom shear stresses under random waves with a current superimposed. *Proc. 24th Conf. on Coastal Eng.*, ASCE, Kobe, Japan, Vol. 1, pp. 565-578.

Soulsby, R.L., Davies, A.G., Fredsøe, J., Huntley, D.A., Jonsson, I.G., Myrhaug, D., Simons, R.R., Temperville, A. & Zitman, T., 1993a. Bed shear-stresses due to combined waves and currents. Book of Abstracts, MAST-2 G8M Coastal Morphodynamics, Overall Workshop, Grenoble, pp. 2.1-1/2.1-4.

Soulsby, R.L., Hamm., L., Klopman, G., Myrhaug, D., Simons, R.R., & Thomas, G.P., 1993b. Wave-current interaction within and outside the bottom boundary layer. *Coastal Eng.*, 21:41-69.

Sveshnikov, A.A., 1966. *Applied Methods in the Theory of Random Functions*, Pergamon Press, New York, N.Y.

Zhao, Y. & Anastasiou, K., 1993. Bottom friction effects in the combined flow field of random waves and currents. *Coastal Eng.*, 19(3,4):223-243.

4 Stochastic hydrology

Stochastic Hydraulics'96, Tickle, Goulter, Xu, Wasimi & Bouchart (eds) © 1996 Balkema, Rotterdam. ISBN 90 5410 817 7

Probabilistic estimation and simulation of water resources system due to global warming

T. Kojiri
Gifu University, Japan

ABSTRACT: The purpose of this paper is to evaluate the impact of global warming to water resources systems under the conditions of limited scenarios through probability simulation methodologies. Hydrographs are generated obeying the transition probabilities between classified patterns and probability density functions. The basin run-off model includes the utilization processes of water demand for municipal and agriculture use with function of temperature and precipitation. Then the concept of equivalent variation is introduced by combining the occurrence probability of drought and the expected utility level in future.

1 INTRODUCTION

Water resources systems have to be planned by considering the uncertain input discharge and water demand in future. Though GCMs under the condition of CO_2 doubling in 2050 denote 3° C temperature rise and 10% precipitation increase, the observation results at the specified point reported the gradual decrease in precipitation in Japan. Moreover, to plan the water resources system, not only the uncertain input but also the increase of population and change of water utilization must be considered.

In this paper, firstly yearly hyetographs with monthly unit are classified into several patterns by considering their futures and volume. Future hyetographs will be generated with the transition probabilities between patterns and probability density functions for identified factors under the assumption that the coefficient of variation is steady. Secondly the mesh-typed multi-layer run-off model is introduced to get the yearly hydrographs with daily units. The precipitation and temperature at the arbitrary observation point according to the different geographical conditions in the basin are estimated with linear regression analysis. This basin run-off model includes the utilization processes of water demand for municipal and agriculture use. Finally the drought situation should be analyzed. Three factors such as reliability, resiliency and vulnerability are evaluated. From an economical viewpoint, the concept of equivalent variation is applied by combining the occurrence probability of drought and the expected utility level in future.

These methodologies are applied into the real basin for their verifications.

2 SIMULATION PROCEDURES

The meteorological or hydrological data such as temperature or precipitation does not only posses the unique characteristics but also show several typical features in substance. Pattern classification might be applied to classify the input data based on simulation purpose. According to the temperature patterns, the precipitation characteristics are analyzed. Then the daily discharge will be estimated through mesh-typed run-off model and the utilized situation on water supply will evaluated under the limited scenarios of global warming from a viewpoint of both side of physical and economical aspect. The flow chart of these procedures is shown in Fig. 1.

2.1 Temperature and precipitation Model

In natural, the temperature and precipitation patterns are classified into several representative ones. The ISODATA algorithm is introduced for classification (Kojiri et al., 1994) with the following objective function as;

$$\text{Dis} = \frac{|X_i - Z_j|}{Z_j} \qquad (2.1)$$

where Xi denotes the ith observed data sequence (pattern vector) and Zj does the j-th representative sequence (reference vector or cluster center). The

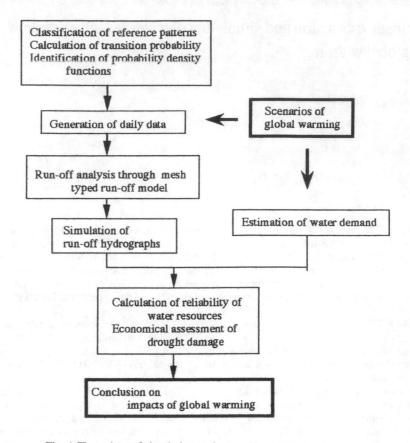

Fig. 1 Flow chart of simulation and assessment procedures

Table 1 Transition probabilities on classified patterns

	Pattern 1	Pattern 2	Pattern 3	Pattern 4	Pattern 5
Pattern 1	0.529	0.176	0.118	0.118	0
Pattern 2	0.143	0.429	0.143	0.143	0.14
Pattern 3	0.5	0	0.25	0.25	0
Pattern 4	0.75	0.25	0	0	0
Pattern 5	1	0	0	0	0

transition probability from the reference vector j to m is expressed as follows;

$$P_{m|j} = \frac{NP_{jm}}{NP_j} \qquad (2.2)$$

where NP_{jm} is the number of observed data sequences which move from the reference vector j to m and NP_j is the number of data belonging to the reference j.

The parameters of daily temperature and precipitation are identified among the classified pattern vectors. The normal distribution, the log-normal distribution with two parameters and the log-normal distribution with three parameters are checked for their accuracies.

The monthly data, which takes a role of input into the basin simulation, is generated with following procedures;

i)Firstly the current representative sequence is extracted among the classified patterns based on it's characteristics.
ii)The following future sequence with monthly data are simulated after denoting the number of reference vector with the transition probability.
iii)The daily data is generated around the simulated monthly data obeying the identified probability density function.
iv)The monthly pattern is also analyzed according to the classified ones after averaging the daily data.
v)Go back to the step ii), otherwise the iteration should stop.

The current situation and three scenarios of global warming, which are recognized as the common outcomes in the popular GCMs, are applied as listed below (Takara & Kojiri, 1991);
(i) Case 1 - no temperature rise and no precipitation increase
(ii) Case 2 - 3°C temperature rise and no precipitation increase
(iii) Case 3 - 3°C temperature rise and 10% precipitation increase
(iv) Case 4 - 3°C temperature rise and -10% precipitation increase
Additionally the temperature rise will affects only the evapotranspiration in run-off process. The land use in future is assumed not be changed and keep same hydrological and geological characteristics in space. The parameters of probability distribution function on daily temperature are shifted according to the following equations;

$$\mu' = \mu + \Delta T \qquad (2.3)$$

$$\sigma'^2 = \frac{\mu + \Delta T}{m} \sigma^2 \qquad (2.4)$$

Besides the probability density function, the duration of continuous precipitation, and non-precipitation, and the total amount of one precipitation event are analyzed for designated levels of averaged monthly temperature. After generating the temperature sequence, the precipitation is simulated by using the coincident and modified parameters through eq.(2.3) and (2.4). Concretely, the duration of non-precipitation and precipitation, and total amount of one precipitation event are generated through random variables obeying the classified frequency distributions. Then, the daily precipitation intensities are also generated through the probability functions. If statistical analysis checks to satisfy the rationality of the monthly temperature against the scenario, the precipitation is handled as the input data for run-off process.

2.2 Basin run-off model

In the case of four layers, the continuous equations of the mesh-typed run-off model (Kojiri, 1993) are formulated as follows;

First layer;
$$dSA/dt = f + r + QAH_{in} - QAH_{out} \qquad (2.5)$$
$$- QAV_{out} - EVA$$

Second layer;
$$dSB/dt = QBH_{in} + QBV_{in} - QBH_{out}$$
$$- QBV_{out} \qquad (2,6)$$

Third layer;
$$dSC/dt = QCH_{in} + QCV_{in} - QCH_{out} \qquad (2.7)$$
$$- QCV_{out}$$

Forth layer;
$$dSD/dt = QDH_{in} + QDV_{in} - QDH_{out}$$
$$- QDV_{out} \qquad (2.8)$$

where, $S\bullet$ is the storage volume at the considered mesh, f is the infiltration discharge at surface, r is the run-off discharge from the upstream surface, $Q\bullet Hin$ is the horizontal inflow from upper stream mesh, $Q\bullet Hout$ is the horizontal outflow to the down stream mesh, $Q\bullet Vin$ is the vertical infiltration discharge at the considered mesh, $Q\bullet Vout$ is the vertical outflow, and EVA is the evapotranspiration, respectively. On the other hand, the kinematic equation is represented as follows;

$$dS/dt = QI_b - QO_b \qquad (2.9)$$

$$QO_b = 1/\kappa * S \qquad (2.10)$$

where κ is the ttransmissivity, and QI_b and QO_b are the total amunt of input and output discharge at the considered mesh, respectively.

Three assumptions are set to ascertain the water movement in a mesh as follows;
i)In the horizontal coordinates, the discharge comes into from three dirsection and comes out to one direction according to the gradients. It's volume is steady for the considered calculation time stage.
ii)The flowing velocity in a mesh is estimated through Darcy's law of $v=\kappa i$, where i is the gradient of mesh. The discharge flows down to the mesh according to the maximum distance with it's velocity for the calculation time stage.
iii)The maximum storage capacity at a mesh is set

and the excess discharge at the end of calculation time stage is allotted to the upstream meshes from which the inflow came with following criteria;

$$\Delta Q_u = \Delta Q A_u v_u / \Sigma A_u v_u \qquad (2.11)$$

where ΔQ_u is the total volume of excess discharge at the considered time stage, A_u is the u-th section area and v_u is the velocity passing the u-th section. ΔQ_u is defined as the leakage discharge from the u-th mesh.

iv) The leakage discharge is not decided with water pressure because the velocity is handled as the mainly dominant factor.

The infiltration volume f is calculated through the Horton' equation. The evapo-transpiration is also calculated through the Hamon' equation.

$$Ev = 0.14 Do^2 Pt \qquad (2.12)$$

where, Ev is the evapotranspiration converted into EVA in eq.(2.5), Do is the hourly sunlight periods at the considered month and point, and Pt is the absolutely saturated humidity.

3 ASSESSMENT OF WATER USAGE

To assess the water resources environment due to the climate change in future, the systematic evaluation for drought must be introduced. The daily discharge as the feasible water supply is analyzed with the simulated results of water demand. The drought should be evaluated through not only the occurrence probability but also the duration or severity, because the damage becomes different from each other depending on the drought patterns. Assessment of the economical damage is, off course, important. From a physical aspect, the following three indices are proposed (Kojiri & Ikebuchi, 1989).

(i) Reliability: The non-occurrence probability for drought situation is represented as follows;

$$REL = 1 - \frac{\text{the number of drought days}}{\text{total days}} \qquad (3.1)$$

(ii) Resiliency: The ability to recover the drought back to the normal situation is equal to the reciprocal number of expected drought periods follows;

$$RES = \frac{1}{\dfrac{\text{the drought periods}}{\text{the occurence number}}} \qquad (3.2)$$

(iii) Vulnerability: The physical damage is evaluated through the drought volume against the water demand representing the severerity as follows;

$$VUL = \frac{\Sigma(1 - \dfrac{\text{the discharge}}{\text{the water demand}})}{\text{the number of drought events}} \qquad (3.3)$$

From an economical aspect, the increase of drought occurrence leads to change the land price. The damage per one household is assumed to be represented as the Equivalent Variation (Morisugi & Oshima, 1985). The household wants to keep the same expected utility level E{Up(I)} with probability p after it is changed into E{Up'(I)} with probability p' as follows;

$$E\{U_p(I+EV)\} = E\{U_{p'}(I)\} \qquad (3.4)$$

where,

$$E\{U_p(I)\} = PU(x,1) + (1-P)U(x,0)$$

$$hU(x,n|n = 1 or 0) = -(h/\rho)[\ln\{W_1 x_1^{-\rho} + W_2 x_2^{-\rho}$$

$$+ W_3 (a_3 - x_3)^{-\rho} + W_4 (a_4 + x_4)^{-\rho} + W_5 (a_5 - x_5)^{-\rho}$$

$$+ W_6 (a_6 + x_6)^{-\rho} + W_7 (a_7 + n x_7)^{-\rho}\}]$$

$$(3.5)$$

where h is the error variance, W. is the weighing factors, ρ is the parameter, x_1 is the net income at one household, x_2 is the house space, x_3 is the commutation hours for working, x_4 is the sunshine hours, x_5 is the transportation for shopping, x_6 is the effectiveness of public service, and x_7 is the insecurity against the drought. Consequently the value of EV is equal to the economical damage.

4 APPLICATION OF PROPOSED THEORIES INTO THE REAL BASIN

4.1 Basin simulation

The proposed theories are applied into the real basin whose basin area is about 1,000 km2 consisting of 117 meshes of 3km2. The basin divided into three zones such as mountain, agriculture and urban by using the cct values of AVHRR-2 (channel 2) in the satellite NOAA. After decision of discrimination criteria through field survey against cct, all of meshes are characterized by their cct values.

The temperature at each mesh was calculated through the regression function as follows;

$$Ti = -0.785 + 0.990TA - 0.088Hi \qquad (4.1)$$

$$R^2 = 0.966 \quad \text{(correlation coefficient)}$$

where Ti is the temperature at the considered mesh, Ta is the temperature at the gauging station and Hi is the elevation of the considered mesh. There are three gauging station in the basin, only one station were available to estimate the temperature because

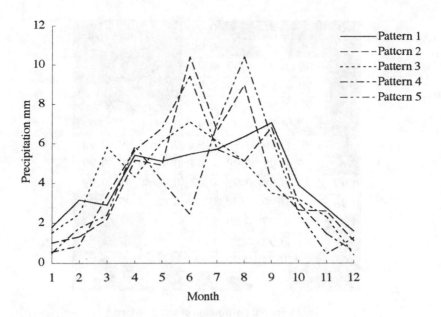

Fig. 2 Classified precipitation patterns

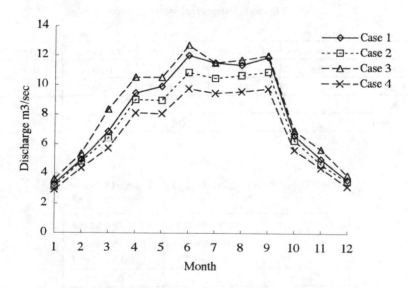

Fig. 3 Simulated discharge sequences among for assumed scenarios

of their locations. The probability density functions at every month are identified among normal distribution, log-normal one with two parameters and log-normal one with three parameters, individually.

Table 1 shows the transition probability at the reference point Furu. Thirty year data were handled in this study and through ISODATA algorithm the monthly hyetographs are classified into five patterns

as shown in Fig. 2. This figure denotes the effectiveness of pattern classification and objective function because the peculiar patterns are extracted, in other words, especially from June to September, the difference between patterns become distinguished. As the probabilities in some section is kept zero, the one pattern is accompanied by his related one.

To avoid the unusual value for simulated

				0.99			
		0.99		0.99			
	0.99	0.99	0.99	0.99			
	0.99	0.99	0.99	0.99			
	0.99	0.99	0.99	0.99			
		0.99	0.99	0.99	0.99	0.99	
	0.99	0.99	0.99	0.99	0.99	0.99	0.99
	0.99	0.98	0.99	0.99	0.99	0.99	0.99
0.99	0.99	0.88	0.92	0.99	0.99	0.99	0.99
0.99	0.99	0.99	0.87	0.89	0.98	0.99	
	0.99	0.97	0.89	0.99	0.99		
	0.84	0.92	0.99	0.99	0.99		
		0.94	0.99	0.99			
		0.99					

Fig. 4 Spacial distribution of soil moisture

Table 2 Drought indices for assumed scenarios

	Frequency	Duration	Reliability	Resiliency	Vulnerability
Case 1	24	42	0.934	0.571	0.545
Case 2	25	44	0.932	0.568	0.55
Case 3	21	35	0.942	0.6	0.508
Case 4	27	55	0.929	0.382	0.603

Table 3 Drought economical damage for assumed scenarios

	Yearly damage per household x1,000 Yen/H/Y	Monthly damage per household x1,000 Yen/H/M
Case 2	2.7	0.23
Case 3	-8.4	-0.7
Case 4	5.5	0.46

precipitation, the averaged monthly precipitation over or under the half of standard deviation around the statistical mean one was dismissed in calculation. The duration of non-precipitation or precipitation without global warming (case 1) were ascertained to keep the same characteristics with confidence interval of 95%. Actually through the detail checking, the amount of simulated precipitation around 10 mm seemed to increase and the amount more than 30 mm was not changed.

Case 2 with the global warming denotes the increase of precipitation amount of 20 to 40 mm and decrease of precipitation amount under 10 mm or 50 mm. Yearly precipitation decrease from 1922 to 1848 mm. To match the yearly amount of precipitation to the global scenarios, the averaged monthly one had to be shifted up to 14.5% against 10% increase and down to 6.4% against 10% decrease. The monthly precipitation shows the decrease tendency and large variance according to the temperature rise.

The parameters of meshes were identified on the basis of minimizing the mean square error as follows;

$$SF = \Sigma \frac{(Qo(t) - Qc(t))^2}{Qo(t)} \quad -> \min \quad (4.2)$$

where $Qo(t)$ is the observed discharge at time t and at the downstream reference point and $Qc(t)$ is the calculated discharge.

At the forth layer the vertical transmissivity was dismissed because it is assumed to lie down on the impermeable zone.

After identification with the hydrological data sequence in 1989, the parameters were decided to be as follows;

$$\begin{array}{ll} \kappa AH=0.80 & \kappa AV=0.16 \\ \kappa BH=0.01 & \kappa BV=0.02 \\ \kappa CH=0.001 & \kappa CV=0.008 \\ \kappa DH=0.00003 & \end{array}$$

Fig. 3 shows the comparison of monthly discharge sequences among scenarios. The observed one means the statistical average from 1962 to 1992. From February to May, more discharge is represented in the case 1, because the precipitation of the range 30 to 50 mm happened to be generated. The available discharge for water resources is more affected by the precipitation increase and pattern than the temperature rise.

Fig. 4 shows the ratio of the soil moisture in the case 2 to that of the case 1. It seems to decrease at the first layer in all meshes, In summer there is the big descrease of 16%. However, other layers has no remarkable difference.

4.2 Assessment of droughts

Daily water demand is estimated by the function of temperature and precipitation in the previous days. For each purpose such as the agricultural use, the industrial one and the municipal one, the parameters of following regression equation is identified through the step-wise approach;

$$\begin{aligned} Q.(t+1) &= d_0 + a_{0}r(t) + a_{1}r(r-1) + \ldots + a_{N}T(r-N) \\ &\quad + b_{0}T(t) + b_{1}T(t-1) + \ldots + b_{N}T(t-N) \\ &\quad + c_{0}Q.(t) + c_{1}Q.(t-1) + \ldots + c_{N}Q.(t-N) \\ &\quad\quad (4.3) \end{aligned}$$

The equations for water demand are identified as follows;

Agricaltural use

$$\begin{aligned} Qa(t+1) &= -0.03 - 0.01r(t-4) + 0.02T(t-4) \\ &\quad + 0.83Qa(t) + 0.14Qa(t-2) \\ &\quad\quad (4.4) \end{aligned}$$

Industrial use

$$\begin{aligned} Qf(t+1) &= 0.10 + 0.67Qf(t) - 0.10Qf(t-1) \\ &\quad + 0.10Qf(t-2) + 0.14Qf(t-3) \\ &\quad\quad (4.5) \end{aligned}$$

Municipal use

$$\begin{aligned} Qc(t+1) &= 0.20 + 0.017T(t-4) + 0.63Qc(t) \\ &\quad + 0.09Qc(t-1) + 0.09Qc(t-2) \\ &\quad\quad (4.6) \end{aligned}$$

The water demand for agriculture depends on all of factors of temperature, precipitation and the previous water demand. However the water demand for industry is independent on temperature, because the re-use process in the factory is well developed and the necessary water demand is mainly decided with the production plan.

The water demand increases with the temperature rise in all cases. Table 2 sows the drought indices including the reliability, the resiliency, the vulnerability, the drought frequency and the drought duration. There are not so big difference in the reliability because the only drought frequencies are analyzed. As the drought duration and deficit amount are changed, the resiliency and vulnerability denote the great difference between in the case 4 and others.

Equivalent Variation is calculated with the occurrence probability which is regarded as one minus reliability. The values of parameters are identified as follows;

$\alpha_1=0, \alpha_2=0, \alpha_3=81, \alpha_4=1, \alpha_5=20,$
$\alpha_6=3, \alpha_7=2,$
$W_1=96.88, W_2=49.56, W_3=8.195,$
$W_4=13.02, W_5=5.759, W_6=50.08,$
$W_7=26.77,$
$\rho=0.062, h=42.54.$

The calculated damage per household are explored as in Table 3, where in the case 3 the benefit was estimated because of enough water supply. In the whole basin, the drought damage was 490.8 in the case 2, -1526.9 in the case 3 and 999.8 in the case 4 (million Japanese yen per year) after being multiplied by the number of households of 181,774.

5 CONCLUSIONS

In this paper, I discussed the basin simulation and assessment through the probabilistic approach due to the global warming. To sum up, the following results were obtained;
i)The daily precipitation was generated by combining the classified monthly precipitation pattern, the transition probability between their patterns and the steady characteristics of daily precipitation specified to th classified pattern.
ii)The run-off discharge was calculated with mesh-typed run-off model according to the catchment conditions of mountain, agriculture and urban zone.
iii)The water usage was evaluated by considering the generated hydrograph and water demand which are carried on with scenarios of global warming and simulated precipitation and temperature.
iv)The impact of global warming to the considered water resources system was evaluated through physical indices of the reliability, the resiliency and the vulnerability, and moreover the economical index of Equivalent Variation with the expected utility.

REFERENCES

Kojiri, T. & Ikebuchi, S. 1989. Optimal Modeling in Water Resources Management Systems Based on Probabilistic Matrix Method. Proc. Hydrology and Water Resources Symposium "Comparison in Austral Hydrology": 279-283
Kojiri, T. 1993. Impact Analysis on Water Resources System due to Global Warming Through Classified Input Patterns and Mesh-typed Run-off Model, Proc. XXV Congress of IAHS. Technical Session A. Volume I: 377-384
Kojiri, T., Unny, T.E. & Panu, U.S. 1994. Cluster Based Pattern Recognition and Analysis of Streamflows. Stochastic and Statistical Methods in Hydrology and Environmental Engineering K.W. Hipel et al. (eds.). Vol.3: 363-380
Morisugi, H. & Oshima, N. 1985. A Method for Evaluating Household Benefits of Reduction in Water-shortage Frequency. Proc. JSCE. No.359 / IV-3: 91-98
Takara, K & Kojiri, T. 1991. A Simulation Study on Catchment Response Change due to Global Warming. Proc. The Int. Symp. on Environmental Hydraulics: 1451-1456

Stochastic Hydraulics'96, Tickle, Goulter, Xu, Wasimi & Bouchart (eds) © 1996 Balkema, Rotterdam. ISBN 90 5410 817 7

Regional effects of climate change on the operational reliability of multi-purpose reservoirs

M.A. Mimikou
Hydraulic & Maritime Engineering, Department of Water Resources, Faculty of Civil Engineering, National Technical University of Athens, Greece

ABSTRACT: Climate changes due to the expected increase of the annual temperature (the "greenhouse effect") will affect the design, construction and operation of multi-purpose reservoirs. By using plausible climate change scenarios the sensitivity of the guaranteed annual fresh water and energy supply levels is being evaluated. It is shown that reservoirs designed and operated under current climatic conditions, are in general affected by the climate changes examined.

1 INTRODUCTION

The global climate change (i.e. increase of the annual temperature), which is likely to occur within the next few decades (the "greenhouse effect"), caused by the increase in the concentrations of carbon dioxide and other trace gases in the atmosphere, will definitely affect the hydrological cycle as well as the quantity and quality of water resources. Recent research strongly suggests that significant temporal and spatial alterations will be caused in the various forms of water resources and associated water management works, resulting in the reexamination of issues, such as design criteria, operational reliability and others (Gleick, 1990; Mimikou, 1995).

Because existing reservoirs have long lifetimes, they may be subjected to climatic conditions for which they were never designed. Reservoirs are designed and constructed to be robust and resilient. However, extreme meteorological events such as damaging floods or droughts still occasionally cause failures. Those extremes are most often manifested during unusual flows or altered timing of flows, such as seasonal floods, persistent droughts, major storms and early or delayed onset of a range of meteorological events, such as spring snowmelt.

For new plants, early planning concerning the hydrologic impact of climate changes can prevent their failure when conditions change and thus save expensive redesign or reconstruction. The appropriate response to this fact is to incorporate into new designs those climate changes that will probably occurred.

In Greece, the purpose of constructing a reservoir is mainly to satisfy irrigation, water supply and power generation needs.

The 16% of electricity production in Greece is generated by hydropower facilities. Although this percentage is low, existing hydropower facilities are valuable because their operating costs are very low relative to the operating costs of other forms of electric generation (thermo-electric plants, etc.). However, hydropower production is particularly vulnerable to climate changes resulting in either possible reduced average annual runoff and, increased variability, and therefore an increased frequency of drought episodes, or in given limited storage capacity, which could be expressed as seasonality of streamflows.

In this paper, the regional climate change impacts on critical water management issues, such as reservoir storage and hydroelectric production, are presented. A methodology using a reservoir simulation model and based on risk analysis of annual firm reservoir quantities for energy production and for water supply, is developed. Plausible hypothetical climate change scenarios are used for equilibrium changes in monthly rainfall, temperature and potential evapotranspiration.

The study area comprises four connected multi-purpose reservoirs in Central Greece, where the huge project of the Acheloos river diversion to the Thessalia plain is under construction. The sensitivity of the guaranteed annual fresh water and energy supply levels was evaluated under conditions of altered runoff due to the climate change scenarios. It was proved that a dramatic increase of the risk associated with the annual guaranteed quantity of water supply and energy production would occur if greenhouse warming were accompanied by reduction in annual precipitation. Significant increases of storage volume would then be needed to maintain existing water and energy yields at tolerable risk levels.

2 THE STUDY REGION

The study area lies in the central mountainous region of Greece, about the $39°$ 30' N parallel. It comprises four drainage basins with the following characteristics:

Table 1. Characteristics of the study basins.

Drainage basin	River	Area (km^2)	Mean elevation (m a.m.s.l.)
Mesohora	Upper Acheloos	633.0	1390
Sykia	Upper Acheloos	1173.0	1299
Pyli	Portaikos	134.5	800
Mouzaki	Pliouris	140.5	575

A major agricultural development plan for the Thessalia Plain has been initiated by the Greek Government. The main purpose is to transfer water from the humid parts of Greece to irrigate the fertile but semiarid Thessalia plain. This will be accomplished be diverting a significant portion of the Upper Acheloos water to the Mouzaki reservoir. Hypsometric differences *en route* will be exploited for power production. The basic scheme comprises four multipurpose reservoirs (for irrigation and power generation) at the outlets of the respective basins, a diversion tunnel from the Sykia reservoir to the Mouzaki reservoir, a tunnel connecting the Pyli reservoir with the Mouzaki reservoir, and four power plants. The study area along with the associated hydrometeorological stations and the water development scheme are depicted in Figure 1.

The Mesohora reservoir, receiving runoff from the upstream drainage basin, is designed to guarantee annual energy production (E=231 GWh). The Mesohora power plant is situated at the end of the Sykia reservoir. The latter receives

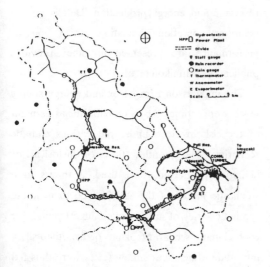

Figure 1. General plan of the study area.

Thessalia plain of W=41 m^3 s^{-1}. The basic operation and design characteristics are given in Table 2.

Table 2. Design characteristics of the Acheloos diversion scheme.

Reservoir	Storage capacity (10^6 m^3)	Hydroelectric power plant	Design head (m)	Primary energy (GWh year^{-1})	Firm water supply (m^3 s^{-1})
Mesohora	228	Mesohora	200	231	-
Sykia	500	Sykia	137	48.8	5
		Pefkofyto	155.5	300	32
Pyli	47	-	-	-	Variable to Mouzaki reservoir
Mouzaki	530	Mouzaki	138.4	370	41

runoff from the intermediate drainage basin between the Sykia and Mesohora dam sites as well as the outflow from the Mesohora reservoir.

A minor quantity of water from the Sykia reservoir is released through the Sykia power plant situated at the toe of the dam to produce primary energy (E=48.8 GWh annually).

The respective firm water flow downstream into the Acheloos River is W= 5 m^3 s^{-1} is diverted from the Sykia reservoir through a tunnel into the Mouzaki reservoir on the western mountain sides of the Thessalia plain. The diverted flow just before entering the Mouzaki reservoir passes through the Pefkofyto power plant. The primary energy generated there amounts to E=300 GWh per annum. The Mouzaki reservoir is also enriched from the Pyli drainage basin through a tunnel connecting it with the Pyli reservoir situated at the outlet of the Pyli basin.

Finally, the Mouzaki reservoir, also receiving runoff from the upstream basin, is designed to produce primary energy (E=370 GWh) annually and a firm water supply to the downstream

3 REGIONAL WATER RESOURCES UNDER CLIMATE CHANGE

The assessment of climate change impacts on regional water resources was based on the use of a monthly water-balance model initially presented in Mimikou et al., (1991a).

The hypothetical climate change scenarios used are based on the most widely accepted estimation regarding the effects of equivalent CO_2 doubling on temperature, which predicts that the mean annual global temperature will rise by 3±1.5° C. The amount of rise is found to increase with increasing latitudes. However, neither the regional nor the temporal distribution of this rise are predictable yet. Therefore, for reasons of compatibility with the above estimated value and comparability with similar efforts, three scenarios were chosen with 1°, 2°, and 4° C increases. Further, the simplest, yet plausible, assumption of uniform increase over the months of the year was made, taking as a basis the historical temperature time-series of each basin. Although increased rates of global evapotranspiration and precipitation are anticipated after greenhouse warming, little is known on the regional distribution of precipitation

313

changes ranging from -25% to 25% (Nemec and Schaake, 1982; Gleick, 1986) in order to conduct rather a sensitivity analysis of the pertaining hydrologic variables and water-management systems. This approach was also maintained here. Five hypothetical scenarios were considered, with changes -20%, -10%, 0%, 10%, 20%. Furthermore, the simplest, yet plausible at any rate, assumption of uniform change over the months of the year was again made, taking as a basis the historical precipitation time-series of each basin. All temperature and precipitation perturbations mentioned previously combine into 3x5=15 climatic scenarios. By using these scenarios as inputs to the previously mentioned water-balance model, calibrated to each one of the four study basins, climatically affected series of runoff and soil moisture were obtained as outputs from the model. Especially, the estimated runoff series for zero change, referred to as the base run, was used for comparison purposes. The selection of the base runs as basis of comparison was made for uniformity reason and it was accepted since each base run was found to be similar to the historical record for all study basins.

More specifically, for each of the 15 pairs of temperature and precipitation scenarios, climatically affected runoff time series I_t for the basins of the study area were produced (Mimikou and Kouvopoulos, 1991). The 15 monthly time series I_t were finally used as inputs to the reservoirs at the outlet of each basin in order to check their sensitivity to the expected climate changes.

The operation of the system of four reservoirs shown in Figure 1 is simulated by a monthly water balance equation under various constraints concerning storage volume, outflow from the

reservoir and energy production. Details on the reservoir operation simulating model and its performance are presented in other previous publication (Mimikou et. al., 1991b).

The impacts on firm water and energy supply levels were estimated by a probabilistic assessment of the risk involved under the various climatic scenarios. The monthly flow I_t, historical or climatically affected, were used as inputs to a stochastic model of the AR(2) type to generate synthetic series of inflows spanning 50 years. For each climatic scenario (current or hypothetical) a multitude of synthetic inflow series were thus fed into the system operation simulating model and the relative frequencies of monthly failures, or risk values, were estimated for various combinations of primary energy E and firm water supply W. A failure was considered to occur when either of the produced quantities failed to meet the preselected demand.

The risk analysis using the historical series revealed that the probability of failure at the examined reservoirs, associated with the design values for W and E, was well below 1% for all of them.

The results of the analysis on the probability of failure associated with the design values of E for the Mesohora reservoir and W and E for the Sykia and Mouzaki reservoirs under the 15 climatic scenarios are presented in Table 3 (Mimikou et al., 1991b). The Pyli reservoir was not analysed since its operation is only secondary.

From this table it is deduced that, under the changed climatic conditions considered, reservoir failure is affected rather by precipitation reduction than by temperature increases. Under the same precipitation reduction, downstream reservoirs exhibit larger risk levels than upstream ones.

Table 3. Risk (%) for the design values of energy and water yield under climate change.

Climatic scenarios	Mesohora HPP	Pefkofyto HPP	Mouzaki HPP
Current-climatic conditions	0.35	0.04	0.33
ΔT= 1°C			
ΔP=-20%	21.43	26.99	51.40
-10%	7.02	5.92	13.77
0%	1.5	0.88	1.35
10%	0.56	0.03	0.16
20%	0.33	0.03	0.04
ΔT= 2°C			
ΔP= -20%	21.92	27.28	49.37
-10%	9.01	6.95	15.12
0%	3.77	1.05	1.56
10%	1.74	0.31	0.38
20%	0.61	0.07	0.03
ΔT= 4°C			
ΔP= -20%	25.29	32.99	53.65
-10%	11.15	9.90	20.54
0%	4.43	1.38	2.47
10%	2.10	0.30	0.28
20%	11.07	0.02	0.04

This is explained by the cumulative effect of upstream failures transferred downstream.

Regardless of the temperature increase, precipitation increases do not significantly improve risk levels, while precipitation reductions significantly and disproportionately raise the risk levels. Regarding temperature increases alone, the Mesohora reservoir appears to be relatively more sensitive than the other two, owing to the intensification of the already high seasonality of runoff in conjunction with its comparatively smaller storage capacity.

When temperature increases combine with precipitation reductions then the risk levels for water and energy supply were shown to be unacceptably high.

However, increasing net storage of the reservoirs compensates for this effect of precipitation reduction.

To obtain an idea of how much should the net storage volume be increased in order the design (W, E) values be maintained at tolerable risk levels, the upstream Mesohora reservoir was examined under precipitation reduction scenarios. Results are shown in Figure 2.

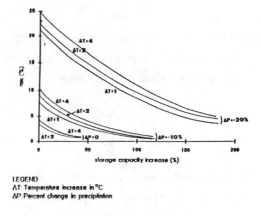

LEGEND
ΔT: Temperature increase in °C
ΔP: Percent change in precipitation

Figure 2. Risk of guaranteed annual production as a function of storage volume increase for the Mesohora reservoir (adapted from Mimikou et al., 1991b; Mimikou, 1995).

As can be seen in Figure 2, radical increases of net storage volume sometimes greater than 100% are required for the design energy production to be met with tolerable risk levels. From this aspect, precipitation reductions play again the important role, the effect of warming appearing to be marginal as compared to that of precipitation reductions.

4 CONCLUSIONS

The main conclusions drawn from this paper are the following:

(1) The climatically affected runoff causes serious increases of the risk associated with the annual firm water and energy production from the system of reservoirs in the study area, when precipitation decreases are involved. The effect of temperature rise on risk levels is marginal. Downstream reservoirs appear to be more sensitive than upstream ones.

(2) Radical increases of reservoir storage volume are required to compensate for the effect of precipitation reduction on risk levels.

5 ACKNOWLEDGEMENTS

The author wishes to thank the Public Power Corporation of Greece for providing the hydrometeorological and other data used in the study.

REFERENCES

Gleick, P,H, 1986. Regional water availability and global climatic change: The hydrologic consequences of increases in atmospheric CO_2 and other trace gases. PhD Dissertation, Univ. of California, Berkley, California, 615 pp.

Gleick, P,H, 1990. Vulnerability of water systems. In: Waggoner P (ed) climate change and U.S. Water Resources John Wiley and Sons, New York, pp 223-240.

Mimikou, M.A. & Kouvopoulos, Y.S. 1991. Regional climate change impacts: I. Impacts on water resources. Hydrol. Sc. J., 36,3,6: 247-258.

Mimikou, M.A., Kouvopoulos, Y.S., Cavadias, G. & Vayiannos, N. 1991a. Regional hydrological effects of climate change. J. Hydrol., 123: 119-146.

Mimikou, M.A., Hadjissavva, P.S., Kouvopoulos, Y.S. & Afrateos, H. 1991b. Regional climate change impacts: II. Impacts on water management works. Hydrol. Sc. J., 36,3,6: 259-270.

Mimikou, M.A. 1995. Climatic change. Chapter 3 of the Book "Environmental hydrology". Kluwer academic publishers, Singh VP (ed) pp. 69-106.

Nemec, J. & Schaake, J. 1982. Sensitivity of water resource systems to climate variations Hydrol. Sc. J., 27(3): 327-343.

Stochastic Hydraulics'96, Tickle, Goulter, Xu, Wasimi & Bouchart (eds) © 1996 Balkema, Rotterdam. ISBN 90 5410 817 7

A statistical model of rainfall and temporal patterns

Martin Lambert
Department of Civil and Environmental Engineering, University of Newcastle, N.S.W., Australia

George Kuczera
Department of Civil Engineering and Surveying, University of Newcastle, N.S.W., Australia

ABSTRACT: A simple and parsimonious point rainfall model capable of representing the inter-event time, storm duration, average event intensity and temporal distribution is presented that can be used to simulate storm rainfall events in hydrological risk assessment. Pluviograph information from Sydney has been used to obtain values for the model parameters and a comparison is made between the simulated Intensity-Frequency-Duration distribution and that observed in the original data set. This is a rigorous test of the modelling approach and the model produced acceptable results. The proposed model has several novel features which include a conditioning of the rainfall intensity on the rainfall duration and a temporal model of the rainfall event which uses a constrained random walk in the dimensionless depth-time space to represent the random nature of rainfall in each storm event.

1 INTRODUCTION

Rainfall is the main input to all hydrological models which predict peak catchment outflow discharges, flood levels, water yield, ground water levels and soil moisture states. A useful engineering model of the rainfall process would find application in many areas of hydrological risk assessment. For instance, such a model could be used as an alternative to the design techniques given in Australian Rainfall and Runoff (Institution of Engineers, Australia, 1987) when various parts of the hydrological system (such as initial rainfall loss, different initial storage reservoir levels, groundwater levels or ocean water level) interact to produce joint probabilities which are difficult to assess. The design techniques currently presented in Australian Rainfall and Runoff are difficult to apply in such situations whereas design techniques based on simple but continuous rainfall models may enable significant insights into the behaviour of the hydrological system. This paper describes a stochastic rainfall model which could fulfil this role.

Eagleson (1978) stated, when discussing his rainfall model, that "an engineering model of this process must retain those natural features which are important to the problem at hand, while for economy of computation and for clarity of behaviour it should omit those features deemed unessential." Eagleson (1978) extended earlier work by Todorovic and Yevjevich (1967) and Ison et al. (1971) to produce a crude but useful model of rainfall with focus primarily on the derivation of analytical expressions for the frequency of annual rainfall for use in long-term water budget analysis. Valdes et al. (1990) used Eagleson's approach to derive a probability

density function for the initial soil moisture in a catchment. They argued that the approach could be improved if the assumption that the precipitation follows rectangular pulses was removed. They suggested the use of more realistic models of point precipitation, such as the Neyman-Scott and Barlett-Lewis point process models (Rodriguez-Iturbe et al., 1987). These models are capable of sometimes replicating a variety of rainfall statistics such as the mean, variance and lag autocorrelation over a large range of rainfall aggregation levels from 1 to 200 hours. They are typically calibrated on a monthly time interval and give reasonable, but by no means perfect, estimation of the monthly extreme values (see Entekhabi, 1989). Unlike the model being presented herein, the random pulse models are essentially calibrated to the Intensity-Frequency-Duration (IFD) characteristics of the data so the above result is to be expected.

Since the work by Rodriguez-Iturbe et al. (1987), the original ideas of Eagleson (1978) seemed to have lost favour over the cluster point process models described by Cox and Isham (1980). While a considerable amount of work has been undertaken using these models (Entekhabi, 1989, Onof and Wheater, 1993, 1994, and Velghe, 1994), we feel that the markedly increased complexity and lack of a simple physical basis has moved much of this work outside the realm of the useful engineering model described by Eagleson (1978). We feel that some relatively simple modifications to Eagleson's approach may significantly improve the model.

It is Eagleson's (1978) approach which is significantly extended in this paper. Our focus has been on the examination of water level frequency distributions in coastal river systems which are jointly affected by fluvial floods and ocean water levels (Lambert and Kuczera, 1996). For this work a model of storm rainfall was needed which could closely replicate both the storm events and the period between these events. Another criterion was that the storm events be temporally distributed in such a way that the Intensity-Frequency-Duration data (obtained using a fixed duration window that progressively moves through the year) was also well produced. This work has led to the simple point rainfall model described in this paper. We believe that this model will prove to be a useful tool for examining the long-term and extreme event behaviour of hydrological systems which are dependent on rainfall input.

2 THE RAINFALL GENERATION MODEL

Figure 1 describes the Eagleson (1978) model of the precipitation event series. He assumed that the rainfall events followed a Poisson arrival process which leads to an exponential distribution for the storm inter-arrival time, t_a, where the cumulative distribution function is given as

$$F_T(t_a) = 1 - e^{-t_a/\mu_{ta}} \qquad (1)$$

and μ_{ta} is the mean storm inter-event arrival time.

The storm duration, t_d, was also found by Eagleson (1978) to be exponentially distributed according to

$$F_T(t_d) = 1 - e^{-t_d/\mu_{td}} \qquad (2)$$

where μ_{td} is the mean storm duration. The original proposal by Eagleson was that the storm intensity was also exponentially distributed as

$$F_I(i) = 1 - e^{-i/\mu_i} \qquad (3)$$

where μ_i is the mean of the average storm intensity. This distribution for rainfall intensity was not used by Eagleson because the resulting distribution for rainfall depth $h = i\, t_d$ is not self preserving which was not satisfactory for producing an analytical expression for the distribution of annual rainfall. Eagleson (1978) instead chose to use a gamma distribution for the rainfall depth (h).

With Eagleson's model of rainfall as a guide we investigated the distribution of rainfall events using 75 years of Sydney pluviograph information that was available from the Bureau of Meteorology in 6 minute time increments. A dry interval of $t_{dry} = 2$ hours was adopted as the criterion for distinguishing independent consecutive events. This is an arbitrary threshold but is consistent with the criterion used by Eagleson (1978). As an initial comparison the interstorm time (t_a), storm duration (t_d) and average storm rainfall intensity (i) were plotted on exponential probability paper. Figure 2 shows the interstorm time distribution while Figures 3 and 4 show the storm duration and the average storm intensity distributions respectively. If the random variables are exponentially distributed then the probability plots shown in Figures 2 to 4 should indicate straight lines. Clearly, none of the random variables are exactly exponentially distributed as suggested by Eagleson, but this fact detracts little from the approach.

The distribution of interstorm time (t_a) shown in Figure 2 is well approximated by the two-parameter gamma distribution presented in Eq. 4 with mean $\mu_{ta} = 37.43$ hours and standard deviation $\sigma_{ta} = 69.44$ hours determined by the method of moments.

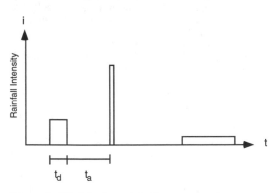

Figure 1. Model of precipitation event series.

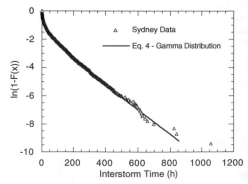

Figure 2. Distribution of time between storms at Sydney.

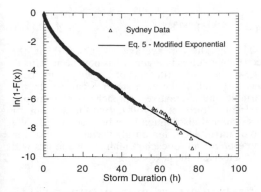

Figure 3. Distribution of storm durations at Sydney.

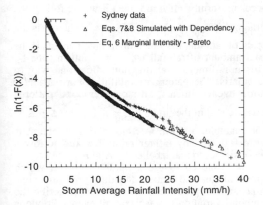

Figure 4. Distribution of average storm rainfall intensities at Sydney.

$$f_{ta}(t) = \frac{x^{\alpha-1} e^{-t/\beta}}{\beta^\alpha \Gamma(\alpha)}, \quad t>0 \qquad (4)$$

where $\alpha = \left(\dfrac{\mu_{ta}}{\sigma_{ta}}\right)^2$ and $\beta = \left(\dfrac{\mu_{ta}^2}{\sigma_{ta}}\right)$ Eq. 4 is plotted along with the data in Figure 2 and shows good agreement. It should be noted that when the shape parameter (α) is less than one, the maximum likelihood estimates of the gamma distribution parameters can produce misleading results. The method of moments was found to produce superior results in this study.

The distribution of the storm duration shown in Figure 3 is also approximately gamma distributed with mean $\mu_{td} = 4.13$ h and $\sigma_{td} = 6.18$ h. However, an improved fit to the storm duration data can be obtained by using a modified exponential distribution of the form

$$F_{td}(t) = 1 - e^{-\alpha t^\beta} \qquad (5)$$

where $\alpha = 0.48$ and $\beta = 0.67$.

The combination of the random variables for t_a and t_d can now provide the locations of the pulses shown in Figure 1. What remains to be determined is the intensity or height of the pulse for each storm event.

The distribution of storm intensities in Figure 4 shows a more pronounced deviation from an exponential form than either the interstorm time or the storm duration. In this case neither the gamma distribution nor the modified exponential distribution could be used to represent the marginal average rainfall intensity. A distribution which provided a reasonable fit to this data is the Generalised Pareto Distribution (GPD) (see Rosbjerg et al., 1992) shown below

$$F_I(i) = 1 - \left(1 - \alpha\frac{i}{\beta}\right)^{1/\alpha} \qquad (6)$$

where $\alpha = \dfrac{1}{2}\left(\dfrac{\mu_i}{\sigma_i}\right)^2 - \dfrac{1}{2}$, $\beta = \dfrac{\mu_i}{2}\left[1+\left(\dfrac{\mu_i}{\sigma_i}\right)^2\right]$ with

the mean $\mu_i = 1.72$ mm/h and $\sigma_i = 2.22$ mm/h. Figure 4 indicates that Eq. 6 is a useful, but not exact, probability model for the marginal average rainfall intensity.

However, Eq. 6 cannot be used in the rainfall model because, being a marginal distribution, it can not make allowance for the dependence of storm intensity on storm duration so clearly displayed in Figure 5. A consequence of this dependency is the increase in the mean and standard deviation of the intensity as storm duration decreases. What is required is the capability to simulate the higher intensity short duration events while maintaining the observed marginal distribution of average storm rainfall intensity. One way to proceed is to assume that the average rainfall intensity is Pareto distributed but the mean and standard deviation of the distribution are functions of the storm duration (t_d). The following power law relationships have been calibrated

$$\mu_i = 1.772\, t_d^{-0.063} \qquad t_d \leq 20 \text{ h} \qquad (7a)$$

$$\sigma_i = 2.22\, t_d^{-0.136} \qquad t_d \leq 20 \text{ h} \qquad (7b)$$

with the four numerical constants determined using the method of maximum likelihood.

The simulated marginal distribution that results from sampling from Eq. (5) and the conditional distribution given by Eqs (6) and (7) is also shown in Figure 4. Although the fit is similar to that of Eq. (6) the dependence between intensity and duration is better preserved. The power law relationships used in Eq. 7 were decided upon after an examination of the variation of the mean and standard deviation

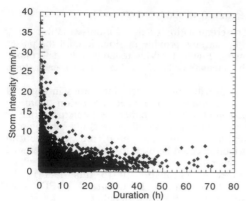

Figure 5. Dependence of average storm rainfall intensity on storm duration.

with storm duration. While other more complicated expressions can be used to describe the variation, little improvement in the maximum value of the likelihood function was obtained. It is envisaged that when the model is applied to other geographic locations the exponents in Eq. 7 will be relatively invariant.

Figure 6 presents the number of storms recorded per hour of storm duration and shows that from about the 10-hour duration onwards comparatively little information is available about the mean or the standard deviation of average rainfall intensity. In particular the region $t_d > 20$ hours has significantly fewer than 50 storms in each hour of increasing duration. Moreover, there are only 49 storm events with a duration greater than 40 hours. As a result there is limited information available to deduce the variations of the mean and standard deviation in the region $t_d > 20$ hours. Therefore, the mean and standard deviation given in Eq. 7 were held at the values corresponding to $t_d = 20$ hours for storm durations greater than 20 hours giving

$$\mu_i = 1.772 * 20.0^{-0.063} \qquad t_d > 20 \text{ h} \quad (8a)$$

$$\sigma_i = 2.22 * 20^{-0.136} \qquad t_d > 20 \text{ h} \quad (8b)$$

The height (or average intensity i) of the rectangular pulses in Figure 1 has now been defined. One of the major criticisms of Eagleson's approach was that rainfall during a storm event was highly variable and alternative approaches (for example, Rodriguez-Iturbe et al., 1987, 1988) to account for this temporal distribution have been advanced. We have introduced temporal variation into the Eagleson model for each rainfall event and this will be discussed in the next section.

3 TEMPORAL DISTRIBUTION OF RAINFALL

Rainfall during a storm event does not usually occur uniformly. If it did, then it would follow the straight line $\frac{d}{d_o} = \frac{t}{t_o}$ shown in Figure 7 where d is the depth of rain that has fallen up to time t and d_o is the total amount of rainfall which will fall in time t_o. This non-dimensional diagram offers one way to summarise the pattern of rainfall for a particular storm event. In fact, all storm temporal patterns must lie within the limits $0 \leq \frac{d}{d_o} \leq 1$ and $0 \leq \frac{t}{t_o} \leq 1$. For example, a storm in which the initial burst is large followed by lighter rainfall would follow a curve similar to that labelled by A in Figure 7.

It is proposed to use a random process which approximates the traces of rainfall in this non-dimensional space (or mass curve) to describe the temporal variation of each rainfall event. Previous descriptions of some rainfall mass curves for Sydney are presented in Pilgrim et al. (1969) but these are for extreme rainfall bursts rather than complete storm events.

We have examined all the rainfall events of 1 hr and greater duration in the 75-year Sydney pluviograph record and found that the temporal pattern could be well described by the following

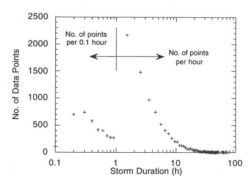

Figure 6. Number of data points per hour (or per 0.1hour if $t_d <$ 1hour) of duration.

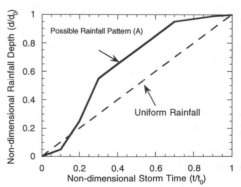

Figure 7. A non-dimensional description of the rainfall temporal pattern.

stochastic process. The dimensionless time space $\left(\dfrac{t}{t_o}\right)$ is divided into N equal intervals giving the N+1 time locations $\left\{\dfrac{t_i}{t_o}, i = 0,..,N\right\}$. A constrained random walk jumps from one time location to the next. Let $f\left(\dfrac{d_i - d_{i-1}}{d_o}\Big|\dfrac{d_{i-1}}{d_o} \le \dfrac{d_i}{d_o} \le 1\right)$ be the conditional probability density function of jumping from $\left(\dfrac{t_{i-1}}{t_o}, \dfrac{d_{i-1}}{d_o}\right)$ to $\left(\dfrac{t_i}{t_o}, \dfrac{d_i}{d_o}\right)$, $i=1,..,N-1$. For the Sydney data the random jumps were well approximated by truncated log-normal distributions. As i approaches N, the mean and standard deviation of the jump were observed to decrease as one would expect given the constraint $\dfrac{d_{i-1}}{d_o} \le \dfrac{d_i}{d_o} \le 1$. This relationship was preserved by parameterising the jump mean and standard deviation on $\dfrac{d_{i-1}}{d_o}$. The following equations were derived from the Sydney data for N equal to 10. The equations for μ_{ji} and σ_{ji} have been generalised to N intervals recognising that the expected jump, μ_{ji}, should be proportional to the size of the jump interval $\dfrac{1}{N}$, whereas the jump variability, represented by σ_{ji}, should be independent of the jump interval.

For the first jump from the origin the distribution is a truncated log-normal with mean and standard deviation given by

$$\mu_{j1} = \frac{0.108}{N} \qquad (9a)$$

$$\sigma_{j1} = 0.102 \qquad (9b)$$

and the range of a valid jump is $0 \le \dfrac{d_1}{d_o} \le 1$.

For subsequent jumps from $\left(\dfrac{t_{i-1}}{t_o}, \dfrac{d_{i-1}}{d_o}\right)$ to $\left(\dfrac{t_i}{t_o}, \dfrac{d_i}{d_o}\right)$, the jump distribution is also a truncated log-normal with mean and standard deviation given by

$$\mu_{ji} = \frac{1}{N}\left(1 - \frac{d_{i-1}}{d_o}\right)\left(0.1 + 0.33\frac{d_{i-1}}{d_o}\right)$$

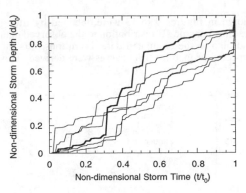

Figure 8. An example of 10 generated storm temporal patterns.

$$\sigma_{ji} = 0.126\left(1 - \frac{d_{i-1}}{d_o}\right), \qquad i=2,..,N-1 \quad (10)$$

with the jump range being $\dfrac{d_{i-1}}{d_o} \le \dfrac{d_i}{d_o} \le 1$.

The current rainfall model subdivides non-dimensional time into 50 intervals. Ten simulated storm traces are illustrated in Figure 8. This stochastic model of the rainfall temporal pattern is relatively simple but we show that it manages to capture the essential elements of the process.

4 RAINFALL MODEL SIMULATIONS

The temporal pattern generator can now be coupled with the rainfall event generator to produce simulated rainfall. This model will, through its very nature, preserve many of the dominant features of the long-term rainfall record. One of the most rigorous tests of the model's ability to simulate the temporal nature of rainfall involves a comparison of the Intensity-Frequency-Duration (IFD) curves derived from the simulated data and from the Sydney data. The IFD curves are obtained by moving windows of fixed duration incrementally through each year and determining the annual maximum rainfall depths for each of the windows. The shorter duration IFD curves (say 1, 3 and 6 hours) will provide a good test of the rainfall model's ability to replicate the random bursts that occur during rainfall events. For example, the maximum 1-hour duration rainfall may not be produced by a 1-hour storm but could (more likely) be produced by a burst within a longer duration storm.

Figures 9 to 11 compare the observed and predicted annual maximum probability distributions for the 1, 3 and 6 hour bursts, while Figures 12 and 13 present the IFD comparison for the longer bursts of 12 and 72 hours respectively. The IFD curves shown in Figures 9 to 13 were obtained by fitting the Log-Pearson III distribution to the observed data and also to the simulated data. From these fits the expected probability IFD curves were derived. This is consistent with the probability distributions used in Australian Rainfall and Runoff for IFD data and excellent fits were obtained in every case. This procedure has also allowed the 90% probability limits to be obtained for the Sydney data and these as shown on all the plots. It was also decided to plot the IFD rainfall intensity versus log(ARI) to exaggerate the tail of the probability distribution. Simulations of 1000 years in length were used for the comparisons with the observed data.

It can be seen from Figures 9 to 11 that the rainfall model gives good predictions of the observed data and these lie within the 90% probability limits obtained from the Sydney data.

However, examination of the longer IFD durations of 12 and 72 hours in Figures 12 and 13 indicates that the model is under-predicting the average IFD rainfall intensity for these events. We feel that this shortcoming in the current model is due to inadequate modelling of the rainfall intensity that is to be expected in long (> 20 hours) duration events. Figure 14 shows that very short duration high intensity (see Region A) events do not occur.

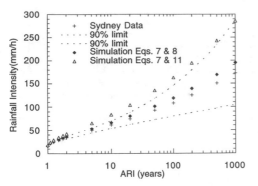

Figure 9. Annual maximum distribution of rainfall intensities for the 1 hour duration IFD.

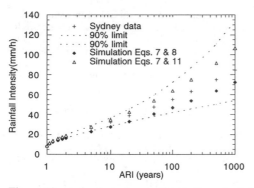

Figure 10. Annual maximum distribution of rainfall intensities for the 3 hour duration IFD.

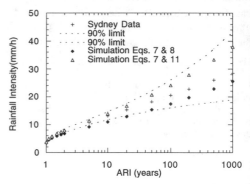

Figure 12 Annual maximum distribution of rainfall intensities for the 12 hour duration IFD.

From these considerations the long duration

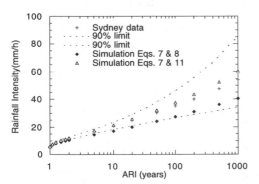

Figure 11. Annual maximum distribution of rainfall intensities for the 6 hour duration IFD.

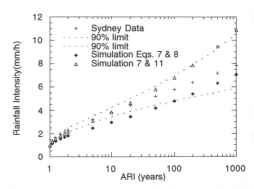

Figure 13 Annual maximum distribution of rainfall intensities for the 72 hour duration IFD.

Similarly low intensity long duration events (see Region B) are also absent from the data record. Moreover, for $t_d > 20$ hours, the mean intensity reverses its declining trend and starts to increase. Events (> 20 hours) should have their intensity generating parameters adjusted to compensate for the inadequacies of Eqs (8) and (9). It was found that an increase of 35% in the mean and standard deviation given by Eq. (9) resulted in a general over-prediction of the IFD results as shown by the open triangles in Figures 9 to 13. This proposed modification to Eq. (9) results in

$$\mu_i = \left(1.772 * 20.0^{-0.063}\right) *1.35 \, , \ t_d > 20\,\text{h} \quad (11a)$$

$$\sigma_i = \left(2.22 * 20^{-0.136}\right) *1.35, \quad t_d > 20\,\text{h} \quad (11b)$$

Figure 15 shows a comparison of the mean annual maximum intensity from the simulation with the adjusted parameters given by Eq. 11 for each duration from 1 hour up to 72 hours and the observed values. The simulated results were quite close to observed data but were seen to over-predict the IFD data for the more extreme events.

The 35% increase in the storm intensity parameters when $t_d > 20$ hours had almost no impact on the marginal distribution of intensity as these storms represent less than 3% of all the storms that will be generated. It can be concluded that some increase of the long duration ($t_d > 20$) rainfall intensity generating parameters would be beneficial in providing the best possible simulation of the longer duration IFD rainfall bursts. It should also be noted that the 35% increase in intensity generating parameters that was adopted for the long duration storms (Eq. 11) produced a significant rise in the extreme values of the short IFD bursts. This is a good indicator of the effectiveness of the temporal pattern generator at producing significant short duration bursts rainfall bursts within long duration storm events.

5 SEASONALITY

At no time has the seasonality present in the Sydney data been considered in this model. The view has been taken that the seasonality will require additional parameters to be added to the model to improve the calibration. For example, two seasons gives twice as many parameters. Our aim has been to keep our approach as simple as possible to try to achieve a useful design model. We feel that this has been achieved but we observe that some seasonality (shown as mean monthly rainfall for the 75 years of data in Figure 16) is present in the Sydney data and, at times, the mean monthly intensity produced by the simulation lies outside the 95% probability limits of the observed monthly mean rainfall depth. A two season approach would improve the rainfall model.

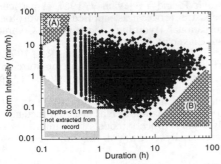

Plot of all Storm Intensities versus Duration

Figure 14 Log Intensity versus Log Duration Plot showing the paucity of data for describing long storms and the lack of long duration small intensity storms.

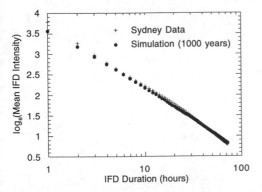

Figure 15 Comparison of the observed mean IFD rainfall intensities and the simulation.

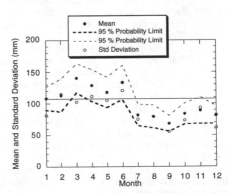

Figure 16. Seasonality of annual precipitation at Sydney.

6 CONCLUDING REMARKS

A simple and parsimonious rainfall model has been presented which is capable of replicating the general characteristics of rainfall at a variety of time scales. While we have followed Eagleson's (1978) original approach we have incorporated several improvements to the model. Two notable additions to the model are:

1) The addition of a dependence relationship between storm event duration and average event intensity.
2) A temporal pattern generator which uses a constrained random walk through the dimensionless time and depth space (or mass curve) for each rainfall event.

The good replication of the observed Intensity-Frequency-Duration relationship was considered to be of paramount importance if the model is to be used for hydrological risk assessment. It was shown that it is possible to produce a good replication of the IFD relationship even though the model has not been calibrated to this information. The use of observed distributions for inter-storm time and storm duration will ensure that the longer term rainfall characteristics are also well preserved. No account has been taken of the seasonality observed at Sydney but the extension of the model to incorporate seasonal variation is straightforward.

7 ACKNOWLEDGMENTS

Funding for this work was provided by the Australia Research Council.

8 REFERENCES

Australian Rainfall and Runoff 1987. A guide to flood estimation, D. Pilgrim (ed.), Institution of Engineers, Australia.

Cox D.R. and Isham V. 1980 Point Processes. London, Chapman and Hall.

Eagleson P. 1978. Climate, Soil, and Vegetation: 2. The distribution of annual precipitation derived from observed storm sequences. Water Resources Research, Vol. 14, No. 5, pp. 713-721.

Entekhabi D., Rodriguez-Iturbe I. and Eagleson P. 1989. Probabilistic representation of temporal rainfall process by a modified Neyman-Scott rectangular pulses model: parameter estimation and validation, Water Resources Research, Vol. 25 No. 2 pp. 295-302.

Ison N.T. Feyerherm A.M. and Bark L.D. 1971, Wet period precipitation and the gamma distribution. Journal of Applied Meteorology, Vol. 10, No. 4, pp. 658-661.

Lambert M. and Kuczera G. 1996. Approximate joint probability analysis of extreme water levels in coastal catchments, this issue.

Onof C. and Wheater H.S. 1993. Modelling of British rainfall using a Barlett-Lewis rectangular pulse model. Journal of Hydrology, Vol 149, pp. 67-95.

Onof C. and Wheater H.S. 1994. Improvements to the modelling of British rainfall using a modified random parameter Barlett-Lewis rectangular pulse model. Journal of Hydrology, Vol 157, pp. 177-195.

Pilgrim D., Cordery I. and French R. 1969. Temporal patterns of design rainfall for Sydney, Civil Engineering Transaction, Institution of Engineers, Australia, April, pp. 9-14.

Rodriguez-Iturbe I., Cox D.R. and Isham V. 1987. Some models for rainfall based on stochastic point processes. Proceedings of the Royal Society, London, A410, pp. 269-288.

Rodriguez-Iturbe I., Cox D.R. and Isham V. 1988. A point process model for rainfall: further developments. Proceedings of the Royal Society, London, A417, pp. 283-298.

Rosbjerg R. and Madsen H., 1992. Prediction in Partial Duration Series with generalised Pareto-Distributed Exceedences. Water Resources Research, Vol. 28, No. 11, pp. 3001-3010.

Todorovic P. and Yevjevich V. 1967. A particular stochastic process as applied to hydrology, International Hydrology Symposium, Colarado State University, Fort Collins.

Valdes J., Diaz-Granados M. and Bras R.L. 1990. A derived PDF for the initial soil moisture in a catchment. Journal of Hydrology, Vol. 113, pp. 163-176.

Velghe T., Troch P., De Troch P. and Van de Velde J. 1994. Evaluation of cluster-based rectangular pulses point process models for rainfall. Water Resources Research, Vol. 30, No. 10, pp.2847-2857.

Stochastic Hydraulics '96, Tickle, Goulter, Xu, Wasimi & Bouchart (eds) © 1996 Balkema, Rotterdam. ISBN 90 5410 817 7

Rainfall forecasting using neural networks

T. S. V. Ramesh & P. P. Mujumdar
Department of Civil Engineering, Indian Institute of Science, Bangalore, India

ABSTRACT : Neural network models are developed for one step ahead forecasting of ten daily rainfall, in a semi arid command in South India. Two network architectures, the Back Propagation and the Radial Basis Functions are used to develop two different models for the same sets of inputs and outputs. Both the architectures use a basic learning network consisting of the input, hidden and the output layers. Training is conducted through a series of learning sets consisting of rainfall and the associated time periods. The neural network models showed good predictive capabilities. Architecture using Radial Basis Functions is suggested for real-time applications in which the network must re-learn as and when new data becomes available.

1 INTRODUCTION

Rainfall forecasting is one of the most challenging problems in hydrology. This is mainly due to a very large range of variability of the rainfall over space and time, a highly random nature of the rainfall process and a large number of climatic variables affecting the process. The forecasting problem becomes even more difficult for shorter time periods of less than a month. However in many practical applications, reliable forecasts for shorter duration are essential. Typical examples of such applications include (i) flood control, for which at least hourly and preferably even shorter period rainfall forecasts are necessary and (ii) irrigation scheduling where weekly or ten daily rainfall forecasts are useful. Because of such a need, a number of quantitative models have been developed for short term rainfall forecasting (e.g. Georgakakos and Kavvas 1987). The applicability of simple regression-based techniques and large scale numerical models to various situations and their limitations are discussed by French et. al (1992). They indicate the advantage of using neural network (NN) models for short term rainfall forecasting.

The present paper addresses the problem of ten daily rainfall forecasting using neural networks. A semi-arid region in South India is used as a case study.

1.1 Neural networks

Neural networks (NN) are mathematical models that attempt to mimic the massive parallel architecture of the human brain. They offer potential tools for solving problems involving pattern recognition, data compression, function approximation and optimization. Discussion of network architectures suitable to various classes of problems with supervised and unsupervised learning algorithms is available in literature (e.g., Lippmann 1987, Wasserman 1990). Flood and Kartam (1994) provide a basic introduction to neural networks with their applications to civil engineering problems.

Recent interest in the use of neural networks for solving problems related to hydrology and water resources has led to many beneficial applications. French et. al (1992) demonstrated the capability of neural networks for short term rainfall forecasting. One hour lead forecasts were obtained by them using input and output pairs for training generated from a complex simulation model. The input and output values were the current step rainfall field and the one hour ahead rainfall fields respectively. They concluded that a neural network is capable of learning complex and non-linear relationships of rainfall events and performed well in multi-site forecasting.

In the present study neural network models are developed for one step ahead ten day rainfall forecasting. The inputs presented to the NN models are the rainfall in the current time period and the time period itself. The ten day rainfall of the next period forms the output. Sets of input-output pairs are used for training. Using time period as one of the inputs has helped the network to learn the relationship between the time period and rainfall values. This is an important feature of NN models developed in the present study, and has not been addressed in the other neural network models for hydrologic forecasting. Also, actual historic data on ten day rainfall values has been used. In the absence of actual data, synthetically generated data may be used but with a major disadvantage of the network acquiring the limitations of the simulation model. Simulated data may help to demonstrate the strength of the network in solving a problem of function approximation but will be of little practical use. The use of data from a simulation model (e.g. French et.al 1992) provides the neural network with an established, well defined relationship between the various input and output variables. This helps the network to obtain the function approximation as it can easily mimic the implicit relationship. Thus the network is bound to perform well if the network architecture is chosen appropriately. However in such a case, the NN model would be good only to the extent that the original simulation itself is good and therefore would be of no additional use in real-time forecasting. On the other hand if the network is provided with actual data, it can capture the actual non-linear functional relationships between various input and output variables, and thus would provide more realistic forecasts.

In the present work, two network models using two different architectures, namely the Back Propagation (BP) and the Radial Basis Functions (RBF) are developed. The former architecture is widely used in literature as it has proved to be effective in learning and generalization. It derives the name from an iterative training algorithm "Back Propagation". One major drawback of this architecture is the long training time it requires for most problems. Various architectures have been tried by many researches to overcome this drawback. RBF networks have provided a good alternative to the BP networks in regard to training times. They have been used in the areas of speech pattern classification (Renals 1989) and time series modeling (Chen et.al 1991). Renals (1989) compared the training times of BP and RBF networks and concluded that RBF networks trained much faster than the BP networks. Chen et.al (1991) demonstrated the applicability of RBF network to a problem of modeling time series data. RBF networks are proved to be good at function approximation problems with good generalization and faster training times. The RBF network is especially useful when the network has to be trained repeatedly for re-learning, as is the case in real-time forecasting. The BP network is not advisable in such cases as it requires considerably long training time for an acceptable solution. Both Back Propagation and Radial Basis Function architectures are used in the present study and their performance in real-time rainfall forecasting is compared.

The following paragraphs give a brief introduction of the network architectures used in the present study. Mathematical details of training algorithms is avoided as they are available in literature elsewhere (Lipmann 1987, Tsoi, 1989, Wasserman 1990).

2 NEURAL NETWORK ARCHITECTURES

The neural network models developed in the present study adopt a general structure of input, hidden and output layers. The BP network has two hidden layers and the RBF network has only one hidden layer. The number of hidden layers used in case of BP network is specific to the present study. It may be possible to achieve a better performance with a different set of hidden layers and hidden layer neurons. Only two hidden layers are used in the BP network as they provided an acceptable solution. Learning is achieved by presenting the input-output pairs and training the network with a learning algorithm. A brief outline of each of the network architectures is presented here. Details of training and testing are presented next.

2.1 Back propagation network

The network has a general structure consisting of three layers of input, hidden and output neurons. The number of neurons in the input and output layers are decided based on the number of input and output values presented to the network for training. The number of hidden layers and neurons are decided by trial and error procedure as no exact method is available in literature. First the number of hidden layers and neurons are fixed arbitrarily and the network is trained. Once the network shows an acceptable performance, the number is fixed or else the number is decreased or increased. For the present study two hidden layers, two neurons in the input layer and one in the output layer are used. Neurons in the input layer are fixed as the inputs are the current period's rainfall and the time period. Figure 1 gives the structure of the BP network used in the present study.

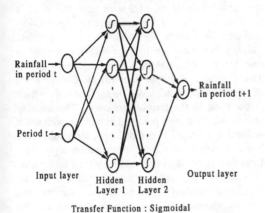

Fig. 1 Network architecture using Back Propagation

The network is fully connected in the sense that each of the neurons are successively connected to the neurons in the next layer. The output neurons receive connections from the hidden layer which in turn is connected to the input layer. Interconnection weights exist in between each layer. Excepting the input layer neurons all other neurons provide a

transformation to the inputs received from the previous layer neurons. The transformation at each hidden layer neuron and the output neuron is given by

$$T(X) = \frac{1}{1 + e^{-\psi(X-\lambda)}} \qquad (2.1)$$

where X is the weighted sum of inputs from the previous layer neurons, $T(X)$ is the output, ψ is the gain and λ is the bias. $T(X)$ is often referred to as sigmoid transfer function or logistic function. The values of ψ and λ are specified for each layer. In the present study the values are taken to be same for all the neurons in a particular layer following French et.al (1992). Training comprises of obtaining the interconnection weights using Back Propagation algorithm. Discussion on this algorithm may be found in Wasserman (1990).

2.2 Radial basis function network

Network architecture using Back Propagation has several drawbacks, a crucial one being the extremely long training time required for most of the problems. In order to avoid the problems of long and uncertain training time, a RBF network is experimented in the present study. Applications involving speech recognition using this architecture are available in the literature (Renals 1989). RBF networks have shown a good promise of being better than the networks using Back Propagation in regard to training times. They are equally good at generalization capabilities when compared to that of BP networks. As the training time required is very less, these networks are ideally suited for applications involving real-time forecasting. A complete derivation of RBF network is given by Tsoi (1989). The network is a three layer network with one input, one hidden and one output layers. The number of neurons in the input and the output layer are decided based on the number of input and output variables respectively. The structure of the RBF network used in the present study is presented in Figure 2. The transfer function (Radial Basis Function) for the hidden layer neurons is given by

$$f_h = e^{\left(-\sum_{k=1}^{2}(x_{kh}-c_{kh})^2/\sigma^2\right)} \qquad h=1...j \qquad (2.2)$$

where f_h represents the output of the neuron h in the hidden layer given the input x_{kh} from k th neuron in the input layer to the h th neuron in the hidden layer and j is the number of hidden layer neurons. Here c_{kh} represent the centers and σ indicates the spread of the Radial Basis neurons (units). Each output layer neuron (in the present case, only one) receives a weighted input from all the neurons in the hidden layer of the network.

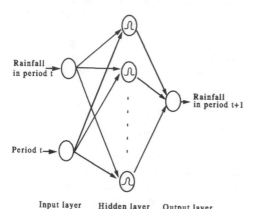

Fig. 2 Network architecture using Radial Basis Functions

The hidden layer neurons are completely connected to the neurons in the output layer. The value of the i th output neuron is given by

$$y_i = \sum_{m=1}^{i} W_{mi}(f_m) \qquad i = 1...n \qquad (2.3)$$

where n indicates the number of output layer neurons, j indicates the number of hidden layer neurons and W_{mi} represents the interconnection weight from the m th hidden layer neuron to the i th output layer neuron. In the present case, since there is only one output variable (or neuron) the value of n is equal to 1. The value, y_1 of the output layer neuron resulting from the transformations (Eq. 2.3) is infact the forecasted value of the rainfall in the next period. The learning algorithm of the RBF network essentially calculates the centers of the Radial Basis neurons and the interconnection weights between hidden and output layer. The learning algorithm is discussed by Chen et.al (1991).

3 TRAINING AND TESTING

Training comprises of presentation of input and output pairs to the network and obtaining the interconnection weights in case of BP network and the values of centers and weights in case of RBF network using a training algorithm. In case of Back Propagation, the network is first initialized by assigning arbitrary values in the interval of 0 and 1 to the interconnection weights and defining the transfer function parameters. Table 1 gives the network parameter values used in the Back Propagation architecture. The input and output pairs are normalized to fall within the interval of 0 and 1. Training involves two phases : a forward pass and a backward pass. Presenting inputs to the network and evaluating the output constitutes a forward pass. The difference between the actual output and the desired output is calculated and is termed as error. In the backward pass, this error is used to modify the interconnection weights by moving backwards. Presentation of complete training data sets once to the network is termed as one epoch. The network is presented with all the sets of input and output pairs till the desired level of performance is achieved on the test data which is not used in the training process. The training may require many epochs. The rate at which the weights are altered is decided by a term called the learning rate in the Back Propagation algorithm. Usually a network is trained first with a high value of learning rate and the rate is reduced as the training progresses. Also to improve the training time a momentum term which determines the weight adjustment is used. Range of the values of learning rate and momentum used in the present study are given in Table 1. Generally the network is presented with the input and output pairs till the error summed over all the training pairs is below an acceptable value set a priori. This value, called the system error, may be set to any constant value to give the desired network performance. Once an acceptable performance is achieved, training is stopped and the interconnection weights are stored. Using the interconnection weights and the transfer function for the hidden and output layers (Eq. 2.1) the output value for the known input values are calculated.

Table. 1 Back Propagation network parameters

Parameter	Value / range of values
Hidden layer neuron gain	0.01-0.1
Hidden layer neuron bias	0-1
Learning rate	0-1
Number of training sets	58
Momentum term	0-1

In case of RBF network the input and the output sets remain the same as those in Back Propagation. The training algorithm (Chen et.al 1991) needs the value of spread (σ) and the error value to be specified. Various values of σ are used for training and the one which results in the best performance on both training and test data is selected. The spread values used in the present study lie between 0 and 1. Once the training is complete the network stores the values of interconnection weights between hidden layer and output layer and the centers of Radial Basis Function units.

Actual rainfall for a period of 88 years related to Malaprabha command area, South India, is used for the present study. The ten daily Thiessen weighted values are used for training and testing. Two thirds of the data is used for the training and the rest is used for testing.

The training data is presented in the following manner. Starting from any year (an year comprising of 36 ten day periods), the two inputs current period's rainfall (R_t), and current time period (t) and the output, the next period's rainfall value (R_{t+1}) are presented to the network. For the inputs corresponding to last period (t=36), the output is equal to the next year's first period's rainfall. After the each presentation of input and output values the training algorithm is used to change the interconnection weights. This is continued till all the training data are exhausted. Similar procedure is followed for testing also.

4 RESULTS AND DISCUSSION

Neural network model using the Back Propagation architecture is developed using FORTRAN and was run on IBM RISC 6000 machine. The model using RBF architecture is developed using Neural Network Tool Box available under MATLAB 4.2c under IBM RISC 6000 environment. The use of the Tool Box facilitated the model development with only a moderate time investment. The RBF network uses the training algorithm proposed by Chen et.al (1991). Details of the selected BP network are given in the Table 2. Parameter values for RBF network are given in Table 3.

Table 2 Details of the BP network

Parameter	Value
Input layer neurons	2
First hidden layer neurons	13
Second hidden layer neurons	12
Output layer neurons	1
Hidden layers and output neuron gain	0.17
System Error	1.79

Table 3 Details of the RBF network

Parameter	Value
Input layer neurons	2
Hidden layer neurons	50
Output layer neurons	1
Spread (σ)	0.018
System Error	1.0

For testing, the developed models are presented with the input values of current period's rainfall and the time period to obtain the one step ahead forecast. The forecasted rainfall values obtained using both the network models are compared with the actual rainfall values. Figure 3 gives the comparison of actual and forecasted rainfall values using the BP network. Similar comparison is given in the Figure 4 with the values obtained from RBF network. Two different sets of data are used in figures 3 and 4. It can be concluded from the figures 3 and 4 that both the networks performed fairly well. The comparisons also indicate that the

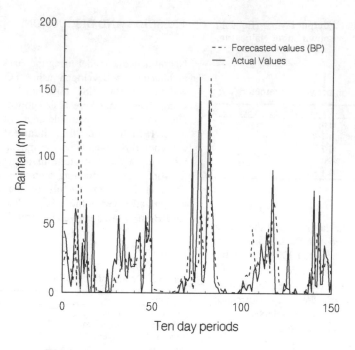

Fig. 3 Observed versus forecasted rainfalls by BP network

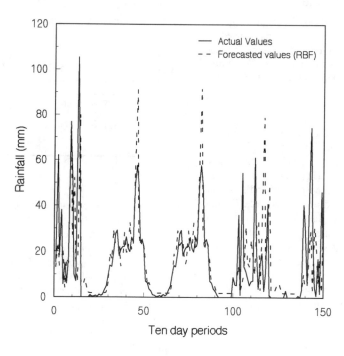

Fig. 4 Observed versus forecasted rainfalls by RBF network

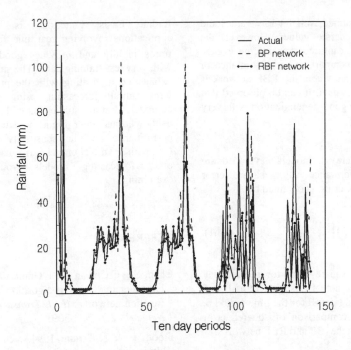

Fig. 5 Comparision of rainfalls forecasted by two networks

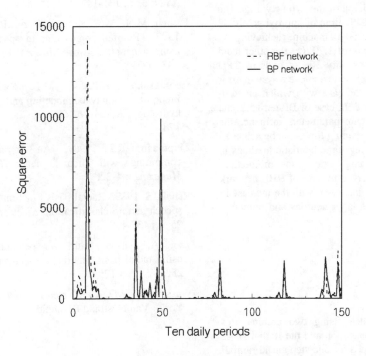

Fig. 6 Square errors of two networks

331

networks have performed well for low values and fairly good for moderate values. However, the networks have not captured the extreme events. Figure 5 gives the comparison of performance of RBF and BP networks where the RBF network is trained with very less data. It can be observed that the RBF has shown good generalization with very less data.

The two neural network models developed are compared for performance using square error criterion. The square error is obtained by

$$E(p) = [A(p) - O(p)]^2 \qquad (4.1)$$

Here E(p) is the square error for the ten day period p, A(p) is the actual observed rainfall and O(p) is the forecasted rainfall for the time period p. Figure 6 gives the comparison of squared error values resulting from the BP and RBF networks.

It is interesting to compare the computational time required for training of the two networks. The BP network took a CPU time of approximately 32 hours on IBM Risc 6000 machine achieving an acceptable system error of 1.79. On the other hand the RBF network took only 300 seconds of CPU time on the same machine to achieve a system error of 1. This indicates the ease with which an RBF network can be trained. In case of BP network, the learning and momentum parameters influence the training time. Lower training times can be achieved with BP networks using some heuristic methods in selecting the learning rate and momentum parameters. It is observed in case of RBF network that the training time increased with the increase in the number of hidden layer neurons and reduction in the system error.

5 CONCLUSIONS

Neural network models using two architectures namely the Back Propagation and the Radial Basis Functions are developed for one step ahead rainfall forecasting. The networks showed good predictive capabilities. Actual data available from a command area in South India is used. The architecture using

Radial Basis Functions is suggested for applications involving real-time forecasting as it trains rapidly and provides good generalization with very less training data. The study is in no way exhaustive in dealing with the problem of short term rainfall forecasting using neural network approach. Future studies should concentrate on using various other inputs to study the network performance and its sensitivity to the inputs. Comparison of NN model performance with that of other forecasting models would be worth examining.

REFERENCES

Chen, S., C.Cowan & P.Grant 1991. Orthogonal least squares learning algorithm for radial basis function networks. *IEEE Transactions on Neural Networks* 2,2 : 302-309.

Flood, I & N.Kartam 1994. Neural networks in civil engineering : Systems and applications. *Journal of Computing in Civil Engineering, ASCE* 8, 2 : 131-147.

French, M.N., W.F. Krajewski & R.R.Cuykendall 1992 . Rainfall forecasting in space and time using a neural network. *Journal of Hydrology.* 137 : 1-31.

Georgakakos, K.P. & M.L. Kavvas 1987. Precipitation analysis, modeling and prediction in hydrology. *Review of Geophysics.* 25(2) :163-178.

Lippmann, R.P. 1987. An introduction to computing with neural nets. *IEEE ASSP Magazine.* : 4-22.

Renals, S. 1989. Radial basis function network for speech pattern classification. *Electronics Letters,* 25, 7 : 437-439.

Tsoi, A.C. 1989. Multilayer perceptron trained using radial basis functions. *Electronics Letters,* 25, 19 : 1296-1297.

Wasserman, P.D. 1990. *Neural computing.* New York : Van Nostrand Reinhold.

Stochastic Hydraulics'96, Tickle, Goulter, Xu, Wasimi & Bouchart (eds) © 1996 Balkema, Rotterdam. ISBN 90 5410 817 7

Synthetic generation of missing run-off data for the White Nile, River Sobat and Bahr el-Jebel

M.Y. Elwan
Nile Water Sector, Giza, Egypt

Mamdouh Shahin & M.J. Hall
International Institute for Infrastructural, Hydraulic and Environmental Engineering, Delft, Netherlands

ABSTRACT: Owing to the security situation in the southern Sudan, streamflow data for certain key stations in the Nile River Basin have not been available since 1983. An investigation into the infilling of the missing data using statistical techniques has demonstrated the utility of·both multiple regression and ARIMA modelling for this purpose. The results obtained have indicated the importance of having one more stream gauging station in the Equatorial Lakes, upstream of the insecure area.

1 PHYSICAL SETTING

The River Nile is the source of life for Egypt and a large part of the Sudan. The Nile has two main sources of supply: the Equatorial Lakes Plateau (basins of lakes Victoria, Kioga, George, Edward and Albert, and of Rivers Kagera, Victoria, Albert, Bahr el Jebel and Bahr el Ghazal), and the Ethiopian Plateau (basins of Rivers Sobat, Blue Nile, Dinder, Rahad and Atbara).

The basins of Bahr el Jebel, Bahr el Ghazal and the Sobat are partly occupied by the Sudd (swampy) area and the Machar swamps respectively. Excessive amounts of water are lost in these swamps, causing a reduction in the flow of the White and the Blue Niles as well as the Main Nile.

2 STATEMENT OF PROBLEM AND AIM OF STUDY

Most of the flow measurements in the ·Nile Basin began around 1905. Collection and analysis of these data are essential for flood forecasting, management of existing storage reservoirs and planning future conservation schemes.

Since 1983, the flow measurements at the gauging stations in the basins of the Bahr el Jebel and the Sobat have ceased due to security reasons in southern Sudan. The present work, therefore, aims at firstly, investigating and presenting the main features of the annual and monthly historical flow data; and secondly, searching for the best approach for infilling the missing flow data.

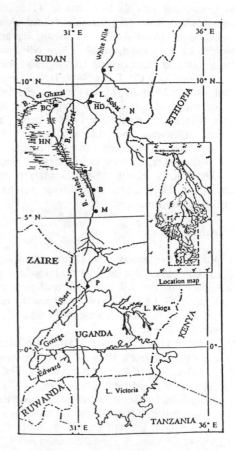

Figure 1. Map of the Nile Basin showing the locations of key stream gauging stations used in the present study.

Table 1. Basic statistics, trend and homogeneity of the annual flow series.

Basic Statistics	Mongalla on Bahr el-Jebel			Malakal on White Nile			Hillet Doleib on the Sobat		
	1905-1983	1905-1961	1962-1983	1905-1983	1905-1961	1962-1983	1905-1983	1905-1961	1962-1983
period, yr	79	79	22	79	57	22	79	57	22
$\bar{x}$, 10^9 m³/yr	32.98	26.59	49.53	29.62	27.64	34.76	13.68	13.47	14.22
C_v	0.38	0.27	0.13	0.18	0.13	0.15	0.19	0.19	0.20
C_s	0.73	1.70	0.56	1.59	2.31	1.56	1.10	1.56	0.25
X_{mx}, 10^9 m³/yr	64.00	55.80	64.00	48.60	44.40	48.60	23.10	23.10	19.40
X_{mn}, 10^9 m³/yr	15.30	15.30	39.10	22.60	22.60	28.20	8.29	9.10	8.29
Trend	Yes	Yes	Yes	Yes	Yes	Yes	No	No	Yes
consistent mean	No	No	No	No	No	No	Yes	Yes	No
consistent variance	No	No	Yes	Yes	No	No	Yes	No	Yes

3 STREAMFLOW DATA

The annual and monthly flow data at 11 stream gauging sites have been examined. Of these stations, three (Jebel Aulia, JA; Melut, T; and Malakal, L) are located on the White Nile, two (Nasser, N; and Hillet Doleib, HD) on the Sobat, five (Buffalo Cape, CP; Hillet Nuer, HN; Jonglei, J; Bor, B; and Mongalla, M) on the Bahr el Jebel, and one station (Pakwach or Panyango, P) on the Albert Nile near the outlet of Lake Albert. The different tributaries and the location of the respective gauging stations are shown in Figure 1.

The period of record used varies from 23 years for Pakwach to 79 years for Mongalla, Malakal and Hillet Doleib. The last three stations were therefore regarded as the primary stations.

3.1 Analysis of data

The plots of the annual flow series at the key stations show clearly that the flow at Mongalla and, to a less extent, that at Malakal suffered from an abrupt rise in 1961-62, followed by alternating fall and rise, though to a smaller scale, until the end of the series (see Figure 2).

This state of affairs prompted separate statistical analysis of the full series, 1905-1983, the pre-jump series, 1905-1961, and the post-jump series, 1962-1983 (Shahin 1985 and 1990; Elwan 1995).

The analysis included compilation and comparison of the basic statistics for each time series (mean $\bar{x}$; standard deviation, s; coefficient of variation, C_v; coefficient of skewness, C_s; and the maximum, X_{mx}, and minimum, X_{mn}, values); application of the Spearman Rank Correlation test for absence of trend; and split-record tests on the consistency and

homogeneity of means and variances of each time series divided into both two and three subsets of equal size. Both the Spearman and the split-record tests were carried out using the DATSCR package (Dahmen and Hall, 1990). The results obtained are summarised in Table 1.

In addition, the independence of each time series was evaluated by applying the Anderson (1942) test to the serial correlation coefficients, again using the facilities of the DATSCR package. As an example, Figure 3 shows the correlogram for the 79 years of annual flows at Mongalla along with the 95 per cent confidence limits by the Anderson test. The strong serial dependence of these data is well illustrated.

Further analyses were carried out of the cross correlations between the time series at selected sites. For example, Table 2 shows the correlation coefficients between the annual flows at Malakal and the remaining sites.

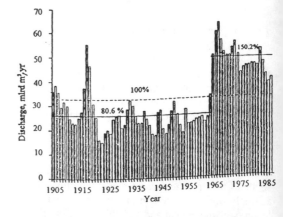

Figure 2. The annual flow series at Mongalla on the Bahr el-Jebel for the period 1905-1983.

Table 2. Correlation coefficient of the annual flow between Malakal and the remaining stations (Elwan 1995).

Period of record at Malakal, yr	T	JA	HD	N	M	B	J	HN	BC
1905-1983	0.98	0.97	0.68	0.68	0.84	0.68	0.39	0.77	0.72
(79)	(36)*	(35)	(79)	(29)	(79)	(41)	(39)	(34)	(34)
1905-1961	0.95	0.93	0.93	0.86	0.74	0.66	0.33	0.27	0.04
(57)	(14)	(13)	(57)	(24)	(57)	(35)	(29)	(25)	(25)
1962-1983	0.97	0.96	0.50	0.19	0.87	0.86	0.33	0.19	0.31
(22)	(22)	(22)	(22)	(5)	(22)	(6)	(10)	(9)	(9)

* Figures in parentheses are the number of years in common between Malakal and the other stations.

Of special importance is the finding (not shown in Table 2) that the available records at Pakwach for the period 1948-1970 are very strongly correlated with the annual flows at Mongalla for the same period. The coefficient of determination, R^2, was 0.96.

The cross correlations between the monthly flows at the different stations was also computed in order to determine the degree of dependence between the flows at the different sites. An illustration of the cross correlogram of the monthly flow between Mongalla station and both Malakal on the White Nile and Hillet Doleib on the Sobat is shown in Figure 4. Both stations appear to be lagging one to two months behind Mongalla, the station above the swamps.

3.2 *Generating and infilling missing data*

Table 1 has shown that the majority of the records in the Nile River Basin are neither stationary nor consistent and homogenous. The reasons for this behaviour have been well-documented elsewhere (Piper et al, 1986; Kite, 1981; 1982). The water level of the equatorial lakes in the Nile Basin rose

2-2.5 m between 1961 and 1964. Following a slight fall it began to rise again in 1978 and has yet to fall back to the pre-1961 level. As a result the average outflow of Lake Victoria jumped from 20.8×10^9 m³/yr for 1900-1960 to 41.2×10^9 m³/yr for 1962-1978. The corresponding jump of the river flow is clearly shown in Figure 2. The rise in water levels and river flows in the Upper Nile system can be attributed to the increase of rainfall in the wet years, and the subsequent improvement of the runoff coefficient due to the reduction of initial losses. The 1961-1964 rise is not unique, and similar fluctuations have occurred in the past. The available level and discharge series show that, except for a few short periods of 10-20 years each, the river flow was never "at rest".

Since the available algorithms for data infilling are strictly only applicable to stationary, consistent and homogeneous series, the subsequent analyses of annual flows was confined to their first differences.

In addition, the monthly flow series were standardised using the observed monthly means and

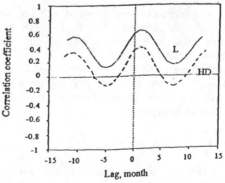

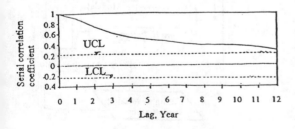

Figure 3. The serial correlogram of the annual flow series at Mongalla station.

Figure 4. Cross correlation between the mean monthly flow of Mongalla and Malakal and of Mongalla and Hillet Doleib.

335

standard deviations for the period of record prior to analysis. Transformations were also applied in order to normalise the data sets. Two basic data infilling algorithms were investigated, involving stochastic modelling and multiple linear regression analysis, and a flowchart of the procedure is presented in Figure 5.

The stochastic models applied to the differenced series are: AR (p), MA (q), ARMA (p, q) and ARIMA (p', d, q'), where p = 1, 2 and 3, q = 1, 2 and 3, and p', d and q' = 1 and 2.

The regression models used for generating annual flows belong to the linear (2 parameter), 2nd order polynomial (3 parameter), and the logarithmic (2-parameter) types of models. The missing monthly flows were generated using the 'best set' multiple regression type, which contained up to a maximum of 13 parameters, as given by

$$\hat{Y} = \beta_o + \beta_1 X_1 + \beta_2 X_2 + ... + \beta_j X_j + ... + \beta_{12} X_{12} ...$$
$$j = 1, 2, ..., 12 \qquad (1)$$

where $\hat{Y}$ is the response variable, Xj is the predictor variable and βj's are the model parameter. A popular algorithm for selecting the best subsets of the predictor variables, which has been used in the present study, has been developed by Furnival and Wilson (1974).

Other than filling in the missing monthly flow series directly, data can be generated indirectly from the generated annual flows using either the monthly pattern of flow distribution (using historical data) or through the application of disaggregation models.

3.3 Estimation of parameters and testing goodness-of-fit

Model parameters can be estimated by means of one or more of three methods. These are the method of moments, the least squares method and the maximum likelihood method.

The goodness of fit of the stochastic models was based upon the sum of square (SS) and the mean square (MS) of the residuals for each of the models used. The SS is expressed as

$$SS = \sum_{i=1}^{N-d} (y_i - \hat{y_i})^2 \qquad (2)$$

where d is the number of differences and N is the number of observations.

The adequacy of the fit of autoregressive-moving average models to series of observed discharges was tested using the Modified Box-Pierce Portmanteau

lack-of-fit test (Ljung and Box, 1978). This test gives the statistic $\bar{Q}(f)$ as

$$\bar{Q}(f) = N(N+2) \sum_{k=1}^{L} \hat{r}^2(k)/(N-k) \qquad (3)$$

where $\hat{r}(k)$ is the serial correlation coefficient of the residuals left after fitting the model to the series at lag k, and L is the maximum lag. The quantity $\bar{Q}(f)$ for large N is distributed as Chi-square with L-m degrees of freedom, where m is the number of parameters.

The goodness-of-fit of the multiple-regression models was based upon the adjusted coefficient of determination, adjR², to be computed from

$$adj\ R^2 = 1 - \left[1 - \left(\sum_{i=1}^{N} (\hat{y_i} - \bar{y})^2 / \sum_{i=1}^{N} (y_i - \bar{y})^2 \right) \right] (N-1)/(N-m) \qquad (4)$$

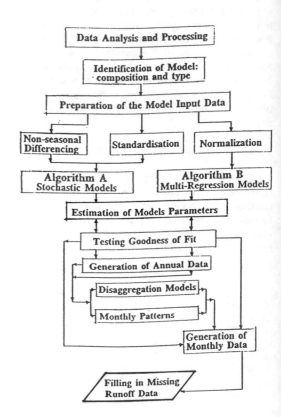

Figure 5. Generation and filling in missing annual and monthly runoff data using stochastic and multi-regression models.

336

Table 3. The results obtained from the ARMA (1, 2) model applied to the first runoff differences.

| Station | Parameters | | | Constant | Box-Pierce Test |
| | AR | MA | | | |
	p_1	q_1	q_2		
Mongalla	0.770	0.489	0.483	0.056	OK
Hillet Doleib	-0.177	0.425	0.533	-0.004	OK
Malakal	0.511	0.472	0.581	0.047	OK
Melut	0.718	0.575	0.530	0.021	OK

4 RESULTS AND DISCUSSION

4.1 Generated annual flows

4.1.1 Stochastic models - The models that gave the least MS of the residuals for the first runoff differences were ARMA (1, 3) for Mongalla and Hillet Doleib stations and ARMA (1, 2) for Malakal and Melut stations. Since the MS for ARMA (1, 2) for the first pair is only slightly larger than the MS given by the corresponding ARMA (1, 3), the former model can conveniently be adopted for all four stations. The results obtained are given in Table 3.

4.1.2 Regression models - The different physiographic and hydrologic characteristics of the study area justify its subdivision into four sub-catchments for each of which a model can be established and calibrated.

The first sub-catchment extends from Pakwach on Albert Nile just downstream of the exit to Mongalla. Since the distance between the two stations is short and the slope is steep, no delay has been taken into account in modelling this reach.

The second sub-catchment extends from Mongalla to Malakal. The approach that was used here was based on the simple regression relation between the two stations. Another approach was investigated, based on computing the difference between the flow at Malakal, Q_L (t), and the flow at Hillet Doleib, Q_{HD} (t) and denoting the difference, i.e. the flow below the swamps, by Q_s (t). For modelling this reach the net losses in the swamps were regressed on the flow at Mongalla. As such, the net losses can be written as

$$NL(t) = Q_M(t) - Q_s(t) = Q_M(t) - Q_L(t) + Q_{HD}(t) \qquad (5)$$

The net losses were also regressed on the flow at

Malakal. Unfortunately, the introduction of a time delay to allow for the residence of water in the swamps did not yield any substantial improvement in fitting the regression line.

The third and fourth models were regression relations between the flow at Hillet Doleib and Malakal and between the flow at Malakal and Melut respectively. This division agrees with the division of the Nile system into sub-catchments for elaborating the constraint linear system model adopted by Fahmy et al (1982).

The second-order polynomial regression proved to be the most adequate type of regression model for the annual flow. The adj R^2-values for the different reaches were: 0.954 for P-M, 0.739 for M-L, 0.636 for NL-L, 0.952 for M-NL, 0.475 for HD-L, and 0.984 for L-T.

4.2 Generated monthly flows

4.2.1 Stochastic models - The results obtained from the application of the stochastic models to the standardised monthly flow series showed that the ARMA (3, 1), (1, 2), (1, 3), (2, 2), (2, 3) and (3, 2) models give the least MS of the residuals compared to other models. However, none of them passed the Box-Pierce test except the flow series at Melut and Hillet Doleib. At the other two stations, Mongalla and Malakal, ARIMA (2, 1, 2) gave the second smallest MS of the residuals and in addition passed the Box-Pierce test (see Table 4).

4.2.2 Monthly distribution patterns and disaggregation models - The annual data generated from stochastic or regression models can be distributed between the months of the year using the distribution patterns obtained from the historical data. These are included in Table 5.

The disaggregation model developed by Lin (1990) was applied to the log-transformed data of the three stations: Pakwach, Hillet Doleib and Melut. The average value of the generated flows for each month divided by the sum of the twelve months gives another distribution pattern. A comparison between this pattern and the one obtained from the historical data can be seen in Table 5. Underlined figures in parentheses are the ratios obtained from the disaggregation model, and the remaining figures are those obtained from historical data. The ratios obtained from the disaggregation models, except from February to May for Hillet Doleib, are shown to be in good-to-excellent agreement with the corresponding ratios obtained from the historical record.

4.2.3 Regression models - The missing streamflows were predicted using multiple-regression models reach-wise: Pakwach from Mongalla, P-M reach; Mongalla from Malakal, M-L reach; Hillet Doleib from Malakal, HD-L reach and Melut from Malakal, M-L reach. The model in its general form is already expressed by Equation 1. Although models based upon this equation underestimate the variance of the dependent variable (Matalas and Jacobs, 1964) and introduce spurious correlations (Valadares Tavares, 1977), equations with high coefficients of determination can sometimes be identified that have some utility.

The results obtained from the multiple-regression analysis show that the number of B_j parameters needs not to be 12. In fact, the 48 models (4 stations x 12 months) have parameters between 2 and 9. Additionally there is no similarity between any two models, even for the same station. The models for the four sub-catchments taking the month of January as an example are:

$$P\text{-}M : \quad \hat{Y} = -0.52 + 1.75X_1 - 1.44X_2 + \\ + 0.72X_5 - 0.51X_6 + 0.43X_7$$

$$M\text{-}L : \quad \hat{Y} = -1.14 - 0.92X_2 + 1.27X_3 + \\ + 1.62X_4 + 0.69X_5 - 1.54X_9 + 1.30X_{11}$$

$$HD\text{-}L : \quad \hat{Y} = 0.17 + 0.98X_1 - 0.54X_4 + 0.52X_5 + \\ + 1.09X_6 - 0.86X_7$$

$$L\text{-}T : \quad \hat{Y} = -0.34 + 1.09X_1 - 0.43X_2 + \\ + 0.71X_{11} - 0.41X_{12}$$

$$(6)$$

The irregularity of the multiple-regression pattern extends to all other months of the year.

The goodness of fit of these models was judged by the adjusted coefficient of determination, adj R^2, as already given by Equation 4. The results obtained can be summarized as follows:

Reach	adj R^2_{min}	adj R^2_{max}	adj R^2_{mean}
P-M	0.919	0.965	0.954
M-L	0.641	0.795	0.734
HD-L	0.362	0.838	0.570
L-T	0.892	0.977	0.943

The lowest coefficients of determination are for the reach HD-L. This result was to be expected since the hydrological characteristics of the River Sobat, on which HD is located, and the White Nile, on which Malakal is situated are not similar. The

Table 4. The results obtained from the stochastic models ARIMA (2, 1, 2) and ARMA (1, 2) applied to the standardised monthly flows.

Station	Parameters				Constant	Box-Pierce test
	AR		MA			
	p_1	p_2	q_1	q_2		
Mongalla	0.193	0.285	0.248	0.424	0.0001	OK
Malakal	0.513	0.232	0.391	0.505	0.0001	OK
Melut	0.955		-0.166	-0.0012	-0.0012	OK
Hillet Doleib	0.760		-0.332	-0.0007	-0.0007	OK

Table 5. Monthly distribution patterns of stream flow as obtained from historical data and from the disaggregation models.

Sta-tion	Average flow in a month divided by the sum of monthly flows											
	Jan.	Feb.	Mar.	Apr.	May	Jun.	Jul.	Aug.	Sep.	Oct.	Nov.	Dec
P	.086	.075	.083	.080	.081	.079	.082	.084	.084	.087	.088	.091
	.099	.062	.076	.085	.079	.087	.082	.082	.096	.085	.079	.090
M	.077	.066	.070	.071	.083	.079	.087	.099	.098	.098	.089	.083
HD	.072	.033	.022	.019	.030	.064	.096	.118	.130	.146	.144	.126
	.077	.042	.035	.035	.035	.063	.092	.113	.127	.134	.127	.120
M	.083	.059	.056	.051	.057	.069	.086	.098	.105	.116	.113	.107
T	.093	.067	.062	.055	.057	.065	.080	.092	.099	.111	.110	.109
	.098	.071	.064	.052	.055	.067	.080	.092	.098	.110	.107	.104

Underlined figures were obtained from disaggregration models.

moderate to moderately high values of adj R² for the M-L reach can be attributed to the excessive losses in the Sudd area (swamps) separating these two stations.

5 CONCLUSIONS AND RECOMMENDATIONS

5.1 *Conclusions*

- Statistical analyses of the available historical data have shown that the streamflow series at all stations considered, except for HD on the Sobat are neither stationary nor homogeneous.
- The advantage of using stochastic models for data generation is that they can help produce synthetic series of any required length. The generated values, however, cannot be used in place of the missing data. For the non-stationary series the ARIMA model (2, 1, 2) produced satisfactory results, whereas the ARMA (1, 2) proved to be good enough for the stationary series at HD.
- The different characteristics of the study area justify its subdivision into subcatchments to each of which a certain multiple-regression model can be established and calibrated. The multiple regression models have proved to be the models most suitable for infilling the monthly flow data in these sub-catchments. Furthermore, among these models the second-degree polynomial was found to provide the best fit.

5.2 *Recommendations*

- The station at Malakal on the White Nile is the only station with a complete record. These data are strongly recommended for use in filling in the missing annual as well as monthly flows at Melut the same river. Malakal also produces acceptable results regarding the missing data at Hillet Doleib on the Sobat, but not for the stations on the Bahr el Jebel, and therefore cannot be recommended for this purpose.
- The only station that helps fill in the missing data at Mongalla and probably other stations in the Bahr el Jebel subcatchment is Pakwach. As Pakwach is a local station in Uganda it would be worthwhile to take up with the administration of Uganda the matter of upgrading the status of this station to serve the interests of the hydrology of the Nile Basin.

REFERENCES

Anderson, R.L. 1942. Distribution of serial correlation coefficient, *Ann. Math. Stat.* 13:1-13.

Dahmen, E.R. & Hall, M.J. 1990. Screening of hydrological data: tests for stationarity and relative consistency, *ILRI Pub. 49*. Wageningen: ILRI.

Elwan, M.Y., 1995. Synthetic generation of missing runoff data of the White Nile, River Sobat and Bahr el Jebel, *M.Sc. Thesis*. Delft: IHE.

Fahmy, A., Panattoni, L. & Todini, E. 1982. A mathematical model of the River Nile. In: Engineering applications of computational hydraulics (editors: M.B. Abbott & A.J. Cunge), 1:111-130, London: Pitman.

Furnival, G.M. & Wilson, R.W. 1974. Regression by leaps and bounds. *Technometrics* 16: 499-511.

Kite, G.W. 1981. Recent changes in level of Lake Victoria. *Bul. Hydro. Sci.* 26:233-243.

Kite, G.W. 1982. Analysis of Lake Victoria levels.. *Bul. Hydro. Sci.* 27:99-110.

Lin, G.F. 1990. Parameter estimation for seasonal to subseasonal disaggregation. *J. Hydro.* 120:65-77.

Matalas, N.C. & Jacobs, B. 1964. A correlation procedure for augmenting hydrologic data, *US Geol. Survey Prof. Pap.* 434-E.

Piper, B.S., Plinston, D.T. & Sutcliffe, J.V. 1986. The water balance of Lake Victoria. *Bul. Hydro. Sci.* 31:25-37.

Shahin, M.M.A. 1985. *Hydrology of the Nile Basin*. Development in Water Sciences, 21:575. Amsterdam, Elsevier.

Shahin, M.M.A. 1990. Annual flow variation in the Nile River system. *Proc. Int. Symp. on Hy. and Hydro. of Arid Zones (organized by ASCE)*:7-13. San Diego.

Valadares Tavares, D. 1977. A reconsideration and a generalisation of Lawrance and Fiering two-station models. *J. Hydro.* 33:185-190.

Stochastic Hydraulics '96, Tickle, Goulter, Xu, Wasimi & Bouchart (eds) © 1996 Balkema, Rotterdam. ISBN 90 5410 817 7

Numerical solutions of stochastic differential equations in rainfall-runoff modeling

Gwo-Fong Lin & Yu-Ming Wang
Department of Civil Engineering and Hydraulic Research Laboratory, National Taiwan, China University, Taipei, Taiwan, China

ABSTRACT: Rainfall-runoff processes are mathematically described by stochastic equations because of their stochastic nature. In this paper, the theory of stochastic differential equations (SDEs) is used to study a nonlinear hydrologic system with a stochastic input. A SDE model having the form of Itô equation is developed. The differential equations for moments of the solution process are derived. The probability density function of the solution process in steady state is also derived by the Fokker-Planck equation. Furthermore, the stochastic Euler scheme is used to approximate the solution process. The accuracy of the numerical scheme is examined through examples. Applications to a linear and a nonlinear hydrologic systems are also provided to demonstrate the applicability of the numerical scheme. The proposed methodology is particularly useful when the solution process of the SDE model does not have an explcit form.

1 INTRODUCTION

Rainfall-runoff processes occurring over the watershed are stochastic in nature and, hence, mathematical descriptions of the processes result in stochastic equations. From the stochastic viewpoint, the rainfall-runoff simulation for a hydrologic system is a mathematical problem of solving a stochastic differential equation (SDE) which is a differential equation containing stochastic processes or random variables. Unny and Karmeshu (1984), Unny (1984), Bodo and Unny (1987), and Unny (1987) employed the theory of stochastic differential equations (SDEs) in stochastic rainfall-runoff studies. In their studies, the SDEs have the form of Itô equation, and the corresponding solution processes have no explicit forms. The lack of explicit solution prevents one from determining the realizations (sample functions) of the solution processes immediately. Recently, the development of the numerical solutions of SDEs (Kloeden and Platen 1992; Kloeden et al. 1994) makes the determination of realizations possible when the solution process does not have an explicit form.

In this paper, the watershed is taken as a nonlinear hydrologic system and the input is considered to be stochastic. Then the hydrologic system model is developed using the theory of SDEs.

The developed stochastic model has the form of Itô SDE, and the solution process is Markovian. The differential equations for moments of the solution process are derived. The probability density function of the solution process in steady state is also derived using the Fokker-Planck equation. Furthermore, the stochastic Euler scheme is investigated and applied to approximate the solution process. Examples are given to demonstrate the applicability and accuracy of the numerical scheme.

2 STOCHASTIC MODEL

For a hydrologic system, the continuity equation is written as

$$\frac{dS(t)}{dt} = I(t) - Q(t) \tag{1}$$

where $I(t)$ is the input to the system and $Q(t)$ is the output from the system, and $S(t)$ is the storage. We assume that the storage $S(t)$ and the output (outflow) $Q(t)$ are related by the following nonlinear equation

$$Q(t) = aS(t)^b \tag{2}$$

where a is a coefficient and b is an exponent.

Consider that the input $I(t)$ is a stochastic processes. Furthermore, we express stochastic process $I(t)$ in terms of its mean and a zero-mean perturbation, i.e.

$$I(t) = \bar{I}(t) + I'(t) \qquad (3)$$

where $\bar{I}(t)$ denotes the mean of the stochastic process $I(t)$ and $I'(t)$ refers to the perturbation (the zero-mean process). Substituting (2) and (3) into (1) gives

$$\frac{dQ(t)}{dt} = CQ(t)^{1-1/b} \left(\bar{I}(t) - Q(t) + I'(t) \right) \qquad (4)$$

where $C = a^{1/b} b$.

The $I'(t)$ in (4) is often regarded as a Gaussian white noise. Because the formal derivative of a Wiener process has the same statistical properties of a white Gaussian noise (Soong 1973), we can further write $d\beta(t) = I'(t)dt$, where $\{\beta(t), 0 \leq t \leq T\}$ is the Wiener process. Hence, (4) can be rewritten as

$$dQ(t) = f(Q(t), t) \, dt + g(Q(t), t) \, d\beta(t) \qquad (5)$$

where

$$f(Q(t), t) = C\bar{I}(t)Q(t)^{1-1/b} - CQ(t)^{2-1/b} \qquad (6)$$

and

$$g(Q(t), t) = CQ(t)^{1-1/b} \qquad (7)$$

Theoretically, the solution of (5) can be written as

$$Q(t) = Q_0 + \int_0^t f(Q(\tau), \tau) \, d\tau$$
$$+ \int_0^t g(Q(\tau), \tau) \, d\beta(\tau) \qquad (8)$$

The first integral on the right-hand side of (8) is a mean square Riemann integral and the second integral is a stochastic integral. In general, two ways are often used to evaluate the stochastic integral. One is the Itô integral and the other is the Stratonovich integral (Jazwinski 1970).

In reality, white noise processes are idealizations of real colored noise processes. In such cases, the appropriate interpretation of a stochastic differential equation is in the Stratonovich sense (Wong and Zakai 1965). For the hydrologic system considered herein, the stochastic process I' is not a pure white noise process, but a colored noise processes. Thus, the Stratonovich interpretation of the SDE of (5) is appropriate. That is, the stochastic integral in (8) should be considered as the symmetrically defined Stratonovich integral.

For convenience in analysis and simulation, the Stratonovich SDE can be converted into an equivalent Itô SDE (Jazwinski 1970; Soong 1973). The equivalent Itô SDE of (5) is derived as

$$dQ(t) = \left(f + \frac{1}{2} g \frac{\partial g}{\partial Q} \right) dt + g \, d\beta(t) \qquad (9)$$

which is the proposed SDE model for the nonlinear hydrologic system considered herein.

2.1 Equations for moments of the solution process

Eq. (9) is the nonlinear Itô SDE. The solution of (9) is expected to be nonlinear and time-varying. It is difficult to obtain the exact explicit solution of (9). Using the Taylor expansions, the general equation for moments of the solution process can be derived as

$$\frac{dm_k(t)}{dt} = E \left[\left(f + \frac{1}{2} g \frac{\partial g}{\partial Q} \right) kQ(t)^{k-1} \right]$$
$$+ E \left[\frac{1}{2} g^2 k(k-1) Q(t)^{k-2} \right]$$
$$k = 1, 2, \ldots \qquad (10)$$

where $m_k(t) = E[Q(t)^k]$, i.e. the kth moment of $Q(t)$. From (10), equations for the first and second moments are obtained as

$$\frac{dm_1(t)}{dt} = E \left[\left(f + \frac{1}{2} g \frac{\partial g}{\partial Q} \right) \right] \qquad (11)$$

and

$$\frac{dm_2(t)}{dt} = 2E \left[\left(f + \frac{1}{2} g \frac{\partial g}{\partial Q} \right) Q(t) \right] + E[g^2] \qquad (12)$$

Equations for higher moments of the solution process can be obtained in a like manner.

2.2 Probability density function of the solution process

Eq. (9) is the Itô stochastic differential equation whose solution is a Markov process. The evolution of the transition probability density of the Markov process generated by the the Itô equation can be described by the the Fokker-Planck equation or the Kolmogorov forward equation (Jazwinski 1970). The Fokker-Planck equation associated with the Itô equation of (9) is written as

$$\frac{\partial p}{\partial t} = -\frac{\partial}{\partial Q}\left[p\left(f + \frac{1}{2}g\frac{\partial g}{\partial Q}\right)\right] + \frac{1}{2}\frac{\partial^2}{\partial Q^2}(pg^2)$$

$$(13)$$

where $p = p(Q,t \mid Q_0,t_0)$, the probability density of the solution process $Q(t)$. From (13), the probability density function of $Q(t)$ can be found. The solution to (13) is subject to the initial condition

$$p(Q,t|Q_0,t_0) = \delta(Q - Q_0) \qquad (14)$$

where $\delta(Q - Q_0)$ is the Dirac delta function and the boundary conditions

$$\lim_{Q\to\pm\infty} p(Q,t) = 0 \qquad (15)$$

$$\lim_{Q\to\pm\infty} \frac{\partial p(Q,t)}{\partial Q} = 0 \qquad (16)$$

In addition, the solution is required to satisfy the normalization condition

$$\int_{Q_{min}}^{Q_{max}} p(Q,t)dQ = 1 \qquad (17)$$

In general, the Fokker-Planck equation is very difficult to solve. Only in very special cases, an explicit time-dependent solution can be found. In the application of the Fokker-Planck equation to non-linear problems, a less ambitious goal is the determination of the stationary solution (Soong 1973). Setting $\partial p/\partial t$ in (13) equal to zero, we can derive the stationary solution

$$p(Q) = \frac{C_0}{B(Q)}\exp\left[2\int_{Q'}^{Q}\frac{A(\xi)}{B(\xi)}d\xi\right] \qquad (18)$$

where C_0 is the integration constant; $A(Q) = f + \frac{1}{2}g(\partial g/\partial Q)$; $B(Q) = g^2$; and Q' is the integration lower limit. The stationary probability density function is independent of t.

3 NUMERICAL SOLUTION

The numerical method used herein for solving (9) is the stochastic Euler scheme which is based on the stochastic Taylor expansion (Platen and Wagner 1982). Let $\{Y(t), t = t_0, t_1, \ldots, t_m\}$ be the numerical approximations of the solution $\{Q(t), t = t_0, t_1, \ldots, t_m\}$. For a given discretization $t_0 = \tau_0 < \tau_1 < \ldots < \tau_i < \ldots < \tau_n = t_1$, of the

time interval $[t_0, t_1]$, the Euler scheme for the Itô equation of (9) is

$$Y_{i+1} = Y_i + \left[f(Y_i) + \frac{1}{2}g(Y_i)\frac{\partial g}{\partial Y}(Y_i)\right]\Delta_i$$
$$+ g(Y_i)\Delta\beta_i$$
$$i = 0,1,2,\ldots,n-1 \qquad (19)$$

with the initial value $Y_0 = Q(t_0)$. In (19), $Y_i = Y(\tau_i)$; $\Delta_i = \tau_{i+1} - \tau_i$ refers to the step size (time increment); and $\Delta\beta_i = \beta_{\tau_{i+1}} - \beta_{\tau_i}$ denotes the increment of the Wiener process on the interval $[\tau_i, \tau_{i+1}]$. $\Delta\beta_i$ is a $N(0,\Delta_i)$ Gaussian process. It should be noted that $Y_n = Y(t_1)$. Based on (19), one can obtain a set of realizations of the process $Y(t)$ through the simulation of $\Delta\beta_i$. Then, the statistical properties of the solution process $Q(t)$ can be estimated from realizations of $Y(t)$.

4 ILLUSTRATIVE EXAMPLES

Four illustrative examples have been performed to test the accuracy of the stochastic Euler scheme described above. Due to the limitation of space, we present here only one nonlinear example in which the Itô equation is of the form

$$dQ(t) = dt + 2\sqrt{Q(t)}d\beta(t) \qquad (20)$$

with the initial value $Q(0) = 100$. The explicit solution process of (20) is

$$Q(t) = (\beta(t) + \sqrt{Q(t_0)})^2 \qquad (21)$$

The value at $t = 1$ is to be calculated using the stochastic Euler scheme. In the numerical experiments, we have tried 5 different numbers of subintervals on the time interval $[0, 1]$, namely $n = 10$, 20, 50, 100, and 200 or, equivalently, step size $\Delta = 0.1$, 0.05, 0.02, 0.01, and 0.005. In addition, we consider four different numbers of batches of simulations, namely, $M = 10$, 20, 50, and 100. In each batch, 1000 trajectories are simulated, i.e., $N = 1000$. In total there are 20 combinations of n and M in the experiments. We use the absolute error

$$\epsilon = E[\,|\,Q(1) - Y(1)\,|\,] \qquad (22)$$

as a measure of the pathwise closeness at the end of the time interval $[0, 1]$ (Kloeden and Platen 1992).

From the 1000 trajectories in a batch, an estimate of ϵ, denoted as $\hat{\epsilon}$, can be obtained. Subsequently, M batches of 1000 trajectories give M

Table 1. Summary of the error analysis.

M	n	$\widetilde{\widehat{\epsilon}}$	$\sigma^2_{\widehat{\epsilon}}$	$\delta\widehat{\epsilon}$
10	10	0.3437	0.0013	0.0196
10	20	0.2491	0.0009	0.0165
10	50	0.1561	0.0001	0.0058
10	100	0.1104	0.0000	0.0029
10	200	0.0797	0.0000	0.0026
20	10	0.3549	0.0010	0.0124
20	20	0.2532	0.0006	0.0096
20	50	0.1565	0.0001	0.0041
20	100	0.1106	0.0000	0.0027
20	200	0.0793	0.0000	0.0023
50	10	0.3553	0.0007	0.0065
50	20	0.2527	0.0006	0.0058
50	50	0.1566	0.0001	0.0024
50	100	0.1113	0.0000	0.0017
50	200	0.0802	0.0000	0.0016
100	10	0.3556	0.0007	0.0047
100	20	0.2528	0.0003	0.0032
100	50	0.1568	0.0001	0.0019
100	100	0.1113	0.0000	0.0012
100	200	0.0824	0.0000	0.0011

estimates of ϵ from which error analyses are performed. The results of the error analyses are summarized in Table 1 in which $\widetilde{\widehat{\epsilon}}$, $\sigma^2_{\widehat{\epsilon}}$ and $\delta\widehat{\epsilon}$ refer to the estimated mean, variance and half length of 90% confidence interval of the absolute error, respectively. The 90% confidence interval is $[\widetilde{\widehat{\epsilon}} - \delta\widehat{\epsilon}, \widetilde{\widehat{\epsilon}} + \delta\widehat{\epsilon}]$. Fig. 1 shows that the confidence interval of ϵ decreases as the number of batches M increases. Fig. 2 indicates that ϵ increases with increasing time increment Δ. The line in Fig. 2 has a slope of about 0.5 which means that the stochastic Euler scheme has a 0.5 order of strong convergence.

In what follows, we consider two examples regarding the hydrologic system. First we assume $b = 1$ and $a = 0.2$ in (2), i.e., the hydrologic system is linear. The input process I has a mean function $\bar{I}(t) = 10$ mm/hr for $t > 0$ and a variance function $D(t) = 20\Delta$ mm^2/hr^2 for $t > 0$. In addition, we use $n = 20$, 50 and 100 (equivalently, $\Delta = 0.05$, 0.02 and 0.01) on the interval $[t, t+1]$ and the initial value $Q(0) = 1.0$. For each case of n, we simulate 1000 trajectories ($N = 1000$) from which the simulated mean and second moment of the output (outflow) can be calculated. On the other hand, from (11) and (12), the equations for the first and second moments for this example are

written as

$$\frac{dm_1(t)}{dt} = aI(t) - am_1(t) \qquad (23)$$

and

$$\frac{dm_2(t)}{dt} = 2aI(t)m_1(t) - 2am_2(t) + a^2 \qquad (24)$$

From (23) and (24), we can obtain the first and second moments which are referred to as "theoretical" moments hereafter. Fig. 3 shows a realization of the outflow process for the case $n = 100$ as well as the theoretical mean. Comparison of the mean outflow hydrographs for various values of n is shown in Fig. 4. Probably because the time increment for the three cases studied herein is small enough, the mean outflow hydrographs are all very close to the theoretical mean as shown in Fig. 4. Fig. 5 shows that the simulated second moment of the outflow for the case $n = 100$ also agrees well with the theoretical second moment. Comparison of Figs. 4 and 5 indicates that the simulated second moment has a larger fluctuation than the simulated mean.

Next, we consider the example for a nonlinear hydrologic system with $b = 0.5$ and $a = 0.2$ in (2). Unlike the previous example of linear hydrologic system, the theoretical moments for this example cannot be found because the equations for moments are nonlinear (see (10)). For instance, the equation for the first moment, (11), now contains $E[Q(t)^{-1}]$ and $E[Q(t)^{-3}]$. Such a drawback always exists whenever the governing differential equation is nonlinear (Soong 1973). Another drawback is that the explicit solution process for this example cannot be obtained either. This prevents us from finding realizations of the solution process directly. Here we use the aforementioned stochastic numerical method to overcome the drawbacks.

The layout of the numerical experiments is the same as that in the previous example of liner hydrologic system. In a like manner, Fig. 6 shows a realization of the outflow process for the case $n = 100$ as well as the simulated mean. The mean outflow hydrographs for $n = 20$, 50, and 100 are compared in Fig. 7. These three n values have almost the same performance in obtaining the simulated mean. The theoretical mean is not provided for comparison because it cannot be found.

5 SUMMARY AND CONCLUSIONS

In this paper, a nonlinear hydrologic system with a stochastic input is considered. A SDE model for

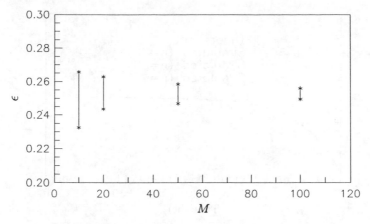

Fig. 1. The 90% confidence interval of ϵ for $n = 20$.

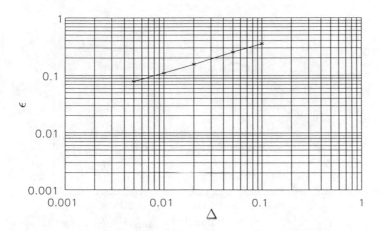

Fig. 2. $\log\Delta$ versus $\log\epsilon$ for $M = 20$.

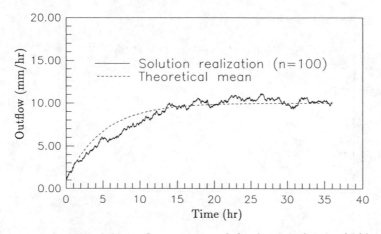

Fig. 3. A realization of the outflow process and the theoretical mean ($Q(t) = 0.2S(t)$).

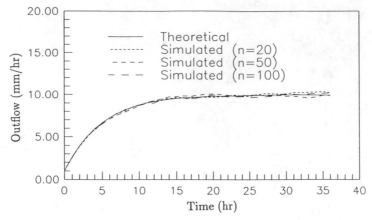

Fig. 4. Simulated and theoretical mean outflow hydrographs ($Q(t) = 0.2S(t)$).

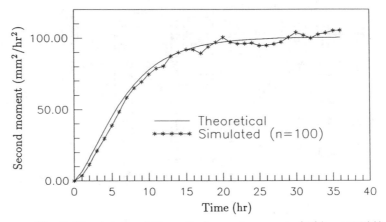

Fig. 5. Simulated and theoretical second moments ($Q(t) = 0.2S(t)$).

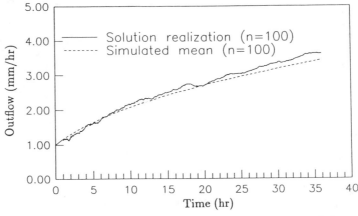

Fig. 6. A realization of the outflow process and the simulated mean $\left(Q(t) = 0.2S(t)^{0.5}\right)$.

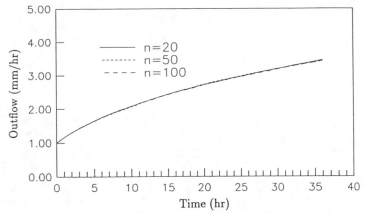

Fig. 7 Simulated mean outflow hydrographs $(Q(t) = 0.2S(t)^{0.5})$.

the hydrologic system has been developed based on the theory of SDEs. The developed SDE model has the form of Itô equation. The differential equations for moments of the solution process are derived. The probability density function of the solution process in steady state is also derived using the Fokker-Planck equation. Finally, to facilitate the rainfall-runoff simulations, the stochastic Euler scheme is used to determine the numerical solution of the solution process. The accuracy of the stochastic Euler scheme is examined through examples. The results show that the confidence interval of absolute error decreases as the number of simulation batches increases and the absolute error increases with increasing time increment. The reliability and applicability of the numerical solutions are also demonstrated through applications to a linear and a nonlinear hydrologic systems. For the linear hydrologic system example, the simulated first and second moments of the output agree well with the theoretical. The advantage of the proposed numerical solution is clearly demonstrated in the nonlinear hydrologic system example for which an explicit solution cannot be found.

ACKNOWLEDGEMENT

This paper is based on research supported by the National Science Council of the Republic of China under Grants NSC 83-0410-E002-012 and 84-2211-E002-048.

REFERENCES

Bodo, B.A. and T.E. Unny 1987. On the outputs of the stochasticized Nash-Dooge Linear Reservoir Cascade. In I.B. MacNeill and G.J. Umphrey (eds), Stochastic Hydrology: 131-147.

Jazwinski, A.H. 1970. Stochastic Processes and Filtering Theory. New York: Academic Press.

Kloeden, P.E. and E. Platen 1992. Numerical solution of stochastic differential equations. Berlin: Springer-Verlag.

Kloeden, P.E., E. Platen and H. Schurz 1994. Numerical solution of SDE through computer experiments. Berlin: Springer-Verlag.

Platen, E. and W. Wagner 1982. On a Taylor formula for a class of Itô process. Prob. Math. Statist. 3: 37-51.

Soong, T.T. 1973. Random differential equations in science and engineering. New York: Academic Press.

Unny, T.E. and Karmeshu 1984. Stochastic nature of outputs from conceptual reservoir model cascades J. Hydrol. 33: 161-180.

Unny, T.E. 1984. Numerical integration of stochastic differential equation in catchment modeling. Water Resour. Res. 20(3): 360-368.

Unny, T.E. 1987. Solutions to nonlinear stochastic differential equation in catchment. In I.B. MacNeill and G.J. Umphrey (eds), Stochastic Hydrology: 87-110.

Wong, E. and M. Zakai 1965. On the relation between ordinary and stochastic differential equations. Int. J. Eng. Sci. 3: 213-229.

Stochastic Hydraulics'96, Tickle, Goulter, Xu, Wasimi & Bouchart (eds) © 1996 Balkema, Rotterdam. ISBN 90 5410 817 7

Prediction of rare inundation volumes of runoff

Arie Ben-Zvi
Israel Hydrological Service, Jerusalem, Israel

ABSTRACT: River constrictions interefere with streamflow and cause an accumulation of water on their upstream side. When approaching discharge exceeds the design discharge of the constriction, the excessive volume of water might cause a damage in excess of that taken care of in the design. Traditional methods for estimation of the excees volume are incapable of assigning recurrence intervals to it. The present work proposes a method for determination of runoff volumes beyond design discharges from standard hydrologic records, and for probabilistic analysis of such volumes. The method applies distributions fitted to partial duration series of peak discharges and of excess volumes beyond relatively low threshold discharges. Subsequently, it synthesizes probability distribution for excess volumes beyond high threshold discharges. In an example for the application of the method, threshold values are so selected as to enable best fit of the distribution to the samples and adherence to the trend of upper tails of the sereis.

1 THE PROBLEM

River constrictions such as bridges and culverts interefere with the flow of run-off and cause an accumulation of water at their upstream side. This accumulation is taken into consideration as long as approaching discharges are lower than the design discharge of the constriction. Excessive accumulation, which might cause damage to the site and its neibourhood, arises when the approaching discharge is higher than the design discharge. The magnitude of this damage depends upon the volume of accumulating waters, as well as the geometry, hydraulic and physical properties of the site. Estimation of such volumes is an important objective of hydrologic work.

Conventional hydrologic design of river constrictions provides a relationship between peak discharges and their recurrence intervals. In cases where consideration of volumes is required, entire hydrographs are prepared and routed through planned conditions of the sites of interest. The peak, or total volume, of such a hydrograph can be associated with a chosen recurrence interval. Yet, there is no method for assigning such an interval to volume of runoff in excess of the design discharge of the constricted site. Consequently, the excess volume obtained by the routing is not associated with a recurrence interval of its own, and the risk of damage cannot be properly estimated. This study proposes a method for direct association of recurrence intervals to runoff volumes accumulated upstream from river constrictions.

2 THE PROPOSED METHOD

The method considers a river site which is constricted to a given design discharge, Q_d, associated with the recurrence interval T_d. Flow events of diverse discharges approach this site. The events whose discharges are lower than Q_d would not generate any excessive accumulation of water. Yet, on average, once in T_d, there arises an event, i, in which, for a certain duration, T_i, discharges exceed Q_d. The excess water accumulates upstream from the constricted site. Assuming that the site's conveyance is little affected by this accumulation, and that the water does not find a new flow path (i.e. over-topping, piping, etc.), the maximum volume accumulated during the event, V_i, can be computed by:

$$V_i = \int_{T_i} (Q_i(t) - Q_a)dt \qquad (1)$$

in which $Q_i(t)$ is the discharge of event i at time t, and dt ϵ T_i. Determination of this volume is illustrated in Fig. 1.

In case that the site's conveyance, including overtopping, can be related to the accumulating volume, i.e. Q_a is a known function of V, Eq. 1 can be modified to:

$$V_i = \int_{t=0}^{T_i} (Q_i(t) - Q_a(t))dt \qquad (1a)$$

and

$$Q_a(t) = \int_{\theta=0}^{t} (Q_i(\theta) - Q_a(\theta))d\theta \qquad (1b)$$

with a numerical solution of Eq.'s 1a and 1b through methods developed for estimation of reservoir outflow.

Upon application of Eq. 1 to a number of events with sufficiently high dis charges, a series of excess volumes is obtained. This series has a peculiar property. Its magnitudes are of volumes, while the frequency of its occurrences is that of partial duration series of peak discharges associated with threshold value equals to Q_a.

Having defined the series, and trying to fit a probability distribution to its magnitudes, the hydrologist encounters an extreme shortage of data. Ordinary design discharges of river constrictions are rarely exceeded, thus leaving an inadequate margin for recording a sufficient number of events of excess volume. The only way remains for proceeding with a fit is through extrapolation of available data. A method for such an extrapolation is proposed below.

The method involves a statistical analysis, carried out in three stages. The first stage assigns recurrence intervals to peak discharges. It commences with collection of momentary data for a site of interest and determination of a complete duration series of peak discharges. Next, a probability distribution is fitted to

the entire series or to a partial duration series associated with a low threshold value (i.e. having a recurrence interval of 0.5 to 2 years). The particular distribution and the fitting method which may be selected for this purpose do not constitute an integral part of the proposed method, and therefore those selected as examples for the present work will be described later. The fitted distribution associates discharges and recurrence intervals to each other through the formula:

$$T(Q) = (\underline{n}P(Q))^{-1} \qquad (2)$$

where $T(Q)$ is recurrence interval of the discharge Q (in years), $\underline{n}$ is mean number of events per year in the fitted series (year^{-1}), and $P(Q)$ is exceedance probability of Q in the fitted distribution. The recurrence interval of a discharge lower than the threshold value is assumed as:

$$T(Q) = (\underline{n}(Q))^{-1} \qquad (3)$$

where $\underline{n}(Q)$ is recorded mean number of times per year when Q is exceeded (year^{-1}).

The second stage fits probability distributions to series of excess volumes. For this purpose, a number of relatively low design discharges are arbitrariliy selected, and a series of excess volumes is prepared for each design discharge. This is accomplished by means of the technique formulated in Eq. 1, or 1a and 1b, and illustrated in Fig. 1. A distribution is fitted to each series, preferabely using those methods applied in determining peak discharges.

Similarly to Eq. 2:

$$T_Q(V) = (\underline{n}_Q P_Q(V))^{-1} \qquad (4)$$

where subscript Q refers to the design discharge with respect to which the series has been determined, $T_Q(V)$ is the recurrence interval of the excess volume V (years), $\underline{n}_Q$ is the mean number of events per year in the fitted series, and $P_Q(V)$ is exceedance probability of V. Following the fit, a list of excess volumes associated with given recurrence intervals is prepared.

The third stage performs the required extrapolation. This is done by means of a

350

Table 1. Watersheds and records.

#	Stream	Station	Area km²	Rain mm	N	n
1	Kziv	Gesher Haziv	131	830	39	130
2	Hillazon	Yavor	158	677	41	203
3	Alexander	Ma'aba˜ & Elya˜	505	606	39	207
4	Lahish	Yavne Rd.	992	451	36	271
5	Shikma	Bror Hayil	378	348	35	195
6	Sheva	Nevatim	405	175	24	91
7	Bsor	Re'im	2630	192	20	66
8	Zin	Zin Valley	660	80	22	70

Legend: # is station number, Area is catchment area for the station, Rain is mean annual depth of precipitation over the catchment, N is number of years on record, n is number of events in these years, ˜ denotes abbreviations of Ma'abarot and Elyashiv.

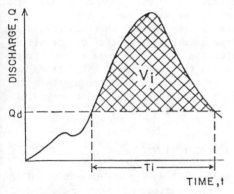

FIG. 1: ILLUSTRATIVE SKETCH

graph as shown in Fig. 2. Volumes abstracted in the course of the second stage are plotted against their recurrence intervals. For each design discharge, a curve is drawn through the relevant data points. Noting that when $Q = Q_d$ no excessive volume is accumulated, each curve may be extrapolated down to its $(T_d, 0)$ point determined by the relationship derived in the first stage. The curves for the different design discharges constitute a family to which additional members can be adopted.

When an estimation is requested for the excess volumes which would accumulate upstream from a river constriction of a given design discharge, Q_g, an appropriate curve is added to the family. This curve should pass through point $(T_g, 0)$, where T_g is the recurrence interval of Q_g. Assuming that the curves in the family look consistent, the shape of the new curve can be so drawn as to adhere to the previously drawn ones. This adherence allows for a minimum of freedom in subjective drawing of the adopted curve.

3 REFLECTIONS ON ORDINARY EXTRAPOLATIONS

The foregoing extrapolation utilizes some ordinary extrapolations of data by involving fitted distributions. Unfortunately, the results of these extrapolations are sensitive to the distribution and thre-

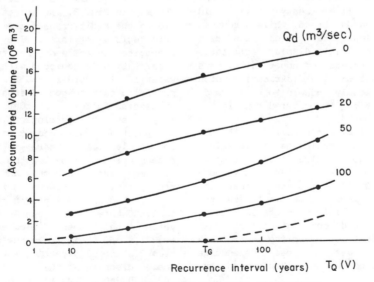

FIG. 2: EXTRAPOLATION TECHNIQUE AND EXAMPLE

shold value selected for the fit. This selection is currently accomplished by means of subjective considerations, about which opinion varies among hydrologists. The opinions reflected in the example of the present work have already been published (Ben-Zvi 1993, 1994, 1995), but their essential features are repeated here for a better understanding of the example.

Since the recurrence intervals, which are of interest in many hydrologic predictions, exceed the lengths of records to which the selected distributions are fitted, certain researchers prefer distributions which adhere to trends in the upper tails of recorded data (e.g. Bryson 1974; Picands 1975; Smith 1987). Bryson (1974) defined the trend as heavy when the observed density distribution approaches zero more slowly than does the negative exponential one, and as light when the distribution approaches zero more rapidly. He tested the rapidity of approach through a variable called conditional mean exceedance, CME_x, defined by:

$$CME_x = E(X - x \mid X \geq x) \qquad (5)$$

where X is a considered variable and x is a pre-selected value of X.

Bryson (1974) demonstrated that CME_x results in an increasing trend for heavy tails and a decreasing trend for light tails. Smith (1987) proposed a simple graphical display in which CME_x is plotted against x. In this display, an unbounded thick (i.e. heavy) tail would render an upwards trend, while a bounded thin (i.e. light) tail would render a downwards trend. This display facilitates detection of trends of upper tails, thus enabling selection of a distribution which would more closely adhere to this trend. Such a selection is carried out in the example below.

Another problem, encountered with the use of partial duration series is selection of threshold value. This is usually accomplished by means of subjective considerations. Obviously, the higher the threshold the fewer the events in the selected series. This results in reduced data handling and computational load, an increased effect of sampling variations and, in many cases, a lesser dependence between selected events (Bhuiya and Yevjevich 1968).

Conventional statistical selections of threshold values are based upon the occurrence frequency of events in the samples (e.g. Langbein 1949; Chow 1953; Cunnane 1973; Taesombat and Yevjevich 1978; Ashkar and Rousselle 1983a, 1987; Ben-Zvi 1991), while some others find appropriate physical considerations (e.g. Kavvas 1982; Ben-Zvi and Cohen 1983). Another approach, based upon goodness of fit, has recently been proposed (Ben-Zvi 1993, 1994, 1995). In this approach, the selected distribution is fitted to nested sub-samples, and the threshold associated with the best fit is selected. The first sub-sample is composed of the uppermost five peaks. The following sub-samples are composed of the uppermost six, seven, eight... peaks; such that each new sub-sample is one member larger than its preceding sub-sample. The process of testing the fit to new sub-samples is terminated when the series is exhausted, or the parameters of the fitted distribution no longer suit the trend determined for the upper tail. It is believed, though not demonstrated, that this approach could contribute to the credibility of predictions.

4 CASE DESCRIPTION

The above methodology has been applied to data series abstracted from records of eight hydrometric stations located on ephemeral streams in Israel, as shown on the map in Fig. 3. The selected watersheds lie in the Mediterranean semi-arid and the arid regions of Israel. They enjoy winter precipitation and a dry hot summer season. Each station is equipped with a continuous water level recorder, and its rating curve providing fair to good discharge accuracy. Processing follows standard techniques adopted by the Hydrological Service. Sufficiently accurate and complete records for at least 20 years are available for each station. The lengths of records, catchment areas, and mean annual depths of precipitation are presented in Table 1.

Watershed areas involved range from 131 to 2,630 km^2, record lengths from 20 to 41 years, total number of flow events from 66 to 271, and mean frequency of occurrence from 2.2 to 7.5 year^{-1}. For each stream, design discharges, Q_d's, of 0, 20, 50, 100, 150 and 200 m^3/sec have been sele-

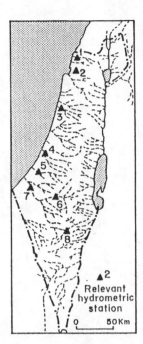

FIG. 3: MAP OF STUDY AREA

cted. Series of excess volumes have been prepared for each stream and with respect to each design discharge. The series with respect to nil design discharge are essentially the complete duration series of runoff volumes.

The data series have been abstracted from files of momentary discharges of complete runoff records. A determination criterion of zero flow for at least 24 hours has been chosen to distinguish between a multi-peaked event and a number of distinct events. Application of such a criterion, requiring a minimal time interval between distinct event occurrences, might interefere with the fit of Poisson distribution (as probably with others) to occurrence frequency (Ashkar and Rousselle 1983b). None the less, this eliminates subjective decision-making regarding multi-peaked records, be they out-of-phase contributions of various tributaries or results of distinct rainfall events. In addition, it precludes application of a model for runoff separaration which might insert an additional source of error into the results.

5 ANALYSIS OF DISCHARGES

For each stream, a graphical display was prepared for CME_q vs q, where q is peak discharge of a runoff event. Based upon these displays, decisions about tail thicknesses were made with respect to trend directions of the upper tails. Conclusions about tail thickness could easily be reached for six of the eight streams. Therefore, it can be concluded that the upper tails of peak discharges of runoff events in the present case study are generally thin. Yet, the possibility of a thick tail or of a stable trend should not be ruled out. Consequently, if a hydrologist wishes to fit the same distribution to the entire case study, one should be selected which is flexible with respect to tail thickness. The generalized Pareto distribution (symbolized by GP) does have this flexibilty (e.g. Hosking and Wallis, 1987), and therefore it is selected for the description of magnitudes in the present case study. This selection does not mean to acknowledge, however, that the GP is the only distribution which possesses this flexibility.

The selected distribution was fitted to all the upper sub-samples composed of at least five members and maintaining the trend which had been determined for their upper tail. The goodness of fit of the distribution was determined through the Anderson-Darling test, modified with respect to sample size, and concluded for the case where the parameter values are estimated from the sample (Stephens, 1986). This test is selected since according to Stephens (1986, p. 167) "it tends to lead the others because it is effective in detecting departures in the tails". The significance levels of 0.25 and 0.01 for rejecting the fit are obtained when the test statistic, A^{2*}, attains values of 0.474 and 1.038, respectively.

The test results indicate that for seven of the eight streams, there have been found a number of sub-samples for which $A^{2*} \leq 0.474$. This means that for each such stream there exist a number of partial duration series to which the GP is well fitted. The multitude of good fits allows for preparation of a number of good, and probably varying, predictions as to the recurrence interval of rare events at a station. This situation may have some

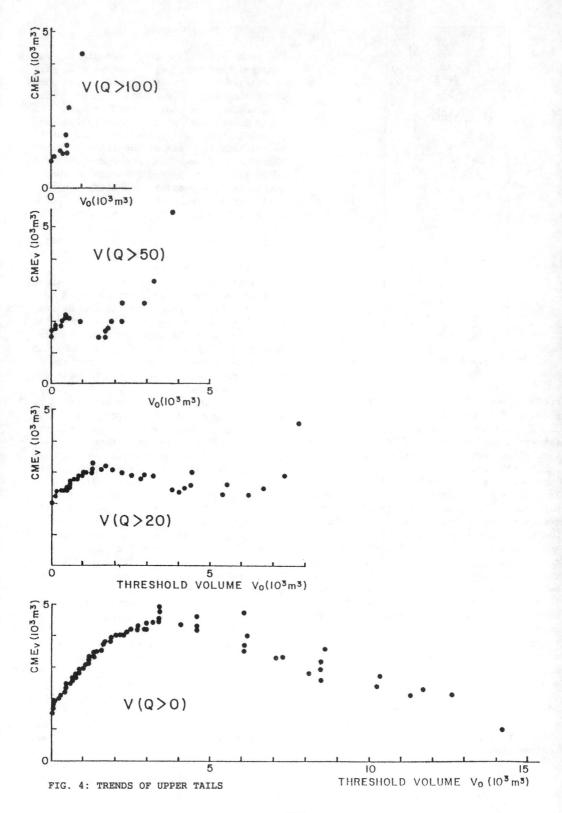

FIG. 4: TRENDS OF UPPER TAILS

354

scientific merits, but might also cause an ambiguous application.

The best fit approach has been chosen here as an objective means for selection of one of these sub-samples. The parameter values and threshold discharge associated with the sub-sample for which A^{2*} attains its minimal value are selected for prediction of rare discharges. The extrapolation was carried out by use of a display of Q (on a linear scale) vs 1 - F (on a log scale) with plotting positions as recommended by Hosking and Wallis (1987):

$$F(i) = (i - 0.35)/n \qquad (6)$$

in which F(.) is non-exceedance frequency, i is rank of peak discharge in an increasing order, and n is sample size.

6 ANALYSIS OF EXCESS VOLUMES

The statistical analysis of excess volumes applies the same procedures as those used for peak discharges. For each stream and design discharge, the conditional mean exceedance of volumes was plotted against its corresponding threshold volume, as shown in Fig. 4. The displays indicate that, in general, samples related to low design discharges render an upwards trend in their lower tails and a downwards trend in their upper tails, while samples related to high design discharges render continuous upwards trends. Such trends are, however, not always clear and consistent. A consistent and parsimonious description of the entire case study requires application of a distribution which can follow the trends found for all upper tails. The GP distribution, which has this property, is again selected.

The GP distribution has been tested for fit to all of the upper sub-samples of excess volumes containing at least five members and maintaining the trend concluded for their respective upper tails. Ranges of good fit (i.e. having significance levels of rejection larger than 25%) are found for each one of the streams and for almost all of the design discharges. The best fits cover the upper 50 to 90% (with an average of 75%) of the proper ranges of excess volumes. It is consequently concluded that the GP distribution is suitable for the description of

magnitudes of flood volumes beyond design discharges, at least for cases having hydrologic properties similar to the present ones.

The best fit of the GP distribution to excess volumes for each station and design discharge, has been displayed in a form similar to that described for the peak discharges. From these displays, predictions were made for excess volumes accumulating at recurrence intervals of 10, 20, 50, 100, and 200 years. These volumes were plotted as shown in the example on Fig. 2. In this example, an extrapolation curve is drawn for the volumes associated with the given design discharge, T_d, whose recurrence interval is 50 years. This curve reaches the 200-year recurrence interval at an excess volume of 2×10^6 m³. Thus, it is estimated that for that site, a volume of at least 2×10^6 m³ would accummulate, on the average, once in 200 years, at the upstream side of a constriction designed for a 50-year peak discharge.

7 DISCUSSION AND CONCLUSIONS

The proposed method applies two kinds of extrapolation, one, the ordinary style of extension of fitted probability distributions, and the other, the generation of a synthetic curve. The results depend upon the selection of the distributions, the goodness of fit, and upon the subjective considerations in the drawing of the synthetic curve. Both kinds involve a high degree of uncertainty. Therefore, the accuracy of the results cannot be too high.

Noting this low accuracy, a question can be raised about the merits of the proposed method. The principal ones are the simple definition and determination of excess volumes beyond design discharges, and the assignment of recurrence intervals to these volumes. This relationship can serve as an input to physical, hydraulic, and economic considerations, and to estimation of risk. All other methods assign recurrence intervals to peak discharges, to total volumes of direct runoff, or to causative rainfall depth (or intensity). None of these recurrence intervals is identical, or directly related, to that of excess volumes. Therefore, the conventional methods cannot relate risk of possible damage to accumulated volume of water.

It seems that the proposed method should provide a proper solution to a hydrologic engineering need. The accuracy of this method will be improved when accuracy of ordinary extrapolations is raised, and when more accurate techniques for drawing synthetic curves are developed.

REFERENCES

Ashkar, F. & Rousselle, J. 1983a. Some remarks on the truncation used in partial duration series models. Water Resour. Res.19:477-480.

Ashkar, F. & Rousselle, J.1983b. The effect of certain restrictions imposed on the interarrival times of flood events on the Poisson distribution used for modeling flood counts. Water Resour. Res.19:481-485.

Ashkar, F. & Rousselle, J. 1987. Partial duration series modeling under the assumption of Poissonian flood count. J. Hydrol.90:135-144.

Ben-Zvi, A. 1991. Observed advantage for negative binomial over Poisson distribution in partial duration series. Stoch. Hydrol. Hydraul.5:135-146.

Ben-Zvi, A. 1993. Distribution of flood volumes beyond design discharges. In C. Y. Kuo (ed),Engrg. Hydrol.:97-102. San Francisco, CA, USA: Am. Soc. Civil Engrs.

Ben-Zvi, A. 1994. Fit of probability distributions to upper sub-samples of partial duration series. In K.W. Hipel (ed), Stochastic and Statistical Methods in Hydrol. and Environ. Engrg., Vol. I, Extreme Values: Floods and Droughts:95-107. Dordrecht, Netherlands: Kluwer.

Ben-Zvi, A. 1995. Flood prediction by use of partial duration series. In V.P. Singh (ed), Proc. Intl. Conf. Hydrol. and Water Rresour. 1993: (in press). Dordrecht, Netherlands: Kluwer. Water

Ben-Zvi, A. & Cohen, O. 1983. Maximal discharges in the lower Yarkon River. Rep. 4/83, Israel Hydrol. Service, Jerusalem, Israel (Hebrew).

Bhuiya, R.K. & Yevjevich, V. 1968. Effects of truncation on dependence in hydrological time series. Hydrol. Paper 31, Colorado State Univ., Fort Collins CO, USA.

Bryson, M.C. 1974. Heavy-tailed distributions: properties and tests. Technometrics 16:61-68.

Chow, V.T. 1953. Frequency analysis of hydrologic data with special application to rainfall intensities. Bull. 414, Engrg. Exp. Station, Univ. of Illinois, Urbana IL, USA.

Cunnane, C. 1973. A particular comparison of annual maxima and partial duration series methods of flood frequency prediction. J.Hydrol.18:257-271.

Hosking, J.R.M & Wallis, J.R. 1987. Parameters and quantile estimation for the generalized Pareto distribution. Technometrics 29:339-349.

Kavvas, M.L. 1982. Stochastic trigger model for flood peaks, I. Development of the model. Water Resour. Res. 18:383-398.

Langbein, W.B. 1949. Annual floods and the partial duration flood series. Trans. AGU 30:879-881.

Pickand, J. 1975. Statistical inferences using extreme order statistics. Annals Statis.3:119-131.

Smith, J.A. 1987. Estimating of the upper tail of flood frequency distributions. Water Resour. Res.23:1657-1666.

Stephens, M.A. 1986, Tests based on EDF statistics. In R.B. D'Agostino & M.A. Stephens (eds.) Goodness-of-fit Techniques:97-103. New-York NY USA: Dekker.

Taesombut, V. & Yevjevich, V. 1978. Use of partial flood series for estimating distribution of annual flood peaks. Hydrol. Paper 97, Colorado State Univ., Fort Collins CO, USA.

Stochastic Hydraulics '96, Tickle, Goulter, Xu, Wasimi & Bouchart (eds) © 1996 Balkema, Rotterdam. ISBN 90 5410 817 7

Statistical evaluation of monitoring networks in space/time dimensions

Sevinc Ozkul, Okan Fistikoglu & Nilgun B. Harmancioglu
Dokuz Eylul University, Turkey

Vijay P. Singh
Louisiana State University, La., USA

ABSTRACT: Application of the informational entropy principle in assessing the technical (temporal, spatial, and combined temporal/spatial) features of an existing hydrometric network provides satisfactory results as the benefits of alternative monitoring practices can be evaluated quantitatively as a function of information. The study presented develops an entropy-based methodology in evaluating combined space/time frequencies of water quality sampling networks and demonstrates its application on a case study.

1 INTRODUCTION

The current status of existing water quality monitoring networks reflects significant shortcomings although design methodologies have become sophisticated in time (Harmancioglu and Singh, 1991). Basically, specific technical design features related to what variables to measure, where, when and how long still remain unresolved. In essence, there are no definite criteria yet established to decide upon the optimum solutions for these problems. Within this respect, the major difficulty underlying both the design and the evaluation of monitoring systems is the lack of an objective criterion to assess: (a) the efficiency, and (b) cost-effectiveness of a network.

Efficiency is related to objectives of monitoring such that the latter delineates "information-expected" from monitoring and the former describes "information produced" by a network. Furthermore, the "information produced" is a function of the technical features of a network related to variables sampled, spatial and temporal sampling frequencies, and the duration of sampling. It is plausible then to define efficiency as the "informativeness" of a network. Yet, current design and evaluation procedures lack a precise definition of information. They often express it indirectly in terms of other statistical parameters like standard error or variance criteria (Sanders et al., 1983). Tirsch and Male (1984) propose a measure of monitoring precision as a function of sampling locations and time frequencies. Such a measure is defined by the

corrected regression coefficient of determination obtained by a multivariate linear regression model. Some researchers (Schilperoort and Groot, 1983) relate efficiency to the statistical power of a network, which is often analyzed by variance (ANOVA) techniques. Then, optimization methods are used to maximize network's statistical power (MacKenzie et al., 1987; Reinelt et al., 1988). These approaches are reasonable and have their merits; yet there is still the question of how one relates such statistical criteria to the value of data. The result is that evaluation of network efficiency remains vague unless a quantitative measure of information is provided.

A similar difficulty is encountered in assessment of cost-effectiveness of a network, where costs are easy to estimate, but benefits are often described indirectly in terms of other parameters, using optimization techniques, Bayesian decision theory and regression methods (Schilperoort et al., 1982; Tirsch and Male, 1984; Attanasi and Karlinger, 1979; Moss, 1976). Thus, a realistic evaluation of benefit/cost considerations cannot be achieved since benefits are not directly quantified. Actually, benefits of monitoring can only be measured in terms of the information conveyed by collected data; that is, they are a function of the value or worth of data. Since most design methodologies do not provide an objective measure of information, benefits of monitoring networks, i.e., the value of data, still remain as unquantifiable parameters in the decision making process.

This study proposes a statistical procedure

based on the entropy principle to address the assessment of both network efficiency and cost-effectiveness. The concept of entropy, as defined in information theory, is a quantitative measure of information. It may as well be used to quantify the efficiency and benefits of monitoring since it describes the utility or usefulness of data (Langbein, 1979). The approach presented here considers the monitoring network to be an "information system". Efficiency is ensured when information produced by the network is maximized. Cost-effectiveness can be evaluated by comparing costs of monitoring versus information gained via monitoring. The issue is then an optimization problem to maximize the amount of information (benefits of monitoring) while minimizing the accruing costs. The technical features of design, i.e., variables sampled, temporal and spatial sampling frequencies, and duration of sampling, can be evaluated with respect to network efficiency and cost effectiveness. The entropy method enables one to evaluate how much information is risked in an attempt to decrease costs by selecting smaller number of stations or larger time intervals. Similarly, the efficiency of the network can be described by entropy measures in quantitative terms with respect to both sampling frequencies and sampling sites (Harmancioglu and Alpaslan, 1992; Harmancioglu et al., 1994).

The work presented focuses basically on the evaluation of an existing network with respect to combined temporal/spatial design features. An entropy-based assessment of spatial and temporal features on a separate basis is presented on a complementary paper submitted to this conference, where the results are demonstrated in the case of the highly polluted Porsuk River in Turkey. Analysis of combined space-time features was first carried out on the same sets of data; however, the sporadic nature of available Porsuk data has led to significant difficulties in evaluating this multidimensional problem. Consequently, the methodology is tested on the more regularly observed water quality series at 12 stations along the Mississippi River in the USA.

2 METHODOLOGY

2.1 Basic approach

Some network design procedures combine both the spatial and the temporal design criteria to evaluate space-time tradeoffs. The approach in such combined design programs is to compensate for lack of information with respect to one dimension by increasing the intensity of efforts in the other dimension. Combined spatial/temporal design features constitute a problem which is multivariate with respect to both time and space. In this case, the joint entropy of multivariate space/time variables need to be computed, with corresponding transinformations (common information) for alternative combinations of numbers and locations of stations versus different sampling intervals. An increase in the sampling interval decreases the common information between the stations considered, whereas an increase in the number of stations increases the transinformation for a given time frequency. One would look for the best combination with respect to time and space for reduction of the total uncertainty about a water quality variable.

To analyze spatial and temporal frequencies on a joint basis, the best combination of monitoring stations has to be selected first. Next, starting with the first priority station, the number of stations is successively increased by adding to the combination the next station on the priority list. For each number of stations, the temporal frequencies are decreased to identify how much information is provided by those stations at different sampling intervals. Finally, changes in information are plotted on the same graph with respect to both the increases in the number of stations and the decreases in temporal frequencies of sampling. The particular information measure used in this analysis is transinformation which represents redundant information in space and time dimensions. The objective is the selection of a space/time combination that produces the least amount of transinformation. Increases in either the space or the time frequencies implies increases in accruing costs so that one has to compare loss of information due to decreased space/time frequencies versus decreased costs, or vice versa.

2.2 Selection of the best combination of stations

Analysis of spatial frequencies requires the assessment of reduction in the joint entropy of two or more variables due to the presence of stochastic dependence between them. Such reduction is equivalent to the redundant information (transinformation) in the series of the same variable observed at different sites. Thus, the objective in spatial design is to minimize the transinformation by an appropriate choice of the number and locations of monitoring stations. The combination of stations with the least transinformation reflects the variability of the quality variable along the river without producing redundant information.

To realize the above procedure for a particular

water quality variable, the joint entropy (uncertainty) of two or more stations and their transinformations must be computed. The total entropy of M stochastically independent variables X_m (m=1,...,M) is (Harmancioglu, 1981; Harmancioglu and Alpaslan, 1992):

$$H(X_1, X_2, ..., X_M) = \sum_{m=1}^{M} H(X_m) \quad (1)$$

where $H(X_m)$ represents the marginal entropy of each variable X_m in the form of:

$$H(X_m) = K \sum_{n=1}^{N} p(x_n) \log [1/p(x_n)] \quad (2)$$

with K=1 if $H(X_m)$ is expressed in napiers for logarithms to the base e. Eq (2) defines the entropy of a discrete random variable X_m with N elementary events of probability $P_n = p(x_n)$ (n=1,...,N) (Shannon and Weaver, 1949). For continuous density functions, $p(x_n)$ is approximated as $[f(x_n).\triangle x]$ for small $\triangle x$, where $f(x_n)$ is the relative class frequency and $\triangle x$, the length of class intervals (Amorocho and Espildora, 1973). Then the marginal entropy for an assumed density function $f(x_n)$ is:

$$H(X_m; \triangle x) = \int_{-\infty}^{+\infty} f(x) \log [1/f(x)] \, dx + \log[1/\triangle x] \quad (3)$$

If significant stochastic dependence occurs between M variables, the total entropy has to be expressed in terms of conditional entropies $H(X_m/X_1..., X_m)$ added to the marginal entropy of one of the variables (Harmancioglu, 1981; Topsoe, 1974):

$$H(X_1, X_2, ..., X_M) = H(X_1) + \sum_{m=2}^{M} H(X_m \mid X_1, ..., X_{m-1}) \quad (4)$$

Since entropy is a function of the probability distribution of a process, the multivariate joint and conditional probability distribution functions of M variables need to be determined to compute the above entropies (Harmancioglu, 1981):

$$H(X_1, X_2, ..., X_M) = - \int_{-\infty}^{+\infty} ... \int_{-\infty}^{+\infty} f(x_1, ..., x_M).\log f(x_1, ..., x_M).$$

$$dx_1 \, dx_2 ... dx_M \quad (5)$$

$$H(X_M \mid X_1, ..., X_{m-1}) = - \int_{-\infty}^{+\infty} ... \int_{-\infty}^{+\infty} f(x_1, ..., x_m).\log f(x_m \mid x_1, ..., x_{m-1})$$

$$dx_1 \, dx_2 ... dx_m \quad (6)$$

The common information between M variables, or the so-called transinformation $T(X_1, ..., X_M)$, can be computed as the difference between the total entropy of Eq.(1) and the joint entropy of Eq. (4). It may also be expressed as the difference between the marginal entropy $H(X_m)$ and the conditional entropy of Eq. (6). It follows from the above that the stochastic dependence between multi-variables causes their marginal entropies and the joint entropy to be decreased. This feature of the entropy concept is used to select appropriate numbers and locations of monitoring stations so as to avoid redundant information (Harmancioglu and Alpaslan, 1992).

To compute the above entropies, one has to determine the type of probability distribution function which best fits the analyzed processes. If a multivariate normal distribution is assumed, the joint entropy of X (the vector of M variables) is obtained as (Harmancioglu, 1981):

$$H(X) = (M/2)\ln 2 + (1/2) \ln |C| + M/2 - M \ln(\triangle x) \quad (7)$$

where M is the number of variables and $|C|$ is the determinant of the covariance matrix C. Eq. (7) gives a single value for the entropy of M variables. If logarithms of observed values are evaluated by the above formula, the same equation can be used for lognormally distributed variables. In the above formula, the covariance matrix C involves the cross covariances of M different variables.

The calculation of conditional entropies in the multi-variate case can also be realized by Eq.(7) as the difference between two joint entropies. For example, the conditional entropy of variable X with respect to two other variables Y and Z can be determined as:

$$H(X \mid Y, Z) = H(X, Y, Z) - H(Y, Z) \quad (8)$$

Next, the following procedure is applied to select the best combination of stations. First, the marginal entropy of the variable at all stations is computed by Eq. (2). The station with the highest entropy is selected as the first priority station to represent the location of the highest uncertainty about the variable. Next, this station is coupled with every other station to select the pair that leads to the least transinformation. The station that fulfills this condition is marked as the second priority location. The same procedure is continued by considering successively combinations of 3, 4, 5,..., stations and selecting the combination that produces the least transinformation by satisfying the condition:

$$\min \{H(X_1,...,X_{j-1}) - H(X_1,...,X_{j-1}|X_j)\} =$$

$$\min \{T((X_1,...,X_{j-1}),X_j)\} \qquad (9)$$

where X_1 is the 1st priority station and X_j is the station with the jth priority. Increasing the number of stations by the above procedure ensures he selection of those locations where the uncertainty about the variable is the highest and redundant information is the lowest.

2.3 Redundant information in the time domain

The marginal entropy of a single process that is serially correlated is less than the uncertainty it would contain if it were independent. If the values that a variable assumes at a certain time t can be estimated by those at times t-1, t-2,..., the process is not completely uncertain because some information can be gained due to the serial dependence present in the series. In this case, stochastic dependence again acts to reduce entropy and causes a gain in information (Harmancioglu, 1981). This feature is suitable for use in the temporal design of sampling stations. Sampling intervals can be selected so as to reduce the redundant information between successive measurements (Harmancioglu, 1984).

For a single process, the marginal entropy as defined in Eq.(3) represents the total uncertainty of the variable without having removed the effect of any serial dependence. However, if the ith value of variable X, or x_i is significantly correlated to values x_{i-k}, k being the time lag, knowledge of these previous values x_{i-k} will make it possible to predict the value of x_i, thereby reducing the marginal entropy of X. To analyze the effect of serial correlation upon marginal entropy, the variable X can be considered to be made up of series X_i, X_{i-1},..., X_{i-k}, each of which represents the sample series for time lags k=0,1,...,K and which obey the same probability distribution function. Then conditional entropies such as $H(X_i|X_{i-1})$, $H(X_i|X_{i-1}, X_{i-2},..., X_{i-k})$ can be calculated. If X_{i-k} (k=1,...,K) are considered as different variables, the problem turns out to be one of the analysis of K+1 dependent multi-variables; thus, formulas similar to Eq.(6) can be used to compute the necessary conditional entropies (Harmancioglu, 1981):

$$H(X_i|X_{i-1},...,X_{i-K}) = -\int_{-\infty}^{+\infty}...\int_{-\infty}^{+\infty} f(x_i,...,x_{i-K}).\log f(x_i|x_{i-1},...,x_{i-K}).$$

$$dx_i...dx_{i-K} \qquad (10)$$

For a serially correlated variable, the relation:

$$H(X_i) \geq H(X_i|X_{i-1}) \geq ... \geq H(X_i|X_{i-1},...,X_{i-K}) \qquad (11)$$

exists between the variables X_{i-k} (k=0,...,K). Thus, as the degree of serial dependence increases, the marginal entropy of the process will decrease until the condition:

$$H(X_i|X_{i-1},...,X_{i-k}) - H(X_i|X_{i-1},...,X_{i-(k-1)}) \leq \epsilon \qquad (12)$$

is met for an infinitesimally small value of ϵ. It is expected that the lag k where the above condition occurs indicates the degree of serial dependence within the analyzed process (Schultze, 1969; Harmancioglu, 1981).

To compute the above joint entropies and conditional entropies for lag-k series of a single variable, Eqs.(7) and (8) may again be used. This time, the C matrix of Eq.(7) includes the autocovariances as a measure of the serial dependence within the process and has the dimensions (K+1)x(K+1) (Harmancioglu, 1981).

The analysis of of redundant information in the time domain is based on the assessment of reduction in the marginal entropy of the variable due to presence of serial dependence. This analysis has to be carried out separately for each sampling site in case of the particular variable analyzed. The marginal entropy of the variable is computed by Eq. (2). Next, successive reductions in uncertainty at time lags k=1,2,3,... are determined by Eq. (10) and transinformations by:

$$T(X_i,...,X_{i-k}) = H(X_i) - H(X_i|X_{i-1},...,X_{i-k}) \qquad (13)$$

In the above, time lags k=1,2,3,... refer respectively to sampling intervals of $\triangle t$=2,3,4,.. months. Accordingly, if the redundant information between successive observations, $T(X_i,...,X_{i-k})$, or the ratio $T(X_i,...,X_{i-k}) / H(X_i,...,X_{i-k-1})$ is found to be high at any lag k, the frequency of sampling may be decreased to reduce such redundancy.

2.4 Assessment of combined space/time frequencies

Combined spatial/temporal design features constitute a problem which is multivariate with

360

respect to both time and space. In this case, the joint entropy of Mx(K+1) variables need to be computed, with corresponding transinformations for alternative combinations of numbers and locations of stations versus different $\triangle t$ sampling intervals. The computation of joint and conditional entropies are again realized by Eqs.,(7) and (8); this time, the C matrix of the former equation includes both the auto and cross covariances to represent temporal and spatial dimensions of the problem.

As noted earlier, the approach here is to increase the number of stations one by one and to investigate transinformations at different sampling intervals for each number of stations. The different sampling intervals are represented by time lags k. Accordingly, for a particular combination of stations (e.g., X, Y, Z,..), transinformations such as $T(X_i,Y_i,Z_i,...)$ (lag k=0 representing the existing sampling interval), $T(X_{i-1},Y_{i-1},Z_{i-1},...)$ (lag k=1 representing sampling interval decreased 1 time unit), or $T(X_{i-2},Y_{i-2},Z_{i-2},...)$ (lag k=2 representing sampling interval decreased 2 time units) are computed. Such transinformations are obtained by:

$$T(X_{i-k},Y_{i-k},Z_{i-k},...) = H(X_i) - H(X_{i-k}|Y_{i-k},Z_{i-k},...) \quad (14)$$

where the conditional entropy is;

$$H(X_{i-k}|Y_{i-k},Z_{i-k},...) = H(X_{i-k},Y_{i-k},Z_{i-k},...) -$$
$$H(Y_{i-k},Z_{i-k},...) \quad (15)$$

When these transinformations are obtained for all combinations of stations, the final graph obtained shows the changes in transinformations with respect to both the number of stations and varying sampling intervals.

3 APPLICATION

The above methodology is applied to the case of Mississippi River basin in Louisiana, USA, where the water quality monitoring network comprises 12 stations. The total available record at these sampling locations cover a period of 27 years with monthly observed values of water quality variables. Essentially, almost all data series have pretty regular observations which permit entropy computations in space/time dimensions. The results of the application are demonstrated below for a few selected variables.

The analysis of sampling locations is applied to available DO, EC, Cl⁻, TSS, COD, P and NO_3-N data from 12 sampling stations. All variables are assumed to be lognormally distributed except for

DO where the normal distribution gives a better fit. Joint entropies are computed by Eq. (7) for M=2,...,12, which can be used to determine the conditional entropy by Eq. (8). The number of stations is increased by fulfilling the condition in Eq. (9). Next, transinformations are computed for M=2,...,12. Results of such computations are shown in Fig.1 for all variables. As it can be observed from this figure, for variables such as COD and P, the transinformations are low even with 12 stations. For DO, a higher transinformation is obtained with a combination of only 8 stations, where the percentage of redundant information (ratio of total traninformation to total joint entropy of 8 stations) is 7%. The same number of stations produce only 3% redundant information for Cl⁻, which increases to 12% for 12 stations.

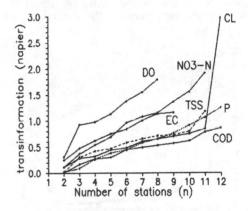

Figure 1. Increases in redundant information for selected water quality variables with respect to number of monitoring stations.

Results of computations for analysis of temporal frequencies, defined by Eqs. (10) through (13) have indicated that even the first order serial dependence within the analyzed processes are pretty low. Accordingly, reduction of the sampling frequency from monthly to bimonthly observations causes a loss of information in the order of at least 80%. This loss increases further at larger sampling intervals. Figure 2 shows this result for DO and Cl⁻ at selected stations, where even the highest percentage of redundant information is less than 20% at a sampling interval of $\triangle t=2$ months. This percentage is reduced further as the sampling interval is increased, indicating that, if the monthly sampling frequency is decreased to bimonthly or beyond, one must expect at least 80% of information loss. Thus, it is concluded here that the current practice of monthly observations should be continued.

Next, transinformations are computed for different numbers of stations and different time frequencies. The changes of redundant information in case of DO are shown in Fig.3(a) with respect to both the space and the time domains. The transiformation curves obtained clearly indicate how much redundancy each sampling site adds to the network at particular sampling intervals. Combinations of 5,6,7 and 8 stations basically increase redundancy to the same level; thus, one may consider here to select only one of the stations to reduce the number of stations from 10 to 7 as shown in Fig.3(b). On the other hand, for a particular number of stations, the redundant information reduces slightly as sampling frequencies are decreased. Thus, in the two dimensional (space/time) case, the space dimension seems to be more dominant in describing the variability (or the uncertainty) of

the variable; that is, the spatial variability of the variable appears to be a more significant factor in producing the required information about DO.

A similar result is obtained for Cl⁻ as shown in Fig.4(a & b). Fig.4(a) indicates that the number of stations can be decreased further as some groups of monitoring sites produce the same level of redundancy. Accordingly, 11 stations are reduced to 7 without changing the total transinformation for the basin as in Fig.4(b). Here again, the spatial variability of Cl⁻ seems to be more effective than temporal variability in production of information.

To make final decisions about the selection of space/time frequencies, the transinformations in Figs. 3 and 4 can be rated against the total uncertainty about the variable in the basin. Each space/time alternative gives a particular percentage of redundant information. If one prefers to reduce this redundancy by either decreasing the number of stations or increasing

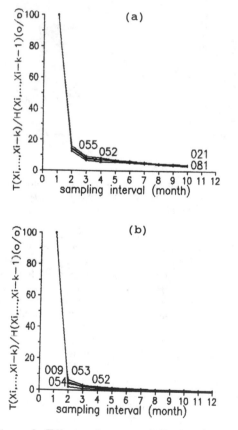

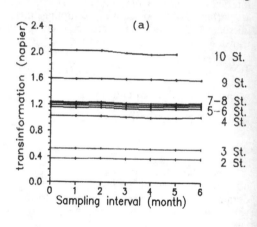

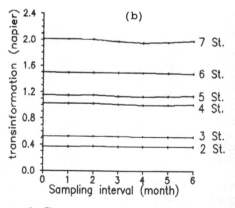

Figure 2. Effects of temporal frequencies upon information gain about: (a) DO, (b) Cl⁻.

Figure 3. Changes in redundant information on DO with respect to space/time dimensions for: (a) all 10 stations; (b) reduced number of stations.

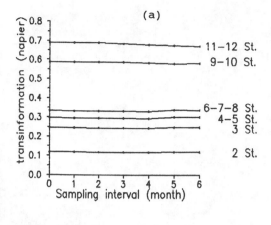

(a)

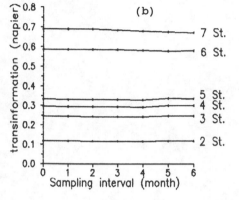

(b)

Figure 4. Changes in redundant information on Cl⁻ with respect to space/time dimensions for: (a) all 10 stations; (b) reduced number of stations.

the sampling interval, monitoring costs at decreased frequencies must be evaluated against information loss. Here, the amounth information loss may be represented by the joint multivariate entropy for a particular space/time alternative rated against the total uncertainty represented by all stations operating at the most frequent sampling intervals.

4 CONCLUSION

The work presented shows an application of the entropy principle to assessment of a water quality monitoring network in combined space/time dimensions. The results of the case study indicate that the methodology proposed can be used effectively to quantify network efficiency in terms of information provided by the network. Cost-effectiveness may also be evaluated quantitatively if cost components can be incorporated into the

method, to be compared with benefits of monitoring (information obtained) for each alternative sampling design.

It must be noted that multivariate applications of the method require sufficiently long and regularly observed data series. The use of sporadic data often lead to difficulties particularly in the assessment of time frequencies.

The application presented shows that the entropy principle is a promising method in hydrologic network design problems since it permits quantitative assessment of efficiency and benefit/cost parameters. Some mathematical difficulties pointed out earlier by the authors (Harmancioglu et al., 1994; Harmancioglu and Alpaslan, 1992) need to be solved as part of future research efforts to elaborate further on its application to network design problems.

REFERENCES

Amorocho, J. and B. Espildora, 1973. Entropy in the Assessment of Uncertainty of Hydrologic Systems and Models. *Water Resources Research*, 9(6):1511-1522.

Attanasi, E.D. and M.R. Karlinger, 1979. Worth of Data and Natural Disaster Insurance. *Water Resources Research*, 15(6):1763-1766.

Harmancioglu, N. 1994. "An Entropy-based approach to station discontinuance", in: *Stochastic and Statistical Methods in Hydrology and Environmental Engineering, vol.3: Time Series Analysis and Forecasting* (eds: K.W.Hipel & I. McLeod, Kluwer Academic Publishers, Dordrecht, the Netherlands, pp.163-176.

Harmancioglu, N.B.; Alpaslan, N.; Singh, V.P. 1994. "Assessment of the entropy principle as applied to water quality monitoring network design", in: *Stochastic and Statistical Methods in Hydrology and Environmental Engineering, vol.3: Time Series Analysis and Forecasting* (eds: K.W.Hipel & I. McLeod, Kluwer Academic Publishers, Dordrecht, the Netherlands, pp.135-148.

Harmancioglu, N.B., Alpaslan, N. 1992. "Water quality monitoring network design: a problem of multi-objective decision making", AWRA, *Water Resources Bulletin*, Special Issue on "*Multiple-Objective Decision Making in Water Resources*", vol.28, no.1, pp.1-14.

Harmancioglu, N.B. and V.P. Singh, 1991. An Information-based Approach to Monitoring and Evaluation of Water Quality Data. In: *Advances in Water Resources Technology*, G. Tsakiris (Editor). Proc. of the European Conference ECOWARM, Balkema, pp. 377-386.

Harmancioglu, N. 1981. "Measuring the information content of hydrological processes by the entropy concept", Centennial of Ataturk's Birth, *Journal of Civil Engineering*, Ege University, Faculty of Engineering, pp.13-38.

Langbein, W.B., 1979. Overview of Conference on Hydrologic Data Networks. *Water Resources Research*, 15(6):1867-1871.

MacKenzie, M., R.N. Palmer and S.T. Millard, 1987. Analysis of Statistical Monitoring Network Design. *J. of Water Resources Planning and Management*, 113(5):599-615.

Moss, M.E. 1976. "*Decision theory and its application to network design*", Hydrological Network Design and Information Transfer, World Meteorological Organization WMO, no.433, Geneva, Switzerland.

Reinelt, L.E., R.R. Horner and B.W. Mar., 1988. Nonpoint Source Pollution Monitoring Program Design. *J. of Water Resources Planning and Management*, 114(3):335-352.

Sanders, T.G., R.C. Ward, J.C. Loftis, T.D. Steele, D.D. Adrian and V. Yevjevich, 1983. *Design of Networks for Monitoring Water Quality*. Water Resources Publications, Littleton,Colorado, 328pp.

Schilperoot and Groot, 1983. *Design and Optimization of Water Quality Monitoring Networks*. Waterloopkundig Laboratorium, Delft Hydraulics Lab. No. 286, Delft, the Netherlands.

Schilperoot, T., Groot, S., Wetering, B.G.M., Dijkman, F. 1982. *Optimization of the sampling frequency of water quality monitoring networks*, Waterloopkundig Laboratium Delft, Hydraulics Lab, Delft, the Netherlands.

Schultze, E., 1969. *Einfuhrung in die Mathematischen Grundlagen der Informationstheorie*. Berlin, Springer-Verlag, Lecture Notes in Operations Research and Mathematical Economics, 116 pp.

Shannon, C.E. and Weaver, W. 1949. *The Mathematical Theory of Communication*, The University of Illinois Press, Urbana, Illinois.

Tirsch, F.S., Male, J.W. 1984. "River basin water quality monitoring network design", in: T.M. Schad (ed.), *Options for Reaching Water Quality Goals, Proceedings of 20th Annual Conference of AWRA*, AWRA Publ., pp.149-156.

Topsoe, F. 1974. *Informationstheorie*. Stuttgart, B.G. Teubner, 88p.

Stochastic Hydraulics'96, Tickle, Goulter, Xu, Wasimi & Bouchart (eds) © 1996 Balkema, Rotterdam. ISBN 90 5410 817 7

Analytical loop rating curves of flood waves

Ching-Ruey Luo & Suphat Vongvisessomjai
Department of Engineering, Asian Institute of Technology, Bangkok, Thailand

ABSTRACT A rating curve of steady uniform flow for a nonerodible section is unique but it is a loop for an erodible section when the flow is nonuniform. For a gradually varied flow in the presence of flood wave, a rating curve does exist but it is more complicated.

An analytical solution for subsidence of flood wave with unsteady state is derived from diffusion equation, a simplified form of St. Venant equations, which is later used to derive for the corresponding discharge from continuity equation in order to obtain the loop rating.

The analytical solution is finally compared with numerical solution based on the full form of St. Venant equations with weighting factor P=0.5 and P=0.7 in order to show its applicable range.

1. INTRODUCTION

There are three relevant groups of researchers on rating curves working independently on i) stage-time or subsidence and speed of flood waves, ii) their discharge-time and iii) rating curves. Emphasis will be made in this study on analytical approach.

Hayami (1951) solved diffusion equation, analytically for subsidence of flood wave with a step function in prismatic channel with no lateral inflow. Sutherland and Barnett (1972) showed how Hayami's solution could be used in natural river channels which varied in width and slope and into which lateral inflows might occur. Tingsanchali and Manandhar (1985) added backwater effects in the Hayami's solution.

Based on St. Venant equations, a general linear solution was solved by Dooge and Harley (1967) for discharge per unit width in semi-infinite uniform channels subject to a delta function inflow at upstream. The diffusion was shown to be a simplified form of the St. Venant equaltions. Ponce and Simons (1977) applied the theory of linear stability to determine propagation characteristics of various types of shallow water waves in open channel flow which were found to depend on Froude number and the dimensionlewss wave number of the unsteady component of the motions. Furthermore they assessed the applicability of the kinematic and diffusion models which were found to depend on the shape and period of waves while the dynamic model was shown to have markedly strong dissipative tendencies.

Henderson (1966) investigated characteristics of flood waves in prismatic channels for their rates of subsidence and speeds of the subsiding flood waves; loop rating of unsteady flow or the Jones formula was demonstrated. Jolly and Yevjevich (1971) treating loop rating of gradually varied flows as an amplifying or attenuating wave further investigated this phenomenon by a numerical integration of the St. Venant equations.

The above past developments will be integrated and applied in order to solve the simplified form of St. Venant equations analytically for loop rating of flood wave having a shape of sinusoidal function. The validity of the solution will be verified by a numerical model.

2. THEORETICAL CONSTIDERATION

The differential equations (St. Venant) governing for flow in a wide rectagular channel without lateral flow can be expressed as follows:

$$\frac{\partial q}{\partial x} + \frac{\partial y}{\partial t} = 0 \tag{1}$$

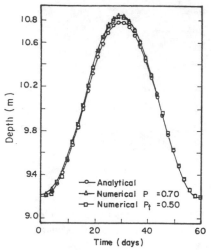

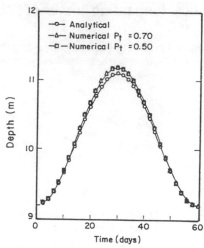

Fig. 1   Comparison of Analtical and
Numerical Results at Station
$X = 60$ km, $S_0 = 10^{-5}$, $T = 60$ days,
$P_t = 0.70$ and $P_t = 0.50$

Fig. 2   Comparison of Analytical and
Numerical Results at Station
$X = 60$ km, $S_0 = 1.55 \times 10^{-3}$,
$T = 60$ days, $P_t = 0.70$ and
$P_t = 0.50$

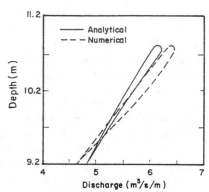

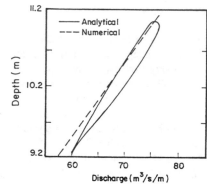

Fig. 3   Comparison of Analytical and
Numerical Results (Depth-
Discharge) at $X = 60$ km,
$S_0 = 10^{-5}$, $T = 60$ days and
$P_t = 0.70$

Fig. 4   Comparison of Analytical and
Numerical Results (Depth -
Discharge) at $X = 60$ km,
$S_0 = 1.55 \times 10^{-3}$, $T = 60$ days
and $P_t = 0.70$

and

$$(gy^3\text{-}q^2)\frac{\partial y}{\partial x} + 2yq\frac{\partial q}{\partial x} + y^2\frac{\partial q}{\partial t} = gy^3[S_0 - \frac{q^2}{C_z^2 y^3}] \quad (2)$$

in which x=distance along the channel; t=time;
g=gravitational acceleration; y=stage; q=unit discharge; S_0=slope of channel; and C_z = the Chézy's coefficient.

By perturbation method

$$q = q_0 + q' \quad (3)$$

and

$$y = y_0 + y' \quad (4)$$

in which q_0=unit discharge of uniform flow; q'=unit

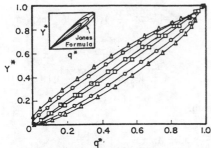

Fig. 5 Rating Curves of the Analytical
Results between the Dimensionless
Depth and the Dimensionless
Discharge for $\mu = 241,500 \text{ m}^2/\text{s}$
Periods 30 days and $S_0 = 10^{-5}$

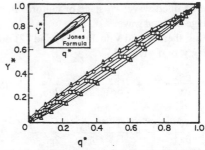

Fig. 6 Rating Curves for the Analytical
Results between the Dimensionless
Depth and the Dimensionless
Discharge for $\mu = 241,500 \text{ m}^2/\text{s}$
Periods 60 days and $S_0 = 10^{-5}$

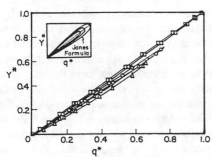

Fig. 7 Rating Curves of the Analytical
Results between the Dimensionless
Depth and the Dimensionless
Discharge for $\mu = 19,500 \text{ m}^2/\text{s}$
Periods 30 days and $S_0 = 1.55 \times 10^{-3}$

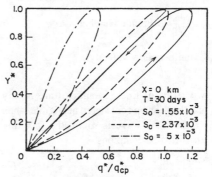

X = 0 km
T = 30 days
$S_0 = 1.55 \times 10^{-3}$
$S_c = 2.37 \times 10^{-3}$
$S_0 = 5 \times 10^{-3}$

Fig. 8 The Loop-Rating Curves for
Different Bed-Slope with
X = 0 km, T = 30 days

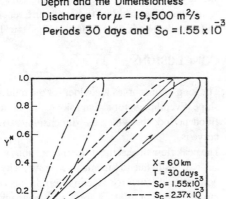

X = 60 km
T = 30 days
$S_0 = 1.55 \times 10^{-3}$
$S_c = 2.37 \times 10^{-3}$
$S_0 = 5 \times 10^{-3}$

Fig. 9 The Loop-Rating Curves for
Different Bed-Slope with
X = 60 km, T = 30 days

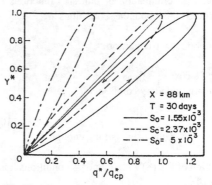

X = 88 km
T = 30 days
$S_0 = 1.55 \times 10^{-3}$
$S_c = 2.37 \times 10^{-3}$
$S_0 = 5 \times 10^{-3}$

Fig. 10 The Loop-Rating Curves for
Different Bed-Slope with
X = 88 km, T = 30 days

discharge of flood wave; y_0=stage of uniform flow; and y'=stage of flood wave.

Substituting these q and y in Eqs. (1) and (2) and eliminating y' or q' yield

$$\mu \frac{\partial^2 q'}{\partial x^2} = C_0 \frac{\partial q'}{\partial x} + \frac{\partial q'}{\partial t} + \frac{y_0 F_0^2}{S_0} \frac{\partial^2 q'}{\partial x \partial t} + \frac{u_0}{2gS_0} \frac{\partial^2 q'}{\partial t^2} \quad (5)$$

or

$$\mu \frac{\partial^2 y'}{\partial x^2} = C_0 \frac{\partial y'}{\partial x} + \frac{\partial y'}{\partial t} + \frac{y_0 F_0^2}{S_0} \frac{\partial^2 y'}{\partial x \partial t} + \frac{u_0}{2gS_0} \frac{\partial^2 y'}{\partial t^2} \quad (6)$$

in which

$$C_0 = \frac{3}{2} U_0 = \frac{3q_0}{2y_0} = \text{celerity of flood wave;}$$

$$m = \frac{q_0 |1 - F_0^2|}{2S_0} = \text{diffusivity coefficient of flood wave; and}$$

$$F_0 = u_0/\sqrt{gy_0} = \frac{q_0}{y_0\sqrt{gy_0}} = \text{Froude Number}$$

$$u_0 = C_z\sqrt{y_0 S_0} = \text{uniform flow velocity}$$

The last two terms of Eqs. (5) and (6) are highly nonlinear and they can be considered negligible when the slope of the flood wave is small; the resulting equations are diffusion equations:

$$\mu \frac{\partial^2 q'}{\partial x^2} = C_0 \frac{\partial q'}{\partial x} + \frac{\partial q'}{\partial t} \quad (7)$$

and

$$\mu \frac{\partial^2 y'}{\partial x^2} = C_0 \frac{\partial y'}{\partial x} + \frac{\partial y'}{\partial t} \quad (8)$$

Considering a dimensionless wave number and a wave shape of a sinusoidal function, which can be treated as the combination of unit step functions in Hayami (1951), the bounday conditions are:

Upstream boundary condition

$$y'(0,t) = y_p[1 - \cos(\frac{2\pi t}{T})]; \quad (9)$$

Downstream boundary condition

$$y'(\ell, t) = 0 \quad (10)$$

if ℓ=the length of the reach is long enough.

The solution of Eq. (8) with boundary conditions Eqs. (9) and (10),

$$\frac{y(x, t) - y_0}{y_p} = (1 - \cos\frac{2\pi t}{T})[1 - \text{erf}(\frac{\pi_3}{2})] = Y^* \quad (11)$$

Solving for q in Eq. (1) using the above y(x,t) gives,

$$\frac{q(x,t) - q_0}{y_p C_0} = q^* = (1 - \cos\frac{2\pi t}{T})\{(\frac{1}{4\sqrt{\pi}})(\frac{\pi_1}{\pi_2})\exp(-\pi_3^2)$$

$$+1/8\exp(\frac{-4\pi_2 + -\pi_3^2}{4\pi_1^2})[\text{erf}(\frac{1}{\pi_1} - \frac{\pi_3}{2}) + \text{erf}(\frac{\pi_3}{2})]$$

$$-(\frac{2\pi t}{T})(\frac{x}{\ell})(\frac{1}{\pi_2})(\frac{\sin(\frac{2\pi t}{T})}{1 - \cos\frac{2\pi t}{T}})[1 - (\frac{\pi_3^2}{2\sqrt{\pi}})]\} \quad (12)$$

in which:

$$\pi_1 = \frac{\sqrt{ut}}{\ell}; \quad \pi_2 = c_0 \ell t; \quad \pi_3 = \frac{x}{\sqrt{ut}}; \quad \ell \geq \frac{C_0 t}{4}$$

3. COMPARISONS AND DISCUSSION

The comparisons of water depth and discharge between analytical and numerical results with different period, diffusivity, bed slope, and time weighting factor for a given position are presented in Figs. 1 to 4, Good agreement was obtained. Here, the characteristic of diffusivity were considered and the results gave the effect due to wave dispersion, which showed the Y*-q* loop curves different from the ones of Jones Formula. The analytical results really could express the river boundary and wave dispersion effect for the flood wave characteristics as in Figs 5 to 7. The diffusivity would be equal to 0 for the critical flow. The rating curve of Y*-q* for subcritical flow could be below Y*=q*, which was for the centerline of critical flow. When the supercritical flow happens, the Y*-q* could be above Y*=q*. The loop rating curves for subcritical and supercritical flow with different positions for a given period are shown in Figs. 8 to 10.

4. CONCLUSIONS

1. The loop rating curves are wider for the milder bed slope. Eventhough for a given slope, the shorter the flood period, the wider the dimensionless rating curve is.

2. The peak flow time at the different locations will be shifted a little due to kinematic wave velocity, and when the diffusivity is considered, the rating curves, not complete a loop like peacock tail feather.

3. The analytical model can be used without any sophisticated computing machine. In fact, a simple desk calculator and a table of error function are sufficient in carrying out the computation.

REFERENCES

1. Dooge, J.C.I. and Harley, B.M., "Linear routing in uniform open channels", Proceedings, International Hydrology Symposium. Fort Collins, Colorado, Septimder 6-8, Vol. 1, paper 8, 1967.
2. Hayami, S., "On the propagation of flood waves, " Disaster Prevention Research Institute, Kyoto Universitg, Bulletin 1, Japan, 1951.
3. Henderson, F.M., "Flood waves in prismatic channels" J. of the Hydraulic Division ASCE. Vol. 89, No. Hy 4, 1966.
4. Jolly, J.P. and Yevjevich, V., "Amplification criterion of gradually varied simple-peaked Waves", Hydrology Paper, No. 51 Colorado State University, Fort Collins, 1971.
5. Ponce, V. M. and Simons, D. B., "The propagation of dynamic waves in open channel flow", Congress of IAHR Vol. 2, Baden-Baden, 1977.
6. Sutherland, A.J. and Barrentt, A.G., "Diffusion solution to flows with upstream control", J. of He Hydraulic Division, ASCE, No. Hyll, 1972.
7. Tingsanchali and Manandhar, "Analytical diffusion model for flood routing with downstream effect", J. of Hydraulic Division, ASCE, Vol. 111, 1985.

Stochastic Hydraulics'96, Tickle, Goulter, Xu, Wasimi & Bouchart (eds)© 1996 Balkema, Rotterdam. ISBN 90 5410 817 7

Overtopping risk analysis for the earth dam of Qinghe Reservoir

Qijun Li
Beijing Institute of Hydraulic Research, China

Zhaohe Chen
North China Institute of Water Resources & Hydropower, Beijing, China

Lisha Liu
Songhuajiang & Lao River Conservancy Commission, Changchun, China

ABSTRACT: Risk analysis against overtopping of an earth dam during a flood was applied to the management of water resources for a large reservoir—Qinghe Reservoir located in Liaoning Province, China. In this study, the flood, wind, volume of reservoir and discharge capacity are all considered as random variables. Then a stochastic differential equation is established for the earth dam to analyze its risk of overtopping against the design series of flood events accompanied with the runup generated by wind wave during flood period. On the basis of risk analyses, some conclusions and recommendations were made.

1 INTRODUCTION

In order to improve the management of Qinghe Reservoir, Liaoning Province, China, a risk model is developed for its earth dam to analyze its risk of overtopping against the design series of flood events accompanied with the runup generated by wind wave during flood period. Taking 10^{-6} as acceptable risk, or 99.999% as acceptable reliability, the risk analysis shows that the earth dam is very reliable against the simultaneous actions of flood and effective wind events. On the basis of risk analysis, we recommended that its original limiting level before flood, say elevation 127m above Dalian datum, may be safely raised to an elevation of 129m. The correspondent increase in water volume of storage should bring an annual direct benefit of about 5.0 million RMB to Qinghe Reservoir Administration along with a great deal of social benefits.

2 RESERVOIR CHARACTERISTIC

Qinghe Reservoir located in Qinghe District, Tieling City, Liaoning Province. It controls a catchment area of 237km² of Qinghe River which is a tributary of Liaohe River in the Northeast China. It has an earth dam with an inclined clay diaphram. The maximum height of the dam is 39.4m. Crest length of the dam is 1622m and crest elevation is 138.1m above Dalian datum. Top elevation of the parapet on the dam crest is 139.25m. The spillway has six spans, each 10m wide with crest elevation of weir at elevation 126m.

The maximum discharge capacity of spillway is 4210 c.m.s. The outlet work is a circular conduit of 5.5m diameter with maximum discharge of 389 c.m.s. The maximum storage capacity of the reservoir is $9.71 \times 10^{8} m^{3}$. All of these makes the resrvoir operate as a multipurpose reservoir for overyear storage.

3 HYDROLOGICAL CHARACTERISTICS

In the design of the reservoir, its design flood has a frequency of 0.1% and the extreme flood has a frequency of 0.01%. The design hydrograph is magnificated on the basis of 7 days flood volume and flood peak under the same frequency from the measured flood hydrograph in 1953 shown in Table 1, which is selected as representative hydrograph.

4 UNCERTAINTIES

In our study, the flood, wind, volume of reservoir and discharge capacity are all considered as random variables.

4.1 Flood

It is well known that flood of any frequency is random event. Generally, it follows P — III type of distribution, which is very familiar with us. For Qinghe Resevoir the probabilistic density functions for the design flood peak, $f_1(Q)$, and 7 days flood volume, $f_2(V)$, may be expressed as follows:

$$f_1(Q) = 0.030031064(Q - 171.8)^{-0.673469388}$$
$$\times exp(-4.7516096 \times 10^{-4}$$
$$\times (Q - 171.8))$$
$$for \quad 171.8 \leqslant Q \leqslant 17400 \qquad (4.1)$$
$$f_2(V) = 1.036852V^{0.524158}exp^{(-0.946682V)}$$
$$for \quad 0 \leqslant V \leqslant 13.662 \qquad (4.2)$$

For our case, we use only the equation (4.2) due to the data presented by Qinghe Reservoir Administration.

Table 1. Representative hydrograph selected —measured flood hydrograph in 1953

Date (M. D. H)	typical flow (m³/s)	Date (M. D. H)	typical flow (m³/s)
8. 15. 24	14.7	8. 20. 3	4720
16. 24	19.6	4	4760
17. 24	62.5	5	4730
18. 24	35.4	6	4660
19. 2	46.5	7	4550
4	71.2	9	4260
8	425	14	3490
11	575	18	2900
12	668	21	2560
13	679	22	2480
14	668	23	2430
16	592	21. 1	2060
18	592	2	2040
20	640	6	1250
21	820	8	1020
22	1200	10	982
24	2870	12	860
20. 1	3800	14	675
2	4490	20	515
		22. 5	368

4.2 Wind

It is well known also that wind of any magnitude from any direction is a random event. The wave height and runup generated by wind is then random event too. As for the overtopping risk of the earth dam, the wind toward the dam during flood period is major concern and is defined as "effective wind for overtopping " by us. For Qinghe dam, the effecfive directions of wind are E, NE and SE. The maximum wind velocities during 1955 – 1990 were collected from Tieling Bureau of Meteorology and the pertinent effective winds as well as their maximum velocities at the water surface were determined. It was shown that the maximum velocities of effective winds in our case follow the extreme Type I distribution. Its distribution function may be expressed as follows:
$$F(x) = exp(-exp(-(x - 7.34)/1.95)) \quad (4.3)$$

The mean value is W=8.46m/s and the standard error σ is 2.5.

4.3 Reservoir area and storage volume

Although the reservoir area and storage volume were traditionally considered as deterministic quantities, they are actually random variables. The contour lines plotted by different surveyors for a given reservoir topography may be different. Under the same contours, the calculated reservoir area and storage volume at a given water level may be different due to variety of computation philosophy and the instruments used. Moreover, the sediment transportation during and after flood may change the underwater topography and thereby the reservoir area and storage volume unless the underwater contours be surveyed after each flood immediately.

Relationship between water level and reservoir area $F(Z)$ is shown in the last literature listed in References of this paper.

A normal distribution is assumed for it with a standard error $\sigma = 0.1F(Z)$.

4.4 Discharge Capacities

The discharge capacities of Qinghe Reservoir are consisted of two parts: spillway and outlet work. The uncertainties of each part arise from many sources, such as the simplification of 3–D flow to 1–D flow model, measurement errors and roughness estimate. We may consider the random nature of discharge coefficient of either spillway or outlet work as the main uncertainty source. It was shown that discharge coefficient follows approximately the normal distribution. We take the relationship between discharge coefficint of either spillway or outlet work as mean value curve. Their standard error are taken as $\sigma = 0.05S$ where S is the mean value of discharge coefficient.

5 RISK MODEL

The risk model of overtopping was developed and may be expressed as follows:
$$\overline{R} = P(Z_0 + H_{max} + e + R_p \geqslant Z_c) \qquad (5.1)$$
where $\overline{R}$=risk against simultaneous actions of a series of flood and effective wind events; Z_0 = initial reservoir level; H_{max} = reservoir level increase due to flood; e=wind setup; R_p =wave runup on slope of the dam; Z_c =predetermined critical elevation, such as the dam crest elevation or top elevation of parapet.

In the case of concurrence of flood event $[Q_{i-1}, Q_i]$ and wind event $[W_{j-1}, W_j]$, the risk P_{ij} is

$$P_{ij} = P(Z_0 + H_{max} + e_{ij} + R_{ij} \geq Z_c) \qquad (5.2)$$

Then, the total risk may be expressed as follows:

$$\bar{R} = \sum_{i=1}^{\infty} f(Q_i) dQ_i [\sum_{j=1}^{\infty} f_w(W_j) dW_j P_{ij}] \qquad (5.3)$$

In the above equations, the mean value of wind setup may be calculated by the following equation:

$$\bar{e} = (K^2 D/2gh) cos\beta \qquad (5.4)$$

where W=velociy of wind at 10m above reservoir level (m/s); D=fetch length(m); H=the average depth of the reservoir along the fetch length; K=a coefficient, its value varied in the range of $(1.5 \sim 5.0) \times 10^{-6}$, and may be taken as 3.6×10^{-6}; β=angle between the wind direction and water body fetched, and may be take as 0° on the safe side. Because variables W follow extreme Type I distribution, the vairables e do so too. The standard error for e may be determined as

$$\sigma_e = (KWD/(gH)) \cdot \sigma_W \qquad (5.5)$$

where σ_W=standard error of the variable W.
It was well known that wind velocities follow Rayleigh distribution, and wave heights do so too. Its distribution function and probablility density function are:

$$F(x) = 1 - e^{-x^2/(2\mu^2)} \quad for \ x \geq 0 \qquad (5.6)$$

$$f(x) = (x/\mu)e^{-x^2/(2\mu^2)} \quad for \ x \geq 0 \qquad (5.7)$$

Mean value M(x) and standard error σ(x) are related with distribution parameter μ as follows:

$$M(x) = \sqrt{0.5}\mu \quad and$$

$$\sigma(x) = \sqrt{0.5(4 - \pi)}\mu \qquad (5.8)$$

As for wave runups on the dam slope, they follow Rayleigh distribution because the coefficient of correlation between wave runup and wave height is 1. The mean value of wave runups may be computed by the following equations:

$$\bar{R}_p = (K_\Delta K_w / \sqrt{1 + m^2})(\bar{h} \bar{\lambda})^{0.5} \qquad (5.9)$$

$$\frac{gh}{W^2} = 0.13th[0.7(\frac{gH}{w_2})^{0.7}] \times$$

$$th\{\frac{0.0018(gD/W^2)}{0.13th[0.17(gH/W^2)^{0.7}]}\} \qquad (5.10)$$

$$\bar{\lambda} = \frac{g\bar{T}^2}{2\pi} \qquad (5.11)$$

$$\bar{T} = 4.0(\bar{h})^{0.5} \qquad (5.12)$$

where K_Δ=coefficient considering the roughness and permeability of dam slope; K_w=empirical coefficient as a function of dimensionless parameter $W/\sqrt{gH}$; m = ctgα and α= slope angle $\bar{h}$= average wave height (m); $\bar{\lambda}$=average wave length (m); W=wind velocity (m/s); H=water depth before the dam (m); D= fetch (m); g=gravitational acceleration (m/s²); T= average wave period (s).

Because wind events are stochastic, the variables R_p are composite random variables. It may be shown that the probability P_{wi} of occurence of R_p when the wind velocity is within the interval $[W_{i-1}, W_i]$ is

$$P_{wi} = P(R_p < x)$$

$$= \int_0^x f_{R_p}(R_p) dR_p \qquad (5.13)$$

where f_{R_p} is the probability density function of Rayleigh distribution for the interval $[W_{i-1}, W_i]$. Therefore, the distribution function of R_p may be expressed as

$$F(x) = \sum_{i=1}^{\infty} f_w(W_i) P_{wi} dW_i$$

$$= \sum_{i=1}^{\infty} f_w(W_i) dw_i \int_0^x f_{R_p}(R_p) dR_p \qquad (5.14)$$

In order to compute P_{ij}, it is necessary to normalize at R_p due to Rayleigh distribution of R_p and do iteration by means of the AFOSM method.

H_{max} is a function of flood flow Q, reservoir area F and discharge capacity S and may be determined through flood routing. A stochastic differential equation for flood routing may be established as follows:

$$Q(t) - S(t) = \frac{d\bar{V}}{dt} = F(Z)\frac{dZ}{dt}$$

$$= F(Z)f(t, Z) \qquad (5.15)$$

Because the variable $Z_{f\ max} = Z_0 + H_{max}$ did not expressed as analytic function, we have to solve eq. (5.15) numerically by means of Integration—AFOSM method.

6　CRITICAL　MODE

Two modes were analyzed. The first taking the crest elevation of the earth dam, i. e. in Qinghe case 138. 1m above Dalian datum, as the critical elevation Z_c. The second, taking the top elevation of parapet, i. e. 139. 25m as Z_c.
For the first mode, the performance formula is

$$g(\cdot) = 138.1 - Z_f - e \qquad (6.1)$$

whereas for the second one, the performance formula is

$$g(\cdot) = 139.25 - Z_f - e - R_p \qquad (6.2)$$

7　RESERVOIR　REGULATION

Two alternatives of regulation of Qinghe Reservoir were studied and finally, due to the requirements of simultaneous operation of four reservoirs (including Qinghe Reservoir) in the Liaohe River system and staggering the flood peaks in time, one of the alternatives was adopted and described herein. The adopted regulation was given as follows: If the reservoir level is within the range of 127. 00 ~ 131. 60m, then according to the inflow, the maximum discharge be limited to 300 c. m. s; during reservoir level is 131. 60 ~ 133. 90m, 300 c. m. s. be released through the outlet conduit and simultaneously water be discharged through two openings of spillway gates; during reservoir level 133. 90 ~ 134. 4m, 300c. m. s. be released through the outlet conduit and simultaneously water be discharged through four openings of spillway gates:

during the reservoir level is higer than 134. 40m, the outlet conduit release 300 c. m. s. again but the six spillway gates are all opened to discharge the inflow. If the flood peak with frequency of 0. 1% occurs, it will not be controlled by any water level in the reservoir but only by the fully—opened spillway gates while the outlet conduit releases still 300 c. m. s.

It should be noted that the accuracy of forecasting on flood volume for the Qinghe River watershed is 87. 7% whereas the accuracy of forecasting on flood peaks for the watershed is 80% with a foresight of 7 hours.

8 RESULTS OF RISK ANALYSES

For simplicity, only the results of risk analyses for the first and second critical modes under the adopted alternative of reservoir regulation are listed in Table 2.

It is traditional practice that the reservoir level, during flood period, should be limited to such a specified elevation that the actual water level in the reservoir before the next flood coming into the reservoir is equal to or lower than the specified one. So the specified elevation is defined as "limiting reservoir level befor coming flood" or simply "limting level before flood". If we take a risk of order of magnitude of 10^{-6} as an acceptable risk, it can be seen from Table 2 that the limiting reservoir level before coming flood originally suggested by designer, say 127m above Dalian Datum, may be safely raised to elevation 129. 15m.

Table 2.   Overtopping risk under the finally adopted alternattive of reservoir regulation

Initial Reservoir Level (m)	Risk $\overline{R}_1$ under the 1st critical mode	Risk $\overline{R}_2$ under the 2nd critical mode
128. 50	0. 120382×10^{-6}	0
128. 80	0. 271038×10^{-6}	0. 376499×10^{-22}
129. 00	0. 339837×10^{-6}	0. 629230×10^{-18}
129. 15	0. 403396×10^{-6}	0. 259276×10^{-15}

9 RECOMMENDATIONS AND CONCLUSIONS

On the basis of risk analysis against overtopping for Qinghe dam, it was recommended that its limiting reservoir level before the coming flood originally suggested by designer, say 127. 0m above Dalian datum, may be raised to an elevation of 129. 0m much more on the safe side which corresponds to a risk of the order of magnitude of 10^{-6} against overtopping under concurrence of flood series and wind events during flood period. The increase of volume of storage corre-

sponding to the suggested increment of limiting level, 2m, will be 78. 5×10^6m³, which may bring an annual direct benefit of about 5. 4 million RMB to Qinghe Reservoir Administration. Besides the direct benefit, a great deal of social benefit will be gained due to the corresponding reduction of the possibility of damage caused by waterlogging in the farm land downstream of the reservoir, as well as the improvement of reservoir water quality and aquatic environment.

The risk analysis presented herein may be applied to similar case where the earth or rockfill dam should never be overtopped. Besides, it may be applied to serve the decision — making of reservoir regulation with scientific quantitative data. During the enlargement or strengthening of existing dam, the risk analysis presented herein may be applied to provide pertinent guidelines.

REFERENCES

CHEN Zhaohe, LI Qijun et al 1991 Research Report—Increasing the economic benefits of Douhe Reservoir through overtopping risk analysis (in Chinese).

LI Qijun 1991 Theory and application of overtopping risk analysis of earth dams. Master thesis. Supervisor: Prof. CHEN Zhaohe and Assoc. Prof. YE Shouzhong. Beijing Postgraduate School, North China Inst. of Water Resources & Hydropower.

LI Qijun 1992 A Study of computation of risk, Information on Science & Technology, Vol. 12, No. 1, March, pp. 1——8 (in Chinese).

CHEN Zhaohe, LI Qijun & YE Shouzhong 1992 Stochastic numerical simulation of reservoir flood regulation , Information on Science & Technology, Vol. 12, No. 3, Sept. , pp. 1——6 (in Chinese).

LI Qijun & CHEN Zhaohe et al 1991 Risk identification and analyses of accidents for earth dams , Information on Science & Technology, Vol. 11, No. 3, Sept. , pp. 1——8 (in Chinese).

LI Qijun, YE Shouzhong & CHEN Zhaohe 1995 Application of risk analysis in management of water resources for a large reservoir, Advances in Hydro — science and — engineering, Vol. I , Part A, pp. 1561~1568

LI Qijun, YE Shouzhong & CHEN Zhaohe 1995 Application of risk analysis in management of water resources for Zhuzhuang Resevoir, Proc. of National Conference on Hydraulics, Wulumuqi, China. (in Chinese)

Qinghe Reservoir Administration 1991 Fundamentals of Qinghe Reservoir.

Stochastic Hydraulics'96, Tickle, Goulter, Xu, Wasimi & Bouchart (eds) © 1996 Balkema, Rotterdam. ISBN 90 5410 817 7

The effect of initial state in flood estimation model utilizing Kalman filter algorithm

Chang-Shian Chen
Department of Hydraulic Engineering, Feng Chia University, Taichung, Taiwan, China

Shaohua Marko Hsu & Jung-Kun Lin
Graduate Institute of Hydraulic Engineering, Feng Chia University, Taichung, Taiwan, China

ABSTRACT: Prior to practical application of Kalman filter algorithm, some initial state values have to be determined beforehand. When the initial values are not set appropriately, Kalman filter algorithm may not improve the accuracy of its subsequent simulated result. The derivation of Kalman filter algorithm may be divided into three categories which are (a) both the variances of system error (R_k) and of measurement error (Q_k) being time-correlated; (b) Q_k being time-correlated while R_k being time-independent; (C) both Q_k and R_k being time-independent. This study investigates the influence of different derivations and the selection of initial state values on the flood estimation model. Some suggestions are proposed, to enhance the practicality of Kalman filter algorithm.

For illustration, the Pa-Chung River Basin in Taiwan was used as the study area. The results from the study revealed that the flood estimation model using, the Muskingum differential model, along with Kalman filter algorithm with wide range of initial values for the state vector and using time-independent derivation for Q_k and R_k (Q_k equals to 0 and R_k is smaller than 1000) may achieve the most stable and accurate results. In other words, if Q_k and R_k were given the representative values adequately, the accuracy of the initial values for the state vector is not very crucial.

1. INTRODUCTION

Since Kalman filter technique first introduced in 1960 , it has been successfully applied in different fields , such as electrical , automatic , chemical, nuclear , hydraulic and hydrologic engineering etc. In general, the derivation of Kalman filter algorithm can be divided into three categories : (1) both the covariance of system error (Q_k) and covariance of measurement error (R_k) are time-correlated (Wood 1978 , Simous 1978) ; (2) Q_k is time-correlated while R_k is time-independent (Kuo 1988 , Wilson 1978); and (3) both Q_k and R_k are time-independent (Chen 1994 ; Ngan 1986 ; Van Geer 1991) . The initial values of the state vector can be chosen from the averaged state vector of historical records (Wood 1978 ; Ngan 1986). However in most case, the initial values are determined by trial and error . In order to improve the practicality of Kalman filter

algorithm in flood routing model , the effects of the initial values of the state vector on the routing results are studied under different derivations . Then, the most suitable derivation and the most convenient approach for setting the initial values are suggested.

2. THE FLOOD ESTIMATION MODEL

Kalman filtering is a recursive process based on the principle of minimizing the variance of error. Parameters of the model is then updated and the system output can be estimated.

The governing equations involve:

(1) System equation

$$X_k = \Phi_{k-1}X_{k-1} + W_{k-1} \qquad (2.1)$$

(2) Measurement equation

$$Z_k = H_kX_k + V_k \qquad (2.2)$$

where k is the index for time step, **X** is the state vector, Φ is the state transition matrix, **H** is the

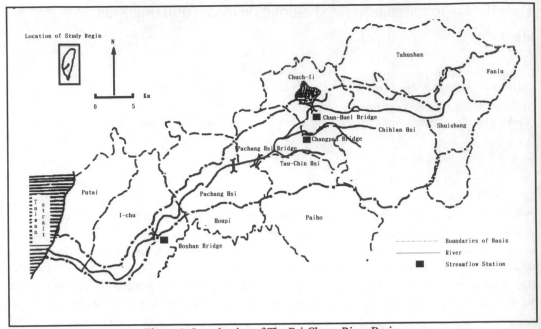

Figure 1. Introduction of The Pai-Chung River Basin

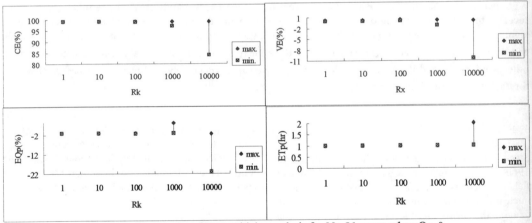

Figure 2. The results of R_k sensitivity analysis for No.81 event when $Q_k=0$

Table 1 : Typhoon-flood events used in this study

No.	Date	Peak discharge(cms)
70	1981.9.01	1360
74	1985.8.23	500
75	1986.9.19	633
76	1987.7.27	610
77	1988.8.13	1020
78	1989.9.12	1750
79	1990.8.19	848
80	1991.6.22	530
81	1992.7.06	820

Table 2 : Values of evaluating indicators in the case of Q_k and R_k being time-correlated

Typhoon-flood event			No.77				No.81			
C0	C1	C2	CE%	VE%	EQp%	ETp	CE%	VE%	EQp%	ETp
0.1	0.1	0.8	92.18	3.32	1.21	1	81.00	-3.89	0.83	1
0.1	0.2	0.7	-149.46	8.56	204.69	-27	98.11	-0.25	0.94	1
0.1	0.3	0.6	96.60	3.22	29.57	-9	77.54	1.77	64.64	16
0.1	0.4	0.5	26.18	7.98	153.18	-4	91.61	1.16	14.52	3
0.1	0.5	0.4	44.96	-18.19	3.70	4	74.14	-1.35	49.17	-1
0.1	0.6	0.3	64.81	-3.76	38.12	22	62.84	1.85	126.91	7
0.1	0.7	0.2	77.79	0.81	24.22	0	-1184.79	-5.74	803.86	-36
0.1	0.8	0.1	-696.79	40.42	302.39	-18	-121.49	-0.88	161.30	-30
0.2	0.1	0.7	-7.44	5.13	159.03	-13	-78.80	-1.89	134.26	16
0.2	0.2	0.6	84.20	2.13	44.10	-11	-0.56	-6.53	31.47	-9
0.2	0.3	0.5	98.55	-1.17	0.70	2	92.29	0.71	28.39	-2
0.2	0.4	0.4	98.91	-0.20	0.96	1	91.08	0.34	7.45	5
0.2	0.5	0.3	94.94	-0.07	1.07	1	96.26	-0.35	1.70	0
0.2	0.6	0.2	-194.66	-37.81	41.18	-16	-32.36	7.80	183.99	-54
0.2	0.7	0.1	-57.63	10.08	190.28	0	-219.13	5.57	404.94	11
0.3	0.1	0.6	98.59	-0.55	0.74	1	80.23	-3.73	0.46	1
0.3	0.2	0.5	61.82	-2.26	23.92	-12	96.51	-1.06	0.50	1
0.3	0.3	0.4	97.49	-0.14	0.53	1	82.82	1.46	23.47	32
0.3	0.4	0.3	-0.40	2.90	151.97	22	74.25	-1.92	17.73	6
0.3	0.5	0.2	55.26	6.96	96.27	-3	84.06	1.91	24.83	7
0.3	0.6	0.1	91.56	-0.28	18.94	-6	-369711	891.25	9067.02	5
0.4	0.1	0.5	99.21	-0.89	1.32	-1	27.05	6.59	211.83	1
0.4	0.2	0.4	96.65	-0.20	19.84	5	94.37	2.11	30.45	-3
0.4	0.3	0.3	89.25	-1.25	9.71	1	85.53	1.59	57.39	18
0.4	0.4	0.2	52.92	4.74	34.53	22	-905.64	11.22	682.15	-24
0.4	0.5	0.1	25.45	7.05	165.69	-1	61.76	3.21	92.36	15
0.5	0.1	0.4	97.61	1.18	8.85	-3	83.39	4.06	10.42	18
0.5	0.2	0.3	94.16	1.23	17.86	-7	94.45	-0.86	0.48	0
0.5	0.3	0.2	95.11	-0.75	-0.01	1	-23138.4	81.49	3822.60	21
0.5	0.4	0.1	22.63	7.69	104.75	12	-13626.5	112.02	1976.91	-24
0.6	0.1	0.3	93.17	2.42	25.76	6	94.90	0.15	0.60	1
0.6	0.2	0.2	93.90	-1.45	2.29	1	-4046.34	6.69	1581.73	-4
0.6	0.3	0.1	-83.94	11.31	257.01	12	69.25	-0.12	30.31	3
0.7	0.1	0.2	93.15	2.58	7.77	21	-16777.9	186.69	1970.40	-27
0.7	0.2	0.1	-3591.74	71.94	845.06	28	88.32	1.43	51.15	-7
0.8	0.1	0.1	-2.01	-5.22	23.58	11	68.35	4.47	72.93	8

measurement matrix, $\mathbf{W}$ is the system error vector, $\mathbf{Z}$ is the measurement vector, and $\mathbf{V}$ is the measurement error vector. Note that $\mathbf{W}_k$ is normal distributed with mean zero and variance $\mathbf{Q}_k$ $(\mathbf{W}_k \sim N(0,\mathbf{Q}_k),\ \mathbf{V}_k \sim N(0,\mathbf{R}_k))$, and the expected value

$$E\left[\mathbf{W}_i \cdot \mathbf{W}_j^T\right]=0,\ =E\left[\mathbf{V}_i \cdot \mathbf{V}_j^T\right]=0 \text{ if } i \neq j, \text{ in which}$$

$$E\left[\mathbf{V}_i \cdot \mathbf{W}_j^T\right] = 0 \text{ for any } i, j.$$

In this study, we chose the Muskingum model to estimate the discharge hydrograph of a flood-event in a catchment . The differential model is

$$O_{k+1}=C_0 I_{k+1}+C_1 I_k+C_2 O_k \qquad (2.3)$$

where O is outflow, I is inflow, C_0, C_1, C_2 are the parameters of the model. In the framework of Kalman filter algorithm, the state vector, measurement vector, and measurement matrix are, respectively, as follow:

Table 3 : Values of evaluating indicators in the case of $Q_k=0$, R_k time-correlated

Typhoon-flood event			No.77				No.81			
C0	C1	C2	CE%	VE%	EQp%	ETp	CE%	VE%	EQp%	ETp
0.1	0.1	0.8	85.39	14.91	13.43	1	98.10	-3.66	-7.75	1
0.1	0.2	0.7	82.59	15.34	13.98	-3	98.07	-3.65	-7.88	1
0.1	0.3	0.6	81.14	15.50	15.77	-3	98.07	-3.65	-7.90	1
0.1	0.4	0.5	80.45	15.56	16.52	-3	98.06	-3.64	-7.92	1
0.1	0.5	0.4	78.51	15.47	17.43	-3	98.05	-3.59	-7.98	1
0.1	0.6	0.3	-10551.7	137.16	1161.34	-11	98.02	-3.44	-8.08	1
0.1	0.7	0.2	78.74	15.41	16.57	-3	97.98	-3.02	-8.07	-1
0.1	0.8	0.1	65.96	15.98	22.21	-11	94.42	1.24	-9.58	1
0.2	0.1	0.7	85.11	15.01	13.48	1	98.10	-3.72	-7.70	1
0.2	0.2	0.6	82.42	15.41	14.93	-3	98.07	-3.69	-7.85	1
0.2	0.3	0.5	80.93	15.55	17.04	-3	98.06	-3.67	-7.91	1
0.2	0.4	0.4	80.41	15.52	17.75	-3	98.05	-3.61	-7.98	1
0.2	0.5	0.3	-2133.62	48.85	517.63	-10	98.02	-3.48	-8.08	1
0.2	0.6	0.2	-405.11	29.45	232.38	-10	97.98	-3.09	-8.08	-1
0.2	0.7	0.1	66.50	16.10	20.48	-11	94.67	1.07	-9.50	1
0.3	0.1	0.6	85.03	15.04	13.50	1	98.11	-3.79	-7.64	1
0.3	0.2	0.5	82.25	15.44	15.40	-3	98.08	-3.74	-7.82	1
0.3	0.3	0.4	80.73	15.56	17.62	-3	98.05	-3.66	-7.94	1
0.3	0.4	0.3	79.82	15.35	18.53	-3	98.02	-3.51	-8.07	1
0.3	0.5	0.2	-6931.85	107.52	987.24	-10	97.97	-3.14	-8.10	-1
0.3	0.6	0.1	68.51	16.09	20.18	-3	94.90	0.92	-9.43	1
0.4	0.1	0.5	84.84	15.09	13.50	1	98.10	-3.81	-7.63	1
0.4	0.2	0.4	81.89	15.51	15.97	-3	98.07	-3.71	-7.86	1
0.4	0.3	0.3	80.56	15.53	17.90	-3	98.02	-3.54	-8.05	1
0.4	0.4	0.2	-984.74	37.22	361.56	-10	97.97	-3.16	-8.11	-1
0.4	0.5	0.1	69.70	16.20	20.35	-3	94.99	0.88	-9.40	1
0.5	0.1	0.4	84.26	15.24	13.46	1	98.10	-3.73	-7.71	1
0.5	0.2	0.3	81.42	15.62	16.63	-3	98.04	-3.54	-8.01	1
0.5	0.3	0.2	75.73	15.79	22.48	-3	97.98	-3.16	-8.09	-1
0.5	0.4	0.1	65.60	16.81	24.98	-3	94.87	1.01	-9.43	1
0.6	0.1	0.3	83.12	15.50	13.37	1	98.07	-3.52	-7.88	1
0.6	0.2	0.2	82.36	15.40	14.55	-3	97.99	-3.13	-8.04	-1
0.6	0.3	0.1	33.87	19.03	50.55	-10	94.53	1.29	-9.53	1
0.7	0.1	0.2	83.92	15.31	13.70	1	98.01	-3.06	-7.91	-1
0.7	0.2	0.1	29.42	19.04	46.44	-10	93.99	1.65	-9.70	1
0.8	0.1	0.1	27.89	18.72	47.41	-11	93.33	2.04	-9.68	1

$$X_k = \left[C_0^k, C_1^k, C_2^k\right]^T$$

$$Z_{k+1} = O_{k+1}$$

$$H_{k+1} = \left[I_{k+1}, I_k, O_k\right]$$

3. MODEL ROUTING AND RESULT ANALYSIS

The Pa-Chung River Basin in Taiwan was selected as a study area . A map of the Pa-Chung River Basin is shown in Figure 1. The gage station at the upstream is Chun-Huel Bridge with a drainage area around 122 km² .The gage station at the downstream is Ichu station , and the drainage area is around 441 km² . In this study , 9 typhoon-flood events during 1981 to 1992 are chosen and they are listed in Table 1

Four indicators are used to evaluate the accuracy of the routing model :

(1) Coefficient of efficiency:

Table 4 : Values of evaluating indicators in the case of Q_k=0.1, R_k time-correlated

Typhoon-flood event			No.77				No.81			
C0	C1	C2	CE%	VE%	EQp%	ETp	CE%	VE%	EQp%	ETp
0.1	0.1	0.8	92.78	-1.38	3.12	1	-297.78	-59.53	6.47	16
0.1	0.2	0.7	48.46	-4.11	52.2	3	-425.88	6.71	543.84	18
0.1	0.3	0.6	95.32	-1.8	1.46	2	86.81	-0.81	0.59	1
0.1	0.4	0.5	41.79	-1.98	40.45	30	89.54	-0.37	0.74	1
0.1	0.5	0.4	84.72	-2.31	1.26	1	84.86	-0.22	53.2	26
0.1	0.6	0.3	79.61	0.2	5.38	2	92.93	-0.42	0.67	1
0.1	0.7	0.2	79.67	-1.4	9.75	-3	76.14	0.25	19.2	-23
0.1	0.8	0.1	71.58	1.27	25.58	25	48.34	3.18	137.35	9
0.2	0.1	0.7	10.94	-17.62	4.39	2	-13.1	-19.17	0.75	1
0.2	0.2	0.6	68.78	-4.47	59.34	1	87.06	0.69	8.82	10
0.2	0.3	0.5	79.36	-4.49	4.34	2	89.92	-0.74	0.69	1
0.2	0.4	0.4	87.9	-2.67	2.8	2	90.56	0.37	0.7	1
0.2	0.5	0.3	5.22	5.25	123.82	26	-1313.5	44.58	330.93	1
0.2	0.6	0.2	-1224.06	14.07	738.07	23	71.65	-5.88	-0.41	1
0.2	0.7	0.1	-3494.06	27.79	1260.09	18	72.93	4.1	7.89	28
0.3	0.1	0.6	73	-5.39	2.67	-2	-843.9	15.79	718.06	31
0.3	0.2	0.5	21.7	-15.34	4.37	2	90.33	-1.09	0.55	1
0.3	0.3	0.4	-31.42	-23.4	-18.48	-6	94.88	0.69	0.58	1
0.3	0.4	0.3	91.44	-1.61	3.36	0	93.31	-1.17	0.61	1
0.3	0.5	0.2	97.92	-0.93	1	1	86.73	-2	0.65	1
0.3	0.6	0.1	50.99	-6.39	41.96	14	64.51	-9.65	4.45	-1
0.4	0.1	0.5	-21.56	-17.88	38.23	0	14.63	2.59	154.35	-23
0.4	0.2	0.4	27.11	0.63	93.72	-9	30.9	-14.18	18.58	22
0.4	0.3	0.3	-297.42	-72.06	-39.06	-14	80.22	-2.07	0.62	1
0.4	0.4	0.2	41.49	4.67	157.95	5	61.67	4.56	64.25	13
0.4	0.5	0.1	-3156.62	10.98	1226.92	9	93.6	-0.33	0.5	1
0.5	0.1	0.4	24.94	2.52	138.74	-11	88.47	-2.93	0.55	1
0.5	0.2	0.3	79.66	1.91	43.51	-11	89.42	0.21	0.65	1
0.5	0.3	0.2	45.08	-13.7	21.86	-4	27.26	-12.7	7.86	-2
0.5	0.4	0.1	57.52	-6.94	7.88	-3	78.74	-0.82	11.6	-23
0.6	0.1	0.3	71.28	-0.51	16.11	-4	-1326.43	12.38	1635.65	13
0.6	0.2	0.2	34.31	7.33	82.39	-3	86.15	-1.45	0.63	1
0.6	0.3	0.1	89.72	-0.07	2.27	1	79.57	-0.31	44.43	21
0.7	0.1	0.2	71.13	4.07	101.72	5	55.76	1.33	116.16	21
0.7	0.2	0.1	83.07	2.93	82.98	2	87.89	0.11	0.5	1
0.8	0.1	0.1	-151.16	10.86	350.58	2	-234.45	9.57	428.78	10

$$CE = 1 - \frac{\Sigma \left(Q_{obs} - Q_{est} \right)^2}{\Sigma \left(Q_{obs} - \overline{Q}_{obs} \right)^2} \quad (3.1)$$

where Q_{est} is the estimated flood discharge(cms), Q_{obs} is the observed flood discharge (cms) , and $\overline{Q}_{obs}$ is the mean value of the observed flood discharge . The closer the CE value to 1, the higher the accuracy is ,which indicates that the closer of the simulated results to the recorded data.

(2) Volume error:

$$VE = \frac{\Sigma Q_{est} - \Sigma Q_{obs}}{\Sigma Q_{obs}} \quad (3.2)$$

where VE is the percentage error of the accumulated discharge. When VE value is closer to zero, the better of volume is conserved by the model.

(3) Peak discharge error:

$$EQ_p = \frac{Q_{pest} - Q_{pobs}}{Q_{pobs}} \quad (3.3)$$

Table 5 : Values of evaluating indicators in the case of $Q_k=0$, $R_k=100$

Typhoon-flood event			No.77				No.81			
C0	C1	C2	CE%	VE%	EQp%	ETp	CE%	VE%	EQp%	ETp
0.1	0.1	0.8	98.53	3.51	3.61	1	99.09	0.11	-0.96	1
0.1	0.2	0.7	98.42	3.51	3.61	1	99.07	0.12	-1.00	1
0.1	0.3	0.6	98.39	3.50	3.61	1	99.07	0.13	-1.01	1
0.1	0.4	0.5	98.37	3.51	3.61	1	99.06	0.14	-1.01	1
0.1	0.5	0.4	98.36	3.52	3.60	1	99.06	0.15	-1.02	1
0.1	0.6	0.3	98.34	3.53	3.60	1	99.05	0.18	-1.03	1
0.1	0.7	0.2	98.29	3.51	3.57	1	99.02	0.27	-1.08	1
0.1	0.8	0.1	98.04	3.26	3.46	1	98.84	0.41	-1.35	1
0.2	0.1	0.7	98.49	3.53	3.63	1	99.08	0.15	-0.96	1
0.2	0.2	0.6	98.33	3.56	3.63	1	99.07	0.13	-1.01	1
0.2	0.3	0.5	98.28	3.56	3.63	1	99.06	0.13	-1.01	1
0.2	0.4	0.4	98.26	3.56	3.62	1	99.06	0.13	-1.02	1
0.2	0.5	0.3	98.23	3.57	3.62	1	99.05	0.16	-1.03	1
0.2	0.6	0.2	98.18	3.55	3.59	1	99.02	0.23	-1.08	1
0.2	0.7	0.1	97.91	3.28	3.47	1	98.84	0.38	-1.35	1
0.3	0.1	0.6	98.48	3.54	3.63	1	99.08	0.16	-0.96	1
0.3	0.2	0.5	98.31	3.58	3.63	1	99.07	0.14	-1.01	1
0.3	0.3	0.4	98.25	3.58	3.63	1	99.06	0.13	-1.02	1
0.3	0.4	0.3	98.21	3.58	3.62	1	99.05	0.15	-1.04	1
0.3	0.5	0.2	98.16	3.55	3.60	1	99.02	0.22	-1.08	1
0.3	0.6	0.1	97.87	3.28	3.48	1	98.83	0.36	-1.35	1
0.4	0.1	0.5	98.48	3.56	3.63	1	99.08	0.17	-0.96	1
0.4	0.2	0.4	98.30	3.60	3.63	1	99.06	0.14	-1.02	1
0.4	0.3	0.3	98.22	3.59	3.62	1	99.05	0.15	-1.04	1
0.4	0.4	0.2	98.16	3.56	3.60	1	99.02	0.22	-1.08	1
0.4	0.5	0.1	97.87	3.29	3.48	1	98.83	0.35	-1.35	1
0.5	0.1	0.4	98.47	3.57	3.63	1	99.07	0.18	-0.97	1
0.5	0.2	0.3	98.28	3.61	3.63	1	99.06	0.17	-1.03	1
0.5	0.3	0.2	98.18	3.58	3.60	1	99.02	0.23	-1.08	1
0.5	0.4	0.1	97.88	3.35	3.49	1	98.83	0.36	-1.35	1
0.6	0.1	0.3	98.46	3.59	3.63	1	99.07	0.22	-0.99	1
0.6	0.2	0.2	98.24	3.60	3.61	1	99.03	0.26	-1.08	1
0.6	0.3	0.1	97.90	3.33	3.49	1	98.84	0.38	-1.35	1
0.7	0.1	0.2	98.42	3.58	3.61	1	99.04	0.31	-1.03	1
0.7	0.2	0.1	97.97	3.35	3.49	1	98.85	0.41	-1.34	1
0.8	0.1	0.1	98.18	3.34	3.49	1	98.86	0.47	-1.29	1

where Q_{pest} and Q_{pobs} are the estimated and observed peak discharges , respectively. When EQ_p is positive, the value of the estimated peak discharge is larger than that of the observed one. When EQ_p is negative, the value of estimation is smaller than that of the observation. When the value is closer to zero, the model accuracy for peak discharge is higher.

(4) Time to peak error:

$$ET_p = T_{pest} - T_{pobs} \tag{3.4}$$

where T_{pest} and T_{pobs} are the time to peak discharge of estimation and observation, respectively. Smaller ET_p indicates better prediction of occurring time for peak discharge.

The effects of routing results using different state initial values along with the different derivations are as follow:

3.1 Both Q_k and R_k are time-correlated

Since there is no propagation error initially, so Q_0 and R_0 can be set to zero, and the initial value of error covariance matrix, that is P_0, equal to $E\left[\mathbf{X}_0 \cdot \mathbf{X}_0^T\right]$. The total sum of the elements of the initial state vector, C_0^0, C_1^0, C_2^0, based on model theory, must be equal to 1. One may simply set the values of these three elements between 0.1 and 0.8. This gives us thirty-six sets of initial state vector. To evaluate the accuracy of model simulation , nine typhoon-flood events shown in Table 1 were used. The simulated results of two events, No.77 and No.81, are shown in Table 2 from which one may conclude that, in the case of Q_k and R_k being time-correlated, the different initial state vectors made a substantial difference in routing results.

3.2 R_k being time-correlated while Q_k being time-independent

To use the thirty-six sets of the initial state vector mentioned previously, the model was routed subject to four cases in which Q_k was set to 0,0.001, 0.01, and 0.1. The routing results of the No.78, No.79 and No.81 events in the case Q_k=0 showed that the average of CE can be up to 95%, the VE is smaller than 4%, the EQ_p smaller than 8%, and ET_p is within 2 hrs. Routing of the same thirty-six sets initial state vector of other six flood events, however still produce unstable and variable results. All other cases of Q_k being not equal to zero, the most of the results were not stable. The simulated results for No.77 and No.81 events with Q_k=0 and 0.1 are shown in Table 3 and 4, respectively.

3.3 Both Q_k and R_k being time-independent

We routed the model by using thirty-six sets initial state vector with R_k set equal to 1, 10, 100, 1000, and 10000 when Q_k equal to 0. From the routing results, we concluded that no matter which value of R_k is, the simulated result is of little difference in all nine events by using the thirty-six sets initial state vector. The highest value of averaged CE is 99.2%, the lowest being 84.9%. For VE, the highest is 5.22%, with the lowest being 0.22%. For EQ_p, the

highest is 8.1%, with the lowest being 0.99. In the case of ET_p, the highest is 4.8, and the lowest is 1. The values of R_k between 1 and 1000 showed little influence on the simulated results. Yet, in the sub-case of R_k=10000, some events showed significant difference in the simulated results. The simulated results of event No.77 and No.81 for the case of R_k=100 is shown in Table 5.The results of R_k sensitivety analysis for No.81 event is shown in Figure 2.

4. CONCLUSIONS

Combining Kalman filter algorithm with Muskingum flood routing model, one may pick up any initial state value belonging feasible solution in the case of system error covariance and measurement error covariance are both time-independent, the former is set to be zero and the latter is smaller than 1000. This conclusion makes Kalman filter algorithm earier to be applied in the flood routing model to estimate flood discharge.The accuracy of model forecast is excellent which can be shown by routing result of Pa-Chung River Basin. As for the future study , it is necessary to go further and evaluate the conclusion in different drainage basins. One may also investigate different flood models other than the Muskingum model to see if they have the same behavior as we reported here.

5.ACKNOWLEDGEMENTS

This investigation was made possible by Grand NSC 85-2211-E-035-016, awarded by the National Science Council, Taiwan, the Republic of China.

REFERENCES

Chen, C.S., J.K.Lin, 1994. Application of Kalman Filter Algorithm on the Linear Flood Routing Model. J. of Taiwan Water Conservancy, Vol.42, No.4, pp 56~67

Kuo, J.T., W.S.Chu, N.S.Hsu, Y.J.Lin, 1988. Real-time Operation for Te-Chi Reservoir. Research Report NO.7711, Hydraulics Eng. Division, Dep. of Civil Eng. National Taiwan University, Taipei, Taiwan R.O.C.

Ngan, P. and S. O. Rusell, 1986. Example of Flow Forecasting with Kalman Filter. Journal of Hydraulic Engineering, ASCE, Vol.112, No.9, pp.818~832

Simons, D.B., N. Duong, R.M. Li, 1978. An Approach to Shore-Term Water and Sediment Discharge Prediction Using Kalman Filter. in Applications of Kalman Filter to Hydrology, Hydraulic, and Water Resources, Ed. Chao-Lin Chiu, Univ. of Pittsburgh, Pa., pp.473~481

Van Geer, F.C., C.B.M. Te Stroet, and Zhou Y, 1991. Using Kalman Filtering to Improve and Quantify the Uncertainty of Numerical Groundwater Simulations 1 the Role of System Noise and its Calibration. Water Resources Research, Vol.27, No.8, pp. 1987~1994

Wilson, J. 1978. Stantard Parameter Estimation in Groundwater Model. in Applications of Kalman Filter to Hydralic, and Water Resources, Ed. Chao-Lin Chiu, Univ. of Pittsburgh, Pittsburgh Pa., pp.657~680

Wood, E.F. 1978. An Application of Kalman Filtering to River Flow Forecasting. in Applications of Kalman Filter to Hydrology, Hydraulic, and Water Resources, Ed. Chao-Lin Chiu, Univ. of Pittsburgh, Pittsburgh, Pa., pp.385~408.

Stochastic Hydraulics'96, Tickle, Goulter, Xu, Wasimi & Bouchart (eds) © 1996 Balkema, Rotterdam. ISBN 90 5410 817 7

Geometric characteristics of Arial Khan river in Bangladesh

M. Monowar Hossain
Department of Water Resources Engineering, Bangladesh University of Engineering & Technology, Dhaka, Bangladesh

A. K. M. Ashrafuzzaman
River Research Institute, Faridpur, Bangladesh

ABSTRACT: A study was carried out to investigate the geometric aspects of Arial Khan river, a major tributary of the Padma (lower Ganges) in Bangladesh for a reach length of about 200 km from its offtake. The study includes analysis of cross sectional geometry, development of non dimensional correlation, thalweg and bank line shifting, mean bed level change, specific gauge analysis and determination of meander parameters. The study showed that planform geometry as well as hydraulic geometry of the river have undergone considerable change over a period of about 20 years. The study further revealed that flood discharge showed an increasing trend and mean bed level suffered considerably.

1 INTRODUCTION

Bangladesh is a land of rivers and comprised mostly of sediments brought down by the Ganges-Padma, Brahmaputra-Jamuna and Meghna river system. Channel migration, flooding and erosion-scour of these rivers and of their numerous tributaries and distributaries have resulted a complex geomorphic history and brought enormous sufferings to millions of people. These river systems drain a vast area of about 1.65 million sq.km. spreading over China, Nepal, India and Bangladesh. Out of this area the Ganges-Padma basin covers an area of about 0.9 million sq.km. of which the Arial Khan is a major tributary. It may be mentioned that the area of Bangladesh is about 144000 sq.km i.e. about 9% of the drainage basin of the major river system (Hossain 1992).

Due to very wide variation of water and sediment discharge in the major river system, significant changes to geometry and hydraulic characteristics are not only brought to the main rivers themselves but also to their distributaries in particular. These changes in river geometry play an extremely important role in connection with the river bank stabilization, navigation, flood control and development of water resources projects. The study of geometric characteristics of alluvial stream involves uncertainties to a great extent as their prediction involves random variables. Arial Khan

river flows through one of the highest density region of Bangladesh and several irrigation projects are being developed in the area, it is therefore vital to investigate on the morphological aspects of this river. The present undertaking is directed towards this end.

2 THE ARIAL KHAN RIVER

The Arial Khan originates from the right bank of the Ganges-Padma at a distance of about 70 km. below Ganges-Brahmaputra confluence at Aricha. It is a meandering river having two parts, viz., the Arial Khan Upper(AKU) and the Arial Khan(AK). The distance between the two parts is about 30 km. The length of the AKU and the AK is about 70 km and 125 km respectively. There is a 8 km. long loop between section 2 and 4 for the AKU. A small channel of 0.8 km. in length has developed which connects this loop. Near Shalmabak the AKU and AK meets and follows a winding course. The river flows further down and takes various names such as Nayabanga, Kirtonkhola and finally discharges into the Bay of Bengal as Bishkhali through the Sundarban forest.

The study covers both courses of the river which totals about 200 km and extends from section AKU1 to AKU12 and section AK0 to AK19 (Fig.1).

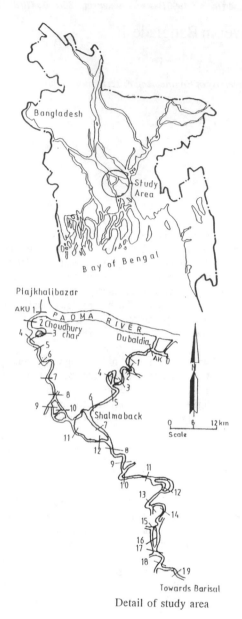

Detail of study area

Fig.1 The Arial khan river

Technology (BUET 1989) studied the Arial Khan in connection with the design of bank protection works near Bhanga-Mawa road. Snowy Mountains Engineering Corporation (SMEC), Australia, in association with Bangladesh Consultants Limited (BCL 1990) studied this river in connection with the same protection works. These reports did not consider any detail analysis of hydrologic, geometric or morphological behavior of the river rather concentrated on protection works necessary at a site. Other studies included those by Bangladesh Water Development Board (BWDB 1988) and Lesleighter (1993).

4 DATA ANALYSIS

Data on rating curve, discharge and water levels etc. were collected from Bangladesh Water Development Board (BWDB) for the years from 1965-66 to 1989-90 and on cross-section for the years 1970-71 and 1987-88. Landsat imageries were collected from Space Research and Remote Sensing Organization (SPARRSO), for the years 1972 and 1991.

Data on cross sections, water areas, average depth etc. were analyzed and their changes were determined over a period. Non-dimensional correlations among parameters such as W/D (shape factor), $[V/\sqrt{(gD)}]$ (Froude number), and $Q/(\nu D)$ (non-dimensional discharge) and the trend of each relationships was determined statistically. ν is the kinematic viscosity (m^2/s) of the fluid and Q,W, D are water discharge, with, depth respectively. Mean bed level (MBL) at various sections and changes in MBL at these sections during 1970-71 to 1987-88 was also determined. The trend analysis on specific gauge was done at Choudhury Char at discharge levels of 100, 500, 1000, 1500 and 2000 cumecs using data from 1965 to 1986. The landsat imageries for the years 1972 and 1991 were utilized to determine the meander parameters of selected bends and the shifting of bankline was determined by overlay techniques during this period.

3 PREVIOUS STUDIES

No significant hydrologic, morphologic or sediment transport studies have so far been undertaken on the Arial Khan. But similar studies have been undertaken on the major rivers of Bangladesh (Coleman 1969, Galay 1980, Hossain 1988, 1989, 1991). Bangladesh University of Engineering and

5 RESULT AND DISCUSSION

5.1 *Cross Sectional shape and Thalweg*

The available cross sections were plotted on a same scale for 1970-71 and 1987-88. From these plots variation of area, width, depth, shifting of thalweg were ascertained. One such typical plot is shown for

AKU6 (Fig. 2). The variation of area, width etc. at different sections are summarized in Table 1. It was observed that average cross sectional area and width increased by about 22% while average depth calculated on the basis of area by width decreased. Again the average water depth calculated on the basis of effective water area and width was found to increase. The thalweg in all sections were found to shift very widely and randomly, the highest shift being at AKU 3, 5 and 8 and of the order of about 500m, 520m and 730m respectively.

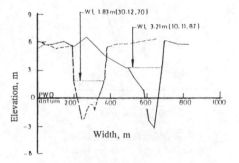

Fig.2 Cross-section for the arialkhan upper (AKU-6)

Table 1. Summary of cross-section, width and thalweg shifting etc. for Arial Khan Upper.

Section No.	Year	Cross-Sectional Area (sq.m.)	Top Width (m)	Thalweg Shifting
1	1970-71	1419	384	
	1987-88	2628	419	120
2	1970-71	980	223	
	1987-88	2309	406	20
3	1970-71	4331	958	
	1987-88	2030	486	500
4	1970-71	2785	625	
	1987-88	3043	597	90
5	1970-71	1430	305	
	1987-88	2820	554	520
6	1970-71	1072	191	
	1987-88	1594	437	395
7	1970-71	1948	591	
	1987-88	5823	984	115
8	1970-71	2546	481	
	1987-81	1866	594	730
9	1970-71	1298	240	
	1987-88	780	255	75
10	1970-71	1262	328	
	1987-88	1656	336	110
11	1970-71	1143	130	
	1987-88	383	114	205
12	1970-71	1119	304	
	1987-88	983	389	95

5.2 Mean Bed Level

Mean Bed Level(MBL) at different sections for the AKU was computed and plotted for the years 1970-71 and 1987-88 as shown in Fig.3. Change in

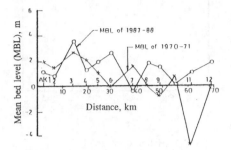

Fig.3 Variation of mean bed level (MBL) for arial khan upper

MBL was also computed to visualize the magnitude of erosion-deposition in the river. It was found that MBL rises at sections AKU3, 5, 6, 8, 9, 11 and 12 and falls at sections AKU1, 2, 4, 7, and 10 in 1987-88 compared to MBL in 1970-71. Assuming linear variation the erosion and deposition was computed and was found to be 7.25 and 27.67 million cubic meters respectively. The issue of aggradation-degradation was cross checked by plotting area-elevation curves at various sections. It was found that the trend obtained for MBL was in agreement with area-elevation relations. A typical area elevation curve is shown in Fig.4.

5.3 Specific gauge analysis

Specific gauge analysis based on discharge data at

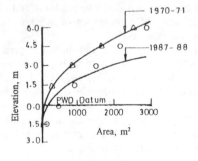

Fig.4 Area-elevation relationship (AKU-4)

385

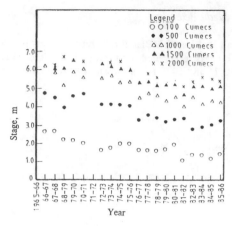

Fig.5 Variation of stage of
the arial khan upper at choudhury
char at different discharge level

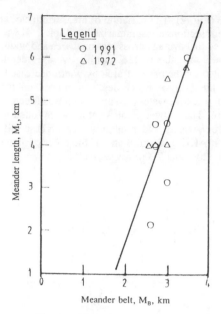

Fig.6 Relation between meander
length (M_L) and meander belt (M_B)

Choudhury Char was conducted for the period
between 1965-66 and 1985-86 as shown in Fig. 5.
The trend of the curves show that stage is
decreasing in a sinuous pattern which characterizes
the general dynamism existing at alluvial beds. The
figure also gives an indication that degradation at
Choudhury Char might have taken place.

5.4 Bend movement

Shifting of various bends during 1972 to 1991 was
found to occur at various bends both laterally and
longitudinally based on satellite imageries. Many
new bends were found to form and old one
abandoned. The confluence of the AKU and AK
was found to shift laterally by about 5 km to right
and longitudinally by about 3 km upstream. Lateral
shifting near Bhanga-Mawa road was about 0.5 km.
The study revealed that a minimum lateral shifting
of 0.25 km and a maximum of 3.5 km occurred
during the study period.

A study on sinuosity and meander parameters
were conducted considering well defined bends of
the Arial Khan river during 1972 and 1991. It was
observed that average sinuosity decreased from 2.16
in 1972 to 1.88 in 1991. This indicates that the AK
is a highly meandering stream (Schumm 1977). A
typical relation between meander length and
meander belt is shown in Fig.6.

5.5 Non dimensional correlations

Use of non dimensional correlations in the study of
alluvial geometry are becoming gradually popular
for many reasons. One potential application of such
correlation is the study of scale models. Keeping
this in mind, correlations between shape(W/D),
flow Froude Number [$V/\sqrt{(gD)}$] and [$Q/\nu D$] were
established. The correlations show a general trend
of decrease of flow Froude Number with the
increase of width-depth ratio as shown in Fig. 7.

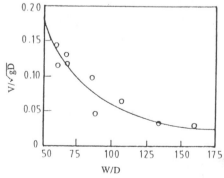

Fig.7 Relation between $V/\sqrt{gD}$ and
W/D at choudhury char (1983)

Relationships were also established for $V/\sqrt{(gD)}$ and $Q/(\nu D)$ and good correlations were found to exist. This correlations however showed a general increasing trend of flow Froude number with increase of non-dimensional discharge $[Q/(\nu D)]$. This relationship is shown in Fig. 8. The regression equations developed for the two figures as shown in Figs. 7&8 may respectively be written as follows:

$$V/\sqrt{(gD)} = 98.753 \, (W/D)^{-1.606} \quad (5.1)$$
$$V/\sqrt{(gD)} = 6.4 \times 10^{-8} \, [Q/(\nu D) \times 10^{6}]^{0.747} \quad (5.2)$$

The coefficient of correlations for Eqs. (5.1) and (5.2) were determined statistically and found to be 0.941 and 0.989 respectively.

The basis of selecting these non-dimensional parameters stems from the findings of previous studies (Hossain 1987). Similar type of correlations were established by Hossain (1989) and Hossain and Das (1995) for some major rivers of Bangladesh.

5.6 Hydrological aspects

The variation of discharge, sediment flow and flood plain characteristics are mainly responsible for bringing the morphological changes in an alluvial stream. The Arial Khan is highly influenced by the flow characteristics of the Ganges-Padma. Thus the magnitude and duration of the discharge and concomitant sediment activity in the parent river as well as in the AK is of vital importance. Due to

Table 2. Comparison of Arial Khan flows width the Padma flows (m³/s).

Year	Arial Khan Maximum	Padma Maximum	Arial Khan % of Padma
1965-66	1520	85000	1.8
1966-67	1130	85200	1.3
1967-68	2280	69200	3.3
1968-69	2230	91000	2.5
1969-70	1910	98200	1.9
1970-71	3540	85200	4.2
1971-72	-	-	-
1972-73	1540	78600	2.0
1973-74	3960	100000	4.0
1974-75	3200	111000	2.9
1975-76	1950	-	-
1976-77	1760	84400	2.1
1977-78	2150	99900	2.2
1978-79	2310	88500	2.6
1979-80	1790	77300	2.3
1980-81	2260	109000	2.1
1981-82	3040	77300	3.9
1982-83	3240	65400	5.0
1983-84	2100	84600	2.5
1984-85	2640	110000	2.4
1985-86	2600	73300	3.5
1986-87	3230	85400	3.8
1987-88	3370	118000	2.9
1988-89	4880	87300	5.6
Average	2564	88841	2.9

this complexity the study and prediction of morphological parameters are still not fully understood and highly dependent on empirical treatments supported by statistical analysis. The distributary entrance to the AK has changed over years, at time through a deep pool, and at other time behind a bar. Accordingly the discharge into the AK varied considerably as shown in Table 2. The table shows that the percent entry of discharge has a slightly increasing trend.

6 CONCLUSIONS

The cross sectional area of the AKU varies widely and randomly. In most places area was found to increase due to increase in width. The study indicated rise of mean bed level at some places while fall in other places. A net deposition was found to occur. However due to increase in discharge water depth was found to increase.

Analysis on the lateral shifting of bankline revealed that all bends shifted to a certain extent during the study period ranging from 0.25 km to about 3.5 km. The river also shifted longitudinally. The average sinuosity decreased from 2.16 to 1.88. At a station specific gauge analysis indicated that

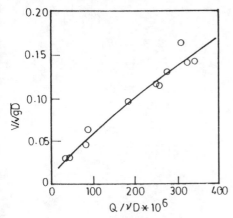

Fig.8   Non-dimensional plot of $Q/\nu D$ vs. $V/\sqrt{gD}$ for the arial khan upper at chowdhury char(1993)

stage has a decreasing trend due to bed degradation at that section. Good correlations were obtained for non dimensional plots from this study. The findings of this study would be highly beneficial for better management of the water resources of this river.

REFERENCES

BCL. 1990. Protection of Bhanga-Mawa highway. *Final Report*:Snowy Mountain Engineering Corporation, Australia & Bangladesh Consultants Ltd., Dhaka.

BUET. 1989. Design of bank protection works of Arial Khan river near Bhanga-Mawa road. *Report*: Bangladesh University of Engineering & Technology, Dhaka.

BWDB. 1988. Morphological survey data of the river Mohananda and Arial Khan. *Report 3*:Bangladesh Water Development Board, Dhaka.

Coleman, J.M. 1969. Brahmaputra River: Channel processes and Sedimentation. *Sedimentary Geology* 3, 2/3: 129-239.

Galay, V.J. 1980. River channel shifting on large rivers in Bangladesh. *Proc. ISRS I*:543-562.Beijing: Tsinghua University.

Hossain, M.M. 1987. A simplified sediment transport function from dimensional considerations. *Journal of the Indian Water Resources Society*. 7,2: 35-44.

Hossain, M.M. 1988. Meander travel of the lower Ganges within Bangladesh. *Proc. ISSRD II*, S2:1-10. Dhaka:Bangladesh University of Engineering & Technology.

Hossain, M.M. 1989. Geomorphic characteristics of the Padma upto Brahmaputra confluence. *Research report R/02/89*:1-158. Dhaka:Bangladesh University of Engg. & Technology.

Hossain, M.M. 1991. Problems of erosion and deposition in the major rivers of Bangladesh. *Proc.ISSPAR*:247-255. Seoul:Korea Institute of Construction Technology.

Hossain, M.M. 1992. Total sediment load in the lower Ganges and Jamuna. *Journal of the Institution of Engineers*. 20,1:1-8.

Hossain,M.M. & S.K.Das 1995. Hydraulic geometry of the Ganges in Bangladesh. *Proc. ICHCE* 2:1954-1961. Beijing:Tsinghua University Press.

Lesleighter, E.J. 1993. Management of morphologically active rivers. *Proc. ISMRF I*:6B1.1-6B1.11. Kuala Lumpur:Dept. of Irrigation and Drainage.

Schumm, S.A. 1977. *The fluvial system*. New York: John Wiley and Sons.

Stochastic Hydraulics'96, Tickle, Goulter, Xu, Wasimi & Bouchart (eds) © 1996 Balkema, Rotterdam. ISBN 90 5410 817 7

The empirical simulation technique – A bootstrap-based life cycle approach to frequency analysis

Norman W. Scheffner
US Army Engineering Waterways Experiment Station, Vicksburg, Miss., USA

Leon E. Borgman
University of Wyoming, Laramie, Wyo., USA

ABSTRACT: This paper describes theory and implementation of the Empirical Simulation Technique (EST), bootstrap-based statistical procedure for simulating multiple time sequences of non-deterministic multiparameter systems such as storm events and their corresponding environmental impacts. Results of the multiple repetitions are subsequently analyzed to compute frequency-of-occurrence relationships for storm effects such as coastal erosion and storm surge. Because multiple life-cycle scenarios are simulated through the EST, mean value frequencies are computed along with error estimates of deviation about the mean. The general methodology is described herein as well as an application of the approach for computing frequency relationships for tropical and extratropical storm induced erosion of coastal berms and dunes.

1 INTRODUCTION

Recreation is a vital industry to many countries of the world with the coast representing one of the primary attractions for both residents and tourists. Unlike many other regions, coastal areas are subject to storm events which can significantly impact the condition of the beach, often rendering it unacceptable for recreational use. As a result, there exists a need to provide an adequate, yet affordable, level of protection against coastal erosion and inundation. Design and construction of coastal structures requires accurate estimates of the frequency and severity of storm events and their respective damages.

This paper describes theory and implementation of the Empirical Simulation Technique (EST), a bootstrap-based statistical procedure for simulating multiple time sequences of non-deterministic multiparameter systems such as storm events and their corresponding impacts to the environment. The general methodology is described herein as well as an application of the approach for computing frequency relationships for tropical and extratropical storm-induced erosion of coastal berms and dunes.

2 EMPIRICAL SIMULATION TECHNIQUE

Many past attempts to assign frequency relationships to hurricane events are based on the assumption that hurricane parameters are independent and that their individual probabilities can be modeled with empirical, or parametric, relationships. The joint probability of occurrence for a particular storm is then computed as the product of the individual storm parameter probabilities via assumed parametric relationships. This assumption is the primary basis of the Joint Probability Method (JPM). Some of the difficulties associated with application of the JPM for storm related damages are discussed in a companion paper presented at this conference (Butler and Scheffner 1996).

The parameters which describe tropical storm events are not independent, but are interrelated in some nonlinear sense. Because they are not independent, joint probability can not be computed as the product of individual parameter probabilities. The EST utilizes observed and/or computed parameters associated with site specific historical events as a basis for developing a methodology for generating multiple life cycle simulations of storm activity and the affects associated with each simulated event. Contrary to the JPM, the technique does not rely on assumed parametric relationships but uses the joint probability relationships inherent in the local data base. Therefore, in this approach, probabilities are site specific, do not depend on fixed parametric relationships, and do not assume parameter independence. Thus, the EST is "distribution free" and nonparametric.

The EST is based on a "Bootstrap" resampling-with-replacement, interpolation, and subsequent smoothing technique in which random sampling of a finite length data base is used to generate a larger data base. The only assumption is that future events will be statistically similar in magnitude and frequency to past events. The EST begins with an analysis of historical events which have impacted a specific locale. The selected data base of events are then parameterized to define the characteristics of the event and the impacts of that event. Parameters which define the storm are referred to as input vectors. Response vectors define storm-related damages such as inundation and shoreline/dune erosion. These input and response vectors are then used as a basis for generating life cycle simulations of storm event activity. Details of the approach follow.

2.1 Input vectors

Input vectors describe the magnitude of the storm event and the location of the event with respect to the area of interest. These values are defined as an N-dimensional vector space as follows:

$$\underline{v} = (v_1, v_2, v_3, \ldots, v_N) \tag{1}$$

For tropical events, appropriate vectors include: 1) the central pressure deficit, 2) the radius to maximum winds, 3) maximum winds, 4) minimum distance from the eye of the storm to the location of interest, 5) forward speed of the eye, and 6) tidal phase during the event. Extratropical event input vectors can include: 1) duration of event as measured by some threshold damage criteria, 2) associated wind wave heights and periods, and 3) tidal phase.

2.2 Response vectors

The second class of vectors describe any response which can be attributed to the passage of the storm. This M-dimensional space is defined as:

$$\underline{r} = (r_1, r_2, r_3, \ldots, r_M) \tag{2}$$

Tropical and extratropical responses include parameters such as maximum surge elevation, shoreline erosion, dune recession.

Although response vectors are related to input vectors

$$v \rightarrow r, \tag{3}$$

the interrelationship is highly nonlinear and involves correlation relationships which can not be directly defined, i.e., a non parametric relationship. For example, in addition to the storm input parameters, storm surge is a function of local bathymetry, shoreline slope and exposure, ocean currents,

temperature, etc., as well as their spatial and temporal gradients. It is assumed however, that these combined effects are reflected by the response vectors.

Response vectors define storm effects such as maximum surge elevation, shoreline erosion, or dune recession. These parameters are usually not available from post storm records at the spatial density required for a frequency analysis. Therefore, response vectors are generally computed via numerical models. For example, if the response of interest is maximum surge elevation, long wave hydrodynamic models are coupled to tropical storm surge models or data bases containing the spatial and temporal distribution of extratropical wind fields. If however, the damage of interest involves storm-related erosion, additional models are used which access the hydrodynamic model surge elevation and current hydrographs to compute, for example, berm/dune erosion. An example will subsequently be given for such a case.

The historical data for storms can thus be characterized as

$$[v_i; i = 1, \ldots, I] \tag{4}$$

where I is the number of historical storm events. For example, let v_i have d_v-components

$$v_i = \Re^{d_v} \tag{5}$$

where $\Re^{d_v}$ denotes a d_v-dimensional space.

Ideally, there are an adequate number of historic event parameters to fill the d_v-dimensional vector space. An adequate number refers to both the number of events and the severity of events as measured by their descriptive parameters. If the number of historic events is sparse, then some event augmentation may be necessary. If the historic population contains many redundant events, i.e., similar events with respect to input and response vector space, then events may be omitted from the set of historic events. The methodology for developing a set of events for input to the EST is described below.

Introduce a separate set of storm events which we term the "training set"

$$[v_j^* j = 1, \ldots, J] \tag{6}$$

The propagation of these events will be modeled to compute event-specific response vectors. The set of v_j^* usually includes the historical events but may include storms which could have occurred. For example, a historical storm with a slightly altered path. The set may not include certain historical events if two events are nearly identical such that both events would produce the same response.

3 EST IMPLEMENTATION

The goal of the EST is summarized as:
1) given the following:
    a) the historical data $[v_i \in \Re^{dv}; i=1,...,I]$
    b) the "training set" data $[v^*_j \in \Re^{dv}; j=1,...,J]$
    c) the response vectors calculated form the training set $[r^*_j \in \Re^{dr}; j=1,...,J]$
    d) additionally, the EST has a capability to interpolate responses for any historical storms not in the training set computed from the responses of the training set $[r^*_k \in \Re^{dr}; k=1,...,K]$
2) produce N simulations of a T-year sequence of events, each with their associated input vectors $v \in \Re^{dv}$ and response vectors $r \in \Re^{dr}$.

Two criteria are required of the T-year sequence of events. The first criteria is that the individual events must be similar in behavior and magnitude to historical events, i.e., the inter-relationships among the input and response vectors must be realistic. The second criteria is that the frequency of storm events in the future will remain the same as in the past. The following sections describe how these two criteria are preserved.

3.1 Storm event consistency

The first major assumption in the EST is that future events will be similar to past events. This criteria is maintained by insuring that the input vectors for simulated events have similar joint probabilities to those of the training set. For example, a hurricane with a large central pressure deficit and low maximum winds is not a realistic event - the two parameters are not independent although their precise dependency is unknown. The simulation of realistic events is accounted for in the nearest-neighbor interpolation, bootstrap, resampling technique developed by Borgman (Borgman, et al 1992).

The basic technique can be described in two dimensions as follows. Let $X_1, X_2, X_3, ..., X_n$ be n independent, identically distributed random vectors (storm events), each having two components $[X_i = \{x_i(1), x_i(2)\}; i=1,n]$.

Each event X_i has a probability p_i as $1/n$, therefore, a cumulative probability relationship can be developed in which each storm event is assigned a segment of the total probability of 0.0 to 1.0. If each event has an equal probability, then each event is assigned a segment s_j such that $s_j \rightarrow X_j$. Therefore each event occupies a fixed portion of the 0.0 to 1.0 probability space according to the total number of events in the training set.

A random number from 0 to 1 is selected to identify a storm event from the total storm popula-

$$[\ 0 < s_1 \le \frac{1}{n}\ ]$$

$$[\ \frac{1}{n} < s_2 \le \frac{2}{n}]$$

$$[\ \frac{2}{n} < s_3 \le \frac{3}{n}] \quad\quad (7)$$

$$[\ \frac{n-1}{n} < s_n \le .\ 1]$$

tion. The procedure is equivalent to drawing and replacing random samples from the full storm event population.

The EST is not simply a resampling of historical events technique, but rather an approach intended to simulate the vector distribution contained in the training set data base population. The EST approach is to select a sample storm based on a random number selection from 0 to 1 and then perform a random walk from the event X_i with x_1 and x_2 response vectors to the nearest neighbor vectors. The walk is based on independent uniform random numbers on $(-1,1)$ and has the effect of simulating responses which are not identical to the historical events but are similar to events which have historically occurred.

3.2 Storm event frequency

The second criteria to be satisfied is that the total number of storm events selected per year must be statistically similar to the number of historical events which have occurred at the area of concern. Given the mean frequency of storm events for a particular region, a Poisson distribution is used to determine the average number of expected events in a given year. For example, the Poisson distribution can be written in the following form:

$$Pr(s;\lambda) = \frac{\lambda^s e^{-\lambda}}{s!} \quad\quad (8)$$

for $s=0,1,2,3,...$ The probability $Pr(s;\lambda)$ defines the probability of having s events per year where λ is a measure of the historically-based number of events per year.

A 10,000 element array is initialized to the above Poisson distribution. The number corresponding to $s=0$ storms per year is 0.7261, thus if a random number selection is less than or equal to 0.7261 on an interval of 0.0 to 1.0, then no hurricanes would occur during that year of simulation. If the random number is between 0.7261 and 0.7261 +

P[N=1] = 0.7261 + 0.2324 = 0.9585, one event is selected. Two events for 0.9585 + 0.0372 = 0.9957, etc. When one or more storms are indicated for a given year, they are randomly selected from the nearest neighbor interpolation technique described above.

4 RECURRENCE RELATIONSHIPS

Estimates of frequency-of-occurrence begin with calculating a cumulative distribution function (cdf) for the response vector of interest. Let X_1, X_2, X_3, ..., X_n be n identically distributed random response variables with a cumulative cdf

$$F_X(x) = Pr[X \leq x] \qquad (9)$$

where Pr[ ] represents the probability that the random variable X is less than or equal to some value x and $F_X(x)$ is the cumulative probability distribution function ranging from 0.0 to 1.0. The problem is to estimate the value of F_X without introducing some parametric relationship for probability. The following procedure is adopted because it makes use of the probability laws defined by the data and does not incorporate any prior assumptions concerning the probability relationship.

Assume that we have a set of n observations of data. The n values of x are first ranked in order of increasing size such that

$$x_{(1)} \leq x_{(2)} \leq x_{(3)} \leq \ldots \leq x_{(n)} \qquad (10)$$

where the parentheses surrounding the subscript indicates that the data have been rank-ordered. The value $x_{(1)}$ is the smallest in the series and $x_{(n)}$ represents the largest. Let r denote the rank of the value $x_{(r)}$ such that rank 1 is the smallest and rank r = n is the largest.

An empirical estimate of $F_X(x_{(r)})$, denoted by $\hat{F}_X(x_{(r)})$, is given by Gumbel (1954) (see also Borgman and Scheffner, 1991 or Scheffner and Borgman 1992).

$$\hat{F}_x(x_{(r)}) = \frac{r}{(n+1)} \qquad (11)$$

for $\{x_{(r)}, r = 1, 2, 3, ..., n\}$. This form of estimate allows for future values of x to be less than the smallest observation $x_{(1)}$ with probability of 1/(n+1), and to be larger than the largest value $x_{(n)}$ also with probability 1/(n+1). In the implementation of the EST, tail functions (Borgman and Scheffner 1991) are used to define the pdf for events larger than the largest or smaller than the smallest observed event so that there is no discontinuity in the pdf.

The cumulative cdf as defined by Equation 11 is used to develop stage-frequency relationships in the following manner. Consider that the cdf for some storm impact corresponding to an n-year return period event can be determined from:

$$F(x) = 1 - \frac{1}{n} \qquad (12)$$

where F(x) is the simulated cdf of the n-year impact. Frequency-of occurrence relationships are obtained by linearly interpolating a stage from Equation 11 corresponding to the cfd associated with the return period specified in Equation 12. Application of this interpolation technique to the EST results obtained for the berm/dune erosion study is presented in the following section.

5 EST APPLICATION TO BERM/DUNE EROSION

The EST provides an effective and accurate approach to developing life cycle simulations of environmental forcing which can be used to develop frequency-of-occurrence relationships for various coastal design parameters. The following example describes application of the EST to establish relationships for tropical and extratropical storm induced horizontal recession of beaches in Brevard County, Florida, located on the lower east coast of the United States. Those relationships were required as input to the economic analysis of storm damages, in a feasibility phase study of the Brevard County Shore Protection Project.

5.1 Training set

The study area is subjected to both tropical and extratropical events. Because the size and frequency of the these types of storms are fundamentally different, a separate EST analysis is required for each. Following each analysis, computed frequency relationships are combined into a single horizontal erosion versus frequency-of-occurrence relationship. The tropical and extratropical training sets for this study were extracted from an existing data base of events historical events.

The data bases were generated through application of a long wave hydrodynamic model covering the computational domain encompassing the east coast of the United States, Gulf of Mexico, and Caribbean Sea (Luettich, Westerink, and Scheffner 1992). Contents of the data bases consist of surface elevation and depth integrated currents corresponding to: 1) 134 historically based tropical events which impacted the east coast and Gulf of Mexico during the period of 1886 through 1989 (Jarvinen et al

1988) and 2) a 16 season (September through March) simulation of wind fields over the full computational domain. Data are archived at 686 discrete nearshore locations. Data from the station nearest to the study area was used to provide appropriate storm events for input to the dune erosion model.

Analysis of the tropical event data base showed that 20 tropical events had impacted the study area. Criteria for inclusion in the data base was that the storm must generate a minimum surge elevation of 0.3048 m (1.0 ft). The surge elevation hydrograph for each event was then linearly combined with the four phases of the tide to produce a training set of 80 events. Each of these events were parameterized by the criteria described above to generate 80 input vector sets. Each of the 80 events were then input to the dune erosion model to generate an 80 event response vector set.

All events of the extratropical data base which generated a surge in excess of 0.216 m (.707 ft) were selected to represent the extratropical event training set. This analysis resulted in the selection of 22 events. As with the tropical storms, each was combined with the four phases of the tide to generate an 88 event response vector set.

5.2 EST simulations

The EST simulation approach was to generate 100 simulations of a 200-year sequence of storms. The number of events per year was specified according to Equation 8 as 0.192 (20/104) for tropical and 1.375 (22/16) for extratropical events. Each 200-year sequence of storm responses is rank ordered with zero specified for years of no storm activity and according to the largest event for years with multiple storms. A cumulative probability density function (pdf) was computed for each ranked series and a frequency-of-occurrence relationship developed for each simulation.

Because the computation is repeated 100 times, 100 individual pdf's and 100 recession frequency relationships are computed. This family of curves is averaged and the standard deviation computed, resulting in the generation of a stage-frequency relationship containing a measure of variability of data spread about the mean value. Figures 1 and 2 show the computed mean value as well as all 100 frequency relationships for the tropical and extra-tropical analysis for a typical station of the study reach.

In order to generate a single damage-frequency relationship, the frequency curves for the tropical and extratropical events were combined using the assumption that the two events are independent.

The following relationship was used to combine results:

$$T_{comb} = \left(\frac{1}{T_{trop}} + \frac{1}{T_{extrop}} \right)^{-1} \qquad (13)$$

where T_c = combined return period, T_{trop} = tropical storm return period, and T_{extrop} = extratropical storm return period. The combined curve for Figures 1 and 2 is shown in Figure 3.

6 CONCLUSIONS

This paper describes an application of a the Empirical Simulation Technique (EST), a bootstrap-based resampling with replacement and parameter interpolation scheme which generates multi-year life cycle simulations of storm events and storm effects. However, the approach is applicable to any periodic event for which some effect can be attributed. In the application presented herein, frequency relationships for berm and dune erosion resulting from the passage of tropical and extratropical events are computed.

The approach exploits the historical data to develop joint probability relationships among the various descriptive input vector storm parameters and response vectors such that multiple repetitions of multiple years of storm activity are generated. Each simulation reflects the joint probability inherent in the original data base of historic events and does not rely on assumed parametric relationships. Therefore, computed probabilities are site specific, do not depend on fixed parametric relationships, and do not assume parameter independence.

EST results have been shown to be accurate and reasonable easy to implement. Many of the difficulties associated with the more traditional Joint Probability Method are overcome by the EST. Additionally, because the approach produces multiple repetitions of simulations, post processing of results gives mean value frequency relationships along with estimates of error bounds. A companion paper in these proceedings (Butler and Scheffner 1996) describes in more detail some of the advantages of the EST.

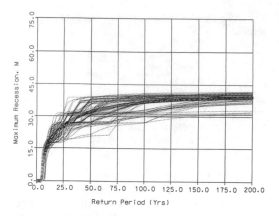

Figure 1 Tropical event frequency relationship

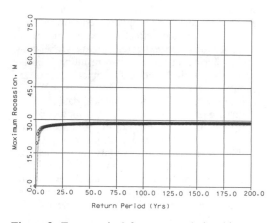

Figure 2 Extratropical frequency relationship

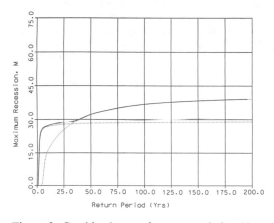

Figure 3 Combined event frequency relationship

ACKNOWLEDGEMENTS

The methodology for developing frequency-of-occurrence relationships for multiple locations described in this paper was developed at the Coastal Engineering Research Center, U.S. Army Engineer Waterways Experiment Station, Vicksburg, Mississippi, USA. The Empirical Simulation Technique was developed by Dr. Leon E. Borgman of the University of Wyoming. Permission was granted by the Chief of Engineers to publish this material.

REFERENCES

Borgman, L.E., Miller, M.C., Butler, H.L., and Reinhard, R.D. 1992. Empirical simulation of future hurricane storm histories as a tool in engineering and Economic Analysis. ASCE Proc. Civil Engineering in the Oceans V. College Station, TX. 2-5 November 1992.

Borgman, L.E. and Scheffner N.W. 1991. The simulation of time sequences of wave height, period, and direction. TR-DRP-91-2. USAE WES. Vicksburg, MS.

Butler, Lee and Scheffner, Norman. 1996. Comparison of stochastics for stage-frequency analysis. Proc. 7th IAHR Int Symp on Stochastic Hydraulics. MacKay, Australia. 29-31 July 1996.

Gumbel, E.J. 1954. Statistical theory of extreme value and some practical application. National Bureau of Standards Applied Math. Series 33. U.S. Gov. Publ. Washington, DC.

Jarvinen, B.R., Neumann, C.J., and Davis, M.A.S. Second Printing October 1988. A tropical cyclone data tape for the North Atlantic basin, 1886-1983: Contents, limitations, and uses. NOAA Tech Memo NWS NHC 22.

Luettich, R.A., Westerink, J.J., and Scheffner, N.W. 1992. ADCIRC: Theory and Methodology. TR-DRP-92-6, Rpt 1. November 1992. USAE WES. Vicksburg, MS.

Scheffner, N.W. and Borgman, L.E. 1992. A stochastic time series representation of wave data. ASCE Journal of Waterways, Ports, Coastal and Ocean Engineering. Vol 118, No. 4. Jul/Aug 1992.

Feasibility of the supervision and steering of a storm drainage system through rainfall scenarios recognition

M. Erlich

Laboratoire d'Hydraulique de France - LHF, Echirolles, France

ABSTRACT: The results of the feasibility study of a new approach to the real-time management of large urban drainage system is described. It aims at the real-time supervision of the overall sewer network with aid of typical, ready-to-apply scenarios (including various strategies of management corresponding to referenced situations), prepared and tested off-line, is based on the automatic selection of appropriate scenario corresponding to current meteorological event.

1 PROBLEM POSITION

A global supervision of steering of a large urban sewer network evolves towards a simultaneous management of multiple risks what in consequence modifies the methods of hydraulic control of flows.

In the case a sewer network monitoring system has a capacity of an automatic control of remote hydraulic equipment (gates, sluices, retention storage reservoirs) while the risks increase with a number of decisions to be taken, the lapse of time left to the network operator for decision making inversely decrease. From the practical point of view one of the most important issues for real-time management of the system is to reduce the time necessary for the decision making and to make, when possible, the appropriate corrections of these decisions.

For the above mentioned reasons the idea of scenario based approach, which could include various strategies of management corresponding to referenced situations, prepared and tested off-line, was considered as the most promising to be tested.

The approach requires a selection of a limited number of typical rainfall events for which management actions of the sewer network facilities are possible, and simulation of the network reaction by the coupled hydrological/hydrodynamic models. The resulting time and space varying physical magnitudes such as discharges, water levels, stored volumes, overflows and commands will be considered as representative responses of the system to a given class of rainfall scenario. In turn these results could be useful to establish a strategy of management to be adopted (time schedule of actions to be undertaken in the real-time) in order to minimise flooding and the pollution rate with respect to the security rules of operation.

The present paper describes the research feasibility study carried on by the LHF on the problem of the automatic selection of appropriate rainfall scenario

corresponding to actual meteorological event. The methodology, which exploits different pattern recognition techniques, was tested through the simulation of real-time conditions of system operation for sewer network of Seine Saint-Denis County (Paris region).

2 PATTERN RECOGNITION APPROACH

2.1 Application context

The past experience of the sewer network management allowed to identify the following three categories of rainfall events:

- a weak intensity rainfall (< 10 mm/hour): below a certain threshold level no serious consequence is expected, so no particular intervention is necessary. This is the most frequent class of events;
- a very intense rainfall (>50 mm/h): over a certain threshold level the rainfall causes the considerable damages and few remedial actions are possible. These events are very rare and their return period is between 5 and 10 years;
- an average and high intensity rainfall: this class of frequent events (2-3 events occur over the County every year) allows some remedial actions which should be rapidly implemented.

The remaining problem consists to recognise in the real-time, using the information furnished, for example, by the meteorological radar and ground rain gauges, an appropriate rainfall scenario for a fixed, 2 hours ahead, time horizon.

2.2 Applicability of the pattern recognition method

The pattern recognition technique as a part of statistical multidimensional data analysis applied in the context of description of dynamic processes, tends to a definition of the state vector being

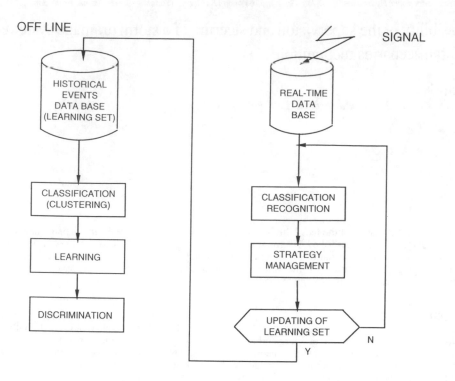

OFF LINE   SIGNAL

Fig. 1 - Overall scheme of pattern recognition approach

representative to any state of a time evolving system. The system is considered as having a finite number of states, each being represented by a base of known vectors. When in real-time the system state updating takes place the application of the appropriate pattern recognition techniques allows to compare a new state vector with a set of known vectors in order todetermine a class of situations a current state of the system belongs to.

To be applied the pattern recognition technique requires the following three elements :
- a set of historical information called learning set;
- a set of decision rules allowing a classification of the elements of the set;
- a discriminator, which using the above mentioned decision rules, allows to select in the real-time the class of events the closest to the current one.

The following scheme presents a general sequence of operations necessary for the application of pattern recognition approach to the diagnosis problem (Fig. 1).

2.3 Construction of the learning set

The set of 20 typical rainfalls was defined by the sewer network management agency of Seine Saint Denis. The hyetograms give the intensity of the rainfall as a function of time sampled every 15 minutes with event maximal duration of 3 hours.

Two approaches to the identification of a set of criteria allowing the best description of each rainfall event were tested. The first consisted in the definition of the following 5 representative variables:

X_1 - duration of the event

X_2 - total depth of the rainfall event

X_3 - time interval of the maximum

X_4 - duration of the maximum

X_5 - depth of the rainfall maximum

The variables X_1, X_3 and X_4 were dimensionless as the instant was indexed as a function of sampling period iT ($i = 1, ..., 12$), T being known, while the variables X_2 and X_5 were in millimetres.The variables $X_1, ..., X_5$ describe perfectly and without redundancy the set of typical rainfalls.

However, it is worth noting, that the variables X_1 and X_3 can furnish the necessary information only when the event is already completed and the variable X_2 is constant for a constant intensity. Consequently, as such they are not useful in the real-time context of rainfall pattern recognition application.

For these reasons the second approach, consisting in the definition of the significant variables at each time step allows a description of each of the selected rainfall events as a vector having for components rainfall intensity rate (mm/h) every 15 minutes for a total duration of 3 hours. The vectors are the rows of the observation matrix $X = (x_{ij})$ while the number of columns correspond to a number of sampling time steps of the homogeneous event length ($i = 1, ..., m$; $j = 1, ..., n$).

2.4 Classification

The goal of the preliminary data analysis to be carried out with the observation matrix is to eliminate from the learning set a redundant information and to reduce the dimension n of the original space R^n to a sub-space W of dimension n' ($n < n'$) with a minimal loss of information. For this purpose the Principal Components Analysis (PCA) was applied in the study (Preisendorfer, 1988). The selected criterion which allows to measure the representativity of a sub-space W is an inertia criterion estimate J of the average square distance of points x_i of the original space from their projections x^{p_i} in a new space W:

$$J = \frac{1}{n}\sum_{i=1}^{n} \| x_i - x^{p_i} \|^2 \qquad [1]$$

where ‖.‖ is the Euclidean norm. From the known properties of the PCA it can be shown that :

$$J = \text{trace } (\Phi)*(1 - \frac{\sum_{k=1}^{n'} \lambda_k}{\sum_{k=1}^{n} \lambda_k}) \qquad [2]$$

where $\Phi = (\Phi_{kl})$ (k,l = 1, ..., n) is the variance-covariance matrix of X, trace $(\Phi) = \sum_{k=1}^{n}\lambda_k$ and λ_k is a k-th eigenvalue of X'X. The quantity :

$$\frac{\sum_{k=1}^{n'} \lambda_k}{\text{trace}(\Phi)}$$

indicates the quality of the representation, closer it is to 1, the sub-space W is more representative for the elements of the initial learning set under study.

In the case of the set of typical rainfall events studied, the original space of dimension 12 can be reduced to the first 6 principal directions, which allow to explain 99.8 % of the total variance of X.

In order to structuralize the learning set of events it has to be clustered into a certain number of classes.

Almost all clustering techniques involve a process of measurement, either of the magnitude of the distance between two objects, or of the magnitude of their similarity to each other, where the objects are described by values of the variables. The knowledge of the system can be useful for a priori classification otherwise mathematical criteria must be used.

A general algorithm of clustering into a maximal number of q classes, satisfying all the criteria, is based on a threshold distance between two elements, which defines the rejection criterion, i.e. decision when two points are not affected to the same class. For each element three possibilities can occur:
- its distance from the element founding a class is smaller than the fixed threshold, the element is consequently affected to the class;
- there is no class fulfilling the threshold condition , the element founds a new class subject to the maximal number of classes q has not been reached ;
- it is not clustered and rejected.

Of course the classification depends on the threshold value and a maximal number of classes. When a threshold is two weak many elements will be rejected, when it is stronger, then a small number of classes will be created. However, this algorithm has an advantage on others as it avoids the determination a priori of q centres of gravity for the classes.

The results of the above classification applied to the typical rainfall events are shown on Fig. 2. The learning set was clustered into 9 classes respecting the differences of events duration, their pattern and the total rainfall depth. The obtained classification can be summarised as follows.
- Class 1 : rainfalls representing no dangerous character;
- Class 2 : short (30 minutes long) rainfalls with intensities > 30 mm/h;
- Class 3 : 1 hour long rainfalls with intensities closed to 20 mm/h
- Class 4 : long rainfalls (duration > 2 hours);
- Class 5 : 1 hour long rainfalls with intensities closed to 30 mm/h;
- Class 6 : rainfalls with high intensities >50 mm/h and duration of 1 hour;
- Classes 7-9 : each class contains a specific single rainfall event.

The result is interesting as these three parameters can be available in the real-time by the radar imagery and qualitative information furnished by the meteorological service.

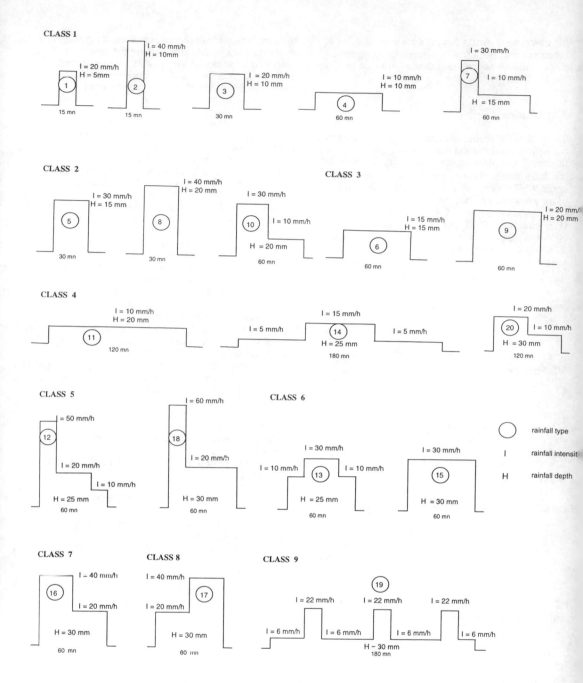

Fig. 2 Clustering results of 20 typical rainfall events

2.5 Discrimination

Among the available discrimination methods (parametric, nonparametric, distance or the frontier distance methods, Späth, 1980) the nonparametric techniques were selected. In the principle, they are based on a probability distribution function estimation at a point x_0 of the vector space. As in the case under study no information was available allowing to establish the sample pdf, the k-Nearest Neighbours method (kNN) was applied. The method consists in searching the k-nearest neighbours (where k is an odd number in order to avoid the ambiguity of indecision when an equal number of neighbours of every class is found) of any point x_0 and in a selection of a class the most representative for them (Fig. 3).

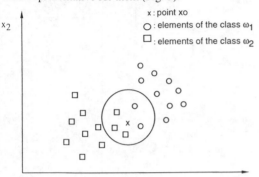

Fig. 3 Example of a discrimination without reject by k-Nearest Neighbours Method (k = 5), x_0 is assigned to the class ω_2

In the case of a discrimination without reject between two classes the following discrimination rule is applied:

$$d(x_0) = \begin{cases} 1, & \text{if } k_1 > k_2 \\ 0, & \text{otherwise} \end{cases} \qquad [3]$$

where $k_1 + k_2 = k$, k_i being the number of neighbours of x_0 belonging to the class ω_i ($i = 1,2$). If k=1, then x_0 is assigned to the class of its nearest neighbour.

The following rejection criteria can be used:

- ambiguity criterion: x_0 is assigned to ω_i if a minimal number k'_i of its k-nearest neighbours belongs to ω_i. In the case of two classes ($i = 1,2$) the decision function is defined as:

$$d(x_0) = \begin{cases} 1, & \text{if } k_1 > k'_1 \\ 2, & \text{if } k_2 > k'_2 \\ 0, & \text{if } k_i < k'_i \end{cases} \qquad [4]$$

in the latter case, x_0 is not assigned to any of these two classes;

- distance criterion: the point x_0 is rejected if an average distance from its k-nearest neighbours is greater than a certain threshold distance.

Although the k nearest neighbours method requires to calculate all the pairwise distances of any point x_0 from all the elements of the learning set, it offers the advantage of simplicity and efficiency for an operational application (Dubuisson, 1990).

2.6 Application in the real-time

For the purpose of feasibility study a simulator of a real-time operation, which integrates the pattern recognition techniques, the available meteorological forecast, and the information about the state of the network was developed (Fig. 4).

The objective consisted in the validation of the overall approach and testing if the available information allows to make a series of optimal decisions until the end of the rainfall event. It is worth mentioning that radar imagery allows to forecast the depth and duration of the rainfall, the meteorological file provides the typical qualitative forecast on the rainfall pattern based on the seasonal variations of the micro climate, expected number of thunderstorms, wind speed etc. The tasks fulfilled by the simulator are marked on the Fig. 4 by the solid line.

The following input data are necessary for the simulation:

- the forecast of the total duration of the rainfall event, which can vary every time step;
- the forecast of the total rainfall depth;
- the rate of the filling up of retention storage reservoir.

Every time step of simulation the vector representing the current event is completed and assigned to an appropriate class using the kNN method, subject to the compatibility of this decision with the forecast of the event duration and rainfall depth. The margin of error for the rainfall depth is fixed at $\varepsilon = 10$ mm around the forecasted value while the selected duration should be located within 25% error bounds of the forecasted duration. Moreover, the filling up of the most important from the management point view retention storage reservoir is simulated through control rule depending on the class of rainfall. At every time step of simulation a control strategy, consisting in opening or closing the reservoir gates depending on the identified rainfall class and the state of the reservoir, to be followed is indicated.

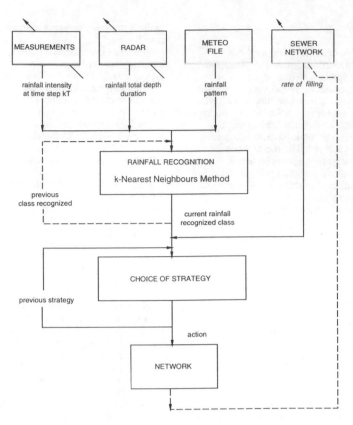

Fig. 4 The functional architecture of the simulator

The following management strategies were tested
with aid of the simulator :

CLASS OF EVENT	MANAGEMENT STRATEGY
CLASS 1	Constant release discharge $Q(t) = 7.0 \text{ m}^3/\text{s}$
CLASS 2	$Q(t) = \begin{cases} 4.0 \text{ m}^3/\text{s, if t} >25 \text{ hours and } V_s < 0.5 \text{ V}_{useful} \\ \\ 5.5 \text{ m}^3/\text{s, if t} > 1,5 \text{ hours and } 0.5 < V_s < V_{useful} \end{cases}$
CLASS 3	Same as far Class 3
CLASS 4	$Q(t) = 5.5 \text{ m}^3/\text{s, if t} > 1$ hour after rainfall event beginning and $V_s < 0.7 \text{ V}_{useful}$
CLASS 5	$Q(t) = 5.5 \text{ m}^3/\text{s, if t} > 2.5$ hours and $V_s < 0.7 \text{ V}_{useful}$
CLASS 6	$Q(t) = 5.5 \text{ m}^3/\text{s, if t} > 2$ hours and $V_s < 0.7 \text{ V}_{useful}$

Classes 7-9 were not tested. Whatever is the class only one strategy modification was allowed during the rainfall event, with V_S - the stored volume and V_{useful} - the useful storage of the reservoir.

In the Tables 1-3 some examples of simulation results obtained are presented.

Table 1

initial storage = 0.0 %   forecasted rainfall = 25.0 mm

Time step	rainfall type	distance	intensity mm/h	duration min.	class	volume stored	release discharge Q m3/s
1	20	16.0	16.0	90	4	6.	7.0
2	20	32.0	16.0	90	4	12.	7.0
3	20	48.0	16.0	90	4	18.	7.0
4	20	64.0	16.0	90	4	24.	5.5
5	20	74.5	16.0	90	4	30.	5.5
6	20	106.	16.0	90	4	36.	5.5
7	20	138.5	0.0	90	4	42.	5.5
8	20	122.7	0.0	90	4	42.	5.5
9	20	122.7	0.0	90	4	42.	5.5
10	20	122.7	0.0	90	4	42.	5.5
11	20	122.7	0.0	90	4	42.	5.5
12	20	122.7	0.0	90	4	42.	5.5

volume of the selected rainfall : 30.0
volume of the input rainfall   : 24.0

Table 2

initial storage = 0.0 %   forecasted rainfall = 20.0 mm

Time step	rainfall type	distance	intensity mm/h	duration min.	class	volume stored	release discharge m3/s
1	5	25.0	35.	30.	2	15.	7
2	8	50.0	35.	30.	2	29.	7
3	8	50.0	0.	30.	2	44.	4
4	8	50.0	0.	30.	2	39.	4
5	8	50.0	0.	30.	2	39.	4
6	5	50.0	0.	30.	2	39.	4
7	8	50.0	0.	30.	2	39.	4
8	5	50.0	0.	30.	2	39.	4
9	5	50.0	0.	30.	2	39.	4
10	5	50.0	0.	30.	2	39.	4
11	5	50.0	0.	30.	2	39.	4
12	5	50.0	0.	30.	2	39.	4

volume of the selected rainfall : 15.0
volume of the input rainfall   : 17.5

Table 3

initial storage = 20.0 %   forecasted rainfall = 20.0 mm

Time step	rainfall type	distance	intensity mm/h	duration min.	class	volume stored	release discharge Q m3/s
1	11	25.0	5.	120.	4	22.	7.0
2	11	25.0	10.	120	4	26.	7.0
3	11	61.0	16.	120	4	32.	7.0
4	11	125.0	18.	120	4	38.	5.5
5	11	135.5	16.	150	4	44.	5.5
6	14	315.0	14.	150	4	50.	5.5
7	14	315.0	15.	150	4	55.	5.5
8	14	329.3	8.	150	4	58.	5.5
9	14	329.0	6.	150	4	61.	5.5
10	14	331.6	8.	150	4	64.	5.5
11	14	333.5	0.	150	4	68.	5.5
12	14	338.3	0.	150	4	68.	5.5

volume of the selected rainfall : 25.0
volume of the input rainfall   : 29.0

The tested events were either selected from the learning set or were created artificially by combination of different depths and durations from the available sample of events. In order to compare the stability of the pattern recognition in the simulated real-time conditions an additional, a posteriori clustering is performed at the end of the rainfall event. In majority of cases (approximately 90 % of simulations) both results coincide, however for some events completely different results were obtained. For example, the rainfall of the type 1 from the Class 4 (Table 1) is not assigned to any class a posteriori. The artificial events are clustered when the rainfall depth is not greater than 30 mm/h (a rainfall of 25 mm/h during 90 minutes, i.e. 37 mm depth was rejected).

The simulator allowed however to test the stability of the method as far as the time evolving changes in recognised rainfall events are concerned. (Rapport LHF/DEA93,1991).

4. CONCLUSIONS

The real-time pattern recognition of the typical rainfall opens the possibility to use the simulation scenarios of physical behaviour of a hydraulic network, which can be prepared and tested in advance and stored in a data base. In turn it can allow to anticipate the effects of flow in some dangerous reference situations using the capacity of a local remote control of valves.

The obtained results confirm that application of the pattern recognition technique can be useful for a purpose of rapid identification of a current rainfall event, when the approximate forecast of rainfall duration and its depth is available, in a well defined class of known typical rainfall events.

Contrariwise, the study did not allow to precise if such an approach could be extended to the multi-dimensional classification problem, consisting in considering the state vector composed of rainfall event, state of the network and applied control decision.

REFERENCES

Dubuisson B. 1990. *Diagnostic et reconnaissance des formes*. Paris : Editions Hermès.
Preisendorfer, R.W. 1988. *Principal Component Analysis in Meteorology and Oceanography*, Developments in Atmospheric Science 17, Elsevier Science Publ. B.V.
Späth, H. 1980. *Cluster analysis algorithms for data reduction and classification of objects*. Ellis Horwood Publishers.

Rapport LHF/DEA93. 1991. Etude de faisabilité des méthodes de reconnaissance de scénarios pour l'aide à la conduite d'un réseau d'assainissement - Rapport final.

Stochastic Hydraulics'96, Tickle, Goulter, Xu, Wasimi & Bouchart (eds) © 1996 Balkema, Rotterdam. ISBN 90 5410 817 7

Effect of rainfall characteristics on runoff in bamboo watershed

Chin-Yu Lee

Department of Soil and Water Conservation, National Pingtung Polytechnic Institute, Taiwan, China

ABSTRACT: With data collected from Jul.'78 to Dec.'94, this study is concerned with the factors of rainfall characteristics affecting runoff of a bamboo watershed. Analyses of the estimative parameters are as follows: rainfalls (P), maximum one-day rainfall (Pm), discharge (Q), peak discharge (Qp), Time peak (Tp), Time lag (TL), infiltration index (Id), and effective rainfall duration (T). As the result of calculation, which used the analysis of frequency distribution and step-wise regression, the following facts were found: Tp, TL and T occurrence four to eight hours, two to six hours and one to four hours, respectively, and the regression equation is Q = -3.89 + 0.28(P) + 673.84 (Pm) + 1.01(Id) + 1.7 (T). The effect of discharge could be arranged as: Pm > T > Id > P, and the application of the result will be offer hydrologic budget of watershed management.

INTRODUCTION

The relation between rainfall and runoff is one of the most important indices for expressing the hydrological behaviour of a watershed. The rainfall-runoff relation is an important subject which has long been studied by many researchers and should be developed intensively as well as extensively in the future. Many studies have been carried out upon the relation between rainfall and runoff within the same period, yearly, seasonal, monthly, daily, hourly, etc. But due to the delay or lag of runoff after rainfall such rainfall-runoff relations are usually meaningless especially for data in short periods (Ogihara, 1967).

Until recently, much knowledge of the relation between runoff, climate and physiography was based on the work of Horton(1945), Strahler(1964), and others, who developed physically based approaches to understanding runoff and its relation to climate and drainage basin characteristics. However, it is now recognized that runoff processes are much more complex than Horton envisioned, and that runoff rates are sensitive to the spatial variation in rainfall, soils, and topography. Results from this more recent work indicate those peak rates of runoff from hillslopes and small (less than 10 km^2) watersheds may vary over several orders of magnitude depending on the mechanism of runoff production. Increasing magnitudes of storm rainfall and urban runoff can be especially problematic in Taiwan because most of the area is encircled by a levee system. Levees were constructed to protect the city from storm surges induced by typhoon.

The purpose of this study was to quantify the factors of rainfall characteristics affecting runoff of a bamboo watershed. Analyses of the estimative parameters are as follows: rainfalls (P), maximum one-day rainfall (Pm), discharge (Q), peak discharge (Qp), Time peak (Tp), Time lag (TL), infiltration index (Id), and effective rainfall duration (T). We tried to use the step-wise regression analysis, and in accordance with the frequency analysis to analyse the hydrologic data in different return periods. The match of the results of fieldwork and the relative errors are also discussed. A detailed survey was performed on the topography data of the bamboo watersheds.

LITERATURE REVIEW

Direct precipitation onto the saturated area was another major contributor of storm flow, storage of water with these source areas was small and travel times out of them were short. Runoff from them was largely controlled by Ran fall intensity. These partial areas contributing quick runoff could expand or contract seasonally or during a storm. The experimental work by Dunne and Black (1970) has thrown doubt on the role of subsurface storm flow from hillsides as a major contributor to storm hydrographs in upland watersheds of northern Vermont. They found that the importance of a hillslope as a producer of storm runoff seems to depend largely on its ability to generate overland flow. Recently, a rapid progress was made in the analysis of rainfall and runoff relationship. Lots of techniques other than the rational formula have been developed and applied. However most of these techniques are unsatisfactory in the analysis of rainfall characteristics if only the available rainfall intensity curve is used (Chang, 1980).

Hawkins(1982) reviewed other published interpretations of rainfall and runoff records which concluded that the proportion of some drainage basins generating overland flow would increase as rainfall intensity increased. Studies of forest management effects on a water yield have also reviewed by Douglas (1983) in the East, Troendle (1983) in the Rocky Mountains, and Hibbert (1983) and Harr (1983) in the West. Forests in relation to streamflow have not been studied in the bamboo watershed region of Taiwan. The present study aims to use available data to evaluate the association of varying proportions of forest area, along with climatic and topographic parameters, with recorded streamflow in a bamboo watershed, and to develop empirical models to estimate stream flows at ungaged watersheds for water resource's management.

Of the 31 physioclimatic parameters analysed, watershed areas, percent forest area, shape index, spring precipitation, and annual temperature were the most significant in affecting streamflow characteristics (Chang & Watters, 1984). Estimates of annual runoff are needed for planning of water resources projects. There is a lack of long-period runoff records in a large number of a watershed in Taiwan. Based on limited data, earlier investigators have developed equations, albeit empirical, relating annual runoff to factors such as annual rainfall, annual average temperature, and watershed characteristics. Since more data are now available, these equations have been found to give results with very large errors. Time series of rainfall, when observed on a scale of days, may be treated as a stationary stochastic process with the property of white noise. However, when it is observed on a scale of hours, it is no longer a stationary stochastic process but belongs to a non-stationary stochastic process (Hino and Hasebe, 1984). The transfer of hydrological data, parameters or relationships obtained from one basin, e.g. from a representative basin, to others, or their estimation for ungaged basins, had obvious practical importance. Several research studies are available in the literature dealing with this subject and especially with the transferability of runoff or of related process characteristics and relationships and with the estimation of the latter by using geomorphological characteristics of the basin. Current knowledge of hydrological and geomorphological processes and the results of published studies indicate that basin size, or some of its related functions like stream length and slope, can be expected to influence runoff significantly in a number of direct and indirect way (Mimikou, 1984).

Kothyari et al., (1985) developed a relationship for mean annual runoff as a function of mean annual rainfall, mean annual temperature, and forest cover. Several recent investigations have shown that the lack of knowledge about the spatial distributions of short-term rainfall is the greatest source of errors in runoff simulation. It has been shown that the spatial variability of rainfall is important even on relatively small urban catchments (Niemczynowicz, 1988). Krajewski and Smith (1989) have developed a statistical model of space-time rainfall for which parameter estimation is based on time-integrated radar rainfall observations from a single radar and time-integrated raingage observations from a network of raingages. The accuracy of the parameter estimators is affected by the estimation procedure, data measurement errors, and data sample size.

Sivapalan et al.(1990) used a partial area runoff generation model and a geomorphologic unit hydrograph (GUH) based runoff routing model to study the impact of hydrologic controls such as runoff production mechanisms, rainfall characteristics and basin scale. Comparisons of estimates of total rainfall, total runoff, peak runoff and time to peak between the hydrographs resulting from the random sampling and interpolation, and that computed from the full resolution radar accumulation field were made (Duncan et al., 1993). In connection with forest decline caused by air pollution and its effects on amount, timing and quality of discharge there is an increasing demand to use sound long-term quantitative and qualitative precipitation and streamflow measurements of forest catchments such as 'bench-mark-catchments'. However, for this purpose the procedures of a climatic calibration and/or watershed modelling must be realised (Brechtel and Fuhrer, 1994).

MATERIALS AND METHODS

The measurements of rainfall and runoff were observed in a small bamboo watershed, and one set of a triangular weir was installed for this study in the campus of National Pingtung Polytechnic Institute (located about 21km north of Pingtung in southern Taiwan). The average annual rainfall of the site was 2600mm, and the rain season generally concentrated from May to October. It covers an area of about 0.56 hectares and has a length of 95 m and an average 48% slope (Lee and Hsu, 1993).

To quantify the effect on runoff, we tried to use a standard step-wise regression selection procedure. It is an attempt to achieve a similar conclusion working from the other direction, that is, to insert variables in turn until the regression equation is satisfactory. The basic procedure is as follows:

1. Calculate the correlations of all the predictor variables with the field data results. Select, as the first variable to enter the regression, the one must

be highly correlated with the response.

2. The overall F-test shows that the regression equation is significant(α=0.05).

3. Calculate the partial correlation coefficients of all variables not in regression with the response.

4. Choose as the next variable to enter into the regression, the one with the highest partial correlation coefficient.

5. The decision is made to either reject or retain the corresponding predictor and the equation is recomputed or the next candidate is sought, respectively.

6. The stepwise method now selects as the next variables to enter, the one most highly partially correlated with the response, and the percentage square of the multiple correlation coefficient, R, must be increased. At this point, the other partials F-values for the variable are examined.

7. The only remaining variable to be considered at this juncture is immediately rejected, the stepwise regression procedure terminated and chooses its best regression equation.

We believe this to be one of the best of the variable selection procedures discussed and recommended its use. It is a more economical use of computer facilities than any methods, and it avoids working with more X's than are necessary while improving the equation at every stage (Lee, 1992).

RESULTS AND DISCUSSION

Four frequency analysis methods were used in this study, and according to the results of correlation coefficient (r) and estimation of standard error (S_E) by regression analysis, the Log-Pearson Type 3 distribution method and Iwai method were found to be the best for frequency analysis. The Gumbel-Chow method was the second, and the Pearson Type 3 distribution method was the third (Table 1).

Table 1. Frequency Eq. of annual Max. one-day rainfall in the Laopi area.

Methods	Regression Equations	R^2	S_E
Iwai Distribution	Y = 127.8 + 220.6 * LOG (T)	0.998	3.7
Pearson Type 3 Dis.	Y = 175.9 + 203.7 * LOG (T)	0.992	8.3
Log-Pearson Type 3 Dis.	Y = 159.5 + 220.4 * LOG (T)	0.999	2.0
Gumbel-Chow method	Y = 180.0 + 194.3 * LOG (T)	0.998	4.0

[Note] T: Return period (yr)

With increasing frequency, forest hydrologists are being asked to provide estimates not only of the total rate of runoff, but to locate the sources of sediment and the processes contributing the sediment to channel at various stages during and after disruptions of forests by major storms, timber harvest, or fire. Reich(1972) reported that mean annual flood series, runoff volume from individual storms, and peak discharge decrease with increasing percentage of woodland area. In a regression analysis of flow characteristics (Lowham, 1976), percent forest cover was not found to be significant enough to warrant inclusion in equations estimating peak flows for 243 gaging stations in Wyoming.

Knowledge of streamflow characteristics is necessary to adequately plan, design, and implement many natural resource management activities. Annual estimates of a water yield are needed for cumulative effects analyses and reservoir design; estimates of peak flow for hydraulic design of culverts, bridges, riprap, check dams, log weirs, and other in stream structures (Evan and Johnston, 1980).

It is shown that of throughfall and relative light intensity of plots, and we obtain the high relationship. Prediction of these sort are useful for understanding watershed yields, improving our ability to anticipate hydrologic budget and to examine conservation strategies. Also, as mentioned above, the engineering geologist's ability to recognize the potential for runoff at a particular site on the basis of local topography, hydrology, and geological materials is an important form of rainfall-runoff prediction, which need to be utilized more frequently in forest management so that particular storm event problems can be avoided.

The soils in the plot are generally shallow and have undeveloped profile characteristics. The silt contents are almost all in the range from 39 to 45%. The surface layer (A horizon) is only 10 cm deep and has medium value of moisture content, 6 to 15%. The specific gravity of the surface layer has a slightly lower value. This might be due to a rather high percentage of organic matter from 0.5 to 5%.

The dry bulk density of the surface layer was, in general, from 1.31 to 1.40 g/cm^3. The clay contents are almost in the range from 25.2 to 30.2%, and the soil texture is silty loam. No great difference in physical properties of the soil in the plot, so that the soil physical conditions could be regarded as homogeneous. Concave hillslope sections accumulated soil moisture due to both saturated and unsaturated lateral flow precesses. In winter, streamflow generation was controlled by return flow, saturation overland flow and throughflow. In summer, postponing, infiltration excess and saturation overland flow dominated. (George and Conacher, 1993).

The amounts of runoff and other parameters were measured of the plot, as the results of calculation of selection of the estimative parameters to runoff, which used the stepwise regression analysis. By using the stepwise regression selection procedures, we calculated the correlations of all the predictor variables with the correlation matrix analysis (Table 2).

Table 2. A correlation matrix of hydrologic parameters from July 1992 to April 1994.

	P	Pm	Q	Qp	Tp	TL	Id	T
P	1.00	0.71**	0.99**	0.81**	0.75**	0.56**	0.03	0.86**
Pm		1.00	0.64*	0.76**	0.35	0.40*	0.53**	0.35
Q			1.00	0.83	0.77**	0.57**	-0.10	0.88**
Qp				1.00	0.50*	0.38*	-0.05	0.62**
Tp					1.00	0.53**	-0.21	0.90**
TL						1.00	0.04	0.43*
Id							1.00	-0.28
T								1.00

P : Rainfall   Pm: Max. one-day rainfall
Q : Discharge   Qp: Peak discharge
Tp : Time peak   TL: Time lag
Id : Infiltration index   T: Effective rainfall duration
**, *: 0.1% & 1% significant level

And then, as with all the procedures discussed, sensible judgement is still required in the initial selection of variables and in the critical examination of the model through examination of residuals. Stepwise regression analysis was performed on all parameters selected to examine the contribution of variations of the regression equation. It is easy to rely the automatic selection performed in the computer. Analyses of the estimative parameters are as follows: rainfalls (P), maximum one-day rainfall (Pm), discharge (Q), peak discharge (Qp), Time peak (Tp), Time lag (TL), infiltration index (Id), and effective rainfall duration (T). As the result of calculation, which used the analysis of frequency distribution and stepwise regression, Fig.1 consisted of the following facts were found: Tp, TL and T occurrence four to eight hours, two to six hours and one to 4 hours, respectively. Those independent variables did not present significant contributions were dropped from the final models. The regression equation was as (Table 3):

$$Q = -3.89 + 0.28(P) + 673.84\ (Pm) + 1.01(Id) + 1.7\ (T)$$

Where Q is the discharge, P is the rainfall, Pm is the maximum one-day rainfall, Id is the infiltration index, and T is the effective rainfall duration, and as an effect on Q in the regression equation could be arranged as Pm > T > Id > P.

Table 3. Values of step-wise regression method.

Step									
	Predictor variables (P.V.) in regression							P.V. not in Eq.	
	P.V.	R	R^2	S_E	F	B	Beta	P.V.	P.C.C.
1st	X_2	0.98	0.97	9.87	1614.17	B_2=785.09	0.99	X_1	0.625
								X_3	0.003
								X_4	-0.039
								X_5	0.164
								X_6	0.763
								X_7	-0.092
2nd	X_6	0.99	0.98	6.45	1925.36	B_2=795.75	0.99	X_1	0.090
						B_6= 1.67	0.13	X_3	-0.071
								X_4	0.195
								X_5	-0.117
								X_7	0.358
3rd	X_7	0.99	0.99	6.08	1443.59	B_2=734.62	0.92	X_1	0.293
						B_6= 1.32	0.15	X_3	0.092
						B_7= 1.24	0.89	X_4	-0.094
								X_5	-0.068
4th	X_1	0.99	0.99	5.88	1160.02	B_2=673.84	0.85	X_3	-0.180
						B_6= 1.01	0.11	X_4	-0.160
						B_7= 1.70	0.12	X_5	-0.030
						B_1= 0.28	0.07		

X_1 : Rainfall (mm)   R : Correlation coefficient
X_2 : Max. one-day rainfall (mm) R^2 : Deterministic coefficient
X_3 : Peak discharge (m³/s) S_E : Estimation of standard error
X_4 : Time to peak (hr)   F : F% points for α=0.05 that the
X_5 : Time lag (hr)   regression is significant
X_6 : Infiltration index (mm/hr)   P.C.C.: Partial correlation coefficient
X_7 : Effective rainfall duration (hr)

Size of annual peak flow in a small watershed in Western Oregon was reduced 32%, and average delay of all peak flows was nearly nine hours following clear-cut logging. Size of annual peak flows caused by rain with snowmelt was reduced 36%, and peak flows resulting from rain with snowmelt were delayed an average of nearly 12 hours following logging (Harr and McCorison, 1979). Changes in runoff are shown to be associated with long-term increases in rainfall amount, intensity, and frequency. Time series analyses of rainfall and watershed runoff demonstrated those changes in rainfall patterns and amounts can mask the benefital impacts of floodwater retarding structures (Fernandez and Garbrecht, 1994).

CONCLUSIONS

From what we have seen so far, various factors play a decisive role in determining the processes affecting in rainfall characteristics on runoff: the configuration of the terrain, vegetation, slope gradient, strength of the native rock, and in particular the greater dynamic properties of the accelerated runoff of heavy rainstorms. Runoff mechanisms on a small hillslope were dependent on the extent and development of variable source areas. The extent of the variable source area and the magnitude of streamflow were due to antecedent soil moisture, rainfall & slope morphology.

In this paper, the author described the present problems on the factors of rainfall characteristics affecting runoff of a bamboo watershed. As described previously, technical solutions may be unable to provide sustainable agriculture where rainfall approaches the point of no return. On suitable sites, we expect technical evolution to provide methods that improve the quantity & quality of crop yields and conserve soil. So, land clearly unsuitable for farming may return to forests.

ACKNOWLEDGEMENTS

The field work reported in this paper was assisted by Y.-L.Chen, C.-W.Lin, and C.-H. Chen. , and was supported by the National Science Council (NSC), grant no. NSC-82-0409-B-020-009.

407

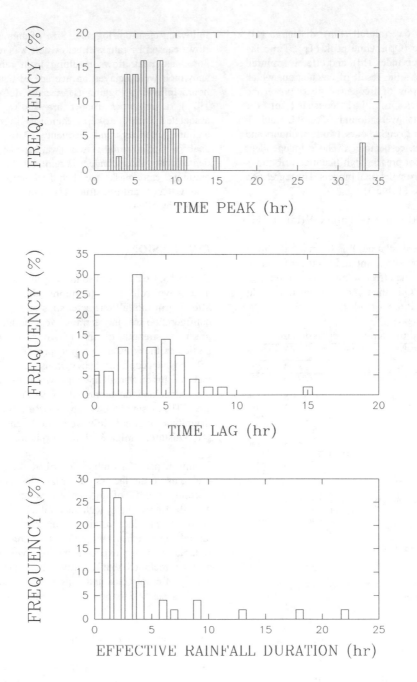

Fig. 1 Tp, TL and T VS. frequency distribution

REFERENCES

Brechtel, H.M. and H.W. Fuhre 1994. Importance of forest hydrological 'benchmark-catchments' in connection with the forest decline problem in Europe. Agric. and For. Meteorol., 72:81-91.

Chang, M. and S.P. Watters 1984. Forests and other factors associated with stream-flow in East Texas.Water Resour. Bull. ,20(5): 713-719.

Chang, Y.-T. 1980. Study on runoff by using hyetograph in the rational formula. Int. Conf. on Water Resour. Development, p.373-382.

Duncan, M.R. et al. 1993. The effect of gage sampling density on the accuracy of streamflow prediction for rural catchments. J. Hydrol. , 142:445-476.

Dunne, T. and R.D. Black 1970. Partial area contributions to storm runoff in a small New England watershed. Water Resour. Res., 6(5): 1296-1311.

Dunne, T. 1978. Field studies of hillslope flow processes. In M.J. Kirkby(Ed.), Hillslope Hydrology, Wiley-Interscience, NY, p.227-293.

Evan, W.A. and B. Johnston 1980. Fish migration and fish passage, A practical guide to solving fish passage problems. USDA For. Service, Washington D.C., 63pp.

Fernandez, G.P. & J. Garbrecht 1994. Effect of trends & long-term fluctuations of rainfall on watershed runoff. Trans., ASAE, 37(6): 1841-4.

George, R.J. and A.J. Conacher 1993. Mechanisms responsible for streamflow generation on a small salt-affected & deeply weathered hillslope. Earth Surface Pro. & Landforms, 18:291-309.

Harr, R.D. and F.M. McCorison 1979. Initial affects of clear-cut logging on size and timing of peak flows in a small watershed in Western Oregon. Water Resour. Res., 15(1): 90-94.

Hawkins, R.H. 1982. Interpretation of source-area variability in rainfall-runoff relationships, edited by V.P.Singh, p.303-324, Water Res. Pub. Fort Collins, Colorado.

Hino, M. and M. Hasebe 1984. Identification & prediction of nonlinear hydrologic systems by the filter-separation autoregression(AR) method: extension to hourly hydrologic data. J. Hydrol. , 68:181-210.

Kothyari, U.C. and R.J. Garde 1991. Annual runoff estimation for catchments in India. J. Water Resour. Planning and Mgmt. ,117(11): 1-10.

Krajewski, W.F. and J.A. Smith 1989. Sampling properties of parameter estimators for a storm field rainfall model. Water Resour. Res., 25(9): 2067-2075.

Lee, C.-Y. 1992. Quantitative study on the landslide risk of reservoir watersheds in Taiwan. Biannual of Taiwan Museum 45(1): 19-27.

Lee, C.-Y. and S.-H. Hsu 1993. Study on the hydrologic characteristics of bamboo watershed. NSC82-0409-B-020-009, 40pp.

Lowham, H.W. 1976. Techniques for estimating flow characteristics of Wyoming streams. USGS, Water Resources Invest. 76-112,83pp.

Mimikou, M. 1984. Regional relationships between basin size and runoff characteristics. J. Hydrol. Sci., 29(1): 63-73.

Niemczynowicz, J. 1988. The rainfall movement - A valuable complement to short-term rainfall data. J. Hydrol. , 104:311-326.

Ogihara, S. 1967. One-day rainfall and its corresponding runoff. In "Int. Symp. on For. Hydrol. , Ed. by W.E.Sopper & H.W.Lull, Pergamon Press, NY, P.523-526.

Pitlick, J. 1994. Relation between peak flows, Precipitation, and physiography for five mountainous regions in the Western USA. J. Hydrol. , 158:219-240.

Reich, B.M. 1972. The influence of percentage forest on design floods. In: Nat. Symp. on Watersheds in Transition. ARWA, Urbana, Illinois, p.335-340.

Sivapalan, M., E.F. Wood. , and K.J. Beven 1990. On hydrologic similarity 3, A dimensionless flood frequency model using a generated geomorphologic unit hydrograph and partial area runoff generation, Water Resour. Res., 26(1): 43-58.

Stochastic Hydraulics'96, Tickle, Goulter, Xu, Wasimi & Bouchart (eds) © 1996 Balkema, Rotterdam. ISBN 90 5410 817 7

Dynamic channel routing for an urbanized basin

Edmond D. H. Cheng
University of Hawaii at Manoa, Honolulu, Hawaii, USA

X. J. Tang
M&E Pacific, Honolulu, Hawaii, USA

ABSTRACT: In order to develop a post flood hazard mitigation plan, it is important to understand the dynamic process of the event. A debris flow simulation procedure for an urbanized small watershed in Hawaii is presented for illustration. During the torrential rains of a more than 100-year storm, numerous debris flow occurred in a highly urbanized watershed in eastern Honolulu. In an attempt to reconstruct the extreme event, the first step is the establishment of a debris bulked capital inflow hydrograph to the urbanized watershed of interest. A dynamic channel routing procedure is thus followed. During the process, a one dimensional unsteady open-channel flow is considered for the watershed.

1 INTRODUCTION

Dynamic debris flow had not received much attention on the island of Oahu, Hawaii, until 1987. December 1987, because of a strong and sharp upper tropospheric trough which stayed over the Hawaiian islands, eastern Oahu had heavy rainfalls. On December 31, another trough moved into the islands and produced a historical storm. In some areas, such as Hahaione Valley in eastern Oahu, the rainfall was even greater than the 100-year frequency. In order to understand the dynamic process of this event, a dynamic channel routing is investigated at Hahaione Valley (Fig. 1).

2 INFLOW HYDROGRAPH

Hahaione Valley is located on the leeward side of southeastern Oahu (Fig. 1). The area of Hahaione watershed is about 2.4 square kilometers; twenty-one percent of the area is urbanized. The land is mostly covered by rock or stony clay with low permeability. The vegetation consists of brush-type plants. This watershed was selected for the following reasons: firstly, this valley is a highly urbanized watershed, it could be used as an example of watershed development in the future; secondly, there are no continuous rainfall records available in the area except for the only confirmed

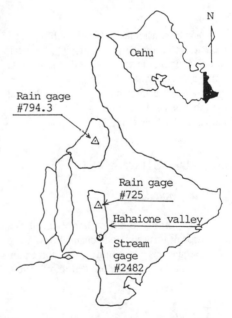

Figure 1. Map of flooded areas, southeastern Oahu, Hawaii.

maximum 24-hour rainfall recorded at Rain Gage No. 725, located at the upper Hahaione Valley (Fig. 1). It is, therefore, possible to create a reasonable synthetic rainfall sequence for Hahaione watershed by transferring rainfall recorded at a nearby Rain Gage No. 794.3 (Fig. 1); finally, some debris data of the 1987 New Year's Eve flood are available for this particular Valley.

Based on the 60 cm rainfall measured at Gage No. 725 a time series of hourly rainfall was established at Gage No. 725. By means of the Nash-Muskingum method (Nash 1959) of routing rainfall excess, a clear water inflow hydrograph for the Hahaione watershed is established.

The basic elements involved in the determination of debris bulked hydrograph are the volume and bulking rate of debris. Debris bulking rate is a time-dependent function, and can only be determined experimentally in the laboratory or in the field. For engineering application, this variable bulking rate, R(t), may be defined by an empirical power equation, which is modified from an empirical equation used by the Department of Public Works, Los Angeles County (LACDPW 1989):

$$R(t) = a(Q_c/Q_{pc})^b \tag{1}$$

where R(t) = dimensionless variable bulking rate; Q_c = clear water discharge in cfs; Q_{pc} = peak clear water discharge in cfs; a = peak bulking rate in %; and b = constant.

With the available information, a debris bulked hydrograph can be developed by distributing the total debris in the flow (Fig. 2).

3 CHANNEL ROUTING

3.1 *Governing equations*

In the channel routing, the flow was assumed as a one dimensional, unsteady open channel flow. The governing equations (continuity and momentum equations) can be written as (Strelkoff 1969; DeLong 1985):

$$\frac{\partial A}{\partial t} + \frac{\partial Q}{\partial s} - q(x) = 0 \tag{2}$$

$$\frac{\partial Q}{\partial t} + \frac{\partial}{\partial s}\frac{(\beta Q^2)}{A^2} + gA(\frac{\partial y}{\partial s} + S_f(s)) = 0 \tag{3}$$

w h e r e t = time; s = channel distance; A= cross-sectional area; Q = volumetric discharge; q

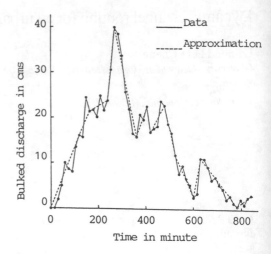

Figure 2. Inflow hydrograph for channel routing.

= lateral inflow per unit of channel length; y = distance of water surface above a given datum; g = gravitational acceleration; ß = momentum coefficient resulting from nonuniform velocity distribution and is defined as

$$\beta = \frac{\int v^2 dA}{V^2 A} \tag{4}$$

where v = velocity and V = mean velocity at a cross section; and S_f = friction slope, and can be represented as

$$S_f = \frac{Q|Q|}{k^2} \tag{5}$$

where K = conveyance.

It is noteworthy that Equations (2) and (3) are a special case of Strelkoff's general unsteady open channel flow. The flow is assumed as one dimensional so that the pressure in the y direction is considered to be hydrostatic, therefore, the streamline curvature and vertical acceleration are ignored. It is further assumed that the momentum coefficient ß sufficiently accounts for nonuniform velocity distribution and that S_f adequately describes channel resistance. Consequently, Equations (2) and (3) may not apply to channels with a large slope. Since the maximum channel slope in this study is only 10.5 percent, Equations (2) and (3) are therefore valid.

Strelkoff used a time dependent natural channel alignment system in his derivations. However, it is more convenient to discuss the problem in a time independent reference system. DeLong transformed equations (2) and (3) into the time independent x-t coordinate system by using Green's theorem (DeLong and Schoellhamer 1988). The governing equations become

$$\frac{\partial(AM_a)}{\partial t} + \frac{\partial Q}{\partial x} - q(x) = 0 \tag{6}$$

$$\frac{\partial(QM_q)}{\partial t} + \frac{\partial}{\partial x}(\frac{\beta Q^2}{A}) + gA(\frac{\partial y}{\partial x}|_A$$
$$+ \frac{\partial y}{\partial A}|_x \frac{\partial A}{\partial x} + \frac{Q|Q|}{k^2}) = 0 \tag{7}$$

In which

$$M_a = \frac{1}{A}\sum_{i=1}^{n} m_i A_i \tag{8}$$

$$M_q = \frac{1}{Q}\sum_{i=1}^{n} m_i Q_i \tag{9}$$

where n = number of subsections; m = subsection sinuosity; A_i = subsection area; and Q_i = subsection discharge.

The second and last terms in Equation (7) are nonlinear in nature. Therefore, it is difficult to solve it analytically. A numerical model was built by DeLong and Schoellhamer (1988) to solve Equations (6) and (7). The model uses the orthogonal-collocation finite-element method (Lapidus and Pinder 1982) in space and a weighted finite difference scheme in time.
The basic strategy of solving the numerical model is to use Hermit polynomials to approximate the dependent variables A and Q over discrete elements.

3.2 Numerical Model

A FORTRAN computer program called HYDRAUX (DeLong and Schoellhamer 1988) was employed to perform the channel routing. HYDRAUX was developed by the U.S. Geological Survey (USGS) based on the numerical model described earlier. The HYDRAUX package consists of five programs: XHYDRP, INTRPH, EXPAND, BNDRAW, and HYDRAUX. The first four programs do preparation work for the last program HYDRAUX (same name as the package). However, HYDRAUX was developed for mainframe computers. Considerable modifications were made on HYDRAUX, so that it can be executed on personal computers, as well as applicable to island-type of hydrologic environments.

3.3 Expert system shell

A user-interface shell was developed by using EXSYS Professional, a generalized expert system development package (EXSYS 1988). The purpose of operating HYDRAUX by this shell is to make the application easier for the user. The user does not have to spend a lot of time learning the details of HYDRAUX, just follow the instructions shown on the screen and enter the required data.
The shell includes a text editor and programs of HYDRAUX package. The text editor, called KE editor, which can be used to create or edit data files and program source files in the HYDRAUX model. If the user is not familiar with a KE editor, the user may use any wordprocessor to work on files but the files have to be saved as text files that can be read by HYDRAUX. After finishing the preparation of the data files, the user may execute the programs listed on the menu screen one by one, that is from XHYDRP to HYDRAUX.
The execution of the shell is controlled by a command file DEBRIS.CMD created in this study. The shell provides an overview about this debris flow simulation model. For the programs XHYDRP, EXPAND and HYDRAUX which need user's input data files, the shell ensures the programs' execution by asking the user whether the data files are ready. The procedure is illustrated in Figure 3.

4 RESULTS

The length of the routing channel section in Hahaione Valley was 4,361 feet, from the junction point of channels I and II to Stream Gage Station No. 2482 (Fig. 4). The routing channel I was divided into nine reaches (nine elements) based on the change of cross section shape, slope or rainfall intensity. The data, such as elevations, sinuosities

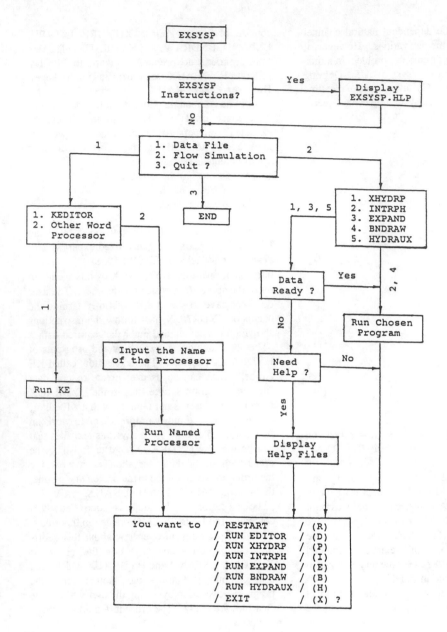

Figure 3. Flow chart of user-interface shell.

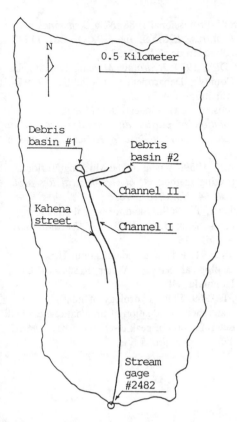

Figure 4. Locations of channels and debris basins in Hahaione watershed, Honolulu, Oahu, Hawaii.

and Manning's n of each cross section are stored in a file called XHD.DAT. For reaches longer than 500 feet, extra cross sections were automatically inserted by a program called EXPAND. The entire process was controlled by a program called CNTRL, which specified that the length of an element was in the range of greater than 200 feet but must be less than 500 feet. The data used for inserted cross sections is interpolated from adjacent known cross sections. The total number of elements, inputted or generated was 20.

4.1 *Adequacy of existing channel*

The channel routing was performed for five-minute intervals. The results indicated that no flow area exceeds the channel cross section areas even without considering the detention functions of Debris Basin #1 and Basin #2 (Fig. 4) in the

routing. Several warning messages were issued for the last two elements from t=9:15 p.m. to t=9:45 p.m. The messages indicate that flow areas at these nodes are larger than the allowed discharge area - 95 square feet, as inputted by users in INTRPH. However, the channel cross section areas at these nodes are 169 square feet and the flow areas are all smaller than 169 square feet. In other words, overflow does not occur along the channel routing section. This proves that the original design of the channel is sufficient and confirms Cheng's conclusion (Cheng et al 1991) that the cause of the 1987 New Year's Eve flood was due to the inadequate design of the Debris Basin #1. The two 7 feet by 10 feet box culverts under Kahena Street (Fig. 4) were more than adequate for draining the clear water flow. Because Debris Basin #1 did not have adequate storage to contain all of the debris generated during the Storm, the debris was washed down and blocked the culverts (Cheng 1992). The water then overflowed the bank and followed its original natural path onto Kahena Street and caused severe damage.

4.2 *Peak discharge*

A peak discharge of 72 cms (Fig. 5) at the last cross section (Stream Gage No. 2482 at t=9:30 p.m., December 31, 1987) obtained from the channel routing simulation is similar to the peak discharge of 75 cms used by USGS (1988). The USGS' 75 cms peak discharge was calculated by using the area-slope method with data obtained from actual high water mark immediate downstream from Stream Gage No. 2482. In addition, the peak discharge about 79 cms predicted by Wu's empirical formula (Wu 1969) is for an 85% runoff coefficient.

It appears that the peak discharge of the 1987 New Year's Eve flood measured at Stream Gage No. 2482 was about 72 to 79 cms.

5 CONCLUSIONS

A debris flow simulation procedure for an urbanized watershed in Hawaii has been presented herein. In the course of reconstructing this more than 100-year storm, a debris-bulked capital inflow hydrograph was established. With the aid of the modified HYDRAUX, a dynamic channel routing

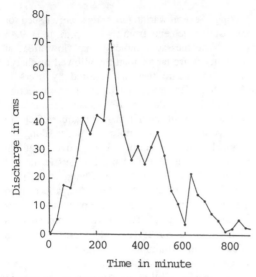

Figure 5. Outflow hydrograph at stream gage #2482 in Hahaione watershed.

process was performed. This process was further applied to evaluate the performance of a debris basin--channel system.

In addition, the expert system shell developed makes HYDRAUX user friendly in a PC environment.

REFERENCES

Cheng, E. 1992. Debris flow prediction for an urbanized watershed. *Proc. of the Sixth IAHR International Symposium on Stochastic Hydraulics*:213-220. Taipei,Taiwan.

Cheng, E., J. Dracup, J. Nigg & T. Schroeder 1991. The new year's eve flood on Oahu, Hawaii, December 31, 1987-January 1, 1988. *Vol. 1, an investigative series of the Committee on Natural Disasters,* National Research Council, Washington, D.C.

DeLong, L. 1985. Extension of the unsteady one-dimensional open-channel flow equations for flow simulation in meandering channels with flood plains. Selected paper in the hydrologic science, *USGS water-supply paper no. 2220.*

DeLong, L. & D. Schoellhamer 1988. HYDRAUX. *USGS water-resources investigation report 88-4226.*

EXSYS-Professional 1988. *New Commands and Features.* EXSYS, Inc., P. O. Box 11247. Albuquerque, NM 87192.

LACDPW 1989. *Hydrology Manual.* Los Angeles County Department of Public Works. Alhambra, CA 91803.

Lapidus, L. & G. Pinder 1982. *Numerical solution of partial differential equations in science and engineering.* New York:John Wiley.

Nash, J. 1959. A note on the Muskingum flood routing method. *J. of Geophysical Research, Vol. 64, No. 8:1053-1056.*

Strelkoff, T. 1969. One-dimensional equations of open-channel flow. *ASCE J. of Hydraulics Div. 95:861-876.*

USGS 1991. Private communication. U.S. Geological Survey, Water Resources Div., Honolulu, HI.

Wu, I. 1969. Flood hydrology of small watershed: evaluation of time parameters and determination of peak discharge. *Trans. ASAE, Vol. 12, No. 5:655-663.*

5 Groundwater hydraulics

Stochastic Hydraulics'96, Tickle, Goulter, Xu, Wasimi & Bouchart (eds) © 1996 Balkema, Rotterdam. ISBN 90 5410 817 7

Uncertainty analysis of unsaturated hydraulic conductivity

Liang Xu & Larry W. Mays
Department of Civil and Environmental Engineering, Arizona State University, Tempe, Ariz., USA

ABSTRACT: Determination of appropriate unsaturated hydraulic conductivity is crucial to the modeling of unsaturated flow for the operation of soil aquifer treatment systems (SAT). Uncertainties of the parameters in the *van Genuchten-Mualem* water retention equation further complicates the modeling. This paper presents an approximate, but practical, uncertainty analysis method by using first-order analysis of uncertainty to estimate the mean and variance of the unsaturated hydraulic conductivity. Different levels of the parameter uncertainties were examined to determine that all the parameters in the water retention equation have significant contributions to the uncertainty of the unsaturated hydraulic conductivity.

1. INTRODUCTION

The basic objective of this paper is to quantify the uncertainty of the unsaturated hydraulic conductivity. This is important in modeling the operation of soil aquifer treatment (SAT) systems. A major portion of the complexity of the unsaturated hydraulic conductivity is the inherent uncertainty exhibited in the parameters of the *van Genuchten-Mualem* water retention equation. Basically the direct field methods to determine the unsaturated hydraulic conductivity are time consuming, expensive, and usually subject to simplifying assumptions. An attractive alternative to extensive direct measurement is the theoretical calculation of the hydraulic conductivity from more easily measured field or laboratory soil water retention data.

Normally an SAT system is operated in a wet-dry cycle operation, in which, the pond is filled and the influent is supplied at a rate equal to the infiltration rate. When the recharge rate reaches an unacceptable value, the operation is stopped. Based on this operation the pressure head of water is changing with time during the cycle. So the uncertainties contributing from every parameter are also changing during the wetting and drying period.

This paper presents a method to assess the uncertainty of the unsaturated hydraulic conductivity due to the parameters based on the *van Genuchten-Mualem* water retention equation. The

methodological framework is based on the first-order analysis of the equation of the unsaturated hydraulic conductivity. The uncertainty of unsaturated hydraulic conductivity due to the uncertainties of the parameters have been studied considering different levels of parameter uncertainties.

2. FIRST-ORDER ANALYSIS OF UNCERTAINTY

The theory and mathematics of first-order uncertainty analysis can be found in *Benjamin* and *Cornell* (1970) and *Mays and Tung* (1992). Essentially first-order analysis provides a methodology for obtaining an estimate for the moments of a random variable which is a function of one or several random variables. It estimates uncertainty in a mathematical model involving parameters which are not known with certainty. By using first-order analysis, the combined effect of uncertainty in a model formulation, as well as the use of uncertain parameters, can be assessed.

First-order analysis considers a random variable, Y, which is a function of N random variables, mathematically expressed as

$$Y = g(\mathbf{X}) \tag{1}$$

where $\mathbf{X}=(X_1, X_2,...., X_N)$ is a vector of uncertain input parameters.

Through the use of a Taylor's expansion, about the means of $\mathbf{X}$ and assuming independence of the uncertain parameters, the random variable Y can be approximated as in a second-order approximation:

$$Y=g(\overline{X})+\sum_{i=1}^{N}\left[\frac{\partial g(\mathbf{X})}{\partial X_i}\right]_{\mathbf{X}=\overline{x}}(X_i-\overline{X}_i)$$

$$+\frac{1}{2}\sum_{i=1}^{N}\sum_{j=1}^{N}\left[\frac{\partial^2 g(\mathbf{X})}{\partial X_i \partial X_j}\right]_{\mathbf{X}=\overline{X}}(X_i-\overline{X}_i)(X_i-\overline{X}_i) \qquad (2)$$

in which $\overline{\mathbf{X}}=\left(\overline{X}_1, \overline{X}_2,...\overline{X}_N\right)$, a vector containing the means of the N parameter random variables; and $\left[\dfrac{\partial g}{\partial X_i}\right]_{\overline{\mathbf{X}}}$ is called the sensitivity coefficient representing the rate of change of function value $g(\mathbf{X})$ at $\mathbf{X}=\overline{\mathbf{X}}$.

The mean (expected value) of Y is approximated as

$$\mu_Y = E[Y] \approx g(\overline{\mathbf{X}}) \qquad (3)$$

It follows that the first-order approximation of the variable of Y, for the X_i's are uncorrelated is

$$\sigma_Y^2 = \sum_{i=1}^{N}\left[\frac{\partial g}{\partial X_i}\right]_{\overline{\mathbf{X}}}^2 \sigma_i^2 \qquad (4)$$

where σ_i^2 is the variance corresponding to random variable X_i.

3. WATER RETENTION EQUATION

For the soil water retention equation, several mathematical models have been developed for soil moisture, θ. The most well-known expression of these equations is the van Genuchten (1980) soil characteristics equation,

$$\theta = \theta_r + \frac{\theta_s - \theta_r}{\left(1+|\alpha h|^n\right)^m} \qquad (5)$$

where θ_r and θ_s are residual and field-saturated volumetric water contents, respectively; h is the

suction head; and α, n, and m are empirical constants. Using the simplifying assumption that $m=1-1/n$, equation (5) can be combined with the predictive conductivity model of Mualem (1976) to yield the van Genuchten-Mualem water retention equation for the unsaturated hydraulic conductivity K_{uns} of the form,

$$K_{uns} = K_s S_e^{\frac{1}{2}}\left[1-\left(1-S_e^{\frac{1}{m}}\right)^m\right]^2 \qquad (6)$$

where the effective saturation is:

$$S_e = \frac{\theta-\theta_r}{\theta_s-\theta_r} = \frac{1}{\left(1+|\alpha h|^n\right)^m}, \qquad (7)$$

and K_s is the saturated hydraulic conductivity.

From the above discussion, the unsaturated hydraulic conductivity is mainly related to the three parameters K_s, α, and n. In this study, the first-order uncertainty analysis was utilized to calculate the uncertainty of unsaturated hydraulic conductivity which is a function of the K_s, α, n, These parameters are treated as random variables.

Based on the first-order analysis described previously, the expected value of the K_{uns} can be expressed as:

$$E[K_{uns}] \approx K_{uns}(\overline{K}_s, \overline{\alpha}, \overline{n}) \qquad (8)$$

Similarly the variance of the K_{uns} can be approximated by the following equations:

$$\sigma_{K_{uns}}^2 = \left(\frac{\partial K_{uns}}{\partial K_s}\right)^2 \sigma_{K_s}^2 + \left(\frac{\partial K_{uns}}{\partial n}\right)^2 \sigma_n^2$$

$$+\left(\frac{\partial K_{uns}}{\partial \alpha}\right)^2 \sigma_\alpha^2 \qquad (9)$$

To analyse the relative contribution of each parameter, equation (9) can be expressed in terms of the coefficient of variation, Ω, by dividing both sides by the mean of unsaturated hydraulic conductivity $\mu_{K_{uns}}$

$$\Omega_{K_{uns}}^2 = \left(\frac{\partial K_{uns}}{\partial K_s}\right)^2\left(\frac{\overline{K}_s}{\mu_{K_{uns}}}\right)^2 \Omega_{K_s}^2 +$$

$$\left(\frac{\partial K_{uns}}{\partial a}\right)^2\left(\frac{\overline{a}}{\mu_{K_{uns}}}\right)^2 \Omega_a^2 + \left(\frac{\partial K_{uns}}{\partial n}\right)^2\left(\frac{\overline{n}}{\mu_{K_{uns}}}\right)^2 \Omega_n^2 \qquad (10)$$

420

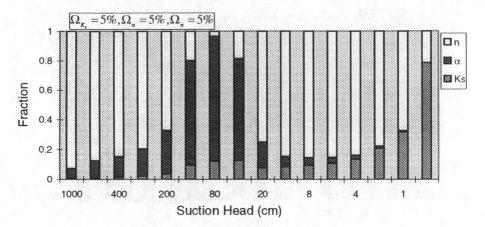

Figure 1 Fraction of Uncertainty Due to n, α, and K_s
for Various Suction Heads

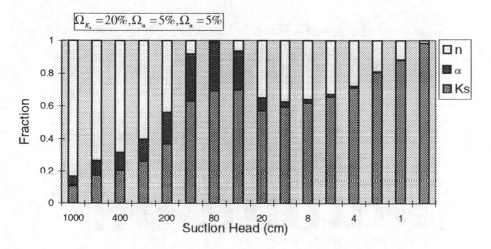

Figure 2 Fraction of Uncertainty Due to n, α, and K_s
for Various Suction Head

where $\Omega_x = \dfrac{\sigma_x}{\overline{x}}$ (11)

The above expression is based on the assumption that K_s, α, and n are independent. Normally the mean and variance of the parameters are known and only elements which need to be calculated are the first derivatives of the unsaturated hydraulic conductivity respect to the three parameters. These derivatives are listed in Appendix A.

4. RESULTS AND DISCUSSIONS

A silt soil is used as an example for the application, the average values for each parameter are $\overline{K}_s$ =16cm/day, $\overline{\alpha}$ =0.033, $\overline{n}$ =1.631.

Figures 1-2 illustrates the effect of uncertainty of unsaturated hydraulic conductivity as a function of the soil suction head for two coefficients of variation 0.05 and 0.1.The results show that K_s, α and n are all important in contribution to the uncertainty of K_{uns}.

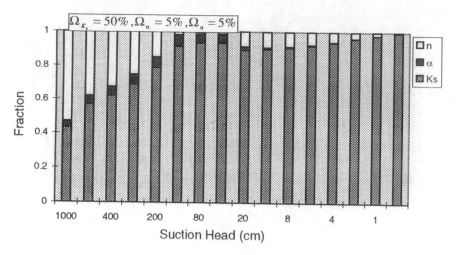

Figure 3 Fraction of Uncertainty Due to n, α, and K_s
for Various Suction Heads

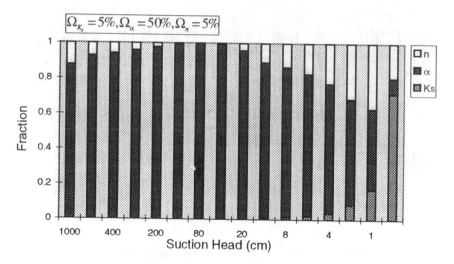

Figure 4 Fraction of Uncertainty Due to n, α, and K_s
for Various Suction Heads

During the wetting time the main contribution is K_s. During drying time the α and n are the main contributions. The uncertainties are increasing with increase of the parameter uncertainties.

Figures 1-4 illustrate the percentage contribution of each parameter in the total uncertainty for its different uncertainty levels. Each parameter is significant to the total uncertainty and the increasing of one parameter increases its percentage in the total uncertainties.

Based on the above analysis, the three parameters K_s, α, n all have effects on the uncertainties of the hydraulic conductivity and have the different relative contributions during wetting and drying period.

REFERENCE:

Benjamin, J. R., and Cornell, C.A. (1970). *Probability, statistics, and decisions for civil engineers*. McGraw Hill, Inc., New York, N.Y..

Mays, L. W. and Y.K. Tung, *Hydrosystems Engineering & Management*, McGraw-Hill, Inc., 1992

Mualem, Y. (1976). A new model for predicting the hydraulic conductivity of unsaturated porous media. *Water Resources Research*, Aug. 12(3), 512-522

Tung, Y. K. and Wade E. Hathhorn, Assessment of probability distribution of dissolved oxygen deficit", *Journal of Environmental Engineering*, Vol. 114, December 1988, 1421-1435

van Genuchten, M.T. (1980). A closed form equation for predicting the hydraulic conductivity of unsaturated soils. *Am. J. of Soil Science Society*, 44, 982-998

APPENDIX A:

The derivatives of the unsaturated hydraulic conductivity respect to the K_s, α, n:

$$K_{uns} = K_s \left(\frac{1}{\left(1+|\alpha h|^n\right)^m} \right)^{\frac{1}{2}} \left[1 - \left(1 - \frac{1}{1+|\alpha h|^n} \right)^m \right]^2 \quad (A1)$$

$$\frac{\partial K_{uns}}{\partial K_s} = \left(\frac{1}{\left(1+|\alpha h|^n\right)^m} \right)^{\frac{1}{2}} \left[1 - \left(1 - \frac{1}{1+|\alpha h|^n} \right)^m \right]^2 \quad (A2)$$

$$\frac{\partial K_{uns}}{\partial a} = K_{uns} \{ -\frac{mn|h|}{2} \left(\frac{1}{\left(1+|\alpha h|^n\right)} \right) * |\alpha h| +$$

$$2mn|h| * \left[1 - \left(1 - \frac{1}{1+|\alpha h|^n} \right)^m \right]^{-1} * \quad (A3)$$

$$\left(1 - \frac{1}{1+|\alpha h|^n} \right)^{m-1} \frac{|ah|^{n-1}}{\left(1+|\alpha h|^n\right)^2} \}$$

$$\frac{\partial K_{uns}}{\partial n} = K_{uns} \{ \frac{\left(1+|\alpha h|^n\right)^m}{2} [\left(-1 + \frac{1}{n} \right)$$

$$\left(1+|\alpha h|^n\right)^{-2+\frac{1}{n}} |\alpha h|^n \ln(|\alpha h|) - \frac{\ln\left(1+|\alpha h|^n\right)}{n^2 \left(1+|\alpha h|^n\right)^m}] -$$

$$2 \left[1 - \left(1 - \frac{1}{1+|\alpha h|^n} \right)^m \right]^{-1} [\left(1 - \frac{1}{n} \right) * \quad (A4)$$

$$\left(1 - \frac{1}{1+|\alpha h|^n} \right)^{-\frac{1}{n}} \frac{|\alpha h|^n \ln(|\alpha h|)}{\left(1+|\alpha h|^n\right)^2} - \frac{1}{n^2}$$

$$\left(1 - \frac{1}{1+|\alpha h|^n} \right)^{1-\frac{1}{n}} * \ln\left(1 - \frac{1}{1+|\alpha h|^n} \right)] \}$$

Stochastic Hydraulics'96, Tickle, Goulter, Xu, Wasimi & Bouchart (eds) © 1996 Balkema, Rotterdam. ISBN 90 5410 817 7

Fractal structure and dynamic scaling of infiltration front

Ray-Shyan Wu & Kuo-Liang Yeh
National Central University, Chungli, Taiwan, China

Jet-Chau Wen
Chung-Yuan Christian University, Chungli, Taiwan, China

Tuso-Rn Wu
Agricultural Engineering Research Center, Chungli, Taiwan, China

ABSTRACT: The study of the transport behavior of pollutants in an unsaturated zone is an important facet in the estimation of the time consumed and the magnitude of pollutants reaching ground water table. In recent years many papers about fingering flow in unsaturated porous media have been published, most of them focusing on the time, the location, the condition and reasons for the occurrence of fingering flow, and factors influencing the shape of fingering flow. This study investigates the characteristics of the interface between the soil granules and the fluid, which would have an important influence on organic reactions during clean up ground water pollution. Experiments are conducted to investigate the dynamic shape of the wetting front. A model is proposed to calculate the interfaced area between the fingering flow and the soil granules by applying fractal geometry analysis. Guidelines for the calculation of the fractal dimension from the Richardson plot, and methods for deriving high precision measurements from low precision measurements are suggested.

1 INTRODUCTION

The unstable interface phenomenon between two fluids in a porous medium has received continued attention since the 1950s (e.g., Taylor, 1950; Staffman and Taylor, 1958; Hill and Parlange, 1972; Homsy, 1987; Baker and Hillel, 1990). In engineering practice, an understanding of how this interface forms is crucial to the successful prediction and monitoring of groundwater pollution and its consequent remediation. A special case of this problem is the shape of the interface that forms when water infiltrates vertically downward into an unsaturated porous medium. Here, the interface is the well known wetting front which forms due to the nonlinear dependence of the hydraulic properties to the moisture content. When the wetting front becomes unstable, fingers form and move down bypassing much of the unsaturated porous medium. Wetting front instability is driven by gravity and stabilized by surface tension which arises through capillaries within the porous media (Glass et al., 1989a).

Analysis of the finger structure of the flow through an unsaturated zone is difficult due to the highly nonlinear behavior of the governing equation, Richard's equation. However, quite a few laboratory experiments have been conducted in Hele-Shaw cells and detailed descriptions of the fingering phenomena have been provided in several papers (Glass et al., 1989a; Baker and Hillel, 1990; Selker et al., 1991; Chang et al., 1994). Glass et al (1989a,b) reported on the relationship between finger width, propagation velocity, moisture content, and flow rate through individual fingers. The experimental observation of the characteristics of the fingers is then related to the properties of the medium and the system's flow rate.

Mandelbrot (1983) first used a fractal concept to provide a mathematical framework for describing the irregular shapes common in natural phenomena. He concluded that if the space, dimension, direction, or shape is self-similar, then it can be described with fractals. The coastline was given as a classic example. Recently this new point of view, differing from ordinary Euclidean geometry, has been widely referred to as an excellent way to describe Nature. Fingering is a manifestation of complex geometry and can also be described in fractals. For instance, Chang et al. (1994) estimated the effective surface tension of fingering from the fractal dimension and the mean pore size.

The research described in this paper, composed of both experiment and theory analysis, explores an engineering approach describing fingering, applying this fractal concept. First, we present an experimental setup that is able to simulate the movement of flows in an unsaturated zone and utilizes an image processing technique in observing flow paths in the soil layers. Through these experiments, we applied fractal geometry analysis to study the finger structure. The fractal structure was analyzed both by divider algorithm and Richardson plot, which identifies the characteristics of fractals within some upper and lower limits, and an efficient dimension is obtained. On the basis of this fractal dimension, we suggest a method to estimate the length of the reaction, defined as the interface between the flow path and the soil matrix.

2 FRACTAL CONCEPT

The shape of the interface between fingering flow and soil granules presents a scale-invariance property that is essential in natural fractal sets, such as coastlines, artery and river networks. One cannot express the total length of the interface properly because, as the scale gets smaller and smaller, the interface shows more and more kinks. It is expected that these meander shapes are intermediate between regular forms and purely random ones. This implies the existence of fractal behavior and attracts a fractal analysis of the shape of the interface.

The most distinguishable feature of a fractal set is its dimension. In mathematics, the fractal dimension is the idea that the observation of the behavior of measurement at scale β, as β approaches zero. By doing this, the irregularities of the scale less than β is ignored. In this paper, the length of the interface, L, can be expressed as

$$L \propto \beta^{1-D} \tag{1}$$

where β is a scale length and D is the fractal dimension.

Let it be noted that it is possible to define the dimension of a set in many ways. Even for the same set, different definitions may result in different values for the dimensions which may also have very different properties. Inconsistent usage of the definition may sometimes cause considerable confusion. Hence, a clarification of the dimension being used is necessary. In most applications, only measurements made in exactly the same fashion can be compared directly. To be sure a set of fractal dimensions are identical, the dimensions measured

each time should be consistent to two and a half digits. The so called dimension is strictly defined in pure mathematics. However, no such proof is conceivable in any natural science. The use of a fractal dimension simply provides a manageable and convenient way to describe Nature.

Clearly, in engineering applications, the fractal features disappear if they are viewed at a sufficiently small or sufficiently large scale, known as the cut-offs. Any real-life object exhibits fractal behavior, or self-similarity only over a finite range of scales. In order words, the scale length, β, should be chosen within a certain interval. The practical measurement of dimension should be altered from its original mathematical consideration. Since the fractal dimension describes how many new pieces of a set are resolved as the resolution scale is decreased, it can therefore be evaluated by comparing properties between any two scales. Furthermore, due to the cut-offs, no natural objects fulfill the definition of fractals in mathematics. Nevertheless, over certain scale ranges bounded by the cut-offs, they do appear much like fractals. Therefore, a revised name "natural fractals" would be proper.

Fractal structure can be analyzed both by a divider algorithm and the Richardson plot. To measure the fractal dimension of the interface, the divider walking algorithm is used here as a tool. To measure the total length of a curve, a divider of a series of fixed spans can be utilized to count the number of steps needed to finish the whole length. Plot the diagram of the dividers open length versus the total length measured in each step, in logarithm. This is called the Richardson plot, following L.F. Richardson's pioneering work, which identifies the characteristic of fractals within some upper and lower limits. An S-shape curve is found in the plot. The divider dimension, D_d, can be obtained as $1-d_s$, where d_s is the slope (negative) of the intermediate section. The applications of the divider walking algorithm are limited. It can only be applied on curves which are one-dimensional. However, the advantage of the divider walking algorithm is that it clearly reveals the fractal nature of curves within the full range of data scales because in practice, each unit length can be taken as a very small variation from the last one. The log-log plot is hence very dense and is beneficial in observing delicate changes of the plot. These delicate changes might indicate important physical meanings. Many researchers noted that the divider walking algorithm provides meaningful results only when applied to self-similar curves (Kaye, 1984; Gilbert, 1989).

The key advantage of the divider walking algorithm is its capability to obtain much information through the full data scale range of a wiggle curve. The algorithm used in this study progresses through a range of divider step sizes, walking each step size along the digitized curve to obtain the number of steps and the remaining partial step at the end. This number is used to calculate an estimate of the curve length.

3 EXPERIMENTAL STUDY

In the experimental study, we observed water infiltrating into a dry soil zone with the aid of an image processing technique. The two-dimensional infiltration chamber, similar to but about twice as large as those used in Glass et al. (1989) and Chang et al. (1994), was 150 cm high, 100 cm wide, and 2 cm thick, constructed out of clear acrylic plastic plates. The chamber was completely sealed to air and water except at its bottom, where two holes were drilled to allow air and water to escape and at the top, which was open. The experimental porous media used were (1) standard Ottawa sand having a mean particle diameter of 0.085 cm passing through a No.10 and N0.30 sieve; (2) natural clay from the Shieh Men Reservoir in northern Taiwan, with a mean pore size of around 0.001 cm, much less than the Ottawa sand.

The sand was first air dried. During packing in the chamber, sand was poured evenly across the top of the chamber in 5-cm layers. Each layer of the sand was packed by tamping the sand with a steel stick. The surface of the each layer was also scarified before another layer of sand was added, to avoid the separation of in-between layers. The flow of water from a reservoir with a constant head was initiated and a thin piece of cotton cloth was placed on top of the sand zone to keep the water depth even across the chamber. At selected intervals, i.e., every 10 seconds, an image was taken using a digital image processing unit consisting of a PC, a Coreco OC-300 image card with 2MB frame buffer, a CCD, and a RGB color monitor. These images were stored and analyzed later. After each experiment the chamber was cleaned and dried in preparation for another experiment.

Eight experiments, all applied with a constant flow rate, were conducted, designated here as; filled with Ottawa sand (Experiments 1-6), Ottawa sand with a top thin layer of Shieh Men clay to emulate top soil in the field (Experiment 7), Ottawa sand

with a slanting band of Shieh Men clay in the zone (Experiment 8).

The image processing technique was used to calculate distance, area, velocity, and moisture content. The interface, i.e., the wetting front, at different infiltration times was also digitized. A relationship between the gray level in the image and water content was established, shown in Figure 1. The saturated water content in the Ottawa sand, measured in the laboratory test was 18%.

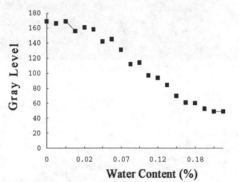

Figure 1: Relationship between gray level and water content

4 RESULTS AND DISCUSSIONS

4.1 *Infiltration flow field*

For each experiment, the results were presented in five types of figures. An overall picture of the development of the flow field, as in Figure 2, reports fingers in every 30 seconds. The increase in the wetted area can be easily calculated by counting the number of pixels that change their gray level, indicating from dry to wet at given time intervals, as in Figure 3. A more detailed analysis can be done with the use of the information in Figure 1 to estimate the dynamic redistribution of the moisture content in the flow field. For each finger, displacement is calculated based on the difference between two images taken at an interval of 10 seconds, as in Figure 4. The velocity of each finger, shown in Figure 5, is then calculated based on the displacements.

Fingering flow, in a homogeneous media, at a constant flow rate generally agreed with previous studies (Glass et al., 1989b; Chang et al., 1994): that is (1) the higher the flow rate, the wider and faster

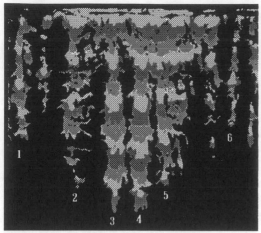

Figure 2: Development of the flow field

the finger; (2) slower movement in the lateral wetting fronts; (3) finger core areas acted as conduits for the majority of the flow. The distribution of the moisture content in the flow field indicates that fingers have a moisture content structure consisting of a saturated wet finger tip and a relatively wet core surrounding by drier media. In other words, Figure 6 shows a higher moisture content with a lower depth in a finger core. This may
only happen in cases where there is a high conductivity in the media and a low flow rate.

4.2 Dynamic behavior of fingering flow

In general, fingering flow develops and moves downward with a constant mean velocity, but

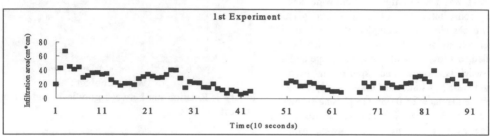

Figure 3: Increase of the wetted area

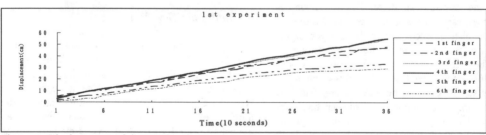

Figure 4: Displacement of the wetting front

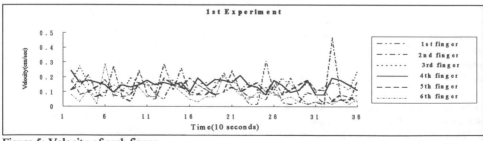

Figure 5: Velocity of each finger

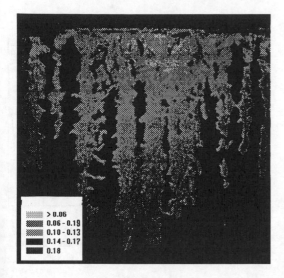

Figure 6: Distribution of the moisture content

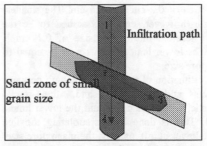

Figure 7: Flow direction when encountering a slanting band of clay, moved in the order 1-2-3-4

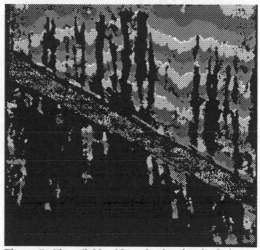

Figure 8: Flow field with a slanting band of clay in the chamber

varying within short time intervals. The separation of a finger was rarely observed. What happens is that a finger can develop into two and after a while merges back into one finger. Different finger widths were observed in the same experimental runs. It was found that the wider the finger, the higher the flow velocity. However, according to Richard's equation, the width should be constant, given some set of boundary conditions (Rajesh, 1991). This is due to two things: (1) homogeneity was not obtained during the packing of the sand or (2) a wider finger actually consisted of several small fingers.

In Experiment 7, the top clay layer became saturated before the infiltration took place. Once a breakthrough point in the sand zone occurred, a finger developed at the spot with an increasing width and depth. Since a constant flow rate was given, the increase in the number of fingers and the width of each finger reached a point when the flow in equaled the flow out of the top layer. A stable flow field was then formed.

In Experiment 8 when flow encountered a slanting band of clay, the direction of the flow gradually paralleled the slant and the slant became partially saturated. However, when flow appeared beneath the slant, the original vertical direction of the finger core was preserved (see Figures 7, 8). This may be due to the fact that the slanting layer was thin and the slope of the slant was small. The slant did not affect the flow path in the sand zone but only the travel time.

4.3 Fractal analysis of the finger structure

The interface between fingers and the dry sand in Experiment 3, as shown in Figure 9, was analyzed with a divider algorithm. It is apparent from Figure 10 that with dividers with a unit length greater than 3 cm, i.e., the log unit length being 0.5, the curve starts fluctuating and a kind of 'beating' phenomenon prevails. The fluctuations are caused by the use of larger dividers that would miss small bends and kinks. At unit lengths greater than 10 cm, a whole finger could be missed when measuring the length. A reasonable upper limit was set, 3 cm. In most of the experiments, the width of a finger was about 10 cm.

To identify the lower limit is a bit more complicated. Precision in laboratory measurement is limited by the quality of the instruments used, in this case the resolution of the CCD. A pixel in the image equals an area of 0.2 cm by 0.25 cm. The smallest scale length, however, can be assumed to be the mean pore size. A mean pore size of 0.1 cm is assumed. The lower limit is then set to be equal to the mean pore size.

For an estimation of the interface length between the fingers and the dry sand, the fractal characteristic, within the upper and the lower limits, is applied. This is done by extrapolating the line toward the lower cutoff, namely, the mean pore size. As an example, the interface lengths at different times are calculated based on extrapolating the curves in Figure 10, also known as Richardson's Plot, and identifying the intersects in the vertical axis denoting a unit length of 0.1 cm (see Table 1). An increase in the interface length is reported in Table 1. This piece of information is essential in the prediction, monitoring an designing of the clean-up of groundwater pollution.

4.4 *Derivation of High Precision Measurements*

The cost in measuring the interface between flow path and soil matrix increases as a higher precision is required. We propose a procedure for the estimation of the interface area by applying fractal geometry analysis based on the self-invariance property.

Traditionally, the quality of the measurement depends on the resolution of the instrument. A fairly high precision microscope is required to identify each soil granule. However, given fractal behavior exists over a range of scales, one can extrapolate the information measured in a low precision to a higher precision. For example, we measured a finger flow in four resolutions that representing different levels of information quality (see Figure 11). The Richardson Plot for all four sets of measures indicates a very similar fractal characteristic. Four regressed lines are very close to each other. Therefore, it is possible to estimate a high resolution measurement based on the low resolution information, say measured in 1/64 accuracy. In this particular example, the length measured in the best accuracy (about the order of 3 mm) is 150 cm. Although a relatively low precision measurement, with an accuracy about the order of 2 cm, is used, one can extrapolate the curve to an estimation of the length as 151 cm. This estimation is within 0.2% difference of the direct measurement. Without the use of the fractal concept, one would obtain a measurement of 130 cm.

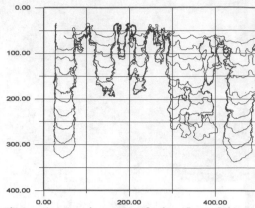

Figure 9: Development of the flow field in Experiment 3 (unit: X-axis : 0.2 cm Y-axis 0.25 cm)

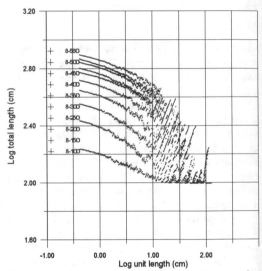

Figure 10: Dependence of the divider on the measured length of the interface between flow path and soil matrix in Experiment 3 over time (from t=100 second to t=560 second)

Table 1: Fractal dimension and the measured interface length in Experiment 3

Time (sec.)	100	150	200	250	300
Fractal Dimension	1.109	1.131	1.126	1.113	1.113
Interface Length (cm)	207	337	446	536	630
Time (sec.)	350	400	450	500	560
Fractal Dimension	1.105	1.115	1.108	1.117	1.105
Interface Length (cm)	703	766	833	888	934

430

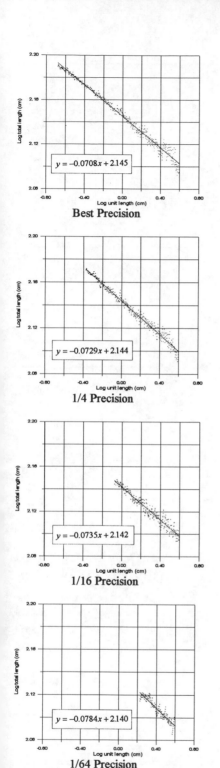

Best Precision

$y = -0.0708x + 2.145$

1/4 Precision

$y = -0.0729x + 2.144$

1/16 Precision

$y = -0.0735x + 2.142$

1/64 Precision

$y = -0.0784x + 2.140$

Figure 11: Richardson plots at four different resolutions representing four levels of information quality

CONCLUSIONS

This research encompassed both experimental and theoretical analysis. We presented an experimental setup that was able to simulate the movement of flows in unsaturated zone utilizing image processing technique to observe flow paths in the soil layers. Through these experiments, we applied fractal geometry analysis to study the finger structure. In theory analysis, we used fractals to analyze the flowing routes in the unsaturated medium, especially at the interface between soil granules and the boundary of fingering flow. Fractal structure was analyzed by a divider algorithm and Richardson's plot. An S-shape curve was found in the Richardson plot and the fractal dimension was obtained. Based on this fractal dimension, we estimated the length of the interface between the flow path and the soil matrix. Furthermore, it is possible to apply this fractal dimension of the fingering flow for the calculation of an efficient surface tension.

Some findings from this study are presented:

(1) The higher the flow rate, the wider and faster the finger is. It is recalled that the wetting front of the finger is dispersed with the large mean pore size (Srivastava and Yeh 1991, Chang et al. 1994). In other words, surface tension plays an important role during fingering development. Therefore, it is conceivable that there is a relationship between fractal characteristics and the surface tension. Exploring of this relationship may be a good future research topic.

(2) For a constant inflow rate at the soil top, a stable flow field can be developed in the soil when an increase in the number of fingers and the width of each finger reaches a point where the flow rate is equal to the inflow rate.

(3) It is found that a slanting clay band laying in an unsaturated sand zone did not affect the flow path in the sand.

(4) It is worth noting that the reasonable upper and lower scale lengths, during the theoretical analysis of the cut-off length, are 3.0 cm and 0.1 cm, respectively.

(5) It is noted in the Richardson plot, that, under an existing fractal behavior over a range of scales, high precision measurements can be obtained from the extrapolation of low precision measurements. In other words, it is possible to estimate a high resolution measurement based on a low resolution measurement. This implies that costs can be reduced by using low precision instruments instead of high precision instruments.

Acknowledgments: The research in this paper was supported in part, by a grant from the National Science Council of the Republic of China on Taiwan through contract NSC 83-0410-E008-002. The authors are grateful for Cheng-Hsung Liu's help in conducting the experiments.

REFERENCES

Baker, R. S., D. Hillel, Laboratory test of a theory of fingering during infiltration into layered soils, Soil Science Society America, Journal, 54, 20-30, 1990.

Chang, W-L, J. W. Biggar, D. R. Nielsen, Fractal description of front instability in layered soils, Water Resources Research. 30(1),125-132, 1994.

Gilbert, L. E., *Are topographic data sets fractal?, in Fractals in Geophysics*, edited by C. H. Scholz and B. B. Mandelbrot, Basel: Birkhauser Verlag, 241-254, 1989.

Hill, D. E., J.-Y. Parlange, Wetting front instability in layered soils, Soil Science Society America, Proceedings, 36, 697-702, 1972.

Homsy, G. M., Viscous fingering in porous media, Annual Review Fluid Mechanics, 19, 271-311, 1987.

Kaye, B. H., Fractal description of fine particle systems, in Particle Characterization in Technology, Vol. 1: Applications and Microanalysis, 81-100, 1984.

Mandelbort, B. B., *The Fractal Geometry of Nature*, W. H. Freeman and Company. New York, 1983.

Srivastava, R., T.-C. Yeh, Analytical solutions for one-dimensional, transient infiltration toward the water table in homogeneous and layered soils, Water Resources Research, 27(5),753-762, 1991.

Staffman, P. G., G. Taylor, The penetration of a fluid into a porous medium or Hele-Shaw Cell containing a more viscous liquid, Proceeding of Royal Society London. A245:312-331, 1958.

Taylor, G. I. 1950. The instability of liquid surfaces when accelerated in a direction perpendicular to their planes, Proceeding of Royal Society London, A201:192-196, 1950.

Selker, J. S., Steenhuis, T. S., Parlange, J. -Y., Theory and laboratory experiments of finger structure in unstable flows, Proceedings of the International Hydrology and Water Resources Symposium, Perth, Australia, October 2-4, 817-1227, 1991.

Stochastic Hydraulics'96, Tickle, Goulter, Xu, Wasimi & Bouchart (eds) © 1996 Balkema, Rotterdam. ISBN 90 5410 817 7

Stochastic optimal control framework for operation of SAT system considering parameter uncertainty and spatial variability of the unsaturated hydraulic conductivity

Liang Xu & Larry W. Mays
Department of Civil and Environmental Engineering, Arizona State University, Tempe, Ariz., USA

ABSTRACT: This paper presents a new methodology for SAT(Soil Aquifer Treatment) operation under uncertainties. The optimal operation of SAT systems is formulated as a discrete-time optimal control problem and solved by interfacing the SALQR (Successive Approximation Linear Quadratic Regulator) optimizer with the MSTS (Multiphase Subsurface Transport Simulator) model. Both deterministic and stochastic formulations have been solved. A chance constrained formulation of the optimization is utilized to account for the uncertainties of water content as a result of the variation of parameter uncertainties. The Monte Carlo method is used to explicitly deal with the influence of the spatial variability of aquifer properties on the SAT operation management. This approach enables one to quantify the uncertainties and automatically account for them in the decision-making process through the SAT management model.

1. INTRODUCTION

The concept of SAT (Soil Aquifer Treatment) systems depends on the infiltration of treated wastewater into the soil and percolation through the vadose zone. Improvements in water quality can occur due to many different mechanisms such as filtration, biological degradation, physical adsorption, ion exchange and precipitation. The efficiencies of these mechanisms are a function of the type of soil, level of pretreatment, loading rates, cycle time, and temperature. Normally, an SAT system is operated by wet-dry cycles, in which, the pond is filled and the influent is maintained at a rate equal to the infiltration rate. When the recharge rate reaches an unacceptable value, the operation is stopped. The pond is allowed to drain and then dry for a period of time. These typical operation procedures are presently done on an ad-hoc basis relying on the limited judgment of the design engineer and/or operator. To date, only limited research has been done on the optimal operation of SAT systems (Tang et al., 1995), however, no one has previously considered the uncertainties in this SAT operation process.

There are many deterministic models for describing the behavior of an unsaturated flow in space and time. The assumption underlying deterministic modeling is that the temporal and spatial behavior of the unsaturated flow is described exactly if the initial conditions, boundary conditions and parameters are precisely known. The soil character and parameters are assumed to be precisely known. However, this case does not exist in the real world, because the exact value of the model parameters and soil variability are not known.

This paper presents a methodology for determining SAT operation under uncertainties which is based upon chance-constrained and Monte Carlo methods. The solution algorithm for this methodology interfaces the SALQR (Successive Approximation Linear Quadratic Regulator) (*Chang* 1992) optimization model with the MSTS (Multiphase Subsurface Transport Simulator) (*Nichols* 1994) simulation model. The first-order, second-moment method quantifies the parameter uncertainties in the predicted water content and the chance-constrained method was used to explicitly deal with the parameter uncertainty in the SAT stochastic optimal management model. The Monte Carlo method is used to explicitly deal with the influence of the spatial variability of aquifer properties on the SAT operation management.

2. SAT DETERMINISTIC OPERATION MODEL

The objective of optimal management of an SAT system is to obtain a maximum infiltration which satisfies various hydraulic and operation constraints,

Maximize Infiltration:

$$F = \sum_{t=1}^{N} F_t(\omega_t, X_t, Z_t, t) \qquad (1)$$

Subject to :

Water content :

$$\omega_{t+1} = T_1(\omega_t, X_t, Z_t) \qquad (2)$$

Infiltration volume:

$$F_t = T_2(\omega_t, X_t, Z_t) \qquad (3)$$

Drainage time:

$$Y_t = T_3(\omega_t, X_t) \qquad (4)$$

Maximum water content:

$$\omega_{t+1} \leq \omega_{max} \qquad (5)$$

Definition of cycle time:

$$X_t + Y_t + Z_t \leq CT_{max} \qquad (6)$$

where

N = number of cycle time;

X_t = application time during which there is inflow to the pond;

Y_t = drainage time during which the water drains after inflow is stopped during cycle time t;

Z_t = drying time of basin i during cycle time t;

CT_{max} = maximum cycle time for the pond; and

ω_{t+1} = average water content at the end of cycle t.

Water content is the state variable and application time X_t and drying time Z_t are the control variables. The objective function of the model is to maximize the infiltration during the operation time for the SAT system. The equality constraints (2)-(4) are the transition equations, which are solved by the simulator MSTS. The inequality constraints (5)-(6) are bound constraints and are considered through penalty functions, discussed later.

In the discrete time optimal control solution process, the optimizer (SALQR) computes the new values of the control variables and passes that information to the simulator (MSTS) to update the corresponding state variables for each iteration, as shown in Figure 1. Basically, the state and control variables are related through the MSTS simulator. In essence, the simulator equations are used to implicitly express the states in terms of the controls yielding a much smaller nonlinear optimization problem.

SALQR differs from DDP (Differential Dynamic Programming) only in that the nonlinear simulation equations are linearized in the optimization step in the SALQR. Therefore, if the simulation equations are linear, SALQR and DDP are identical.

The purpose of the unsaturated flow model is to simulate the water flow in the SAT system in order to solve equations (2)-(4) compute the water content, ponding depth and drainage time for the different wetting and drying time when the SAT system is in operation. The Multiphase Subsurface Transport Simulator (MSTS) model was developed by the Pacific Northwest Laboratory (Nichols and White, 1994) to numerically model the flow of liquid water, water vapor, air and heat through fractured porous media in the one, two or three dimensional domain under different initial and boundary conditions

3. CHANCE-CONSTRAINED SAT MODEL

The uncertainties considered result from the parameters in the *van Genuchten* (1980) soil characteristics equation which can be combined with the predictive conductivity model of *Mualem* (1976) to yield an expression for unsaturated hydraulic conductivity $K(S_e)$ of the form

$$K(S_e) = K_s S_e^{\frac{1}{2}} \left[1 - \left(1 - S_e^{\frac{1}{m}} \right)^m \right]^2 \qquad (7)$$

where the effective saturation, S_e, is

$$S_e = \frac{\theta - \theta_r}{\theta_s - \theta_r} = \frac{1}{\left(1 + |\alpha h|^n \right)^m} \qquad (8)$$

434

K_s is the saturated hydraulic conductivity; θ_r and θ_s are the residual and field-saturated volumetric water contents, respectively; h is the pressure head; and α, n, and m are empirical constants.

The unsaturated hydraulic conductivity is related to the three parameters, K_s, α, and n; and the water content of the soil is related to the unsaturated hydraulic conductivity through the MSTS model

The problem was formulated using a nonlinear stochastic management methodology which accounts for parameter uncertainties. The approach is to incorporate the SAT simulation model as part of a nonlinear chance-constrained stochastic optimization problem. The key transformation is from deterministic optimization assuming no uncertainty in the chance-constrained stochastic optimization. The constraint on water content, which is a function given by the unsaturated flow model, is replaced by a constraint accounting for parameter uncertainty given by the simulation model and the first-order analysis.

Formulation as a chance-constrained model transforms the constraint into a probabilistic constraint. Water content ω_i is defined as a function of parameter uncertainty, so that the water content constraint is satisfied under a specified reliability level p:

$$Prob\{\omega_t \le \omega_{max}\} \ge p \qquad (9)$$

Generally, p can be a vector, and therefore required reliability levels can vary from location to location and from time to time. Here the required reliability levels are treated as a single value.

The water content, as a function of the uncertain parameters, is assumed to be normally distributed, so that constraint (9) becomes

$$Prob\left\{z \le \frac{\omega_{max} - E[\omega_t]}{sd[\omega_t]}\right\} \ge p \qquad (10)$$

where z is a standard normal random variable with mean zero and standard deviation one and $sd[\ ]$ indicates the standard deviation. This is equivalent to

$$\frac{\omega_{max} - E[\omega_t]}{sd[\omega_t]} \ge F^{-1}(p) \qquad (11)$$

where $F^{-1}(p)$ is the value of the standard normal cumulative distribution corresponding to reliability

level p. Rearranging (11) gives the deterministic equivalent of the chance constraint

$$E[\omega_t] + F^{-1}(p)sd[\omega_t] \le \omega_{max} \qquad (12)$$

where $E[\omega_t]$ is the expected value component and $F^{-1}(p)sd[\omega_t]$ is the stochastic component. The mean and variance of the water content are evaluated through the first-order analysis of the MSTS simulator.

For the optimal SAT operation management problem, the final form of the chanced-constrained management model is

$$Max: F = \sum_{t=1}^{N} F_t(\omega_t, X_t, Z_t, t) - R_1$$

$$\left[min\left(0, \omega_{max} - \left(E[\omega_t] + F^{-1}(p)sd[\omega_{,t}]\right)\right)\right]^2 \qquad (13)$$

$$-R_2\left[min\left(0, CT_{max} - (X_t + Y_t + Z_t)\right)\right]^2$$

subject to (2)-(4), (6), and (12).
Using the first-order analysis of uncertainty (*Mays and Tung*, 1992) the mean and variance of the water content are evaluated respectively as

$$\mu_\omega = E[\omega] \approx g(\overline{x}) \qquad (14)$$

and

$$\sigma_\omega^2 = \left(\frac{\partial\omega}{\partial k_s}\right)^2 \sigma_{k_s}^2 + \left(\frac{\partial\omega}{\partial n}\right)^2 \sigma_n^2 + \left(\frac{\partial\omega}{\partial a}\right)^2 \sigma_a^2 \qquad (15)$$

The first-order approximation of the variance of the simulated water content is a function of the variance of the estimated model parameters and the sensitivity of the response to changes in these parameters.

The overall solution procedure is further illustrated through the flowchart in Figure 1. There are two loops in this procedure, with the outer loop determining the penalty weights. The inner loop solves the augmented SAT problem by interfacing SALQR with the MSTS simulator, in which the penalty weights are fixed at the values determined by the outer loop. At this stage when the SALQR optimizer has a new control value, the first-order second-moment analysis is called and the uncertainties of water content due to the parameter uncertainties are calculated. Once an inner loop is finished, the convergence criterion is checked by looking at the bound violations of water content and cycle time. If it is small enough, the procedure terminates, otherwise the procedure returns to the outer loop and updates the dual variables and penalty weights and then goes to the inner loop and solves the new augmented problem again with the updated penalty from the outer loop. This process continues until an optimal solution of the overall problem is found.

4. MONTE CARLO SAT OPERATION MODEL

As a consequence of the geologic processes through which groundwater systems evolve, the hydrogeologic properties of an aquifer vary through space. In fact, the hydraulic conductivity can vary by an order of magnitude or more over short distances. Moreover, there is never sufficient data to completely describe this heterogeneity. To overcome the difficulties associated with characterizing aquifer heterogeneities, a stochastic approach to the unsaturated flow modeling is used. In this approach, the heterogeneous hydraulic conductivity is represented as a random field and described by a probability density function. The true hydraulic conductivity distribution is now considered to be a single realization of the random conductivity field.

A complete characterization of the variability and correlation of this random field would require knowledge of the joint probability density function for conductivities at all points throughout the aquifer.

Generally, the hydraulic conductivity is assumed to follow a lognormal distribution. Support for this assumption can be found in the work by *Freeze* (1975) and *Hoeksema and Kitanidis* (1985). Define the natural logarithm of hydraulic conductivity at point X_i as Y_i, under the assumptions of stationarity, lognormality, and exponential covariance, the mean and covariance function for the random log-hydraulic conductivity field are

$$E[Y_i] = \mu_Y \qquad (16)$$

$$Cov[Y_i, Y_j] = \sigma_Y^2 \, exp\{-|\Delta Xij|/\lambda_Y\} \qquad (17)$$

where $E[\ ]$ denotes expected value, $Cov[\ ]$ denotes covariance, μ_Y is the log-hydraulic conductivity mean, σ_Y^2 is the log-hydraulic conductivity variance,

λ_Y is the log-hydraulic conductivity correlation scale, and $|\Delta X_{ij}|$ is the distance separating Y_i and Y_j.

A single realization of the scaling factor $Y=ln\delta$ distribution over our simulated heterogeneous field was generated by using the turning bands method (*Tompson* et al. 1989) with an anisotropic exponential decay correlation function. A FORTRAN code TURN2D modified by *Tompson* (personal communication, 1995) from his original three-dimensional generator was used for random field generation.

For every single realization of the hydraulic conductivity field, the mathematical model for the Monte Carlo method takes the form of equations (1)-(6).

The procedure of the Monte Carlo SAT operation model is to solve a series of individual deterministic optimization problems for each realization of hydraulic conductivity. To account for the uncertainty associated with the heterogeneous hydraulic conductivity, a representative sample of the hydraulic conductivity field is generated, and for each realization, the nonlinear discrete-time optimization problem is solved. Therefore if there are n hydraulic conductivity realizations, the Monte Carlo management model provides n optimal loading schedules, each corresponding to a different realization of hydraulic conductivity

The results from this Monte Carlo model can be used to characterize the probability distribution of the optimal loading schedule. This data is used to analyze the effects of soil heterogeneity on the optimal loading schedule.

5. MODEL APPLICATIONS

An example problem has been used to illustrate application of the SAT deterministic and stochastic management models. The hydraulic and numerical parameters of the problem are the same as listed in Table 1. The reliability of the water content constraint p is 0.95. As an initial demonstration of the SAT operation management model, the deterministic model was first tested to investigate the efficiency of the model. The objective of the SAT deterministic model is to determine the optimal loading schedule that maximizes infiltration subject to the hydraulic constraints.

Tables 2 lists results assuming no clogging and Table 3 lists results constraining the clogging effects on the optimal operation schedule. If the soil surface clogging effects have been considered in the

436

simulation, the wetting time becomes shorter and drainage time becomes longer for the same wetting time, but there is little effect on the drying time. In reality, the clogging effects should be included in the model. So in the next calculation the model with clogging is considered.

Table 1
Parameters used in the example problem

Parameter	Quantity	Unit
Initial loading rate	30	cm/day
Initial water content	0.1	
Maximum water	0.2	
Maximum cycle time	25	days
Maximum ponding	30	cm
Saturated conductivity	25	cm/day
Soil column depth	5	m
Coefficient on clogging	0.0005	
Concentration of solids	20	mg/L
Concentration of algae	20	mg/L
Number of cycles	10	

If the model parameters are known precisely, then the operation schedule can be determined from the deterministic model with results shown in Tables 2 and 3. But this never happens in the real world. In the following section, the stochastic SAT management model was used to determine the optimal SAT systems loading schedule with the presence of parameter uncertainty under the reliability of 95%, which means that the water content constraint is satisfied at 95% probability.

Table 4 lists the operation schedule if variation of the parameters is 5% and Table 5 is for a variation of 10%. The results show that there is a significant difference in the loading schedule between these two conditions. In order to satisfy the water content constraint with higher uncertainty under a specific certain reliability the drying time must be extended, and then the maximum cycle will affect the wetting time and make it smaller. The drainage time changes

with the wetting time, so that with increasing parameter uncertainty, the drying time will increase, and the wetting and drainage time will decrease.

Table 2
Results from the example without clogging effects

Cycle	X(day)	Y(day)	Z(day)	CT(day)
1	9.47	1.24	14.16	24.87
2	8.63	1.20	14.52	24.35
3-10	8.63	1.20	14.52	24.35

Total Infiltration 2873.4 (cm)
Number of calls to MSTS: 144
CPU: 1693.32 (sec.) (on IBM RS/6000)

Table 3
Results from the example with clogging effects

Cycle	X(day)	Y(day)	Z(day)	CT(day)
1	6.88	4.21	13.65	24.74
2	6.85	4.18	13.81	24.84
3-10	6.74	4.12	14.04	24.90

Total Infiltration 1601.3 (cm)
Number of calls to MSTS: 166
CPU: 1942.20 (sec.) (on IBM RS/6000)

Table 4
Results from SAT chance-constrained model
($\sigma_{Ks}=0.05$, $\sigma_\alpha=0.05$, $\sigma_n=0.05$, $CT_{max}=25$)

Cycle	X(day)	Y(day)	Z(day)	CT(day)
1	5.25	2.63	16.58	24.46
2	5.19	2.54	16.78	24.51
3-10	5.15	2.51	16.82	24.48

Total Infiltration 1473.4 (cm)
Number of calls to MSTS: 178
CPU: 2082.6 (sec.) (on IBM RS/6000)

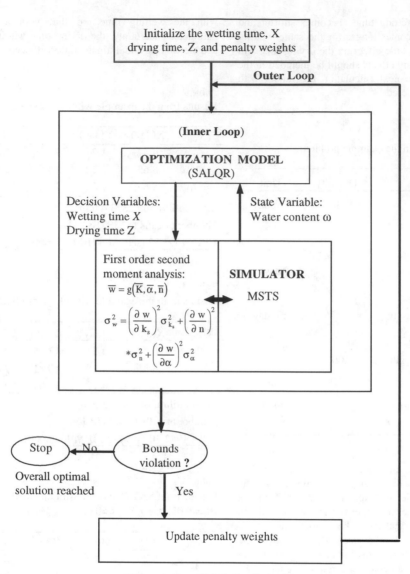

Figure 1 Overall Solution Procedure

Table 5
Results from SAT chance-constrained model
(σ_{Ks}=0.1, σ_{α}=0.1, σ_n=0.1, CT_{max}=25)

Cycle	X(day)	Y(day)	Z(day)	CT(day)
1	2.28	1.07	21.56	24.91
2	2.25	1.06	21.63	24.94
3-10	2.25	1.06	21.63	24.94

Total Infiltration 954.4 (cm)

Number of calls to MSTS: 212

CPU: 2480.4 (sec.) (on IBM RS/6000)

Comparing the results between the deterministic model (Table 3) and the stochastic model (Table 4), it should be noted that there is a difference in every element of the loading schedule. So the loading schedule is different if the parameter uncertainties are considered.

Computational requirements for the above stochastic examples increase with the increased parameter uncertainties. Comparing the deterministic model and the stochastic model, the computational requirements increase 25% percent for the stochastic model.

438

The Monte Carlo operation model was solved for 200 realizations. The results show the loading schedule is in the range, and the mean of wetting is less than the deterministic model about 15%, the drying time is more than 10%.

6. SUMMARY

Soil aquifer treatment (SAT) systems are assumed to operate in order to obtain a maximum infiltration while satisfying the hydraulic constraints. A new methodology is presented to solve the SAT optimal operation problem by interfacing the SALQR with the MSTS model. The MSTS simulator is based on nonlinear partial differential equations of the unsaturated water flow in one, two or three dimensions, however only one dimensional flow is simulated herein.

The solution algorithm for this problem can be briefly summarized in two steps. The first step is to determine the penalty parameter for each hydraulic constraint. The second step consists of the optimizer simulator SALQR-MSTS to solve for the optimum for the given problem from the first step. If the constraints are not met, the problem returns to the first step to update the penalty parameters. The two-steps are repeated until an optimum is found with all constraints satisfied within certain bound criteria.

An example problem has been used to show the application of the SAT deterministic and stochastic management model. The results suggest that although the problem is large, the algorithm interfacing SALQR with MSTS in an optimal control framework holds much promise.

Typically, the matter of uncertainty involving the simulation is treated by employing sensitivity analyses. In this paper the SAT operation models were developed using a stochastic approach to the SAT operation problem in which the chance constrained and Monte Carlo methods were used to deal with parameter uncertainty and heterogeneity of the soil. The techniques show usefulness for the same kind of uncertainty problems.

REFERENCES

Chang, Liang-Cheng, C.A. Shoemaker, and Philip L. F. Liu, "Optimal time- varying pumping rates for groundwater remediation: application of a constrained optimal control algorithm", *Water Resources Research*, 28(12), 3157-3173, 1992

Freeze, R. A., "A stochastic-conceptual analysis of one-dimensional groundwater flow in nonuniform homogeneous media", *Water Resources Research*, 11(5),725-741, 1975

Hoeksema, R. J., and P.K. Kitanidis, " Analysis of the spatial structure of properties of selected aquifers", *Water Resources Research*, 21(4), 563-572

Mays, L. W. and Y.K. Tung, *Hydrosystems Engineering & Management*, McGraw-Hill, Inc., 1992

Mualem, Y. (1976). "A new model for predicting the hydraulic conductivity of unsaturated porous media", *Water Resources Research*, 12(3), 512-522

Nichols W. E., and M. D. White, "MSTS, Multiphase Subsurface Transport Simulator User's Guide and Reference", Pacific Northwest Laboratory, Feb. 1994

Tang, Z, G. Li, and L,W. Mays, "New methodology for the optimal operation of soil aquifer treatment systems:, *Proceedings of ASCE Water Resources Planning and Management annual meeting*, Cambridge, Massachusetts, May, 1995

Tompson, A.F.B., R. Ababou, and L.W. Gelhar, Implementation of the three-dimensional turning bands random field generator, *Water Resources research*, 25, 2227-2243, 1989.

van Genuchten, M.T. (1980). "A closed form equation for predicting the hydraulic conductivity of unsaturated soils". *Am. J. of Soil Science Society*, 44, 982-998

Stochastic Hydraulics'96, Tickle, Goulter, Xu, Wasimi & Bouchart (eds) © 1996 Balkema, Rotterdam. ISBN 90 5410 817 7

A stochastic model for clogging of porous column by sediment

Fu-Chun Wu & Ming-Hsi Hsu
Department of Agricultural Engineering, National Taiwan, China University, Taipei, Taiwan, China

Hsieh Wen Shen
Department of Civil Engineering, University of California, Berkeley, Calif., USA

Diana Y. Ma
Hydraulic Research Laboratory, National Taiwan, China University, Taipei, Taiwan, China

ABSTRACT: A stochastic model for clogging of porous column by sediment is presented. The distribution of deposited sediment within the porous column after clogging stage can be estimated with this model. The two parameters of the stochastic model, λ_1 and λ_2, were found to vary with the medium-sediment size ratio, quantity of sediment introduced, and seepage flow rate through the porous column. Experimental study was conducted to investigate the variabilities of the two parameters with the above-mentioned factors. Two series of experiments with more complicated testing conditions were carried out and the results indicated that the proposed stochastic model is valid for the physical process currently considered.

1 INTRODUCTION

Clogging of a porous medium by fine solid particles has been reported as an important engineering problem in many natural and technical processes such as filtration of industrial and potable waters, artificial groundwater recharge, and soil infiltration into filters (Behnke, 1969; Sakthivadivel and Einstein, 1970; Rice, 1974). In addition, intrusion and deposition of sediment particles into alluvial streambed has also caused adverse effects to natural environment such as pollution of spawning gravel bed substrate (Diplas and Parker, 1985; Alonso et al., 1985; Lisle, 1989) and streambed contamination (Cerling et al., 1990; Berndtsson, 1990). This paper presents a stochastic model for clogging of a gravel column by sediment particles.

1.1 Mechanism of clogging

The term "clogging" of a porous medium is the phenomenon of retaining fine solid particles in the pores of the medium (Sakthivadivel and Einstein, 1970). In the final stage of clogging, the pores are filled with fine particles which prevent further motion of such particles. Sakthivadivel and Einstein (1970) conducted the experimental study with uniform sediment of size d moving downward

through uniform spheres of size D. According to their experiments, the patterns of surface deposition, deep-bed infiltration, and penetration of sediment particles, as illustrated in Fig. 1, occur respectively for the size ratio $D/d \leq 7$, $7 < D/d \leq 14$, and $D/d > 14$. They also found that the characteristic mechanism of sediment deposition within the porous medium for sediment with the size ratio in the range $7 < D/d \leq 14$ is clogging of the pores. Rice (1974) classified factors that result in soil clogging as chemical, biological, and physical. Clogging of a porous medium by sediment is dominated by physical processes (McDowell-Boyer et al., 1986). Physical clogging is the result of suspended solids blocking the pores. Hall (1957) proposed two mechanisms for particle removal from suspension by physical processes, namely, gravitational settling and interstitial straining. When the pore size is larger than the diameter of suspended solids, some particles may penetrate to greater depths. As suspended solids are deposited, however, a restricting layer is built up. Clogging of the fine solids in upper layers inhibited fine particles from further infiltration into the deeper layers. From laboratory observations, hardly any fine solids penetrate through the bottom of the porous column during the clogging stage, with only a few particles falling down from the bottom (Wu, 1993). Sediment infiltration is mainly a deposition process without much erosion to move sediment from a level to a lower level. Therefore, once clogging of the porous medium occurs, the fine particles deposited within the porous medium are claimed to reach the steady state.

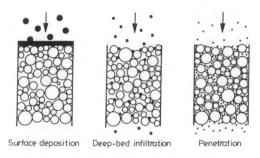

Surface deposition   Deep-bed infiltration   Penetration

Fig. 1 Three patterns of sediment infiltration into porous medium.

1.2 Previous studies

The distribution of deposited sediment particles within the porous medium is interesting to engineers and soil scientists from the viewpoints of practical applications and sediment behavior. Particulate filtration or transport through porous media and clogging of porous media by fine solids have been investigated by many researchers. For example, Sakthivadivel (1966) developed a mathematical model for filtration of non-colloidal particles through the porous column of uniform spheres; McDowell-Boyer et al. (1986) thoroughly discussed the transport of fine particles through the porous media for a broad range of particle size; Joy et al. (1993) presented a stochastic model for the equilibrium sediment transport in a porous medium under non-linear flow condition; Behnke (1969) investigated the clogging in surface spreading operations for groundwater recharge; Rice (1974) tried to understand the soil clogging process that causes infiltration rate reduction.

The present study applies a stochastic model to clogging of a porous column by sediment particles. The distribution of deposited sediment within the porous medium at clogging stage is estimated by this stochastic model with the two parameters, λ_1 and λ_2, varying with sediment size, medium size, quantity of sediment input, and seepage flow rate. The relationships of the two parameters with these

variables were determined with the experimental data (Wu, 1993).

2 STOCHASTIC MODELS

Einstein (1937) proposed the first stochastic model of bed-load transport based on the concept that the movement of sediment particles follows a sequence of alternate step and rest. Following Einstein's pioneering work, Shen and Todorovic (1971) developed a general 1-D stochastic sediment transport model, or the compound nonhomogeneous Poisson process (NHPP) model. Einstein's model was found to be a special case of the NHPP model. The major result of Shen-Todorovic's NHPP model is the 1-D cumulative probability distribution function (CDF), $F_t(x)$, given by

$$F_t(x) = P(X_t \le x)$$

$$= \exp\left[-\int_{t_0}^{t} \lambda_1(s)ds\right] \cdot \exp\left[-\int_{x_0}^{x} \lambda_2(s)ds\right] \cdot$$

$$\sum_{n=0}^{\infty}\sum_{j=n}^{\infty} \frac{\left[\int_{t_0}^{t} \lambda_1(s)ds\right]^n\left[\int_{x_0}^{x} \lambda_2(s)ds\right]^j}{n!\,j!}$$

$$= \exp\left[-\Lambda_1(t)\right] \cdot \exp\left[-\Lambda_2(x)\right] \cdot$$

$$\sum_{n=0}^{\infty}\sum_{j=n}^{\infty} \frac{\left[\Lambda_1(t)\right]^n\left[\Lambda_2(x)\right]^j}{n!\,j!} \tag{1}$$

where

$$\Lambda_1(t) = \int_{t_0}^{t} \lambda_1(s)ds \tag{2}$$

$$\Lambda_2(x) = \int_{x_0}^{x} \lambda_2(s)ds \tag{3}$$

X_t denotes x-direction displacement of a sediment particle at time t. The two parameters of the NHPP model, λ_1 and λ_2, are intensity functions physically representing the inverse of the average rest period and the inverse of the average step length. λ_1 and λ_2 are time-dependent and space-dependent functions respectively.

When the deposited sand particles accumulate within the pore space of the gravel matrix, the sizes of the pore openings for the sand particles to pass through are reduced, as a consequence, the average rest period of the infiltrating sand particles becomes longer due to the difficulty of passage. It has been mentioned in the previous section, the sediment particles remain steady at clogging stage. In other words, λ_1 remains zero after this stage since the average rest period of sediment particles is infinitely long after clogging. If t_c is time to reach clogging stage, then the CDF after t_c, $F_c(x)$, may be expressed as the following:

$$F_c(x) = P(X_t \leq x \text{ for } t \geq t_c)$$

$$= \exp\left[-\Lambda_1^c\right]\exp\left[-\Lambda_2(x)\right]\sum_{n=0}^{\infty}\sum_{j=n}^{\infty}\frac{\left[\Lambda_1^c\right]^n\left[\Lambda_2(x)\right]^j}{n!\,j!} \quad (4)$$

in which Λ_1^c is the Λ_1 value at clogging stage, i.e.

$$\Lambda_1^c = \Lambda_1(t_c) = \int_{t_0}^{t_c}\lambda_1(s)ds \quad (5)$$

The distribution of sediment deposited within the clogging porous column can be estimated with Eq. (4). To apply Eq. (4), the two parameters λ_1 and λ_2 are needed. Wu and Shen (1993) have presented first-order finite-difference approximations to determine λ_1 and λ_2 experimentally. Experimental study for determining λ_1, λ_2, and their variabilities is briefly described in the next section.

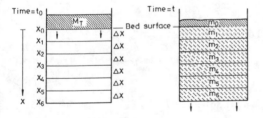

Fig. 2 Gravel matrix configurations at time t_0 and t.

3 EXPERIMENTAL STUDY

The experiments of sediment infiltration into the gravel column with 25cm × 25cm cross-section were conducted in the Hydraulic Engineering Laboratory, University of California at Berkeley, to determine the intensity functions λ_1 and λ_2, and further to estimate the distribution of deposited sediment within the porous column. Details of the experimental setup and procedures were adequately

described by Wu (1993), thus only the testing conditions and the materials used in the experiments are summarized here. For each experimental run, sand of the quantity M_T was initially placed on the surface of gravel matrix, as shown in Fig. 2. After a period of sediment infiltration subject to a constant seepage flow of water through the gravel column, the quantities of sand deposited within the six 5-cm-thick layers were physically measured by detaching the plastic column.

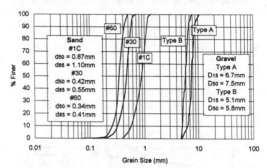

Fig. 3 Size distributions of the materials used.

3.1 Materials

Two types of uniformly graded gravel, types A and B, were used as the porous media packed in the plastic column. Three kinds of uniform sand, #1C, #30 and #60, were used as the fine particles infiltrating into the porous matrix. The size distributions of these materials are summarized in Fig. 3.

3.2 Testing conditions

Based on a series of granular filter experiments (Yim and Sternberg, 1987; Wu, 1993), three experimental variables were found to be the primary factors significantly affecting experimental results. Namely, (1) gravel-sediment size ratio, R_s, defined as D_{15} of gravel / d_{85} of sediment; (2) quantity of sediment introduced into the system; and (3) seepage flow rate through the porous medium. In all, a total of 41 experimental runs with different testing conditions were carried out. The testing conditions included 5 different R_s; sand input, M_T, in the range between 1 kg and 16 kg; seepage flow rate from 63.08×10^{-6} m^3/sec (1 gpm) to 37.85×10^{-5} m^3/sec (6 gpm).

443

3.3 Results

(1) Experimental results

After each experimental run, the quantities of sand on the surface of gravel matrix, and the deposited sediment within the six 5-cm-thick layers (as m_0, m_1, m_2, m_3, m_4, m_5, and m_6 respectively illustrated in Fig. 2), were measured. For the experimental setup and testing conditions considered herein, the cumulative probability distribution at depth x and time t is then calculated by:

$$F_t(x) = P(X_t \leq x) = \frac{M_x(t)}{M_T} \qquad (6)$$

where $M_x(t)$ is the total quantity of sediment above the depth x of the porous column at time t.

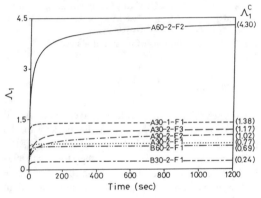

Fig. 4 Λ_1 curves for seven series of experiments.

(2) Λ_1 curve

According to the previous study (Wu, 1993), $\Lambda_1(t_i)$, defined as the integration of $\lambda_1(t)$ from t_0 to t_i, is given by the following:

$$\Lambda_1(t_i) = -\ln\left[\frac{m_0(t_i)}{M_T}\right] \qquad (7)$$

in which $m_0(t_i)$ is the quantity of sand left on the surface of gravel matrix at time t_i. Given $\Lambda_1(t_i)$ for various t_i, the Λ_1 curve is accordingly determined by fitting these $\Lambda_1(t_i)$ points. Fig. 4 shows Λ_1 curves obtained from 7 series of experimental data. The Λ_1 curves approach to the asymptotic Λ_1^c values

indicated in the parentheses. The clogging Λ_1 value, Λ_1^c, was found to vary with testing conditions, the following relationship was obtained by regression:

$$\Lambda_1^c = (2.42 \times 10^{-5}) \cdot R_s^{4.3} \cdot I^{-1.0} \cdot V^{7.0} \qquad (8)$$

in which I is total sand input in kg; V is the ratio of sand particle velocity to fall velocity, where sand particle velocity, defined as the summation of sediment fall velocity and actual seepage velocity through pores, considers the gravitational settling of a sand particle and the flow advection.

(3) λ_2 curve

The first-order finite-difference approximation was used to determine λ_2 (Wu and Shen, 1993). A total of 18 data series were analyzed with this approach (Wu, 1993). In summary, it is found that λ_2 values are decreasing exponentially with the depth x, which may be expressed as:

$$\lambda_2(x) = \lambda_0 \cdot e^{-kx} \qquad (9)$$

The coefficients λ_0 and k also vary with testing conditions. It has been mentioned in section 1.1, the morphology of sediment deposition and infiltration greatly depends on the gravel-sediment size ratio. Therefore, λ_0 and k are in different forms for various categories, the results are summarized in the following:

For $6 \leq R_s \leq 13$,

$$\lambda_0 = (14.33) \cdot R_s^{-1.2} \cdot I^{-0.8} \cdot V^{-3.5}$$
$$(10)$$
$$k = (21.36) \cdot R_s^{-2.1} \cdot I^{-0.4} \cdot V^{-3.9}$$

For $R_s = 16$,

$$\lambda_0 = (0.42) \cdot I^{-0.7} \cdot V^{-0.4}$$
$$(11)$$
$$k = (0.06) \cdot I^{-0.1} \cdot V^{-3.4}$$

Given R_s, I, and V, the distribution of sediment within the gravel column at clogging stage can be predicted with Eqs. (4), (8), (9), (10), and (11).

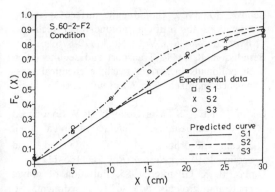

Fig. 5 $F_c(x)$ for the experiments with gravel columns S1, S2, and S3.

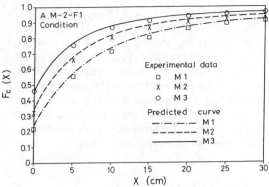

Fig. 6 $F_c(x)$ for the experiments with sand mixtures M1, M2, and M3.

4 APPLICATIONS

In the following sections, two series of experiments with more complex testing conditions were conducted to examine the applicability of the preceding results and verify the predicted results.

4.1 Gravel column of two media

Three experiments were performed with gravel columns of coarse-to-fine media, as described in the following:

Gravel column S1:
4 layers (20 cm) of type A gravel on top of 2 layers (10 cm) of type B gravel.

Gravel column S2:
3 layers (15 cm) of type A gravel on top of 3 layers (15 cm) of type B gravel.

Gravel column S3:
2 layers (10 cm) of type A gravel on top of 4 layers (20 cm) of type B gravel.

In each run, the sand input was 2 kg of #60 sand, the seepage flow rate was set to 2 gpm (126.16×10^{-6} m^3/sec).

With the Λ_1^c values and λ_2 curves obtained through Eqs. (8) to (11), the distribution of sediment after clogging can be evaluated with Eq. (4), details of the calculations refer to Wu (1993). The predicted $F_c(x)$ curves and the experimental data points with gravel columns S1, S2, and S3 are shown in Fig. 5.

It appears that the predicted curves are in satisfactory agreement with the experimental data.

4.2 Sand mixture

Three experiments with sand mixture infiltrating into the gravel column of uniform medium were carried out. The sand mixtures used in these experiments are as follows:

Sand mixture M1:
1.5 kg of #60 sand (75%) and 0.5 kg of #30 sand (25%).

Sand mixture M2:
1 kg of #60 sand (50%) and 1 kg of #30 sand (50%).

Sand mixture M3:
0.5 kg of #60 sand (25%) and 1.5 kg of #30 sand (75%).

The d_{50} and d_{85} sizes for sand mixture M1 are 0.35mm and 0.47mm, for M2 are 0.36mm and 0.50mm, for M3 are 0.39mm and 0.54mm respectively. In each run, type A gravel was used as the porous medium in the column, the seepage flow rate was set to 1 gpm (63.08×10^{-6} m^3/sec).

To use Eq. (4), λ_2 curves for sand mixtures are needed. Assuming the superposition of step length in sand mixture leads to:

$$\frac{1}{\lambda_2^M(x)} = \frac{m_A}{m_A + m_B} \frac{1}{\lambda_2^A(x)} + \frac{m_B}{m_A + m_B} \frac{1}{\lambda_2^B(x)} \quad (12)$$

445

in which $\lambda_2^A(x)$, $\lambda_2^B(x)$, and $\lambda_2^M(x)$ are the intensity functions of sands A, B, and the mixture respectively; m_A and m_B are mass components of sands A and B. The $\lambda_2^M(x)$ curves obtained with Eq. (12) were used to calculate the sediment distribution after clogging (Wu, 1993). The predicted $F_c(x)$ curves and the experimental data points with sand mixtures S1, S2, and S3 are shown in Fig. 6. It also appears that the predicted curves are in satisfactory agreement with the experimental data.

5 CONCLUSIONS

Clogging of porous media by fine particles is of important engineering concern in many natural and technical processes. Clogging of porous media by sediment is dominated by the physical processes of gravitational settling and interstitial straining. Sediment infiltration into the porous medium is mainly a deposition process and thus the deposited sediment particles remain steady after clogging. In this paper, a stochastic model based on the Shen-Todorovic's NHPP model is applied to the clogging of gravel column by sediment.

The distribution of deposited sediment particles within the porous column at clogging stage is estimated by the stochastic model with the two parameters varying with gravel-sediment size ratio, quantity of sediment input, and seepage flow rate. Experimental study was conducted to determine the relationships of the two parameters with these variables. Two series of experiments with more complex testing conditions, one with the gravel column of two media and the other with sand mixture, were carried out to examine the applicability of the stochastic model and verify the predicted sediment distributions. The results appear to indicate that the predicted curves are in satisfactory agreement with the experimental data and the stochastic model presented is valid for the experimental setup and the physical processes currently considered.

REFERENCES

Alonso, C.V., G.Q.III.Tabios & C.Mendoza 1985. Sediment Intrusion (SEDINT) Model. Proj. Report. Fort Collins, Colorado.

Behnke, J.J. 1969. Clogging in surface spreading operations for artificial ground-water recharge. Water Resour. Res. Vol.5 No.4: 870-876.

Berndtsson, R. 1990. Transport and sedimentation of pollutants in a river reach: a chemical mass balance approach. Water Resour. Res. 26: 1549-1558.

Cerling, T.E., S.J.Morrison & R.W.Sobocinski 1990. Sediment-water interaction in a small stream: adsorption of ^{137}Cs by bed load sediments. Water Resour. Res. 26:1165-1176.

Diplas, P. & G.Parker 1985. Pollution of gravel spawning grounds due to fine sediment. St. Anthony Falls Hydraulic Laboratory Proj. Report 240. University of Minnesota, Minneapolis, Minnesota.

Einstein, H.A. 1937. Bed load transport as a probability problem. Dr. Sc. Thesis, Federal Institute of Tech., Zurich, Switzerland. W.W. Sayre (trans). In H.W.Shen (eds), Sedimentation (Einstein). Water Resources Publication, 1972.

Hall, W.A. 1957. An analysis of sand filtration. J. Sanit. Eng. Div. ASCE. 83(SA3), paper 1276.

Joy, D.M., W.C.Lennox & N.Kouwen 1993. Stochastic model of particulate transport in porous medium. J. Hyd. Eng., ASCE, 119(7): 846-861.

Lisle, T.E. 1989. Sediment transport and resulting deposition in spawning gravels, north coastal California. Water Resour. Res. 25: 1303-1319.

McDowell-Boyer, L.M., J.R.Hunt & N.Sitar 1986. Particle transport through porous media. Water Resour. Res. 22: 1901-1922.

Rice, R.C. 1974. Soil clogging during infiltration of secondary effluent. J. WPCF. Vol.46 No.4: 708-716.

Sakthivadivel, R. 1966. Theory and mechanism of filtration of non-colloidal fines through a porous medium. HEL 15-5, Hydraulic Engineering Lab., University of California, Berkeley, California.

Sakthivadivel, R. & H.A.Einstein 1970. Clogging of porous column of spheres by sediment. J. Hyd. Div. ASCE. 96(HY2): 461-472.

Shen, H.W. & P.Todorovic 1971. A general stochastic model for the transport of sediment bed material. In C.-L.Chiu (eds), Stochastic Hydraulics. Proc. 1st International Symposium on Stochastic Hydraulics.

Wu, F.-C. 1993. Stochastic modeling of sediment intrusion into gravel beds. Ph.D. Thesis, Civil Engineering, University of California, Berkeley.

Wu, F.-C. & H.W.Shen 1993. An application of nonhomogeneous Poisson process in sediment infiltration into gravel bed. In H.W.Shen, S.T.Su & F.Wen (eds), *Hydraulic Engineering '93. Proc. 1993 National Conference on Hydraulic Engineering and International Symposium on Engineering Hydrology, ASCE.*

Yim, C.S. & Y.M. Sternberg 1987. Development and testing of granular filter design criteria for stormwater management infiltration structures (SWMIS). Maryland Dept. of Transportation, State Highway Administration Research Report FHWA/MD-87/03.

6 Dispersion and diffusion

Characteristics of turbulent shear stress applied to bed particles in an open channel flow

Alireza Keshavarzy & James E. Ball
Department of Water Engineering, School of Civil Engineering, The University of New South Wales, Sydney, N.S.W., Australia

ABSTRACT: The entrainment of sediment particles is influenced by many characteristics of the flow and particularly the turbulence which produces stochastic shear stress fluctuations at the bed. Recent studies of the structure of turbulent flow have recognised the importance of bursting processes for description of the turbulent processes. Of these process, the sweep event has been suggested to be the most important turbulent event for entrainment of particles from the bed. It imposes variable forces in the direction of the flow and assist movement of particles by rolling, sliding and occasionally saltating. In this study, the characteristics of bursting processes and, in particular, the sweep event were investigated in a flume with a rough bed. The velocity fluctuations of the flow were measured in two-dimensions using a small electromagnetic velocity meter and the turbulent shear stresses were determined. The instantaneous shear stress applied to sediment particles on the bed resulting from sweep events depends on the magnitude of the turbulent shear stress and its probability distribution. From a statistical analysis of the experimental data, it was found that the magnitude of the turbulent shear stress in sweep and ejection events had a normal distribution after application of a Box-Cox transformation. This enabled determination of the mean shear stress, angle of action and standard error of estimate for the bursting events. It was found that the mean turbulent shear stress for a sweep event was 40 percent higher than for the time-averaged shear stress.

1. INTRODUCTION

The initiation of motion or threshold of movement of sediment particles at the bed of open channels is an important component of the management of river systems. Among the many processes which influence the availability and entrainment of sediment particles, the turbulence of the flow and the associated coherent structures are important but, as yet are not completely understood. In this regard, Nezu and Nakagawa (1993) pointed out that the mechanism of turbulence in flow over rough beds in rivers required investigation. Associated with turbulence and momentum transfer are the processes by which sediment particles are entrained and transported with the flow.

The concept of the bursting phenomenon was introduced by Kline *et al.* (1967) as a means of describing the transfer of momentum between the turbulent and laminar regions near the boundary. The phenomenon was considered to consist of four separate events which are defined by the zone in a phase where the event occurs; shown in Figure 1 is a

phase diagram with the zone of each event indicated. As illustrated in Figure 1 the sweep event results in the shear stress being angled towards the bed of the channel.

Recently, the contributions of coherent structures, such as the sweep and ejection events, to momentum transfer have been extensively studied by quadrant analysis or probability analyses based on two-dimensional velocity information. Studies by Nakagawa and Nezu (1978) and Grass (1982, 1971) have indicated that just above the channel bed, the sweep event is more responsible than the ejection event for transfer of momentum into the bed layer. Nakagawa and Nezu (1978), Thorne et al. (1989) and Keshavarzy and Ball (1995) all concluded that sweep and ejection event occurred more frequently than outward and inward interaction events.

The four types of bursting events identified earlier have different influences on the rate, and mechanisms of sediment entrainment in a turbulent flow. In studies of sediment transport by, for example, Thorne *et al.* (1989) and Williams (1990) it has been found that sediment entrainment occurs from the bed more frequently during sweep events and occasionally during outward interaction events whereas transport of suspended sediment depends on the ejection event. Keshavarzy and Ball (1995) also pointed out that the magnitude of the instantaneous shear stress during sweep event is greater than that which occurs during outward and inward interactions events. Bursting events impose a rapid and significant pressure fluctuation on the bed; it is these fluctuations that are considered to have a significant influence on the entrainment of sediment particles from the bed with a resultant variable temporal rate of sediment entrainment. Ball and Keshavarzy (1995) pointed out the effect of instantaneous shear stress on incipient motion and the consequent reasons for the difficulty in defining the sediment entrainment function. These instantaneous pressure fluctuations lower the local pressure near the bed and hence particles may be ejected from the bed by hydrostatic pressure. Raudkivi (1990) postulated that even sheltered particles can be entrained by this mechanism. Despite the importance of the bursting events in sediment transport, the statistical characteristics of bursting events have not been investigated in sufficient detail during previous studies.

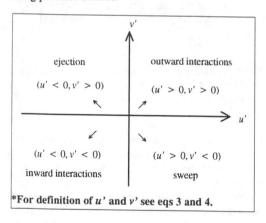

*For definition of u' and v' see eqs 3 and 4.

Figure. 1: Bursting events and their associated regions

In the study reported herein the statistical characteristics of the bursting phenomenon in an open channel flow over a rough bed will be investigated. The magnitude of the shear stress, the frequency and the angles of all events will be determined from experimental data. The variation of shear stress, frequency and angle for each individual event with depth will be presented also.

2. EXPERIMENTAL APPARATUS

The experiments undertaken during this study were carried out in a non-recirculating tilting rectangular flume of 0.61 m width, 0.60 m height and 35 m length. The side walls of the flume were made of glass, making it possible to observe and record the flow. The bed was constructed with movable concrete sheets covered by sand particles of $D_{50}=2mm$ for this study. As a result, it is possible to perform experiments with different bed roughness simply by changing the concrete sheets.

The longitudinal and vertical components of the instantaneous velocity were measured using a small electromagnetic velocity meter (henceforth referred to as EMC). Use of a trolley mounting device enabled the EMC to be moved to any location for measuring of the velocity fluctuations. The measurements were performed along centerline of the flume at a location of 7 m downstream from the inlet of the flow.

These recorded velocities were analysed to calculate the time averaged velocity in the horizontal and vertical directions, the overall mean shear stress, turbulent velocity fluctuations, the mean shear stress for each event, and to count the number of bursting events during the sample period. The analysis also determined the angle from the horizontal for the sweep and ejection events.

Shown in Table 1 are details of flow conditions for a number of experimental tests. The velocity at different point of flow normalised by cross sectional mean velocity (U) for some tests and shown in Figure 2 as a normalised depth (d/H). As expected, the normalised velocity profile are similar suggesting that the effect of variation of flow conditions are minimal.

3. ANALYSIS OF EXPERIMENTAL DATA

As previously noted, data obtained from the experiments were analysed to calculate the following characteristics of the flow: time-averaged velocity, velocity fluctuations, root mean square and the Reynolds shear stress. Relationships used were:

- For the time averaged mean velocity of the longitudinal and vertical components

$$\overline{U} = \frac{1}{N}\sum_{i=1}^{N} u_i \qquad (1)$$

Table 1: The hydraulic conditions of the flow in the experiments

Test	E	F	G	H	J	K	L	M	N	O	P	Q
Q(l/s)	63.7	76.3	52	58	73.6	40.2	61	30.3	22	48.5	79.8	97.5
H(mm)	355	154	276	120	145	120	212	154	70	166	230	265
T(°C)	15	15	15	15.5	14	13.5	13.5	13	12.2	13	13	12.6
U(m/s)	0.29	0.81	0.31	0.79	0.83	0.55	0.47	0.32	0.51	0.48	0.57	0.60
U_{max} (m/s)	0.35	0.14	0.38	0.86	0.99	0.65	0.58	0.40	0.56	0.38	0.38	0.37
Fr	0.16	0.66	0.19	0.73	0.7	0.51	0.33	0.26	0.62	0.38	0.38	0.37

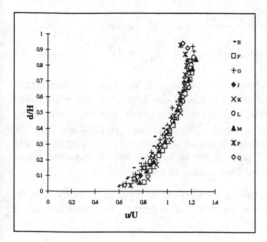

Fig. 2. Normalised velocity profile of the experimental runs

$$\overline{V} = \frac{1}{N}\sum_{i=1}^{N} v_i \qquad (2)$$

where u_i and v_i are instantaneous velocities, U and V are the local temporal mean velocities in the longitudinal and horizontal directions, respectively and N is the number of samples.

- For the velocity fluctuations in two components about the mean value were given by

$$u'_i = u_i - \overline{U} \qquad (3)$$
$$v'_i = v_i - \overline{V} \qquad (4)$$

where u'_i and v'_i are the velocity fluctuations in the longitudinal and vertical directions, respectively.

- For the root mean square value of the velocity

$$u_{rms} = \sqrt{\overline{u'^2}} = \sqrt{\sum_{i=1}^{N} \frac{(u_i - \overline{U})^2}{N}} \qquad (5)$$

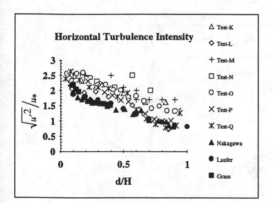

Fig. 3. Turbulence intensity in horizontal direction

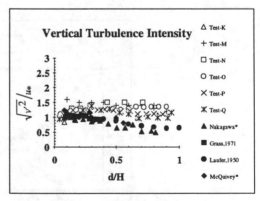

Fig. 4. Turbulence intensity in vertical direction

* Nakagawa et al. 1975, Grass, 1971, Laufer, 1950, McQuivey & Richardson 1969

$$v_{rms} = \sqrt{\overline{v'^2}} = \sqrt{\sum_{i=1}^{N} \frac{\left(v_i - \overline{V}\right)^2}{N}} \qquad (6)$$

- For the Reynolds shear stress

$$\tau = -\rho \overline{u'v'} \qquad (7)$$

$$\overline{u'v'} = \frac{1}{N}\sum_{i=1}^{N} u_i' v_i' \qquad (8)$$

The experimental data were employed also to calculate the turbulence intensity of flow in the horizontal and vertical directions. The turbulence intensities in the horizontal and vertical directions normalised by the flow shear velocity ($u*$), are shown in Figures 3 and 4 as a function of the normalised depth. To validate the data collected, the turbulence intensities in the present experiments were compared with previously published data. As shown in Figures 3 and 4 close agreement was obtained.

The Reynolds shear stress normalised by the shear velocity is shown in Figure 5. Values obtained during this study were compared with previously published data and a good agreement was found in this parameter also.

4. RESULTS AND DISCUSSION

The analysis of the experimentally determined data resulted in the characteristics of the bursting events differing with the normalised flow depth. Important aspects of the analysis are presented in the following discussion.

4.1 Box-Cox transformation

The magnitude of the mean instantaneous shear stress for each category of event is different, and it differs from the overall mean shear stress for the flow. The mean normalised shear stress for each event was determined for alternative normalised flow depths.

Prior to undertaken any statistical analysis, a Box-Cox transformation was used to transform the original data set into a normally distributed data set. As presented by Box and Cox (1964), the transformation is defined by

$$\begin{cases} \dfrac{(z_t + c)^{\lambda} - 1}{\lambda} & \text{if} \quad \lambda \neq 0 \\ \ln(z_t + c) & \text{if} \quad \lambda = 0 \end{cases} \qquad (9)$$

where z_t are the original data, and C is a constant. If all of the values in the time series are greater than zero, then the constant C usually is set to equal zero (Box and Cox, 1964). Application of the transformation to the data set comprising more than 50 tests of sweep and 50 tests for ejection events resulted in a value of λ equal to 0.28.

If the transformed data is represented by C_2, then the reverse transformation from C_2 to the original data is defined as

$$z_t = e^{\left(\ln\left(\overline{C_2}.\lambda + 1\right)/\lambda\right)} = e^K \qquad (10)$$

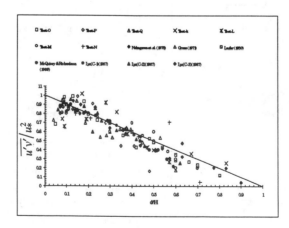

Fig. 5. The normalised Reynolds shear stress compared with previous data.

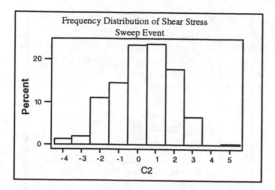

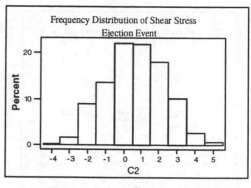

Fig. 6: Frequency Distribution of shear stress in sweep event

Fig. 7: Frequency Distribution of shear stress ejection event

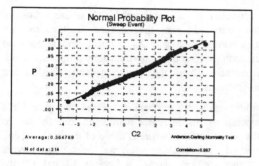

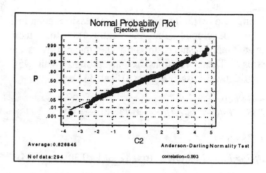

Fig. 8: Probability Distribution of shear stress in sweep event

Fig. 9: Probability Distribution of shear stress in ejection event

where

$$K = \frac{\ln(\overline{C_2}.\lambda + 1)}{\lambda}$$ (11)

The frequency distributions after transformation of shear stress in the sweep and ejection events are shown in Figures 6 and 7. The probability distribution of the transformed shear stress for sweep and ejection events is shown also in Figures 8 and 9.

4.2 Instantaneous shear stress

The mean values of the transformed data (C_2) were calculated and reverse transformation applied. It was found that the magnitude of the shear stress for a sweep event was approximately 1.4 times the mean shear stress close to the bed. Shown in Figures 10 and 11 are the variation of the normalised mean shear stress for each event in relation to normalised depth

of flow for all of the experimental runs. The normalised mean shear stress for each event is determined from equation 12. It can be seen that the mean shear stress for an individual event increases with depth from the bed to the free surface.

$$RSS_i = \frac{\tau_i'}{\overline{\tau}} = \frac{u'v_i'}{u'v_{total}'},$$ (12)

where RSS is the ratio of shear stress in the event.

A focus of this study was analysis of the characteristics of the bursting events and particularly, the sweep event in the region close to the bed. This focus resulted from a need to consider the influence of the shear stress magnitude during a sweep event on particle entrainment. Due to the instantaneous shear stress being greater than the mean shear stress, it is expected, that particles which would not move at the mean shear stress are able to move during a sweep event due to the higher induced shear force.

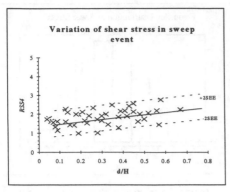

Fig. 10. Normalised shear stress in sweep event

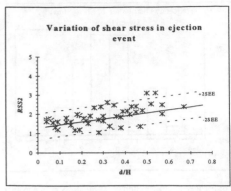

Fig. 11. Normalised shear stress in ejection event

Using the experimental data, regression equations were developed for the ratio of the mean instantaneous shear stress to the mean flow shear stress. These relationships as plotted with 95% confidence limits in Figures 10 and 11, are

- Sweep

$$RSS_4=1.48+1.13(d/H), \quad (r=0.73, SEE=0.36) \qquad (13)$$

- Ejection

$$RSS_2=1.35+1.83(d/H), \quad (r=0.81, SEE=0.38), \qquad (14)$$

where RSS is shear stress ratio and r is correlation coefficient.

4.3 Frequency of the events

The instantaneous shear stress data was divided into different groups based on phase of the bursting process. The frequency of each event was determined by

$$P_k = \frac{n_k}{N} \quad \text{and} \quad N = \sum_{k=1}^{4} n_k \qquad (15)$$

where P is the probability of an event occurring, n_k is the number of event in each category, and N is total number of events.

The probability of each category of event was calculated for all of the experimental data and is shown as a function of normalised depth of flow in Figures 12 and 13. It can be seen that the probability of an individual event varies with depth. From these figures, it can be seen also that the occurrence of the sweep and the ejection events, particularly close to the bed of an open channel, are higher than outward and inward interactions. Close to the bed, the sweep and the ejection events have a probability of

occurrence of approximately 30%, whereas those of the outward and the inward interactions are approximately 20%. As a result, the sweep and ejection events occur approximately 30 percent more frequently than the outward and the inward interactions.

Additionally, the probability of the sweep and the ejection events decreases with depth from the bed to the water surface, whereas those of the outward and inward interactions increase with the depth. Furthermore, as shown in Figures 12 and 13, the probability of the events approaches a value of 25% near the free water surface which suggests that all events are equally probable.

The relationship between the probability of the event and the normalised depth was investigated also. A linear regression was performed between these two variables resulting in , for a

- Sweep event
  $$Y=31.9-7.2X \qquad (r=0.77) \qquad (16) \text{ and}$$
- Ejection event
  $$Y=31.3-8.9X). \qquad (r=0.83) \qquad (17)$$

4.4 Angle of the events

The applied force on the bed particles depends on the inclination angle of the sweep force to the bed. The instantaneous angle of the sweep and the ejection events was determined from the experimental data using equations 18 and 19. The mean angle of the events was calculated by taking the average after original data transformed by Box-Cox transformation at a particular depth within the flow and a period of time.

456

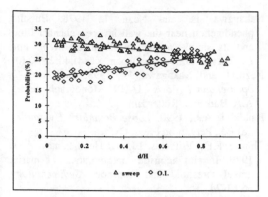

Fig. 12. frequency of the Sweep and O.I. events

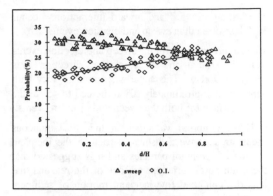

Fig. 13. frequency of the ejection and I.I. events

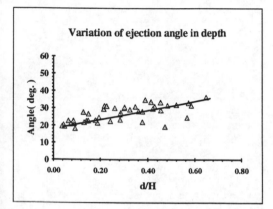

Fig.14. Variation of the sweep angle in depth

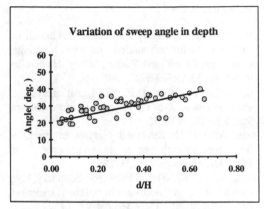

Fig. 15 Variation of ejection angle with depth

$$\theta\ (sweep) = \tan^{-1}\left[\frac{v'(sweep)}{u'(sweep)}\right] \qquad (18)$$

$$\theta\ (ejection) = \tan^{-1}\left[\frac{v'(ejection)}{u'(ejection)}\right] \qquad (19)$$

where θ is the angle of the event from the horizontal direction (clockwise).

Angles determined from the experimental data are shown in Figure 14 and 15 in relation to the normalised depth of flow for all experimental runs. It can be seen that the angle of the sweep event increases with depth from the bed to the free water surface. The angle of the sweep event is about 20° close to bed and increases to 45° at the free water surface. In a similar manner, the angle of ejection events is about 20° near the bed and increases to approximately 45° at the free water surface.

5. RESULTS AND CONCLUSIONS

From the analysis of the bursting process, the following observations were made;

• The magnitude of instantaneous shear stress in an event differed from the mean shear stress. In the sweep event the shear stress was approximately 1.4 times of the mean shear stress in the region near the bed and increased to approximately twice the mean shear stress near the water surface. For the ejection events, the shear stress ratio is about 1.4 near the bed and increases to approximately 2.5 at the water surface.

• The frequency of each event was determined from the experimental data. It was found that, near the bed, the probability of occurrence for the sweep and ejection events was 30%, whereas that of outward interaction and inward interaction events were 0.20.

Thus, the outward and inward interactions occur 30% less often than sweep and ejection events.

• The average angle for the sweep and ejection events from horizontal in the clockwise direction was determined also. This angle was found to vary with depth from approximately 20° at the bed to 45° at the water surface for both the sweep and ejection events.

• These temporal variations in the instantaneous shear stress, have a distinct impact on the incipient motion of sediment particles and it is suggested that these temporal variations are one of the reasons that it is difficult to define incipient motion of sediment particles.

6. REFERENCES

Ball, J.E. and Keshavarzy, A., 1995. Discussion on Incipient Sediment motion on non-horizontal slopes by Chiew and Parker, *J. of Hydraulics Research*, 33(5):723-724.

Box, G.E.P. and Cox, D.R., 1964. An analysis of transformation, *J. of the Royal Statistical Society*, Series B, 26: 211-252.

Grass, A.J., 1971. Structural features of turbulent flow over smooth and rough boundaries, *J. of Fluid Mechanics*. 50(2):233-255.

Grass, A.J., 1982. The influence of boundary layer turbulence on the mechanics of sediment transport, Proc. Euromech 156: *Mechanics of Sediment Transport*, Istanbul, pp. 3-17.

Keshavarzy, A. and Ball, J.E., 1995. Instantaneous shear stress on the bed in a turbulent open channel flow, *Proc. of XXVI IAHR Congress*, London.

Kline, S.J., Reynolds, W.C., Schraub, F.A. and Runstadler, P.W., 1967. The structure of turbulent boundary layers, *J. of Fluid Mechanics*, 30(4):741-773.

Laufer, J., 1950. Some recent measurement in a two dimensional turbulent channel, *J. of Aeronautical Science*, 17(5):277-287.

Lyn, D.A., 1986. Turbulence and turbulent transport in sediment-laden open-channel flows, *Report KH-R-49, W. M. Keck Laboratory of Hydraulics and Water Resources,* California Institute of Technology, Pasadena, Calif.

McQuivey, R.S., and Richardson, E.V., 1969. Some turbulence measurement in open-channel flow, *J. of Hydraulics Division*, ASCE, 95(HY1):209-223.

Nakagawa, H., Nezu, I. and Ueda, H., 1975. Turbulence of open channel flow over smooth and rough beds, *Proceeding of the Japan Society of Civil Engineers*, 241:155-168.

Nakagawa, H. and Nezu, I., 1978. Bursting phenomenon near the wall in open channel flows and its simple mathematical model, *Mem. Fac. Eng., Kyoto University*, Japan, XL(4),40:213-240.

Nezu, I. and Nakagawa, H., 1993. *Turbulence in open-channel flows,* IAHR Monograph Series, A.A. Balkema, Rotterdam.

Raudkivi, A.J., 1990. *Loose Boundary Hydraulics*, 3rd Ed., Pergamon Press, Oxford.

Thorne, P.D., Williams, J.J. and Heathershaw, A.D., 1989. In situ acoustic measurements of marine gravel threshold and transport. *Sedimentology*, 36:61-74.

Williams, J.J., 1990. Video observations of marine gravel transport, *Geo. Mar. Lett.*, 10:157-164.

Stochastic Hydraulics'96, Tickle, Goulter, Xu, Wasimi & Bouchart (eds) © 1996 Balkema, Rotterdam. ISBN 90 5410 817 7

Prediction of longitudinal dispersion coefficient in natural streams

I.W. Seo & T.S. Cheong
Seoul National University, Korea

ABSTRACT: A new simple method for predicting dispersion coefficient by using hydraulic and geometric data easily obtained in natural streams has been developed. Dimensional analysis was implemented to select physically meaningful parameters that are to be included in the new equation for predicting longitudinal dispersion in natural streams. One-step Huber method, which is one of the nonlinear multi-regression methods, was applied to derive a new dispersion coefficient equation. Fifty-nine measured data that were collected in twenty-six streams in the United States were used in the derivation of the new equation. The new dispersion coefficient equation was proven to be superior in explaining dispersion characteristics of natural streams more precisely compared to the existing equations.

1. INTRODUCTION

Contaminants and effluent discharged into a river undergo stages of mixing when they are transported downstream in a flowing water. They spread in the longitudinal, the transverse, and the vertical directions by the advective and dispersive transport processes. After the cross-sectional mixing is complete, the process of longitudinal dispersion is the most important mechanism erasing any longitudinal gradients of concentration (Fischer et al., 1979). In that case, the one dimensional (1-D) Fickian-type dispersion equation derived by Taylor (1954) has been widely used to get a reasonable estimate of the rate of the longitudinal dispersion. The 1-D dispersion equation is

$$A\frac{\partial C}{\partial t} = -UA\frac{\partial C}{\partial x} + \frac{\partial}{\partial x}\left(KA\frac{\partial C}{\partial x}\right) \qquad (1)$$

in which A = cross-sectional area; C = cross-sectional average concentration; U = cross-sectional average velocity; K = dispersion coefficient; t = time; and x = direction of the mean flow.

Analytical solutions of Eq. (1) are easy to obtain with given initial and boundary conditions in such case when the river flow is uniform and the dispersion coefficient is constant with river reach. However, the use of the 1-D dispersion equation is limited to locations far downstream from the source at which the balance between advection and diffusion assumed by Taylor is achieved. Fischer et al. (1979) reasoned that during the early period of the transport process, the advective transport due to the velocity distribution is dominant. During this so called "initial period" advection and diffusion are not in balance, and Taylor's analysis cannot be applied. He also reasoned that because of the dominant effect of the velocity distribution during the initial period, the longitudinal distribution of the cross-sectionally averaged concentration is highly skewed, with a steep gradient in the downstream direction and a long tail in the upstream direction. The variance of the longitudinal concentration distribution increases non-linearly with time during the initial period, and during the later or so-called "Fickian (Taylor) period," the variance increases linearly for steady, uniform flow. So, the 1-D dispersion equation should be applied only after Fickian period is arrived.

When the 1-D dispersion model is applied to predict concentration variations of pollutants in natural streams, selection of a proper dispersion coefficient is the most important and difficult task. It is a rather easy task to use measured dispersion coefficient if it is known in some streams. However, in streams of which mixing and dispersion characteristics are unknown, dispersion coefficient should be estimated by using theoretical or empirical equations. Theoretical method to predict longitudinal dispersion coefficient was first proposed by Taylor (1954) and expanded by Elder (1959) who derived an equation to compute the longitudinal dispersion coefficient for a uniform flow in an infinitely wide open channel assuming a logarithmic velocity profile. Since then, based on experimental and field data, many empirical equations for predicting dispersion coefficient have been proposed by various investigators. However, because most works have been done based on specific assumptions and channel conditions, behavior of each equation varies widely for the same flow condition and the same stream.

In this study, a new simple equation that can predict dispersion coefficient by using hydraulic data that can be easily obtained in natural streams has been developed. Dimensional analysis was done to select physically meaningful parameters in the mechanisms of natural dispersion. One-step Huber method, which is one of the nonlinear multi-regression, was applied to derive a regression equation of dispersion coefficient using 59 measured data collected in 26 streams in the United States.

2. PREVIOUS WORKS

Taylor (1954) first introduced a concept of the longitudinal dispersion coefficient for longitudinal mixing in a straight circular tube in turbulent flow. Taylor derived his equation theoretically as follows

$$K = 10.1rU_*$$ (2)

in which r = pipe radius; and U_* = shear velocity which is given as

$$U_* = \sqrt{gRS}$$ (3)

in which g = gravitational acceleration; R = hydraulic radius; and S = slope of the energy grade line.

Elder (1959) extended Taylor's method for uniform flow in an infinitely wide open channel. He derived dispersion equation assuming a logarithmic velocity profile and that the mixing coefficients for momentum transfer and mass transfer in the vertical direction are the same. Elder derived the equation as

$$K = 5.93dU_*$$ (4)

in which d = the depth of flow.

Elder's equation has been widely used because it is simple and has sound theoretical background. However, it has been indicated that his equation does not describe dispersion in real streams (Fischer et al., 1979). Fischer (1966, 1968) showed that Elder's equation is underestimating the natural dispersion in the real streams by far, because Elder's equation does not consider the transverse variation of the velocity profile across the stream. He postulated that in most natural streams, transverse profile of the velocity is far more important in producing longitudinal dispersion than the vertical profile.

Parker (1961) adapted Taylor's turbulent flow equation to open channel by introducing the hydraulic radius in place of the half pipe radius. The resulting equation is;

$$K = 14.28R^{3/2}\sqrt{2gS}$$ (5)

Using the lateral velocity profile instead of the vertical velocity profile, Fischer (1966, 1968) obtained integral relation for dispersion coefficient in natural streams of large width-to-depth ratio. The result is given as

$$K = -\frac{1}{A}\int_0^W u'd\int_0^y \frac{1}{\varepsilon_t d}\int_0^y u'ddydydy$$ (6)

in which u' = deviation of the velocity from the cross-sectional mean velocity; W = channel width; y = Cartesian coordinate in the lateral direction; and ε_t = transverse turbulent diffusion coefficient. In his study he showed that the agreement of measurements and the prediction by Eq. (6) is within a factor of four in nonuniform streams, and within an error of 30% in uniform streams.

Eq. (6) is rather difficult to use because detailed transverse profiles of both velocity and cross-sectional geometry are needed. So, Fischer (1975) developed simpler equation introducing a reasonable approximation of the triple integration, velocity deviation, and transverse turbulent diffusion coefficient. The result is

$$K = 0.011\frac{U^2W^2}{dU_*}$$ (7)

Eq. (7) has the advantage of being simple because it can predict dispersion coefficient by using only the data of cross-sectional mean parameters that are easily obtained in the stream.

Sooky (1969) studied effects of the cross-sectional shape and the velocity distribution on the dispersion coefficient. Assuming a logarithmic velocity profile and power-function velocity profile, Sooky (1969) developed dimensionless dispersion equation as the function of the width-to-depth ratio (= aspect ratio) for a uniform flow in straight open channels of which cross section is triangular and circular. Through the analysis of field data of Godfrey and Frederick (1970), Sooky (1969) showed that the dimensionless dispersion coefficient increases as the width-to-hydraulic radius ratio increases. Sooky's work does not describe the natural dispersion in real streams adequately because his equation was derived assuming a uniform channel cross section. Bansal (1971) reviewed and summarized the empirical and theoretical equations to compute the dispersion coefficient. He also demonstrated that dimensionless dispersion coefficient increases as width-to-hydraulic radius ratio increases using the dispersion data obtained from the U.S. Geological Survey.

McQuivey and Keefer (1974) developed a simple equation of dispersion coefficient using the similarity between the 1-D solute dispersion equation and the 1-D flow equation especially when Froude number is less than 0.5. First they derived the equation relating the longitudinal dispersion coefficient and the flow dispersion coefficient. Then, by the linear least-square regression of the field data, they derived empirical equation of longitudinal dispersion coefficient as

460

$$K = 0.058 \frac{dU}{S} \tag{8}$$

Even though the method suggested by McQuivey and Keefer (1974) is simple, Fischer (1975) criticized that their equation lacks analytical basis because the mechanisms for dispersion of a flood wave and a dissolved contaminant would be quite different

Abd El-Hadi and Daver (1976) tried to relate the longitudinal dispersion coefficient to the form of the bed roughness and the other hydrodynamic characteristics of the channel flow. They performed experiments in a recirculating flume with different roughness arrangements on the bed, and reported that the dimensionless dispersion coefficient is a function of both the relative roughness height and the relative roughness spacing. They also showed that the relation between K and dU_* is clearly nonlinear beyond values of about 0.009 m²/s for dU_*, and the degree of nonlinearity increases with increasing roughness density.

Liu (1977) derived dispersion coefficient equation using Fischer's equation, Eq. (6) taking into account the role of lateral velocity gradients in dispersion in natural streams as

$$K = \beta \frac{U^2 W^2}{dU_*} \tag{9}$$

in which β = parameter that is a function of the channel cross-section shape and the velocity distribution across the stream. He suggested that the parameter, β, can be determined by considering sinuosity, sudden contractions and expansions, and dead zones in natural stream. By least-square fitting to the field data obtained by Godfrey and Frederick (1970) and others, he deduced following expression.

$$\beta = 0.18 \left(\frac{U_*}{U} \right)^{1.5} \tag{10}$$

He postulated that the maximum deviation of the field data from the prediction by his equation is less than six.

Chatwin and Sullivan (1982) investigated the effects of aspect ratio, ratio of channel width to depth, on dispersion coefficient in channels of which cross-section is approximately rectangle. They determined analytically the dispersion coefficient for laminar flow, and expanded it for turbulent flow in a flat-bottomed channel of large aspect ratio. However, it is practically difficult to use their method in predicting the dispersion coefficient because detailed information on velocity profile and cross-sectional geometry are needed to calculate the dispersion coefficient.

Magazine et al. (1988) studied the effect of large-scale bed and side roughness on dispersion by experiments. They derived an empirical predictive equation for the estimation of dimensionless

dispersion coefficient based on roughness parameters of a channel, such as the Reynolds number and the details of boundary size, spacing, etc., of roughness elements to account for blockage effects. Based on the experimental results of their study and an analysis of the available existing dispersion data, they developed the following expression

$$\frac{K}{RU} = 75.86P^{-1.632} \tag{11}$$

in which P = a generalized roughness parameter incorporating the influence of the resistance and the blockage effects due to the roughness elements. For the prediction of dispersion coefficient in natural streams, Magazine et al. (1988) suggested the following equation

$$P = 0.4 \frac{U}{U_*} \tag{12}$$

Asai and Fujisaki (1991) examined the dependence of the longitudinal dispersion coefficient on the channel aspect ratio by using k-e model. They showed that the dispersion coefficient increases as the channel aspect ratio increases up to 20, and that as the channel aspect ratio increases furthermore, the dispersion coefficient tends to decrease. Iwasa and Aya (1991), by analyzing their laboratory data and previous field data collected by Nordin and Sabol (1974) and others, obtained the equation to predict the dispersion coefficient in natural stream and canal. The result is as follows

$$\frac{K}{dU_*} = 2.0 \left(\frac{W}{d} \right)^{1.5} \tag{13}$$

Since natural streams are sinuous, and have sudden contractions, expansions and dead zones of water, the dispersion coefficient of the natural streams tends to increase compared to those of the simple straight open channels. Thus Eq. (13) tends to underestimate because Eq. (13) is derived by using both open channel and natural stream data.

3. DEVELOPMENT OF THE NEW EQUATION

3.1 Selection of meaningful parameters

Major factors which influence dispersion characteristics of pollutants in natural streams can be categorized into three groups: properties of fluid, hydraulic characteristics of the stream, and geometric configurations. The properties of fluid include fluid density, and fluid viscosity, and so on. The cross-sectional mean velocity, shear velocity, channel width, and depth of flow can be included in the category of bulk hydraulic characteristics. The bed forms, sinuosity can be regarded as the geometric configurations. The dispersion coefficient can be

461

related to these parameters as

$$K = f_1(\rho, \mu, U, U_*, d, W, S_f, S_n) \qquad (14)$$

in which ρ = fluid density; μ = fluid viscosity; S_f = bed shape factor; and S_n = sinuosity.

By using dimensional analysis, new functional relationship between dimensionless terms was derived as

$$\frac{K}{dU_*} = f_2\left(\rho\frac{Ud}{\mu}, \frac{U}{U_*}, \frac{W}{d}, S_f, S_n\right) \qquad (15)$$

in which K/dU_* = dimensionless dispersion coefficient; $\rho dU/\mu$ = Reynolds number; W/d = width-to-depth ratio, which is same as the aspect ratio; U/U_* = friction term which can be defined as

$$U/U_* = (8/f)^{1/2}. \qquad (16)$$

in which f = Darcy-Weisbach's friction factor. Bed shape factor, S_f, and sinuosity, S_n are vertical and transverse irregularities in natural stream respectively. These vertical and transverse irregularities cause secondary currents and shear flow that affect the hydraulic mixing processes in streams. However, in this study, these two parameters were dropped because they are not easily collectable parameters in natural stream, and furthermore, the influences of these two parameters can be included in the friction term. For fully turbulent flow in rough open channels, such as natural streams, the effect of Reynolds number is negligible. Thus Eq. (17) reduces to

$$\frac{K}{dU_*} = f_3\left(\frac{U}{U_*}, \frac{W}{d}\right) \qquad (17)$$

The functional relationship given in Eq. (17) indicates that even though this equation is composed of dimensionless parameters, if this equation is converted into a dimensional form, only hydraulic and geometric parameters easily obtained in natural streams are included. These parameters are channel width, W, depth, d, mean velocity, U, and shear velocity, U_*, and these are the same parameters that were used in the Fischer's and Liu's equations.

To test correlation between the dimensionless dispersion coefficient and dimensionless parameters included in Eq. (17), plots of measured dispersion coefficient versus measured hydraulic and geometric parameters were drawn on a log-log paper. Fig. 1 a) presents a plot of K versus dU_* on a log-log graph. This figure shows clearly that a nonlinear relation exists between K and dU_*. This was also confirmed by Abd El-Hadi and Davar (1976) through their experiments. Therefore, from these results, it could be concluded that Elder's equation and Elder-type

equation in which a linear relation between K and dU_* is assumed are not appropriate to explain natural dispersion process in stream.

The plot of dimensionless dispersion coefficient versus Reynolds number is depicted in Fig. 1 b). This figure shows that, for the data collected in natural streams, the Reynolds number has an insignificant effect on the dimensionless dispersion coefficient. This confirms the assumption that, for turbulent flow in rough natural stream, the effect of Reynolds number may be negligible.

The plots of K/dU_* versus U/U_*, and K/dU_* versus W/d are shown in Fig. 1 c) and d). These figures demonstrate that the dimensionless dispersion coefficient appears to have a strong dependency on these two dimensionless parameters. It increases as the friction term and the aspect ratio increase.

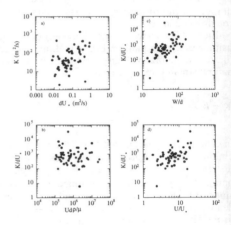

Fig. 1 Plots of dispersion coefficient versus various parameters : a) K vs. dU_*; b) K/dU_* vs. $Ud\rho/\mu$; c) K/dU_* vs. W/d; d) K/dU_* vs. U/U_*

3.2 Regression method

A standard nonlinear multiple model in which dependent variable $\mathbf{Y}$ is related to p unknown independent variables $\mathbf{X}$ can be given as

$$\mathbf{Y} = \alpha\mathbf{X}_1^\beta\mathbf{X}_2^\gamma\mathbf{X}_3^\delta\cdots\mathbf{X}_p^\lambda\varepsilon \qquad (18)$$

in which $\mathbf{X}$ = independent variables which represent the hydraulic and geometric parameters; α, β, γ, ..., and λ = unknown regression coefficients; and ε = independent random residuals which usually follow arbitrary distributions. In Eq. (18), vectors and metrics were noted by bold letters. Taking logarithms of Eq. (18), a linear multiple form can be derived as follows.

$$\ln Y = \ln\alpha + \beta\ln X_1 + \gamma\ln X_2 \cdots +\lambda\ln X_p + \ln\varepsilon \qquad (19)$$

The solution of the transformed model, Eq. (19), is usually obtained by using least square method in which a sum of square of residuals is minimized. However, in case that residuals do not follow the normal distribution, especially none of the residuals is not exceptionally large compared to the estimated standard deviation of the observations, the estimation of regression coefficients includes leverage points that have an overriding influence on the fit.

The Robust estimation minimizes a sum of less rapidly increasing functions of the residuals instead of minimizing a sum of squares. One-step Huber method, developed by Huber (1980), is one of the Robust regression method that estimates reasonably well even in the presence of moderately bad leverage points. The One-step Huber method in which a sum of Robust loss function of the residuals is minimized can be written as

$$\sum_{i=1}^{p} \xi\left(\frac{y_i - \sum x_{ij}T_j}{s} \right) = \min, \qquad j=1,2,3,\cdots,n \qquad (20)$$

in which y_i = measured value of the dependent variable; T_j = initial value of the Robust estimator; s = a known or previously estimated scale parameter; x = Robust loss function. If we let $\psi = \xi'$, which is the first derivative of x, then a necessary condition for a minimum is that the Robust estimator, $\hat{T}$ satisfies the following equation.

$$\sum_{i=1}^{p} \psi\left(\frac{y_i - \sum x_{ij}T_j}{s} \right)x_{ij} = 0 \qquad (21)$$

in which y = weight function.

In general, Eq. (21) is a set of nonlinear equations and thus an iterative method is required to solve this equation. If a starting value, $\hat{T}_o$ is given, then there are various iteration schemes for finding the solution of Eq. (21). Among various iteration schemes, in this study, applied Newton's rule was used, and the result is

$$\hat{T} = \hat{T}_o + s(X^TX)^{-1}X^T\psi\left(\frac{Y - X\hat{T}_o}{s} \right) \qquad (22)$$

To minimize the least absolute residuals estimator, $\hat{T}$, the following equation was used in this study.

$$r_i = r_i(T) = y_i - \hat{y}_i = y_i - \sum x_{ij}T_j \qquad (23)$$

in which $\hat{y}_i$ = predicted value of dependent variable; r_i = residual. It is also necessary to estimate the value of the scale parameter to solve Eq. (22). In this study,

scale parameter was estimated by the following function.

$$s = 1.48\left[\, \underset{i}{med}\left|\left(y_i - x_{ij}T_j\right) - \underset{j}{med}\left(y_i - x_{ij}T_j\right)\right| \right] \qquad (24)$$

In this study, weight function proposed by Huber (1980) was used. That is

$$\psi(r) = r, \qquad\qquad |r| \le a \qquad (25a)$$

$$\psi(r) = a\sin(r), \qquad |r| > a \qquad (25b)$$

in which a = constant which is proposed as 1.5 by Huber.

For the solution of the One-step Huber method, preliminary estimates of $\hat{T}_o$ are assumed with a value computed by least squares method and then values of scale parameter are computed by Eq. (24). Then Eq. (21) is solved approximately for Robust estimates $\hat{T}$ by applying Newton's rule. The iteration method is utilized in which estimates $\hat{T}$ are computed with guessed value, $\hat{T}_o$ until the sum of residual of computed value is minimum.

3.3 New dispersion coefficient equation

In this study, a nonlinear multi-regression equation for predicting the dimensionless dispersion coefficient equation as a function of the friction factor and width-to-depth ratio is derived by using the One-step Huber method. The data sets used in the development of the new dispersion coefficient equation are listed in Table 1. Among fifty-nine data sets, thirty-five measured data sets, which are shaded data sets in Table 1, were selected to derive the dispersion coefficient, and twenty-four measured data sets were used to verify the new dispersion coefficient equation. Separation of total data sets into two groups was done considering that the two groups have similar statistical characteristics. The range of dimensionless values of the measured dispersion coefficient, and hydraulic and geometric parameter are summarized in Table 2.

Table 2. Range of dimensionless hydraulic and dispersion parameters

Dimensionless parameter	Range
Dimensionless dispersion coefficient, K/dU$_*$	6.17 - 35712
Aspect ratio, W/d	13.8 - 157
Friction term, U/U$_*$	1.29 - 20.8
Reynolds number, Udρ/μ (x10^5)	1.10 - 259

The new regression equation derived in this study is given as

463

Table 1. Summary of hydraulic and dispersion data measured at 26 streams in the United States

Streams	Width W (m)	Depth d (m)	Vel. U (m/s)	Slope S	Shear velocity U_* (m/s)	Dispersion coefficient $K\,(m^2/s)$	Reference
Antietam Creek, MD	12.80	0.30	0.42	0.00095	0.057	17.50	Nordin & Sabol (1974)
	24.08	0.98	0.59	0.00135	0.098	101.50	
	11.89	0.66	0.43	0.00095	0.085	20.90	
	21.03	0.48	0.62	0.00100	0.069	25.90	
Monocacy River, MD	48.70	0.55	0.26	0.00050	0.052	37.80	
	92.96	0.71	0.16	0.00045	0.046	41.40	
	51.21	0.65	0.62	0.00040	0.044	29.60	
	97.54	1.15	0.32	0.00045	0.058	119.80	
	40.54	0.41	0.23	0.00045	0.040	66.50	
Conococheague Creek, MD	42.21	0.69	0.23	0.00060	0.064	40.80	
	49.68	0.41	0.15	0.00060	0.081	29.30	
	42.98	1.13	0.63	0.00060	0.081	53.30	
Chattahoochee River, GA	75.59	1.95	0.74	0.00072	0.138	88.90	
	91.90	2.44	0.52	0.00037	0.094	166.90	
Salt Creek, NE	32.00	0.50	0.24	0.00033	0.038	52.20	
Difficult Run, VA	14.48	0.31	0.25	0.00127	0.062	1.90	
Bear Creek, CO	13.72	0.85	1.29	0.02720	0.553	2.90	
Little Pincy Creek, MD	15.85	0.22	0.39	0.00130	0.053	7.10	
Bayou Anacoco, LA	17.53	0.45	0.32	0.00054	0.024	5.80	
Comite River, LA	15.70	0.23	0.36	0.00058	0.039	69.00	
Bayou Bartholomew, LA	33.38	1.40	0.20	0.00007	0.031	54.70	
Amite River, LA	21.34	0.52	0.54	0.00048	0.027	501.40	
Tickfau River, LA	14.94	0.59	0.27	0.00117	0.080	10.30	
Tangipahoa River, LA	31.39	0.81	0.48	0.00061	0.072	45.10	
	29.87	0.40	0.34	0.00069	0.020	44.00	
Red River, LA	253.59	1.62	0.61	0.00007	0.032	143.80	
	161.54	3.96	0.29	0.00009	0.060	130.50	
	152.40	3.66	0.45	0.00009	0.057	227.60	
	155.14	1.74	0.47	0.00008	0.036	177.70	
Sabine River, LA	116.43	1.65	0.58	0.00014	0.054	131.30	
	160.32	2.32	1.06	0.00013	0.054	308.90	
Sabine River, TX	14.17	0.50	0.13	0.00029	0.037	12.80	
	12.19	0.51	0.23	0.00018	0.030	14.70	
	21.34	0.93	0.36	0.00013	0.035	24.20	
Mississippi River, LA	711.20	19.94	0.56	0.00001	0.041	237.20	
Mississippi River, MO	533.40	4.94	1.05	0.00012	0.069	457.70	
	537.38	8.90	1.51	0.00012	0.097	374.10	
Wind/Bighorn River, WY	44.20	1.37	0.99	0.00150	0.142	184.60	
	85.34	2.38	1.74	0.00100	0.153	464.60	
Copper Creek, VA	16.66	0.49	0.20	0.00135	0.080	16.84	Godfrey & Frederick (1970)
Clinch River, VA	48.46	1.16	0.21	0.00085	0.069	14.76	
Copper Creek, VA	18.29	0.38	0.15	0.00332	0.116	20.71	
Powell River, TN	36.78	0.87	0.13	0.00032	0.054	15.50	
Clinch River, VA	28.65	0.61	0.35	0.00039	0.069	10.70	
Copper Creek, VA	19.61	0.84	0.49	0.00132	0.101	20.82	
Clinch River, VA	57.91	2.45	0.75	0.00041	0.104	40.49	
Coachella Canal, CA	24.69	1.58	0.66	0.00010	0.041	5.92	
Clinch River, VA	53.24	2.41	0.66	0.00043	0.107	36.93	
Copper Creek, VA	16.76	0.47	0.24	0.00135	0.080	24.62	
Missouri River, IA	180.59	3.28	1.62	0.00020	0.078	1486.45	Yotsukura et al.(1970)
Bayou Anacoco, LA	25.91	0.94	0.34	0.00049	0.067	32.52	McQuivey & Keefer (1974)
	36.58	0.91	0.40	0.00050	0.067	39.48	
Nooksack River, WA	64.01	0.76	0.67	0.00963	0.268	34.84	
Wind/Bighorn River, WY	59.44	1.10	0.88	0.00131	0.119	41.81	
	68.58	2.16	1.55	0.00133	0.168	162.58	
John Day River, OR	24.99	0.58	1.01	0.00346	0.140	13.94	
	34.14	2.47	0.82	0.00134	0.180	65.03	
Yadkin River, NC	70.10	2.35	0.43	0.00044	0.101	111.48	
	71.63	3.84	0.76	0.00044	0.128	260.13	

$$\frac{K}{dU_*} = 5.915 \left(\frac{W}{d}\right)^{0.620} \left(\frac{U}{U_*}\right)^{1.428} \tag{26}$$

In deriving Eq. (26), the correlation coefficient is 0.75. Eq. (26) can be rearranged as

$$K = 5.915 W^{0.620} d^{0.380} U^{1.428} U_*^{-0.428} \tag{27}$$

4. VERIFICATION

Twenty-four measured data sets that were not used in the derivation of the regression equation were used to verify the proposed equation for predicting dispersion coefficient. For the verification of the proposed equation, the dispersion coefficients predicted by the proposed equation and some existing equations were compared with measured dispersion coefficient. Three existing dispersion equations were selected; they are equations by McQuivey and Keefer (1974), Liu (1977), and Iwasa and Aya (1991).

The comparisons of estimated dispersion equations with measured data are shown in Fig. 2. Fig. 2 shows that the proposed equation predicts quite well whereas McQuivey and Keefer's equation overestimates and Iwasa and Aya's equation underestimates in many cases. Prediction by Liu's equation is generally in good agreement with the measured data. However, for large rivers, Liu's equation is overestimating significantly.

measure that is defined as (White et al. 1973)

$$\text{Discrepancy Ratio} = \log \frac{K_p}{K_m} \tag{28}$$

in which K_p = predicted value of the dispersion coefficient; K_m = measured value of the dispersion coefficient. If discrepancy ratio is 0, the predicted value of the dispersion coefficient is identical to the measured dispersion coefficient. If the discrepancy ratio is larger than 0, the predicted value of the dispersion coefficient is overestimating, and if discrepancy ratio is smaller than 0, it is underestimating. The accuracy is defined as the proportion of number of which the discrepancy ratio is between -0.3 and 0.3 to the total number of data. Discrepancy ratios of each equation for twenty-four field data sets are shown in Fig. 3. Accuracy of each equation is listed in Table 3. The proposed equation predicts better, compared to the existing equations and the discrepancy ratio of new dispersion coefficient equation is ranging from -0.6 to 1. Accuracy of proposed equation is 79%, which is the highest of all. These results demonstrate that the new dispersion coefficient equation developed in this study is superior to the existing equations in predicting dispersion coefficient more precisely in natural streams.

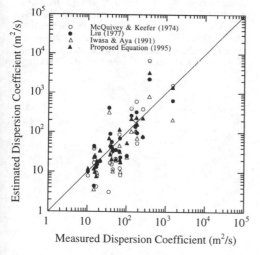

Fig. 2 Comparison of estimated dispersion coefficients with 24 measured data used in verification

To evaluate the difference between measured and predicted values of the dispersion coefficient more quantitatively, discrepancy ratio was used as an error

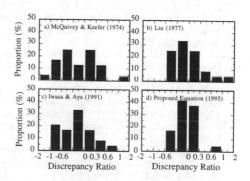

Fig. 3 Comparison of discrepancy ratios of existing and new equations for 24 measured data used in verification

Table 3. Accuracy of existing dispersion coefficient equations and proposed equation

Dispersion coefficient equation	Accuracy (%)
McQuivey & Keefer (1974)	50.0
Liu (1977)	66.7
Iwasa & Aya (1991)	58.3
Proposed equation (1995)	79.2

5. CONCLUSIONS

The results of this study show that the new dispersion coefficient equation, that has been developed in this study is superior in explaining dispersion characteristics of natural streams compared to existing equations. The dispersion coefficient estimated by the proposed equation can be used when the 1-D dispersion model is applied to stream where mixing and dispersion data has not been collected, so that a measured dispersion coefficient is not available.

Acknowledgments

This research work reported herein was partially supported by the 1994 Non Directed Research Fund of the Korea Research Foundation.

References

Abd El-Hadi, N. D., and Daver, K. S. (1976). "Longitudinal Dispersion for Flow over Rough Beds," *Journal of Hydraulics Division,* ASCE, 102(4), 483-498.

Asai, K., and Fujisaki, K. (1991). "Effect of Aspect Ratio on Longitudinal Dispersion Coefficient," *Proceedings of the International Symposium on Environmental Hydraulics,* Hong Kong, 493-498.

Bansal, M. K. (1971). "Dispersion in Natural Streams," *Journal of Hydraulics Division,* ASCE, 97(11), 1867-1886.

Chatwin, P. C., and Sullivan, P. J. (1982). "The Effect of Aspect Ratio on Longitudinal Diffusivity in Rectangular Channels," *Journal of Fluid Mechanics,* 120, 347-358.

Elder, J. W. (1959). "The Dispersion of a Marked Fluid in Turbulent Shear Flow," *Journal of Fluid Mechanics,* 5(4), 544-560.

Fischer, B. H. (1966). "Longitudinal Dispersion in Laboratory and Natural Streams," *Report KH-R-12,* Keck Laboratory of Hydraulic and Water Resources, California Institute of Technology, Pasadena, California.

Fischer, B. H. (1968). "Dispersion Predictions in Natural Streams," *Journal of Sanitary Engineering Division,* ASCE, 94(5), 927-943.

Fischer, B. H. (1975). Discussion of "Simple Method for Predicting Dispersion in Streams," by McQuivey, R. S., and Keefer, T. N., *Journal of Environmental Engineering Division,* ASCE, 101(3), 453-455.

Fischer, B. H., List, E. J., Koh, R. C. Y., Imberger, J., and Brooks, N. H. (1979). *Mixing in Inland and Coastal Waters,* Academic Press, Inc., New York, N.Y.

Godfrey, R. G., and Frederick, B. J. (1970). "Stream Dispersion at Selected Sites," *United States Geological Survey Professional Paper 433-K,* Washington, D.C.

Huber, J. P. (1980). *Robust Statistics,* John Wiley & Sons, New York, N.Y.

Iwasa, Y., and Aya, S. (1991). "Predicting Longitudinal Dispersion Coefficient in Open-Channel Flows," *Proceedings of the International Symposium on Environmental Hydraulics,* Hong Kong, 505-510.

Liu, H. (1977). "Predicting Dispersion Coefficient of Stream," *Journal of Environmental Engineering Division,* ASCE, 103(1), 59-69.

Magazine, M. K., Pathak, S. K., and Pande, P. K. (1988). "Effect of Bed and Side Roughness on Dispersion in Open Channels," *Journal of Hydraulic Engineering,* ASCE, 114(7), 766-782.

McQuivey, R. S., and Keefer, T. N. (1974). "Simple Method for Predicting Dispersion in Streams," *Journal of Environmental Engineering Division,* 100(4), 997-1011.

Nordin, C. F., and Sabol, G. V. (1974). "Empirical Data on Longitudinal Dispersion in Rivers," *United States Geological Survey Water Resources Investigation 20-74,* Washington, D.C.

Parker, F. L. (1961). "Eddy Diffusion in Reservoirs and Pipelines," *Journal of Hydraulics Division,* ASCE, 87(3), 151-171.

Sooky, A. A. (1969). "Longitudinal Dispersion in Open Channels," *Journal of Hydraulics Division,* ASCE, 95(4), 1327-1346.

Taylor, G. I. (1954). "Dispersion of Matter in Turbulent Flow through a Pipe," J. *Proceedings of the Royal Society of London,* Series A., 223, 446-468.

White, W. R., Milli, H., and Crabbe, A. D. (1973). "Sediment Transport : An Appraisal Methods," Vol. 2, *Performance of Theoretical Methods, Station Report,* No. IT119, Wallingford, U.K.

Yotsukura, N., Fischer, H. B., and Sayre, W. W. (1970). "Measurement of Mixing Characteristics of the Missouri River between Sioux City, Iowa, and Plattsmouth, Nebraska," *United States Geological Survey Water-Supply Paper 1899-G,* Washington, D.C.

Stochastic Hydraulics'96, Tickle, Goulter, Xu, Wasimi & Bouchart (eds) © 1996 Balkema, Rotterdam. ISBN 90 5410 817 7

Risk analysis of salinity intrusion in Keelung River after channel regulation

Jan-Tai Kuo & Yao-Cheng Kuo

Department of Civil Engineering and Hydraulic Research Laboratory, National Taiwan, China University, Taipei, Taiwan, China

ABSTRACT: This study evaluates the effect of channel regulation on the salinity distribution in Keelung River. Deterministic and probabilistic/risk analyses are used to compare the difference of salinity distribution before and after channel regulation. The risk analysis uses the method of mean value first-order second moment. A "key parameter" method is adopted to decrease the computation effort while maintain a certain level of accuracy. Simulation of salinity is carried out by coupling a one-dimensional unsteady flow model to a finite segment water quality model, WASP4. Model calibration and validation are extensively performed for both models. The uncertainty in the study considers the influence due to model parameters, such as Manning coefficient n and dispersion coefficient E_{ij}, as well as input data or boundary conditions, such as upstream freshwater flow, lateral inflow, and downstream salinity boundary condition.

1 INTRODUCTION

Keelung River is one of the three major tributaries of Tanshui River (Fig. 1). Tanshui River is the third largest river in Taiwan and flows through the city of Taipei. Keelung River has a drainage area of 501 km² and a total length of 86 km. The low portion of the river, from Hsi-Tzu to Kuan-Tu, is affected by tidal variation. The Keelung River channel regulation project, which includes short-cut for two meanings in the low part, began in 1991 and was completed in 1994. The purpose of the project is mainly for flood control. The project is to resolve the housing problem of 13,000 residents who lived in the original floodplain of 555 hectares. The residents were reallocated to the reclamation land and thereby improving their original poor environmental and living conditions(Fig.2).

This study investigates the influence of the channel regulation project on the salinity distribution in Keelung River. The research works

include developments, calibration, and validation of hydraulic model and water quality model as well as risk analysis for salinity distribution in the river. Hydraulics and salinity concentrations are simulated before and after channel regulation and the differences are compared, analyzed and discussed.

2 HYDRAULIC SIMULATION

2.1 Modeling approach

Low part of Keelung River is subject to tidal effect. Hydraulics of Keelung River are simulated by a one-dimensional unsteady flow model originally developed at Department of Civil Engineering and Hydraulic Research Laboratory, National Taiwan University (Yen and Hsu 1982). The model solves the following De Saint Venant equations:

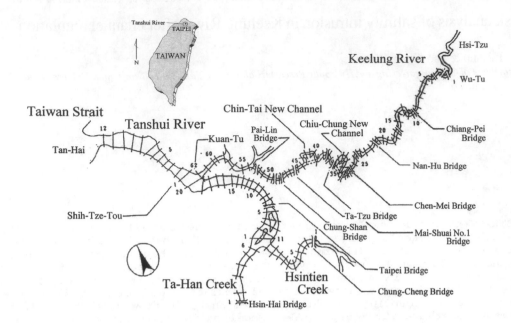

Fig.1. Tanshui River system and model segmentation.

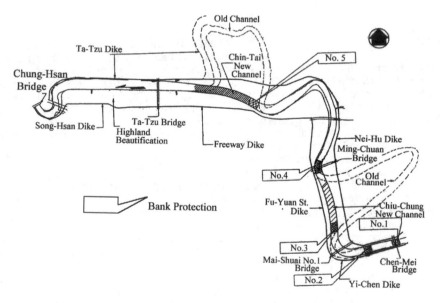

Fig.2. Channel regulation project of Keelung River.

$$\frac{\partial A}{\partial t} + \frac{\partial Q}{\partial x} = q_l \tag{1}$$

$$\frac{\partial Q}{\partial t} + \frac{\partial (Q^2/A)}{\partial x} - gA\left(S_0 - \frac{\partial y}{\partial x} - S_f\right) = q_l V \tag{2}$$

where A=cross-sectional area, Q=flow, q_l=lateral inflow per unit length of river, t=time, x=distance, g=accelerational gravity, S_0=slope of river bed, S_f=energy slope which can be calculated by $(Q|Q|n^2)/(A^2 R^{4/3})$, n=Manning's roughness coefficient, R=hydraulic radius, y=water depth, and V=average velocity component of lateral inflow in x direction.

Equations (1) and (2) are solved by the four point Preissmann implicit scheme. The weighting factor θ in the numerical computation is set to 1.0 in this study.

2.2 *Application to Tanshui River system*

The hydraulic model is applied to the whole Tanshui River system, including Kee- lung River, Ta-Han Creek, and Hsintien Creek. This has the advantage of overcoming the problem of without complete downstream hydraulic boundary conditions at Kuan-Tu in Keelung River. The segmentation of finite difference scheme for the whole Tanshui River system is shown in Fig.1. A total number of 111 segments is used. For Keelung River, it is divided into 62 segments. Upstream boundary conditions are available field streamflow data at Hsin-Hai Bridge (Ta-Han Creek), Chung-Cheng Bridge (Hsintien Creek) and Wu-Tu (Keelung River).

The downstream boundary conditions are measured field data of water surface elevation at river mouth. A time step Δt of 0.5 hour is chosen in the model. Model calibration and validation are extensively carried out using available streamflow data of high flow and low flow. The range of Manning's n after calibration is from 0.025 to 0.0175 with a decreasing value downstream. Fig.3 shows an example result of model validation and the comparison of simulated flows and stages with field data of May 10, 1990 at Pai-Lin Bridge at downstream

Keelung River. In general, the hydraulic model gives favorable simulated results of hydraulics compared to field data.

upstream freshwater = 12.2 cms

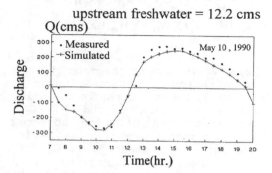

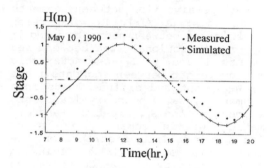

Fig.3 Model validation for hydraulics.

3 SALINITY SIMULATION

3.1 *Modeling approach*

The above hydraulic model is coupled to a water quality model to simulate the salinity distribution in Keelung River. The water quality model selected is WASP4 (Water Quality Analysis Simulation Program) (Ambrose, et al. 1988) developed by Manhattan College and U.S. Environmental Protection Agency. WASP4 is a finite segment (finite difference) model which performs material balance for constituents around the boundary (interfaces) of each segment due to advection and dispersion. The biochemical reaction and transfer (e.g. atmospheric reaeration) have also to be considered in each segment or across the air-water surface.

The generalized mass balance equation of WASP4 for a constituent in segment(section) j can be written as :

$$\frac{\Delta(V_j C_j)}{\Delta t} = \sum_i [-Q_{ij} C_{ij} + R_{ij}(C_i - C_j)]$$

$$+ \sum_L W_{Lj} + \sum_B W_{Bj} + \sum_K V_j S_{Kj} \quad (3)$$

where V_j=volume of segment j; C_j= concentration of the constituent in segment j; t=time; Q_{ij}=advectine flow between segments i and j; C_{ij}= constituent concentration advected between segments i and j, $= \nu C_j + (1-\nu) C_i$ when entering segment j, $= \nu C_i + (1-\nu) C_j$ when leaving segment j; ν =numerical weighting factor, 0-0.5; R_{ij}=dispersive flow between segments i and j, $=E_{ij} A_{ij}/l_{ij}$; E_{ij}=dispersion coefficient between segments i and j; A_{ij}=cross-sectional area between seg-ments i and j; l_{ij}=characteristic mixing length between segments i and j; W_{Lj}=point and diffuse loads into segment j; W_{Bj} = boundary loads into segment j, and S_{Kj}=kinetic transformations within segment j. For simulating the salinity distribution, the last term in equation(3) is neglected since salinity is a conservative substance.

3.2 Application to Keelung River

Keelung River is divided into 62 segments for WASP4 model, same as that in the hydraulic model. Hydraulic model output of Tanshui River system generates the information on stream-flow, velocity, and cross-sectional area to be input to Keelung River's WASP4 model. Upstream boundary con-ditions for salinity at Wu-Tu are set to zero. Downstream boundary con-ditions at Kuan-Tu are based on the available measured salinity data. Dispersion coefficients are varied spatially and are determined from model calibration. WASP4 uses an explicit scheme for time discreti-zation. A small Δt of 12 minutes is chosen in numerical computation. Fig.4 is the simulated salinity concentrations of November 29, 1989 at Ta-Tzu Bridge and Ming-Chuan Bridge,

respectively, for model calibration. Dispersion coefficient is found to be in the range of 0.5 to 8.0 km^2/day in Keelung River, with a increasing value downstream.

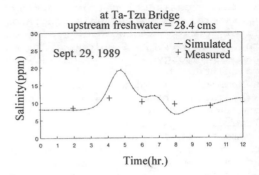

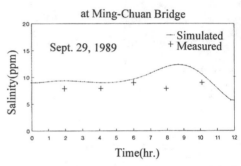

Fig.4. Model calibration for salinity distribution.

3.3 Change of salinity distribution after channel regulation

The change of salinity distribution in Keelung River after short-cut for two meanderings is due to the faster freshwater velocity in new channels and the increasing salinity intrusion potential at high tide because of less resistance at straight channels. The change of river cross-sectional areas also affects the hydraulics as well as salinity concentrations.

New channel geometry(for Chin-Tai and Chiu-Chung new channels) is input to the hydraulic model to obtain hydraulics after channel regulation at varying upstream freshwater flow. Then salinity distribution after channel regulation can be simulated by

470

WASP4 using appropriate boundary conditions .

Fig.5 demonstrates the change of salinity distribution in Keelung River after channel regulation. The simulation is performed under the design freshwater flow condition Q_{75}, a flow with an exceedance probability 0.75. Q_{75} is the low flow criterion currently used in Taiwan for river water pollution control. From Fig.5, it can be seen that the salinity is higher after channel regulation and the difference before and after channel regulation is more significant near the river mouth at high tide. Using 1 ppt as a base for comparison, the length of salinity intrusion is 300 m longer after channel regulation which occurs at about 6 km from river mouth for tidally average condition. At high tide, the length is 200 m longer which occurs at about 8 km from river mouth. The above results can be used for a further study in aquatic ecology and environmental impact analysis.

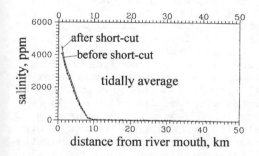

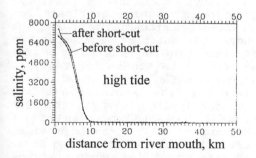

Fig.5. Distribution of salinity in Keelung River before and after channel regulation.

4 RISK ANALYSIS OF SALINITY INTRUSION

4.1 Risk analysis

Risk of a system can be defined as (Yen, et al. 1986; Yoon and Melching 1992):

$$\text{Risk} = P_r[Z < 0] \qquad (4)$$

$$Z = C - L = g(X) \qquad (5)$$

where Z=performance function, C=resistance, L=load, $X = (x_1, x_2, \ldots, x_n)$, representing all n random variables.

Through Taylor series expansion and neglecting the higher order terms, equation(5) can be expressed as:

$$Z \cong g(\overline{X}) + \sum_{i=1}^{n}[\frac{\partial g}{\partial x_i}]_{x=\overline{x}}\,(x_i - \overline{x}_i) \qquad (6)$$

Using the method of mean value first-order second moment (MFOSM) and assuming uncorrelated x_i, the first moment E(Z) and second moment Var(Z) can be written as:

$$E(Z) = g(\overline{X}) \qquad (7)$$

$$\text{Var}(Z) = \sum_{i=1}^{n}\left(\frac{\partial g}{\partial x_i}\right)_{x=\overline{x}}\text{Var}(x_i) \qquad (8)$$

where $(\partial g / \partial x_i)_{x=\overline{x}}$ is the sensitivity coefficient for random variable x_i. If Z is normally distributed, the risk or probability of failure is:

$$P_E = 1 - \phi(\frac{E(Z)}{\sqrt{\text{Var}(Z)}}) \qquad (9)$$

where $\phi(\beta)$ is the cumulative standard normal distribution evaluated at β. β is reliability index and is equal to $E(Z) / [\text{Var}(Z)]$.

4.2 *Application to Keelung River*

Uncertainty from model coefficients n(Manning coefficient) and E_{ij} (dispersion coefficient) as well as upstream freshwater flow at Wu-Tu, lateral inflow, and downstream salinity boundary condition are considered in analysis. The numerical models used in the study are unsteady ones with 62 segments and coupling between hydraulic and salinity computation. If complete, detailed computation for risk is considered, the work load will be very large. To reduce the computation effort and increase the efficiency, a method of using key parameters is used herein (Yoon and Melching 1992; Kuo 1994) to account for the major variability of n and E_{ij} in the river.

Sensitivity(S_N) of n and E_{ij} is computed first by varying the coefficient ±10% and ±20% of the calibrated value. S_N is calculated by:

$$S_N = \frac{\Delta C / C}{\Delta K_i / K_i} \qquad (10)$$

where ΔC=chang of salinity concentration C, and ΔK_i=change of parameter K_i. It is found that segment no.20 to 60 are the most sensitively responded reaches for D_{ij}, and segment no.20 to 50 are the most sensitively responded reaches for n.

To obtain the key parameters, the following equation is used to account for the variance component for each segment i:

$$S_i = \left(\frac{\Delta C}{\Delta K_i}\right)^2 \mathrm{Var}(K_i) \qquad (11)$$

The key parameter segments are obtained when the percentage of the sum of S_i from those segments is at least 90% at a certain location (segment). We also make sure that at least 5 segments are considered as the key parameter segments. The standard deviation of n and E_{ij} for each segment is obtained from previous studies and literatures related to hydraulic modeling in Keelung River. For upstream freshwater flow, lateral inflow and downstream salinity boundary condition, based on data analysis and experience, a coefficient of variation 0.3 is assumed (for all those three variables) in computation.

Equations (8) and (9) are then used to obtain the risk of salinity intrusion under different definition. Fig.6 shows the exceedance probability of salinity intrusion vs. distance in the river at Q_{75} under different base on salinity concentration for comparison. Before channel regulation, using 1,000 ppm as the base, the exceedance probability at distance 5km, 6km and 7.5km is 0.8, 0.15 and 0.0, respectively at tidally average condition. After channel regulation, it is 0.85, 0.6 and 0.0, respectively. The difference for risk of salinity intrusion before and after channel regulation is not significant, but it does exist from modelling analysis.

Fig.7 is the plot for exceedance probability of salinity intrusion after channel regulation at low flow Q_{75} vs. distance for tidally average and high tide condition subject to different frequency of downstream salinity boundary condition. A base of 1,000 ppm is used to define the length of salinity intrusion. From observed salinity data and through statistical tests, a normal distribution is reasonable to describe the frequency of salinity at Keelung River's mouth. From Fig.7, it can be seen that the difference in results due to varying fre-quency of boundary salinity condition is not very significant. For an exceedance probability of 0.8, the length of salinity intrusion is different in a range from 350 m (for T=5,000 yr) to 600 m (for T=5 yr). Various simulation results demonstrate that the length of salinity intrusion is from 8 km (at Chen-Der Bridge) to 9 km (at Chung-Shan Bridge) under high tide. This result provides useful information for Keelung River restoration work and aquatic ecology study.

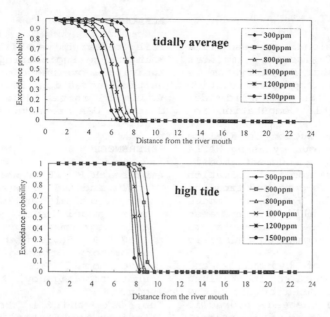

Fig.6. Exceedance probability of salinity intrusion vs. distance in Keelung River.

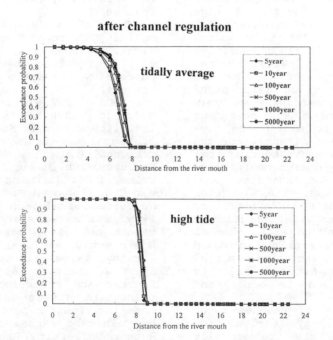

Fig.7. Exceedance probability of salinity intrusion vs. distance after channel regulation at Q_{75} under different frequency of downstream salinity boundary condition.

5 CONCLUSIONS

1. A one-dimensional, unsteady flow model using Preissmann implicit scheme for Tanshui River system is established. After model calibration and verification, the hydraulic model provides favorable computation results.

2. Salinity simulation in Keelung river is carried out by using WASP4 model which is a segment model. Hydrodynamic transport information needed in WASP4 is obtained from the hydraulic model. Extensive model calibration and validation are also performed and WASP4 produces favorable results when compared with field data.

3. The change of hydraulics and salinity distribution in Keelung River is investigated by using the output from the unsteady hydraulic model and WASP4. At low flow, using 1 ppt as a base for comparison, the length of salinity intrusion is 300m longer after channel regulation which occurs at about 6 km from river mouth at tidally average condition. At high tide, the length is 200 m longer which occurs at about 8 km from river mouth.

4. The risk analysis of salinity intrusion before and after channel regulation is carried out by the method of mean value first-order second moment. Sensitivity analysis is performed and a key parameter method is used to reduce the computation time significantly for the uncertainty related to Manning coefficient and dispersion coefficient in both models. The exceedance probility of salinity at certain location in the river or the salinity intrusion length at different tidal condition is estimated before and after channel regulation. The uncertainty in regard to upstream freshwater flow, lateral inflow, and downstream salinity boundary condition is also considered.

5. The influence of channel regulation on the salinity distribulation and its associated risk are generally not significant in Keelung River. However, the results from this study provides useful information for environmental impact analysis on aquatic ecology and river water quality.

ACKNOWLEDGMENTS

This research was funded by the Maintenance Engineering Department, Manicipal Government of Taipei. The helpful suggestions from Prof. Albert Y. Kuo of Virginia Institute of Marine Science, USA are acknowledged.

REFERENCES

Ambrose, R.B., T.A. Wool, J.P. Connolly, and R. W. Schanz, 1988. WASP4, a hydrodynamic and water quality model- model theory, user's manual, and programmer's guide. EPA/600/3-87/039, Environmental Research Laboratory, Office of Research and Development, U.S. Environmental Protection Agency, Athens, Georgia.

Kuo, J.T. and S.L. Lo, 1994. The Influence of Channel Regulation on Water Quality in Keelung River (II). Technical report no. 172, Hydraulic Research Laboratory, National Taiwan University, Taipei, Taiwan. (in Chinese)

Kuo, Y.C., 1994. Risk analysis of salinity distribution in Keelung River. Master thesis, Department of Civil Engineering, National Taiwan University, Taipei, Taiwan. (in Chinese)

Yen, B.C., S.T. Cheng, and C.S. Melching, 1986. First-order reliability analysis. Stochastic and risk analysis in hydraulic engineering. B.C. Yen, ed., Water Resources Publications, Littleton, Colo., 1-36. (in Chinese)

Yen, C.L., and M.H. Hsu, 1982. Numerical simulation of unsteady flow in a river system. Research report no. 7105, Dept. of Civil Engineering, National Taiwan University, Taipei. (in Chinese)

Yoon, C.G. and C.S. Melching, 1992. Sources and reduction of uncertainty in stream water quality modeling. Final report, New Jersey Water Resources Research Institute, Rutgers University, Piscataway, NJ.

Stochastic Hydraulics'96, Tickle, Goulter, Xu, Wasimi & Bouchart (eds) © 1996 Balkema, Rotterdam. ISBN 90 5410 817 7

Optimal control and data assimilation on the sea water circulation in Chinhae Bay

Jong-Kyu Kim, Sun-duck Chang & Cheong-Ro Ryu
Institute of Ocean Hydraulics, National Fisheries University of Pusan, Korea

ABSTRACT: An examples of data assimilation and optimal control for the prediction of sea water circulation in Chinhae Bay are introduced. Numerical modelling was carried out to foresee the reduction of the pollution. The ocean numerical models contain various uncertain elements such as basic equations and boundary conditions, thereby making it difficult to analyze the sea water circulation with a sufficient accuracy. In so far as numerical models are incomplete, therefore a possible method to obtain more realistic results is to make adjustments by replacing predicted values with observed values. In this study, numerical simulations using a data assimilation and optimal control system were conducted in Chinhae Bay areas to examine the effectiveness of data assimilation and optimal control on the prediction system of sea water circulation with fluctuation.

1 INTRODUCTION

Chinhae Bay is an enclosed bay and has surface area of 680 km^2 (Fig. 1). The bay is connected with the southern coastal waters by two narrow and shallow channels. Therefore, the exchange of sea water in the bay with the outer sea is inefficient. Since pollutants produced by urban and industrial complexes are discharged into the bay, significant decrease in the self-purification capacity of sea water in the bay poses a serious environmental and social concern.

Eutrophication and red tide of Chinhae Bay were documented since 1960's (Park and Kim, 1967). Catastrophic red tide of toxic Dinoflagellates in 1981 and the oxygen depletion water bodies caused significant damage to aquaculture. A large scale underwater sewage outfall for citizens of more than one million in Masan and Changwon City is under construction. Massive reclamation and construction projects along the coast are under planning. These civil engineering works have raised the question about the effective management of the bay.

All of these aspects encouraged the oceanographic to develop a hydraulic and numerical model with the data assimilation and optimal control system which allow the temporal and spatial evolution of water mass to be simulated.

Hydrographic Office of the Ministry of Trans-portation set tide gauge in Chinhae Bay after World War II. Environmental sciences team of the National Fisheries Research and Development Agency is monitoring the water quality bimonthly since 1970's (NFRDA, 1989). Lee et al. (1980 82) carried out oceanographic surveys in Chinhae Bay. Sea water circulation in Chinhae Bay were studied by Chang et al. (1984), Chang and Yanagi (1986), and Kim et al. (1986), Chang et al. (1993), Kim et al. (1994), and so on.

These observational data were used to feed the hydraulic and numerical modelling of Chinhae Bay. In spite of the simplifications inherent in the model experiments approach, the numerical algorithm as well as the physical scale model is able to simulate the main pattern of water circulation induced by tidal forcing.

In this study, numerical simulations using a data assimilation and optimal control system were conducted in Chinhae Bay areas to examine the effectiveness of data assimilation and optimal control for the prediction system of sea water circulation with fluctuation. The problem of data assimilation is to use the observations in such a way that: i) the solution of the model is as close as possible (in a sense to be quantitatively defined) from the observations, ii) the solution of the model is in agreement (in a sense to be qualitively defined) with the properties of the actual coastal sea conditions.

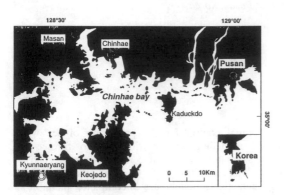

Fig. 1. Maps of Chinhae Bay.

2. DATA AND METHODS

2.1 Oceanographic observations

Seven series of oceanographic observations are carried out in Chinhae Bay as well as in the two channels linked with the outer sea from July 1983 to May 1988. Current speed are measured by DPCM-4 and CM-2 at stations in the eastern, central, northern and western parts of the bay at spring and neap tide.

Drogue tracking at 2m layer below the surface and drift bottle experiment were carried out at stations listed the bay. Water temperature and salinity were also observed at each stations for checking the stratifications.

2.2 Hydraulic model experiment

1) Experimental facility and similarity

Scope of the hydraulic model basin is 34° 44′ ~ 35° 09′ N and 128° 24′ ~ 128° 45′ E (Fig. 2). The dimension of the model basin 24m in length and 12m wide. As the tidal waves are propagated through the Kaduk Channel and the Kyunnaeryang Channel, two water tanks are provided for supplying waters, and an electronic automatic control system of tidal current is employed to regulate phase and amplitude of

water circulation corresponding to those of the tidal current. Electronic water level gauges are incorporated in the control system to monitor and report the water level change.

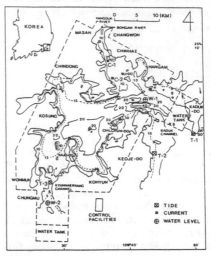

Fig. 2. Schematic diagram of experimental facilities.

Physical scales of parameters are deduced in accordance with the Froude's similarity law, and are listed in Table 1.

Table 1. Physical parameters in the prototype and hydraulic model.

Parameter	Scale	Prototype	Model
Distance	1/2,000	2.0 km	1.0 m
Water depth	1/159	50 m	31.5 cm
Tidal range	1/159	2 m	1.26 cm
Tidal period	1/159	12h 25m	4m 41s
Velocity	1/12.6	1.25 m/s	5 cm/s
Discharge	$1/(4 \times 10^6)$	100 m³/s	25 cm³/s

2) Experiment

As the semi-diurnal tide predominate in the tidal circulations observed in Chinhae Bay, tidal harmonic constants at mean spring tide (M_2+S_2) and mean neap tide (M_2-S_2) are deduced and used in the model experiments (Table 2).

Propeller type current meter were uses to measure flow speed, and plastic drogue tracking were carried out with the aid of 54mm video camera.

Table 2. Tidal harmonic constants used in the hydraulic experimental model.

St.	Type	Amp.(mm)	Period(s)	Phase(°)
	Spring tide (M_2+S_2)			
T-1	Proto.	862	44,700	243.7
	Model	5.4	281	243.7
T-2	Proto.	868	44,700	244.5
	Model	5.5	281	244.5
T-3	Proto.	1,084	44,700	253.4
	Model	6.8	281	253.4
	Neap tide (M_2-S_2)			
T-1	Proto.	214	44,700	243.7
	Model	1.3	281	243.7
T-2	Proto.	220	44,700	244.5
	Model	1.4	281	244.5
T-3	Proto.	279	44,700	253.4
	Model	1.8	281	253.4

2.3 Numerical modelling

1) Observational networks

The information about the state of the atmosphere and ocean are provided by observations using various sensors (Fig. 3).

Surface stations are distributed irregularly over the continents, in generally, their density is relatively high over Western Europe and North America and very low over the oceans especially over the south Pacific. In the best case atmospheric pressure, wind components, temperature and relative humidity are measured every 3 hours. Radiosonde and pilot balloons give information about the vertical structure of the atmosphere. The density of the stations is weaker than the density of surface stations, the densities are roughly homogeneous. Balloons are launched every 12 hours, the variables observed are, in the best case, temperature, pressure, relative humidity and wind components. Satellites provide numerous data about the atmosphere and ocean. Variables are observed and they have to

be transformed into oceanic and meteorological variables. From the polar-orbiting satellites vertical sounding of air temperature can be deduced. From geostationary satellites some estimation of the wind can be deduced from the observation which are avaliable for the numerical prediction are heterogeneous in quality and in density (spatial and temporal).

The ocean circulation is not as well understood as that of the atmosphere because of the relative paucity of observations. There has been little need for oceanic forecasting and hence no real-time network. Most oceanographic observations have been obtained by research oceanogra-

Fig. 3. A schematic components of the observational networks.

phic vessels on anonroutine basis; these vessels often take measurements in long, straight lines called sections. Because of the longer time scales in the oceans, the value of a given observations are taken down to many hundreds of meters using BT. Currents are infrequently measures because of the cost of ocean moorings. Lagrangian tracer techniques that followed the dispersal of tritium from 1950s bomb tests have proved useful. New observing techniques such as XBT, satellite scatterometer measurements of surface winds, satellite infrared imagery from sea surface temperatures, satellite-tracked drifting buoys, and acoustic tomography techniques are also being tested.

Oceanographic data have been objectively analyzed directly, without the aid of an assimi-

477

lating model by using statistical interpolation (Gauss-Markov method, Table 3).

Table 3. Inhomogeneous and incomplete listing of algorithms methods and mathematical topics with a relationship to inverse methods (Wunsch, 1989).

Related Methods and Algorithms
Backus-Gilbert Procedure, Singular Value Decomposition, Objective Mapping, Objective Analysis, A priori filter theory (Kalman form), Control Theory, Optima Estimation Theory, Linear Programming, Quadratic Programming Dynamic Programming, Rank Deficient Regression, Ridge Regression, Assimilation, Regularization, Least-squares (many variations), Gauss-Markov Estimators, Generalized Inverses, Adjoint Methods, Duality Theory, Constrained Estimation, Krigging, Collocation (in geodesy)

2) Oceanographic Data Assimilation

Data assimilation has been studied for a long time in Meteorology where it is essential to make good weather forecasts. In oceanography, assimilation of data into models is only at its beginning (Fig. 4; Robinson and Leslie, 1985). The possible availability of satellite derived measurements that will cover the globe in space and time urge us to use this technique when we want to make profit of all the detailed information satellites can provide.

The analysis and interpretation of oceanographic and meteorological data is generally done in conjunction with a model. This model can for instance be statistical as in optimal interpolation where a linear regression model based on a priori statistical information is used to fill the space between measurements with data. Dynamical information is used e.g. when meteorologists allow sharp gradients in the vicinity of a front while the rest of the field is required to be smooth.

Data assimilation is a process which combines information of measurements with the corresponding

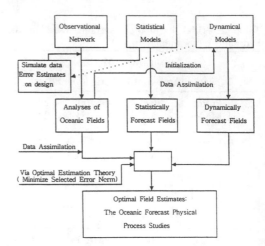

Fig. 4. A schematic of the components of an oceanic descriptive system (Robinson and Leslie, 1985).

field of a dynamical model to give better estimates of the measurements and simultaneously improve the model state. The model is then integrated forward for some time until new measurements are available and the process is repeated. After a number of assimilation cycles all model variables may have felt the influence of the measurements and even the part of the model state that is not observed can be estimated by this method. Time independent flow problems can also be dealt with. The same data are assimilated each cycle until an equilibrium solution has been reached and changes in the model state from cycle to cycle are very small. The weight given to measurements relative to that of model values in the assimilation determines whether the model is closer to observations or to steady state.

In oceanography the sparseness of data both in space and time until now has allowed to make estimates of global fields only for mean values and mean variances and in some cases for average seasonal cycles. Interannual variations or variations on short time scales are regarded as noise for these estimates and their information content is discarded. Optimal interpolation of temperature and salinity can lead to vertically unstable density profiles which of course cannot exist for average fields. To inhibit errors like this we have to use more than statistical information for the process of interpolation. In addition to the statistical requirements we can force the interpolated field to obey constraints. These constraints are either satisfied in a statistical sense (e.g. a least squares solution) or they are required to hold exactly as equalities. The first kind of constraint is denoted as "weak" or "soft" while the latter kind is called a "strong" or "hard" constraint.

The problem of an oceanographic data assimilation is to use the observations in such a way that: First, the solution of the model is as close as possible (in a sense to be quantitatively defined) from the observations. Second, the solution of the model is in agreement (in a sense to be defined) with the oceanic characteristics. In the following we will consider "observation" in the broad sense, it means that we will use physical observations as described above, as well as synthetic observations which can be the results of a numerical model.

3) Mathematical formulation

The equations that physical oceanographers need to solve in order to be able to describe the dynamics of the ocean are known as the equations of motion. The equations governing the evolution of the ocean are deduced from the general equations of thermohydrodynamics. For the sake of simplicity we only consider the various assumptions, makes the equations much more complicated. Thus, these are simply Newton's Second Law of Motion applied to a fluid moving over the surface of the Earth. The system of partial differential equations governing the oceanographic flow is given (Pond and Pickard, 1983):

Equation of motion:

$$\frac{dV}{dt} = -\alpha \nabla P - 2\Omega \times V + g + F \quad (1)$$

Equation of mass conservation:

$$\frac{1}{\rho}\frac{dV}{dt} + \nabla \cdot V = 0 \quad (2)$$

Equation of salinity and temperature

$$\frac{ds}{dt} = K_s \nabla^2 S \quad (3)$$

$$\frac{dt}{dt} = K_T \nabla^2 T + Q_T \quad (4)$$

Equation of state:

$$\rho = \rho(S, T, P) \quad (5)$$

In these equations:

$$V = [u, v, w] \quad (6)$$

represents the velocity, u and v are the horizontal components, w the vertical one, P is the pressure, α is the specific volume, Ω is the magnitude of the angular velocity of rotation of the Earth, g is the gravity, F is the external force, ρ means for the density, S is the salinity and T is the temperature, respectively. K_S and K_T are molecular kinematic diffusivities for salt and for temperature (or heat), respectively. And ∇^2 is a Laplancian, Q_T represents a source function. The total derivative of a scalar function A is:

$$\frac{dA}{dt} = \frac{\partial A}{\partial t} + u\frac{\partial A}{\partial x} + v\frac{\partial A}{\partial y} + w\frac{\partial A}{\partial z} \quad (7)$$

These equations are the general equations of oceanography, according to the scale and the phenomenon studied, an approximation can be used, for instance most of the oceanographic models use the assumption of vertical hydrostatic equilibrium. After discretization in space, we obtain a set of ordinary differential equations. In significant operational models the order of magnitude of the number of discrete variables considered is of the order of 10^6. Therefore one of the encountered difficulty is related to the dimension of the systems under consideration. If an approximation of the system $(1 \sim 5)$ can be written:

$$\frac{\partial X}{\partial t} + AX + B(X) + CX = F \quad (8)$$

in an Hilbert space H where A is a linear, positive definition operator with a compact inverse, B is a quadratic operator, C is a linear operator mapping $D(A)$ into H and if some additional conditions are satisfied then it can be shown, Temam (1989), that the system has an inertial manifold M. This manifold is of finite dimension, it is a lipschitzian manifold exponentially attracting all the trajectories of the system. Furthermore M is an invariant of the motion. The existence of such a manifold is conjectured for oceanographic systems it is of theoretical and practical importance to determine approximations of the inertial manifold.

Shallow water equations are often used as approximation of $(1\sim 5)$, if the normal modes of the linearized equation are computed then it can be seen, Daley (1991), that the spectrum of the eigen-frequencies may be divided into two parts : the so-called Rossby modes associated with typical oceanographic motion and the gravity modes associated with fast motion carrying only few kinetic energy in the actual ocean. The attractor of the oceanographic system might be considered as the state of dynamic equilibrium between Rossby and gravity modes. If the initial condition which is used to integrate the model is not located in the vicinity of the attractor then the trajectory will be contaminated by gravity motion masking the short range forecast.

3 RESULTS AND DISCUSSION

Data assimilation and optimal control for the prediction of sea water circulation in Chinhae Bay are introduced.

Numerical modeling was carried out to foresee the reduction of the pollution. The hydraulic and numerical experiments are able to reproduce the semi-diurnal tidal current and gave an idea on the general feature of the flow pattern of water motions in the bay. The numerical modeling simulated the large and the small scale eddies formed at the central and the eastern part of the bay, respectively. The available data show that the effluents from northern Chinhae Bay area will have to be treated to a certain level before discharge. The less saline water of surface layer moves eastward opposite to the lower layer saline water, especially in the rainy season. The results of the oceanographic observations reveal the density stratification in the northern Chinhae Bay when the fresh water run-off prevails. Tidal transport through the eastern channel, Kaduk channel, during the semi-diurnal tidal period is estimated to be $86\sim 90\%$ of total transport at spring tide, and $61\sim 80\%$ at neap tide.

Fig. 5 and Fig. 6 show the tidal residual currents by the hydraulic and numerical model during the spring and neap tide. Tidal residual currents in the Kaduk and Kyunnaeryang channel and upper part of Chilchundo during spring tide are strong having eddies occur. Tidal residual currents in the western and northern part of the Chinhae Bay during spring tide are very weak. Tidal residual currents rotating clockwise and anticlockwise in the central part of the bay

occur, and the currents rotating clockwise around Budo are generated (Fig. 5 and Fig. 6). Currents in the bay depend strongly on the river flow and wind (Fig. 7).

Tidal residual currents rotating clockwise occur in the central part of the bay. The surface current in the bay depend strongly on the wind and river flow, and it seems to be remarkable during the neap tide (Fig. 7).

Fig. 8 shows the procedure for solving the present system with and optimization strategies. The minimization determines the best fit of the data when the optimal solution is approached. Many different minimization are available.

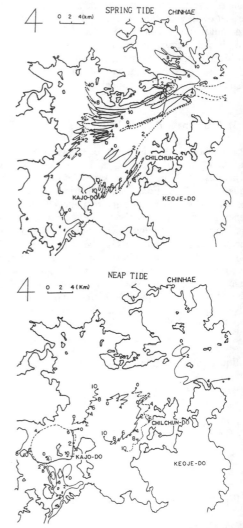

Fig. 5. Tidal residual currents in the hydraulic model during the spring tide and neap tide.

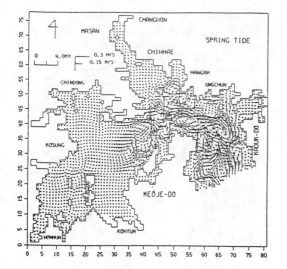

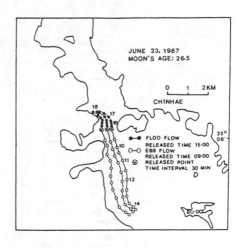

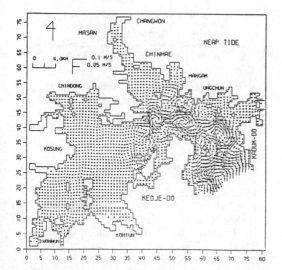

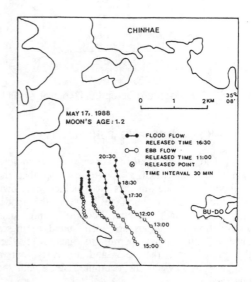

Fig. 6. Computed tidal residual currents during the spring tide and neap tide.

Fig. 7. Drogue tracking the study area during the spring and neap tide.

Fig. 9a shows the values of the cost function, it's gradient, and data misfit versus the number of iterations in the minimization procedure. All values have been normalized by their own initial values to allow a direct comparison of the convergence rate. As can be seen, the cost function drops rapidly in the first couple of iterations. The ratio of the norm of the gradient also experiments a sharp decrease in the first two iterations. The convergence criterion is satisfied after 11 iteration. Fig. 9b shows the variation of the values corresponding to Fig. 9a

with the number of iterations. The cost function decreases to about 53% of its initial value after 11 iterations. The norm of the gradient has a rapid reduction during the first few iterations. After 11 iterations it drops to 5% of its initial value and reaches a steady state.

In this study, numerical simulations using a data assimilation and optimal control system were conducted in Chinhae Bay to examine the effectiveness of data assimilation and optimal control techniques for the prediction of coastal sea conditions

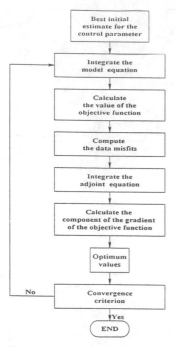

Fig. 8. The procedure of optimal control system.

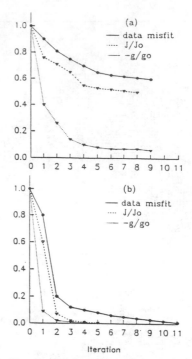

Fig. 9. Results of the (a) variation of the cost function, norm of the gradient, and the data misfit with the number of iterations, (b) real data assimilation corresponding to (a).

4 CONCLUSIONS AND PERSPECTIVES

A field experiment in coastal sea conditions cannot be duplicated a consequence is that the method of modeling has to include its own validation and the interface between the model and the data. Optimal control and data assimilation techniques are well suited to theses purposes, thus, they permit to handle together the model and the data and to carry out a comparison between the computed fields and the observed ones.

An examples of data assimilation and optimal control for the prediction of sea water circulation in Chinhae Bay are introduced.

We will have to update the model results with use of observational data obtained at buoy stations or by satellite in order to carry out the effective data assimilation and optimal control in the numerical model of the coastal sea conditions.

REFERENCES

Chang, S.d. et al., 1984. Sea water circulations in eastern Chinhae Bay. Bull. Fish. Res. Dev. Agency, 32, 7-23.

Chang, S.d. and T. Yanagi 1986. Water circulation in Chinhae Bay. Summary Rept. Oceanogr. Soc., Japan, 325-326.

Chang et al., 1993. Field observations and hydraulic model experiments of tidal current in Chinhae Bay. Bull. Korean Fish. Soc., 26(4), 346-353.

Kim, J.H., S.d. Chang and S.K. Kim 1986. Variability of current velocities in Masan Inlct. Bull. Korean Fish. Soc., 19(3), 274-280.

Kim et al., 1994. Two-dimensional hydraulic and numerical modeling of tidal current in Chinhae Bay. J. Oceanol. Soc., Korea, 29(2), 85-94.

Lee, K.W. et al., 1980-82. A study on the monitoring system for red tides in Chinhae Bay. KORDI.

NFRDA, 1989. A comprehensive study on marine pollution for the conservation of the Korean coastal ecosystem with respect to culture areas and fishing grounds. Tech. Rept. 84, 347p.

Park, J.S. and J.D. Kim 1967. Studies on the red tide in Chinhae Bay. Bull. Fish. Res. & Dev. Agency, 1, 63-79.

7 Hydrological time series

Stochastic Hydraulics'96, Tickle, Goulter, Xu, Wasimi & Bouchart (eds) © 1996 Balkema, Rotterdam. ISBN 90 5410 817 7

Estimation of groundwater recharge rate by application of time series analysis in the Pingtung Plain, Taiwan, China

Cheh-shyh Ting

Department of Civil Engineering, National Pingtung Polytechnic Institute, Taiwan, China

ABSTRACT: Data from previous research on the plain with respect to groundwater recharge and specific yield are obviously inadequate even though there are records extending for more than 30 years of groundwater level, river discharge, precipitation and abstraction. An application of time series analysis for the estimation of groundwater recharge was thus executed. Six observation wells from the present monitoring network were selected via a single input-output, linear, time-invariant, lumped parameter model. The main aim of this work is to attempt to solve the uncertainties concerning hydrological stresses by alternative methods. To estimate the infiltration coefficient with respect to specific yield has been achieved and resulted in the estimated values. This model can predict aquifer water levels for known specific yield values. Not only is it an important parameter in water resources evaluation, but its value also effects the accuracy of model calibration and prediction.

1 INTRODUCTION

1.1 *Current problems related to groundwater resources development in Taiwan*

Large scale groundwater utilization and development in Taiwan began in the 1950s. Investigations and surveys on groundwater resources of Taiwan were initiated in 1956, and at the end of 1961, the areas of Choshui River Alluvial, Chianan Plain, Pingtung Plain and Hualien Taitung Valley (Fig. 1) were completed on a reconnaissance scale. More detailed research was carried out in the Yunlin Plain in 1958 (TPGDB, 1961), and in later years (Willis, 1977, 1984, 1988), these studies included a groundwater flow analysis by numerical methods, salt water intrusion in coastal areas and the development of a groundwater management model. The first study with a mathematical model in the Pingtung Plain is described in Ting (1992,1993). In later years, regional groundwater resources analysis and quality protection studies using the MODFLOW and SEFTRAN models were carried out by Tsao (1991) to simulate alternative groundwater extraction patterns. In both the Ting and Tsao studies, the models are calibrated against the hydrogeological parameters: transmissivity, conductivity and storativity.

LEGEND

1. Taipei Basin
2. Taoyuan Chungli Terrace
3. Hsinchu Miaoli Coastal Area
4. Tai Chung Area
5. Choshui River Alluvium
6. Chianan Plain
7. Pingtung Plain
8. Lanyang Plain
9. Hualien Taitung Valley

Figure 1 Groundwater regions of Taiwan (after WRPC, 1984)

Since the previous studies of groundwater resources use in the plain were based on an ad hoc basis, data are insufficient and imprecise. Only in recent years has the government become aware of the excessive groundwater developments, and their side effects, and initiated a study to improve the groundwater monitoring system in Taiwan (TPWCB, 1989). The first year of the project is aimed at the Pingtung Plain. The objective of the study is to improve the quality of basic hydrogeological information for an efficient management of the groundwater resources. This project is expected to offer sufficient information for evaluation of the groundwater resources in an economic and ecological context, associated with the development and allocation of groundwater supplies of known quality to compete water demands.

Before completion of this project, uncertainty in the hydrogeological parameters will be dealt with through a parameter uncertainty model. To optimize the groundwater level monitoring network for the plain the Kalman filtering technique, kriging and time series analysis should be further applied to extend an existing primary sampling programme (TPWCB, 1989).

1.2 Situation in the Pingtung Plain

The aquaculture industry in Taiwan began in the 1960s. During that time, in the coastal area of Pingtung Plain (such as the Linbien and the Chaton areas, refer to Fig. 2), people raised eels by mixing seawater with freshwater pumped from boreholes. Because of a significantly higher profit from eel culture compared with other agricultural products, the aquaculture industry began to prosper in this area. After intensifying the eel culture industry, the adjacent agricultural land became salty because of the infiltration of brackish water from fish ponds.

Because of the negative effect of this salinisation on agriculture and the higher profit from fish farming, more farmers were stimulated to turn their land into fish ponds. This situation led to over-pumping of the groundwater reservoir and initiated lowering of the hydraulic head in the aquifers that originally showed artesian characteristics. This in turn caused subsurface seawater intrusion along the coast. Another side effect was the reduction in well yields for licensed owners, leading to disputes and legal cases between the fish farmers and other groundwater users. In fact, many of the fish farmers did not possess a license from the groundwater authority for groundwater abstraction.

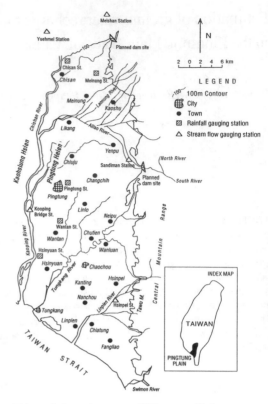

Figure 2 Location map of Pingtung Plain

The problems to be addressed can be summarized as follows: (1) To determine the discharge of the rivers, the capacity of the irrigation system and the amount of groundwater abstraction. (2) To define the optimal conjunctive use of surface water and groundwater resources, such that the operation will not produce an unacceptable increase in salt water intrusion, water level decline, land subsidence, and a deterioration in water quality.

The present paper focuses on application of time series analysis for the estimation of groundwater recharge rate with respect to specific yield was thus executed. Six observation wells (No. 9100, 9150, 9230, 9160, 980, and 9190 from the present monitoring network) were selected for analysis (Fig. 3). The ultimate aim of this study is to attempt to solve the uncertainties concerning hydrological stresses(i.e. specific yield) on the estimation of groundwater recharge rate at the unconfined aquifer zone of the Pingtung Plain. Thus, this results provide the further research on groundwater resources evaluation and management.

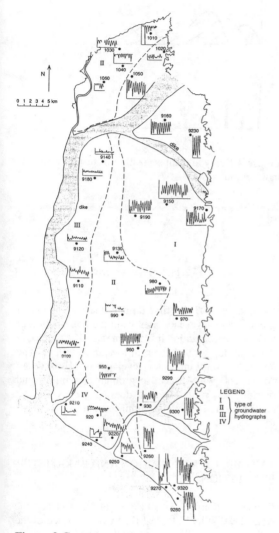

Figure 3 Groundwater hydrographs

2 PRINCIPLES OF THE INPUT-OUTPUT MODEL

Considering a single input-output, linear, time-invariant, lumped parameter system, the linear differential equation can be written as follows: (Box and Jenkins, 1976; Pandit and Wu, 1983)

$$\alpha_n \frac{d^n Y(t)}{d^n(t)} + \alpha_{n-1} \frac{d^{n-1} Y(t)}{dt^{n-1}} + \dots + \alpha_0 Y(t) = b_m \frac{d^m X(t)}{dt^m}$$
$$+ b_{m-1} \frac{d^{m-1} X(t)}{dt^{m-1}} + \dots + b_0 X(t)$$

$$(1)$$

where α_i (i=0, 1, 2, ...n) and b_j (j= 0, 1,2, ...m) are constant differential coefficients; $Y(t)$ and $X(t)$ are output and input respectively. The solution to equation (1), a so-called continuous Autoregressive Moving Average model, can be expressed as following via Laplace transformation

$$Y(t) = \int_{-\infty}^{\infty} X(t-\tau) H(t) d\tau \qquad (2)$$

where $Y(t)$ is the output, $X(t-\tau)$ is the input and $H(\tau)$ is the Green's function, also called the impulse response function, and $t-\tau$ is the time.

It is known from physical reality that any mass must be distributed over a non-zero volume in space. Equation (2) should thus be written as

$$y(t) = \int_0^{\infty} x(t-\tau) h(t) d\tau \qquad (3)$$

For discrete series and setting the time interval $\Delta t = 1$, equation (3) may be re-written in the form

$$y(t) = \sum_{\tau=l}^{n} x(t-\tau) h(\tau) \qquad n \geq t \qquad (4)$$

Given a delay in impulse response, selected terms are removed from the discrete equation i.e., if the delay is infinite, the last unavailable terms n - l should be neglected; l represents the output lag time with respect to input.

When the initial state is zero, i.e., there is either no water in the aquifer before rainfall or no release of stored water occurs, the so-called zero-state response results. This can be expressed as

$$y(t) = \sum_{\tau=l}^{k} x(t-\tau) h(\tau) \qquad 0 \leq k \leq t \qquad (5)$$

Alternatively, when the initial state is non-zero, which means that water can be released and is called the non-zero state response, this equation can be given as

$$y(t)=\sum_{\tau=l}^{k}x(t-\tau)h(\tau)+\sum_{\tau=k+1}^{n}x(t-\tau)h(\tau)\qquad 0\leq k\leq t,\ n\geq t$$

$$(6)$$

If the time lag is less than a time step, i.e., $\tau=0$ and under the non-zero state, response is superimposed on the antecedent time t-i; the relationship between input, output and state can be approximately expressed as the following

$$y_0 \leftarrow \boxed{e_0} \leftarrow x_0$$
$$y_1 \leftarrow \boxed{e_1} \leftarrow x_1$$
$$y_2 \leftarrow \boxed{e_2} \leftarrow x_2$$

where e_i denotes state of time i. From the state system, the above responses y_i can be expressed by the following

$$y_0 = e_0 + T_0X_0;$$
$$y_1 = (e_0 + T_0X_0) + T_1X_1;$$
$$(e_0 + T_0X_0) = e_1$$
$$y_2 = [(e_0 + T_0X_0) + T_1X_1] + T_2X_2;$$
$$(e_0 + T_0X_0) = e_2$$

where T_i is a translation operator.

Hence, in equation (6), the last term describes the initial state at time t which is superimposed from t-i, where i > 0. Fig. 4 shows that the water level hydrograph shows a different decay to that from the precipitation which contributes to the system and is determined by the hydrogeological characteristics at time t.

3 THE TRANSFER FUNCTION MODEL

From a systems point of view, natural groundwater level fluctuations are the response of the groundwater system to recharge. The cause-effect relationship between groundwater level and precipitation can thus be modelled using a simple transfer function of form

$$y(t)=\sum_{\tau=l}^{k}p(t-\tau)h(\tau)+\sum_{\tau=k+1}^{n}p(t-\tau)h(\tau)\qquad(7)$$

Equation (7) expresses the composition of state for previous groundwater levels and increments of response for time t-1 to t.

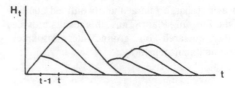
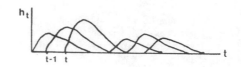
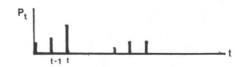

Figure 4 Delay and superposition of response for an input-output system

$$y(t)=\sum_{\tau=l}^{k}p(t-\tau)h(\tau)+\phi\cdot H(t-1)\qquad(8)$$

In equation (8), the first term is the summation of groundwater level increases from the contribution of k months precipitation infiltration rate. The term does not consider groundwater inflow and outflow. The last term is groundwater level decay (H is decay coefficient) with respect to groundwater inflow and outflow for time t-1 and does not consider the contribution of precipitation infiltration from time t-1 to t.

4 MODEL IDENTIFICATION AND PARAMETER ESTIMATION

From equation (8), it is easy to understand that the water level at hydrological time t is composed of the precipitation infiltration from the years t, t-1, t-2, ...etc, i.e., it is composed of the weighted average of precipitation infiltration from time t and earlier. The recharge at time t will contribute partly to the water level at time t and the rest to the subsequent times t+1, t+2,.... etc. The recharge at time t can therefore be estimated from the water level hydrograph through equation (8), which is applied to

488

the unconfined aquifer including O.W. Nos.9160, 9230, 9150 and 9160. This means that the water table at time t is composed of t, t-1, t-2,.... water level contributions and infiltration from the unsaturated zone. If the water level in the aquifer had been stable, these remain constant in the aquifer for a certain lag time past the input pulse. The lag time is a measure of the time required for the first water to reach the well recorder via the unsaturated zone. Therefore, by summing the monthly precipitation times a characteristic factor (or decay coefficient), ϕ $\times$ H(t-1), one can derive the water level resulting from previous contributions and current precipitation, viz,

$$\Delta H = \sum_{i=1}^{12} \Delta h_i = \sum_{i=1}^{12} \sum_{\tau=l}^{k} P(t_i - \tau) h(\tau) \qquad (9)$$

$$R = P \times \alpha \times A \times 1000 \qquad (10)$$

where in equation (10), R is the recharge rate (m^3/year), P is the annual rainfall (mm), α is the infiltration coefficient and A is the contributing area (km^2). By neglecting the discharge q during the time step, α becomes

$$\alpha = \frac{\Delta H \times \mu}{P} \qquad (11)$$

The values for α are thus subject to the accuracy of determining μ (specific yield).

5 RESULTS AND ANALYSIS

The following results were derived from the time series analysis:

O.W.9100: Considering the previous water level H_{t-1} and the contribution from the average infiltration coefficient to be $\alpha = 2.127 \mu$, the average fitting precision s (standard deviation) = 0.42 and r (correlation coefficient) = 0.86. This well reflects an infiltration lag time that does not exceed 3 months.

O.W.9150: This well was not significantly effected by river infiltration though lateral inflow has had an influence on the water level fluctuations. Average infiltration coefficient $\alpha = 3.225 \mu$, and infiltration lag time is 2 months. If the previous water level, lateral inflow and local precipitation infiltration are all considered, then the average infiltration coefficient should be 11.2815μ, with a lag time of 8 months.

O.W.9230: Fitting precision was the worst for the present analysis s is in the range of 3 to 5 m which has the correlation coefficient r in the range of 0.86 to 0.91. This may be explained by the fact that the annual water level fluctuates about 20 meters because of the highly permeable alluvial area. If the monthly water level records represent the instant water level, then this aspect is of lesser consequence. Sampling frequency in space and time may need to be reconsidered.

O.W.9160: The water level of the well, which in a highly permeable unconfined aquifer and at the same location as O.W. 9230, shows infiltration from the river. Plus the previous water level contribution, $\alpha = 5.68 \mu$. The water level fluctuation is around 12 m, s is in the range of 1.28 to 1.43 m and r = 0.95; lag time for precipitation infiltration is 2 months.

O.W.980: considering previous water level contributions and local precipitation infiltration, the average infiltration coefficient is 2.454μ and the lag time is 4 months.

O.W.9190: river infiltration does not effect the water level but lateral inflow does, as does local precipitation infiltration. The average infiltration coefficient is 12.90μ and the lag time is 7 months.

The above results are considered at this stage to represent the hydrogeological characteristics for each subregion and the related coefficients are adopted for initial model calibration.

6 RECOMMANDATIONS

Application of time series analysis to estimate the infiltration coefficient with respect to specific yield has been successfully achieved and resulted in the estimated values. Although only a simple linear system and lumped parameter model, this programme can predict aquifer water levels for known specific yield values. Specific yield is therefore one of the important components of groundwater recharge which is needed to advance further study on the plain. Not only is it an important parameter in water resources evaluation, but also its value effects the accuracy of model calibration and prediction. In addition, time-variant, non-linear, distributed parameter models of time series analysis should be developed and applied to this area in future research.

7 ACKNOWLEDGEMENTS

The financial support of the study was received from the National Science Council (NSC) and National

Pingtung Polytechnic Institute (NPPI), Taiwan, ROC; their support is gratefully acknowledged.

I would like to thank Prof. Dr. Ian Simmers, Prof. Dr. J.J. De Vries and Dr. Y. Zhou who gave fruitful discussions and critical comments. The visiting research hydrogeologists from China University of Geosciences for their assistance with several of the more tedious study aspects was also appreciated.

REFERENCES

Box, G.E.P. and Jenkins, G.M.. 1976. *Time Series Analysis: forecasting and control*, Holden-Day, Oakland, CA, USA, p575.

National Pingtung Polytechnic Institute (NPPI). 1990. 'Evaluation of Water Supply Facilities and Their Effects on Groundwater Condition', Technical report. (in Chinese)

Pandit, S.M. and Wu, S.M.. 1983. *Time Series and System Analysis with Applications*, John Wiley and Sons, New York, p586.

Taiwan Provincial Groundwater Development Bureau (TPGDB). 1961. 'Investigation Report on Groundwater Resources of the Pingtung Plain'. (in Chinese)

Taiwan Provincial Water Conservancy Bureau (TPWCB). 1989. 'Study on the Improvement of the Groundwater Monitoring System in Taiwan: Pingtung Plain'. (in Chinese)

Ting, C.S.. 1992. 'Evaluation of the Dispute among the Water Resource Users in Hsinyuan District of Pingtung Plain, Taiwan', *European Geophysical Society XVII General Assembly*, Annales Geophysicae, part II, p.C321.

Ting, C.S.. 1992. 'Application of a Groundwater Model in the Dispute among Water Users in the Pingtung Coastal Plain, Taiwan', *Proceedings of International workshop on groundwater and environment*, Beijing, China, pp 332-345.

Ting, C.S.. 1993. 'Development of an Optimum Model for Groundwater Resources Management: A Case Study for Pingtung Plain', *European Geophysical Society XVIII General Assembly*, Annales Geophysicae, part II, p.C245.

Ting, C.S.. 1993. *Development of an Optimum Model for Groundwater Resources Management: A Case Study for Pingtung Plain*' Free university, Amsterdam. p87.

Tsao, M.C.. 1991. 'A Study of the Local Effective Groundwater Resources Usage and Pollution Prevention', M.Sc thesis, *National Chung Hsing University*, p128 (in Chinese).

Water Resources Planning Commission (WRPC). Ministry of Economic Affairs. 1984. 'Water Resources on Taiwan'.

Willis, R. and Newman,B.A.. 1977. 'Management Model for Groundwater Development', *Journal of Water Resources Planning and Management Division*, ASCE, Vol. 103, No. WR1, pp 159-171.

Willis, R.. 1984. 'A Unified Approach to Regional Groundwater Management', *Water Resources Monograph Series 9*, American Geophysical Union Washington, D.C. pp 392-407.

Willis, R., Finney, B.A.. 1988. 'Planning Model for Optimal Control of Saltwater Intrusion', *Journal of Water Resources Planning and Management Division*, ASCE, Vol. 114, No. 2, pp 163-178.

Stochastic Hydraulics '96, Tickle, Goulter, Xu, Wasimi & Bouchart (eds) © 1996 Balkema, Rotterdam. ISBN 90 5410 817 7

Ergodic limit for spatial moments in nonuniform groundwater flow

I. Butera & M.G. Tanda
D.I.I.A.R. Politecnico di Milano, Italy

ABSTRACT: The results of the stochastic theory of transport phenomena can be applied to a single realization only if it evolves under the ergodic hypothesis. In literature the minimum size of a plume necessary to reach the ergodicity for uniform in the average groundwater flows through heterogeneous porous formations has been investigated analytically and numerically. Less attention has been addressed to nonuniform in the average flows.

In this paper an attempt to evaluate, by Monte Carlo simulations, the minimum size of a plume necessary to reach the ergodic limit for spatial moments is carried out in the presence of uniform recharge (both positive and negative). The results, obtained for different degree of flow nonuniformity, are discussed and compared with the case of uniform in the average flows.

1. INTRODUCTION

In the stochastic approach groundwater flow and transport phenomena in heterogeneous porous formations are studied through the analysis of their ensemble behavior. All the realizations of the ensemble are statistically equivalent.

By a practical point of view, however, the question arises if the real case in study is going to be "well described" by the ensemble behavior; only if it evolves in ergodic manner this is true, otherwise the difference between the ensemble behavior and the single realization -the real case- could be significant and too high to feasibility purposes.

In the analysis of transport phenomena at not large travel time, the input volume size is determinant to achieve the ergodic limit; this has been investigated analytically, by first order approximation (Dagan, 1990; 1991), and numerically through Monte Carlo simulations (Guadagnini et al., 1992; Fiorotto and Salandin, 1993), for uniform in the average flow fields. The conclusion of these studies is that for a thin initial solute body transversal the flow, its spatial extension should be at least 25λ (where λ is the trasmissivity correlation length).

The target of the present work is to investigate the minimum input size necessary to ensure ergodicity when the mean flow is not uniform because of a evenly distributed recharge, as it occurs for rainfall or for leakage in multilayer aquifers.

The analysis is inferred on the basis of the numerical results obtained by Monte Carlo simulations. The 2D domain has been considered at regional scale, adopting the shallow flow approximations. The findings are related to some features of the velocity field pointed out by the first order analysis (Rubin and Bellin, 1994).

2. STATEMENT OF THE PROBLEM

Dealing with heterogeneous aquifer, the main difficulty is the impossibility to get a detailed knowledge of the conductivity filed $K(\mathbf{x})$ because of its erratic nature; in the stochastic approach it's usual to model $K(\mathbf{x})$ as a spatial random function of prescribed characteristics, computed by the statistical inference from the data available (e.g. Delhomme, 1979). Through the analysis of fields tests, both at regional and local scale, it came out, and it is by now accepted (Dagan, 1989), that trasmissivity -T- in the first case and conductivity in the second one, are characterized by a lognormal p.d.f., hence completely defined by the first and the second statistical moments (Freeze, 1975; Delhomme, 1979). In the stochastic approach, the real heterogeneous porous formation to be studied is considered as one in the ensemble of realizations characterized by the same statistical structure of K (or T).

The randomness of the trasmissivity leads to the definition of the velocity as a spatial random function as well, and so trajectories are subjected to uncertainty.

Using a Lagrangian approach the solute transport phenomena is studied following the particles in time and analyzing their spatial distribution; well known are the following formula (Dagan, 1990):

$$X(t,a) = a + \int_0^t V(X) \, dt$$

$$X = a \quad \text{at} \quad t = 0$$ (1)

for the particle trajectory $X(t,a)$ where $V(X)$ is the Eulerian velocity field;

$$C(x,t) = \int_{A_0} C_0(a) \delta(x - X) \, da$$

$$C = C_0 \quad t = 0 \quad x \in A_0$$ (2)

for the concentration value, and neglecting pore scale diffusion;

$$M = nC_0 A_0$$ (3)

$$R = \frac{1}{M} \int nC \, dx = \frac{1}{A_0} \int_{A_0} X(t,a) \, da$$ (4)

$$S_{ij} = \frac{1}{M} \int n(x_i - R_i)(x_j - R_j) C \, dx =$$

$$\frac{1}{A_0} \int_{A_0} [X_i(t,a) - R_i][X_j(t,a) - R_j] \, da \quad i,j = 1,2,3$$ (5)

for the spatial moments; eq. (3), (4) and (5) defines respectively the total mass M (n is the porosity), the centroid coordinates R and the second spatial moments S_{ij}; the ensemble mean values and variances $<R>$, R_{ij} ; $<S_{ij}>$, $\text{var}(S_{ij})$ depend on the trajectories mean value $<X>$ and covariance $X_{ij}(t,b)$, where b is the lag vector at time t=0.

To apply such results to the movement of a non reactive pollutant plume in a single realization, the question is whether the spatial moments satisfy the ergodic requirement: i.e. all the states of the system are represented in the given realization. If the spatial extent of the plume is large compared to heterogeneity scale, this is true, so that space and ensemble averages can be interchanged; if not, for normal distributions, necessary and sufficient condition for an ergodic behavior is that the coefficient of variation tends to zero (Dagan, 1990). It is well known (Dagan, 1990) that two factors residence time t and input volume V_0 contribute to the achievement of ergodic limit for centroid and spatial moments.

Our target here is to investigate, at small travel time, the influence of V_0 when the flow is not uniform in the mean. It will be paid attention to the behavior of the coefficient of variation of the first and the second spatial moments:

$$CvR_i = \frac{\sqrt{R_{ii}}}{<R_i>}; \quad CvS_{ij} = \frac{\sqrt{\text{var} S_{ij}}}{<S_{ij}>}$$ (6)

for a thin initial solute body transversal the flow direction.

3. THE NUMERICAL EXPERIMENTS

The investigations on the ergodic limit are performed by means of a Monte Carlo process. The procedure has been applied to a 2D heterogeneous trasmissivity field, generated by the FFT algorithm (Guthjahr, 1989). The log-trasmissivity field is characterized by weak stationarity: $Y=\ln T$; $<Y>=\text{cost.}$; $C_Y(x_1,x_2) = C_Y(r)$; $r = x_1 - x_2$; an isotropic exponential correlation function has been adopted:

$$C_Y(r) = \sigma_Y^2 e^{-r/\lambda}; \quad r = |r|$$ (7)

with λ the correlation length.

The flow domain has rectangular shape with 42λ and 60λ sides; it has been discretized in squared blocks of $1/6\lambda$ side; to each of these a value of T is attributed and the flow equation is solved determinig the head value at each node of the same grid. The finite element method with Galerkin formulation and squared elements has been used for the solution. The boundary conditions are deterministic: they consists of fixed head values $H = H_1$ at $x=0\lambda$, $H = H_2$ at $x=60\lambda$ and no flow boundaries at the longer side (Fig. 1). A constant value of recharge is imposed at the nodes (but those of constant head) so that the coming average flow is unidirectional, parallel to the longer side with non constant hydraulic gradient: $J=J(x)$.

The transport phenomena has been simulated injecting, at t=0 in $x_0=5\lambda$, 7500 particles aligned from $y=6\lambda$ to $y=36\lambda$ (250 particles/λ) and following each of them by a particle tracking procedure.

Collecting the particles in different way, plumes of variable size l are studied.

Several replicates (1500) of the process were carried out, so that the spatial moments of the plume in the numerical stuff can be computed with the following relationships:

492

$$< R_i > = \frac{1}{NMC} \sum_{n=1}^{NMC} R_{i,n}; \quad R_{i,n} = \frac{1}{NP} \sum_{p=1}^{NP} X_i \qquad (8)$$

$$R_{ij} = < \left[R_i - < R_i > \right] \left[R_j - < R_j > \right] > =$$

$$= \frac{1}{NMC} \sum_{n=1}^{NMC} R_{ij,n} \qquad (9)$$

$$R_{ij,n} = \left[R_{i,n} - < R_i > \right] \left[R_{j,n} - < R_j > \right]$$

$$< S_{ij} > = \frac{1}{NMC} \sum_{n=1}^{NMC} S_{ij,n}$$

$$S_{ij,n} = \frac{1}{NP} \sum_{p=1}^{NP} \left[X_{i,p} - R_{i,n} \right] \left[X_{j,p} - R_{j,n} \right] \qquad (10)$$

$$\mathrm{var}\, S_{ij} = \frac{1}{NMC} \sum_{n=1}^{NMC} \left[S_{ij,n} - < S_{ij} > \right]^2 \qquad (11)$$

where NMC is the total number of replicates and NP is the number of plume particles.

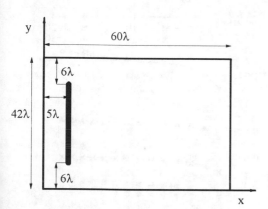

Fig. 1: Flow domain of the numerical experiments

The numerical experiments were made for two values of σ_Y^2 and four recharge intensity. These ones can be described by the parameter $\beta = Q\lambda/T_G J(x_0)$ (Dagan and Rubin, 1987), where Q is the intensity of recharge, λ the transmissivity correlation length, T_G the mean value of T and J the hydraulic gradient. 1500 replicates have been carried out for each case of Table 1

Table 1.

case	Parameters pairs considered	
1	$\sigma_Y^2 = 0.16$	$\beta = 0.067$
2	$\sigma_Y^2 = 0.16$	$\beta = -0.015$
3	$\sigma_Y^2 = 0.16$	$\beta = 0.018$
4	$\sigma_Y^2 = 0.16$	$\beta = -0.009$
5	$\sigma_Y^2 = 0.8$	$\beta = 0.018$
6	$\sigma_Y^2 = 0.8$	$\beta = -0.009$

4. RESULTS

The analysis of the spatial moments values showed in Fig. 2, points out the different trends for positive and negative recharge; in the first case the main flow is accelerated and the longitudinal spread S_{11} of the plume around its centroid position is higher than for a uniform flow, while for a negative recharge it is lower; this is a consequence of the non stationarity of the velocity field induced by the recharge (Rubin and Bellin, 1994). It is usual to plot these values with respect of the dimensionless time $\tau = t V(x_0)/\lambda$, but since $V(x_0)$ depends on β, in order to compare and analyze transport process for different degree of nonuniformity, it is more useful to plot data versus the dimensionless travel length $< R_1 > /\lambda$.

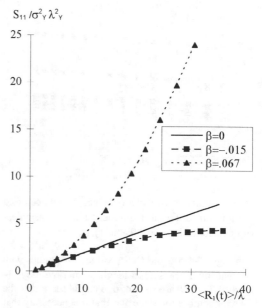

Fig. 2: Longitudinal spread for different recharge intensity $l = 1\lambda$.

493

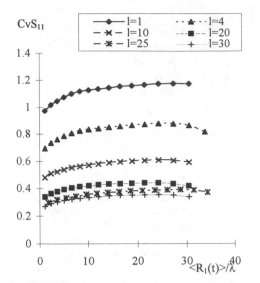

Fig. 3: Coefficient of variation CvS_{11} versus centroid location for different plume extents dimensionless with respect of λ (case 1-Table 1).

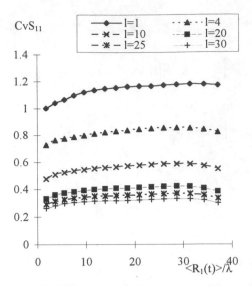

Fig. 5 Coefficient of variation CvS_{11} versus centroid location for different plume extents dimensionless with respect of λ (case 3-Table 1).

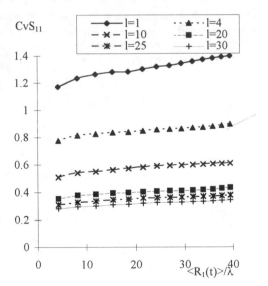

Fig. 4 Coefficient of variation CvS_{11} versus centroid location for different plume extents dimensionless with respect of λ (case 2-Table 1).

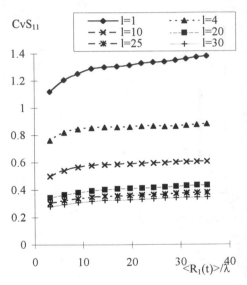

Fig. 6 Coefficient of variation CvS_{11} versus centroid location for different plume extents dimensionless with respect of λ (case 4-Table 1).

In Figs. 3-4-5-6 the CvS_{11} values for plumes with different initial transversal length l and the cases 1-2-3-4 of Table 1 are shown; one can observe that the behavior is quite the same and that it is necessary that l exceeds $25 \div 30\lambda$ to achieve the ergodic limit (if we accept $CvS_{11}=0.2$ as threshold value, otherwise even bigger values of l are necessary); it is worthwise to point out that this value of l ensures ergodicity also for the transversal spread S_{22} (Fig. 7). Change in the trasmissivity variance doesn't affect substantially the matter (Fig.8).

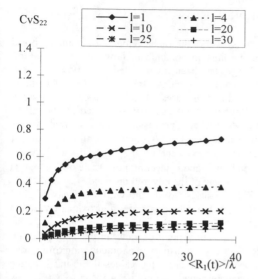

Fig. 7: Coefficient of variation CvS_{22} versus centroid location for different plume extents dimensionless with respect of λ (case 1-Table 1)

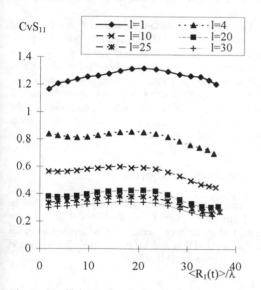

Fig. 8: Coefficient of variation CvS_{11} versus centroid location for different plume extents dimensionless with respect of λ (case 5-Table 1).

The condition of ergodic behavior for the spread process is more strict than for the centroid location. From Fig. 9, where the values of CvR_1 are plotted for the same case of Fig. 3, it appears that almost all the plume centroids after a few traveled distance have a coefficient of variation below 0.2.

The non-dependence of minimum 1 on β, coming out from the analysis of numerical data, agrees with the following considerations developed on the basis of the velocity field features outlined by Rubin and Bellin (1994), with a first order analysis under the hypothesis of the linear theory ($\sigma_Y^2 < 1$): in the presence of a uniform recharge the flow field has been shown to have stationary velocity correlation function (Rubin and Bellin, 1994), just at a few log-trasmissivity integral scales far from the boundary; this implies that the change in the mean velocity due to the recharge has no consequence on the velocity field integral scales I_v (Rubin and Bellin, 1994). These integral scales I_v remain function of the trasmissivity field, the source of randomness, like in the uniform case. The solute spread hence, neglecting pore scale diffusion, depends on the velocity fields and the achievement of ergodic limit depends on I_v. Since I_v isn't function of β the same had to be expected for minimum 1.

Furthermore, the CvS_{11} traveled distance dependence is shown to be rather weak in the spatial domain considered. This one can be regarded quite large compared to extensions dealing with practical applications at regional scale, being λ of order of hundreds or thousands meters. After a few correlation length a costant value of CvS_{11} is reached; this fact suggests to synthetize the data, plotting CvS_{11} with respect of 1 in Fig. 10, referring to its constant value (i.c. at $<R_1> /\lambda = 18$).

One can observe that for initial extension greater than 1λ (for smaller plume may be necessary more replicates of the Monte Carlo process) the data of caese 1-2-3-4-5 collapse in almost the same way, with a trend well represented by the dotted curve of Fig. 10. The data of case 6, relative to the higher value of σ_Y^2, fall on upper position in the figure. This fact can be explained with a not sufficient number of replicates of the Monte Carlo process. In fact for the bigger plume extensions, the data seem to become closer to those of the cases 1÷5.

This diagram, useful and handy, allows one to evaluate the degree of non-ergodicity of a transport phenomena as function merely of the transversal input size: whatever is β value, including hence the uniform in the average flow ($\beta=0$) too. It applies to a width range of traveled distance just greater than a few λ.

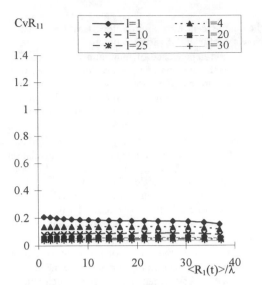

CvR_{11}

Legend:
- $\ell=1$
- $\ell=4$
- $\ell=10$
- $\ell=20$
- $\ell=25$
- $\ell=30$

Fig. 9: Coefficient of variation CvR_{11} versus centroid location for different plume extents dimensionless with respect of λ (case 1-Table 1).

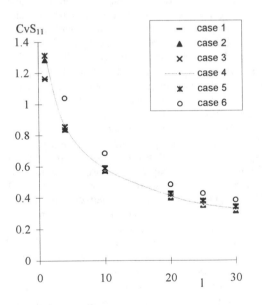

CvS_{11}

Legend:
- — case 1
- ▲ case 2
- × case 3
- case 4
- ✳ case 5
- ○ case 6

Fig. 10: Coefficient of variation CvS_{11} versus different plume extents dimensionless with respesct of λ, injected in $x_0 = 5\lambda$

5. CONCLUSIONS

In this work the achievement of plume spread ergodic limit in nonuniform flow, due to a uniform recharge, has been investigated. The flow has been considered at regional scale and the pollutant is released in a thin area whose extension is transversal the flow direction. The analysis, based on the results of a Monte Carlo process, has been carried out for different degree β of flow nonuniformity; it came out that the achievement of the ergodic limit doesn't depend on β, and neglecting pore scale diffusion, it is function only of the covered distance.

This result is consistent with the velocity field features outlined by first order approximation (Rubin and Bellin, 1994).

By a practical point of view, in planning pollutant recovery or in transport phenomena monitoring, one can refer to the ensemble values if the initial plume extent determines, in Fig. 10, a value of CvS_{11} low enough. According to β, it will lead to different planning spatial extension: the presence of a negative recharge lowers the spread, while the positive one increases it.

These results of certain significance don't exhaust the interest of the problem; remaining into ergodic hypotesis a great importance holds to assume a spatial random recharge, to give reason of variable soil adsorption parameters; nevertheless it would be interesting and useful to study the recharge impact at local scale, considering the vertical component of the flow, and time dependence; more attention, then, should be paid to pore scale diffusion.

6. ACKNOWLEDGMENTS

We are grateful for the financial support provided by the National Research Council of Italy (CNR), by the Group for defence against the hydrogeological disasters (GNDCI) and by the Italian University and Research Ministry (MURST).

REFERENCES

Dagan, G. Solute transport in heterogeneous porous formations. *J. Fluid Mech.* 145, 151-177. 1984.

Dagan, G. & Y. Rubin. Stochastic identification of trasmissivity and effective recharge in steady groundwater flow. *W.R.R.* 23, 1185-1192. 1987.

Dagan, G. *Flow and Transport in Porous Formations.* Spinger. 1989.

Dagan , G. Transport in heterogeneous formations: spatial moments, ergodicity and effective dispersion. *W.R.R.* 26, 1281-1290. 1990.

Dagan, G. Dispersion of a passive solute in non-ergodic transport by steady velocity fields in heterogeneous formations. *J. Fluid Mech.* 233, 197-210. 1991.

Delhomme, J.P. Spatial variability and uncertainty in groundwater flow parameters: a geostatistical approach. *W.R.R.* 15, 269-280. 1979.

Freeze. R. A. A stochastic-conceptual analysis of one dimensional groundwater flow in nonuniform homogeneous media. *W.R.R.* 11, 725-741. 1975.

Guadagnini, A., U. Maione, A. Martinoli & M.G. Tanda. A single realization approach in stochastic modelling of contaminant transport through heterogeneous porous media. Atti del *VI IAHR Int. Symposium on Stochastic Hydraulics*, Taipei, 1992.

Gutjahar, A. L. Fast Fourier transform for random field generation, project report, contract 4-RS of Los Alamos National Laboratory, N. M. Inst. of Min. and Technol., Socorro, 1989.

Rubin, Y. Prediction of tracer plume migration in disordered porous media by the method of conditional probabilities. *W.R.R.* 27, 1291-1308, 1991.

Rubin, Y. & A. Bellin. The effects of recharge on flow nonuniformity and macrodispersion. *W.R.R.* 30, 939-948, 1994.

Salandin, P. & V. Fiorotto. Numerical simulations of non-ergodic transport in natural formations. *XXV Congresso IAHR*, Tokyo, 1993

Stochastic Hydraulics'96, Tickle, Goulter, Xu, Wasimi & Bouchart (eds) © 1996 Balkema, Rotterdam. ISBN 90 5410 817 7

Uncertainty and risk analyses for FEMA alluvial fan method

B. Zhao & L. W. Mays

Civil and Environmental Engineering Department, Arizona State University, Tempe, Ariz., USA

ABSTRACT: Alluvial fans along mountain bases pose interesting problems for highway crossings design and flood insurance studies. In this paper, Rosenblueth's point-estimate is applied to the FEMA (Federal Emergency Management Agency) alluvial fan method to compute the mean and standard deviation for the 100-year discharge at any point on the fan and the mean and standard deviation for the fan arc width. The mean and standard deviation for the 100-year discharge are used to obtain the risk that the 100-year discharge will exceed the discharge capacity of hydraulic structures. The mean and standard deviation for the fan arc width are used to estimate the risk that a given location on the fan is within the hazard flood zone. The HEC-1 rainfall-runoff computer model is used to compute inputs to the FEMA method. The proposed methodology is applied to an alluvial fan in north Scottsdale, Arizona.

1 INTRODUCTION

An alluvial fan is defined as a geomorphic feature characterized by a cone or fan-shaped deposit of boulders, gravel, and fine sediments. The flood hazard areas on an alluvial fan are identified as Zone AO which is defined as the flood insurance rate zone that corresponds to the areas of 100-year shallow flooding (usually sheet flow on sloping terrains) where average depths are between 1 ft (0.3048 m) and 3 ft (0.9144 m) (FEMA, 1985, 1992). The 100-year flood event on an alluvial fan is defined as the flood having a 1 percent chance of being exceeded (at the point at which the definition is being applied) in any given year (FEMA, 1992). If H denotes the event of the point in question being flooded, then the 100-year flood discharge at that point is the q_{100} defined by (FEMA, 1992)

$$0.01 = \int_{q_{100}}^{\infty} P(H|Q=q)\, f_Q(q)\, dq \qquad (1)$$

in which $P(H|Q=q)$ is the probability of the point being flooded, given that a flood with a magnitude of q cfs is realized at the fan apex; and $f_Q(q)$ is the probability density function (PDF) of the discharge Q occurring at the fan apex which follows a log-Pearson Type III or log-normal distribution.

In order to solve Eq. (1), the following three assumptions are typically made (Dawdy, 1979; FEMA, 1992): (1) the probability density function,

$f_Q(q)$, is log-Pearson Type III; (2) the conditional probability, $P(H|Q=q)$, on any contour is equal to the width of the channel carrying the discharge divided by the width of the area subject to flooding measured along the contour; and (3) the width of the "channel" followed by the flood is proportional to the four-tenths power of the flood discharge.

The FEMA alluvial fan methodology can be used to determine the 100-year discharge at any point on an alluvial fan if the 2-year, 10-year, and 100-year discharges at the fan apex are known and the fan arc width associated with the point is known. The FEMA alluvial fan methodology can also be used to estimate the fan arc width to define the hazard flood zone if the 2-year, 10-year, and 100-year discharges at the fan apex are known and the boundaries in terms of total depth of zone AO are given.

Due to the assumptions and many uncertain variables associated with the FEMA alluvial fan procedures, the results from the FEMA alluvial fan methodology are subject to uncertainties. Therefore, uncertainty analysis may be advantageous. In this paper, an uncertainty analysis technique based upon Rosenblueth's point-estimate(Rosenblueth, 1975, 1981; Tung and Yen, 1993) is applied to the combined procedures of FEMA alluvial fan method and HEC-1 rainfall-runoff model to determine the mean and standard deviation for the 100-year discharge at any point on an alluvial fan and the mean and standard deviation for the fan arc width. The mean and standard deviation computed from uncertainty analysis for the 100-year discharge at the point on the fan are then used to compute the risk that the hydraulic structures on the fan fail. The mean and standard deviation computed from the uncertainty analysis for the fan arc width are used in the risk-based delineation of hazard flood zones by which one can estimate the risk that a given location of interest will be within the hazard flood zone.

2 FEMA ALLUVIAL FAN METHODOLOGY

2.1 Procedures for Estimating 100-year Discharge

(1) Use rainfall-runoff model to estimate the $Q_{0.01}$, $Q_{0.10}$, and $Q_{0.50}$ which are flood discharges at the fan apex associated with return periods of 100, 10, and 2 years, respectively.
(2) Compute the skew coefficient, standard deviation, and mean by

$$G=-2.50+3.12\frac{\log_{10}(Q_{0.01}/Q_{0.10})}{\log_{10}(Q_{0.10}/Q_{0.50})} \qquad (2)$$

$$S=\frac{\log_{10}(Q_{0.01}/Q_{0.50})}{K_{0.01}-K_{0.50}} \qquad (3)$$

$$\bar{y}=\log_{10}(Q_{0.50})-K_{0.50}S \qquad (4)$$

in which $K_{0.01}$, $K_{0.10}$, and $K_{0.50}$ are Pearson Type III deviates with skew coefficient G for exceedance probabilities of 0.01, 0.10, and 0.50, respectively.
(3) If G=0.0, then compute log-normal parameters and transformation constant by

$$S_Z=\sigma_Z=S \qquad (5)$$

$$\bar{Z}=\bar{y}+0.92\sigma_Z^2 \qquad (6)$$

$$G_Z=G=0.0 \qquad (7)$$

$$C=\exp(0.92\bar{y}+0.4232\sigma_Z^2) \qquad (8)$$

in which S_Z, $\bar{Z}$, and G_Z are respectively the standard deviation, mean, and skew coefficient of random variable Z; and C is the transformation constant (FEMA, 1985).
(4) If $G\neq 0.0$, then compute the transformation variables, log-Pearson Type III parameters, and transformation constant by

$$m=\bar{y}-\frac{2S}{G} \qquad (9)$$

$$\lambda=\frac{4}{G^2} \qquad (10)$$

$$\alpha=\frac{2}{GS} \qquad (11)$$

$$\bar{Z}=m+\frac{\lambda}{\alpha-0.92} \qquad (12)$$

$$S_Z^2=\frac{\lambda}{(\alpha-0.92)} \qquad (13)$$

$$G_Z=\frac{2}{\lambda^{0.5}} \qquad (14)$$

$$C=(\frac{\alpha}{\alpha-0.92})^\lambda \exp(0.92m) \qquad (15)$$

(5) Compute the exceedance probability of the discharge at a point of interest on an alluvial fan by

$$p=\frac{W}{950AC} \qquad (16)$$

in which W is the fan arc width which corresponds to the contour passing the point, A is avulsion coefficient, and C is from Eq. (8) or Eq. (15). The fan arc width (W) can be measured from the topographic map. The avulsion coefficient is defined as a measure of the "average avulsions per event". A factor of 1.5 is recommended in the absence of data (FEMA, 1985). Eq. (16) is derived from Eq. (1).
(6) Based upon the exceedance probability (p) from Eq. (16) and the skew coefficient value (G_Z) from Eq. (7) or Eq. (14), use Pearson Type III distribution table ("Guidelines" 1982) to find the deviate k_Z by interpolation.
Because Pearson Type III distribution table is not easy to use for computer programming, the following equation can be used to compute the deviate K_Z ("Guidelines" 1982)

$$K_Z=\frac{2}{G_Z}[[(z_{1-p}-\frac{G_Z}{6})\frac{G_Z}{6}+1.0]^3-1.0] \qquad (17)$$

in which z_{1-p} is the standard normal quantile for probability 1-p or $P(Z<z_{1-p})=1-p$.
If p<0.5, the quantile z_{1-p} can be computed by (Abramowitz and Stegun, 1972)

$$z_{1-p}=t-\frac{c_0+c_1t+c_2t^2}{1+d_1t+d_2t^2+d_3t^3} \qquad (18)$$

in which $c_0=2.515517$, $c_1=0.802853$, $c_2=0.010328$, $d_1=1.432788$, $d_2=0.189269$, $d_3=0.001308$, and

$$t=\sqrt{-2\ln(\bar{p})}, \text{ for } p<0.5 \qquad (19)$$

If p>0.5,

$$z_{1-p}=-(t-\frac{c_0+c_1t+c_2t^2}{1+d_1t+d_2t^2+d_3t^3}) \qquad (20)$$

in which the constants c's and d's are the same as defined earlier and

$$t=\sqrt{-2\ln(1-\bar{p})}, \text{ for } p>0.5 \qquad (21)$$

(7) Compute the 100-year discharge q_{100} at the point by

$$\log_{10}(q_{100})=\bar{Z}+K_ZS_Z \qquad (22)$$

in which K_z is from step (6) and $\bar{Z}$ and S_z are from step (3) or (4).

2.2 Procedures for Estimating Fan Arc Width

(1)-(4) Same as (1)-(4) in 2.1.
(5) Input total depth (H) or velocity (V) boundaries for zone AO.
(6) Compute the 100-year discharges at total depth H (or velocity V) zone boundaries by

$$q_{100} = 280 H^{2.5} \qquad (23)$$

or

$$q_{100} = 0.13 V^5 \qquad (24)$$

in which H is the total depth in ft due to pressure head and velocity head and V is velocity in ft/s (FEMA, 1985).
(6) Compute k_z by

$$k_z = \frac{\log_{10}(q_{100}) - \bar{Z}}{S_z} \qquad (25)$$

in which S_z and $\bar{Z}$ are from step (3) or (4).
(7) Based upon the obtained values for k_z and G_z, use Pearson Type III table ("Guidelines" 1982) to find the exceedance probability (p) corresponding to each of the 100-year discharges.

For the ease of computer programming, the following algebraic equations can be used to compute the exceedance probability (p). First, compute the standard normal quantile z_{1-p} by ("Guidelines" 1982)

$$z_{1-p} = \frac{G_z}{6} + \frac{6}{G_z} [(\frac{G_z k_z}{2} + 1.0)^{\frac{1}{3}} - 1.0] \qquad (26)$$

Then, compute the exceedance probability (p) by (Abramowitz and Stegun, 1972)

$$p = \frac{1.0}{\sqrt{2\pi}} \exp(-\frac{z_{1-p}^2}{2})(\sum_{i=1}^{5} b_i v^i) \quad if \ z_{1-p} \geq 0 \qquad (27)$$

$$p = 1.0 - \frac{1.0}{\sqrt{2\pi}} \exp(-\frac{z_{1-p}^2}{2})(\sum_{i=1}^{5} b_i v^i) \quad if \ z_{1-p} < 0 \ (28)$$

in which $b_0 = 0.2316419$, $b_1 = 0.31938153$, $b_2 = -0.356563782$, $b_3 = 1.781477937$, $b_4 = -1.821255978$, $b_5 = 1.330274429$, and $v = 1.0 + 1.0/(1.0 + b_0 |z_{1-p}|)$.
(8) Compute the fan arc width by

$$W = 950 A C p \qquad (29)$$

in which A is avulsion coefficient and C is from step (3) or (4).

3 UNCERTAINTY ANALYSIS BY ROSENBLUETH'S POINT-ESTIMATE METHOD

Methods for performing uncertainty analyses in hydrological and hydraulic applications are well summarized (Tung and Yen, 1993). They are classified into first order second moment estimation method, probabilistic point estimation methods, Monte-Carlo simulation, and integral transformation methods. Rosenblueth's point-estimation method (Rosenblueth, 1975 , 1981; Tung and Yen, 1993) is a probabilistic point estimation method. There are several advantages of using Rosenblueth's method because (1) it is more accurate than first order second moment method; (2) it is more efficient than Monte-Carlo simulation; and (3) it does not require probability distribution assumptions as do Monte-Carlo simulation and integral transformation.

Consider Y=g(**X**) where Y is a scalar and vector **X** = $(X_1, X_2, ..., X_N)$. Each X_i has two concentrated mass points $x_{i,+}$ and $x_{i,-}$ with probability masses $p_{i,+}$ and $p_{i,-}$, respectively. The concentrated points are computed by

$$x_{i,+} = \mu_{X_i} + z_{i,+}\sigma_{X_i}$$
$$x_{i,-} = \mu_{X_i} + z_{i,-}\sigma_{X_i} \qquad (30)$$

in which

$$z_{i,-} = \frac{\gamma_{X_i}}{2} - \sqrt{1 + \left(\frac{\gamma_{X_i}}{2}\right)^2} \qquad (31)$$

$$z_{i,+} = \gamma_{X_i} - z_{i,-} \qquad (32)$$

in which μ_{Xi}, σ_{Xi}, and γ_{Xi} are the mean, standard deviation and skew coefficient of X_i. The associated probability masses $p_{i,+}$ and $p_{i,-}$ are computed by

$$p_{i,-} = \frac{z_{i,+}}{z_{i,+} - z_{i,-}} \qquad (33)$$

$$p_{i,+} = 1.0 - p_{i,-} \qquad (34)$$

By using the sign indicator for sake of simplicity, X_i takes two mass points $x_{i,\delta i}$ with $p_{i,\delta i}$ where δ_i can be either "+" or "-". The probability $p_{i,\delta i}$ for the ith variable is determined as in Eqs. 33-34. Therefore, the random vector **X** can take 2^N possible values $X_{\delta 1, \delta 2, ..., \delta N}$, each of which has the probability mass $p_{\delta 1, \delta 2, ... \delta N}$. This probability mass can be computed by (Tung and Yen, 1993)

$$p(\delta_1, \delta_2, ..., \delta_N) = \prod_{i=1}^{N} p_{i,\delta_i} + \sum_{i=1}^{N-1}\left[\sum_{j=i+1}^{N} \delta_i \delta_j a_{ij}\right] \qquad (35)$$

in which

$$a_{ij} = \frac{\rho_{ij}/2^N}{\sqrt{\prod_{i=1}^{N}\left[1 + \left(\frac{\gamma_{X_i}}{2}\right)^2\right]}} \qquad (36)$$

in which ρ_{ij} is the correlation coefficient between X_i

and X_j. Then, the mth-moment about the origin for Y can be estimated by

$$E(Y^m) \approx \sum g^m(X_{\delta_1,\delta_2,...,\delta_N}) p_{\delta_1,\delta_2,...,\delta_N} \qquad (37)$$

in which the summation is taken over 2^N points (X's) and probability masses (p's). Therefore, mean, standard deviation, or higher moments for Y can be computed.

It may be pointed out that Rosenblueth's point estimation method is only one of the probabilistic point estimation methods. Other methods are Harr's point estimation method (Harr, 1989), a modified Harr's point estimation method (Chang et al., 1995), and Li's point estimation method (Li, 1992; Zoppou and Li, 1993). A detailed discussion on other methods applied to FEMA alluvial fan method is beyond the scope of this paper.

4 UNCERTAINTY ANALYSIS FOR FEMA ALLUVIAL FAN PROCEDURES

To apply Rosenblueth's point-estimate method, one needs to first define Y, g, and X. Y can be defined as the 100-year discharge (q_{100}) or the fan arc width (W). Consider g as the combined procedures of HEC-1 computer model and FEMA alluvial fan method.

The random vector X may consist of the 100-year, 10-year, and 2-year rainfall depths for the watershed upstream of the alluvial fan, random variables in the rainfall-runoff model, and random variables in the FEMA alluvial fan procedures. Mathematically, define the random vector $X=(P_{0.01}, P_{0.1}, P_{0.5}, X_1, X_2)$ in which $P_{0.01}$, $P_{0.1}$, and $P_{0.5}$ are the 100-, 10-, and 2-year rainfall depths, respectively; X_1 of dimension (k_1x1) is the random vector consisting of the random variables in the rainfall-runoff model; and X_2 of dimension (k_2x1) is the random vector consisting of the random variables in the FEMA alluvial fan procedures. Since the runoff computed by a rainfall-runoff model is usually rather sensitive to curve numbers and the curve number is a function of many uncertain variables such as antecedent moisture condition, soil type, land use and treatment, and the condition of ground surface, the curve number of the watershed or curve numbers of sub-basins can be considered as an element or elements in the random vector X_1. Since the assumptions for FEMA alluvial fan method may lead to model errors, a model error correction coefficient (Tung and Mays, 1980) is introduced into Eq. (16) and Eq. (29) as a random variable. Eqs. (16) and (29) become, respectively,

$$p=\lambda_p \frac{W}{950AC} \qquad (38)$$

$$W=\lambda_w 950ACp \qquad (39)$$

The avulsion coefficient and the model error for Eq. (38) can be considered as the elements of the random vector X_2.

For a given fan arc width on an alluvial fan, if the 100-year discharge (q_{100}) at a point on the alluvial fan is of interest, Y is defined as q_{100} or $Y=q_{100}=g(X)$. If the hazard flood zone is to be delineated, then the fan arc width (W) is to be computed. Then Y is defined as W or $Y=W=g(X)$. The following explains the procedures of implementing uncertainty analysis method in order to compute the statistics of the random variable Y.

As discussed in Rosenblueth's point-estimate method, if the dimension of X is (1xN), then 2^N points in N-dimensional space and their probabilities need to be generated based upon the statistics of the random vector X. It may be noted that each generated point is a vector of dimension (1xN). To generate 2^N points, one first needs to compute two values ($X_{i,+}$ and $X_{i,-}$) for each of the elements (X_i) in the random vector X by using Eq. (30). Since each X_i, i=1, 2,..., N, has two values, the random vector X can take 2^N possible points. The 2^N points can be easily found through combinatorial operations after the two values for each of X_i are computed. To compute the probability for each of the 2^N points, one first uses Eqs. (33)-(34) to compute the two probabilities for each X_i. Then, Eqs. (35)-(36) are used to generate 2^N probabilities for the 2^N points.

After the 2^N points and their probabilities are found, each of the 2^N points is substituted into the general function g(X) to obtain Y. Eq. (37) is then used to compute the mth-moment about the origin for Y. Finally, the mean and standard deviation for Y can be found by

$$\bar{Y}=E(Y) \qquad (40)$$

$$\sigma_Y=E(Y^2)-\bar{Y}^2$$

5 RISK ANALYSIS FOR ALLUVIAL FANS

5.1 Risk Analysis for Hydraulic Structures on an Alluvial Fan

The risk for a hydraulic structure associated with the 100-year discharge can be defined as the probability that the loading discharge to the structure is greater or equal to the discharge capacity of the structure. Mathematically,

$$r=Prob(q_{100}>q_C) \qquad (41)$$

in which q_{100} is the 100-year discharge of a point at which the hydraulic structure is located and q_C is the discharge capacity.

For levee system and highway drainage structures such as culverts and bridges, q_C can be computed from using Manning's equation

$$q_C=\frac{1.486}{n}A^{5/3}P^{-2/3}S_f^{1/2} \qquad (42)$$

in which n is the Manning's roughness coefficient, A is the cross-sectional area, P is the wetted perimeter, and S_f is the energy slope. With given n, A, and S_f for the hydraulic structures, the risk can be computed by assuming a probability distribution for

q_{100} and solving Eq. (41) through analytical or numerical integration.

When uncertainties in n, A, and S_f are significant, q_C should be considered as a random variable. If the statistics for n, A, and S_f can be estimated, the Rosenblueth's Point-Estimate uncertainty analysis method can also be used to find the statistics for q_C. Assuming probability distributions for q_{100} and q_C, one can perform analytical or numerical integration for Eq.(41) to obtain the risk. If a log-normal probability distribution is assumed for q_C and the Type I extremal distribution is assumed for q_{100}, the following equations can be derived (Tung and Mays, 1980)

$$r = 1 - \int_0^\infty \frac{1}{\sqrt{2\pi}\, q_C \zeta_{q_c}} \exp\left[-\frac{1}{2}\left[\frac{\ln(q_C) - \chi_{q_c}}{\zeta_{q_c}} \right]^2 \right]$$

$$\exp\left[-\exp\left[-\left(\frac{q_C - \alpha_1}{\alpha_2} \right) \right] \right] dq_C \tag{43}$$

in which $\alpha_2 = \sqrt{6}/\pi\sigma_{q_{100}}$, $\alpha_1 = \bar{q}_{100} - 0.5772\alpha_2$, χ_{q_c} and ζ_{q_c} are, respectively, the mean and standard deviation for $\ln(q_C)$ and can be computed by (Mays and Tung, 1992)

$$\chi_{q_c} = \frac{1}{2}\ln\left[\frac{\bar{q}_C^2}{1 + (\sigma_{q_c}/\bar{q}_C)^2} \right] \tag{44}$$

$$\zeta_{q_c} = \left[\ln\left[\left(\frac{\sigma_{q_c}}{\bar{q}_C} \right)^2 + 1 \right] \right]^{0.5} \tag{45}$$

5.2 Risk-Based Delineation for Hazard Flood Zone on Alluvial Fans

Since fan arc widths computed by the FEMA alluvial fan method for each given depth boundary are subject to uncertainties, the delineation of a hazard flood zone should be probabilistic. Suppose a hazard flood zone of fan shape has been defined by the FEMA alluvial fan method. If there is a point which may be within or beyond the defined zone, the question is to find the probability that the point will fall in the hazard flood zone. Obviously, a point within the computed zone has a relatively high probability that it will be in the zone while a point beyond the zone should have a low probability. The risk for defining the hazard flood zone is proposed to be the probability that a given point on an alluvial fan is within the hazard flood zone. Mathematically,

$$r = Prob(W \geq w) \tag{46}$$

in which fan arc width W is a random variable whose mean and standard deviation are obtained

from uncertainty analysis and w is the fan arc width corresponding to the point of interest. Assuming a probability distribution for the random variable W, one can compute the risk by analytical or numerical integration of Eq. (46).

6 EXAMPLES FOR UNCERTAINTY AND RISK ANALYSES

Two numerical examples are conducted by applying uncertainty and risk analyses to an alluvial fan in north Scottsdale, Arizona. The first example is associated with the culvert design on the alluvial fan while the second example is concerned with the risk-based delineation of hazard flood zone on the alluvial fan. The watershed upstream of the alluvial fan is divided into 12 sub-basins whose areas range from 1.9422 to 0.0438 sq. miles (Cella Barr Associates, 1988). Eight of the 12 subareas have an area approximately equal to or larger than 20% of the largest subarea (1.9422 sq. miles). Two subareas have an area approximately equal to 10% of the largest subarea. The smallest subarea with an area of 0.0438 sq. miles has the lowest curve number (73) among the 12 subareas, having high infiltration rates. The antecedent moisture condition II (AMC II) curve numbers for the 12 sub-basins ranges from 73 to 86. Sensitivity analysis indicates that the discharge at the fan apex computed by the HEC-1 computer model is found to be much more sensitive to the curve number for the largest sub-basin than the curve numbers for the other sub-basins.

The function g(**X**) for the numerical examples is defined as combined procedures of HEC-1 and FEMA alluvial fan procedures. For the first example, Y is defined as the 100-year discharge (q_{100}) at the point near the culvert structure. For the second example, Y is defined as the fan arc width (W). Define **X**=($P_{0.01}$, $P_{0.1}$, $P_{0.5}$, X_1, X_2) in which $P_{0.01}$, $P_{0.1}$, and $P_{0.5}$ are the 100-, 10-, and 2-year rainfall depths for the watershed upstream of the alluvial fan, respectively; X_1 is a vector consisting of curve numbers for the sub-basins; and X_2 is a vector consists of avulsion coefficient and model error correction for the equation used in FEMA alluvial fan procedures.

X_1 is reduced to a scalar by defining X_1 to be the curve number for the largest sub-basin, since the uncertainty in the curve number for the largest sub-basin is much more important than that for other sub-basins to the runoff computed from HEC-1 computer model. Therefore, **X** is reduced to a (1x6) vector as ($P_{0.01}$, $P_{0.1}$, $P_{0.5}$, CN, A, λ_m) in which CN is the curve number for the largest sub-basin, A is the avulsion coefficient, and λ_m is the model error correction term.

Based upon NOAA Atlas 2, Volume III-Arizona for precipitation-frequency (1973), the accuracy of the estimates obtained from the maps of this Atlas varies from a minimum of about 10 percent for shorter return periods in relatively nonorographic regions to 20 percent for the longer return periods in the more rugged orographic regions. Therefore, the coefficients of variation for $P_{0.01}$, $P_{0.1}$, and $P_{0.5}$ are assumed to be 13 percent, 12 percent, and 10 percent, respectively. Based upon the soil type, land use and treatment, and ground surface condition, the AMC II curve number for the largest sub-basin is 85 (Cella Barr Associates, 1988). In the numerical examples, curve number (CN) is assumed to follow a triangular probability distribution. The curve numbers for AMC I, AMC II, and AMC III are considered to be the lower limit, mode, and upper limit of the triangular probability distribution, respectively. Based upon SCS (1978, 1986), the lower and upper limits for AMC II CN=85 for the region are 70 and 94, respectively. Therefore, the mean, coefficient of variation, and skew coefficient for CN can be directly computed as 83, 0.06, and - 0.2358 (Tung and Mays, 1980). Since FEMA (1985) recommended the use of 1.5 for avulsion coefficients (A) in the absence of data, the mean of avulsion coefficient is assumed to be 1.5. The coefficient of variation for A is assumed to be 10 percent in the numerical examples. The mean of the model error correction term λ_m is assumed to be 1.0 with the coefficient of variation being 10 percent.

6.1 Uncertainty for q_{100} and Risk-Analysis for Culvert Design on Alluvial Fan

On the alluvial fan, there is an existing development area. The boundary of the development area is within fan arc width ranging approximately from 0 to 3600 ft (1097.3 m). Suppose there is a culvert at a point whose fan arc width is 2000 ft (609.6 m). Applying uncertainty analysis method to the combined procedures of the HEC-1 computer model and the FEMA alluvial fan method can give the mean and standard deviation of q_{100} at the culvert.

Since $\mathbf{X}=(P_{0.01}, P_{0.1}, P_{0.5}, CN, A, \lambda_m)$, 2^6 points need to be generated. Each of the 2^6 points is the input to the HEC-1 and FEMA alluvial fan procedures. Theoretically speaking, HEC-1 computer model and FEMA alluvial fan procedures should be performed 2^6 times. Computer programs using MATLAb code (Mathworks, 1989) have been developed for FEMA alluvial fan procedures and Rosenblueth's point estimate uncertainty analysis method. As stated above, the HEC-1 computer model should be run 2^6 times; however, since the curve number with only one of the three rainfall depths is used for each run of the HEC-1 computer model, the number of running HEC-1 can be reduced to 12, which is significantly smaller than 2^6.

After the HEC-1 computer model is run 12 times, the computed discharges together with the generated avulsion coefficient and model error correction term are used to form 2^6 points which are the inputs to FEMA alluvial fan procedures. One run of HEC-1 takes about 12 seconds on a 486-DX-33MHz PC. Then, Eq. (22) is used to find q_{100} for each of the 2^6 points. Finally, the mean and standard deviation of q_{100} can be computed by using Eq. (40). The mean

and coefficient of variation of q_{100} are found to be 10390 cfs and 26.6 percent, respectively.

Consider the discharge capacity (q_C) as a random variable. The mean and coefficient of variation for q_C can be obtained by applying an uncertainty analysis method such as Rosenblueth's point-estimate or the First Order Second Moment to Manning's equation (Tung and Mays, 1980) if the uncertainties in Manning's n, energy slope and wetted perimeter are given. Herein, for purposes of illustration, the mean and coefficient of variation for q_C are assumed to be 9351 cfs and 10 percent. It is also assumed that q_{100} follows a Type I extremal distribution and q_C follows a log-normal distribution. Then, Eq. (43) can be used to compute the risk or the probability that q_{100} is greater or equal to q_C, which is found to be 0.4.

If the discharge capacity is considered deterministic, then Prob($q_{100}>q_C$) can be directly computed by (Tung and Mays, 1980)

$$r=1.0-\exp[-\exp(-\frac{q_C-\alpha_1}{\alpha_2})] \tag{47}$$

in which $\alpha_2=\sqrt{6}/\pi\sigma_{q_{100}}$ and $\alpha_1=\overline{q}_{100}-0.5772\alpha_2$. With the assumed value of q_C=9351, the risk that that q_{100} is greater or equal to q_C is found to be 0.607.

6.2 Uncertainty for W and Risk-Based Delineation of Hazard Flood Zone on Alluvial Fan

As defined by FEMA (1985), zone AO is the flood insurance rate zone that corresponds to the areas of 100-year shallow flooding (usually sheet flow on sloping terrains) where average depths are between 1 foot (0.3048 m) and 3 ft (0.9144 m). The zone AO can be divided into several separate sections with similar flooding depths between 1 foot (0.3048 m) and 3 ft (0.9144 m). Given a depth, the FEMA alluvial fan method can be used to compute the corresponding fan arc width which delineates the hazard flood zone. The uncertainty analysis method with HEC-1 and FEMA alluvial fan procedures can be used to compute the mean and standard deviation of the fan arc width for a given depth. Suppose the depth is 3 ft (0.9144 m). Therefore, the mean and coefficient of variation for the fan arc width are computed to be 10223 ft (3115.97 m) and 52.2 percent, respectively.

Suppose there is a point of interest, which is within 10223 ft (3115.97 m) on the contour corresponding to the given depth. The corresponding arc width (w) is assumed to be 9201 ft (2804.5 m). Assume W follows log-normal distribution, implying log(W) follows normal distribution. Then the mean and standard deviation of ln(W) can be computed by (Mays and Tung, 1993)

$$\mu_{\ln(W)} = \frac{1}{2}\ln\left(\frac{\mu_W^2}{1+\Omega_W^2}\right) \tag{48}$$

504

$$\sigma^2_{\ln(W)} = \ln(\Omega^2_W + 1) \tag{49}$$

in which μ_W and Ω_W are the mean and coefficient of variation for W, respectively. The risk or the probability that the point falls in the hazard flood zone can be computed by

$r = Prob(W \geq w)$
$\quad = Prob(\ln(W) \geq \ln(w))$

$$= Prob \left[\frac{\ln(W) - \mu_{\ln(W)}}{\sigma_{\ln(W)}} \geq \frac{\ln(w) - \mu_{\ln(W)}}{\sigma_{\ln(W)}} \right] \tag{50}$$

$$= Prob \left[Z \geq \frac{\ln(w) - \mu_{\ln(W)}}{\sigma_{\ln(W)}} \right]$$

in which Z is the standard normal deviate. By using standard normal probability table or approximation equations (Eqs. 27-28), the risk is found to be 0.488. The CPU time on a 486-DX-33MHz is 0.28 seconds. Suppose there is another point beyond 10223 ft (3115.97 m) on the same contour and the arc width (w) is assumed to be 11245 ft (3427.5 m). Using the same procedures for w = 9201 ft (2804.5 m), one can obtain the risk that the point falls in the hazard flood zone, which is 0.33. It may be noted that the risk for w = 11245 ft (3427.5 m) is smaller than that for w = 9201 ft as expected (2804.5 m).

7 CONCLUSIONS

Rosenblueth's point-estimate uncertainty analysis method has been applied to the combined procedures of the rainfall-runoff model (HEC-1) and the FEMA alluvial fan procedures to find the mean and standard deviation associated with 100-year discharge at any point on an alluvial fan for a given fan arc width. The mean and standard deviation associated with the fan arc width were also found for a given total depth or velocity of zone AO. The results from the uncertainty analysis were then used to compute the risk that the 100-year discharge would exceed the discharge capacity associated with the hydraulic structures on the fan. The results from the uncertainty analysis were used in the risk-based delineation of hazard flood zones by which one could estimate the risk that a location of interest would be within the hazard flood zone. The proposed uncertainty and risk analyses for the FEMA alluvial fan methodology were applied to an alluvial fan in north Scottsdale, Maricopa County, Arizona. By applying uncertainty and risk analyses to the combined procedures of HEC-1 rainfall-runoff model and the FEMA alluvial fan procedures, one can obtain more judicious results useful to the risk-based engineering design and flood insurance studies.

REFERENCES

Abramowitz, M. & I.A. Stegun 1972. Handbook of Mathematical Functions with Formulas, Graphs and Mathematical Tables. 9th Edition. New York: Dover Publications, Inc.

Chang, C.-H., Y.K. Tung, & J-C. Yang 1995. Evaluation of probability point estimate methods. Applied Math. Modelling, 19: 95-105.

Cella Barr Associates 1988. Hydrologic analysis of Scottsdale alluvial fans 1-6, Maricopa County. Report prepared for Federal Emergency Management Agency.

Dawdy, D. R. 1979. Flood Frequency Estimates on Alluvial Fans. Journal of the Hydraulics Division, 107(3): 1407-1413.

Federal Emergency Management Agency (FEMA) 1985. Flood insurance Study-guidelines and specifications for study contractors Washington, D. C.

Federal Emergency Management Agency (FEMA) 1992. Flood insurance Study-guidelines and specifications for study contractors Washington, D. C.

"Guidelines for determining flood flow frequency." 1982. Bulletin 17B., U. S. Water Resources Council, Washington.

Harr, M.E. 1989. Probabilistic estimates for multivariate analysis. Applied Math. Modelling, 13(5): 313-318.

Li, K.S. 1992. Point-estimate method for calculating statistical moments. J. of Engineering Mechanics, 118(7): 1506-1511.

Mathworks 1989. MATLAB User's Guide, South Natick: The MathWorks, Inc.

Mays, L. W. & Y.K. Tung 1992. Hydrosystems Engineering and Management, New York: McGraw-Hill.

NOAA 1973. Precipitation-frequency atlas of the Western United States. In NOAA Atlas 2, Silver Spring, MD.

Rosenblueth, E. 1975. Point estimates for probability moments. Proc. Nat. Academy of Science, 72(10): 3812-3814.

Rosenblueth, E. 1981. Two-point estimates in probabilities. Applied Math. Modelling, 5(5): 329-335.

Soil Conservation Service 1978. Soil survey for Aquila-Carefuree Area, Parts of Maricopa and Pinal Counties, Arizona.

Soil Conservation Service 1986. Urban hydrology for small watershed, technical release 55.

Tung, Y. K. & L.W. Mays 1980. Risk analysis for hydraulic design. J. Hydraulic Engineering, 106(5): 893-913.

Tung, Y. K. & B.C. Yen 1993. Some recent progress in uncertainty analysis for hydraulic design. In Reliability and Uncertainty

Analyses in Hydraulic Design, B.
C. Yen and Y. K. Tung, ed., ASCE,
New York, New York.
Zoppou, C. & K.S. Li 1993. New point   estimate
method for water resources modeling. J. of
Hydraulic Engineering, 119(11):
1300-1307.

Stochastic Hydraulics'96, Tickle, Goulter, Xu, Wasimi & Bouchart (eds) © 1996 Balkema, Rotterdam. ISBN 90 5410 817 7

A methodology for forecasting non-Gaussian hydrological time series

Gwo-Hsing Yu & Wen-Cheng Wen
Department of Water Resources & Environmental Engineering, Tamkang University, Taipei, Taiwan, China

ABSTRACT: Janacek and Swift (1991) proposed a method to forecast the non-Gaussian time series. However, the marginal distribution of data must be known in advance. This basic assumption restricts its use in practice, because the distribution of real data are usually unknown. The distribution-free plotting position formula proposed by Yu and Hwang(1992) is employed to overcome the restriction. Furthermore, a scheme for generating the correlated data which obeys any specified non-Gaussian distribution is proposed in the present study. The results indicate that the accuracy of forecasting for the proposed method is higher than the naive method and lower than linear Z_t method of Janacek and Swift. However, the proposed method does overcome the restriction of Janacek and Swift method.

1. INTRODUCTION

The two major functions of hydrological time series models are to generate synthesis data and to forecast. In traditional models such as ARIMA (p,d,q) model of Box-Jenkins, the basic hypothesis is that the data and noise obey normal distribution while the majority of the hydrological time series do not pass the normality test. The reason may be that the hydrological time series are non-linear model with Gaussian noise, linear model with non-Gaussian noise, or non-linear model with non-Gaussian noise. Lawrence and Lewis (1981), Newbold and Granger(1976) had done a systematic research on linear model with non-Gaussian noise.

Newbold-Granger hypothesizes that time series Z_t is a normal distribution, and assume unknown time series as instantaneous transformation function of Z_t, $Y_t = T(Z_t)$. This could be expressed by Hermit Polynomial, with this, a method to forecast could be proposed. Janacek-Swift(1988) applies this assumption to propose a method for non-Gaussian time series model which is used for forecasting. The transformed stationary non-Gaussian time series Y_t to Z_t and through the selection from optimal model and estimation of model parameter to forecast the future Gussian time series Z_{t+1} to the prediction value Y_{t+1}. However, in the method of Janacek-Swift, the marginal distribution of data must be known in advance.

As we all know, the distribution of real data are usually unknown. Moreover, there is no good standard to test the distribution of non-Gaussian time series. The above restricts the application of Janacek-Swift method.

This paper employs the distribution free plotting position formula proposed by Yu and Hwang (1992) to indicate the cumulative frequency in order to overcome the restriction mentioned above. As for the seasonal time series, this paper intends to transform seasonal non-Gaussian time series to non-seasonal non Gaussian series, and analyze it with the method proposed. The monthly steam flow series in Taiwan were analyzed to ascertain whether the proposed method can be suitably applied for those series.

2. THEORETICAL ASPECTS

2.1 Time series model

Most of hydrological data are seasonal. They could be analyzed by ARIMA (p,d,q)*(P,D,Q) model of time series $\{ Z_t \}$ given in (1) :

$$\phi(B)\Phi(B^s)\left[(1-B^s)^D(1-B)^d(Z_t-\mu)\right] = \theta(B)\Theta(B^S)a_t \qquad (1)$$

where,
s= the length of the season, i.e. months, s=12.
D= order of seasonal difference.

d= order of non-seasonal difference.
u=mean of Z_t

$$\Phi(B^S) = 1 - \Phi_1 B^S - \Phi_2 B^{2S} ..... - \Phi_P B^{PS} ;$$

P=order of seasonal autoregressive operator ;

$$\Theta(B^S) = 1 - \Theta_1 B^S - \Theta_2 B^{2S} ..... - \Theta_Q B^{QS} ;$$

Q=order of seasonal moving average operator ;

$$\phi(B) = 1 - \phi_1 B - \phi_2 B^2 ..... - \phi_p B^p ;$$

p=order of non-seasonal autoregressive operator ;

$$\theta(B) = 1 - \theta_1 - \theta_2 B^2 ..... - \theta_q B^q ;$$

q=order of non-seasonal moving average operator.

For example, ARIMA(1,1,1)*(0,1,1)$_{12}$ model is

$$(1 - \phi_1 B)(1 - B)(1 - B^{12})(Z_t - \mu) = (1 - \theta_1 B)(1 - \Theta_1 B^{12})a_t$$

2.2 The detrended model

Traditionally, if non-Gaussian time series is seasonal, the detrended model is employed to transform data to normal time series. Then, the series is analyzed with normality model. The steps to establish the detrended model are:
  1.The normalization is carried out in terms of the standardized series

$$Z_{v,h} = \frac{Y_{v,h} - \bar{Y}_h}{S_h} \tag{2}$$

where
V=1,2,......,M(year) ;
h=time interval of the year, i.e. month(h=1,2,.....,12);
$\bar{Y}_h$ and S_h are the estimated periodic mean and standard deviation, respectively, that is,

$$\bar{Y}_h = \frac{1}{M} \sum_{V=1}^{M} Y_{v,h}$$

$$S_h^2 = \frac{1}{M-1} \sum_{V=1}^{M} \left(Y_{v,h} - \bar{Y}_h \right)^2$$

  2. The AIC(Akaike Information Criterion) is used to select the optimal model of $Z_{v,h}$. This paper uses method of moments to estimate the model parameters.
AIC is defined in (3):

$$\text{AIC}(L) = n \ln \hat{\sigma}_a^2 + 2L \tag{3}$$

where L= number of parameters in the model; n=sample size; and $\hat{\sigma}_a^2$ = variance of the noises .

  3. Predict the future data $Z_{v,h}{}^1$ and then transform it back to $Y_{v,h}{}^1$ as

$$Y_{v,h}{}^1 = \bar{Y}_h + Z_{v,h}{}^1 * S_h \tag{4}$$

2.3 Janacek-Swift method

Suppose that we have a stationary time series { Y_t } where the marginal density has a continuous probability density function (p.d.f.) f(y). It is known from standard theory that for any f(y) and corresponding cumulative distribution function (c.d.f.) F(y), F(y) is uniform on (0,1). In addition, $\Phi(Z)$, where Φ is the standard normal c.d.f., is uniform on the same interval. Thus, for any time period t we can identify

$$F(Y_t) = \Phi(Z_t) \tag{5}$$

which gives

$$Y_t = F^{-1} \{ \Phi(Z_t) \} = T(Z_t) \tag{6}$$

or

$$Z_t = \Phi^{-1}(F(Y_t)) \tag{7}$$

and our non-linear transformation T is identified. In this way, it regards the original series {Y_t} as a transform of a normal series {Z_t}. A schematic representation is given in Figure 1.

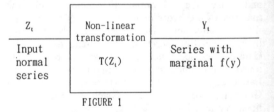

Z_t	Non-linear transformation	Y_t
Input normal series	$T(Z_t)$	Series with marginal f(y)

FIGURE 1

$T(Z_t)$ can be computed by eq.(8) (New-Granger(1976)):

$$T(Z_t) = \sum_{j=0}^{\infty} \alpha_j H_j(Z_t) \tag{8}$$

where α_j are constants, and $H_j(Z_t)$ is Hermite polynomial of order j. We used the definition

$$H_j(Z_t) = \frac{(-1)^j \varphi^{(j)}(Z_t)}{\varphi(Z_t)} \tag{9}$$

where $\varphi(Z)$ is the standard normal p.d.f. . Thus, $H_0(Z_t)=1$, $H_1(Z_t)=Z_t$, $H_2(Z_t)=Z_t^2-1$, ... They follow the recurrence

$$H_{j+1}(Z_t) = Z_t H_j(Z_t) - j H_{j-1}(Z_t) \tag{10}$$

By orthogonality, the α_j can be evaluated by

$$\int_{-\infty}^{\infty} \phi(Z_t)T(Z_t)H_j(Z_t)dZ_t = \alpha_j j!$$ (11)

After non-Gaussian time series $\{Y_t\}$ is transformed to normal series $\{Z_t\}$, the method of moment is used to identify the model and estimate its parameters. Then, the model can be used to forecast. Assume $Z_t(1)$ is the prediction value of Z_{t+1}

$$\hat{Z}_t(l) = E[Z_{t+1}|h_t]$$ (12)

where $h_t = \{Z_t, Z_{t-1}, ...,$ and model parameters$\}$. The prediction value $\hat{Y}_t(l)$ of Y_{t+1} can be computed by the following methods.
(1) Naive method

$$\hat{Y}_t(l) = T\left(\hat{Z}_t(l)\right) = \sum_{j=0}^{\infty} \alpha_j H_j\left(\hat{Z}_j(l)\right)$$ (13)

(2) Optimal linear method
Suppose $\hat{Y}_t(l)$ and Z_t shows a linear relationship in the form of $\sum_{j=0}^{\infty} C_j Z_{n-j}$. The coefficients C_j could be chosen by minimizing the MSE. Whittle(1983) shows that

$$\hat{Y}_t(l) = \alpha_0 + \alpha_1 \hat{Z}_t(l)$$ (14)

2.4 The proposed method

Because Janacek-Swift method can be used only when the distribution of data is known, this paper uses the distribution-free plotting position formula proposed by Yu(1992) which is shown in equation (15) to assume the cumulative probability. If $\{Y_t\}$, $t=1,2,....,n$, is arranged in decreasing order, it is $Y_{(1)} \leq Y_{(2)}... \leq Y_{(n)}$. The cumulative probability $Y_{(t)}$ can be computed by using position formula.

$$F(Y_{(t)}) = P(Y \leq Y_{(t)}) = \frac{t-0.326}{n+0.348}$$ (15)

Putting the result into $Z_t = \phi_1^{-1}[F(Y_{(t)})]$,$\{Y_b\}$ can be transformed to standard normal $\{Z_t\}$ then use AIC to select the optimal model.

To get α_j factor, this paper uses the least squares method, the estimation is as the following. If

$$Y_t = \sum_{j=0}^{\infty} \alpha_j H_j(Z_t)$$ (16)

then, $K = \sum_{t=1}^{m} e_t^2 = \sum_{t=1}^{n}\left(Y_t - \sum_{j=0}^{m} \alpha_j H_j(Z_t)\right)^2$ (17)

where m=Truncated Value. To differentiate individual parameters by equation(15), then

$$\sum_{t=1}^{n} (H_0(Z_t))^2 .... \sum_{t=1}^{n} H_i(Z_t)H_0(Z_t) \qquad \alpha_0$$

$$\sum_{t=1}^{n} H_1(Z_t)H_0(Z_t)...... \sum_{t=1}^{n} H_i(Z_t)H_i(Zt) \qquad \alpha_1$$

$$\cdot \qquad\qquad \cdot \qquad\qquad \cdot$$

$$\sum_{t=1}^{n} H_j(Z_t)H_0(Z_t)...... \sum_{t=1}^{n} (H_j(Z_t))^2 \qquad \alpha_j$$

$$= \begin{matrix} \sum_{t=1}^{n} H_0(Z_t)Y_t \\ \sum_{t=1}^{n} H_1(Z_t)Y_t \\ \cdot \\ \cdot \\ \sum_{t=1}^{n} H_j(Z_t)Y_t \end{matrix}$$ (18)

or $A\alpha=B$

get $\alpha = A^{-1} B$ (19)

The computational scheme of the proposed method is given in Figure 2.

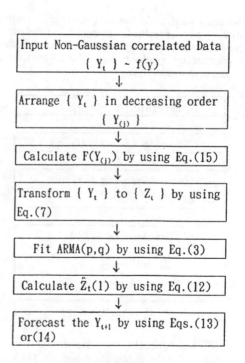

FIGURE 2 Computational scheme of the proposed method

509

If time series is seasonal and non-Gaussian, it must be transformed to Gaussian time series. This paper uses the following method :

If $\{Y_t\}$ time series has a cycle p, t=1,2,...,n, this means $Y_{v,h}$,v=1,2,3,.......M(year), h=1,2,3,.........,p. The$\{Y_t^*\}$ corrected time series is non-Gaussian . $Y_{v,h}$ the related expression is:

$$Y_{n,h^*} = \frac{Y_{v,h}}{F_h} \qquad (20)$$

where, $F_h = \dfrac{\bar{Y}_h}{\bar{Y}}$

$$\bar{Y} = \frac{1}{n}\sum_{t=1}^{n} Y_t \qquad ,and \quad \bar{Y}_h = \frac{1}{M}\sum_{V=1}^{M} Y_{v,h}$$

Now, use(Eqs.(13) or (14))to compute the prediction value $\hat{Y}_t$ (1) of $Y_{t,h}^*$ and the prediction value $\hat{Y}_t$ (1) of Y_{t+1} can be calculated from eq. (21)

$$\hat{Y}_t(1) = F_h * Y_t^*(1) \qquad (21)$$

where   h is the month at time t+1

2.5 Generation of correlated data which obey
      the specified marginal distribution

This study wants to discuss the correlated data which satisfy any specified marginal distribution. Therefore, this paper proposes the following method to generate the synthetic data. The computational scheme is shown in Figure 3.

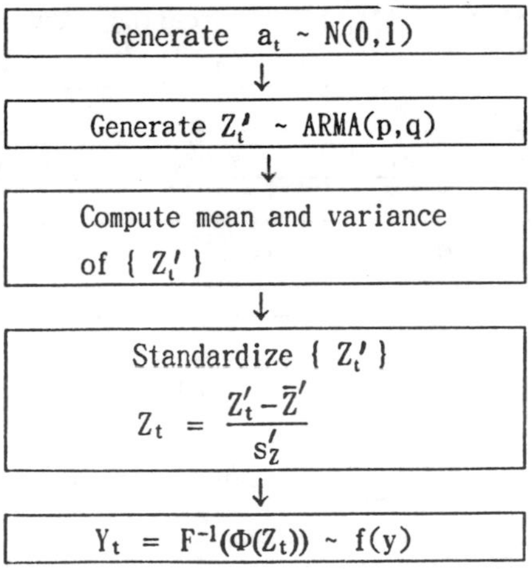

FIGURE 3 Generate the correlated
             data which obey p.d.f. f(y)

2.6 Assessment of the goodness of fit

(1) Q test (or Portmanteau test):

To test whether noise $\{a_t\}$ is independent, it realizes if the ARIMA (p,q) model is correct, then the distribution of quantity Q is approximatedly distributed Chi-Square distribution with degree of freedom k-p-q

$$Q = n\sum_{t=1}^{k} r_{a^2(\tau)} \qquad (22)$$

where,
r_a=autocorrelation function of noise.
k=the maximum lag.
n=number of data.

(2) Cumulative Periodgram Test:

If data is seasonal, one must examine whether the noise is periodical or not. This paper uses Cumulative Periodgram Test. (details in Box and Jenkins(1976),pp.294-295)

(2) Forecasting errors:

The comparative indexes in this paper include MSE (mean square errors) and MAE (mean absolute errors) given in Eqs.(23) and (24)

$$MSE = \frac{1}{k}\sum_{t=n}^{t=n+k-1} \left\{Y_{t+1} - \hat{Y}_t(1)\right\}^2 \qquad (23)$$

$$MAE = \frac{1}{k}\sum_{t=n}^{t=n+k-1} \left|Y_{t+1} - \hat{Y}_t(1)\right| \qquad (24)$$

where,

Y_{t+1}=observed value at time t+1;
$\hat{Y}_t(1)$ =prediction value of Y_{t+1} ;
K=number of forecasting .

The less indexes are, the better forecasting ability indicates.

3.DATA USED IN THE STUDY

Two sets of data were used in this study. The first set comprises the synthetic data generated by four different AR models :

$$Z_t=-0.6Z_{t-1}+a_t \qquad (25)$$
$$Z_t=0.6Z_{t-1}+a_t \qquad (26)$$

510

$Z_t=1.1Z_{t-1}-0.3Z_{t-2}+a_t$ (27)
$Z_t=0.9Z_{t-1}+0.01Z_{t-2}-0.105Z_{t-3}+a_t$ (28)

In each of these models, $\{a_t\}$ is a zero-mean uncorrelated normally distributed with variance unity. For each model, 100 synthetic series of 120 and 360 observations were generated. Then, they were transformed to obey Gamma, Cauchy, Double, Exponential and Extreme value Type I distributions. Thus, 4,000 Synthetic series are analyzed in the present study.

The second set of data consists of 22 monthly streamflow data from Taiwan. Information about these real data is given in Table 1.

Table 1 Characteristics of real data used in the study

Data No	period	n	mean	variance	skewness
F030002	1957-1988	384	17.5	379	3.04
F030004	1957-1988	384	7.0	41	2.85
F030005	1957-1988	384	30.1	1,164	3.71
F030021	1957-1988	384	8.7	58	2.47
F030026	1952-1988	444	18.0	236	2.73
F100013	1950-1988	468	62.8	4,108	2.34
F180001	1957-1988	372	6.0	53	4.32
F250004	1957-1988	372	14.1	166	2.08
F250006	1957-1988	372	8.7	77	3.57
F270014	1956-1986	372	54.3	4,904	2.83
F290009	1934-1983	600	35.8	1,312	3.01
F290035	1967-1993	324	133.2	12,604	1.54
F290042	1955-1988	408	19.2	772	2.40
F330022	1958-1993	420	25.3	1,006	2.11
F340011	1959-1988	360	20.2	437	2.85
F400016	1957-1988	384	36.8	2,010	2.39
F400020	1949-1988	480	98.7	16,807	2.39
F390011	1961-1988	336	8.5	183	2.67
F510010	1959-1988	360	71.1	8,057	2.30
F510031	1959-1986	336	40.6	3,936	2.53
F510036	1961-1988	336	239.3	10,572	2.16
F550004	1962-1988	324	23.3	1,577	2.20

4.RESULTS AND DISCUSSIONS

4.1 Synthetic data

(1)Selection of M value:
Prediction value of the non-Gaussian distribution can be computed by using Hermite Polynomial. The accuracy depends on the selection of m used in (17). There are 1/5 of data used to compare the forecasting ability, while 4/5 of data are used to simulate model. Once a certain m value is set, MSE can be computed form the prediction value and the actual value.

This paper discusses AR(1) model with parameter ϕ_1 =0.6. The results show that the average value of MSE is the smallest in Gamma, Exponential and Extreme value type I distributions when m=2. The value are 0.469,3.027 and 4.675 respectively. The average value of MSE is the smallest in Cauchy and Double exponential distributions when m=15 and m=3. The value are 259.76 and 8.66. Meanwhile when M15, MSE do not significantly improve for each distribution. Therefore, m value are set to be between 2 and 15in the present study.

(2) Estimation of parameters:
The results show that Janacek-Swift and the proposed method are better than the detrended method in Gamma distribution for AR(1) models. The difference between these three methods is diminutive for AR(2) and AR(3) models. When data satisfy Cauchy distribution for AR(1),AR(2) and AR(3) models, Janacek-Swift method and the proposed method are better than the detrended method. The accuracy of the three methods are very close in Double exponentail distribution for AR(1), AR(2) and AR(3) models. However, the accuracy of the estimated parameter in Double exponential is superior to Cauchy distribution when the detrended method is used. In Exponential and Extreme value type I distributions for AR(1) model, Janacek-Swift method and the proposed method are better than the detrended method.

To sum up, the detrended method has the worst result in Cauchy distribution. The accuracy of these three methods are all similar in Double exponential and extreme value type I distributions. When the series are with the same model parameters and the sample size, no matter kind of distribution the data satisfy, Janacek- Swift method and the proposed method has the identical parameter estimated value.

Table 2 shows the comparison of the estimated parameters from these three methods in Gamma distribution.

(3)Comparisons of forecasting ability:
This paper maintains 1/5 of data to conduct the comparison of the forecasting ability, and uses the rest of 4/5 of data to simulate the model. The results indicate that for AR(1) model of the Gamma and Cauchy distributions, Janacek-Swift method and the proposed method are better than the detrended method. The proposed method is not as good as the other two for AR(1) model of Double exponential distribution when sample size is 120; however, the difference is little. The values of MSE of the proposed method and the naive method and optimal linear Z_t of Janacek-Swift are 7.98,7.87 and 7.89, respectively. Overall, the proposed method is much better to be applied to forecast than the others in Cauchy distribution. Majority of MSE values for the proposed method is between those of the naive method and optimal linear Z_t . Therefore, the proposed method can obtain in good accuracy of forecasting for unknown distribution.

511

Table 2 Comparisons of the estimation of parameters(Gamma distribution)

Model			AR(1) $\phi_1 = -.6$	AR(2) $\phi_1 = 1.1$ $\phi_2 = -.3$	AR(3) $\phi_1 = .9$ $\phi_2 = .01$ $\phi_3 = -.105$
Marginal parameter			0.5	0.6	0.8
Sample Size	Method *	Estimated parameter			
120	(1)	$\hat{\phi}_1$	-0.62	1.07	0.87
		$\hat{\phi}_2$		-0.34	-0.04
		$\hat{\phi}_3$			-0.11
	(2)	$\hat{\phi}_1$	-0.57	1.01	0.83
		$\hat{\phi}_2$		-0.26	0.03
		$\hat{\phi}_3$			-0.12
	(3)	$\hat{\phi}_1$	-0.34	0.94	0.79
		$\hat{\phi}_2$		-0.24	-0.01
		$\hat{\phi}_3$			-0.08
360	(1)	$\hat{\phi}_1$	-0.53	1.13	0.96
		$\hat{\phi}_2$		-0.34	-0.10
		$\hat{\phi}_3$			-0.06
	(2)	$\hat{\phi}_1$	-0.59	1.07	0.85
		$\hat{\phi}_2$		-0.29	0.02
		$\hat{\phi}_3$			-0.13
	(3)	$\hat{\phi}_1$	-0.32	1.00	0.84
		$\hat{\phi}_2$		-0.27	0.01
		$\hat{\phi}_3$			-0.09

(1)J-S Method
(2)The proposed method
(3)The detrended method

Table 3 Comparisons of Janacek-Swift Method, the proposed method and detrended method (Cauchy , n=360)

Models		AR(1) $\Phi_1 = -.6$	AR(2) $\Phi_1 = 1.1$ $\Phi_2 = -.3$	AR(3) $\Phi_1 = .9$ $\Phi_2 = .01$ $\Phi_3 = -.105$
mariginal parameters		0.6 0.6	0.6 0.6	1. .1.
mean		0.82	0.6	0.68
variance		12401	2762	9200
skewness		-0.88	-0.51	-0.45
Janacek-Swift method	MSE(Naive)	6640	381	1326
	MSE(Linear Z_t)	6597	499	1620
The proposed method	M value	15	10	10
	MSE	6646	426	1403
MSE(Detrended method)		6685	492	1524

Table 4 MSE of forecasting errors (AR(1) model)

Distribution	Method *	Sample Size	
		120	360
Gamma	(1)	.36	.39
	(2)	.34	.37
	(3)	.35	.38
	(4)	.42	.46
Cauchy	(1)	1,401	6,640
	(2)	1,427	6,597
	(3)	1,403	6,646
	(4)	1,625	6,685
Double Exponeatial	(1)	7.98	8.41
	(2)	7.89	8.21
	(3)	7.98	8.27
	(4)	7.90	8.25
Exponeatial	(1)	70.0	72.4
	(2)	67.4	70.5
	(3)	68.5	71.2
	(4)	78.5	82.9
Extrame Value type I	(1)	6.72	6.84
	(2)	6.63	6.77
	(3)	6.70	6.82
	(4)	7.04	7.25

*(1) J-S Method, Naive
(2) J-S method, Linear Z_t
(3) The proposed method
(4) The detrended method

Table 3 shows the comparisons of forecasting ability of these three methods when n=360 and data obey Cauchy distribution. Table 4 shows that the mean MSE of 100   synthetic series for the Janacek-Swift method, the proposed method and the detrended method when the data were generated by AR(1) model. The results indicate that the proposed method and Janacek-Swift method are better than the detrended method in forecasting.

4.2 Real data

There are 22 monthly stream flow series analyzed in Taiwan. In order to check the forecasting ability, this paper maintains the last 5 years data for forecasting, and uses the rest to simulate the model. Due to the unknown distribution of the real data, Janacek-Swift method is not applicable. This paper discusses the forecasting ability by using the proposed method and the detrended method.   The results are given in Table 5.

Table 5 Comparision of forecasting ability for real data

Data No.	The detrended method		The proposed method		
	MAE	MSE	M value	MAE	MSE
F030002	11.4	282	2	10.3	233
F030004	3.5	26	2	3.5	24
F030005	17.4	667	2	16.4	576
F030021	5.0	48	2	4.9	45
F030026	7.8	155	2	8.0	163
F100013	20.2	890	6	18.5	883
F180001	2.4	14	2	2.6	17
F250004	6.0	72	2	6.5	74
F250006	3.6	27	7	3.2	23
F270014	24.0	1,353	6	20.7	1,071
F290009	16.8	737	2	16.1	686
F290035	53.7	7,345	11	49.7	7,042
F290042	10.7	290	5	9.9	282
F330022	13.8	517	8	13.0	506
F340011	10.7	331	2	10.2	298
F400016	21.8	1,061	11	19.3	921
F400020	51.1	5,676	5	46.0	5,774
F390011	4.4	65	5	4.2	63
F510010	28.8	2,236	8	24.5	1.599
F510031	18.8	1,000	5	14.8	710
F510036	104.4	33,193	4	89.7	28,697
F550004	13.6	552	9	12.5	508

For F-290009(Wu-Chieh Station), the MAE and MSE of the detrended method are 16.82 and 737.27. Overall, the MSE and MAE of the proposed method are smaller than those of obtained by the detrended method. Therefore, the proposed method is actually superior to the detrended method for monthly stream flow in Taiwan.

5.CONCLUSIONS

The proposed method has better accuracy for forecasting than the naive method of Janacek and Swift, though it is a little worse than the linear Z_t method. The proposed method has solved the problem when using Janacek and Swift method that is the marginal distribution of data must be known in advance. Therefore, the method proposed shows its superiority.

ACKNOWLEDGMENT

This work was supported by the National Science Council of the Executive Yuan, grant NSC 84-2211-E-032-04. Furthermore, the referee's invaluable comments are highly appreciated.

REFERENCES

Box, G.E.P. and D.R. Cox, 1964. An Analysis of Transformation., J. of Roy. Stat. Soc., Ser. B., vol. 26, pp. 211-252.

Box, G.E.P. and Jenkins, G.M. 1976. Time Series Analysis Forecasting and Control., 2nded.,Holden-Day, San Francisco.

Fernandez, B. and Sale, J.D. 1986. Periodic Gamma AR Process for Operational Hydrology. Water Resources Research, vol. 22, no, 2, pp. 311-22.

Granger, C.W. and Newbold, P. 1976. Forecasting Transformation Series. J. of Roy. Stat. Soc., Ser. B., vol. 38, no. 2, pp. 311-22.

Granger, C.W.J. and Newbold, P.1987., Forecasting Economic Time Series, Academic Pressm, New York.

Hopwood, W.S., Mackeown, J.C. and Newbold, P. 1982.Time Series Forecasting Model Involving Power Transformation. J. of Forecasting, vol. 3, pp. 57-61.

Janacek, G.J. and Swift, A.L.1988. A Class of Models for Non-normal Time Series. J. of Time Series Analysis, vol. 1, pp. 19-32.

Janacek, G.J. and Swift, A.L. 1991. Forecasting Non-normal Time Series. J. of Forecasting, vol.10, pp.501-520.

Lawrance, A.J. and Lewis, P.A.W. 1981. A New Autoregressive Time Series Model in Exponential Variables, Adv. Appl. Prob., vol. 13, pp.826-45.

Yu, G.H. and Lin, Y.C.1991. A Methodology for Selecting Subset Autoregressive Time Series Models. J .of Time Series Analysis, vol.12, no. 4, pp 363-373

Yu, G.H. and Hwang, C.C. 1994. A Distribution Free Plotting Position on Probablity Paper.Proceeding of Ninth Congress of APD-IAHR, Vol.1, pp.193-200, Singapore.

8 Stochastic hydraulics

Stochastic Hydraulics'96, Tickle, Goulter, Xu, Wasimi & Bouchart (eds) © 1996 Balkema, Rotterdam. ISBN 90 5410 817 7

Studies on statistical characteristics of the pressure fluctuation amplitude in outlet works

Ding Zhuoyi
Yangtze River Scientific Research Institute, Wuhan, China

ABSTRACT: The pressure fluctuation amplitude is very important for the load design of structures and to research flow cavitation, Using 182 sample records of fluctuation pressure obtained from model tests and prototype observations for many kinds of outlet works, this paper studies statistical characteristics of fluctuating pressure amplitude in outlet works. The researched result indicate that the probability of fluctuating pressure amplitude in the outlet works obeys that of Gaussian normal distribution, and the verification of prototype observation data from Gezhouba project further proves that the conclusion is true.

1 PRELIMINARY DEMONSTRATION ON STATISTICAL CHARACTERISTICS OF FLUCTUATING PRESSURE AMPLITUDE

The time process of pressure fluctuation is assumed to be stationary random process with ergodicity. So, for every measured point of fluctuating pressure, it is only necessary to measure one sample record with enough recording length so as to calculate its digital characteristics. Let $x(t)$ is a sample record of fluctuating pressure with recording length T and $x(k \triangle t)(k = 0,1,2, \cdots\cdots N)$ is the sampled value of sample record $x(t)$, where $N = T/\triangle t$ is sample size, $\triangle t = 1/2f_c$ is sample interval and f_c is Nyquist sampling frequency. The deviation coefficient C'_s and kurtosis coefficient C'_E of $x(k \triangle t)$ can be expressed as follows:

$$C'_s = \{\frac{1}{N}\sum_{k=0}^{N-1}[x(k \triangle t) - \overline{x}]^3\}/\sigma_x^3 \qquad (1)$$

$$C'_E = \{\frac{1}{N}\sum_{k=0}^{N-1}[x(k \triangle t) - \overline{x}]^4\}/\sigma_x^4 \qquad (2)$$

where $\overline{x}$ is the mean value of $x(k \triangle t)$ and σ_x is root of mean square of $x(k \triangle t)$.

On the basis of model tests and prototype observations, 182 sample records of fluctuating pressure at 182 measured points are obtained, which are measured from various outlet works including sluices, stilling basin (their aprons and side — walls), spillway tunnels, bottom outlets in the dam , baffle sills, dividing piers at the tunnel exit and galleries in the ship lock etc. At the locations of the measured points, there are various flow patterns including free flow, pressure flow, supercritical flow, subcritical flow, rolled flow in the hydraulic jump, vortex with vertical axis, flow under the gate and flow in the bucket etc. Taking samples from 182 sample records and making statistical calculation according to equation (1) and (2) (in which $f_c = 100HZ$, $\triangle t = 0.05s$, $N = 2^{10} = 1024$), 182 values of deviation coefficient C'_s and 182 values of kurtosis coefficient C'_E are obtained and listed in Table 1 (from prototype observations) and Table 2 (from model tests). Because $x(k \triangle t)$ in equation (1) and (2) is a random variable, C_s, C_E are all statistics (C_s, C_E is deviation coefficient and kurtosis coefficient respectively for ensemble), i. e. C_s, C_E are also random variable. So, 182 values

Table. 1

C'_s	C'_E	C'_s	C'_E	C'_s	C'_E	C'_s	C'_E	C'_s	C'_E
0.239	3.37	0.160	1.89	−0.074	2.79	−0.096	3.38	−0.456	2.61
0.231	2.42	0.347	2.81	−0.510	2.38	−0.472	3.84	0.057	2.49
−0.104	3.18	0.008	2.57	−0.424	4.04	−0.668	4.44	0.495	2.85
−0.077	2.72	−0.091	2.81	0.322	3.19	0.407	2.34	0.040	2.96
0.287	2.33	0.435	2.56	0.396	2.85	−0.498	2.77	0.053	2.40
0.228	2.47	−0.151	2.08	−0.099	2.48	0.095	3.03	0.170	4.13
0.381	2.93	0.007	1.94	0.499	2.62	0.140	3.54	−0.563	2.66
−0.159	2.04	0.014	2.96	0.416	2.67	0.245	3.91	0.263	2.38
0.335	2.92	−0.008	3.38	−0.260	4.15	0.466	3.28	−0.319	2.17
0.052	2.33	0.036	2.62	−0.295	3.73	−0.352	2.69	−0.557	3.36
0.512	2.44	−0.423	2.80	−0.469	2.73	−0.459	2.86	0.397	2.29
0.435	2.43	0.596	2.75	0.253	2.43	−0.012	1.99	−0.614	2.88
0.270	3.42	0.118	2.87	0.196	3.59	0.540	2.60	0.143	2.22
−0.830	4.27	0.571	3.52	0.214	2.18	−0.254	3.34	0.304	3.77
0.656	2.95	−0.005	3.10	−0.027	2.46	0.081	1.75	0.267	2.20
−0.084	2.73	0.170	4.16	−0.098	2.88	0.141	2.81	0.034	1.98
−0.448	3.92	0.326	3.30	−0.267	2.43	0.038	2.33	0.282	2.59
−0.179	2.24	−0.556	2.44	−0.253	2.77	0.451	2.39	−0.067	1.88
−0.314	2.20	0.568	2.99	−0.477	2.61	−0.411	3.41	0.455	3.31
0.093	2.28	0.163	2.59	0.797	2.55	−0.346	3.30	0.177	3.11
0.568	2.63	0.033	2.52	0.002	2.93	0.089	3.05		

Table. 2

C'_s	C'_E	C'_s	C'_E	C'_s	C'_E	C'_s	C'_E	C'_s	C'_E
0.190	3.15	−0.300	3.35	0.160	2.53	0.080	3.38	−0.578	3.55
−0.260	2.77	0.280	2.75	0.250	3.59	0.060	3.70	0.237	3.70
0.180	3.17	0.040	3.01	0.030	3.57	0.540	3.22	0.067	2.63
−0.070	3.63	0.030	3.43	0.210	3.41	0.400	2.96	−0.093	3.05
0.076	4.34	0.660	4.49	−0.250	2.60	0.310	2.93	−0.435	3.13
0.200	2.94	0.140	2.96	0.300	3.11	0.010	2.84	0.777	4.06
0.070	4.14	0.070	2.89	0.250	2.60	−0.160	3.94	−0.005	3.24
0.420	3.73	0.060	3.51	0.300	3.25	−0.033	2.84	0.000	3.22
0.280	3.32	0.130	3.01	0.270	3.66	−0.231	3.61	−0.372	3.63
0.520	3.89	−0.300	3.74	0.010	2.84	0.646	3.20	0.044	3.18
−0.190	2.31	0.350	2.60	0.230	3.12	0.153	4.50	0.244	4.28
0.170	2.92	0.010	3.11	−0.270	3.80	0.384	4.04	−0.900	4.10
0.150	3.75	−0.270	3.10	−0.140	3.69	0.162	3.58	0.336	3.61
−0.036	2.93	0.370	4.02	−0.130	2.60	−0.098	3.20	0.505	3.26
−0.550	4.48	−0.100	2.91	0.080	3.30	0.040	2.80		
−0.330	3.55	0.180	3.83	−0.060	3.20	−0.197	3.44		

of C'_s, C'_E in Table 1, 2 are from a sample of C_s, C_E respectively with sample size n = 182. Using the values in Table 1, 2, the sample mean value $\overline{C'_s}$, $\overline{C'_E}$ of C_s, C_E can be calculated as follows:

$$\overline{C'_s} = \frac{1}{N}\sum_{i=1}^{n}C'_{si} = 0.044 \qquad (3)$$

$$\overline{C'_E} = \frac{1}{N}\sum_{i=1}^{n}C'_{Ei} = 3.055 \qquad (4)$$

It can be seen from equation (3) and (4) that $\overline{C'_s} \approx 0$, $\overline{C'_E} \approx 3$. It is known that $C_s = 0$, $C_E = 3$ for the random variable with Gaussian normal distribution. Owing to the fact that the values in Table 1.2 are obtained from various outlet works with various flow patterns, the preliminary conclusion can be stated: the probability of fluctuating pressure amplitude in outlet works obeys that of Gaussian normal distribution.

2 HYPOTHESIS TEST FOR NORMALITY OF PROBABILITY DISTRIBUTION

The preliminary conclusion stated above is obtained from one sample with sample size n = 182 (Listed in Table 1.2). So, it is necessary to make hypothesis test for normality of probability distribution of fluctuating pressure amplitude, i. e. to test whether the hypothesis of $C_s = 0$ and $C_s = 3$ is accepted or not ($\overline{C}_s$, $\overline{C}_E$ is mean value of deviation coefficient and kurtosis coefficient respectively, for ensemble).

In order to test the mean value $\overline{C}_s$, $\overline{C}_E$ of ensemble, it is necessary to know the probability distribution of random variable C_s, C_E. Using sample values C'_s, C'_E in Table 1.2, χ^2 testing method yields (when sample size n > 50):

$$\chi'^2 = \sum_{i=1}^{m}\frac{(f_i - f'_i)^2}{f'_i} \qquad (5)$$

where m is the total of dividing C'_s, C'_E into groups; χ'^2 is χ^2 distribution with degree of freedom $\gamma = m - 3$; f_i is the appearing number of C'_s, C'_E in group i; f'_i is the appearing number in theory corresponding to Gaussian normal distribution, in group i. Taking the level of significance α

= 0.05 and looking up χ^2 distribution table, $\chi^2_{0.05}$ = 9.488 for C'_s and $\chi^2_{0.05} = 12.592$ for C'_E are obtained. At the same time, $\chi'^2 = 7.444$ for C'_s and $\chi'^2 = 9.540$ for C'_E are calculated from equation (5). These results give that $\chi^2_{0.05} = 9.488 > \chi'^2 = 7.444$ for C'_s and $\chi^2_{0.05} = 12.592 > \chi'^2 = 9.540$ for C'_E. The testing result shows that the samples C'_s and C'_E come from the ensemble of Gaussian normal distribution, i. e. the probability of C_s, C_E obeys that of Gaussian normal distribution. Now, it is only necessary to make a hypothesis test for the mean value of normal ensemble. Because the ensemble variance is unknown, t testing method is used. Taking the level of significance $\alpha = 0.05$, the following hypothesis should be tested:

$$H_1 : \overline{C}_s = 0$$
$$H_2 : \overline{C}_E = 3$$

where H_1, H_2 mean "hypothesis". According to t testing method , the rejection region is :

$$|\overline{C'_s} - \overline{C}_s| > \frac{\sigma'_{cs}}{\sqrt{n}}t_{\alpha/2}(n - 1) \qquad (6)$$

$$|\overline{C'_E} - \overline{C}_E| > \frac{\sigma'_{cE}}{\sqrt{n}}t_{\alpha/2}(n - 1) \qquad (7)$$

where σ'_{cs}, σ'_{cE} is mean square root of sample C'_s, C'_E respectively; $t_{\alpha/2}$ is $100\alpha\%$ point (two sides) of t distribution with degree of freedom n − 1; n is sample size. looking up t distribution table, $t_{\alpha/2}$(n −1) = $t_{0.025}$(182−1) = 1.96 is obtained. According to the data of C'_s, C'_E in Table 1, 2, $\sigma'_{cs} = 0.323$ and $\sigma'_{cE} = 0.609$ are calculated. Substituting all data into equation (6), (7), the following results are given:

$$|\overline{C_s} - \overline{C}_s| = 0.044 < \frac{\sigma'_{cs}}{\sqrt{n}}t_{0.025}(n - 1) = 0.047 \qquad (8)$$

$$|\overline{C_E} - \overline{C}_E| = 0.055 < \frac{\sigma'_{cE}}{\sqrt{n}}t_{0.025}(n - 1) = 0.089 \qquad (9)$$

The equations (8) and (9) indicate that the symbols of inequality in expressions (6) and (7) must be inverse, i. e. the hypothesis of $\overline{C}_s = 0$, $\overline{C}_E = 3$ sould be accepted. So, the following conclusion can be finally stated: the probability distribution of the amplitude of fluctuating pressure in outlet works obeys that of Gaussian normal distribution.

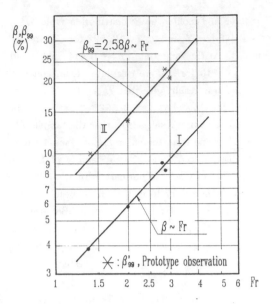

Fig.1 Relation of β, β_{99} and Fr

project are also plotted in Fig. 1 (see symbol "*"). It can be seen from Fig. 1 that β'_{99} is in agreement with β_{99}.

The verification of fluctuating pressure data obtained from the prototype observation in the sluice No. 2 of Gezhouba project further proves that the conclusion mentioned above is true.

3 CONCLUSION

The researched result in the paper can be applied to many outlet works. So long as the root mean square σ_x is measured, the amplitude of fluctuating pressure in many outlet works, correspoding to any probability, can be calculated. It will provide a great convenience for research and application of pressure fluctuation in outlet works.

This conclusion gives a convenience for calculating the amplitude of fluctuating pressure corresponding to any probability P. For example, the amplitude corresponding to probability $P = 99\%$ is 2. 58 σ_x, σ_x is root mean square of the amplitude.

Now, let the fluctuation intensity is defined as follows

$$\beta = \sigma_x / \frac{V_1^2}{2g} \qquad (10)$$

where β is fluctuation intensity; σ_x is root mean square of fluctuating pressure in units of head; V_1 is mean velocity at the initial section of hydraulic jump; g is gravity acceleration. According to the values of σ_x obtained from prototype observation in sluice NO 2 of Gezhouba project, corresponding to different Froude number F_r, the values of β are calculated by Equation (10) and plotted in Fig. 1, and a straight line is obtained (Line I in Fig. 1). Because of $\beta_{99} = 2. 58\beta$ for the probability distribution of Gaussian normality (β_{99} is fluctuation intensity corresponding to probability $P = 99\%$), another straight line $\beta_{99} = 2. 58\beta \sim F_r$ is given (Line II in Fig. 1). The values of β'_{99} obtained from prototype observation in sluice NO. 2 of Gezhouba

REFERENCES

Lin Shaogong, 1978, Basis Probability and Mathematical statistics, Education Press, China.

Ding Zhuoyi, 1981, Data Processing and Calculating Method for Fluctuating Pressure, Yangtze River Scientific Research Institute, China.

Ding Zhuoyi, 1980, Prototype Observation of Fluctuating Pressure for Sluice No. 2 of Gezhouba Project. Yangtze River Scientific Research Institute, China.

Ding Zhuoyi, 1979, Studies on Characteristics of Flow Pressure Fluctuation, Journal of Yangtze River, No. 3, China.

Stochastic Hydraulics'96, Tickle, Goulter, Xu, Wasimi & Bouchart (eds) © 1996 Balkema, Rotterdam. ISBN 90 5410 817 7

Uncertainty in estimations of excess shear stress

Peggy A. Johnson
University of Maryland, College Park, Md., USA

ABSTRACT: Excess shear stress is used to determine sediment transport capacity, determine bed forms that will occur under various conditions, assess the potential for scour at bridges, and evaluate channel adjustment processes. In this paper, Monte Carlo simulation is used to determine the uncertainty in estimating the excess shear stress, based on the uncertainty in both the average boundary shear stress and the critical shear stress.

1. INTRODUCTION

Shear stress plays an important role in river mechanics. The magnitude of the average boundary shear stress above some threshold, typically the critical shear stress above which sediment movement begins, has been used to determine sediment transport capacity, determine bed forms that will occur under various conditions, assess the potential for scour at bridges, and evaluate channel adjustment processes. One of the problems with using boundary and critical shear stresses is that there is considerable uncertainty in estimating the values. Therefore, assessments based on the shear stress or excess shear stress are also uncertain.

In this paper, Monte Carlo simulation is used to determine the uncertainty in estimating the excess shear stress, based on the uncertainty in both the average boundary shear stress and the critical shear stress. The uncertainty will be described in terms of the coefficient of variation and the associated probability distribution. An example is developed.

2. SHEAR STRESS AS A MEASURE OF CHANNEL PROCESSES

The average boundary shear stress is defined as

$$\tau_o = \gamma RS \tag{1}$$

where γ = the specific weight of water, R = the

hydraulic radius, S = the channel or friction slope, depending on whether the flow is uniform or nonuniform, respectively. The flow depth y is often substituted for R since y is more readily available from gaging records than R. However, this substitution is strictly valid only for wide channels.

The critical shear stress τ_c is evaluated using a variety of methods. The most common method is Shields diagram which gives the dimensionless critical shear stress as a function of the particle Reynolds number for uniform sediments. For nonuniform sediments, other methods are available (e.g., Wiberg and Smith 1987). Wiberg and Smith presented a diagram similar to Shields diagram, except that the dimensionless critical shear stress is a function of both a dimensionless particle diameter and the nonuniform particle size. A dimensionless parameter D/k_s is used to describe the nonuniformity. The critical shear stress thus obtained corresponds to a particle size D in a bed characterized by a roughness length k_s (Wiberg and Smith 1987). Andrews (1983) used data from three gravel and cobble bed rivers to develop a regression equation which gives the dimensionless critical shear stress as a function of the ratio of the particle size to the median particle size in the subsurface bed material. Other studies have shown that the critical shear stress for either uniform or nonuniform sediments is probabilistic rather than deterministic. Based on laboratory and geophysical evidence, Lavelle and Mofjeld (1987) determined that the probability that sufficient instantaneous shear stress exists to provide the force needed to cause particle movement is nonzero even for very small mean shear stress and that sediment transport must be viewed as a probabilistic phenomenon. Buffington et al. (1992) also found that critical shear stress is best described by a probability distribution rather than as a deterministic value.

The shear stress in excess of the critical shear stress, excess shear stress, can be defined as either

$$\tau_e = \frac{\tau_o}{\tau_c} \quad (2)$$

or

$$\tau_e = \tau_o - \tau_c \quad (3)$$

depending on the intended use. τ_e has been used in many ways to predict, either quantitatively or qualitatively, the movement of bed material at a given discharge. For example, it has been used in the prediction of scour (Smith 1994; Raudkivi 1990), bedforms in sand beds (van Rijn 1984; Simons and Richardson 1966), sediment transport capacity (Yang 1973), and channel adjustments (Chang 1988) and to describe channel processes (Pitlick 1992; Dinehart 1989; Andrews 1983; Parker and Klingeman 1982).

3. UNCERTAINTY IN SHEAR STRESS

It is clear that there is considerable uncertainty in attempting to determine a critical shear stress due to an inability to describe the exact point at which particle movement begins. There is also considerable uncertainty in estimating the average boundary shear stress due to changes in the specific weight of water transporting sediment, and estimation of the hydraulic radius and slope. In the following discussion, the uncertainty in the average boundary shear stress and critical shear stress is described using the coefficient of variation (standard deviation divided by the mean) for each parameter and the associated probability distribution. This information is required input to determine the probability that the excess shear stress will exceed specified values.

3.1 Average Boundary Shear Stress τ_o

All three parameters in Eq. 1, γ, R, and S, contribute to the uncertainty in τ_o. The uncertainty in γ is due to the change in the amount of sediment being carried. For clear water, $\gamma = 9800$ N/m³, representing

a minimum value. Assuming that a maximum value can be represented by an additional 10 percent increase, the maximum value is 1.1(9800) = 10780 N/m³. Assuming a triangular distribution which is skewed toward $\gamma = 9800$ N/m³ (see Figure 1), the mean $\bar{\gamma} = 10127$ N/m³ and the coefficient of variation $\Omega_\gamma = 0.023$ (Ang and Tang 1984).

The flow depth y is often substituted for the hydraulic radius R in Eq. 1 since it is more readily available. In that case the actual hydraulic radius may be considerably less than the flow depth, depending on the width of the stream and the flood level. The probability distribution may be described by an asymmetric triangular distribution, shown in Figure 2, with the maximum value given by the flow depth. The coefficient of variation in R, Ω_R, can be computed from (Ang and Tang 1984)

$$\Omega_R = \frac{1}{\sqrt{2}} \left(\frac{y - R_l}{2y + R_l} \right) \quad (4)$$

If a value of Ω_R can be assumed, then Eq. 4 can be solved for R_l. One way of determining an appropriate value of Ω_R is to determine the bias between the R and y for a simple rectangular channel of various width to depth ratios. The coefficient of variation in the hydraulic radius can be assumed to be the same as the coefficient of variation in the bias. Taking width to depth ratios varying from 1 to 100, computing the average and standard deviation of the resulting bias, and computing a coefficient of variation yields $\Omega_R = 0.32$.

The uncertainty in slope can be determined based on previous findings (Johnson 1995) which gives $\Omega_S = 0.25$ and a lognormal distribution. This assumption is based on estimating slope as channel slope from a topographic map.

3.2 Critical Shear Stress τ_c

As discussed previously, τ_c can be computed from a host of different equations, providing a wide range of estimates. The lower limit for τ_c can be assumed to be zero (Lavelle and Mofjeld 1987) and the upper limit can be taken as the value given by Buffington (1992). Assuming that these limits cover approximately ± 2 standard deviations from the mean and that τ_c is normally distributed, the mean τ_c is (Ang and Tang 1984)

$$\bar{\tau_c} = \frac{1}{2}(\tau_{c_l} + \tau_{c_u}) \quad (5)$$

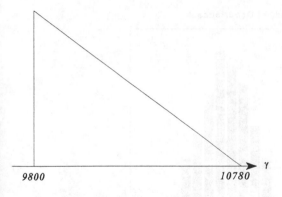

Figure 1. Asymmetric distribution for γ.

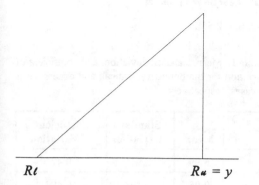

Figure 2. Asymmetric distribution for R.

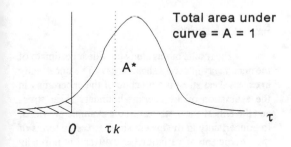

Total area under
curve = A = 1

Figure 3. Truncated normal distribution

where the subscripts u and l represent upper and lower values, respectively, and the coefficient of variation is (Ang and Tang 1984)

$$\Omega_{\tau_c} = \frac{1}{2}\left(\frac{\tau_{c_u} - \tau_{c_l}}{\tau_{c_u} + \tau_{c_l}}\right) \qquad (6)$$

As an example, for $d_{50} = 4$ mm, Andrews gives $\tau_c = 1.9$ N/m², Shields gives $\tau_c = 3.9$ N/m², Wiberg and Smith gives $\tau_c = 4.5$ N/m², and Buffington gives $\tau_c = 6.5$ N/m². From Eq. 5, $\overline{\tau}_c = 3.25$ N/m², which is approximately the value given by Shields, and $\Omega\tau_c = 0.5$ from Eq. 6.

The maximum value of τ_c is unknown, but is given by the tail of the normal distribution. From the lower tail of the distribution, it is possible to obtain a negative value of τ_c; to avoid this problem, the normal distribution can be truncated at zero. To ensure that the area under the truncated distribution curve is one, the area remaining after truncation is normalized as follows. Figure 3 shows a normal curve for a random variable τ_c for which the total area A = 1. If the curve is truncated at $\tau_c = 0$, the area under the remaining curve A* is less than one. However, if A* is divided by itself, the normalized area is again one. A* is equal to the probability that $\tau_c > 0$, so

$$A^* = p(\tau_c > 0)$$
$$A^* = p\left(z > \frac{0 - \overline{\tau}_c}{S_{\tau_c}}\right)$$
$$A^* = 1 - p\left(z < \frac{0 - \overline{\tau}_c}{S_{\tau_c}}\right) \qquad (7)$$

where S is the standard deviation and z is the standard normal variate. For the truncated distribution then, the probability that $\tau_c < k$, where k is a specified value of τ_c, is equal to the probability based on the standard normal distribution multiplied by 1- A*.

4.0 UNCERTAINTY IN EXCESS SHEAR STRESS

Based on the uncertainty in the average boundary shear stress and the critical shear stress, the uncertainty in the excess shear stress can be determined using a simulation program. First the shear stress and critical shear stress are randomly generated N times as follows. All variables in Eq. 1 (γ, R, and S) were treated as random variables with coefficients of variation and probability distributions as described in the previous section. Values of each variable were obtained during a particular simulation cycle by generating a uniform random number and then transforming that number using the inverse transform method (Ang and Tang 1984). Assuming no or minimal correlation between the variables, τ_o was computed from Eq. 1. τ_c was computed also using the inverse transform method, assuming a

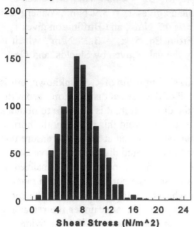

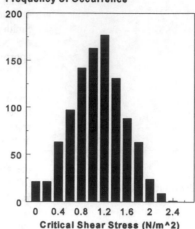

Figure 4. Distribution of τ and τ_c for N = 1000 and flow depth = 2.5 m.

truncated normal distribution and a mean value from Shields diagram.

The process of generating τ_o and τ_c is repeated for N simulation cycles for a specified slope, flow depth, and particle size. The mean and standard deviation of both τ_o and τ_c were then computed. This information can then be used to determine the uncertainty in the excess shear stress. If both τ_o and τ_c are normally distributed, then calculation of the coefficient of variation in the excess shear stress is straightforward. For excess shear stress equal to the difference between τ_o and τ_c, the mean excess shear stress is

$$\overline{\tau_e} = \overline{\tau_o} - \overline{\tau_c} \qquad (8)$$

and the variance is

$$S_{\tau_e}^2 = S_{\tau_o}^2 - S_{\tau_c}^2 \qquad (9)$$

As an example, assume that a particular stream cross section has a slope of 0.0004 and a median sediment size of 1 mm. We are interested in determining the uncertainty in the excess shear stress. Figure 4 shows the results of 1000 simulation cycles for a flow depth of 2.5 m. Both histograms appear to be normally distributed. A chi-square goodness-of-fit test showed that both τ_o and τ_c are normally distributed at all levels of significance. The means, standard deviations, and coefficients of variation are given in Table 1. This yields a coefficient of variation in the excess shear stress of 0.49.

Table 1. Mean, standard deviation, and coefficient of variation for the boundary, critical, and excess shear stresses, respectively.

	Mean	Standard Deviation	Coefficient of Variation
τ_o	6.99	2.91	0.42
τ_c	0.95	0.47	0.49
τ_e	6.04	2.95	0.49

5.0 CONCLUSIONS

The results of this study provide an estimate of the uncertainty in excess shear stress. The uncertainty in excess shear stress is a function of the uncertainty in the estimation of the average boundary shear stress and the critical shear stress. In the example provided, the uncertainty in the excess shear stress, in terms of the coefficient of variation, is 0.49 and is normally distributed. The significance of this relatively high uncertainty is that in using excess shear stress to evaluate sediment transport or bed forms, it should be kept in mind that there is a fairly high uncertainty and that the predicted quantity may vary significantly from the actual value due, in part, to the uncertainty in estimating these important parameters.

REFERENCES

Andrews, E.D. (1983). Entrainment of gravel from naturally sorted riverbed material. Geological Society of America Bulletin, 94, 1225-1231.

Ang, A.H-S., and Tang, W.H. (1984). Probability Concepts in Engineering Planning and Design: Vol. II. Wiley and Sons, New York.

Buffington, J.M., Dietrich, W.E., and Kirchner, J.W. (1992). Friction angle measurements on a naturally formed gravel streambed: implications for critical boundary shear stress. Water Resources Research, 28(2), 411-425.

Chang, H.H. (1988). Fluvial Processes in River Engineering. Krieger Publication Co., Melbourne, Florida.

Dinehart, R.L. (1989). Dune migration in a steep, coarse-bedded stream. Water Resources Research, 25, 911-923.

Johnson, P.A. (1995). Uncertainty of hydraulic parameters. Journal of Hydraulic Engineering, ASCE (in press).

Johnson, P.A., and Ayyub, B.M. (1992). Assessing time-variant bridge reliability due to pier scour. J. Hyd. Eng., ASCE, 118(6), 887-903.

Lavelle, J.W., and Mofjeld, H.O. (1987). Do critical stresses for incipient motion and erosion really exist?. J. of Hydraulic Engineering, ASCE, 113(3), 370-385.

Parker, G., and Klingeman, P.C. (1982). On why gravel-bed streams are paved. Water Resources Research, 18, 1409-1423.

Pitlick, J. (1992). Flow resistance under conditions of intense gravel transport. Water Resources Research, 28(3), 891-903.

Raudkivi, A.J. (1990). Loose Boundary Hydraulics. Pergamon Press, New York.

Simons, D.B., and Richardson, E.V. (1966). Resistance to Flow in Alluvial Channel. U.S. Geological Survey Professional Paper 422-J.

Smith, S.P. (1994). Preliminary Procedure to Predict Bridge Scour in Bedrock. Colorado Department of Transportation, Report No. CDOT-R-SD-94-14.

van Rijn, L.C. (1984). Sediment transport, III, bed forms and alluvial roughness. J. Hydraulics Division, ASCE, 110(HY12), 1733-1754.

Wiberg, P. L., and Smith, J.D. (1987). Calculations of the critical shear stress for motion of uniform and heterogeneous sediments. Water Resources Research, 23(8), 1471-1480.

Yang, C.T. (1973). Incipient motion and sediment transport. J. Hydraulics Division, ASCE, 99(10), 1679-1704.

Stochastic Hydraulics'96, Tickle, Goulter, Xu, Wasimi & Bouchart (eds) © 1996 Balkema, Rotterdam. ISBN 90 5410 817 7

Effect of bed profile on reliability of flood levee design

Mohammad Mafizur Rahman, Nobuyuki Tamai & Yoshihisa Kawahara
University of Tokyo, Japan

ABSTRACT: Variation of the river cross sectional shape is a natural process. In the master plan, though the river bed and the water surface stage are shown to have a smooth profile and the safety measures are taken accordingly with the added assumption of the cross-section to be of idealized shape but in reality the case is not exactly so. 1-D steady flow model was used. The natural variation of the bed level from such smoothness in the longitudinal direction in a 13km reach of Kinu river was analyzed. The effect of natural change of the cross sectional shape was also taken care of by the comparison of the water level stage for it with that for the idealized sections. The idealization of the cross sectional shape was performed from two different point of views. The probability of flooding was observed to be near 1% which should also be taken into account in the reliability analysis of flooding.

1 INTRODUCTION:

The river master plan considers the longitudinal river bed to have a smooth profile. Such master plan also shows water surface stage to have a smooth longitudinal profile. Based on such planning assumptions the levee height is designed. The designed levee usually maintains the uniform freeboard along the longitudinal reach. But in reality, the river bed is under the process of natural variation and it hardly follows such smooth profile. The present study analyses the behavior of such variation with its subsequent effect on the safety considerations of the flood control planning. Fong (1989) and many other researchers have also analyzed the reliability of levee but mostly from the point of view of the discharge uncertainty.

The effect of the change of the cross-sectional shape on the safety measures was also analyzed in the study. The idealization of the existing channel section was done. The 1-D flow simulation was performed to simulate the water surface for the natural sections and each of the idealization types. The effect of change was observed through the difference in the water level stage found in case of the natural river section with that for the two idealized sectional types.

The investigation area covers a river reach of 13.0 kilometers of the Kinu river, in Japan. The downstream and upstream sections of the investigation reach are respectively 45.0 km. and 58.0 km. upstream of the Tone river confluence. (Note: hereafter, section 45 will refer to the most downstream section of the reach and 58 will refer to the most upstream reach and so will be true for the rest of the sections also).That means the reach is selected such that there is no branching of the river within this portion. Another aspect to be observed regarding the location of the experimental reach is that in the upstream side- beyond the reach, the bed slope is steep (ranging between 0.002 to 0.005) and that in the downstream direction beyond the reach is flatter(ranging between 0.0004 to 0.0006). That means this portion of the Kinu river is in a zone of transitional slope.

2 RIVER GEOMETRY DATA:

For Kinu river only the annual river cross sectional geometry data is available. The cross-sectional river geometry map of the Kinu river for the years 1977,1981,1983 and 1989 were obtained. The pitch was 0.5 km. and hence there were 27 river cross-sectional geometry data sets for each of the four years. The number of digitized data points per cross-section ranges from 54 to 201 with an average of 110. The bed actual slope within the reach varies between 1m to 2m per km. A typical section as shown in Fig.1. is for the section 45 surveyed in 1989.

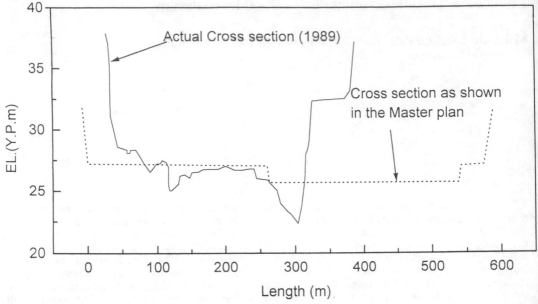

Fig.1 The actual and design cross sectional profile for section 45.

3.MATHEMATICAL MODEL:

Water surface profile for the reach for each year was computed by a 1-D model using the standard step (step-profile) method (Burnham,1990)(Samuel,1995)(Fong,1989)for steady-flow condition. The method solves the steady-flow equations by applying a cross section to cross section, step by step procedure (Chaudhry,1993) (Chow,1959). Along with the continuity equation the energy equation solved in the 1-D model was:

$$F(y_2) = y_2 + \frac{Q^2}{2gA_2^2} + \frac{1}{2}S_{f_2}(x_2 - x_1) + z_2 - H_1 + \frac{1}{2}S_{f_1}(x_2 - x_1) = 0 \quad (1)$$

where the initial approximation of the consecutive

upstream section water depth was done by

$$\frac{dy}{dx} = \frac{S_0 - S_f}{1 - \frac{Q^2 B}{gA^3}} \quad (2)$$

where the suffices 1 and 2 refer to the present section (downstream) and the next section (upstream) respectively in the iteration. In equations (1) and (2) the notations used are as follows:
S_f = the slope of energy grade line at any section,

S_0 = the longitudinal bed slope, H = the total head at any section, y = depth of flow, x = the longitudinal distance of the river reach (positive in the downstream direction), z = elevation of the channel bottom above a specified datum, A = cross sectional area of the river.

Energy coefficient (α) in all cases was assumed to be 1.0. The Newton-Raphson's method was followed for the iteration.

As input data, the bed slope, Manning's roughness coefficient (n), water depth at the downstream section and flow rate (Q) were given. The Manning's roughness coefficient (n) was assumed to be constant (=0.035) for all the simulation cases. This value of n was suggested in the master plan of Kinu river. The actual bed slope data was also obtained.

As the down stream boundary condition the Manning's uniform flow equation was used for a range of flow depths to generate the respective discharge(Q) value. For example, for the year 1989 the actual digitized cross-section(of section 45) was used and for a range of depth of flow- the Manning's equation gave the respective discharge value. Later, from the selection of the desired value of Q the respective depth of flow for section 45 was selected.

As input data, the number of total digitized data for the respective section along with the number of initial and the final points where the calculation should start and terminate were given. These two are actually the highest points on each side of the thalweg within which the water is bounded.

The program calculates the geometrical parameters such as the area of the cross section, wetted perimeter (the hydraulic radius), the top width of a natural river cross section of any arbitrary shape. As output it gives the depth of flow, the longitudinal variation of water surface stage which in other words is produced from its ability to calculate the longitudinal thalweg profile.

4 IDEALIZATION OF SECTIONS

The naturally existing sections were simplified through some idealization techniques. For idealization of the sections two different aspects were considered. In each of the two cases the total cross-sectional area was kept the same as that of the naturally existing section. In order to do this numerical integration was performed on the digitized data within the peaks of the levees on both sides of the channel.

Firstly the natural sections were idealized to the equivalent rectangular ones as shown in Fig.2. Secondly, the idealization was performed to an equivalent shape which resembles more or less to the sectional shapes found in the master plan of the Kinu river Fig.2. This idealization was accomplished for each of the sections for each of the four years.

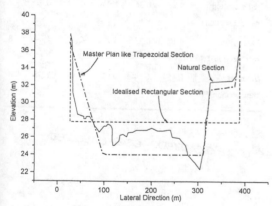

Fig.2 The natural section (section 45) with the equivalent idealized sections.

The bottom point (thalweg) of all the sections were shifted to the (three point) moving average line found by averaging the existing longitudinal profile. This was done to maintain the master plan like smoothness of the bed profile in order to exclude the effect of bed undulation once again which has already been taken care of in the first part.

5 NUMERICAL EXPERIMENTATION

The design depth of flow as shown in the master plan is 8.6m at section 45. The respective design discharge is 5,300 m^3/s. For this value of discharge and the corresponding depth of flow of section 45 for the digitized section in each of the four years were used in the simulation. These gave the water surface profile for the design discharge in each of the four years corresponding to the actual cross sectional shape and hence the actual longitudinal profile.

Fig.3 shows the natural rugged behavior of the longitudinal bed profile for four years(1977,1981,1983 and 1989) compared to the smooth bed profile shown by the master plan. The water surface stage profile for the year 1989 is also shown in Fig.4 with the corresponding longitudinal bed profile.

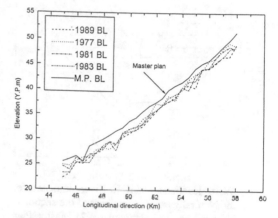

Fig.3 Longitudinal thalweg profile of the studied 13.0km reach of Kinu river as per Master plan and in 1977,1981, 1983 and 1989.

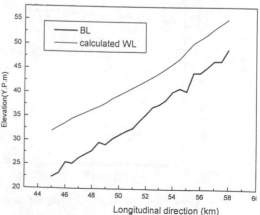

Fig.4 The calculated longitudinal profile of water surface and the bed of Kinu river in 1989.

The similar method was followed for each of the two idealized types of sections. For each of the four years, the normal depth for the section 45 of each of the types were found for the discharge of 5300m³/s. Then the 1-D flow simulation was performed using this discharge and the associated normal depth of water for section 45 as input data. The simulated water surface for the natural section and those of the idealized sections for the year 1989 are presented in Fig.5.

longitudinal water surface stage profile was determined from that of the moving averaged smooth profile at each of the longitudinal locations where the cross sectional data was obtained. In the similar way the deviation at the same locations were determined for the river bed, also for all the four years.

The simulated water surface found for each of the idealized sectional types were compared with those of the natural sections, for all the four years.

6 ANALYSIS:

The longitudinal water surface stage profile and the longitudinal bed profile for the existing sections were made smooth by the moving average method over three points. Then the deviation of the

7 DISCUSSION AND RESULT:

The longitudinal water surface stage profile and the longitudinal bed profile were smoothed because the purpose of the study was to analyze their natural

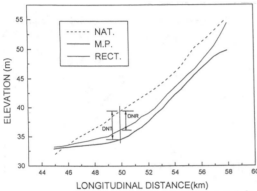

Fig.5 Water level stage of Kinu River simulated for the Naturally existing section and the sections idealised as 1) Rectangular, 2) Master Plan likeTrapezoidal shapes.

Table 1. Statistics of the variations of bed and water surface deviation.

Deviation for the profile of	Mean (m)	Std. Dev. (m)	Max. (m)	λ	ζ	$\chi2$
Water stage (natural section)	0.09	0.08	0.29	-2.67	0.76	0.27
Bed elevation (natural section)	0.28	0.36	1.0	-1.74	0.92	7.75
Water stage (rect. section)*	0.17	0.23	3.1	-2.3	1.04	9.8
Water stage (trap. section)**	0.15	0.34	0.93	-2.8	1.35	8.3

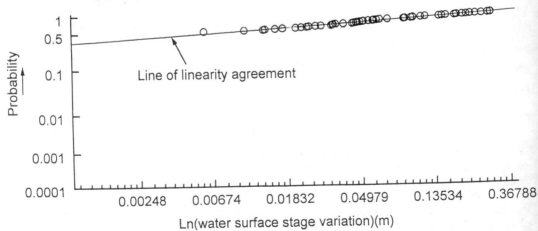

Fig.6 Water surface stage variation about the moving average line for the natural section.

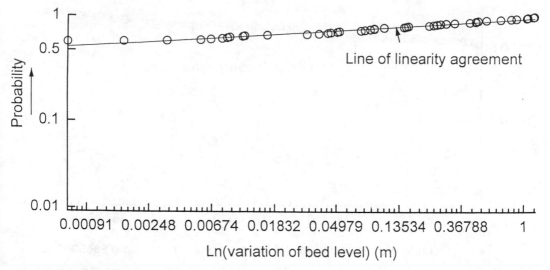

Fig.7 The bed level variation about the moving average line for the natural section.

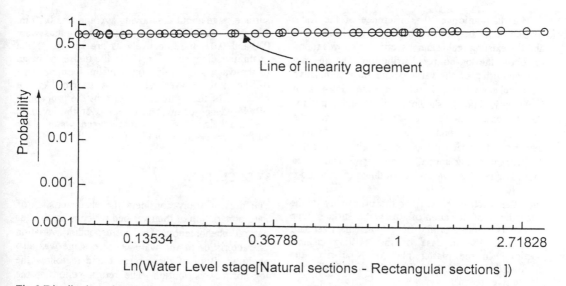

Fig.8 Distribution of Variation of Water Level simulated for the Natural sections and that of the Rectangular sections for Kinu-River.

variation with respect to that of the smoothed one like in the master plan.

The distribution of the deviations of the longitudinal profile of both bed and water stage for the natural sections follows the log-normal distribution. The linearity of relationship of the variates and their cumulative frequencies when plotted in the log-normal probability paper is shown in Fig.6 and Fig.7 for both the cases. The linear best fit line is drawn in the figures. The maximum, mean and the standard deviation of the deviations are shown in Table 1. Respective log-normal parameters(λ and ζ)and the chi-squared(χ^2) values of goodness of fit test are also given. As goodness of fit , Chi-square test(Ang and Tang,.1975) was done for the confidence level of 95% (=1-α). The respective percentile values (Ang and Tang, 1984) of the Chi-square distribution are 77.9.

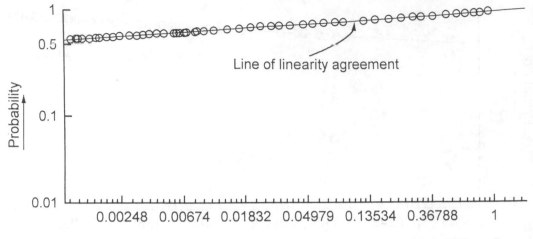

Fig.9 Distribution of Variation of Water Level simulated for the Natural sections and that of the Master plan like trapezoidal sections for Kinu-River.

The distribution of the difference of the water surface stage between each of the idealized types and the existing real channel sections also was found to follow the log-normal distribution. The linearity of relationship of the variates and their cumulative frequencies when plotted in the log-normal probability paper is shown in Fig.8 and Fig.9 for both the cases. The linear best fit line is drawn in the figures. The maximum, mean and the standard deviation of the deviations are also shown in Table 1. Respective log-normal parameters(λ and ζ)and the chi-squared(χ^2) values of goodness of fit test are also given.

For the reach of Kinu river the probability of the exceedence of deviation of the water surface stage above its smooth profile by 1.5 m (the freeboard height) is 1% in case of the existing channel sections. In the master plan of Kinu river the freeboard was maintained 1.5m uniformly for the whole river reach.

The difference of the water level for the rectangular sections* from that of the natural sections were analyzed and the corresponding probability of overstepping was 0.5%. The mean and the standard deviations were 0.17m and 0.23m respectively.

The difference of the water level between the master plan like trapezoidal sections** and the natural sections were analyzed and the corresponding probability of flooding was found to be 1%. The mean and the standard deviations were 0.15m and 0.34m respectively.

As shown in Table 1, for the natural channel sections, the mean of the deviation of the water

surface stage about the moving average was 0.09 m. The 95% confidence upper and lower bounds were 0.1 and 0.01 m respectively as are shown by the qualitative diagram in Fig.10. In the figure the mean of the water stage deviation only in the upper part of the moving averaged line is shown. As the objective of the study is to highlight the probability of exceedence of water surface over the levee so the line below the moving averaged water stage surface was excluded in Fig.10.

8 CONCLUSION:

The present study considers the characteristics of natural variation of the river bed which is an obvious natural phenomenon. The corresponding variation characteristics of the water surface stage was also analyzed. Both of these were found to follow the log-normal distribution. The return period of the exceedence of the water stage above the levee height is almost 100 years. This sort of risk retained from the natural uncertainty of the bed is important and should be taken into consideration by the river planners. This scope, in another way is a source of uncertainty. For an existing river, such consequences should be taken into consideration for the detailed study of the reliability analysis.

It was found that if the master plan would suggest rectangular cross-sections for the Kinu river and maintain the same uniform freeboard of 1.5m as present situation then the possible flooding would occur once in 200 years whereas due to the

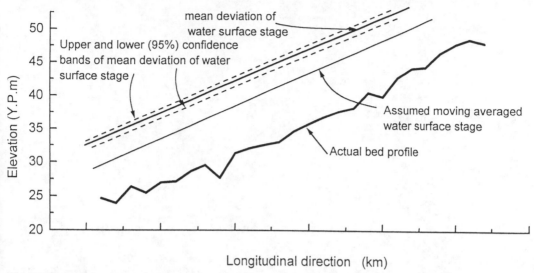

Fig.10 The average of variation of the water surface stage profile about the moving average is shown with the 95% confidence bands in a qualitative diagram (For convenience the vertical scale has been exaggerated).

presently suggested trapezoidal like sections the expected return period is 100 years. Though this may be indicative for the river planners regarding their decisions for the sectional shapes but the flood plains have many practical implications and their importance can not be ignored. The flood plain keeps the levees or dikes safe from the sudden natural destruction caused by the quick migration of the thalweg point due to the sectional shape change. This provides some time to take necessary safety measures when some risk is imminent.

A close observation of the final results will show that the return period is the same for the trapezoidal shaped idealization case and that of the moving average technique done for the natural sections. This, in other words, can be concluded saying that in this type of reliability analysis the simple moving averaging technique can sufficiently produce the desirable results instead of the tedious job of idealizing the trapezoidal like sections.

Acknowledgment:

The authors wish to express thanks to the Shimodate branch office of Ministry of Construction of Japan- for their co-operation to provide the Kinu river cross sectional geometry annual survey maps of four years for the studied area and the master plan.

REFERENCES:

ANG A.H.S. and TANG W.H. : *Probability Concepts in Engineering Planning and Design*, Vol 1, 409p, John Wiley and Sons, (1975).

ANG A.H.S. and TANG W.H. : *Probability Concepts in Engineering Planning and Design*, Vol 2, 562p, John Wiley and Sons, (1984)

BURNHAM M.W. and DAVIS D.W.: *Effects of Data Errors on Computed Steady Flow Profiles*, Journal of Hydraulic Engineering, ASCE, Vol. 116, No.7, 914-929,(1990).

CHAUDHRY M. HANIF: *Open Channel Flow*,483p, Prentice Hall,Inc.(1993)

CHOW, V. T. : *Open Channel Hydraulics*, 680p, McGraw-Hill Book Company (1959).

FONG S. I.: *Risk Analysis on Flood Levee Design*, Dissertation Paper of M.Engg., Department of Civil Engg., University of Tokyo,108p,(1989).

SAMUEL P.G.: *Uncertainty in Flood Levee Prediction*, Hydra 2000(Vol. 1),567-572,Thomas Telford, London(1995).

Stochastic Hydraulics '96, Tickle, Goulter, Xu, Wasimi & Bouchart (eds) © 1996 Balkema, Rotterdam. ISBN 90 5410 817 7

Hydrodynamic characteristics and applications of compound-channel flows

Ching-Ruey Luo
Department of Engineering, Asian Institute of Technology, Bangkok, Thailand

ABSTRACT: In the natural rivers, eventhough man-made canals, the flow situation of compound channels is very common and obviously important. From the view points of high-water regulation, low-water quality controls and flood-plain reclamation and utilization, the compound-channel planning and design will be the most suitable choice for the future river hydraulic study. In this paper, the velocity profiles with different depth-ratio, width-ratio and roughness-ratio, for both center-line of cross-section and lateral direction are solved by statistically semi-empirical method first. Secondly, the turbulent viscosity obtained by introducing a dimensionless turbulent viscosity coefficient is presented. Thirdly, the analytical results of turbulent shear stress and total discharge of compound-channel flows are obtained from velocity profiles and turbulent viscosity, which form the apparent shear stress. Finally, the comparisons of velocity profiles, total discharge with different depth-ratio, width-ratio, and roughnes-ratio among analytical, numerical and experimental results are made.

1. INTRODUCTION

Movement of water during flood period occurs under the conditions of complicated interaction of the main-channel and flood-plain streams. The principal reason of this is the difference in hydraulic resistances of the main-channel and flood-plain. This conditions considerable velocity gradients near the channel edges, formation of eddies, transverse water masses exchange and increased turbulence. This results in the increase of the stream's energy losses, reduction of stream velocity in the main-channel and of its discharge capacity.

River channel sections are normally compound-section shapes composed of the main channel and the berm sections. The latter ones are often hydraulically much rougher than the former one. The difference in depth and roughness under the overbank flow conditions leads to a slower movement of water in the berm section. The slower moving water retards the faster one, thus, creating a shear layer which extends laterally to a considerable extent both into the main-channel and the berm-section as in Fig. 1. In the case of rivers with flood-plains, when the depth of flow ex-

ceeds the one of the main-channel, the flood plains carry a part of total discharge. When D/d is much larger than unity, and it is a very common case in the natural rivers, in Fig. 2, the general practice is to divide the compound flow section into a number of supposedly homogeneous subareas by the introduction of shear-free vertical boundaries as extensions of the banks of the main channel and to compute the flow for each subsection by generally using Manning's equation, and sum up to obtain the total quantity.

In fact, there exists the apparent shear stresses at the edges of main-channel. The vertical apparent shear force increased rapidly for low relative depths and large berm width. Therefore, the flow could be characterized by four dimensionless parameters namely,

D_r, B_r, n_r, and $H_r = \dfrac{(B/2)}{(D-d)}$, where the first two are used to show the geometrical properties of the whole channel; the third one for channel physical property and the last one is for the flow condition of main channel a two or three dimensional flow. Velocity profiles in centerline of compound channel were derived, then the velocity profile U(y,z) were obtained;

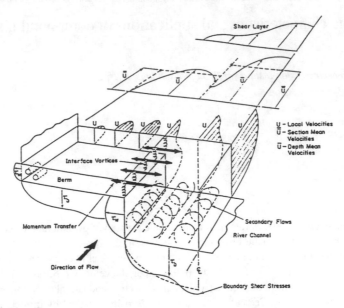

Fig.1 Hydraulic Aspects of Overbank Flow

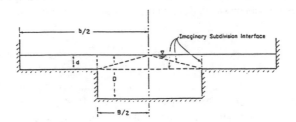

Fig.2 Definition Sketch of Compound Cross-Section

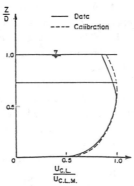

Fig.3 Calibration of Isovels in Centerline
of Compound Section Channel
(Data See Knight, D.W., 1984, Fig. 9)

finally the turbulent viscosity were derived.

2. THEORETICAL CONSIDERATIONS

Velocity profiles in compound-section channels

The following two relationships of the velocity profiles both at the centerline of the compound channels and at any positions respect to the corresponding velocity-value in the centerline are derived and regressed from experimental data. The ranges of each dimensionless parameter are as the follows:

$$D_r=0.197\sim0.505; \quad B_r=1.73\sim4.01; \quad n_r=1.00\sim3.23;$$
and $\quad H_r=1.00\sim6.90$

The proposed equations are:

A. Primary velocity profile at the centerline $U_{C.L}(z)$

$$\frac{U_{C.L}}{U_{C.L.M}} =Exp[-0.71(\frac{z}{D} - 0.65)^2] \tag{1}$$

where $U_{C.L.M}$ is the maximum velocity in the centerline of compound channel.

B. Velocity profiles of primary flow to their corresponding horizontal-plane position $U(y,z)$

$$\frac{U}{U_{C.L}} =1.177(D_r)^{0.2004}(n_r)^{-0.2300}(\frac{y}{(b/2)})^{0.2545}(\frac{z}{D})^{-0.3556} \tag{2}$$

2. Turbulent viscosity of compound section channels

The suggested DEV model from Fischer (1979).

$$\upsilon_t=C_\upsilon Hu_* \tag{3}$$

and regressing the data sets from Ogink, (1985), we have

$$C_\upsilon=0.2(b/B)-0.15 \quad \text{for } b/B\geq2.0$$
$$=0.25; \quad \text{for } b/B\leq2.0 \tag{4}$$

Hanxiang (1980) derived an analytical solution for $\overline{\varkappa}_{xy}$ at steady uniform flow, which is

$$\overline{\upsilon}_t(y)= \frac{C_h gH}{C_c} \overline{U}(y) \tag{5}$$

in which, $C_H=4.8$; $M=6+0.6C_C$; Comparing Eqs. (4) and (5) the new form of the eddy viscosity is,

$$\overline{\upsilon}_t=(\frac{n*h*\sqrt{g}}{R*^{1/6}}) [0.2(\frac{b}{B})-0.15].U(y.z_c) \tag{6}$$

Apparent shear stress Due to the difference of the water depth and resistance coefficient from the main-channel and berm-section, the apparent shear stress resulting from the lateral momentum transfer will form.

The turbulent shear stress in the berm-section, near to the edge of the main-channel, say $y=y_0$, and $z=d_f$, then

$$\frac{\overline{\varkappa}_{f.p}}{P} =(\frac{n_f d_f \sqrt{g}}{R_f^{1/6}})[\frac{\partial U(y,z)}{\partial y}][0.2(\frac{b}{B})-0.15]U(y_0 z) \tag{7}$$

where $$R_f=\frac{[\frac{1}{2}d_f(b-B)]}{[2d_f+\frac{1}{2}(b-B)]} \tag{8}$$

d_f: measure from bed of flood plain to the position of $z=z_c$

The turbulent shear stress in the main-channel, near to the intersection dividing surface, also, $y=y_0$ and $z=z_c$, then

$$\frac{\overline{\varkappa}_{m.c}}{P} =(\frac{n_{m.c} z_c \sqrt{g}}{2R_{m.c}^{1/6}})[\frac{\partial U(y,z)}{\partial y}][0.2(\frac{b}{B})-0.15]U(y_0 z_c) \tag{9}$$

where $$(\frac{BZ_c}{2Z+B})=R_{m.c} \tag{10}$$

z_c: measured from bed of main channel to the position of $z=z_c$.

The net turbulent apparent shear stress is,

$$\frac{\overline{\varkappa}_t}{\rho} = \frac{\overline{\varkappa}_{m.c.}}{\rho} - \frac{\overline{\varkappa}_{f.p.}}{\rho} \tag{11}$$

if $\frac{\overline{\varkappa}_{m.c.}}{\rho}$ is greater than $\frac{\overline{\varkappa}_{f.p.}}{\rho}$, the exflux momentum from main-channel to flood-plain; conversely, the influx momentum from flood-plain to the main-channel.

4. Discharge of compound-channel flow

Many studies derived the relationship between water level and discharge for uniform flow. In uniform flow, the energy gradient of the main-channel is assumed to be equal to that of the flood-plain, and energy gradient, S_f, is equal to the slope of the channel, S_0. However, it is well documented that this energy gradient is not totally consumed as a driving force for discharge. Because the apparent shear stress exerts on the channel flood-plain interface, effective energy gradient becomes nonhomogeneous between the main channel and the flood plain. In this study S_{fm} and S_{ff} are defined as the effective energy gradients of the main-channel and flood-plain, respectively. Due to this nonhomogeneity of the energy gradient, S_{fe} is introduced to explain this phenomenon as follows:

$$S_{fm}=S_f-S_{fe}; \; S_{ff}=S_f+S_{fe} \qquad (12)$$

The relationship between the apparent shear stress and the head loss, Δx, is given by

$$\rho g A . \overline{\Delta h_e} = 2 \bar{\varkappa}_a d \Delta x \qquad (13)$$

that is

$$\frac{\Delta h_e}{\Delta x}=S_{fe}= \frac{2\bar{\varkappa}_a D}{\rho g A} \qquad (14)$$

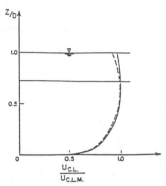

Fig.4 Verification of Isovels in Centerline of Compound Section Channel
(Data See Tominaga , A. 1988 , Fig.4)

3. COMPARISONS AND DISCUSSION

1. Comparisons of velocity profiles The comparisons, including calibrations and verifications, of the velocity profiles on both centerline of the cross section and the corresponding horizontal-plane distributions between analytical methods in Eqs. (1) and (2) and the experimental data obtained from Wormleaton (1982), Knight (1984), Murota (1990), and Tominaga (1988) are plotted in Figs. 3 to 6.

2. Comparisons of dimensionless eddy viscosity The form of dimensionless eddy viscosity, c_v in Eq. (4) gives the mean value of 0.59 by comparing the re-

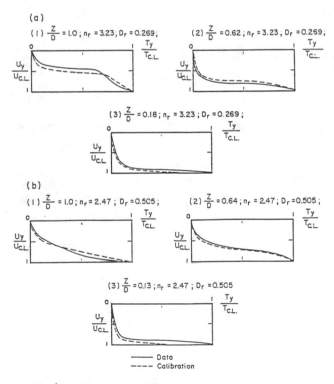

Fig.5 Calibration of Velocity Profiles in Horizontal-Plane of Cross-Section for Compound Channels
(Data See Knight, D.W. 1984. Figs. 9 and 10)

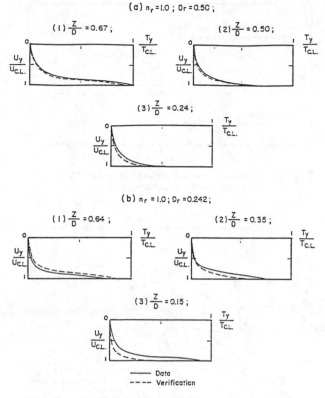

$$(a) \ n_r = 1.0 \ ; \ D_r = 0.50 \ ;$$

$$(1) \frac{Z}{D} = 0.67 \ ; \qquad\qquad (2) \frac{Z}{D} = 0.50 \ ;$$

$$(3) \frac{Z}{D} = 0.24 \ ;$$

$$(b) \ n_r = 1.0 \ ; \ D_r = 0.242 \ ;$$

$$(1) \frac{Z}{D} = 0.64 \ ; \qquad\qquad (2) \frac{Z}{D} = 0.35 \ ;$$

$$(3) \frac{Z}{D} = 0.15 \ ;$$

——— Data
---- Verification

Fig.6   Verification of Velocity Profiles in Horizontal- Plane
of Cross - Section for Compound Channels
(Data See Tominaga , A; 1988, Fig. 4)

sult given by Shiomo, and Knight, (1991), c_v=0.50, the range of B_r=2.2 to 6.6. The value of this dimensionless eddy viscosity suggested by ASCE Hydraulic Committee (1988) is, c_v=0.60. The nice agreement is shown in Fig. 7.

3. Comparisons of discharge after the bankful water level The apparent shear stress causes the additional head loss due to the momentum transfer at the interface between main-channel and flood-plain. The effective slope of energy formed in Eq. (12) was used to calculate the effectvie discharge by taking into account the effect of shear stress. The predicting total discharge is formed as follows:

$$Qe = K_m S_{fm}^{1/2} + K_f S_{ff}^{1/2} \tag{15}$$

where

$$K_m = A_m R_m^{2/3} n_{m.c.}^{-1}; \ K_f = A_f R_f^{3/3} n_{f.p.}^{-1} \tag{16}$$

In order to obtain Q_e in Eq. (15), n, S_{fm}, $\bar{x}_{mc}$ and $\bar{x}_{f.p}$ were calculated. The S_{fm} and S_{ff} were Eq(12). while $\bar{x}_{m.c}$ and $\bar{x}_{f.p}$ from Eqs. (4), (7), and (9). Substituting these results in Eqs. (15) and (16) by considering the interface as the perimeters, the analytical results of Q_e are obtained. By comparing the analytical results with Eddy Model and experimental data from Tamai (1982), good agreement is obtained and plotted in Fig. 8.

When the water-depth just rised over the bankful of main-channel, the mean velocity in main-channel is quite larger than the one in flood-plain. Meanwhile the flux, ρQU, flows from main-channel to flood-plain, too. When the water-depth is high enough, or say d/D is large enough, the ρQU in flood-plain is greater than the one in main-channel; at this moment, the flux flows from flood-plain to main-channel.

The strong main-channel vortex with expanding spanwise flood-plain vortex happens when the water-depth just rises over the bankful or the ratio of d/D is

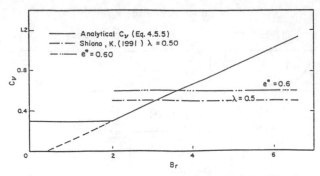

Fig. 7   Relationship between Dimensionless Turbulent Viscosity
Coefficient and the Width-Ratio

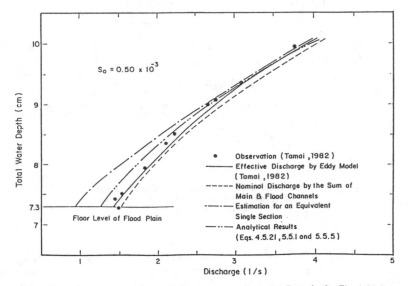

Fig. 8   Variation of Discharge after the Water Level Exceeds the Flood Plain

very small. This is because the very strong flux from main-channel to flood-plain. When the water-depth is high enough or d/D is large enough, the strong flood-plain vortex with weak main-channel occurs, and this strong flood-plain vortex or flux destroys the eddies in the lower part of main-channel.

4. CONCLUSIONS AND APPLICATIONS

1. The analytical velocity profiles for both vertical direction of centerline in compound-channel and the horizontal direction along the lateral, or cross-section, were derived. Good agreements were obtained by comparing the analytical velocity profiles with the experimental data.

2. The new regression form of the dimensionless viscosity coefficient, c_v, was expressed in function of B_r. The mean value of c_v showed acceptable result.

3. By comparing the discharge of compound channel among analytical, experimental and numerical results, good predictions by analytical methods were obtained.

4. From the view of point of flood-control, increasing the conveyance of channel within flood seasons is very important. From the view point of river bed management in dry seasons, creating new-born land to improve the living quality by horizontal-plane utilization, such as tennis court, golf field,

and greens etc., is very necessary for developing and developed countries. The compound-channel flow can be used to achieve this purpose.

REFERENCES
1. KNIGHT, D.W. and MAMED, M.E. (1984) "Boundary shear in symmetrical compound channel", Journal of the Hydraulics Division, ASCE, Vol. 110, No. 10, pp.
2. MUROTA, A.(1990) " Effects of channel shape and flood-plain roughness on flow structure in compound cross-section", J. of Hydroscience and Hydraulic Engg., Vol. 8, No. 2, pp. 39-52.
3. SHIONO, K., and KNIGHT, D. (1991) "Turbulent open channel flows with variable depth across the channel", J. of Fluid Mech., Vol. 222, pp. 617-646.
4. TAMAI, N. and KAWAHARA, Y. (1982) "Resistance law of a flow in composite cross sections", Proceedings of the 3rd Congress of the Asian and Pacific Regiona Division of IAHR. Vol. B, pp. 42-51.
5. THE ASCE TASK COMMITTEE ON TURBULENCE MODELS IN HYDRAULIC COMPUTATION (1988). J. of Hydraulics ASCE, Vol, 114, No. 9. pp 970-1073.

6. TOMINACA, A., NEZU, I. and EZAKI, K. (1988), "Experimental study on secondary currents in compound open-channel flows", IAHR XXIII Congress, pp. A. 15-A. 22.
7. WORMLEATON, P.R., ALLEN, J. and HADDJIPANOS, P. (1982) "Discharge assessment in compound channel flow", Journal of the Hydraulics Division. ASCE, Vol. 109, HY9, p. 975-994.

Stochastic Hydraulics'96, Tickle, Goulter, Xu, Wasimi & Bouchart (eds) © 1996 Balkema, Rotterdam. ISBN 90 5410 817 7

Comparison of stochastics for stage-frequency analyses

Lee Butler & Norman Scheffner
US Army Engineer Waterways Experiment Station, Vicksburg, Miss., USA

ABSTRACT: Two different approaches for computing stage frequency along the south coast of Long Island, New York, are compared. Over a decade ago a state-of-the-art hydrodynamic model was used to compute storm surge impacting the New York Bight and Long Island coast. The stochastic method of choice was the Joint Probability Method (JPM), as in many similar studies conducted at the time. These methods and how they were applied were somewhat dictated by the available computer resources. In the present study a new, more resolved hydrodynamic model as well as an emerging efficient stochastic procedure for life-cycle analysis, based on a bootstrap resampling-with-replacement and nearest neighbor interpolation is applied. The paper compares both hydrodynamic and stochastic approaches, delineating weaknesses in the older technology and suggesting research to strengthen confidence in the new results.

1 INTRODUCTION

The U.S. Army Engineer District, New York, initiated a project reformulation study for the south shore of Long Island, New York, in 1980. The focus of the study included the coastal reach from Fire Island Inlet to Montauk Point and has been referred to as the FIMP study. Its purpose was to assist efficient coastal protection design by providing accurate predictions of the likelihood and effects of destructive storms which shape the coast, damage property and structures, and endanger lives. The FIMP study was completed in the early 1980s and resulted in flood frequency estimations along the coast and within bays of southern Long Island. Now, more than a decade later, we have a unique opportunity to revisit this work and apply state-of-the-art hydrodynamic and stochastic technologies.

Flood protection for several areas along the FIMP coastal reach has significantly been reduced by erosion of beaches and dunes relative to 1980 protection levels. Hence, new calculations are warranted, even if the result is a validation of previous results. This paper begins by presenting the two hydrodynamic approaches. The new hydrodynamic model utilizes an unstructured grid finite element model which is capable of accurate and efficient computation over very large domains while still providing high resolution in areas of interest. This is in contrast to the late-1970s technology which employed a finite difference technique applied on stretched orthogonal grids over the domain. A major issue in the earlier study was lack of computer resources; most of the computations were made on a CDC 7600. However, today we have a CRAY C-90 at our disposal with vastly increased speed and memory.

The original FIMP study used JPM techniques to compute storm-induced stage-frequency relationships. The JPM assumes storm event parameters are basically independent and can be modeled with fixed parametric relationships. This approach is somewhat lacking in technical validity and requires experience in applying to avoid creation of an impossible, or extremely unlikely, event. The new approach is the Empirical Simulation Technique (EST) which utilizes parameter inter-relationships of historic storm events to generate a database of storm parameters and corresponding environmental responses such as the computed surge. The EST then generates multiple life-cycle scenarios of storm activity from which frequency and error estimates can be made. A companion paper on the EST technique is presented in this conference.

2 HYDRODYNAMICS

2.1 Structured dual-grid application (1980)

The hydrodynamic model applied in the original

FIMP study (Butler and Prater 1986) was the WES Implicit Flooding Model (WIFM). The numerical and hydrodynamic features of WIFM are discussed in Butler (1980) and the application to coastal studies is documented in numerous papers and reports. WIFM solves the primitive vertically integrated time-dependent shallow water wave equations of fluid motion in Cartesian coordinates. The governing equations are solved using an alternating direction implicit finite difference algorithm on a stretched rectilinear grid. The code allows subgrid barriers to be exposed, submerged, or overtopped, as well as individual cells over low-lying terrain to flood or dry during a simulation.

It was intractable to construct a single grid extending from beyond the continental shelf to the nearshore and back-bay areas of interest with sufficient resolution to capture inlet conveyance, bay effects, and barrier island overtopping. Thus a coarse global grid (Figure 1) composed of over 3500 cells was constructed to cover the New York Bight from a point south of Atlantic City, New Jersey, to beyond Cape Cod, Massachusetts (and includes New York Harbor and Long Island Sound). The purpose of the global grid was to model large-scale tidal and storm events, providing results for input to a nearshore domain, high-resolution, 4100 cell grid of the main study area. The high-resolution grid stretches from near Jones Inlet to beyond Shinnecock Inlet (including all back-bay and channel systems) and has variable grid size resolution of 200 to 1200 m. Even with this resolution it was difficult to obtain a good calibration of the model to measured water levels and velocities.

The decision to use a dual-grid approach also influenced the choice of stochastic procedure for computing stage frequency within the nearshore gridded area. This will be discussed in a later section of the paper.

2.2 Unstructured single-grid application (1995)

A finite element based hydrodynamic model, called ADCIRC-2DDI (Luettich, et al. 1992), was applied on an unstructured grid for all storm event simulations. The model is the depth integrated option of a system of two- and three-dimensional hydrodynamic codes (Luettich, et al. 1992). The models are based on the Generalized Wave-Continuity Equation (GWCE) solved in conjunction with the primitive form of the momentum equations. Governing equations reflect incompressibility, Boussinesq, and hydrostatic pressure approximations. The model has the capability of wetting and drying computational cells and uses the standard quadratic parameterization for bottom stress.

One problem often encountered in the modeling of nearshore flow hydrodynamics is that computational boundaries of the model are not well removed from the area of interest. For example, the continental shelf can substantially affect the amplitude and phase of a storm surge propagating from open water onto the shelf. If boundary conditions are specified on the shelf, errors are introduced in the solution because the assumed boundary conditions are posed in a dynamic flow region, i.e., the transformation of the flow field over rapidly changing bathymetry.

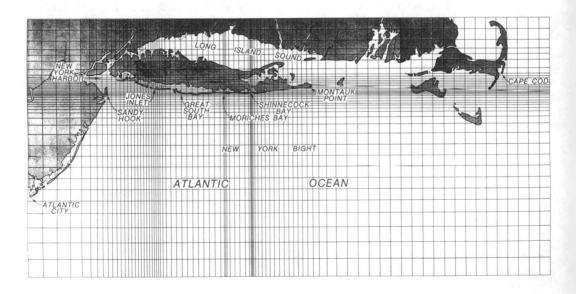

Figure 1 WIFM global computational grid

A major advantage of an unstructured grid is the ability to specify sparse resolution in areas of deep bathymetry and high resolution in shallow areas or near complex boundaries. Figure 2 displays the gridded computational domain for the revised study. In this 15,000 node grid, the offshore boundary to the east is placed at 60° west longitude while the southern boundary is placed at approximately 26° north latitude. An enlargement of Fire Island Inlet in the FIMP study area shown in Figure 2 demonstrates the ability to provide high grid resolution where necessary.

All back bay areas all highly refined and inlet features are fully resolved with grid node spacing as small as 50 m. The computer system used for the calculations was the CRAY C-90. This machine has 16 parallel processors, each of which has a performance rating of over 1 gigaflop. The previous finite difference model was run a CDC 7600 which had a performance rating below 10 megaflops resulting in a gain of over 1000 in performance with the new machines. The lack of computer power in 1980 highly influenced selection of a dual-grid approach in the earlier study.

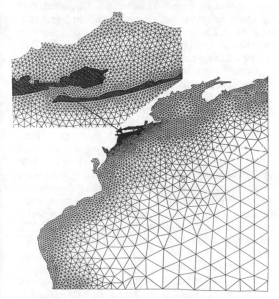

Figure 2. ADCIRC computational grid

3 STOCHASTICS

3.1 Joint probability method

The JPM approach adopted for hurricanes in the FIMP study involved defining probability distributions for each of five descriptive hurricane parameters (central pressure deficit, P, radius to maximum winds, R, forward speed of the storm eye, F, and the storm track direction and landfall point). Review of historical data led to choosing an ensemble of storms represented by the various combinations of storms developed from 6 P's, 3 R's, 3 F's, and five landfall points with 3 track directions (2 tracks for bypassing storms). This selection gives a total ensemble of 918 hurricanes, each with its own probability of occurrence. With the use of sensitivity studies, the actual number of simulations was reduced to 306 hurricanes by simulating storm combinations with extremes of the central pressure probability distribution and interpolating for intermediate values.

Because of the difficulty in parameterizing extratropical storms and the availability of data from frequent storm events, a semi-historical approach was adopted. It was determined that a 41-year period of events (1940 to 1980) was sufficient to adequately cover the area of dominance on the combined event stage-frequency curve (lower than 50-year flood level events). Historical data from 101 storms over the 41-year span that produced at least a 0.7 m surge at the entrance to New York Harbor were used to develop a partial duration exceedence (PDE) curve. Twenty seven representative storms from this set were selected and assigned probability masses in proportion to the amount of PDE distribution each storm was to represent.

Inclusion of the tidal effect was a complicating factor. In the open coast area the tide may be superimposed on the storm surge without significant loss of accuracy. Hence all the storm simulations were made on the global grid without tidal influence. Tides were incorporated through the convolution of average seasonal tidal values with output of the surge-only numerical simulations. This procedure resulted in tens-of-thousands of surge-tide combinations which were rank ordered by elevation and plotted to produce stage-frequency relationships at desired open-coast locations throughout the study area.

To properly inundate the shoreline, including overtopping and breaching of the barrier islands, and subsequent flooding of the back-bay areas, it is necessary to include the tide directly in the computation along with potential increase of water level due to wave effects. To accomplish this feat it was intractable to run hundreds or thousands of additional storm simulations with all tide possibilities. Therefore in simulating storm response on the nearshore grid, it was necessary to represent the statistics of the huge population of global surge-tide events with a small ensemble. This was achieved by taking a random sample (storm-tide event) within every 15 cm change in total flood level at a nearby open-coast location for which a global stage frequency was developed, repeating the process three times. The idea was to develop three separate sets of data from the same population which could be analyzed separately and together (yielding confidence limits on the results) to arrive at a frequency curve for a given location.

For hurricanes the flood level range sampled was 2.5 m resulting in three sets of 17 storm-tide events or a total population of 51 events to be simulated on the nearshore grid. Weaker events were not emphasized due to the dominance of extratropical storms for return periods below 50 years. It was demonstrated that using flood levels from such a sampling procedure was capable of reproducing the open-coast result within an acceptable error band. For extratropical events the range was less and a total of 39 events were randomly selected and simulated on the nearshore grid.

Probabilities of each nearshore event to be simulated were established according to that portion of the nearby open-coast stage-frequency curve (generated via the global grid calculations) each event represented. Stage frequencies at interior locations were developed from these results.

The JPM method was popular a decade ago for use in estimating frequency of flood level occurrence. However, several problems are inherent in the method for use in the New York area. First of all the probability exceedence curves for each storm parameter are based on a very sparse data set (10 storms at the time the first study was conducted). Next each parameter in turn is varied while holding the other four parameters fixed. This results in a large population of storms, many of which are probably unrealistic. There is also a question on the interdependence of parameters. In the original study a relationship between the central pressure and track was determined and probabilities were adjusted accordingly. These weaknesses are overcome by the approach presented below.

3.2 Empirical simulation technique (EST)

The EST statistical "life-cycle" simulation procedure is based on a "bootstrap" resampling-with-replacement, nearest neighbor interpolation, and subsequent smoothing technique, in which a random sampling of a finite length database is used to generate a larger database. The only assumption is that future events will be statistically similar in magnitude and frequency to past events. The EST begins with the development of a database of storm parameters, then uses these parameters as a basis for simulating multiple repetitions of multiple years of storm activity. Details of the EST can be found in a companion paper presented at this conference (Scheffner and Borgman 1996).

The EST requires specifying a set of parameters which describe the dynamics of some physical system, tropical and extratropical storms in this application. These parameters, which must be descriptive of both the process being modeled and the effects of that process, are defined as a N-dimensional vector space. The parameters which describe only the physical attributes of the process are referred to as input vectors. For tropical storms, pertinent input vectors include the same parameters used in the JPM approach. Extratropical event input vectors are limited to the wind generated 1) wave height, 2) wave period, and 3) tidal phase.

The second class of vectors involve some selected response resulting from the input-vector parameterized storm. For the FIMP study, the response of interest is maximum water surface elevation, defined as storm surge plus tide. These values are computed via the ADCIRC model and archived at each specific location corresponding to each particular historical or simulated event of the total set of tropical and extratropical events used in the study.

The tropical storm data base of the National Hurricane Center's HURricane DATa (HURDAT) data base (Jarvinen, et al. 1988) is the source of historical events from which a tropical storm input vector data base was constructed. Documentation of event selection can be found in Scheffner, et al (1994). The extratropical event data base was selected as the 1977-1993 year portion of the U.S. Navy's Fleet Numerical Mete-orological and Oceanography Center wind field database.

The final selected set of events for simulation via the ADCIRC model was 16 historical and 3 hypothetical tropical events and 10 extratropical events. Storm events used to generate input to the EST are referred to as the "training set" of storms. Four separate M_2 tidal phase boundary conditions were specified for each storm because each event is independent of tidal phase. Because tropical events produce extreme surge elevations which can overtop the barrier islands, the tropical events were simulated concurrently with a tidal elevation time series open boundary condition. Therefore, a total of 76 tropical events were simulated with the ADCIRC model.

Extratropical events rarely produce surges which overtop the barrier islands, therefore, only 10 extratropical events were simulated with ADCIRC. A separate simulation of the M_2 tide resulted in an M_2 tidal amplitude at each station location. The four phases of the tidal amplitude were then linearly added to the maximum surge computed for the extratropical event. This combination of tide and surge produced a data base of 40 extratropical events.

The EST utilizes input and response parameters associated with site specific historical events as a basis for developing multiple simulations of future responses. These responses are subsequently used to determine frequency-of-occurrence relationships inherent in the historical data base. It is not incumbered by reliance on an assumption that storm parameters are independent or on assumed parametric relationships. On the other hand this method uses the joint probability relationships inherent in the historical data base.

The number of simulations required to achieve results similar to a JPM approach is far fewer. In this application the reduction in the number of computer simulations was approximately one tenth. Only one grid was required, thus reducing validation efforts and, with fewer simulations, data management requirements were far less.

4 NUMERICAL RESULTS

4.1 Dual-grid application (1980)

Both global and nearshore models were validated for tide, hurricane, and extratropical events. Error estimates on all stage-frequency results were computed (Prater, et al. 1986). Figure 3 displays example stage-frequency curves for Sandy Hook at the entrance to New York Harbor and for Montauk Point at the eastern end of Long Island. Also displayed are Sandy Hook water level measurements analyzed for up to a 25-year return period. Similar comparisons at other historical gage sites give a significant degree of confidence in the older method for open coast simulations.

Prior to the earlier study a single accepted stage-frequency curve was used for the entire Long Island coastline. The 1980 study demonstrated the error in such an assumption. At the 100 year return period there is over a meter difference in projected water surface elevation. This fact has had a significant influence on the design of coastal projects for this coastal reach and has dramatically reduced costs of earlier overly-conservative designs.

4.2 ADCIRC application (1995)

The ADCIRC model was validated for the M_2 tidal constituent and for several historical tropical and extratropical events. The three hypothetical events used in the data training set represent the 1938 hurricane with two altered paths and hurricane Gloria with an increased radius to maximum wind value at landfall. Results presented in this paper are limited to the tropical event analysis in which the 16 historical events and 3 hypothetical events are used to generate an array of response vectors for each location at which stage-frequency relationships are desired.

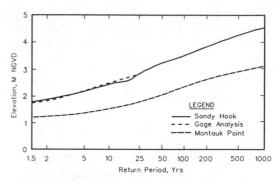

Figure 3   WIFM/JPM generated stage-frequency relationships with comparison to measurements

Figure 4 displays the tropical event stage-frequency relationships computed for the Sandy Hook gage. Because the EST is a multiple life-cycle simulation approach, multiple stage-frequency relationships are generated, each corresponding to a separate simulation. In the present analysis, 100 simulations of a 200 year simulation of storm activity were generated.

Mean value relationships are computed with error estimate bounds defined as plus/ minus one standard deviation. These values are indicated in Figure 5 which displays the analyzed stage frequency curve for Sandy Hook. Also included on the figure are points representing the tropical event curve from calculations in the 1980 study. As shown, the EST generated relationship is below that generated in the original FIMP study, however, the original results are approximately within the error limits of the EST results.

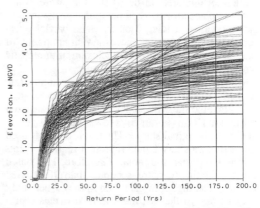

Figure 4   EST stage-frequency relationships for Sandy Hook, New Jersey

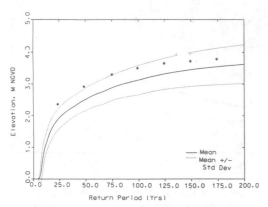

Figure 5 Stage-frequency curve for Sandy Hook

5 DISCUSSION, CONCLUSIONS AND RECOM-MENDATIONS

While the methodology used in the earlier study was innovative more than a decade ago, there were significant drawbacks and problems encountered. Successful simulations of open coast surge were made but it was difficult to accurately model hydro-dynamics in back bay areas as well as barrier overtopping. The study was unable to determine the influence of the JPM assumptions on the slope of the exceedence curve for tropical events. How important was the influence of "unrealistic storms" included of necessity in the storm ensemble? What impact did ignoring the interdependence of tropical storm parameters have? These questions were never answered satisfactorily.

The hydrodynamic and stochastic modeling approaches used in the revisited FIMP study represent a great improvement over the techniques employed in the original study. As a result, the major difficulties encountered in the first study were eliminated: the need for two computational grids; inaccurate modeling of inlet, back-bay, and barrier over-topping hydrodynamics; necessary assumptions of parameter independence and inclusion of unrealistic storm events.

Research on the new approach and the application to the New York area has not been completed. It is noted that the stage frequency relationship generated for Sandy Hook (Figure 5) is lower than that generated in the original study. The tropical events used in the present study should represent realistic events which impact the study area. Therefore, EST-based results should be more accurate than JPM-based results. The higher levels found in the earlier study may be due to the fact that the JPM events represent combinations of storm parameters which are not truly representative of the study area. Since the extratropical events dominate the lower end of the curve (lower return periods), this impact may not have been consequential.

Further research is required to define sufficiency of the training set of storms (input response vectors) as well as how to best define life-cycle scenarios. How far do you go in breaking down probabilities associated with historical events by including slight perturbations of these events (altered track or another appropriate parameter)?

Final conclusions of the study are that the new technology is certainly more accurate and easier to apply than the older approaches. As a result, surge analyses such as the present project can be accomplished in less time and with more accuracy than was possible a decade ago.

ACKNOWLEDGEMENTS

The research described was sponsored by the US Army Engineer District, New York, and permission to publish the material was granted by the Chief of Engineers.

REFERENCES

Butler, H.L. 1980. Evolution of a numerical model for simulating long-period wave behavior in ocean-estuarine systems. Estuarine & Wetland Processes with Emphasis on Modeling. Marine Science Series. Vol. 11. New York: Plenum Press.

Butler, H. Lee and Prater, Mark D. 1986. Innovative determination of nearshore flood frequency. Proceedings. 20th ICCE. ASCE. Taipei, Taiwan. 9-14 Nov 1986.

Jarvinen, B.R., Neumann, C.J., and Davis, M.A.S. 1988. A tropical cyclone data tape for the north Atlantic basin, 1886-1983: Contents, limitations, and uses. NOAA Tech Memo NWS NHC 22.

Luettich, R.A., Westerink, J.J., and Scheffner, N.W. 1992. ADCIRC: Theory and methodology. TR DRP-92-6. Rpt 1. USAE WES. Vicksburg, MS.

Prater, M.D., Hardy, T.A., Butler, H. Lee, and Borgman, L.E. 1984. Estimating error of coastal stage frequency curves. Proceedings. 19th ICCE. ASCE. Houston, TX. 3-7 Sept 1984.

Scheffner, N.W., Mark, D.J., Blain, C.A., Westerink, J.J., and Luettich, R.A. 1994. A tropical storm data base for the east and Gulf of Mexico coasts of the US. TR DRP-92-6. Rpt 5. USAE WES. Vicksburg, MS.

Scheffner, Norman W. & Borgman, Leon E. 1996. The empirical simulation technique, a bootstrap-based life cycle approach to frequency analysis. Proc. 7th IAHR Int Symp on Stochastic Hydraulics. MacKay, Australia. 29-31 Jul.

Stochastic Hydraulics'96, Tickle, Goulter, Xu, Wasimi & Bouchart (eds) © 1996 Balkema, Rotterdam. ISBN 90 5410 817 7

Hydraulic characteristics of stepped spillway

Pan Ruiwen & Xu Yiming
Yunnan Polytechnic University, Kunming, China

ABSTRACT: The hydraulic characteristics of flow on a stepped spillway, over a high sill broad crested weir, are systematically studied by experiment. It is seen that the skimming flow formed on the stepped spillway face makes a lot of air entrained into the discharge. Beneath the main flow are stable vortices with air bubbles on the steps. Flow depth and velocity along the spillway probably remain unchanged after the flow is fully aerated. A great amount of flow energy is dissipated by the stepped spillway. The hydraulic conditions in the downstream can be greatly improved and cavitation damage to spillway face can be prevented properly. The stepped spillway is an effective dissipator.

1 INTRODUCTION

For a dam with a conventional spillway, the energy in the discharge is almost transported to the toe of the spillway and is dissipated concentratedly by a way of under current, jet flow or surface current and so on. Whereas energy dissipation of stepped spillway is different. It has a series of steps cut into the spillway face downstream. Flow energy is dissipated by the steps. The hydraulic conditions after the toe of dam can be greatly improved. Hence the energy dissipator at the toe can be simplified or even omitted. Besides, it has the advantage of stepped spillway keeping continuity for using building material and dam geometry shape in construction process. So, with the technique of roller compacted concrete(RCC) dams appearing, a new type, a stepped spillway has been gradually practiced in some hydraulic projects in the world. But until now, not much of the flow behaviour on a stepped spillway has been understood.[1] We have done a hydraulic model experiment of stepped WES spillway with four different step heights in the condition of free overfall.[3] For the spillway project, located at Bajiacun, Kunming, China, five stepped spillway models were constructed to study flow behaviors in the conditions of free overfall and sluice flow.[2] The findings indicate that reasonable steps can ensure forming typical flow on the stepped spillway-skimming flow. Kinetic energy of flow can be dissipated along the steps and the potential for cavitation damage to spillway face is eliminated. The stepped spillway is an effective dissipator.

2 EXPERIMENTAL APPARATUS AND TESTING

The model includes a high sill broad crested weir, sluice gate and spillway. They are all made of plexiglass. Sluice is 9.6m wide. A reservoir is connected with the upstream sluice gate. A steep glass flume links with dam toe (Fig .1).

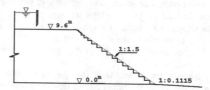

Fig. 1 Stepped Spillway Profile

Table 1. Step dimension

Model No.	Transition step		Regular step	
	Height(m)	Number	Height（m）	Number
I	0.32	1	0.64	12
	0.48	2		
II	0.40	1	0.80	10
	0.60	1		
III	0.48	1	0.96	8
	0.72	1		
IV	0.32	1	0.64	3
	0.48	1	0.80	3
			0.96	4
V	0.32	1	0.64	8
	0.48	2	0.80	2
			0.96	1

Note: All step heights are prototype values

This is a scaled 1:16 model based mainly on Froude similarity law as well as on resistance similarity law. Six spillway models were constructed. Model 0 is an unstepped smooth spillway; Model I , II ,III are stepped spillways with unchanged regular step heights; Models IV and V are stepped spillways with different step heights. For all stepped models, to prevent the underside of discharge away from spillway face, there are 2~3 smaller transition steps before regular steps. The heights of transition steps are 1/2~3/4 of the regular (Table 1). The slope of unstepped spillway profile is 1.5H :1V and all external edges of step on models I to V are fitted to this profile. Piezometric tubes were set up on the face of spillway and steps. Both free overfall and sluice flow are conducted in these models. The range of discharge rate is q=1.72, 3.02, 8.18, 10.94 m³/s/m (prototype values) for free overfalls and q=2.20, 4.08, 5.68, 7.40 m³/s/m (prototype values)for sluice flow. About 50 runs for six spillway models were conducted to study flow conditions and hydraulic parameters.

3 FLOW CONDITIONS

The test results indicate that:

(1) In each test run for model I to V, the flow passing on the spillway contacts well with the stepped spillway face. Two flow conditions appeared from the crest to the toe (Fig. 2). The fore-reach of the spillway is in no air entrainment. The flow speeds up downstream along the flow course, forming gradually a certain turbulent boundary layer. The zones between main flow and steps are filled partly at first and then with water cushion and with rolling vortices. The free surface of flow is smooth and well defined. The post-reach is in air entrainment. Turbulent layer develops to the free surface and air entrainment occurs at some point and flow is swelling after aeration. Instead of water cushion on steps is stable rolling vortices trapped air bubbles. The overlying flow moving on the spillway is supported by the vortices and the tips of step. The aerated region comes to the toe of the spillway and the free surface is tumbling violently.

(2) Because the underside of flow jet on the spillway is broken by the steps, the turbulence of flow is violent and the boundary layer is fully developed. A typical flow regime-- skimming flow of the stepped spillway is formed. The skimming flow is aerated after a certain point. The zone between main flow and steps is filled by the stable horizontal axis vortices with air bubbles(Fig. 2). Flow depth increases obviously due to aeration and then seems to be constant. The free surface probably changes its position around the mean location for each testing discharge rate. The kinetic energy of flow is dissipated by flow dispersing, aeration, the momentum transferring to the recirculating fluid, the shear stress between main flow and vortices, and violent turbulence. Therefore the velocity is down and then tends to stability.

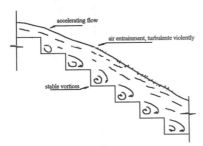

Fig. 2 Skimming flow-typical flow regime
of stepped spillway

(3) With the increasing of discharge rate q , the position of air entrainment and stable vortices move progressively down in the case of specified steps. For example, for model I in sluice flow, the stable vortex occurs at 4# step in the case of q=2.20m³/s/m, at 7# step and 10# step in the case of q=4.08m³/s/m and q=5.68m³/s/m, respectively. The positions of air entrainment are probably the same to that of the stable vortices. After discharge rate increasing to some limit (such as q=7.40), vortices on steps and air entrainment tend to disappear. Free overflow contacting closely with the stepped spillway face appears. Flowing water moves down along the "hydrodynamically smooth surface" formed on the tips of steps.

(4) For skimming flow, the step dimension is proportional to the discharge rate. That is, small steps are suitable for small discharges and vice versa. The combination of different size steps is suitable for keeping better flow regime in the case of a wide range of discharge, especially in middle and small-size hydraulic projects.

4 HYDRAULIC CHARACTERISTICS

4.1 Discharge capacity

For all stepped spillway models, both in free overfall and in sluice flow, the discharge capacity is the same to that of unstepped spillway. For example, the tested synthetic discharge coefficient m of free overfall is 0.351~0.362 and m of sluice flow (relative gate opening 0.2) is 0.488~0.570.

4.2 Flow depth on stepped spillway

The measurement of depth is difficult, since the flow is aerated. Vertical mean water depth measured at the tips of selected steps for model I is listed in Table 2.

Table 2. Mean depth,model I (m)

q (m³/s/m)		1.72	3.02	4.08	5.68	7.40	10.94
Step	0#	0.37	0.56	0.48	0.74	1.07	1.52
	2#	0.35	0.51	0.43	0.61	0.99	1.44
	4#	0.38	0.45	0.40	0.58	0.93	1.31
	6#	0.45	0.51	0.38	0.50	0.86	1.20
	8#	0.45	0.61	0.56	0.61	0.78	1.12
	10#	0.45	0.61	0.56	0.68	0.72	1.01
	12#	0.45	0.61	0.56	0.68	0.69	0.96
	14#	0.45	0.61	0.56	0.68	0.66	0.91

Note: Prototype values

The test results show that the aerated flow depth on the stepped spillway increases by 150%~200% compared with the unstepped spillway.[2] For lower discharges(such as q=5.68m³/s/m or below), the skimming flow appeared on stepped spillway. Air entrainment occurs at certain flight of the spillway and the flow depth tends to be stable after the point of air entrainment. For relatively big discharge (such as q=10.94m³/s/m or above), the free overflow without vortex between steps and main flow appears and flow depth decreases from the crest to the toe of spillway.

4.3 Pressure distribution along spillway face

The results of the experiments showed that pressure on the horizontal face of step is higher than that on the unstepped spillway face. For example, on horizontal face of $7^\#$ step of model V. For four test runs (q=1.72, 3.02, 8.18, 10.94m^3/s/m) in free overfalls, the pressure is 8.13, 10.98, 14.11 and 16.95kN/m^2, respectively. For four test runs (q=2.20, 4.08, 5.68, 7.40m^3/s/m) in sluice flow, the pressure is 9.60, 11.37, 12.25 and 13.33kN/m^2, respectively. Whereas at the corresponding point of the unstepped spillway in the same flow conditions, the pressure is only 0.98, 1.86, 4.41, 5.29, and 1.18, 2.45, 3.33, 3.27kN/m^2, respectively. Minus pressure only appears in vertical faces of step and the value is very small.

4.4 Velocity of skimming flow

The flow depth of stepped spillway tends to be stable after the flow is fully aerated, the velocity also seems to become constant.

Table 3. Discharge, upstream head and toe velocity

		q (m^3/s/m)	2.20	4.08	5.68	7.40
	O	H	0.33	0.63	1.01	1.81
		V_C	12.48	13.16	13.58	14.10
	I	H	0.34	0.64	1.02	1.83
		V_S	9.14	10.32	10.96	11.62
	II	H	0.34	0.64	1.01	1.81
Model		V_S	9.12	10.26	10.88	11.42
	III	H	0.34	0.64	1.03	1.82
		V_S	9.10	10.14	10.72	11.23
	IV	H	0.34	0.64	1.03	1.83
		V_S	8.68	9.52	10.41	10.86
	V	H	0.34	0.64	1.03	1.87
		V_S	8.45	9.24	10.02	10.54

Note: All scaled to prototype values, H: m; V_C, V_S: m/s

The measured values of water head H above the crest and the flow velocity at the toe in the condition of sluice flow are listed in Table 3. V_C(for unstepped spillway) and V_S(for stepped spillway) are the mean velocity in vertical central line in the toe cross section and were measured by a digital propeller velocity meter. It can be seen that V_S decreases by 17.6%~32.3% compared with V_C in the same condition. For each stepped spillway, the reduction of V_S is affected by the discharge rate and the aerated flow length in the case of specified step height.

5 ENERGY DISSIPATION

The energy dissipation rate η_s caused by flow on a stepped spillway can be estimated as follows: If $E_C(E_S)$ is the kinetic

energy per unit weight of liquid at the toe cross section of smooth spillway without steps (stepped spillway).

$$E_c = \frac{V_c^2}{2g}$$

$$E_s = \frac{V_s^2}{2g}$$

$$\eta_s = \frac{E_c - E_s}{E_c} = \frac{V_c^2 - V_s^2}{V_c^2}$$

where, V_C, V_S are flow velocity at the toe cross section of smooth spillway and stepped spillway respectively. The test results are listed in Table 4.

Table 4. Energy dissipation rate (%)

(a) Sluice flow

		q (m^3/s/m)	2.20	4.08	5.68	7.40
		I	46.4	38.5	34.9	32.1
		II	46.6	39.2	35.8	34.4
Model		III	46.8	40.6	37.7	36.6
		IV	51.6	47.7	41.2	40.7
		V	54.2	50.7	45.6	44.1

(b) Free overfall -Model V

q (m^3/s/m)	1.72	3.02	8.18	10.94
η_s (%)	57.7	54.4	45.2	42.8

From Table 4. it can be seen that, either for sluice flow or for free overfall, the energy dissipation rate η_s is affected by q and step height Δh. With the increasing q, η_s decreases in the case of specified Δh, whereas with the increasing Δh to some extent, η_s increases in the case of specified q. The models with mixing steps are superior to the others. Within the testing range of q, η_s is 32%~58%. It is less than the results of reference [5] and approximately agrees to the results of reference [3]. In the present test, η_s is relatively small. The main reasons are that the tested weir height is lower, spillway face slope is mild and the length of flow course is shorter (so the number of steps is reduced). The conditions of forming and keeping skimming flow regime were affected by the model boundaries.

6 CAVITATION CHARACTERISTICS

In order to find out cavitation damage to stepped spillway face, the flow cavitation characteristics have been analyzed. In general, the cavitation number K is used to judge whether cavitation would occur or not. K can be written as follow :

$$K = \frac{p - p_v}{\gamma v^2 / 2g}$$

where, p,v are absolute pressure and mean velocity at the given step, respectively, p_v is the vapor pressure, varying with temperature, for t=25°C, p_v=3.16kN/m^2, γ is specific weight for water and g is gravitational acceleration. The results of cavitation number K for model V at 7$^\#$ step are listed in Table 5. The value of K is 1.72~17.4 for the testing range of discharge rate.

If the step can be considered as a small drop on the spillway face, then the cavitation number at initiation $K_i \approx 1.1$. It is evident that cavitation would not occur, if $K \gg K_i$.

Table 5. Cavitation number at step 7$^\#$, model V

q (m^3/s/m)	p (kN/m^2)	v (m/s)	k
1.72	106.1	3.44	17.4
2.20	107.6	4.15	12.2
3.02	108.9	4.87	8.93
4.08	109.4	6.91	4.45
5.68	110.2	7.68	3.63
7.40	111.3	9.49	2.41
8.18	112.1	10.1	2.14
10.9	114 9	11.4	1.72

Under the condition of skimming flow, it can be considered that step broke up the underside of flow, air bubbles, caused by the steps, are trapped by the flow coming into the vortices. On the flow surface, much air gets in the flow due to turbulence. The air density in the water increases. Pressure and velocity on the step face and fluid property are changed. So the possibility of cavitation damage to spillway face is reduced or even avoided. The prototype observation of M'Bali dam is presented in reference [6]. It is a stepped spillway of 24m high with 25 steps(each step height is 0.8m), the design head is 4m and maximum discharge is 960m^3/s. After the first operation of the spillway, no deterioration of spillway face was found by visual inspection.

7 CONCLUSIONS

(1) In comparison with the conventional spillway without steps, the effectiveness of energy dissipation of stepped spillway is more considerable. A stepped spillway can reduce the erosion and simplify the works for energy dissipation. It can be used in weir with steep slope of spillway in high dam as well as in broad crested weir with mild slope of spillway in low dam for certain discharge rate. A stepped spillway is an economical, practical energy dissipator.

(2) The key for dissipating energy on stepped spillway is to form the flow regime of skimming flow on stepped spillway. Within the aerated region of skimming flow, the flow depth and velocity tend to be stable. The choice of configuration of stepped spillway depends mainly on effectiveness to forming and keeping the regime of skimming flow. The spillway with different step size is more suitable for different discharges.

(3) The efficiency of energy dissipation of stepped spillway is based on dam height, discharge rate, step height, slope of spillway face and the length of flow course on spillway face. Energy dissipation rate decreases with the increasing of discharge rate in the case of specified dam height and step height, whereas with increasing of step height, step numbers and the flow length, the rate increases in the case of specified discharge rate. The suitable discharge rate is in the range of 10~12m^3/s/m. For lower dam with mild slope of spillway face (such as 1.5H : 1V or even milder), the energy dissipation rate of stepped spillway is only 30%~60%.

(4) Because the flow on stepped spillway is broken by the steps, a lot of air bubbles are trapped into the vortices on steps. Flow property is changed. The possibility of cavitation damage to spillway face is greatly reduced. The reasonable type of stepped spillway is also an effective measure to avoid cavitation erosion.

ACKNOWLEDGMENT

This study was supported by Yunnan applied and basic science fundation.

REFERENCES

1 Pan Ruiwen, Xu Yiming, "Stepped dissipator on spillway face", Journal of Yunnan Institute of Technology, 1994(1).

2 Pan Ruiwen, "Model test report on energy dissipation over stepped spillway for Bajacun dam", Yunnan Polytechnic University, 1994,9.

3 Ru Shuxun, Pan Ruiwen,et al., "Energy dissipation on stepped spillway with curved weir", Symposium on Discharge Works and High- velocity Flow, Chengdu University of Science and Technology Publishing House, Chengdu,China, 1994.

4 Wu Chigong, "Hydraulics", Higher Education Publishing House, China, Second edition, 1983.

5 Houston K.L. , Richardson A.T. , "Energy Dissipation Characteristics of a stepped spillway for an RCC Dam ", Proceeding of The International Symposium on Hydraulics for High Dams, Beijing, 1988.

6 Bindo M., Gautier J. and Lacroix F. , "The stepped Spillway of M'Bali dam ", Water Power & Construction, January 1993.

Stochastic Hydraulics'96, Tickle, Goulter, Xu, Wasimi & Bouchart (eds) © 1996 Balkema, Rotterdam. ISBN 90 5410 817 7

Experimental study on the pressure transfer function

Chia-Shan Huang
Union-Tech. Engineering Consult, Taiwan, China

Heng-Haur Chow & Chia-Chuen Kao
Department of Hydraulics and Ocean Engineering, National Cheng Kung University, Tainan, Taiwan, China

ABSTRACT: The aim of the present study is to obtain a practical model for the pressure type wave gauges to transform the bottom pressure data into surface wave information. A series of measurements both in regular and irregular waves were carried out in the laboratory. The data analysis shows that the pressure response function given by the linear wave theory is valid in the shallow water region. In deep water region the linear theory needs to be modified. A modified model has been developed, which is a one-parameter semiempirical formula based on the linear theory. The parameter was determined from the measured data. The comparison results show that the error of the modified formula is within 6.3%.

I. Introduction

Wave measurements are essential for study and engineering purposes. The wave measurement methods commonly include direct ways such as acoustic instruments and step-resistance types, indirect ways such as buoys, radar and the pressure type gauges. Among the measurement ways the buoy is floating with the water surface and the others need a fixed structure. The selection of the measurement method depends on user's requirement and the economic condition. In many cases the pressure type gauges are used for its advantages:

1. The pressure type gauges are placed at the bottom of the sea. They are easy to install and no further structure for holding and supporting the gauges in necessary.
2. The process of measuring does not interfere and block the activities of the vessel nearby, especially in the near shore areas.
3. They have low costs compared to the other types of gauges.
4. Because of placing at the bottom of the sea, they are more immune to the attacks of the sea rollers. Hence, the failure rate of the pressure type wave gauges is lower than the others.
5. The pressure type gauges can provide the information of the mean water level variation resulted from the tides and the storm surges while the moored surface gauges, such as pitch-and-roll buoys, can not.

The pressure-type wave gauges use the information via indirect measurement, which can only measure the subsurface pressure variation, to predict the changes of the surface water level. According to small amplitude wave theory the relationship between the surface water level and the subsurface pressure is linear. In the shallow water area the linearity relation becomes a non linear relation for the influence of forced wave and sea current. Building the correct relationship between the surface water level and the subsurface pressure becomes the most important topic on the applications of the kind of gauges.

The nonlinear characteristics of the waves in the shallow water area result in the errors of the measurement of the pressure type gauges with linear theory. Some relevant researches can be found in Folsom[1947][1], Lee[1984][2], Gabrial[1986][3], and Kao[1983][4], etc.

The researches of the measurement using pressure type gauges have concentrated on the analysis of pressure correction factor and the development of the transfer functions between the wave pressure and the surface water level. In this paper, a practical transfer model which based on the small amplitude wave theory cooperated with the experience function of pressure correction factor was developed. This transfer model was also provided for the purpose of calculating the surface elevation spectra from the bottom pressure spectra.

II. Theoretical Considerations

In small amplitude wave theory, the relation between the surface water elevation η and the underwater dynamic pressure p can be derived from the small amplitude wave theory as

$$\frac{p}{\rho g} = \eta \frac{\cosh k(h+z)}{\cosh kh} \tag{1}$$

Where ρ is the water density, g is the gravitational acceleration, h is the water depth, z is the depth of the gauge position, and k is the wave number.

When the pressure gauge is positioned at the bottom of the sea, i.e. z=-h, Equation(1) could be simplified as

$$\frac{p}{\rho g} = \eta \frac{1}{\cosh kh} \tag{2}$$

From Equation(2), the relation between the underwater dynamic pressure and the surface water variation is linear, and the constant of proportionality is the pressure response function K_p defined in the small amplitude wave theory. The pressure response function K_p is only a function of the relative depth of water.

The pressure response function can be expressed as

$$K_p(h/L) = \frac{1}{\cosh kh} \tag{3}$$

For $\cosh kh$ being greater than 1, the greater h/L, the smaller K_p becomes. It means that the pressure, measured from the equation(2), will be smaller when the waves of the same amplitude but of bigger value of h/L.

To discuss the validity of the pressure response function in linear wave theory, the paper used the pressure correction factor N proposed by Folsom and Seiwell[1947][1]. This pressure correction factor is described in terms of h/L as

$$N(h/L) = \frac{K_p(h/L)_{theory}}{K_p(h/L)_{measure}} \tag{4}$$

where $K_p(h/L)_{theory}$ is the theoretical pressure response function, and $K_p(h/L)_{measure}$ is the ratio of the measured pressure to the amplitude of the surface wave.

When N approaches 1, which means the linearity holds true for the pressure response function, the water surface elevation can be calculated from the pressure response function. However, if N deviate from 1, the pressure response function is not appropriately suitable for the transformation between the wave pressure and the wave elevation.

In practice, since the sea waves can be treated as irregular waves, the wave spectrum is often used to represent the energy distribution. Under the assumption of the linear superposition, each frequency component of wave can be treated as a single component wave. Therefore, whether the linear wave pressure formula can

be applied to all the frequency domain becomes a problem which needs to be investigated in more details.

In order to discuss the relation between the wave pressure and h/L related to the change of surface wave, the measured wave pressure to the surface wave transfer function of each component in the irregular wave $H_p(h/L)$ is defined:

$$H_p(h/L) = \sqrt{\frac{S_{pp}(h/L)}{S_{\eta\eta e}(h/L)}} \tag{5}$$

$S_{pp}(h/L)$ is the measured bottom pressure head spectrum. $S_{\eta\eta e}(h/L)$ is the measured surface wave energy spectrum. Since this paper discusses the relation between the wave pressure and h/L. The frequency of each component wave in the energy spectrum can be transformed to h/L based on the linear wave theory.

The pressure correction factor of the irregular wave, N, can be shown as the ratio of the theoretical pressure response function to the measured response function under each frequency component:

$$N(h/L) = \frac{K_p(h/L)}{H_p(h/L)} \tag{6}$$

As the mentioned above, when N=1 the linear wave theory is applicable under the specified frequency to calculating the relative depth. Whenever $N \neq 1$, some modifications must be considered.

From Equation (5) and (6), the surface wave spectrum $S_{\eta\eta e}(h/L)$ calculated from the pressure spectrum can be expressed as

$$S_{\eta\eta e}(h/L) = [\frac{N(h/L)}{K_p(h/L)}]^2 \cdot S_{pp} \tag{7}$$

III. Experiment Setup and Procedure

A. Experiment Instrument and Configuration

The experiments were carried out in the wind-wave flume of the Department of Hydraulics and Ocean Engineering of the National Cheng Kung University (see Figure 3-1). The size of the flume is 27m × 1m × 1.4m. A flap type wavemaker is installed at the front end of the flume and a coarse surface slope with a tilt slope of 1/10 placed at the back for the water breaking purpose. Both sides of the flume are the transparent glass plates such that the motion of wave can be clearly observed. The stainless plates were placed at the bottom of the flume. The wavemaker uses a personal computer sending out the voltage signals via D/A converter and servo control system to the hydraulic press such that the wave paddle can move back and forth to produce waves. Meanwhile, the surface waves are measured by the capacitance-type wave gauge and the data transmitted back to the computer for analysis and storage.

In the experiments, a capacitance-type wave gauge was placed at a position from the wave paddle by 9.4m. Right below the wave gauge a pressure gauge is mounted at the bottom to measure the surface wave and the pressure variation synchronously. The measured signals of the surface waves and the bottom wave variations are amplified and transmitted to the computer via A/D converter for further processing.

B. Regular Wave Pressure Experiments

To examine the consistency of the relation between the theoretical pressure response functions and the practical experimental bottom-surface wave in any environments, the first step is to carry out the experiments of measuring the regular wave pressure and wave level. In the process, the water depth was kept at 70 cm, also the pressure gauge was positioned at the depth of 70 cm from the surface. Totally 141 waves of different conditions, including 13 deep water waves, 120 intermediate depth waves and 8 shallow water waves were tested. The periods of the wave ranged from 0.7 to 9.5 seconds and the wave heights were from 1.3 to 29.3 cm. The quiescent water levels were measured prior to each sample being tested. Thirty seconds of the surface wave and the sampling rate of 20 Hz were used for determining the average value of the data set. Forty samples of each wave period were obtained to be analyzed. The sampling rate was 20 Hz, too.

C. Irregular Wave Pressure Experiments

This paper also undertook the irregular wave experiments to investigate the phenomena of the wave pressure and levels. Two kinds of spectra, JONSWAP and P-M, were adopted in the experiment. They are explained as follows:

1. JONSWAP Spectrum:

$$S(f) = \frac{3.28}{C_1^2 \cdot C_4^2} (\frac{H_s}{T_s^2}) f^{-5}$$
$$\cdot \exp[-\frac{5}{4}(C_2 f T_s)^{-4}] \cdot r^{\exp(\beta)} \tag{8}$$

where

$$\beta = -\frac{1}{2\sigma_0}(C_2 f T_s -)_1^2. \tag{9}$$

In Equation (8), the parameters are adopted from the results of practical measured waves of Taiwan Strait from Qu[1977][5]:

$$C_1 = 3.8,$$
$$C_2 = 1.13,$$
$$r = 2.08,$$

$$\sigma_0 \begin{cases} \sigma_a = 0.07 & f \le f_p, \\ \sigma_b = 0.09 & f > f_p. \end{cases} \tag{10}$$

2. Pierson-Moskowitz Spectrum:

$$S(f) = 0.121(\frac{H_s^2}{T_s^4}) f^{-5}$$
$$\cdot \exp(-0.485(T_s f)^{-4}) \tag{11}$$

$$T_p = 1.266 T_s. \tag{12}$$

In the irregular wave experiments, 97 JONSWAP spectra and 114 P-M spectra of different peak frequencies and significant wave heights were calibrated under the conditions of water depths of 70, 60, and 50 cm, peak frequency varying from 0.50 to 0.95 Hz, and significant wave height varying from 4 to 14 cm. Totally 211 wave spectra are tested in these bottom wave pressure experiments.

The quiescent water was measured for 30 seconds and sampled at a rate of 20Hz before the experiments. During the process of experiment, sampling rate was kept at 20 Hz while the surface wave and the pressure were sampled every other 204.8 second. Therefore, 4096 data sets for surface wave and wave pressure were acquired.

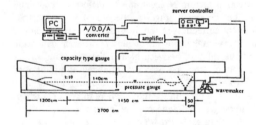

Figure 3-1 The layout of the experimentation setup

IV. Data Analysis and Discussion

A. Verification of the Regular Wave Experiments and Linear Pressure Response Functions

In the regular wave experiments, there were 13 deep water waves, 120 intermediate depth waves, and 8 shallow water waves being tested. The pressure response function measured from the experiments are depicted in Figure 4-1. The lower horizontal axis, upper horizontal axis, and the vertical axis are presented as h/L, frequencies, and the transfer function values, respectively. In the figure, for the convenience, the solid line, the theoretical solution of linear wave, is compared with the measured values. Figure 4-2 illustrates the variation of the pressure correction factor N with respect to the change of h/L. From Figure 4-1 and 4-2, when h/L is less than 0.35, the measured values were approximately equal to the theoretical functions. When h/L is greater than 0.35, the values of the theoretical function become smaller than the measured ones. The greater h/L is, the greater deviation generates.

A conclusion can be reached from the above results: when h/L < 0.35, the relation between the surface wave and the bottom pressure can be estimated for the intermediate depth and shallow water waves. So the theoretical pressure response function can be properly applied to calculate the surface wave for a signal wave form. For the deep water wave, the theory will overestimate the level variations. In addition, when h/L >0.5, the measured and the theoretical K_p both are smaller than 0.1. It shows that the surface waves have little influence on the bottom wave pressure under the deep water conditions.

B. Validity of the Theoretical Pressure Response Functions for Irregular Waves

1. Discussion of N variation
The surface spectrum $S_{\eta\eta}(h/L)$ and the pressure spectrum $S_{\eta\eta}(h/L)$ can be acquired from the Fast Fourier Transform (FFT) of the measured surface elevation and bottom pressure signals. When those spectra are substituted in Equation (5) and (6), the correction factors N according to different frequencies can be obtained. Figures 4-3 to 4-5 demonstrate different correction factors in different wave conditions. In order to make comparison among different data set to find the relation between N and the spectrum energy distribution, the spectra are normalized and multiplied by two after 256-point window smoothing. Finally they are shown with N in the figure where the vertical axis is N, upper axis is frequency, and lower axis is h/L.

From Figures 4-3 to 4-5 one can observe that the N is uniformly distributed around the significant frequency component but it scatters at the low frequency and gathers together at the high frequency area.

Figures 4-3 and 4-4 give the N plotted against h/L for $f_p = 0.71Hz, Hs = 7.5cm$ and $f_p = 0.91Hz, Hs = 7.95cm$, respectively. In Figure 4-3, when f =0.45 to 1.05Hz, i.e., when h/L =0.15 to 0.5, N approaches unity. If f is beyond this range (h/L > 0.5), N becomes smaller gradually. This means that the values of theoretical pressure response function are smaller than the measured values. Although this result shows

consistently in the results of regular wave experiments in qualitative, the displacement is not the same for each h/L in quantity. When the frequencies are low (f, 0.45), the energy for both the surface wave and the pressure is small and the N's are randomly distributed, but it does not mean that the linear wave theory is invalid for the estimation of surface wave. So we can see when f >1.05Hz, N starts decaying, however, N still remains steady in the area where the most energy is located. But in Figure 4-4, although in the significant energy distribution area, when frequency is greater then 1.05 Hz (h/L > 0.5), an unsteady phenomenon of N shows up. So from the above, N is dependent on h/L, not wave condition. When h/L > 0.5, N is deviated from unity and thus the linear wave theory is not suitable applied in this region.

Figures 4-4 and 4-5 have different water depth h=70cm and 50cm. The Ns in Figures 4-4 and 4-5 are decaying starting from frequency 1.05, 1.25 Hz which all corresponds to the same h/L=0.5. Although water depths are different, N approximately equal to 1 in h/L < 0.5 and main energy distribution region, when h/L > 0.5, N starts to decay. So N depends on h/L, not water depth. Note that especially in the shallow water region, when h/L > 0.5, N decays more slowly than in other regions, and thus the range for the valid linear wave theory is larger. One can conclude that the more the pressure gauge is near the surface, the more accurately

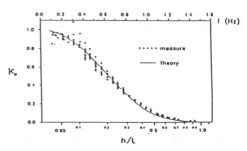

Figure 4-1 The comparison of the measured pressure response factor with linear wave theory.

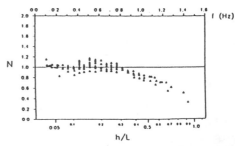

Figure 4-2 The relationship of the pressure correction factor and the relative water depth in the experiments of the regular wave.

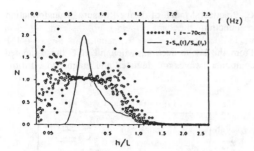

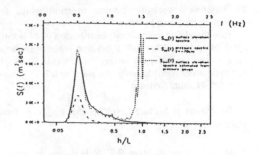

Figure 4-3 The relationship of the pressure correction factor and the relative water depth in the experiments of the irregular wave. Where fp=0.71 Hz, Hs=7.5 cm and h=70 cm

Figure 4-6 The comparison of linear pressure response function transferred spectrum and measured surface wave spectrum. Where fp=0.51 Hz, Hs=6.55 cm and h=70 cm

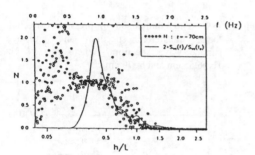

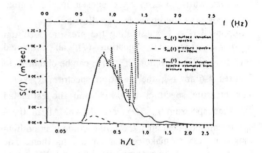

Figure 4-4 The relationship of the pressure correction factor and the relative water depth in the experiments of the irregular wave. Where fp=0.90 Hz, Hs=7.95 cm and h=70 cm

Figure 4-7 The comparison of linear pressure response function transferred spectrum and measured surface wave spectrum. Where fp=0.83 Hz, Hs=8.95 om and h=70 cm

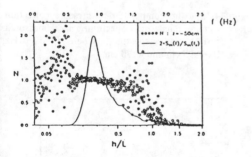

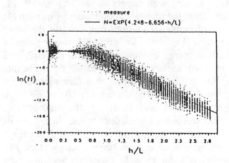

Figure 4-5 The relationship of the pressure correction factor and the relative water depth in the experiments of the irregular wave. Where fp=0.90 Hz, Hs=7.64 cm and h=50 cm

Figure 4-8 The pressure correction factor N regressing with the relative depth h/L.

the theoretical pressure response function works for estimating the wave pressure.

From the above analysis, the distribution of N is only relevant to h/L. When h/L < 0.5 and within the range where the significant energy is located, N is approximately equal to unity. As soon as h/L is greater than 0.5, N starts to decay.

2. Derivation of Surface Level Spectra $S_{\eta\eta e}\,(h/L)$ from the Pressure Spectra

When h/L is greater than 0.5, N becomes to smaller than 1. This means that the theoretical pressure response function is not correctly applied in this situation. To realize the effect of N variation on the high frequency energy and to try to obtain the limitation of the pressure response function applied in the practical case, a JONSWAP spectrum with $f_p = 0.51Hz$, and a P-M spectrum with $f_p = 0.83Hz$ were chosen in the bottom wave pressure experiments in the following.

Supposed N = 1, substituting the pressure spectrum $S_{pp}\,(h/L)\,S_{pp}\,(h/L)$ into Equation (7), an estimated surface elevation spectrum $S_{\eta\eta e}\,(h/L)$ can be obtained. In Figures 4-6 to 4-7, the elevation spectrum $S_{\eta\eta}\,(h/L)$, The pressure spectrum $S_{pp}\,(h/L)$, and the estimated elevation spectrum $S_{\eta\eta e}\,(h/L)$ are shown. All of these spectra are depicted after 256-point window smoothing thus it is easy to make comparison among them. The lower horizontal axis represents h/L and the upper horizontal axis is for frequency. It is shown in Figure 4-6 that when h/L > 0.8 (f > 1.3Hz), the estimated elevation spectrum becomes magnified apparently. The reason for this is that the N is smaller when h/L > 0.5 (theoretical pressure response is less than measured pressure response) and the pressure response function is abruptly decaying with respect to the increase of h/L. Therefore, after the transformation through Equation (7), there is a phenomenon of amplification. But this error amplification has little to do with the region of signification energy distribution and still the estimated and measured spectra are quite consistent with this region. So the linear wave theory transformation can get a very good approximation when the h/L of the peak frequency is 0.16 and most of the concentrated component waves are intermediate waves. In Figures 4-7, the h/L of the peak frequency is 0.35 which represent that the spectra have large portion of energy located in the deep water wave range. And the error becomes large enough to influence the correctness of the transformation in the main energy region of spectrum. The error between the estimated and the measured spectrum is increasing gradually when h/L > 0.5 and even becomes exponentially magnified when h/L >0.8. This shows that the limitation for the applicable theoretical estimation is h/L = 0.5. When the wave components approach the deep water waves, there will be a huge error existing if the linear wave pressure response function is used to estimate the spectrum.

C. Practical Model of Wave-Pressure Transformation Function

From the above section, when h/L>0.5 the estimated component waves are deviated and magnified from the

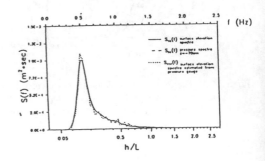

Figure 4-9 The comparison of pressure correction factor transferred spectrum and measured surface wave spectrum. Where fp=0.51 Hz, Hs=6.55 cm and h=70 cm

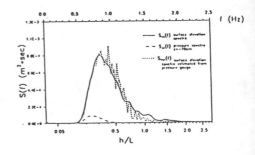

Figure 4-10 The comparison of pressure correction factor transferred spectrum and measured surface wave spectrum. Where fp=0.83 Hz, Hs=8.95 cm and h=70 cm

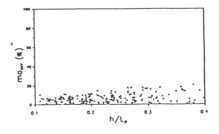

Figure 4-11 The spectral energy deviation of pressure correction factor transferred spectrum from that of measured surface wave spectrum.

practically measured data. Therefore, it is necessary to make some modifications of the theoretical pressure response function in the region of interest for the precise estimation of the surface wave from the bottom wave pressure.

This paper investigated 211 linear irregular bottom wave pressure experiments and calculated the pressure correction factor N of each frequency to find the relation to each h/L. All the results are placed in a single figure for comparison. From Figure 4-8, when h/L is small, N is almost equal to 1 and the theory holds in this case. But when h/L becomes larger, N is different from unity and starts to decay. Although the trend of N decaying is consistent with Figure 4-2, the starting point of h/L for N to decay is not the same. Since the number of the sampling points of the deep water area in Figure 4-2 is not enough for analyzing, the data points in Figure 4-8 are chosen to describe the variation of N. In Figure 4-8, although the N is deviated from unity when h/L > 0.5, all of the N's distributed linearly in the semi-logarithmic plot which can be expressed as

$$N = \exp(a + b \cdot h/L) \qquad (13)$$

By using the iteration of the experimental results, total 211 linear lines of a+b · h/L can be calculated when h/L > 0.5. The average of a and b are 4.248 and -2.656. The standard deviation of a and b are 0.353 and 0.214 respectively. Equation (13) then can be represent as

$$N = \begin{cases} 1 & h/L < 0.64 \\ \exp(4.248 - 6.656h/L) & h/L \geq 0.64 \end{cases} \qquad (14)$$

From the Figure when h/L is between 0.5 and 0.6, N is approximated to 1. Using Equation (14), N=1 corresponds to h/L =0.64. Therefore, in the practical application, when h/L is less than 0.64, N can be used as unity; when h/L is greater or equal to 0.64, N can be calculated from Equation(14).

To obtain the errors using the practical model applied to 211 experiment results, Equation (14) is used to calculate the elevation spectra from pressure spectra and the correctness is investigated qualitatively and quantitatively. The calculation results are shown in Figures 4-9 and 4-10 which are the transforms of Figure 4-6to 4-7 in order. From Figure 4-9, it can be clearly illustrated that there is no magnification in the high frequency area (h/L > 0.8) and meanwhile in the low frequency (h/L < 0.5) they still keep the same consistence. From Figure 4-10, not only the error of magnification (h/L > 0.8) were improved, but also the estimated spectra is consistent with the measured spectra. The ratios of the estimated spectra to the measured spectra are 0.935, 0.991 and 0.953.

The relative energy error which is defined as the error between the estimated elevation spectrum energy and the measured elevation spectrum energy can be expressed as

$$m_{oerr} = \frac{|m_o - m_{op}|}{m_o} \cdot 100\% \qquad (15)$$

Figure 4-11 shows 211 total energy relative errors with respect to h/Lp. (Lp is the wave length according to peak frequency). The average relative error is 6.3% and the maximum relative error is 20.8% . This justifies the validity of the modified of the pressure-elevation transformation.

V. Conclusion

The experimental investigation from 141 regular waves and 211 irregular waves concludes:

1. The validity region of the linear pressure response function depends on the function of relative depth h/L. As h/L > 0.35 in regular wave and h/L > 0.5 in irregular waves the surface wave predicted by linear pressure response function becomes smaller than the real wave. The linear pressure response function needs to be modified.

2. The modified pressure correction factor can be presented as follows,

$$N = \begin{cases} 1 & h/L < 0.64 \\ \exp(4.248 - 6.656h/L) & h/L \geq 0.64 \end{cases}$$

3. As using modified N to transfer the bottom pressure spectral to be surface elevation spectrum in 211 results shows that it can effectively reduce the relative error as compare to the results measured by capacity type wave gauge.

VI. References

Folsom, R. 1947 Subsurface pressure due to oscillatory waves. Transaction on American Geophysics Union, pp. 875-881

Qu, S. 1977 Paramatric Determination of Wave Statistics and Wave Spectrum of Gravity Waves. PhD thesis, Cheng Kung University

Kuo, Yi-Yu 1983 The pressure Characteristics of Random Water Wave. Journal of the Chinese institute of Engineers vol. 6, no. 1, pp.1-8

Lee, D.Y. Wang, H. 1984 Measurement of Surface wave from Sub-surface Gauge. Eng. Conf., ASCE, vol. 1, pp. 271-286

Gabriel and Hedges 1986 Effects of currents on interpretation of sub-surface pressure spectra. Coastal Engineering, vol. 10, pp. 309-323

Nielsen, P. 1989 Analysis of natural waves with local approximations. Journal of Waterway Port, ASCE, vol. 115, No. 3, pp. 384-396

Stochastic Hydraulics'96, Tickle, Goulter, Xu, Wasimi & Bouchart (eds) © 1996 Balkema, Rotterdam. ISBN 90 5410 817 7

The occurrence of critical and supercritical velocities in natural channels

Kenneth L. Wahl & Jeffrey E. Miller
US Geological Survey, Lakewood, Colo., USA

ABSTRACT: Under channel-control conditions, the Froude number (F) for a cross section can be approximated as a function of the ratio $R^{2/3}/d^{1/2}$ and $S^{1/2}/n$, where R is the hydraulic radius, d is the hydraulic (average) depth, S is friction slope, and n is the Manning roughness coefficient. On high-gradient streams, these ratios increase with increasing depth and discharge; thus, F can also be expected to increase with depth. Summary data for more than 1,100,000 current-meter measurements made at about 22,000 locations in the U.S.A. were reviewed, and F was computed for the measurements using average properties of the measurement cross section. About 400 locations were identified with at least one measurement having $F > 0.8$ and at which F generally increases with depth of flow. In some cases, F approaches 1 as the discharge approaches the magnitude of the median annual peak discharge. The data indicate that few actual current meter measurements have been made at the large discharges where velocities can be supercritical.

1 INTRODUCTION

Controversy persists over the behavior of flow velocity in the channels of natural rivers. Some investigators believe that upper regime (supercritical) flows cannot be sustained in natural channels (Jarrett, 1984, 1987; Trieste, 1992). They cite the lack of actual observations of supercritical flow as evidence in support of this theory. However, hydraulic relations indicate that the Froude number for cross sections having particular geometric properties will increase with increasing flow depth (Wahl, 1993). Under those conditions, supercritical flow can occur at large discharges if flow approaches critical depth at moderate discharges (Wahl, 1993, 1994; Simon and Hardison, 1994).

The data needed to settle this controversy have not been forthcoming for several reasons. Velocity is known to positively correlate to stream slope, but most streamflow data are collected on streams of only moderate slope. Also, actual velocity measurements of large floods are difficult to obtain; such floods are commonly of short duration, and their measurement is not without personal risk. Consequently, most measurements of major floods for large-gradient streams are done by indirect methods after the flood waters have receded.

The U.S. Geological Survey (USGS) collects streamflow data at approximately 7,000 continuous-record streamflow gaging stations. Collectively, more than 60,000 current-meter measurements of discharge are made each year at these gaging stations (Wahl et al., 1995). Summary data for most current-meter discharge measurements made at U.S. Geological Survey gaging stations since about 1985 and selected measurements made before 1985 are available in computer files. These files presently contain data for more than 1,100,000 individual current-meter measurements at about 22,000 locations, most of which are at streamflow gaging stations. These measurements provide a data set that can be used to compute Froude numbers for the measurements, if the correction for nonuniform distribution of velocities in the cross section is assumed to be 1.

The authors are presently screening these measurements to identify stream reaches that warrant more rigorous investigation. Wahl (1993, 1994) examined data from Colorado streams and for selected stations in four western States. This paper examines data from the entire U.S.A. and reports the results of initial phases of that examination. In addition to presenting the results of this screening, the paper demonstrates the kinds of analyses that can now be made routinely using large data sets.

2 REVIEW OF HYDRAULIC RELATIONS

The average velocity, V, of a river can be described by the Manning equation if channel control prevails. Although the Manning equation is empirical, the applicability of the relation is widely accepted for steady flow conditions. The metric form of that equation is:

$$V = \frac{1}{n} R^{2/3} S^{1/2} , \qquad (2.1)$$

where n = Manning roughness coefficient, R = hydraulic radius (cross-sectional area divided by wetted perimeter), and S = friction slope.

Similarly, the flow regime of a river can be described by the Froude number, F. Chow (1959, p. 43) gives the equation for F as:

$$F = \frac{\sqrt{\alpha} V}{\sqrt{gd}} , \qquad (2.2)$$

where α = the kinetic-energy correction factor, g = acceleration due to gravity, and d = hydraulic (average) depth. Liggett (1993) has suggested that α should be replaced in equation 2.2 by the momentum coefficient, β.

Recognizing that $1/\sqrt{g}$ is a constant, C_1, equations 2.1 and 2.2 can be combined and simplified to yield:

$$F = \sqrt{\alpha} C_1 \frac{R^{2/3}}{d^{1/2}} \frac{S^{1/2}}{n} . \qquad (2.3)$$

If the behavior of the individual terms is known, the behavior of F can be predicted. Although α is often treated as equal to 1 for a prismatic channel cross section, α is actually greater than 1 unless velocities at all points of the section are equal. However, unless cross-section shape changes significantly as depth increases, α will remain relatively constant or will perhaps increase slightly with increasing depth. Therefore, the change in F with changing depth and discharge depends on the changes in the ratios $R^{2/3}/d^{1/2}$ and $S^{1/2}/n$. The change in $R^{2/3}/d^{1/2}$ is a function of the shape of the channel cross section and the width to depth ratio, W/d. For a rectangular-shaped channel, that ratio will always increase with increasing depth (see Figure 1).

Studies of high-gradient streams in the United States (Barnes, 1967; Limerinos, 1970; Jarrett, 1984) and New Zealand (Hicks and Mason, 1991) have shown that, in general, n decreases and friction slope increases with increasing depth and discharge.

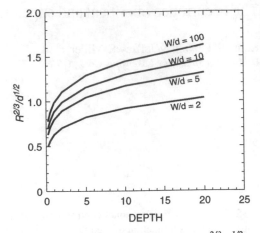

Figure 1. Relation between the ratio $R^{2/3}/d^{1/2}$, hydraulic depth, d, and width-to-depth ratio, W/d, for a retangular cross section.

Roughness elements are submerged as the water depth increases, and channel control becomes fully developed. Bank effects that increase with stream depth, such as heavy or overhanging vegetation, will alter this relation. The apparent change in slope may be an artifact of the relatively short reaches used in those studies. Over a long reach, the friction slope is nearly constant, but pool-and-riffle sequences produce local variations in slope.

In this paper, the measured discharges (Q) have been divided by the discharge of the median annual peak discharge (Qm) for the specific site to produce a dimensionless discharge (Q/Qm), herein referred to as the relative discharge. No significance is attached to the median annual peak discharge; it simply serves as a convenient and easily defined index discharge that provides perspective about the relative magnitudes of the measured discharges and facilitates comparison between streams.

The ratio $S^{1/2}/n$ plotted against the relative discharge, Q/Qm, for selected locations studied by Jarrett (1984) and Hicks and Mason (1991) shows that $S^{1/2}/n$ generally increases with relative discharge (see Figure 2). The relations for other high-gradient sites in those studies followed those general patterns. For high-gradient streams with cross sections that are approximately rectangular, equation 2.3 and Figures 1 and 2 show that F can be expected to increase with increasing depth and discharge.

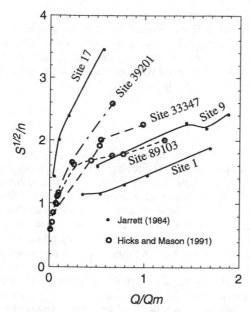

Figure 2. Relation between $S^{1/2}/n$ and Q/Qm for selected streams in Colorado, U.S.A., and New Zealand.

3 DATA ANALYSIS

The primary purpose of this analysis was to define potential locations for later, more detailed and definitive data collection and analysis. At the outset, it was evident that field visits to individual sites were not possible in the current study. Because individual sites could not be visited and because of the size of the data base, screening was done at several levels. A screening process was used because the purpose was to find sites that are suitable for future study, not to find all such sites. Many potential sites were omitted at the beginning because any gaging stations that were discontinued before the 1985 inception of the computer files already have been omitted from consideration. Therefore, omitting sites that may fit the physical criteria but were not readily available for computer analysis was not considered to be a major problem.

Because of the criteria used to locate USGS gaging sites, current-meter measurements at a specific cross section may reflect hydraulic conditions over a reach of channel. Measuring sections are generally located where flow is as uniform as possible. Measurements at low to moderate stages are generally made by wading and

are not always made at exactly the same location along the stream. When stages are too high for wading, measurements are generally made at a fixed location from a bridge or cableway. Cableway sections usually are located in uniform reaches of channels, but bridges may constrict the flow.

3.1 *The data base*

The primary data set used in this analysis is a compilation of summary data from measurements made at streamflow gaging stations. These measurement data currently reside on minicomputers in the various State offices of the USGS and are used primarily to evaluate stage-discharge relations at the individual gaging stations. Summary data for the measurements are in up to 28 fields containing information including date of the measurement, the parties making the measurement, conditions and equipment used, in addition to the top width (W) of the measuring section, the cross-sectional area (A), the stream stage (height above a specified datum), average velocity (V), and discharge (Q). Although the hydraulic depth (d) is not recorded, it can be computed from $d = A/W$. These data are sufficient to compute F for the measurements if α is assumed equal to 1. Both the energy coefficient, α, and the momentum coefficient, β, are greater than 1 for natural channels because velocity is not distributed uniformly in the cross section. Therefore, the actual Froude numbers will be at least as great as the calculated values whether the correction is based on α, as shown in Chow (1959, p. 43) or on β, as suggested by Liggett (1993).

These data were retrieved and combined into a single tab delimited data set on a workstation. The composite data set contained more than 1,100,000 measurements that were made at about 22,000 locations. Many of these measurements were made at low-gradient streams that were not of interest for the current study. Thus, mechanisms were needed to screen this large data set to eliminate measurements for sites that were not pertinent to this study.

3.2 *The screening process*

Three levels of screening were done. The first level of screening was done simply to eliminate data that might contain errors. Because the measurement data had primarily been used to define and evaluate stage-discharge relations, only the entry of stage and discharge fields in the computer files had been subjected to standard USGS checking and review procedures. Therefore, several calculations were made on each measurement in the data set to identify possible erroneous entries of data. The area and velocity were multiplied together and compared to the reported discharge. Measurements in which the product AV differed from Q by more than 10 percent were eliminated from the data base. Similarly, measurements with a calculated mean depth of less than 3 cm were omitted as the extremely shallow depths were indicative of errors in either width or area.

After all measurements with easily defined errors were removed, the hydraulic depth, d, and Froude number, F, were computed. Each measurement in the data set was then tested for F, keeping data only for those locations for which at least one measurement had $F > 0.8$. This step eliminated most of the low-gradient locations. The value of 0.8 was used because calculation of F based on average velocity (and assuming $\alpha = 1$) is known to be only an approximation of the true F for the section. At the conclusion of this second-level screening, the data set had more than 120,000 measurements from 980 locations. This data set contained all sites with at least one measurement with $F > 0.8$, regardless of the relative discharge.

Large Froude numbers made at low-water measurements are recognized to often be indicative only of hydraulic conditions in the immediate vicinity of the measuring site. To eliminate locations for which only low-water measurements gave large F, the third-level screening retained only those sites for which at least one measurement with $F > 0.8$ also had $Q/Qm > 0.25$. Testing against $Q/Qm > 0.25$ had another desirable effect; because Qm is not defined for man-made canals, the measurements made at gaging stations on canals were eliminated by this test. Flow variability, and thus the range in the relative discharges, generally increases in the U.S.A. from the east to the west. The ratio $Q/Qm > 0.25$ was arbitrarily chosen and approximates the 80th percentile of all measurements made in the U.S.A, but exceeds the 90th percentile of all measurements made in the eastern U.S.A. .

To conclude the third-level screening, F was plotted versus the relative discharge (Q/Qm) for each of the remaining sites. The plot of F versus Q/Qm was examined to see if the trends of that relation were consistent with F generally increasing with increasing relative discharge. If measurements appeared to define a consistent relationship, the location was retained. If, however, the measurements with large values of F appeared to be

anomolous, that site was eliminated from the data set. At the conclusion of this step, the data set has data from 418 locations with about 61,800 measurements, of which more than 2,200 have $F > 0.8$. This is the data set that is discussed in the next section of this paper.

3.3 Results

The locations of the measuring sites that remained after the three-level screening process are shown on Figure 3. Now that the candidate sites have been reduced to a managable number, the next step in this procedure will be to closely examine each of these remaining sites to determine if (1) the measurements actually represent natural-channel conditions, and (2) the measuring conditions are representative of conditions in a long reach. Site-by-site examination undoubtedly will show that the measuring sections at some of these sites are not representative of natural conditions. Some sites will have been measured in man-made channels; others will have been made at bridges that constrict the flow. Until the site-by-site examination is made, however, all sites that passed the screening tests remain in the data set.

As expected, the locations shown in Figure 3 are not uniformly distributed across the U.S.A. Although there is no readily available data to define reach characteristics, some generalizations can be made from the locations of the sites and from cursory evaluation of the gaging station data for the sites shown in Figure 3. The heaviest concentration of sites is in the mountains of the western one-half of the USA. These will no doubt prove to be locations with large gradients, and many locations will have relatively stable streambeds with cobbles and boulders. However, measurements with large F are evident in many of the States. Cursory review of the locations shows that many of the non-mountain sites are on sand-bed streams. That review also shows that few of the gaging stations have drainage areas of greater than 500 km^2; most locations have drainage areas of less than 100 km^2. The requirement that sites have at least one measurement with $Q/Qm > 0.25$ and $F > 0.8$ eliminated some of the eastern sites, perhaps unnecessarily. For sites with comparable drainage areas in the U.S.A., generally flows are less variable and median peak discharges are larger in the east than in the west. Thus, the ratio Q/Qm is generally less for measurements in the east.

The relation between F and Q/Qm at selected representative sites is shown in Figure 4. These data are consistent with the data presented by

Figure 3. Map of conterminus United States showing locations at which Froude numbers of greater than 0.8 have been measured by current-meter for relative discharges of greater than 0.25 ($Q/Qm > 0.25$). [An additional 36 sites are located in Alaska.]

Wahl (1993, 1994). Simple, linear correlation coefficients between F and Q/Qm for the entire final data set confirm that F generally increases with increasing discharge (Figure 5). The boxplot of Figure 6 gives the distribution of both Q/Qm and F for all 61,800 measurements in the data set. Measurements are infrequently made at values of $Q/Qm > 1$ where the largest Froude numbers should occur. More than 90 percent of the measurements were made at $Q/Qm < 0.5$, and the median ratio for the measurements is less than 0.05. The histogram of Froude numbers (Figure 7) shows the contrast between the distribution of Froude numbers for all measurements and for only those measurements with $Q/Qm > 1.0$.

The data for these example streams (Figure 4) are typical of data for many gaging stations examined in that few current-meter measurements are available for discharges greater than the median annual peak discharge. Although the relations suggest that the Froude numbers will equal or exceed 1 for very large floods, data are not available to support extrapolation to larger discharges.

4 DISCUSSION

Streams are known to adjust their channels to accommodate flows. Some investigators believe that this adjustment process makes critical and supercritical flows unlikely in natural channels. Those adjustments, however, require that flows be sustained for a sufficient time for the channel to reach equilibrium and that the channel has complete freedom to adjust. Unless both conditions are satisfied, equilibrium might not be reached. Thus, the mechanisms can exist to produce supercritical velocities during large discharges of short duration. The scatter shown in Figure 4 and in comparable illustrations shown by Wahl (1993, 1994) suggests that the channels in these example stations do change, but the data also show that Froude numbers greater than 1, indicating supercritical flow, might occur for the infrequent large discharges. The data also show that relatively few actual current-meter measurements are available for discharges of greater than the median annual flood.

The relations shown in Figure 4 are not isolated cases. This study has identified more than 400 locations where current-meter measurement data

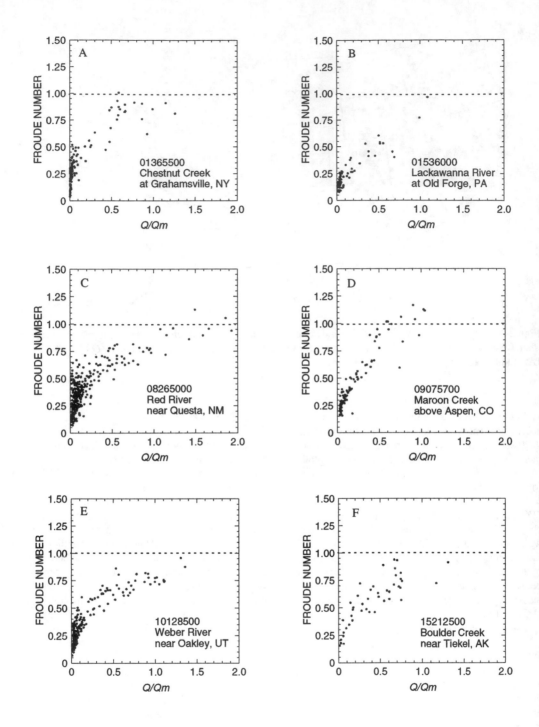

Figure 4. Relation between Froude number and relative discharge for current-meter measurements of selected streams in (A) New York, (B) Pennsylvania, (C) New Mexico, (D) Colorado, (E) Utah, and (F) Alaska.

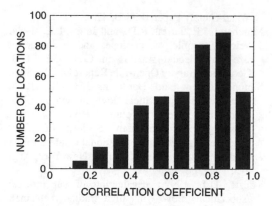

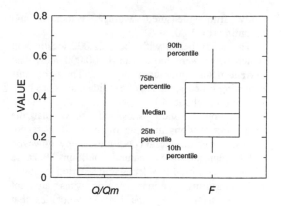

Figure 5. Distribution of correlation coefficients between Froude number and relative discharge for the locations in the final data set.

Figure 6. Boxplots of the distributions of relative discharge and Froude number for measurements in the final data set.

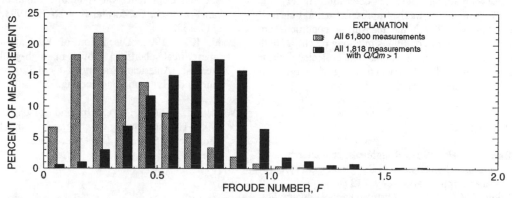

Figure 7. Distribution of Froude number for all measurements in the final data set and for those with relative discharges greater than 1.

include a measurement with $F > 0.8$ and with Froude number increasing with discharge; these sites have a potential for supercritical flow at large discharges. Because measuring sections are generally located in reaches where flows are as uniform as possible, the flow regime in the measurement cross section might be representative of a longer reach. Without additional data, however, nothing is known about the length of reach that exhibits the large Froude numbers. The results of the data screening and analyses now being conducted should serve to identify gaging sites where additional data can be collected. Such data could provide answers to the questions being raised.

Chow (1959) noted that α has been found to vary from about 1.03 to 1.36 for fairly straight prismatic channels. Marchand and others (1984) computed α based only on the vertical variation in velocity at individual locations in cross sections for the Blue River near Dillon, Colorado, and found that α ranged from 1.37 to 1.79. Wahl (1993) computed α using only the lateral variation in velocity from the actual current-meter measurements for the four largest measurements made at Blue River near Dillon, Colorado; those values of α averaged 1.15. Given the possible range in α, one can reasonably assume that the actual values of F would be at least 10 percent larger than the values computed in this study from measured average velocities and assuming $\alpha = 1.00$.

This study began with about 22,000 locations at which there were more than 1,100,000 individual current-meter measurements. Through this screening process, the list of candidate sites has been reduced to about 400. Without a doubt, close examination of these sites will show that the measuring sections at some of these sites are not representative of natural conditions; many, however, will be representative of natural conditions. Because of the criteria used to select gaging-station locations and measuring sections, gaging sites are not expected to represent the highest velocities that might occur in high-gradient streams. And yet this study has shown $F > 0.8$ at about 400 gaged locations. This suggests that even larger Froude numbers may occur on these streams at locations other than at the gaged locations. The next step in this procedure will be to closely examine each of these remaining sites to determine if (1) the measurements actually represent natural-channel conditions, and (2) the measuring conditions are representative of conditions in a long reach? If both conditions are met, additional data collected at these sites would add to the understanding of the behavior of high-gradient streams.

REFERENCES

Barnes, H.H., 1967, Roughness characteristics of natural channels: U.S. Geological Survey Water-Supply Paper 1849, 213 p.

Chow, V.T., 1959, Open channel hydraulics: McGraw-Hill, New York, 680 p.

Hicks, D.M., and Mason, P.D., 1991, Roughness Characteristics of New Zealand Rivers: Water Resources Survey, Kilbirnie, Wellington, New Zealand, 329 p.

Jarrett, R.D., 1984, Hydraulics of high-gradient streams: American Society of Civil Engineers Journal of Hydraulic Engineering, Vol. 110, No. 11, pp. 1519-1539.

Jarrett, R.D., 1987, Errors in slope-area computations of peak discharges in mountain streams: Journal of Hydrology, Vol. 96, pp. 53-67.

Liggett, J.A., 1993, Critical depth, velocity profiles, and averaging: American Society of Civil Engineers, Journal of Irrigation and Drainage Engineering, Vol. 119, no. 2, pp. 416-422.

Limerinos, J.T., 1970, Determination of the Manning coefficient from measured bed roughness in natural channels: U.S. Geological Survey Water-Supply Paper 1898-B, 47 p.

Marchand, J.P., Jarrett, R.D., and Jones, L.L., 1984, Velocity profile, water-slope, and bed-material size for selected streams in Colorado: U.S. Geological Survey Open-File Report 84-733, 82 p.

Simon Andrew, and Hardison, J.H., III, 1994, Critical and Supercritical flows in two unstable, mountain rivers, Toutle River system, Washington: American Society of Civil Engineers Proceedings of National Conference on Hydraulic Engineering, Buffalo, New York, Aug. 1-5, 1995, pp. 743-746.

Trieste, D.J., 1992, Evaluation of supercritical /subcritical flows in high-gradient channel: American Society of Civil Engineers Journal of Hydraulic Engineering, Vol. 118, No. 8, pp. 1107-1118.

Wahl, K.L., 1993, Variation of Froude number with discharge for large-gradient streams: American Society of Civil Engineers Proceedings of National Conference on Hydraulic Engineering, San Francisco, California, July 26-30, 1993, pp. 1517-1522.

Wahl, K.L., 1994, Discussion of Evaluation of supercritical/subcritical flows in high-gradient streams: American Society of Civil Engineers, Journal of Hydraulic Engineering, Vol. 120, no. 2, pp. 270-272.

Wahl, K.L., Thomas, W.O., Jr., and Hirsch, R.M., 1995, The stream-gaging program of the U.S. Geological Survey: U.S. Geological Survey Circular 1123, 22 p.

Stochastic Hydraulics'96, Tickle, Goulter, Xu, Wasimi & Bouchart (eds) © 1996 Balkema, Rotterdam. ISBN 90 5410 817 7

Operational guidance for reliable water supply

Chengchao Xu

Department of Civil Engineering and Building, Central Queensland University, Rockhampton, Qld, Australia

ABSTRACT: An analytical model for assessing the risk of service reservoir operation in a water distribution system with respect to the failure of critical supply components is proposed. The proposed model can be used to determine the minimum service reservoir storage required to safeguard the bulk supply against failures of critical supply components in compliance with a prescribed level of reliability. A sensitivity analysis procedure is described to determine the impacts of various maintenance and operating strategies on the bulk supply reliability. A simple example is presented to illustrate the proposed model.

1 INTRODUCTION

Many water supply systems are equipped with significant storage capacity in the form of service reservoirs or/and water tanks within the distribution network. There are generally two main functions of these reservoir storages:

1. To provide buffers of water for maintaining continued supply during a failure or shutdown of treatment plants, pumping stations and trunk mains etc. leading to the storage reservoirs or/and for meeting other emergency demands, e.g., firefighting; and

2. To balance the fluctuating demands and maintain an adequate pressure within the distribution network, permitting more flexible and cost effective operation of pumping stations.

Traditionally, operation of these reservoirs has been focused on maintaining the security of water supply by essentially keeping the reservoir storage near its full capacity at all the times. In recent years, however, the requirement to operate water supply systems more efficiently has led to more dynamic and comprehensive use of the storage capacity to permit maximum advantage to be taken of electricity tariffs. Specifically, there is a tendency to allow service reservoirs to deplete during the day and refill them during the night when electricity costs are lower. For practical operation purposes, it is therefore very important to know the risk to which the system is being exposed when determining an operational policy which has the

objective of achieving a cost saving. This paper describes an analytical model for assessing the reliability, and providing safety guidance for service reservoir operation in a water distribution system with respect to the failure of critical supply components.

Many methods for reliability assessment of water distribution systems have been developed in recent years. Goulter (1995) has recently presented a comprehensive review of various reliability measures and techniques with an extensive list of references. Most of these works have, however, been focused on the reliability of distribution networks with very little consideration of the impacts of reservoir storage capacity. The reliability of water supply systems with consideration of the reservoir storage issue has been analysed using Monte-Carlo simulation, modified frequency and duration or Markov chain methods, e.g., Damelin et al. (1972), Beims and Hobbs (1988), Hobbs and Beims (1988), Duan and Mays (1990) and Fujiwara and Ganesharajah (1993). However, these methods are usually concerned only with the expected performance of the water supply system using the average frequency and duration of component failures, and therefore, oriented towards the system planning and design stage.

In examining the exact impacts of reservoir storage on the reliability and security of supply, it is, however, equally important to have the knowledge of information other than the averaged performance of the system. This need was demonstrated by

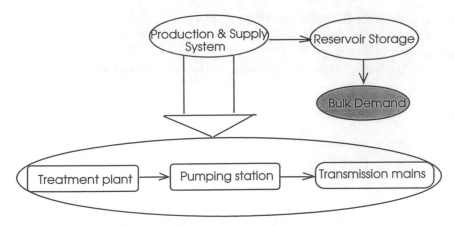

Figure 1 Schematic of the bulk supply system

Germanopoulos et al. (1986) in a practical study investigating the needs for emergency storage. Their approach combined an extended period simulation of the system under failure of a critical component with a probabilistic model to determine the likelihood of the occurrence of such failure events. The model for probability analysis used the probability distributions for the random occurrence of failure events and for the repair time to evaluate the reliability of the system in terms of the occurrence of failure events at different frequencies with various repair durations rather than in terms of the average conditions. Warren et al. (1989) used the same technique to determine the minimum permissible reservoir storage required to maintain specified levels of service in terms of the likelihood of failures of a particular supply component.

However, the probabilistic model for these earlier approaches evaluates the reliability of supply against the failure and repair of one particular component only. For a system in which many failure events may lead an interruption of supply to the service reservoir, such an approach will be optimistic in determining the potential risk. This paper presents a probabilistic model to analyse the reliability of a bulk supply system in which an interruption of supply to the service reservoir may be caused by a failure from one or more critical components.

2 MODEL DEVELOPMENT

The water supply system considered in this paper includes the production and supply system, reservoir storage and bulk demand as shown schematically in

Figure 1. The production and supply system consists of major components the failure of which may interrupt the source supply to the service reservoir. Typical examples of these critical components are the treatment plant, pump station and the major transmission mains delivering purified water to the service reservoir. These critical components, which can usually be identified by close inspection of the supply system structure or for more complex system undertaking detailed simulation study (e.g. Jowitt and Xu 1993), are, for the purposes of reliability analysis, considered as serial components within the production and supply system as illustrated in Figure 1. Due to the presence of reservoir storage, the occurrence of a failure in the production and supply system may, or may not, lead to an interruption of the bulk supply depending on the reservoir storage, the bulk demand and time required to repair the failure. If the failed component is repaired and put back into operation before the reservoir storage is effectively depleted, the bulk supply can be continuously maintained. Since occurrence of a failure and the repair duration for the failed component are random in nature, the risk of interruption of the supply to the bulk demand associated with a given storage level in the service reservoir can only be determined through a probability analysis. (In this paper, the terms of 'reliability' and 'security' of bulk supply are used interchangeably and are defined as the probability that normal supply to the bulk demand can be maintained over a given period, e.g., a year.)

The following assumptions are made in the development of the probabilistic model for reliability analysis of bulk supply:

1. Failures of individual components within the production and supply system are independent of each other; and time to failure of each individual component and the repair duration are exponentially distributed;

2. If a component fails, no other component within the production and supply system can fail until full service has been restored; and

3. Bulk demand is known.

It should be noted that some of these assumptions may slightly deviate from practical situations but as is the case for many probabilistic analyses, these approximations are necessary in order to render the problem analytically tractable.

The key component in the probability analysis of bulk supply reliability is the derivation, from the performance of individual components, of the probability distributions for the failure patterns and for the repair durations of the production and supply system. These probability distributions are obtained through the use of random variable theory and probability theory.

Let the random variable, K, denote the total number of failures occurring per year within the production and supply system. For a serial system, the total number of failures per year is equal to the sum of the number of failures for each component, i.e.,

$$K = \sum_{i=1}^{N} k_i \qquad (1)$$

where

k_i – random variable denoting the number of failures per year for component i;

n = total number of components included in the production and supply system.

Employing the earlier assumption that time to failure for each component is exponentially distributed, it can be shown that k_i follows a Poisson distribution. The quantity on the right hand side of Equation (1) is thus the sum of independent Poisson distributed random variables. It can be proved by random variable theory (Shooman 1968) that the random variable K also has a Poisson distribution with an average value of failure rate, λ_s, equal to the sum of failure rates, λ_p, of all the components involved, i.e.,

$$\lambda_s = \sum_{i=1}^{n} \lambda_i \qquad (2)$$

The probability of a given number of failures per year, $P(K)$, for the production and supply system can now be evaluated easily since the Poisson distribution is completely defined by the single parameter λ_s, i.e.,

$$P(K) = \frac{e^{-\lambda_s} \lambda_s^K}{K!} \qquad (3)$$

Similarly, the probability distribution for the duration of repair to the production and supply system also depends on the probability distributions of repair durations for the individual components. Let $f_i(r)$, $i \in [i,n]$ and the repair duration $r \in [0,\infty]$, be the probability density function for the repair duration of component i. According to probability theory, the marginal density function of the repair duration for the production and supply system can be expressed as follows (Billinton and Allan 1989):

$$f_s(r) = \sum_{i=1}^{n} \frac{\lambda_i}{\lambda_s} f_i(r) \qquad (4)$$

The cumulative distribution of the repair duration for the production and supply system is thus given by

$$P(r \le t) = \int_0^t f_s(y)dy \qquad (5)$$

Recall the earlier assumption that the repair duration of individual components has an exponential distribution, i.e.,

$$f_i(r) = \frac{1}{m_i} e^{-\frac{r}{m_i}} \qquad (6)$$

where

m_i = average repair duration for components i.

Substituting Equations (4) and (6) into (5) and performing the integration gives the following cumulative distribution of the repair duration for the production and supply system:

$$P(r \le t) = 1.0 - \frac{1}{\lambda_s} \sum_{i=1}^{n} \lambda_i e^{-\frac{t}{m_i}} \qquad (7)$$

Based on the probability distributions of the failure patterns and the repair durations of the production and supply system and the future bulk demand, the reliability of bulk supply can be assessed. Assume that the reservoir storage is able to meet the bulk demand for the next T hours. The probability that no failure events with duration greater than T will occur in a year in the production and supply system can be derived by:

$$P(t \leq T) = \sum_{K=0}^{\infty} P(K)\left[P(r \leq T)\right]^K$$

$$= \sum_{K=0}^{\infty} \frac{\lambda_s^K e^{-\lambda_s}}{K!}\left(1.0 - \frac{1}{\lambda_s}\sum_{i=1}^{n}\lambda_i\, e^{-\frac{T}{m_i}}\right)^K$$

$$= e^{-\lambda_s}\sum_{K=0}^{\infty}\frac{\lambda_s^K}{K!}\left(1.0 - \frac{1}{\lambda_s}\sum_{i=1}^{n}\lambda_i\, e^{-\frac{T}{m_i}}\right)^K$$

$$= e^{-\lambda_s}\, e^{\lambda_s\left(1.0 - \frac{1}{\lambda_s}\sum_{i=1}^{n}\lambda_i\, e^{-\frac{T}{m_i}}\right)}$$

$$= e^{\left(-\sum_{i=1}^{n}\lambda_i\, e^{-\frac{T}{m_i}}\right)} \tag{8}$$

which has the form of a Gumbel distribution.

The reliability of bulk supply for a given storage and future demand is thus related to the failure patterns and the repair duration of each individual component through Equation (8). Moreover, if a level of reliability is given, it is possible to solve the inverse problem of Equation (8) to obtain the supply duration T and the necessary storage to achieve the targeted reliability. By varying the targeted levels of reliability and using the diurnal bulk demand profile and reservoir characteristics, a set of time-varying minimum reservoir storage levels, which can serve as 'control curve' in the specification of the system operational safety guidelines, can be derived.

If the current level of reliability for bulk supply is found unsatisfactory, three options are available for reliability enhancement: increasing the reservoir storage, taking effective measures to reduce the number of failures and improving the efficiency of the repair work. The impacts of each of these strategies on the bulk supply reliability can be evaluated through sensitivity analysis by taking the first order partial derivative of Equation (8) with respect to each decision variable as follows:
For increasing reservoir storage

$$\frac{\partial P}{\partial T} = P(t \leq T)\left(\sum_{i=1}^{n}\frac{\lambda_i}{m_i}e^{-\frac{T}{m_i}}\right) \tag{9}$$

For reducing the failure rate

$$\frac{\partial P}{\partial \lambda_s} = P(t \leq T)\left(-\sum_{i=1}^{n}e^{-\frac{T}{m_i}}\right) \tag{10}$$

For speeding up the repair process

$$\frac{\partial P}{\partial m} = P(t \leq T)\left(\sum_{i=1}^{n}-\frac{\lambda_i T}{m_i^2}e^{-\frac{T}{m_i}}\right) \tag{11}$$

These sensitivity information can be used to make informed decisions related to the maintenance and operation of bulk water supply.

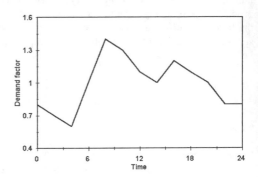

Figure 2 Diurnal demand profile

Table 1 Failure rates and repair durations of components

Component	Failure rate (times/year)	Average repair time (hours)
Pumping station	0.4	8.0
Treatment plant	0.2	12.0
Trans. main	0.1	15.0

3 NUMERICAL EXAMPLES

The application of the above techniques to assess the reliability of bulk water supply and furthermore to

572

derive the operational strategies for the service reservoir is demonstrated by application to the simple system as shown in Figure 1. The production and supply system in this simple example consists of three components, namely, the treatment plant, the pumping station and transmission mains. The failure rates and expected repair durations of each individual component are given in Table 1. In practice, such data can be estimated from the historical operating records on the failure and repair of system components.

The diurnal bulk demand profile with average demand equal to 500 m³/h is given in Figure 2. The service reservoir is assumed to have uniform cross-sectional area of 1000 m².

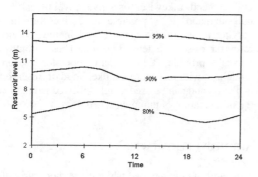

Figure 3 Reservoir 'control curves'

By applying the Equation (2) to the data in Table 1, it was found that the failure rate of the production and supply system is 0.7 times/year, i.e., a supply interruption to the service reservoir is expected to occur on average two times in three years. The probability that a given number of supply interruptions to the service reservoir occur in a year can be easily estimated from the Poisson distribution function of Equation (3). Likewise, the cumulative probability that no more than a specified number of supply interruptions to the service reservoir occur can also be deduced. For instance, the probability of no occurrence of supply interruptions to the service reservoir is 49.7% and the probability of the occurrence of no more than one supply interruption to the service reservoir is 84.4%.

The relationship between reliability of bulk supply and the necessary reservoir storage can be derived from Equation (8). If bulk demand is supplied directly from the sources (i.e., if the service reservoir is effectively empty), the reliability of bulk supply is 49.7%. Any additional reservoir storage

would enhance the reliability of the bulk supply. For instance, if the current reservoir storage is sufficient for the demand over the next 11 hours, it can be calculated from Equation (8) that the level of bulk supply reliability is about 80%, namely, the likelihood of the bulk supply being fully maintained through the year is about 80%. Similarly, if the reservoir storage is increased so that it can supply bulk demand over the next 19 hours, the reliability of the bulk supply is improved to about 90%.

Conversely, the relationship between the reliability of the bulk supply and the required reservoir storage can be used to derive an operating control curve for the service reservoir which complies with a prescribed level of reliability. If the water company wants to pursue a reliability level of 95%, the minimum reservoir storage will have to be maintained at all times at a level sufficient to supply the consumer demand for a period of about 27 hours ahead. By combining this required supply duration with diurnal demand variations, such as those shown in Figure 2, and the reservoir characteristics, it is possible to derive the required service reservoir level at any time for a given level of reliability. Figure 3 shows the minimum levels of reservoir storage for the example problem for levels of reliability at 80%, 90% and 95% respectively. These `control curves' which are associated with different levels of supply reliability can be used operationally for decision support, particularly in the pump scheduling role.

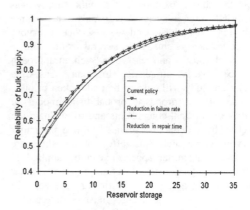

Figure 4 Reliability enhancement with different policies

Information from probability analysis of supply failures can also be used to determine alternative maintenance and operating strategies. In addition to the increase of service reservoir storage, reliability of the bulk supply can be enhanced by either improving the reliability of the supply components through the increase of effort on component maintenance or by increasing the efficiency of repair process. Figure 4 demonstrates how two improved maintenance strategies affect the reliability of the bulk supply. In the first instance, the focus of maintenance is directed at reducing the number of failures within the production and supply system by 10%. In the second case, more effort is directed at improving the efficiency of the repair team and reducing the average repair duration by 10%. As shown in Figure 4, both efforts clearly improve reliability of the bulk supply.

It is interesting to note, however, that when reliability of the bulk supply is low, reliability is more effectively improved by reducing the failure rates of the production and supply system, whereas in the region of higher reliability, improved reliability is best achieved by speeding up the repair process. Improvements in the maintenance and repair of the production and supply system also provide the opportunity to further drawdown the service reservoir storage without deteriorating the reliability of bulk supply. Of course, such an improvement can only be achieved at additional maintenance costs. The most appropriate strategies taken to meet a desired security level must therefore be determined on the rational consideration of the economies for the case in question.

In the above example, the bulk supply was simplified into a system with only one source supplying the bulk demand via a service reservoir. For complex systems with several zones and a number of sources and reservoirs, such an approach may be overly simplistic. In many cases, a simulation study will have to be undertaken prior to the use of the proposed probabilistic model to identify the critical components and to evaluate the effective bulk demand associated with a particular reservoir. Once such cause-effect relationships are identified, a similar approach can be applied to analyse the potential risk of a failure of a particular supply route leading to a particular reservoir. In such cases, the model results must be interpreted accordingly to reflect the practical situations.

4 CONCLUSION AND DISCUSSION

Service reservoir storages within water distribution networks play a very significant role in both the reliable and economic operation of a water supply system. Traditionally, the storage required to ensure security of supply is specified in terms of maximum permissible drawdown. This maximum allowable drawdown is usually fixed arbitrarily based on the experience of the operator, without a detailed quantitative analysis of the risks involved. In this paper, a method for the explicit quantitative assessment of the reliability of a bulk supply system in the development of control curves for optimal service reservoir operation is proposed. A procedure determining the time-varying minimum reservoir levels was also developed. Based on the failure patterns and the probability distribution of the durations of interruptions of the production and supply system, the bulk supply reliability can be estimated for a specified reservoir storage and future bulk demand. Likewise, for a given level of risk and bulk demand, it is possible to derive the time-varying minimum storage required in the service reservoir over an extended period. These `control' curves' provide the system operator secure guidelines with respect to the likelihood of reservoir depletion and can be used as reliability constraints in the optimization of pumping schedule.

ACKNOWLEDGMENTS

This work was completed while the author was with Brunel University. I would like to thank Dr. R.S. Powell for his supports and encouragement during the course of this research. Useful comments and suggestions by Prof. I. Goulter and Mr. F. Bouchart on the earlier version of the paper is gratefully acknowledged.

REFERENCES

Beim, G.K. and Hobbs, B.F.(1988). Analytical simulation of water system capacity reliabiity, Part 2: A Markov-chain approach and verification of the models. *Water Resources Research*, 1988, 24(9),1431-1458.

Billinton, R. and Allan, R.N. (1989). *Reliability assessement of large electric power systems*. Kluwer Academic Publishers, Boston.

Damelin, E., Shamir, U. and Arad, N. (1972). Engineering and economic evaluation of reliability of water supply. *Water Resources Research*, 8(4), 861-877.

Duan, N. and Mays L.W.(1990). Reliability analysis of pumping systems. *Journal of Hydraulic Engineering, ASCE*, 116(2), 230-247.

Fujiwara, O. and Ganesharajah, T. (1993). Reliability assessment of water supply systems with storage and distribution networks. *Water Resources Research*, 29(8), 2917-2924.

Germanopoulos, G., Jowitt, P.W. and Lumbers, J.P.(1986). Assessing the reliability of supply and level of service for water distribution systems. *Proc. ICE.*,80, 413-428.

Goulter, I. (1995). Analytical and simulation models for reliability analysis in water distribution systems. In *"Improving efficiency and reliability in water distribution systems"* Ed. by E. Cabrera and A. F. Vela, Kluwer Academic Publishers, London, 235-266.

Hobbs, B.F. and Beim, G.K.(1988). Analytical simulation of water system capacity reliability, Part 1: Modified frequency duration analysis. *Water Resources Research*, 24(9), 1431-1458.

Jowitt, P.W. and Xu, C. (1993). Predicting pipe failure effects in water distribution networks. *Journal of Water Resoures Planning and Management, ASCE*, 119(1), 18-31.

Shooman, M.L. (1968). *Probabilistic reliability: an engineering approach*. McGraw-Hill, New York.

Warren, R.P., Cook, S.C., Lumbers, J.P. and Fanner, P. (1989). Optimisation of water supply system: risk analysis and reliability of supply. *Journal of Institution of Water and Environmental Management*, 3(1), 22-26.

Stochastic Hydraulics'96, Tickle, Goulter, Xu, Wasimi & Bouchart (eds) © 1996 Balkema, Rotterdam. ISBN 90 5410 817 7

Statistical approaches to assessment of treatment plant efficiencies

Necdet Alpaslan & Filiz Barbaros
Dokuz Eylul University, Turkey

ABSTRACT: Statistical approaches may be effectively used in environmental engineering system design and operation, where the processes encountered are random by nature. A particular area of application covers water and wastewater treatment plants. Although the inputs and outputs of such systems are significantly variable, this variability is often inadequately accounted for in practice, and mean values (or maximum values) of the input and output processes are used in either designing the TP or evaluating their operational performance. The study presented considers the use of statistical measures in evaluating the variability of input/output processes and in assessing the efficiency of a TP with respect to such variability. The approaches used cover the analysis of basic statistical parameters of inputs and outputs, the classical correlation and regression methods, and the entropy measures of statistical information. These statistical approaches are further demonstrated in case of an existing TP of a paper factory in Turkey, for which input and output data on BOD, COD, and TSS concentrations are available. These results show that each statistical measure produces a different kind of information for the planners.

1 INTRODUCTION

The inputs and outputs of water and wastewater treatment systems fluctuate significantly with time so that the proper design and operation of treatment plants (TP) require an assessment of such variability. However, it is not easy to accurately quantify these random fluctuations in both the inputs and the outputs of TP. This is often due to poor knowledge of the various sources of variability that can affect such processes. Consequently, in practice, the random nature of inputs and outputs is inadequately accounted for, and mean or maximum values are used in either designing the TP or in evaluating their operational performance.

With respect to design of TP, the use of mean or maximum values to describe the inputs, which is often the case in practice, may lead to overdesign or underdesign. At this point, one needs to properly assess how variable or how steady the inputs are so that the design parameters can be selected accordingly. With respect to the operation of TP, the performance of the treatment system has to be evaluated by defining the "operational efficiency" of the TP in terms of both the inputs and the outputs. Efficiency is simply described as:

$$E(\%) = (S_i - S_e)/S_i \qquad (1.1)$$

where E is the efficiency, and S_i and S_e are the time variant random input and output (effluent) concentrations, respectively. In practice, the TP system is assumed to reach a level of steady state conditions. That is, no matter how variable the inputs are, the TP is expected to produce an effluent of uniform and stable characteristics. This implies a reduction in input variability to obtain smooth effluent characteristics. The assumption of steady state conditions is made for the sake of simplicity both for design purposes and for assessment of operation. The above definition of efficiency is given either for a single time point or by using the means of S_i and S_e data within the total period of observation. In reality, the operation of a TP reflects a dynamic character; yet, due to the assumptions of steady state conditions, there exists no definition of efficiency that accounts for the dynamic or random character of input/output processes of the TP.

There are only a few studies that focus on the variability of input/output processes of TP in evaluating its performance. Weber and Juanico (1990) describe in statistical terms that the effluents from a TP may be as variable as the input raw water or sewage. Ersoy (1992) and Tas

(1993) arrived at similar results by comparing various statistical parameters to describe random fluctuations of these processes. They also investigated the relationships between input and output variables to express operational efficiency so as to account for the dynamic character of TP. However, none of the above studies have arrived at a precise relationship between the variability of input/output processes and the efficiency of the TP. Tai and Goda (1980 and 1985) define "thermodynamic efficiency" on the basis of thermodynamic entropy and show that the efficiency of a water treatment system can be described by the rate of decrease in the entropy of polluted water.

The study presented considers the use of statistical measures in evaluating the variability of input/output processes and in assessing the efficiency of a TP with respect to such variability. Basic statistical parameters are analyzed first in describing the random fluctuations of TP inputs and outputs. Next, the classical correlation and regression methods are employed to investigate input/output relationships and to define TP efficiency within this respect. Finally, the use of the entropy concept is introduced in assessing the uncertainty of input and output processes of TP within an informational context. Furthermore, the entropy concept is also proposed here to define a "dynamic efficiency index, DEI" as the rate of reduction in input uncertainty or entropy to arrive at a minimum amount of uncertainty in the outputs.

The statistical approaches described above are demonstrated in case of an existing TP of a paper factory, for which input and output data on BOD, COD, and TSS concentrations are available. The results of the application indicate the need for further investigations on the subject, particularly in relating the informational entropy concept to the dynamic processes of treatment.

2 ANALYSIS OF INPUT/OUTPUT VARIABILITY BY STATISTICAL PARAMETERS

As stated earlier, the inputs to a TP show a highly variable character, and the TP is expected to reduce this variability so that the outputs produced are of a more stable nature. The TP is considered efficient if maximum reduction in input variability is realized. Accordingly, the efficiency of TP can be evaluated in relation with input/output variability.

In statistical terms, variability of a random process X can be measured by two parameters: standard deviation, S_x, and the coefficient of variation, C_v. At the first thought, one expects

these two parameters to be high for TP inputs and low for the outputs. However, it is also known that the output process may be as variable as the inputs due to various operational factors that produce "local born variability" in the outputs (Weber and Juanico, 1990). Thus, it is possible that both parameters S_x and C_v assume high values for TP outputs.

There is also the question of which of the above two statistical parameters to use for assessment of input/output variability. The standard deviation is a statistic that describes the variability of a random process in absolute terms expressed in the same units as the variable. The values of this parameter depend not only on the variability of the data but also on the magnitude of the observed variable. Thus, it is not appropriate to use the standard deviation in comparing TP input and output variabilities as the output values of treatment are usually much lower than the input values. Accordingly, the non-dimensional coefficient of variation C_v, described as:

$$C_v = S_x / X_m \qquad (2.1)$$

X_m being the mean value, is a better statistic which allows comparisons between input and output variability (Weber and Juanico, 1990).

The operation of a TP changes both the mean values and the standard deviation of the inputs. The treatment process has to reduce the magnitude of the inputs to produce low values of the output. Thus, for an efficient operation, the TP has to produce a high reduction in the inflowing values of raw water. In fact, this reduction is more significant than the reduction in standard deviation. In statistical terms, this indicates a reduction in the mean value, or X_m. Accordingly, if X_m is reduced significantly, no matter how S_x is changed, the result of efficient treatment will produce outputs with a higher C_v than that of the inputs (Ersoy, 1992; Tas, 1993).

The above considerations are demonstrated in the case of Seka Dalaman Paper Factory treatment plant in Turkey, for which input and output data on BOD (biochemical oxygen demand), COD (chemical oxygen demand), and TSS (total suspended solids) concentrations are available. The Seka treatment plant, with its physical and biological treatment units, is designed to process wastewaters from the factory at a capacity of 4500 m³/day. Daily monitored data on BOD, COD, and TSS concentrations were obtained for the period between January 1989 and October 1991. The input variables were monitored on the main entrance canal to the treatment plant, and the output concentrations were observed at the outlet of the biological treatment lagoon.

Input/output processes of the three variables are analyzed first for their statistical properties. Table 1 shows these characteristics together with the classical efficiency parameter computed by Eq.(1.1) using the mean values of the inputs and the outputs. According to these computations, the TP appears to process TSS more efficiently than the other two variables; in fact, the efficiency drops down to 75% for COD. It is also interesting to note that both the input and the output processes of TSS reflect more uncertainty than BOD and COD if coefficient of variation, C_v, is considered as an indicator of variability. Furthermore, for high levels of efficiency, the C_v of outputs is higher than that of the inputs as in the case of BOD and TSS. For COD, where the efficiency is lower with respect to the other two variables, the C_v of outputs is lower than that of the inputs. This result essentially stems from the rate of reduction of the mean value.

3 ASSESSMENT OF CORRELATIVE RELATIONSHIPS BETWEEN INPUTS AND OUTPUTS

The second requirement for effective treatment is recognized as the insensitivity of effluents with respect to the inputs. That is, the correlation between the inputs and the outputs is required to be minimum for a reliable treatment system. Such a requirement may also be considered as an indicator of TP efficiency.

Table 1: Statistical properties of input/output data of Seka treatment plant.

Variable	Sample size N	Mean (mg/lt)	Standard Deviation (mg/lt)	Coefficient of Variation	Efficiency (%)
BOD in.	843	170.95	54.11	0.32	
BOD out.	843	15.83	9.37	0.58	91
COD in.	819	760.17	343.33	0.45	
COD out.	819	188.17	52.39	0.28	75
TSS in.	880	458.58	304.09	0.66	
TSS out.	880	19.26	14.41	0.73	96

On the other hand, the efficiency of a TP can be described in terms of the relative change between inputs and outputs as in Eq.(1.1) or as:

$$E\ (\%) = 1 - (\text{output/input}) \qquad (3.1)$$

The output/input ratio in the above can be determined by using a simple linear least squares regression of input and output loads (Martin, 1988; Ersoy, 1992; Tas, 1993). In this approach, the slope of the regression line represents the output/input ratio so that the efficiency becomes:

$$E\ (\%) = 1 - \text{regression slope} \qquad (3.2)$$

Furthermore, the intercept can be used as a criterion for assessing compliance with effluent standards. If the intercept is found to be below the critical value of the variable set as the standard, it may be stated that the TP produces outputs that do not violate the standard. Accordingly, if the results of the regression analysis produce a small intercept and a small slope, these two factors indicate that the TP has an efficient operation.

The same data described in the previous section are investigated for correlative relationships between the inputs and the outputs and the results are given in Table 2.

Table 2. Correlative relationships between inputs and outputs of the three variables for Seka TP.

Variable	Correlation Coefficient (mg/lt)	Intercept (mg/lt)	Slope	Efficiency (%)
BOD	0.051	10.4	0.032	99.8
COD	0.260	156.4	0.04	96.0
TSS	0.009	19.6	0.0004	99.96

The statistical significance tests show that the correlation coefficients for TSS and BOD are not significantly different from zero, whereas for COD, the correlation coefficient of 0.26 is significantly different from zero for the available sample size. These results show that, for BOD and TSS, the inputs and outputs are independent of each other, indicating the reliability of the treatment system with respect to these variables. For COD, this reliability is slightly decreased due to the small but nonnegligible correlation between the inputs and outputs. It is interesting to note here that the efficiencies obtained by Eq.(3.2) are higher than those of Table 1, which were computed by Eq.(1.1).

4 ASSESSMENT OF VARIABILITY AND EFFICIENCY USING THE ENTROPY PRINCIPLE

4.1 The informational entropy concept

Entropy is a measure of the degree of uncertainty of random hydrological processes. Since the reduction of uncertainty by means of making observations is equal to the amount of information gained, the entropy criterion indirectly measures the information content of a given series of data (Shannon and Weaver, 1949; Harmancioglu, 1981).

The marginal entropy or the information content of a discrete random variable X with N elementary events of probability $P_n = p(x_n)$ $(n=1,...,N)$ is defined in information theory as (Shannon and Weaver, 1949):

$$H(X) = K \sum_{n=1}^{N} p(x_n) \log [1/p(x_n)] \qquad (4.1)$$

with K=1 if H(X) is expressed in napiers for logarithms to the base e. When two random processes X and Y occur at the same time, stochastically independent of each other, the total amount of uncertainty they impart is the sum of their marginal entropies (Harmancioglu, 1981):

$$H(X,Y) = H(X) + H(Y) \qquad (4.2)$$

When significant dependence exists between variables X and Y, the concept of "conditional entropy" has to be introduced as a function of the conditional probability of X with respect to Y (or of Y with respect to X):

$$H(X|Y) = -K \sum_{n=1}^{N} \sum_{n=1}^{N} p(x_n,y_n) \cdot \log p(x_n|y_n) \qquad (4.3)$$

where $p(x_n,y_n)$ $(n=1,...,N)$ define the joint probabilities and $p(x_n|y_n)$ the conditional probability of the values x_n and y_n. The conditional entropy $H(X|Y)$ defines the amount of uncertainty that still remains in X, even if Y is known.

If the variables X and Y are stochastically dependent, the total entropy is expressed as (Harmancioglu, 1981):

$$H(X,Y) = H(X) + H(Y|X) \qquad (4.4)$$

The total entropy H(X,Y) of dependent X and Y will be less than the total entropy if the processes were independent:

$$H(X,Y) < H(X) + H(Y) \qquad (4.5)$$

In this case, H(X,Y) represents the joint entropy of X and Y and is a function of their joint probabilities:

$$H(X,Y) = K \sum_{n=1}^{N} \sum_{n=1}^{N} p(x_n,y_n) \log [1/p(x_n,y_n)] \qquad (4.6)$$

The difference between the total and the joint entropy is equal to another concept of entropy called "transinformation":

$$T(X,Y) = H(X) + H(Y) - H(X,Y) \qquad (4.7)$$

which represents the amount of information that is repeated in X and Y (Amorocho and Espildora, 1973; Harmancioglu, 1981). By replacing the term H(X,Y) in Eq.(4.7) with its definition given in Eq. (4.4), transinformation can be formulated as:

$$T(X,Y) = H(X) - H(Y|X) \qquad (4.8)$$

For continuous density functions, $p(x_n)$ is approximated as [f(x_n). x] for small x, where $f(x_n)$ is the relative class frequency and x, the length of class intervals (Amorocho and Espildora, 1973). Then the marginal entropy for an assumed density function $f(x_n)$ is:

$$H(X; \ x) = \int_{-\infty}^{+\infty} f(x) \log [1/f(x)] \ dx + \log[1/ \ x] \qquad (4.9)$$

and the joint entropy for given bivariate density function f(x,y) is:

$$H(X,Y; x, y) = \int_{-\infty}^{+\infty} \int_{-\infty}^{+\infty} f(x,y) \log[1/f(x,y)] dxdy + \log[1/ x \ y] \qquad (4.10)$$

Similarly, the conditional entropy of X with respect to Y is expressed as a function of f(x,y) and the conditional probability distribution f(x|y):

$$H(X|Y; \ x) = \int_{-\infty}^{+\infty} \int_{-\infty}^{+\infty} f(x,y) \log[1/f(x|y)] dxdy + \log[1/ \ x] \qquad (4.11)$$

The transinformation T(X,Y) is then computed as the difference between H(X; x) and H(X|Y; x) (Harmancioglu and Alpaslan, 1992).

4.2 Application of the concept to assessment of TP input/output variability

The variability and therefore the uncertainty of input and output processes of a TP can be measured by the entropy method in quantitative terms. This can be realized by computing the marginal entropy of each process by Eq.(4.9) under the assumption of a particular probability density function. Representing the input process by X_i and the output process by X_e, marginal entropies $H(X_i)$ and $H(X_e)$ can be expressed in specific units to describe the uncertainty prevailing in such processes.

As stated earlier, a treatment system is expected to reduce the variability of inputs to produce an effluent with stable characteristics. This implies that the most efficient operation of a TP is realized when maximum reduction in the uncertainty of the inputs is achieved to arrive at a minimum amount of uncertainty in the outputs. In entropy terms, the uncertainty of the effluents X_e, $H(X_e)$, should be minimum (Alpaslan, 1993). Accordingly, a "dynamic efficiency index", or DEI, can be defined as the rate of reduction in uncertainty (entropy) of the inputs, $H(X_i)$, so that it approaches a minimum value for the entropy of the outputs, $H(X_e)$. Such a measure can be expressed in (%) as:

$$DEI = H(X_i) - H(X_e) \ / \ H(X_i) \qquad (4.12)$$

In essence, the above definition involves the first requirement for efficient treatment, namely the difference between input/output variability, which actually is an indicator of treatment capacity. If this difference reflects low efficiency, then the process or design parameters of the TP may have to be changed to increase the DEI.

The above approach to assessment of efficiency appears to comply with the thermodynamic efficiency definition given by Tai and Goda (1980 and 1985). Their description of efficiency refers to the decrease in the thermodynamic entropy of polluted water, where the treated media moves from a state of disorder to order. The terms "order" and "disorder" are analogical in both the thermodynamic and the informational system considered. In the former, "disorder" refers to thermodynamic disorder or pollution, to be measured by the thermodynamic entropy of the system. In the latter, "disorder" indicates high variability in the system, again quantified by entropy measures albeit in an informational context. Accordingly, the two efficiency definitions, one given by Tai and Goda (1980 and 1985) and the other presented here, are similar in concept; the major difference between them is that former is given in a thermodynamic framework, whereas the latter is presented on a probabilistic basis.

The second requirement for effective treatment is the insensitivity of effluents with respect to the inputs, i.e. the correlation between the inputs and the outputs is required to be minimum for a reliable treatment system. Entropy measures can again be employed to investigate the relationship between input/output processes. In this case, conditional entropies in the form of $H(X_e/X_i)$ have to be computed as in Eq.(4.11). The condition $H(X_e/X_i)=H(X_e)$ indicates that the outputs are independent of the inputs and, consequently, that the TP is effective in processing the inputs.

Another entropy measure of correlation between the input and the output processes is transinformation $T(X_i,X_e)$. If transinformation between the two processes is zero, this indicates that they are independent of each other.

The same data used in the previous sections are investigated by means of entropy measures to assess their variability and the efficiency of the TP. Table 3 shows the marginal entropies $H(X_i)$ and $H(X_e)$ of the input/output processes, the conditional entropies $H(X_e/X_i)$ of the effluents with respect to the inputs, transinformations $T(X_i,X_e)$, joint entropies $H(X_i,X_e)$, and finally the DEI of Eq.(4.12) for each variable. These entropy measures are computed by Eqs.(4.9) through (4.11), assuming normal probability density function for each variable. It may be observed from this table that the input TSS has the highest variability (uncertainty or entropy) followed by COD and BOD. With respect to the outputs, COD still shows high variability whereas the uncertainty of BOD and TSS are significantly reduced. Likewise, the joint entropy of inputs and outputs is the highest for COD. These results show that the treatment process applied produces the highest reduction in input uncertainty in case of TSS and the lowest in case of COD. This feature is also reflected by the dynamic efficiency index which reaches the highest value for TSS and the lowest for COD.

It is interesting to note here that when the classical efficiency measure of Eq.(1.1) gives values in the order of 96%, 91%, and 75% respectively for TSS, BOD, and COD, the efficiencies defined by entropy measures on a probabilistic basis result in the respective values 50%, 36%, and 25%. Although the two types of efficiencies described do not achieve similar values, their relative values for each variable are in the same order, i.e. both types of efficiencies are the highest for TSS and the lowest for COD.

The entropy measures shown in Table 3 also reveal the relationship between inputs and outputs for each variable. For TSS, the conditional entropy of outputs $H(X_e/X_i)$ is equal to the marginal entropy $H(X_e)$, which indicates that the outputs are independent of the inputs. The same result is shown by the transinformation $T(X_i, X_e)$ as it is very close to zero. Accordingly, it is observed here that the output TSS of the treatment process is insensitive to the inputs and that the operation of the TP can be considered

Table 3: Assessment of input/output variability by entropy based measures.

Entropy Measures (in napiers)	BOD	COD	TSS
$H(X_i)$	5.4424	7.2110	7.1357
$H(X_e)$	3.4837	5.4264	3.5662
$H(X_e/X_i)$	3.4586	5.3996	3.5660
$H(X_i, X_e)$	8.9011	12.6106	10.7017
$T(X_i, X_e)$	0.0251	0.0268	0.0002
DEI (%)	35.9	24.8	50.0

reliable in case of TSS. The level of dependence increases although slightly for BOD and COD, indicating some correlation between the inputs and the outputs. Thus, one may state that the reliability of the TP decreases for BOD and COD.

5 CONCLUSION

The study presented proposes the use of statistical methods in the assessment of input/output uncertainty of TP. First, the input/output variability of a TP can be identified by the use of statistical parameters such as S_x and C_v. The results of the study reveal that C_v^x is a better representative statistic since it permits comparisons between input/output variability in a non-dimensional form. Furthermore, it is also disclosed here that, as the efficiency of the TP increases, the C_v of the outputs also increases. This is because, the treatment process is expected to significantly reduce the magnitude of the inputs so that, as a result, the C_v of the outputs is increased.

The study presented also uses the entropy measures of information in the assessment of input/output uncertainty of TP. Such measures describe the variability of inputs and outputs in quantifiable terms, i.e. in specific units. They also permit comparisons between input/output uncertainty and those between different variables.

With respect to the efficiency of TP operation, three definitions may be used: (a) the classical one given by Eq.(1.1); (b) regression-based definition of Eq.(3.2); and (c) the DEI defined on the basis of entropy measures. The first definition gives equal weight to each incoming load in the data set; thus, it may produce biased results in evaluating efficiency. Furthermore, it does not provide any information about changes in input data, nor does it provide statistics concerning the input/output data series.

The efficiencies based on regression slopes are higher than the classical efficiency measures of Eq.(1.1). This is basically due to the nature of the method used, i.e. large input loads are given more weight than small loads in the computation of efficiency. The regression slope represents the transformation rate of inputs into outputs, and unity minus this rate is defined as efficiency. Here, the efficiency actually describes the goodness of this transformation rate. Furthermore, the regression-based approach assumes that the efficiency of the TP is the same for all loads. Thus, it provides a measure of overall efficiency of TP, indicating efficiency consistency.

Next, the operational efficiency of TP is also assessed by means of the "dynamic efficiency index, DEI", which represents the rate of reduction in the uncertainty of inputs to produce minimum amount of entropy or uncertainty in the outputs. In essence, two requirements are foreseen for an effective and reliable treatment system: (a) the highest reduction in input uncertainty must be obtained; (b) the outputs must be insensitive to the inputs; that is, the condition of independence must be satisfied. The entropy measures, can be effectively used to assess whether these two requirements are met by the operation of the TP.

The DEI definition proposed in the study is parallel to the thermodynamic efficiency description of Tai and Goda (1980 and 1985), who express efficiency in terms of the rate reduction in thermodynamic entropy within the process of treatment. It is claimed here that further investigations are needed on the subject to disclose the relationship between the informational entropy measures and the dynamic (physical, thermodynamic, etc.) processes of treatment.

In conclusion, it is observed that the three types of efficiencies defined in the text are similar to each other; that is, their relative values for different variables are in the same order. However, each efficiency measure produces a different kind of information for the planner.

582

Thus, water quality regulation objectives must be evaluated to decide which of these efficiency measures are appropriate for use in different situations.

It must also be noted that the entropy measures do not produce information different from that obtained by the other two approaches. The use of informational entropy measures may have merit once they are related to thermodynamic entropy involved in the process of treatment. Thus, further investigations are required to develop this relationship before one attempts to assess treatment efficiency via entropy measures.

REFERENCES

Alpaslan, N. 1993. Assessment of treatment plant efficiencies by the entropy principle. In *Int. Conf. on Stochastic and Statistical Methods in Hydrology and Environmental Engineering,* Waterloo, Canada, June 21-23, 1993.

Amorocho, J. and Espildora, B. 1973. Entropy in the assessment of uncertainty of hydrologic systems and models. *Wat. Resour. Res.,* 9(6), 1551-1522.

Ersoy, M. 1992. *Various approaches to efficiency description for wastewater treatment systems (in Turkish).* Graduation Thesis (dir. by N.Alpaslan), Dokuz Eylul University, Faculty of Engineering, Izmir (Turkey).

Harmancioglu, N. 1981. Measuring the information content of hydrological processes by the entropy concept. *J. of Civil Eng. for Centennial of Ataturk's Birth,* Ege University , Faculty of Engineering, Izmir, Turkey, 13-38.

Harmancioglu, N.B. and Alpaslan, N. 1992. Water quality monitoring network design: a problem of multi-objective decision making, *Water Resour. Bull. Special Issue on "Multiple-Objective Decision Making in Water Resources",* 28, (1), 179-192.

Martin, E.H. 1988. Effectiveness of an urban runoff detention pond - wetlands system. *J. of Envir. Eng.,* ASCE, 114, (4), 810-827.

Shannon, C.E. and Weaver, W. 1949. *The Mathematical Theory of Communication.* The University of Illinois Press, Urbana, Illinois.

Tai, S. and Goda, T. 1980. Water quality assessment using the theory of entropy. In *River Pollution Control,* M.J. Stiff (ed.), Ellis Horwood Publishers, ch.21, 319-330.

Tai, S. and Goda, T. 1985. Entropy analysis of water and wastewater treatment processes. *Int. J. of Envir. Studies,* Gordon and Breach Science Publishers, 25, 13-21.

Tas, F. 1993. *Definition of dynamic efficiency in wastewater treatment plants (in Turkish).* Graduation Thesis (dir. by N.Alpaslan), Dokuz Eylul University, Faculty of Engineering, Izmir (Turkey).

Weber, B. and Juanico, M. 1990. Variability of effluent quality in a multi-step complex for wastewater treatment and storage, *Water Res.,* 24, (6), 765-771.

Stochastic Hydraulics'96, Tickle, Goulter, Xu, Wasimi & Bouchart (eds) © 1996 Balkema, Rotterdam. ISBN 90 5410 817 7

Stochastic hydraulic reliability of a water distribution network

Arun Kumar, M.L. Kansal & Sanjay Kumar
Delhi College of Engineering, India

ABSTRACT: In a water distribution network (WDN), water supply and demand both are driven by the variability of climatic conditions such as rainfall and temperature. Traditionally, a WDN is designed according to certain guidelines which represent reliability considerations. Most of the present studies related to hydraulic reliability of a WDN have assumed that supply is always greater than or equal to demand. In developing countries, water supply to the consumers is quite less than the actual requirement and is also negatively correlated. Even, when the supply available is more then the demand, system may not satisfy the requirement because of the pressure or capacity constraint. In the present study, an analytical approach has been suggested to compute the probabilistic reliability measures for a WDN under these drought induced failure conditions. The methodology is illustrated with the help of an example.

1 INTRODUCTION

In developing countries like India, the urban water supply systems are traditionally designed on the basis of "standard" design criterion or the "rules of thumb" that attempt to provide a hundred percent reliable water service (Indian water supply manual, (1991)). In real terms, the reliability of a water distribution network (WDN) is low, continues to decrease with ageing of the network and is dependent on a large number of factors. It is therefore important to undertake a proper reliability analysis while planning a WDN. Generally, water distribution system engineers are concerned about two types of reliability viz. mechanical and hydraulic. Mechanical reliability assesses the connectivity of the various demand nodes with the source, whereas, the hydraulic reliability measures the ability of the network to provide desired amount of water at required pressure. Extensive studies have been reported on mechanical reliability (Walski et al (1982), Wagner et al (1986), Mays et al (1989), Quimpo et al (1991) and Kansal et al (1995)). The hydraulic reliability aspect has been studied by Lee (1980), Wagner et al (1986), Mays et al (1989) and Quimpo et al (1991,93).

Presently, all the above cited studies related to hydraulic reliability, assume that the supply is greater than or equal to the demand. However, the network has a capacity constraint with the result the residual pressure at demand nodes may be less than the desired one. The other part of the problem is that of supply insufficiency, particularly, when the supply and demand are inversely correlated (when demand increases supply decreases). Al-Weshah and Shaw (1994) have suggested an elaborate simulation technique for estimating the system reliability under these conditions. This is particularly, the major problem in countries like India where vast region of the country experiences a hot dry summer season for two to three months when per capita demand increases manifold and the supply sources start drying up simultaneously.

The objective of the present study is to compute hydraulic reliability of a WDN taking into consideration the stochasticity of the supply and demand. The series of supply and demand are assumed to be normally distributed with a negative correlation among themselves. (If the data is not normally distributed, it is transformed into normal distribution using a suitable transformation.)

2. MATHEMATICAL FORMULATION FOR RELIABILITY ANALYSIS

Theoretically, the probability of failure in a WDN can be evaluated using the joint probability distribution of supply and demand. If the demand 'D' and supply 'S' are randomly distributed over a period of time, the expression for static system reliability 'R_s' can be represented as

$$R_s = \Pr\left[(S-D) > 0\right] \tag{1}$$

$$= \int_0^\infty f_s(S)\left[\int_0^S f_d(D)dD\right]dS \tag{2}$$

where,
$\Pr[(S-D)>0]$ = probability of (S-D) greater than zero.
$f_s(S)$ = probability density function for supply;
$f_d(D)$ = probability density function for demand.

As the joint probability density functions are either unknown or difficult to derive, the solution of equation (2) is not possible. If the supply 'S' and demand 'D' are normally distributed and are correlated (negatively) with a coefficient of correlation 'ρ', then the Reliability (R_s) can be computed from (e.g. Kapur & Lamberson (1977))

$$R = \phi\left(\frac{\mu_s - \mu_D}{\sqrt{\sigma_s^2 + \sigma_D^2 - 2\rho\sigma_s\sigma_D}}\right) \tag{3}$$

where,
ϕ = cumulative density function of standard normal distribution ($N(0,1)$);
μ_s = mean value of the supply series;
μ_D = mean value of the demand series;
σ_s = standard deviation of the supply series;
σ_D = standard deviation of the demand series;
ρ = zero order cross-correlation between supply and demand series.

Equation (3) is valid for estimating the reliability of the system when the supply is greater than the demand. Since, the supply and demand are varying with time, it is possible that during the periods of increased demand, the residual pressure at various demand nodes may go lower than the required service head. This may reduce the nodal reliability and hence the system reliability. This reduction in system reliability can be computed if the demand is bifurcated in two parts i.e. $D < D_0$ and $D > D_0$. Here, D_0 is the capacity of the system, meaning thereby, that the nodal pressures at all the demand nodes will be greater than the service head when D is equal to D_0. Thus, the reliability expression given in equation (1) can be re-written as

$$R_s = \Pr\{(S-D)>0, D<D_0\} * R_{hl} + \Pr\{(S-D)>0, D>D_0\} * R_h \tag{4}$$

where, D_0 = demand truncation level for which the system is designed.i.e. the residual pressure at various nodes are greater than the

586

service head at demand D_o ; R_h = network hydraulic reliability when D is greater than D_o. R_{hl} is the network hydraulic reliability when $D < D_o$ and is equal to 1 as the network is designed for D_o.

Equation (4) requires the computation of joint probability estimates between (S-D) and D series. Since S and D are negatively correlated, having correlation coefficient 'ρ', (S-D) and D will also be correlated with a serial correlation which can be expressed as

$$\rho_1 = \frac{\rho\sigma_s - \sigma_D}{\sqrt{\sigma_s^2 + \sigma_D^2 - 2\rho\sigma_s\sigma_D}} \tag{5}$$

In the first component of equation (4), since the demand is less than the capacity of the system, the computations can be carried out analytically as at any node of the system the discharge and pressure criterion will be satisfied. However, for the computation of second component, hydraulic simulation will be necessary as some of the nodes may or may not receive water with sufficient pressure head. The steps for computation of this component of hydraulic reliability are taken up subsequently in the section.

If (S-D) series is denoted by X series and demand series by Y, then the bivariate probability estimation based on normal distribution can be computed as

$$\Pr\{X \le h, Y \le k ; \rho_1\} =$$
$$\frac{1}{2\pi\sqrt{1-\rho_1^2}} \int^h \int^k \exp\left[-\frac{1}{2}\frac{x^2 - 2\rho_1 xy + y^2}{1-\rho_1^2}\right] dxdy \tag{6}$$

$\Pr\{X \le h, Y \le k; \rho_1\}$ can be expressed in terms of the T-function (Kumar, A.(1980) and Owen (1962)) as

$$\Pr\{X \le h, D \le k; \rho_1\} = \tfrac{1}{2}\Pr(h) + \tfrac{1}{2}\Pr(k)$$
$$-T(h,a_h) - T(k,a_k) - \left\{\begin{matrix} 0 \\ 1/2 \end{matrix}\right. \tag{7}$$

where,

$$a_h = \frac{k}{h\sqrt{1-\rho^2}} - \frac{\rho}{\sqrt{1-\rho^2}}$$

$$a_k = \frac{h}{k\sqrt{1-\rho^2}} - \frac{\rho}{\sqrt{1-\rho^2}}$$

$$T(h,a) = \frac{1}{2\pi}\int_0^a \frac{\exp[-h^2(1+x^2)/2]}{1+x^2} dx$$

the upper choice is made if h*k > 0 and if h*k = 0 but h+k ≥ 0; the lower choice is made otherwise.

The values of the T-function are tabulated in Owen (1962). Alternately equation (6) can be computed numerically.

Also, in the second component of equation (4), the demand is as a continuous function. For the purpose of simplicity, this continuous demand function above D_0 is divided in 'm' discrete demand intervals and is expressed as

$$\Pr\{(S - D) > 0, D > D_0 ; \rho_1\} * R_h =$$
$$\sum_{i=1}^{m} \Pr\{(S-D) > 0; D_{i-1} < D < D_i ; \rho_1\}*(R_h)_{i-1, i} \tag{8}$$

where, 'm' number of discrete intervals depends upon the accuracy of results desired.

For any two normally distributed random variables (correlated or not), which will fall in any region bounded by a polygon, the bivariate normal probability can be computed using T-functions. For example,

$$\Pr\{(S-D) > 0; D_0 \le D \le D_1 ; \rho_1\}$$
$$= \Pr(D \le D_1) - \Pr(D \le D_0)$$
$$- \Pr\{(S-D) \le 0; D < D_1 ; \rho_1\}$$
$$+ \Pr\{(S-D) \le 0; D < D_0 ; \rho_1\} \tag{9}$$

587

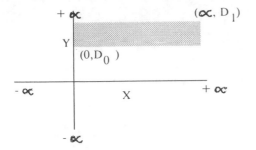

Hydraulic reliability $(R_h)_{i-1, i}$ is computed for 'm' demand intervals. For simplicity, each interval is represented by D_k , the average of D_{i-1} and D_i. Hydraulic simulation of the WDN having 'n' demand nodes is then carried out for this D_k and corresponding pressure heads at all the demand nodes are computed. In reality, the water supplied to a jth node will depend on the head attainable at that node. For each node, two head limits must be given :

1) a minimum head (H_{min}), and
2) a service head (H_s).

The system is said to be performing satisfactorily only when, for each node, all the imposed demands can be met with heads above the service limit (H_s). If however, at some nodes, the available head (H) is below H_s but above H_{min} , it is assumed that the system can not supply the full demand but a reduced supply at that node. Many relationships (Wagner et al (1986), Kumar & Kansal (1995)) are available for estimating this reduced supply. In this study, reduced supply is computed using Equation (10). However, no supply is possible if the pressure head at the node is below H_{min} . The rationale for Equation (10) is that hydraulic laws for flow through pipes show that the flow is proportional to the square root of the pressure head. The supply reduction from normal to no flow is related to the square root of the available head.

$Q_{j, supplied} = Q_{j, reqd.}$ for $H \geq H_s$

$Q_{j, supplied} = [(H - H_{min}) / (H_s - H_{min})]^{1/2} * Q_{j, reqd.}$

for $H_{min} \leq H < H_s$

$Q_{j, supplied} = 0$ for $H < H_{min}$

(10)

The nodal hydraulic reliability (R_j) for the representative demand is taken as

$$R_j = \frac{Q_{j, supplied}}{Q_{j, reqd.}}$$ (11)

The network hydraulic reliability for the representative demand can then be computed by taking either an arithmetic mean or the weighted average of the nodal reliabilities. In the present study, all the nodes have been assumed to be equally important and the network hydraulic reliability is computed by taking simple arithmetic mean of all nodal reliability as shown below :

$$R_h = \frac{\sum_{j=1}^{n} R_j}{n}$$ (12)

Equation (12) evaluates the network hydraulic reliability for a particular discrete demand interval of D_{i-1} and D_i . Since there are 'm' discrete demand intervals, the overall network hydraulic reliability taking stochasticity of supply and demand is computed using Equations (4) and (8).

3 ILLUSTRATIVE EXAMPLE

For illustration of the suggested methodology, a typical real-type WDN as shown in Fig.1 has been considered. The hydraulic equivalent of Fig. 1 with pipe and node details is shown in Fig. 2. The network is then analysed for pressures and flows in various pipelines for a given demand interval by any of the hydraulic simulation algorithm.

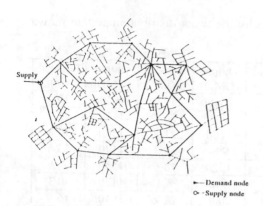

Supply

●—Demand node
○ -Supply node

Fig.1 An illustrative water
distribution network

In the present study, the network has been analysed using the well-known WADISO algorithm which is based on the node method suggested by Walski (1984).

In the network shown in Fig. 2, there are 17 nodes connected by 21 pipelines. The head available in the reservoir is 30 m and the service head for all the demand nodes has been assumed to be 16 m. However, the

minimum head for all the demand nodes has been assumed as 12 m. This means that at any demand node, full demand can be satisfied if it receives water at a pressure of 16 m, the node will receive reduced supply if the pressure of water lies between 12 and 16 m and there will be no supply if the pressure is below 12 m. In the present illustrative example, the following data have been used :

Mean of supply series (μ_s) = 17,000 cmd standard deviation (σ_s) = 2,000 cmd.
Mean of demand series (μ_D) = 15,000 cmd standard deviation (σ_D) = 3,000 cmd.

The coefficient of correlation between supply and demand series (ρ) = - 0.4.

For the computation of system reliability as shown in equation (4), first the value of joint probabilities for various intervals of demand subjected to the condition that supply is more than the demand are computed. These values have been computed using the T-functions and are reported as follows:

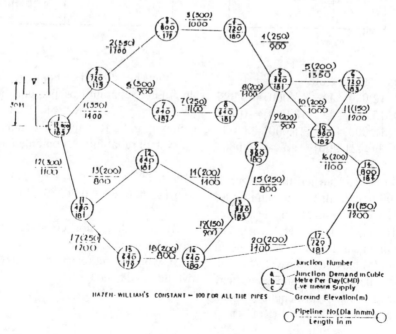

HAZEN-WILLIAM'S CONSTANT = 100 FOR ALL THE PIPES

Junction Number

Junction Demand in Cubic
Metre Per Day(CMD)
(-ve means Supply

Ground Elevation(m)

Pipeline No(Dia in mm)
Length in m

Fig.2 Line diagram of water distribution network shown in fig.1

Table 1 Pressure and flow data at various nodes for the water distribution system shown in Figure 2.

Node No.	Elev. (m)	Output (CMD)	Pr.Head (m)	Output (CMD)	Pr.Head (m)	Output (CMD)	Pr.Head (m)	Output (CMD)	Pr.Head (m)
1	185	- 12000	30.0	-13500	30.0	-16500	30.0	-19500	30.0
2	179	720	30.5	810	29.1	990	26.0	1170	22.4
3	179	800	28.2	900	26.3	1100	21.9	1300	16.8
4	180	720	26.3	810	24.2	990	19.3	1170	13.6
5	181	960	22.7	1080	19.9	1320	13.6	1560	6.2
6	183	720	18.9	810	15.7	990	8.3	1170	-0.3
7	182	640	26.5	720	24.9	880	21.2	1040	16.9
8	181	640	25.7	720	23.7	880	19.0	1040	13.6
9	180	960	23.2	1080	20.3	1320	13.7	1560	5.9
10	182	960	19.4	1080	16.1	1320	8.5	1560	-0.4
11	181	480	29.8	540	28.8	660	26.4	780	23.7
12	181	640	26.0	720	24.0	880	19.5	1040	14.3
13	183	960	20.9	1080	18.2	1320	12.0	1560	4.8
14	178	800	23.0	900	19.5	1100	11.7	1300	2.5
15	179	640	28.6	720	26.8	880	22.7	1040	17.8
16	180	640	24.0	720	21.4	880	15.2	1040	8.1
17	181	720	20.8	810	17.6	990	10.2	1170	1.5
Network hydraulic reliability (R_h)			1.0		0.9976		0.6362		0.439

$Pr\{(S-D) > 0;\quad D < 12,000\}\quad = 0.158$
$Pr\{(S-D) > 0;\ 12,000 < D < 15,000\} = 0.325$
$Pr\{(S-D) > 0;\ 15,000 < D < 18,000\} = 0.192$
$Pr\{(S-D) > 0;\ 18,000 < D < 21,000\} = 0.002$
$Pr\{(S-D) > 0;\ D > 21,000)\qquad \simeq 0.000$

As shown in Table 1, when the demand is less than 12,000 cmd, all the demand nodes will receive water at more than the service head (i.e. 16 m). However, when the demand is more than 12,000 cmd, some of the demand nodes may receive water in full, some may in reduced mode and some may even not receive water at all depending upon the availability of pressure head (Ref. Table 1 for details). The results of pressures at various nodes for different sets of demand

can then be utilised to compute the network hydraulic reliability as per equation (12) as shown in last row of Table 1. After computing the R_h values, the numerical value of system reliability is computed as

$R_s = 0.158 * 1.0 + 0.325 * 0.9976$
$+ 0.192*0.6362 + 0.002*0.439$
$= 0.605$

If one does not consider the capacity constraint, the value of system reliability (R_s), which is represented by $Pr\{(S-D) > 0\}$, would have been 0.683.

4 CONCLUSIONS

This study has shown that certain reliability measures defined on general networks can be calculated analytically. These analytical methods can provide a fast initial assessment of the reliability of a WDN. Since the supply and demand are random variables which depend upon various climatic conditions, the availability of supply and demand is a basic step towards the reliability computation of a WDN. Furthermore, even for a particular system demand, it may not be always possible to provide the required demand at desired pressure because of the certain capacity constraints. In such like situations, hydraulic simulation can play a major role towards the computation of hydraulic reliability of a WDN.

In the present study, a probabilistic method combined with simulation study has been suggested to compute the hydraulic reliability of a WDN under the stochastic conditions of supply and demand. The methodology has been illustrated with the help of an illustrative example.

5 REFERENCES

Al-Weshah, R.A.A. & Shaw, D.T. 1994 Performance of integrated municipal water system during drought *J Water Res. Plang. and Management, ASCE*, Vol 120, No 4, July/August, pp 531-545.

Kansal, M.L., Kumar, A. & Sharma, P.B. 1995 Reliability analysis of water distribution systems under uncertainty, *Reliability Engg. & System Safety*, Vol 50, No.1, 51-59.

Kapur, K.C. & Lamberson, L.R. 1977 Reliability in engineering design *John Wiley and Sons, New York, N.Y.*

Kumar, A. 1980 Prediction and real time hydrological forecasting *Ph.D. Thesis, IIT Delhi*, India.

Kumar, A. & Kansal, M.L. 1995 Reliability analysis of water distribution systems - Discussion *J Env. Engg., ASCE*, September, pp 674-675.

Lee, S.H., 1980 Reliability evaluation of a flow-network *IEEE Trans Reliability*, Vol R-29, April, pp 24-26.

Mays, L.W. 1989 New methodologies for reliability-based analysis and design of water distribution systems, Technical report CRWR-227, *Centre for Research in water resources*, Austin, Texas, USA.

Ministry of Urban Development 1991 *Indian manual on water supply and treatment*, New-Delhi.

Owen, D.L. 1962 Handbook of statistical tables *Addison Wesley Publication Co. Inc., USA.*

Quimpo, R.G. and Shamsi, U.M. 1991 Reliability-based distribution system maintenance *J. Water Res. Plang. and Management, ASCE*, Vol 117, No 3, May/June, pp 321-339.

Wagner, J.M., Shamir, U. & Marks, D.H. 1986 Reliability of water distribution systems *Report No. 312, MIT Cambridge, Massachusetts.*

Walski, T.M., 1984 Analysis of water distribution systems *Van Nostrand-Reinhold, New York.*

Walski, T.M. and Pelliccia, A.,1982 Economic analysis of water main breaks *J. of the American Water Works Asso.*, Vol 79, No 3, March, pp 140-147.

Wu, S.J., Yoon, J.H. and Quimpo, R.G., 1993 Capacity weighted water distribution system reliability *Reliability Engg. and System Safety*, Jan., pp 39-46.

Stochastic Hydraulics'96, Tickle, Goulter, Xu, Wasimi & Bouchart (eds) © 1996 Balkema, Rotterdam. ISBN 90 5410 817 7

Reliability analysis on the stability reduction of armor blocks due to their uneven posture inclination on an artificial mound

A. Kimura
Faculty of Engineering, Tottori University, Koyama Minami, Japan

K. Kobayashi & H. Satoh
TETRA, Tsuchiura, Japan

ABSTRACT: Reliability analysis is applied for the estimation of an armor block stability placed on an artificial reef (wide submerged offshore breakwater). Statistical properties of basic variables such as a block inclination, lift coefficient and etc. which affect the block stability are experimentally investigated. A failure probability of the block is estimated applying the method by Hasofer and Lind [?]. The final result shows that an adoption of larger block does not always bring about a larger safety if there are considerable uncertainty of the physical properties of the blocks such as uneven posture inclination.

1 Introduction

Artificial reefs have been increasingly adopted in Japan for beach protections without appearance changes on coasts. The reef is usually constructed in a shallow water region. Its structure is composed of an inside rubble mound and a surface armor block layer. Several equations such as Hudson's which are mainly directed to give a required weight of an ordinary wave absorbing block, have been used even for the armor blocks on the reef for convenience. Uda et al. [?] developed an experimental equation to give a required weight of rubble mound of a reef. This equation is extensively applied for the calculation of a required block weight. Failures of the blocks are sometimes taking place. Fluctuations of experimental data which induce an uncertainty in the equation may be one of the reasons. For example, if the armor blocks have uneven surface inclinations, different fluid forces may act on the block from those for the blocks placed in a regular order and horizontally. In a practical situation, due to an insufficient finishing of mound surface construction work and/or an uneven subsidence of the rubble mound over a long period, uneven block inclinations are inevitable.

To estimate an effect of the uneven block inclination on its stability, reliability analysis which

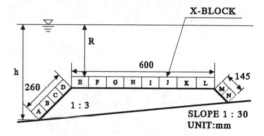

Figure 1: Model artifical reef

can cope with the uncertainty of basic properties in the design equation is applied in this study. The failure probabilities of the block under some prototype conditions are calculated.

2 Experimental measurements of uncertainty for basic variables

2.1 Wave forces at different positions on a reef

Before the measurement on an overall uncertainty of the physical armor block properties, wave forces acting on the blocks at different positions on the reef are experimentally measured. Experimental set up is shown in Fig.1.

An artificial model reef is placed on a 1/30 uniform slope. Measurement are made at the positions A, B, C, ... along the model reef surface. L

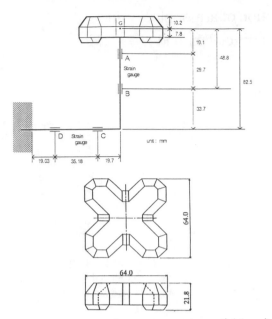

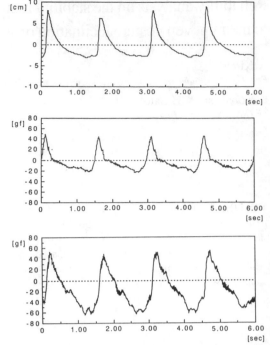

Figure 2: L shape bar spring equipment (above) and the model block (below)

Table 1: Experimental conditions

$H_0(cm)$	6.5, 7.5, 8.5, 9.5, 10.5
$T(s)$	1.0, 1.5, 2.0
$R(cm)$	2.0, 3.0, 4.0

shape bar spring type equipment with the block on its top (Fig.2) is used to measure horizontal and vertical fluid forces simultaneously. Plan and lateral views of the model block are shown in the same figure (below).

The equipment is installed in the central row of totally 14(direction of wave propagation) × 7(transverse direction) blocks placed in a grid on the acryl reef model (impermeable). The other blocks around the equipment are adhered on the model reef with small gaps so that no lateral friction acts on the equipment. The wave tank is $0.5m$ wide, $0.75m$ deep and $29m$ long. The model block is a 1/50 model of $20t$ concrete block. Experimental conditions are shown in Table 1. H_0, T and R are the deep water wave height, wave period and water depth above reef crown respectively. Wave

Figure 3: Water surface elevation(top), horizontal(center) and vertical(bottom) forces

forces acting on blocks C, D, E and F are larger than those placed on other positions of the reef. Figure 3 shows an example of the time record of the water surface elevation (top), and the horizontal (center) and vertical (bottom) wave forces acting on the block at D.

Figure 4 shows the relation between horizontal F_H (above) and vertical F_L (below) wave forces and H_0/L_0 for the blocks at C and D, $(R = 3cm)$. F_H and F_L are their maximum values over one wave period respectively, H_0/L_0 is the deep water wave steepness, $W' = W - B$ is the block weight in fluid, α is the slope angle of the reef ($\tan\alpha = 1/3$). Figure 5 also shows the relations between F_H (above) ,F_L (below) and H_0/L_0 for the blocks at E and F $(R = 3cm)$. Both F_H and F_L take maxima when $H_0/L_0 = 0.02 \sim 0.03$ irrespective of R. Summarizing the results, the maximum wave force along the reef may act on the block at E or F. Therefore the stability of the blocks at E and F are treated in the following sections.

594

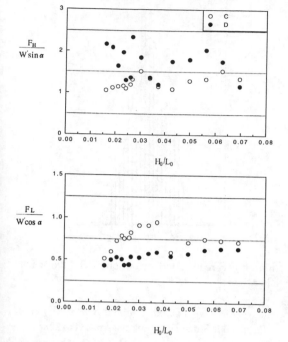

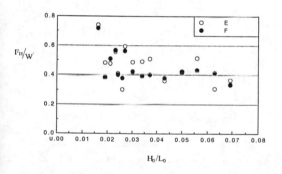

Figure 4: Relations between F_H (above), F_L (below) and H_0/L_0: C($\circ$), D($\bullet$)

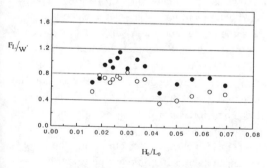

Figure 5: Relations between F_H (above), F_L (below) and H_0/L_0: E($\circ$), F($\bullet$)

2. 2 Visual observation

Experiments on the failure modes of armor blocks are also made. Model blocks are only placed on the acryl model reef. The experimental set up and the conditions are the same as those in section 2.1. Block failures take either of the following modes.

- upward movement

- rotation around the block end

A failure of the very first block on the reef starts following the first mode very frequently. The successive block failures after the first one usually take the second mode.

3 Equilibrium equation of forces

If the initial block movement at the position E or F is concerned, and if the block moves taking the first failure mode, the equilibrium equation of the block becomes

$$F = (W - B) - F_L + F_F, \qquad (1)$$

where F_L and F_F are the lift force and the lateral friction force respectively. Since gaps are usually observed between the next blocks when the initial block movement starts, lateral frictions may be neglected in eq.(1). The critical situation for the block is given by

$$f(W, B, F_L) = (W - B) - F_L = 0. \qquad (2)$$

The lift force is usually expressed as

$$F_L = \frac{1}{2}\rho_w C_L A u^2, \qquad (3)$$

in which ρ_w is the fluid density, C_L is the lift coefficient, A is the projected area which comes above the upstream block surface, on the normal plane in the direction of u, If there is no uncertainty on the basic variables C_L, A, u, and $W - B$ in eqs.(2) and (3), and if moderate values are used for these under the design value of u, no block failure takes place. However due to the physical uncertainties induced by the uneven block inclinations for a long time, those values used in the design may change even under the same design conditions of external forces.

Those properties are approximately realized in the experiment and their effects on the block stability are measured and analyzed. However, neither the values ρ_w, W and B in eqs.(2) and (3) nor eq.(3) itself may not change due to the uneven block properties over long period. C_L, A and u become to be the basic variables, in the present reliability analysis for the blocks on the reef.

3. 1 Uncertainty of the block surface inclination

The basic variable A in eq.(3) changes depending on the unevennesses and inclinations of two adjacent blocks: the concerned and its upstream ones. Placing the blocks on a mound of crushed stones, these properties are approximately realized. The armor block usually has a hole in itself or forms a hole with the next block when it is placed in a grid. The size of stone is selected so that no stone comes out from the hole. Since the hole size of the present model block is about $14mm$, crushed stones with the representative diameter from $16mm$ to $19mm$ (average $17.5mm$) are used. A very wide stone mound of $5cm$ in thickness is prepared on a wide flat table outside the wave tank. Its surface is not finely flattened but only the stones which form remarkable local unevennesses are removed. Their postures are measured with a level using a rule and graduator from two directions at right angles separately. Figure 6 shows a frequency distribution of 1,000 block inclinations in degree.

Its mean and standard deviation are $\mu(\theta) = 1.01°$ and $\sigma(\theta) = 4.36°$ respectively. The solid curve shows the Gaussian distribution of the above $\mu(\theta)$ and $\sigma(\theta)$. The area A for the block at E is larger than that at F since the upstream block D is not placed on the same plane but on the slope top (Fig.1). Therefore the value of A for the block at E is used in eq.(3) in this study. The vertical position of the offshore side block edge at E rises by $(L/2)\sin\theta$ from its mean position where θ is an angle of block inclination and L is a lateral block length ($64mm$). The vertical position of the onshore side block edge at D shifts by $(L/2)\sin\theta\cos\alpha$ from its mean where α is the angle of reef slope. $\tan\alpha = 1/3$ in the present reef model (Fig.1). Since $\sigma(\theta) = 4.36°$, standard deviation of the rise and down of the block edges at

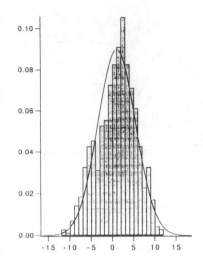

Figure 6: Frequency distribution of the block inclinations

the top of the reef slope are about $3.0mm$ (E) and $2.8mm$ (D). Since the block thickness is ($22mm$), these values are 13.6% and 12.9% of the thickness. Their square sum

$$(0.136)^2 + (0.129)^2 = (0.187)^2, \qquad (4)$$

gives the standard deviation of the area factor C_A where $A = C_A A_*$ and A_* is the whole lateral area of the block. The measured mean value of A for the block at E is $0.2A_*$.

3. 2 Uncertainty of the lift coefficient

Fluctuation of lift forces when blocks have inclinations are also measured experimentally. Figure 7 shows an experimental set up (below). The detailed equipment for the measurement is shown in the same figure (above). A single layer of crushed stones are adhered on a plate of $15cm \times 20cm$.

The stones are the same ones as used in the former section. Two model blocks are placed on the plate in the right direction of the flow. The plate is placed on a force measurement equipment which is installed in a channel bottom so that the stone surface coincides with the bottom level. The plate is exchangeable and 100 different plates with stones are prepared to realize the independent situations of different block inclinations. The channel is $0.5m$

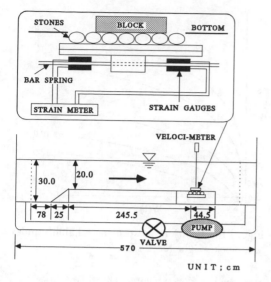

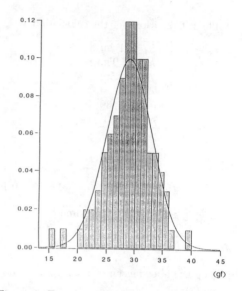

Figure 7: Channel(above) and equipment for the force measurement(below)

Figure 8: Frequency distribution of the lift force

wide, $0.6m$ deep and $6.7m$ long. Two different uniform flows of $30cm/s$ and $20cm/s$ are used. The flow velocity is measured by the velocity meter located $1cm$ above the block. Figure 8 shows the frequency distribution of the lift forces. The mean and standard deviation of C_L are $\mu(C_L) = 2.21$ and $\sigma(C_L) = 0.44$ respectively.

3. 3   Uncertainty of the water particle velocity

Uda [?] made measurements of the maximum water particle velocity on an artificial reef placed on a bottom slope of 1/30. Fore- and rear side slopes of the model reef are both 1/3.

Figure 9 shows the comparison between measured data and the next experimental equation (eq.5: Uda[?]) for the possible maximum water particle velocity u_{max} on the reef.

$$u_{max}/\sqrt{gR} = 8\exp\left(-1.5\frac{H_0}{h} - 2.8\frac{R}{H_0}\right) + 0.2,$$
(5)

in which R and h are water depth at the reef crown and reef. The mean and standard deviation of $u_{max\ (data)}\ /\ u_{max\ (eq.5)}$ are 1.16 and 0.24 respectively.

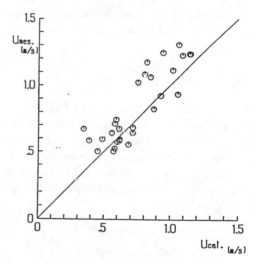

Figure 9: Comparison between the measured and calculated u_{max}, (after Uda et al. :1988)

597

Table 2: Conditions for the calculation

R	1.5m
h	5.95m
H_0'	2.5m
$\tan\alpha$	1/3
$\tan\theta$	1/30
ρ_s	2.3

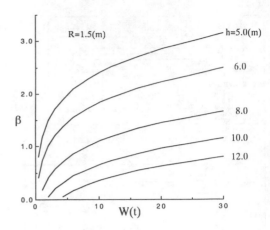

Figure 10: Relation between the reliability index and block weight ($R = 1.5m$)

When

$$H_0 = \left(\frac{2.8}{1.5}Rh\right)^{1/2}, \tag{6}$$

in eq.(5), u_{max} takes the next maximum

$$u_{max} = \sqrt{gR}\{8\exp[-\sqrt{16.8R/h}] + 0.2\}. \tag{7}$$

Multiplying the above mean of $u_{max\,(data)}$ / u_{max} $_{(eq.5)}$ to eq.(7),

$$u_{max}^* = 1.16\sqrt{gR}\{8\exp[-\sqrt{16.8R/h}] + 0.2\}, \tag{8}$$

gives the mean possible maximum u_{max} for the given values of R and h with the standard deviation of $0.24u_{max}^*$.

4 Failure probability of the block

Reliability analysis by Hasofer and Lind [?] is applied to estimate the failure probability of the block. Giving initial values for the basic variables and making iterations following the equations from eq.(10) to eq.(14), a set of the basic variables which makes the reliability index β minimum are obtained.

$$
\begin{aligned}
Z_1 &= [C_L - \mu(C_L)]/\sigma(C_L) \\
Z_2 &= [A - \mu(A)]/\sigma(A) \\
Z_3 &= [u_{max} - \mu(u_{max})]/\sigma(u_{max})
\end{aligned} \tag{9}
$$

$$\beta = \min\left(\sum_{i=1}^{3} Z_i^2\right)^{1/2} \tag{10}$$

In the present study, the physical properties of the prototype situation are used instead of those for models in the experiments. The properties of the prototype armor block (X-block) are also used in the calculation. Those conditions are listed in Table 2, where ρ_s is the specific block gravity.

Instead of Z, γ which is defined by

$$Z_i = \gamma_i\beta, \qquad (i = 1, 2, 3), \tag{11}$$

where

$$\sum_{i=1}^{3}\gamma_i^2 = 1, \tag{12}$$

is used in the iteration. Connecting eqs.(2) and (3), the failure function is obtained as

$$
\begin{aligned}
F &= (W - B) - \frac{1}{2}\rho_w C_L A u^2 \\
&= f(Z_1, Z_2, Z_3) = f(\gamma_1, \gamma_2, \gamma_3 : \beta) \\
&= 0.
\end{aligned} \tag{13}
$$

In the Hasofer and Lind method, new γ is defined with

$$\gamma_i = \frac{\dfrac{\partial f(\overline{Z})}{\partial Z_i}}{\left[\sum_{k=1}^{3}\left\{\dfrac{\partial f(\overline{Z})}{\partial Z_k}\right\}^2\right]^{1/2}}, \qquad (i = 1, 2, 3), \tag{14}$$

in which $\overline{Z} = (Z_1, Z_2, Z_3)$. The convergence of the iteration is fairly good in this method if the moderate initial values for γ and β are used. The failure probability is finally obtained substituting β obtained in the iteration into

$$P_f = \frac{1}{\sqrt{2\pi}}\int_{-\infty}^{-\beta}\exp(-x^2/2)dx. \tag{15}$$

The calculated result of the relation between β and the block weight W is shown in Fig.10. In

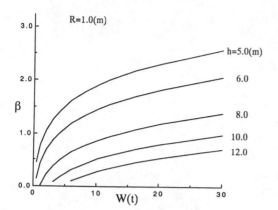

Figure 11: Relation between the reliability index and block weight ($R = 1.0m$)

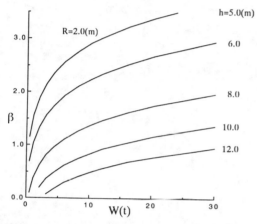

Figure 12: Relation between the reliability index and block weight ($R = 2.0m$)

the ordinary allowable stress method, almost all blocks heavier than 5t satisfy the required weight under the conditions used in the present study. Figures 11 and 12 also show the relation between P_f and β when $R = 1.0m$ (Fig.11) and $R = 2.0m$ (Fig.12) respectively.

5   Concluding remarks

Reliability analysis is applied for the estimation of armor block stability on an artificial reef. Uncertainty of the basic variables brings considerable failure probability of the blocks. If the standard deviations of the fluctuations on A, C_L are very small, this probability becomes negligibly small.

Since the fluctuations on A and C_L are inevitable for a long time, the failure of blocks must be taken place even when the block has the required design weight. Furthermore, we have to recognize that an application of larger block brings unexpectedly small reliability increase for a long time, since increasing block size also brings about larger lateral block area A. We may have to recognize the failure probability even when we use the ordinary allowable stress method in the design.

References

[1] Hasofer, A. M. and N. C. Lind (1974): see Thoft-Christensen, P. and M. J. Baker (1982): Structural reliability theory and its application, Springer- Verlag.

[2] Uda, T., A. Komata and A. Yokoyama (1988): The function and design method of an artificial reef, Tech. Memo. of PWRI, No. 2696, 71p. (in Japanese)

Stochastic Hydraulics'96, Tickle, Goulter, Xu, Wasimi & Bouchart (eds) © 1996 Balkema, Rotterdam. ISBN 90 5410 817 7

Uncertainty in stage-discharge relationships

Gary E. Freeman, Ronald R. Copeland & Mark A. Cowan
US Army Waterways Experiment Station, Vicksburg, Miss., USA

ABSTRACT: The uncertainty of stage-discharge relationships was evaluated for 116 gages located throughout the United States. The uncertainty of stage-discharge relationships is discussed for three differing classes of uncertainty which are described as natural, measurement, and modeling. Relationships are presented for the estimation of stage-discharge uncertainty for ungaged streams and for the estimation of Manning's n values in hydraulic model studies.

1 INTRODUCTION AND DISCUSSION

The U.S. Army Corps of Engineers (Corps) has begun to evaluate flood damage reduction projects using a risk-based analysis. This is a departure from the use of expected values in a deterministic approach for the estimation of project benefits due to the reduction of flooding. This shift in methodology has precipitated a need to understand and quantify the uncertainty associated with the stage-discharge relationship at gaging stations and in ungaged reaches of rivers and streams.

The uncertainty associated with the stage-discharge relationship was identified and classified into three categories. The categories were natural uncertainty, measurement uncertainty, and modeling uncertainty.

1.1 Components of natural stage-discharge uncertainty

The natural uncertainty associated with the stage-discharge curve is a result of many factors. These factors include effects from debris or other obstructions, ice, bed load and suspended sediment transport, bed forms, channel geometry changes during a flood event, changes in vegetation in the channel and overbank areas (winter vs summer), vegetation flattening or removal during high flows,

unsteady flow effects, interaction of flow between the main channel and overbanks in complex channels, and downstream tailwater effects which include storm surge and tidal effects. Aggradational or degradational trends also need to be considered and accounted for in any effort to determine uncertainty in the stage-discharge relationship for a flood damage reduction project.

1.2 Components of measurement stage-discharge uncertainty

The uncertainty associated with the measurement of stream stage and discharge includes effects from measurement errors, accuracies of equipment used in the measurement of stage and discharge, method of measurement, effects of waves on the ability to read stage, and so forth. Standard procedures have been developed to minimize these errors to the extent possible (USGS 1977). Even with the best equipment and an expert team, some error in measurement is inevitable.

1.3 Components of modeling stage-discharge uncertainty

When a numerical model is applied to a river reach, uncertainties in addition to those associated

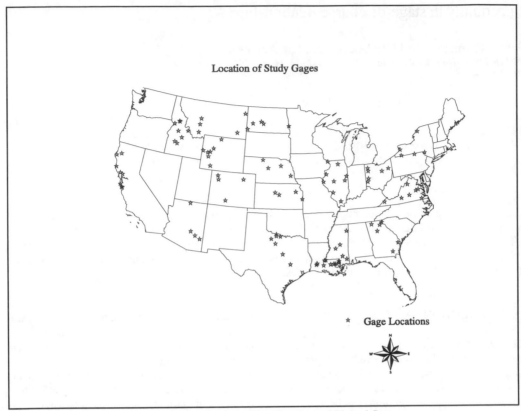

Figure 1. Location of Selected Gaging Station within Continental United States.

with measurement and natural conditions are introduced into the predicted stage-discharge curve. These uncertainties are generated by the inability of the numerical scheme to accurately represent the physical processes, the degree to which the model results are calibrated to observed results, the accuracy of the estimation of model parameters such as hydraulic roughness values (Manning's n values, etc), and the quality of the channel and overbank geometry used in developing the model. These factors have been analyzed previously by researchers at the US Army Hydrologic Engineering Center (HEC) located in Davis California (HEC 1987).

2 STUDY METHODOLOGY

The analysis of stage-discharge uncertainty was accomplished by two methods, one for the natural and measurement uncertainty and one for the uncertainty introduced by a numerical modeling effort.

The first method, used to analyze the combined natural and measurement uncertainty (hereafter called natural uncertainty), was the compilation of period of record data, where available, for 116 gages within the continental United States and Alaska. This data was analyzed for the combination of natural and measurement uncertainty.

A second method of analysis was used to determine stage-discharge uncertainty due to numerical modeling of a stream reach. This effort involved numerous runs of the water surface profile model HEC-2 (HEC 1990).

2.1 Methodology for natural uncertainty (natural plus measurement uncertainty)

In order to estimate the stage discharge uncertainty inherent in rivers and streams within the United States, the period of record data were compiled from 116 gages from through out the United States. Data from four of the gages was not usable resulting in data analysis from only 112 gages. Figure 1 identifies the 107 gage locations within the continental US. Five gages were also included from various Alaskan rivers.

The natural and measurement uncertainty were not separated since any stream gaging effort will include the measurement uncertainty. It was also assumed that the data collection by the USGS and Corps was done in a professional manner so as to reduce to a minimum any collection error. Based on this assumption and the fact that measurement error has been evaluated by the USGS in the analysis of their program and the quality assurance of the gage data (USGS 1977, USGS 1968), the measurement uncertainty included in the gage records was considered a part of the natural variation.

The evaluation of natural and measurement variability involved the determination of uncertainty for two differing conditions. The easier condition was for a river reach where a gage is present. The second and more usual condition is where no gage is present in the river or stream reach and uncertainty must be estimated from the physical characteristics of the river or stream of interest.

The data compiled from USGS and Corps files consisted of records from river gaging activities. Some of this data included the data sheets showing section by section measurements of discharge and the observed stage during the gaging exercise. Most records obtained were summaries (USGS form 9-207) giving the total flow rate, average velocity, stream width, and other basic stream gaging data.

These data were analyzed for changes in gage location, gage datum, or substantial shifts in the rating curves. The record was viewed as a whole except when multiple stage-discharge tracks were apparent. For some rivers it was necessary to break the period of record into two or more sections for analysis. Often the differing stage-discharge tracks were a result of changing the location of the gage. In cases where only the datum was changed, the various datums were shifted over the period of record to produce a single stage-discharge track.

Stage-discharge records were also analyzed for long term rating stability and if aggradational or degradational trends were noticed the records were broken into sections to provide stable periods of the gage records involved. For one gage in this study, located on a stream with high slope (> 0.02), the data was adjusted to remove the effects of a long term degradational trend. This was possible since the degradation at the gage was periodic and occurred during events of approximately 110 m^3/sec (4,000 cfs) or higher. For other gages with a more constant aggradation or degradation rate this method would be difficult and the period of record data may have to be analyzed in short increments or with the aggradational or degradational trends included with the base uncertainty at the gage.

Any long term aggradational or degradational trends must be noted and analyzed to determine long term stability of the channel and ultimate uncertainty for a project. The additional uncertainty due to long term trends should be evaluated separately from the basic stage-discharge uncertainty if possible.

The effects of ice and debris were separated from the data so that a base uncertainty could be established for each stream. The uncertainty from the ice and debris were calculated but data points were normally scarce and may not have been true indicators of the uncertainty of ice and debris effects on stage-discharge relationships. Gaging is also not performed on many streams during periods of heavy ice.

Additional research into the stage-discharge uncertainty associated with ice is currently underway at the US Army Cold Regions Research and Engineering Laboratory in Hanover, New Hampshire.

Subsequent to the data preparation described above, data was analyzed using STATISTICA, a statistical analysis package (StatSoft 1993). This package facilitated the determination of a best fit line through the data and the calculation of the distance from the best fit line to the data point. This distance from the best fit line was taken as a measure of the uncertainty of stage-discharge measurements at a particular gage. The best fit line model that was used was of the form:

$$STAGE = a + bQ^{1/2} + cQ^{1/3} + dQ^{1/4} + eQ^{1/5} + fQ^{1/6}$$

where Q is the observed discharge, and a, b, c, d, e, and f are coefficients determined by the statistical package to produce the best fit line for the stage-discharge data.

The uncertainty was calculated for stage rather

than discharge since stage uncertainty was needed for the other components of the Corps risk based flood damage estimation program.

The distribution of the stage-discharge uncertainty was also needed in order to provide parameters for a Monte Carlo approach to combining the uncertainty from frequency-discharge and stage-damage relationships with the uncertainty obtained from the stage-discharge relationship. The distribution was obtained by breaking the uncertainty into varying numbers of categories and comparing the sample distribution to various distributions. Chi Square and Kolmogorov-Smirnov tests were used to determine the best distribution.

2.2 Methodology for ungaged sites

The physical parameters of the sites used to determine uncertainty at gaged sites were collected and analyzed to determine any correlation with observed uncertainty. The parameters that were collected included bed material, river slope, cross-section information, sinuosity, mean flow data, elevation of floodplains, and a number of other factors that could prove helpful in classifying the streams and rivers. These various physical data were analyzed to determine any correlations between the parameters and the observed uncertainty. Approximately 90 of the sites were visited and photo records were made at each site visited. Field data were collected at many of the sites including surveyed slope, bed material samples, and site characteristics that could be used later to assist in data correlation.

2.3 Model uncertainty methodology

The US Army Hydrologic Engineering Center (HEC) had previously used a set of approximately 100 rivers and streams to analyze the importance of model parameters in the accuracy of calculated water surface profiles. (HEC 1987) These same data sets were used for additional runs to determine the effects of uncertainty in Manning's n value for the rivers and streams.

The HEC study determined that one of the parameters that most influenced model uncertainty was the Manning's n values used in the model. In order to better quantify the uncertainty in the selection of Manning's n values by water resources professionals, a survey was performed as a part of several training courses and in a number of meetings with hydraulic engineers. This survey consisted of presenting a series of slides showing stream reaches. Participants were asked to estimate the Manning's n value for the reach based on the slides and tables of recommended Manning's n values (for example Chow 1959).

The data obtained from these surveys were examined for outliers and evaluated for the mean, standard deviation (or variance), and distribution of the estimates from the mean values. The data were fitted with a best fit line and used to determine the uncertainty associated with a Manning's n value as a function of the mean value of the estimate.

The data thus obtained were used to estimate Manning's n values that were 1, 2, and 3 standard deviations above and below the value used in the calibrated HEC-2 models obtained from HEC.

The HEC study involved the variation of Manning's n value by means of a Monte Carlo method where the value at each cross-section is varied independently from other cross-sections in the model. The HEC method would account for random error introduced in the estimation of Manning's n value. The methodology in this study would simulate bias that might be introduced from either consistently over or under estimating Manning's n values.

This effort required approximately 700 computer simulations with seven runs for each river using the various biased n values. A base run using the Manning's n values as supplied by HEC was used as a basis for comparison. The data from the HEC-2 runs were evaluated both for individual cross-sections and for an average value for the river reach being modeled. The averaging of data allowed comparison between the data from this study and the HEC data.

3 RESULTS

The analysis of data was a complex and time consuming effort. Data on the 116 rivers was stored and analyzed in a spreadsheet environment which consisted of 126 rows and 220 columns. The data from the HEC-2 runs consisted of over 14,000 data points.

3.1 Uncertainty at gaging stations

The data collected from the 116 sites were analyzed for stage uncertainty based on the measured discharge. The best fit line described above was used to determine distribution of uncertainty. The best fit line was usually found using all six terms presented in the above equation. For some rivers it was found that in order to avoid a decreasing stage at flows near or in excess of the flow of record it was advisable to drop some of the higher order terms ($Q^{1/6}$, $Q^{1/5}$, etc.). The dropping of these terms resulted in only minimal reductions in the goodness of fit (R^2) and produced more physically correct stage discharge relationships.

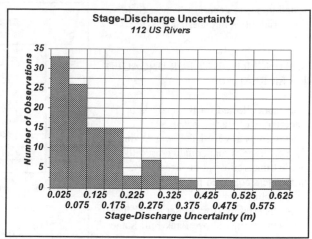

Figure 2. Distribution of Stage-Discharge Uncertainty for Gages Analyzed.

The analysis of the data from the various rivers and streams indicated that the gamma distribution was most applicable to the majority of the data sets. The uncertainty from the calculated line was expressed in terms of the standard deviation. The standard deviation for the gamma distribution is:

$$S_\Gamma = \sqrt{\frac{\kappa}{\lambda^2}}$$

where κ is the shape parameter, λ is the scale parameter, and S_Γ is the standard deviation calculated for the gamma distributed data.

The analysis found that the uncertainty for the rivers varied from a standard deviation of 0.0085 m (0.028 feet) on the New Fork River near Big Piney, Wyoming, (a mountain stream) to 0.624 m (2.05 feet) for the Mississippi at Tarbert Landing, Mississippi and 0.628 m (2.06 feet) for the Red River near Terral, Oklahoma. The Tarbert Landing gage is on the lower Mississippi River in southern Mississippi and is a sand bed river with a small slope.

The uncertainty for the New Fork River was even less than that associated with the Maumee River near Defiance which has a wide weir that controls stage at all flows. The Maumee River has a 0.023 m (0.074 foot) deviation but has a significantly flatter slope (0.0003 vs 0.0009 for the New Fork).

The uncertainties on the lower Mississippi River and the Red River were the largest found in the rivers in the study. The lower Mississippi is noted for bed forms, large sediment loads, and looped rating curves. The Red River at Terral is a sand bed stream with a well defined channel as is the Mississippi River. The gage on the Red River at Burkburnett, Texas, approximately 40 miles farther upstream, was also evaluated as a part of this study. This gage has an uncertainty of 0.311 m (1.02 feet) which is approximately one-half of the uncertainty found for the Terral gage. The difference between the gage sites is that the river cross-section is much wider with a shifting low flow channel at the Burkburnett location but consists of a well defined channel with relatively stable vegetated banks at the Terral site. The Terral site also has a 100 year flood flow nearly double that of the Burburnett site and a total stage range of 1.5 times larger than that of the Burkburnett gage.

The distribution of uncertainty for the rivers analyzed is shown in Figure 2. It can be seen that the uncertainty values are skewed to the lower end of the range with a median uncertainty (standard deviation) of 0.132 m.

Analysis of the data from the 112 rivers indicated a variety of uncertainty distributions. The distribution of uncertainty varied from a distribution centered at the mean which was much narrower than a normal distribution, to highly skewed distributions. The skew in the distribution were found towards both the high and low side of the best fit line. The gamma distribution was

found to be, in general, the best fit for
the variabilities in uncertainty.

3.2 UNCERTAINTY FOR UNGAGED RIVER REACHES

The data collected for the analysis of
uncertainty of stage-discharge
relationships at gaged sites was evaluated
for correlations that could be used to
predict uncertainty at ungaged sites. The
parameters collected for the various gages
are as follows:
1. Uncertainty from Observed Values
2. Uncertainty above Bank Full Stage
(20% exceedence value)
3. Slope from 1:24,000 USGS
Quadrangles
4. Surveyed Slope - from riffle to riffle
where applicable and available
5. Maximum Observed Flow
6. Median Flow
7. Flow Ratio - maximum flow divided by
median flow
8. 100 Year Flood Flow
9. Bed Material Type - Rock, Sand, etc
10. Sediment Classification - D_{50}, D_{84}, and D_{16}
11. Flow Control (Riffle, Weir, etc)
12. Sinuosity
13. USGS Gage Classification
14. Contributing Basin Area
15. Bank Vegetation
16. Ice Uncertainty (limited amount of data)
17. Debris Uncertainty (limited data)
18. Average Basin Rainfall
19. Maximum Observed Stage Range
The individual correlations of these various
parameters with stage-discharge uncertainty
produced results that were not statistically
significant.
The stage discharge uncertainty should also be
affected by the variables that are contained in the
flow equations. These are based on physical
parameters and include:
1. Slope
2. Flow Area
3. Hydraulic Roughness of Channels and
Floodplain
4. Changes in Channel Shape During Events, and
5. Backwater Effects, if any
The effects of backwater were eliminated from the

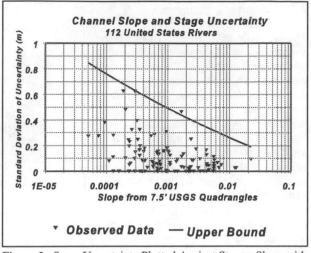

Figure 3. Stage Uncertainty Plotted Against Stream Slope with
Approximate Upper Bound of Data.

analysis where ever backwater effects were
indicated in the USGS data. The changes in
channel shape during events could not be
determined from the data and any affects must be
considered a part of the overall uncertainty. The
change in channel shape during an event are,
however, affected by the bed type and the sediment
type being transported. This combines with the
shape of the channel and floodplain to give the
flow area. In considering streams it also apparent
that steep streams are associated with smaller
floodplains while flat streams are normally
associated with wide floodplains with meandering
channels. Figure 3 is an attempt to view the
relationship between slope and stage uncertainty.
It apparent that at low values of slope there is no
relationship based on slope alone but an upper
bound appears to be possible based on the 112
gages evaluated.

Multiple regressions were also run on the above
data with better results. The results of the multiple
regression identified four factors that were
significant in predicting stage-discharge
uncertainty. These factors were basin area above
gage (contributing area), maximum observed stage
range, bed type identifier, and 100 year flood flow.
The bed type was identified on a simple scale of 0-
4 as shown in Table 1.

The formula for the estimation of stage-discharge
uncertainty is:

$$U_s = [0.07208 - 2.2626x10^{-7}A_{Basin} +$$

$$0.0216H_{Range} + 1.419x10^{-5}Q_{100} + 0.04936\ I_{Bed}\ ]^2$$

Where:

U_S = Standard Deviation of Stage-Discharge Uncertainty

H_{Range} = Maximum Stage Range (m)

A_{Basin} = Basin Area Above Gage (km²)

Q_{Max} = 100 Year Flood Flow (m³/sec)

I_{Bed} = Bed Identifier (Table 1)

This equation has an adjusted R^2 of 0.65.

Table 1. Bed Type Identifiers Used in Correlations.

Bed Identifiers	
Material	Identifier
Rock / Resistant Clay	0
Boulders	1
Cobbles	2
Gravels	3
Sands	4

3.3 UNCERTAINTY FROM PARAMETER ESTIMATION AND MODEL INACCURACIES

The 100 rivers studied previously by HEC (1987) were evaluated for the effect of biases in the estimation of Manning's n values. Before the effects of a bias could be determined it was necessary to understand the uncertainty involved in the estimation of Manning's n values. The uncertainty that could be expected in the estimation of Manning's n values was determined by the development of a series of case studies. The case studies consisted of a series of slides showing different streams through out the United States. The case studies were presented to a series of Corps training classes and hydraulic engineering meetings. A very few data points were eliminated as being outliers by use of the Tukey outlier test. The data thus obtained were used to

estimate the uncertainty associated with n value estimation. The uncertainty obtained from the case studies is presented in Figure 4. The data from this exercise were used to modify the Manning's n values used in rivers and streams used by HEC in their study.

The Manning's n values in the models were modified to represent the 1, 10, 33, 67, 90, and 99 percent values from the distributions presented. This produced changes in water surface elevations from -15 feet to +20 feet depending on the stream and the initial values of Manning's n used in the model cross-sections. Some problems were noted in that the cross-sections did not extend far enough to reach terrain as high as the water surface for the high positively biased n values used for the 90 and 99 percentile tests. In order to eliminate this problem the values for the 33 and 67 percentile values were used to estimate the standard deviation of model uncertainty.

The tests produced over 14,000 data points representing changing water surface elevations based on differing values of Manning's n. The equation presented in section 3.2 could not be used to predict uncertainty since the parameters needed for estimation were not included in the data set obtained from HEC. Additionally the factors used in section 3.2 are not used in the HEC-2 model input and should not affect model output.

The 14,000 data points were evaluated for correlation with channel width, slope, depth, stream power, velocity, hydraulic radius, flow

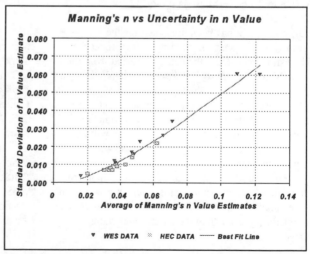

Figure 4. Uncertainty of Manning's n Value Estimates Based on Mean Estimate Values.

area, channel and overbank n values and various combinations of the overbank and n channel values. No correlation could be found between any of the variables used in the attempt to compare Manning's n bias and resulting water surface uncertainty. The data were also evaluated based on each cross-section within the models. The uncertainty obtained from the modeling effort was also plotted against slope as shown in Figure 5. It can be noted that the data points follow the same pattern of reduced uncertainty for increasing slope noted for the observed gage data. The scatter of data in the upper portion of the plot points to the need for careful model calibration in the selection of Manning's n values to ensure accurate model results. The upper bound for the natural data is also plotted in Figure 5. It can be noted that the upper bound for stage uncertainty is significantly lower for the natural uncertainty than for uncertainty associated with the selection of Manning's n value in a modeling effort.

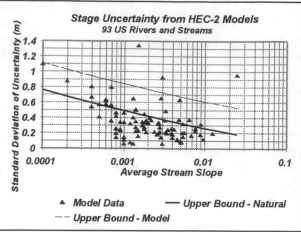

Figure 5. Stage Uncertainty Resulting from Manning's n Value Uncertainty Obtained from HEC-2 Modeling Effort.

4 CONCLUSIONS

The studies described in this paper arrived at the following results:

1. The uncertainty of US rivers at gaging stations ranges from approximately 0.0085 to 0.628 m (0.028 to 2.06 feet) based on the 112 rivers evaluated.

2. A formula has been presented for the estimation of uncertainty in ungaged river reaches. The estimations from the equation varied from the actual values by up to 100% so care should be exercised in applying the equation.

3. There is no easy relationship that can be used to determine stage uncertainty due to estimation errors of Manning's n values. Modeling will be required to determine stage uncertainties through a sensitivity analysis approach.

4. Uncertainty in the estimation of Manning's n values has been assessed and the impact of a bias in the estimation has been evaluated. This impact is, in general, greater than the natural uncertainty for a river reach.

5 ACKNOWLEDGMENTS

Permission has been granted by the Chief or Engineers to publish this research. Funding for this research was obtained from the Risk Analysis of Water Resources Investments Program of the US Army Corps of Engineers. The use of trademarked or program names in no way implies endorsement of the product by the US Government, the US Army Corps of Engineers, or the US Army Waterways Experiment Station.

6 REFERENCES

Chow, Ven Te, 1959. Open Channel Hydraulics, McGraw-Hill Book Company, New York, New York.

HEC, 1987. Accuracy of Computed Water surface Profiles, US Army Hydrologic Engineering Center, Davis, California, USA 95616.

HEC, 1990. HEC-2 Water Surface Profiles, User Manual, US Army Hydrologic Engineering Center, Davis, California, USA 95616

StatSoft, 1993. STATISTICA User's Manual, StatSoft, Inc., Tulsa, Oklahoma, USA 74194

USGS, 1977. National Handbook of Recommended methods for Water Data Acquisition, USGS, Reston, Virginia, USA

USGS, 1968. US Geological Survey Techniques, Water Resources Inventory, Book 3, Chapters A, USGS, Reston, Virginia, USA.

608

Stochastic Hydraulics'96, Tickle, Goulter, Xu, Wasimi & Bouchart (eds) © 1996 Balkema, Rotterdam. ISBN 90 5410 817 7

Uncertainty analysis of water distribution networks

Chengchao Xu & Ian Goulter
Department of Civil Engineering and Building, Central Queensland University, Rockhampton, Qld, Australia

ABSTRACT: The effects of uncertainty and imprecision in nodal demands and in the values of pipe parameters on the output of simulation studies of water distribution networks are examined. Three uncertainty analysis techniques, derived from interval theory, probability theory and fuzzy logic theory respectively, are employed in the examination of the effects of these uncertainties on the results of hydraulic modelling. The hydraulic modelling associated with the uncertainty analysis is performed by an approximate and computationally efficient hydraulic model developed by linearisation of the non-linear hydraulic equations at selected points. Results from a Monte-Carlo simulation study indicate that these linearised models provide very good representations of the nonlinear behaviour of the system provided that the variability of the model parameters is not too large. The approaches are demonstrated by application to an example network.

1 INTRODUCTION

Hydraulic modelling of water distribution networks is a critical component in the planning, design, maintenance and operational control of water supply systems. The principles involved in mathematical modelling of the hydraulic performance of water distribution networks have been well documented and many efficient algorithms and computer programs have been developed, e.g., Boulos and Wood (1990). However, one common feature of these models is that they are inherently deterministic in the sense that the results of the simulation are only valid for the specified set of model input parameters. In practice, whilst much of the model input data can be determined with a reasonable degree of accuracy, some network parameter data such as the nodal demands and pipe roughness are generally much less reliable particularly for existing systems which have been in operation for some times. Although many algorithms have been devised to refine the estimates of these network parameters through model calibrations using field measurement data, e.g., Boulos and Wood (1990), Lansey and Baset (1991) and Liggett and Chen (1994), a great deal of uncertainty in the values of these network

parameters still exists. The situation is made even more complex by the fact that these values are generally changing temporally and spatially within a network.

The imprecision and uncertainty in the values of these input parameters may severely decrease the quality of solution of a deterministic network analysis and degrade the usefulness of simulation results particularly in the operation of an existing water distribution network. It is therefore very important that the level of the uncertainty in the solution of network analysis resulting from the imprecision and uncertainty in the network parameters is appropriately quantified so that the network engineers are clearly informed on not only what is the most likely behaviour of the system but also how that behaviour might vary in some detailed manner. Uncertainty analysis of water distribution networks aims to determine how the uncertainty and imprecision in the model input parameters affect the results of network analysis.

2 PROBLEM DESCRIPTION

The problem of uncertainty analysis in water distribution network model has received relatively little attention in the literature. The main reason for this lack of attention appears to be a result of the difficulty of representing the uncertainty and propagating it through the analysis. Several studies, e.g., Lansey et al. (1989), Bao and Mays (1990) and Bouchart and Goulter (1991) have examined the effects of uncertainty in nodal demand and the pipe capacity on the hydraulic performance of water distribution networks. However most of these efforts were directed at the issue of reliability assessment rather than quantifying the uncertainty associated with the outputs of deterministic network analysis. An important exception is the work by Bargiela and Hainsworth (1989) who attempted to quantify the pressure and flow uncertainty in the system from model input uncertainty. The uncertainties in models including, nodal demands, reservoir heads and some pipe flow measurements, were modelled in that study by an error bound. Three models, including Monte-Carlo simulation, linear programming approaches and sensitivity analysis, were developed to estimate the error bounds on the output of network analysis. The usefulness of techniques was demonstrated by application to problem of optimal design of telemetry monitoring networks. It is important to note that their approach concentrated on the uncertainty in system state which is caused by the imprecision and uncertainty in the nodal demands but ignored the effects of uncertainty in pipe roughness.

This paper takes a broader approach to the consideration of uncertainty and examines the problems of the interrelationships between the uncertainties in the nodal demands and values of the pipe parameters and the uncertainties in network simulation results. Three distinct uncertainty analysis techniques, based on interval theory, probability theory and fuzzy logic theory respectively, are employed in the examination of the effects of these uncertainties on the results of hydraulic modelling. The approach is based upon the principle that by using appropriate information from the deterministic network solution, the uncertainty associated with simulation results can be quantified in terms of bounded values, probability distributions and possibility distributions. In contrast to traditional deterministic network solvers, which produce a single value for each output variable, these uncertainty analysis models provide the network engineers with a range of feasible values for the system state, each associated with a measure of confidence. The potential use of this information to enhance the decision-making in the management of water distribution networks is discussed.

3 HYDRAULIC NETWORK MODEL

Analysis of water distribution networks involves the determination of nodal heads and pipe flow rates which satisfy the two fundamental principles:

1. Continuity: The algebraic sum of the flow rates in all the elements meeting at a junction, together with any external flows, is zero.

2. Energy conservation: The algebraic sum of the head losses in each element, combined with any head generated by pumps, around any closed loop formed by hydraulic components is zero.

There are many alternative formulations for the system governing equations and techniques to solve these equations. By using a nodal formulation which involves taking the unknown nodal heads as system state variables, the network hydraulic model can be expressed as follows:

$$\sum_{j \in \Omega_i} Q_{ji} + D_i = 0 \qquad i = 1, \ldots, N \qquad (1)$$

where Q_{ji} = algebraic flow in the link (pipe) connecting node j to node i (=0 if node j is not connected to node i); D_i = nodal demand; H_i = nodal head; N = total number of nodes with unknown heads and W_i = a set of nodes connected to node i.

The nonlinear function relating the hydraulic loss and flow rate in the pipe connecting nodes i and j is described by the Hazen-Williams equation as follows:

$$Q_{ij} = \alpha \ C_{ij} d_{ij}^{2.63} \left(\frac{H_i - H_j}{L_{ij}} \right)^{0.54} \qquad (2)$$

where C_{ij}, d_{ij} and L_{ij} are the Hazen Williams coefficient, diameter and length of the link (pipe) connecting nodes i and j respectively; α is a constant whose value depends upon the units used in calculation.

Equation (1) can be expressed in more compact form as:

$$\mathbf{F}(\mathbf{H}) + \mathbf{D} = 0 \qquad (3)$$

where $\mathbf{H}$ and $\mathbf{D}$ are vectors of the unknown nodal heads and known nodal demands respectively. The nonlinearity of $\mathbf{F}(\mathbf{H})$ means that the system unknowns cannot be solved directly and an iterative technique such as the Newton-Raphson method must be employed. Iterative linearisation procedures usually start with an initial guess for the system state. This guess is then progressively refined until the nodal continuity equation is satisfied.

The use of such a hydraulic model requires knowledge of the exact values for all the model parameters appearing in Equations (1) and (2). As discussed in the introduction, due to their temporal and spatial variability nodal demands and pipe roughness are particularly difficult to estimate precisely in practice. However, the deterministic network model described above provides no information about how the output will be affected by the imprecision and uncertainty in the model input parameters.

The problem of how imprecision and uncertainty in model parameters is propagated through the network is very complex due to the nonlinearity of hydraulic models. However, if the variability of the input parameters is relatively small, the uncertainty, or more precisely the propagation of uncertainty, can be analysed by using an approximate model derived from linear expansion of the non-linear hydraulic models at appropriate points.

Consider the nodal formulation of water distribution networks rewritten as follows:

$$F(H,C) + D = 0 \qquad (4)$$

By using a first order Taylor series expansion at some values of the nodal heads, demands and pipe roughness, e.g., $\overline{H}$, $\overline{C}$ and $\overline{D}$, Equation (4) becomes:

$$\left\{ F(\overline{H},\overline{C}) + \overline{D} \right\} + J\Delta H + J_c\Delta C + I\Delta D = 0 \qquad (5)$$

where $\Delta H = H - \overline{H}$; $\Delta C = C - \overline{C}$ and $\Delta D = D - \overline{D}$; J and J_c are the Jacobian matrix and sensitivity matrix with respect to change of pipe coefficients respectively; I is a unit matrix and M is the number of pipes in the network.

Note that hydraulic network is in balance at the expansion point. The first term of Equation (5) is therefore zero and Equation (5) can be rewritten as:

$$H = \overline{H} - J^{-1}J_c(C - \overline{C}) - J^{-1}I(D - \overline{D}) \qquad (6)$$

where J^{-1} is the inverse of J.

Defining $A = -J^{-1}$, $B = AJ_c$ and $\overline{\overline{H}} = \overline{H} - A\overline{C} - A\overline{D}$, Equation (6) becomes

$$H = \overline{\overline{H}} + BC + AD \qquad (7)$$

The linearised hydraulic model, represented by Equation (7), forms the basis for the development of three uncertainty analysis models for water distribution networks used in this study and described in the following sections.

4 UNCERTAINTY ANALYSIS

4.1 Interval analysis

Interval analysis represents the uncertainty as unknown but bounded, i.e., the upper and lower limits on the uncertainties are assumed but no assumptions are made about the distribution information. Uncertainty with unknown but bounded characteristic can also be modelled by a so-called grey number (Huang et al. 1995). Mathematically, a grey number $\tilde{a}$ is defined as $\tilde{a} \in [\underline{a}, \overline{a}]$, where $\underline{a}$ is the minimum value and $\overline{a}$ is the maximum values of $\tilde{a}$. The mid-value of a grey number is defined as its mid-value and has the same distance to both its upper and lower bounds, i.e., $0.5(\underline{a} + \overline{a})$.

The use of grey numbers to represent the uncertainty and imprecision in the nodal demands and pipe roughness only requires that the maximum and minimum values that these parameters can possibly take be specified. For the nodal demands, the upper and lower bounds can generally be estimated from the historical demand data or from the factors such as population distribution, type of consumers and level of leakage etc., which dominate the demand conditions. Similarly, the upper and lower bounds of the pipe roughness, more specifically the Hazen-Williams coefficients, can be evaluated based on the age and conditions of the pipe taking into consideration such factors as corrosion of pipes and deposition in pipes. In both cases, the knowledge and experience of network engineers play a significant role in the determination of these bounds. The main task of the uncertainty analysis is generally the estimation of the upper and lower bounds for the nodal heads.

By linearising the nonlinear hydraulic equations at the mid-value points of the grey nodal demands and grey pipe roughness, the grey nodal heads can be expressed as follows:

$$\tilde{H} = \overline{\overline{H}} + B\tilde{C} + A\tilde{D} \qquad (8)$$

where $\tilde{H}$, $\tilde{C}$ and $\tilde{D}$ are vectors of the grey nodal heads, the grey pipe coefficients and the grey nodal demands respectively. For each nodal head, the maximum and minimum values can be obtained by

$$\overline{H}_i = Max(H_i) = \overline{\overline{H}}_i + Max(\sum_{k=1}^{M} b_{ik}\tilde{C}_k + \sum_{i=1}^{N} a_{ij}\tilde{D}_j) \qquad (9)$$

and

$$\underline{H}_i = Min(H_i) = \overline{\overline{H}}_i + Min(\sum_{k=1}^{M} b_{ik}\tilde{C}_k + \sum_{j=1}^{N} a_{ij}\tilde{D}_j) \qquad (10)$$

where a_{ij} and b_{ij} are the elements of sensitivity matrices A and B respectively.

Equations (9) and (10) show that the maximum and minimum values of nodal head are linear functions of the upper and lower bounds of the grey nodal demands and the grey pipe coefficients. The bounds that the grey demands and the grey pipe roughness take in Equations (9) and (10) depend solely on the algebraic sign of the sensitivity coefficient. If the maximum value is desired, the maximum possible values of the grey demands and grey pipe roughness are used if the sensitivity coefficient is positive, and vice versa.

The actual algorithm for analysis of uncertainty in water distribution networks using the interval analysis approach can be summarised as follows:

1. Using the mid-values of the grey nodal demands and the grey pipe coefficients, calculate the mid-values of the grey nodal heads by using a deterministic network solver.

2. From the deterministic solution, obtain the sensitivity matrix $\mathbf{B}$ and $\mathbf{A}$.

3. Calculate the upper and lower bounds for nodal heads based on Equations (9) and (10).

4.2 Probability analysis

Probability analysis considers the uncertainty and imprecision of model input parameters as random processes with known statistics. The statistics of the mean and variance of nodal demands and values of pipe roughness are estimated from historical data or assumed subjectively on the basis of past experience. The main purpose of this type of probabilistic analysis is to derive the probabilistic characterisations of random nodal heads using analytical probability theory.

Provided that the variability of the random demands and values of pipe roughness coefficients is not very large (results from other research work not reported here indicate that the accuracy of linearised hydraulic model is reasonably good when the coefficients of variations for the nodal demands and values of pipe roughness coefficients are less than 0.3 and 0.15 respectively), the random nodal heads can be modelled by the linear Equation (7) which as shown previously were obtained by linearising the nonlinear hydraulic equations around the expected value region. Inserting such a restricted range of random demands and pipe roughness coefficients into Equation (7) gives

$$\hat{\mathbf{H}} = \overline{\overline{\mathbf{H}}} + \mathbf{A}\hat{\mathbf{D}} + \mathbf{B}\hat{\mathbf{C}} \qquad (11)$$

where $\hat{\mathbf{D}}, \hat{\mathbf{C}}$ and $\hat{\mathbf{H}}$ are the vectors of random nodal demands, random pipe coefficients and random nodal heads respectively.

Equation (11) provides the opportunity to compute each random nodal head from a weighted sum of the random nodal demands and random pipe roughness. The weighting factors used in this process are defined as sensitivity coefficients with respect to the random demands and random pipe coefficients. In theory, the distribution function for each random element of nodal heads can always be obtained from a knowledge of the joint distribution of nodal demands and pipe parameters; in practice, however, such a task is often very complicated. However, if the random variables are assumed to be pairwise statistically independent, through application of the central limit theorem, the probability distribution of each nodal head is approximately normally distributed for large or even modest size of networks no matter what the distributions of the random demands and pipe roughness actually are.

The outcome of this step is that the calculations of probability distribution of nodal heads are greatly simplified since only the mean and variance of the random nodal heads need to be estimated. The expected values of nodal heads are obtained from the deterministic solution as described previously. The variances of the nodal heads are estimated using the approach of Yen et al. (1986) by:

$$\sigma^2_{H_i} = \sum_{j=1}^{N} a^2_{ij}\, \sigma^2_{D_j} + \sum_{k}^{M} b^2_{ik}\, \sigma^2_{C_k} \quad i = 1,2...,N \qquad (12)$$

where $\sigma^2_{D_i}$ and $\sigma^2_{C_i}$ are the variances of the random demand at node i and the random pipe coefficient for pipe i respectively.

The actual algorithm for analysis of uncertainty using the probabilistic modelling of water distribution networks can be summarised as follows:

1. Solve, for the estimated mean of nodal heads $\overline{\mathbf{H}}$, a deterministic hydraulic network assuming the nodal demands and pipe coefficients equal to the their expected values $\overline{\mathbf{D}}$ and $\overline{\mathbf{C}}$ respectively.

2. Obtain the sensitivity coefficient matrices $\mathbf{A}$ and $\mathbf{B}$.

3. Compute the variance of random nodal heads from Equation (12).

4.3 Fuzzy sets approach

Fuzzy set theory was introduced by Zadeh (1965) and deals with uncertainty and imprecision by use of fuzzy representations, e.g., fuzzy numbers. A fuzzy number can also be viewed as an extension or generalisation of the concept of the interval of confidence. Instead of considering the interval of confidence at one unique level, it is considered at several general levels between 0 and 1, through the use of a membership function. This membership function can be interpreted as a measure of the degree of belief in the occurrence of certain events. (A detailed description of the concept of fuzzy sets,

612

fuzzy numbers and their arithmetic operations is given in a recent book by Kaufmann and Gupta (1991).)

In this study, the uncertainty and imprecision of the nodal demands and pipe roughness coefficients are modelled by a possibility distribution and represented by a trapezoidal fuzzy interval as shown in figure 1. The possibility distribution described by values D_1, D_2, D_3 and D_4 in that figure is defined to reflect the degree of belief in the occurrence of some values of the nodal demands and pipe parameters. The possibility distribution will have a value of 1 for demand values or pipe roughness coefficients that are highly possible. The possibility distribution will decrease in value from this maximum value of 1 as the possibility of the corresponding value diminishes to the point where a zero possibility is assigned to the values that are rather impossible to occur. These impossible values are located in Figure 1 beyond the two extreme values of D_1 and D_4. In the context of water distribution network parameters, the fuzzy representation of nodal demands shown in Figure 1 can be interpreted as " demand may occur between D_1 and D_4 but it is most likely to be between D_2 and D_3".

Uncertainty analysis based on fuzzy representations aims to derive the possibility distribution for the nodal heads. In common with previous two models, a linearised model at the medium points of possibility distributions of the nodal demands and pipe roughness coefficients is used, i.e.,

$$\hat{H} = \overline{\overline{H}} + A\hat{D} + B\hat{C} \qquad (13)$$

where $\hat{H}$, $\hat{D}$ and $\hat{C}$ are the vectors of fuzzy nodal heads, demands and pipe coefficients respectively.

The actual calculation of fuzzy nodal heads is accomplished in three steps as follows:

1. Perform deterministic network simulation using the medium points of fuzzy demands and pipe parameters.

2. Calculate the sensitivity matrices A and B.

3. Estimate the four critical points, H_1, H_2, H_3 and H_4, of the fuzzy nodal heads from Equation (13) using fuzzy arithmetic.

It is important to note that forms of possibility distributions other than the trapezoidal shape shown in Figure 1 can be used to represent the parameter uncertainty. However, if non-linear membership functions are employed, it may be necessary to evaluate the intervals of nodal heads for selected a-cuts in order to derive the possibility distribution of nodal heads.

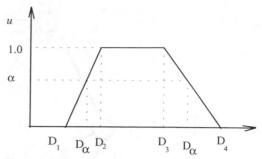

Figure 1 Representation of trapezoidal fuzzy number

5 NUMERICAL EXAMPLES

The models for uncertainty analysis of water distribution networks described above are demonstrated by application to an example network taken from Lansey and Basnet (1991) as shown schematically in Figure 2. The averaged nodal demands in litre per second and the reservoir levels in metres are also given in this figure. Relevant link information is given in Table 1. The Hazen-Williams equation is used to determine the hydraulic losses along each pipe. It should be noted that the Hazen-Williams coefficients given in Table 1 are averaged values.

Hydraulic simulation, results of which are shown in column 2 of Table 2, was performed using a deterministic network solver to calculate the nodal heads under the conditions of averaged demands and mean pipe roughness. This solution forms the basis for the subsequent uncertainty analysis since all three uncertainty analysis models make use of the sensitivity coefficients derived from this baseline solution. By allowing the demands and the values of pipe parameters to vary around the means by 10%, the upper and lower bounds of nodal heads were estimated from the interval analysis model shown in the columns 3 and 4 of Table 2. It can be seen from these nodal head values that the feasible range of nodal heads varies from node to node with the range at node 12 (measured by the ratio of feasible interval to the mid-value) being the biggest (about 24%).

In the same way, the effects of random variation of the nodal demands and the pipe roughness on nodal heads can be evaluated by probability analysis. Column 5 of Table 2 gives the standard deviation for each nodal head assuming coefficients of variation for nodal demands and the pipe roughness

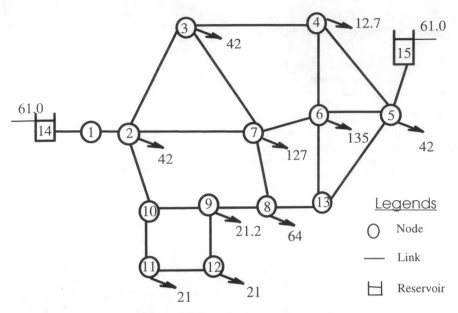

Figure 2 Schematic of example network

equal to 0.1 and 0.05 respectively. The probability that the nodal head at a particular node falls within any specified limits can easily be found from tables of cumulative standard normal distribution or estimated numerically by using approximate functions.

If the only information available on the uncertainty and imprecision of the model parameters is in form of linguistic declaration such as "the change of model parameters is likely to be within 5% range and rarely exceeds 10%", the uncertainty analysis model based on fuzzy sets can be applied to derive the possibility distributions of nodal heads. The most likely range of the nodal heads is given in the columns 6 and 7 of Table 2. These results, together with the interval analysis values in columns 3 and 4, enables the possibility distribution of nodal head to be constructed. This process of constructing the possibility distribution of a nodal head is shown in Figure 3 for node 12. It can be seen in Figure 3 that the nodal head for node 12 may fall anywhere between 40.79 m and 51.98 m but is most likely to be between 43.58 m and 49.18 m. The results shown in Table 2 also indicate that the effects of use of inappropriate values of nodal demand and pipe roughness arising from uncertainty and imprecision in the exact values of these parameters on the predicted nodal heads can be very significant.

Recall that all three models are developed by linearising the hydraulic network models. However, results obtained from the simplified linear model were in very close agreement with the results obtained from Monte-Carlo simulations using non-linear hydraulic model. An example of the closeness of these results is shown in Figures 5 and 6 for the probabilistic analysis of node 12. Similar results are obtained for the other models and nodes but are not given here due to space limits.

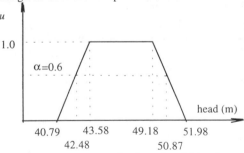

Figure 3 Possibility distribution of nodal head (node 12)

Information from uncertainty analysis can be used in many ways. For instance, they can be used to assess whether a particular operating policy may render the system out of an acceptable 'secure' operating range and to provide some safety guidelines for operation of water distribution networks. They can also be used in the design or expansion of monitoring network for water systems

614

to evaluate the effects of a new measurement station. Both aspects are currently under investigation.

6 DISCUSSION AND CONCLUSION

Analysis of water distribution networks has traditionally been limited to calculation, using hydraulic simulation models, of the exact values of nodal heads and pipe flows from a given set of precise input parameters. In reality the input parameters to the network models can rarely be estimated precisely, particularly for existing systems which have been in operation for some time. Uncertainty analysis provides more information on the system state than the conventional network models. Such information can be used to assist operators to provide a safe and reliable water supply. Of course, the uncertainty and imprecision of input parameters to the network model need to be quantified in order to use these uncertainty analysis tools.

In this paper, three models derived from interval analysis, probability modelling and fuzzy sets respectively are developed and evaluated as a means of estimating how the uncertainty and imprecision in nodal demands and values of pipe parameters affect the solution of network analysis obtained from a particular hydraulic network simulation. The computational procedures for three models are very similar and require a deterministic network simulation, from the associated sensitivity matrices are obtained. All three algorithms were found to be very efficient and require execution time which is only fractionally higher than that required for deterministic network simulation models. The simulation component of the analysis used a linearised approximation of the non-linear hydraulic equation. This approach was found to be sufficient accurate for cases where the variation in the hydraulic parameters around the mean values was not too large.

Each of the three uncertainty representations has its own advantages. When adequate statistical data on the uncertainty, from which the mean and variance of the uncertainty can be estimated reliably, are available, the approach based on probability analysis is most appropriate. However, probabilistic models can be misleading if the uncertainty represents an event that will occur rarely. On the other hand, the

Table 1 Link information of example network

From node	To node	Length (m)	Diameter (m)	Hazen-Williams Coef.
14	1	610.0	0.508	120.0
1	2	240.0	0.457	110.0
2	3	1524.0	0.406	110.0
3	4	1128.0	0.305	100.0
4	5	1189.0	0.381	100.0
5	15	640.0	0.406	110.0
6	4	762.0	0.254	100.0
7	3	945.0	0.254	100.0
2	7	1676.0	0.381	95.0
7	6	1128.0	0.305	100.0
5	6	884.0	0.305	95.0
5	13	1372.0	0.381	100.0
6	13	762.0	0.254	90.0
8	7	823.0	0.254	90.0
2	10	945.0	0.305	100.0
9	10	579.0	0.305	95.0
10	11	488.0	0.203	100.0
12	11	457.0	0.152	110.0
12	9	503.0	0.203	100.0
8	9	884.0	0.203	100.0
13	8	945.0	0.305	95.0

Table 2 Results of uncertainty analysis

Node	Averaged Nodal Head	Interval Analysis Bound		Probability Analysis	Fuzzy Set Interval ($\alpha=1$)	
	(m)	Upper (m)	Lower (m)	Std. dev.(m)	H_1 (m)	H_2(m)
(1)	(2)	(3)	(4)	(5)	(6)	(7)
2	54.59	57.46	51.71	0.59	56.02	53.15
3	50.13	54.63	45.62	0.93	52.38	47.87
4	47.61	52.69	42.53	1.24	50.15	45.07
5	51.68	55.67	47.69	0.95	53.68	49.68
6	46.66	51.98	41.34	1.28	49.32	44.00
7	46.93	52.19	41.66	1.22	49.56	44.29
8	46.70	52.05	41.36	1.18	49.38	44.03
9	47.70	52.85	42.55	0.99	50.28	45.12
11	46.68	52.20	41.16	1.07	49.44	43.92
12	46.38	51.98	40.79	1.08	49.18	43.58

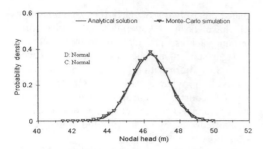

Figure 4 Probability distribution of nodal head (node 12)

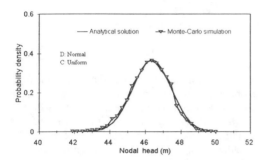

Figure 5 Probability distribution for nodal head (node 12)

model based on fuzzy set theory is best suited for representing the uncertainty caused by vague definition arising from incomplete human knowledge and the imprecision of natural language statements as opposed to uncertainty caused by a truly random condition for the value of the parameters of interest. Interval analysis is a special case of fuzzy modelling. Selection of the most appropriate model for a particular application will depend on the data available for analysis and the objectives of the uncertainty analysis.

ACKNOWLEDGMENTS

The research was jointly supported by an operating grant from the Australian Research Council and a research grant from Central Queensland University.

REFERENCES

Bao, Y. and Mays, L.M. (1990). Models for water distribution system reliability. *Journal of Hydraulic Engineering, ASCE*, 116(9), 1119-1137.

Bargiela, A. and Hainsworth, G. (1989). Pressure and flow uncertainty in water systems. *Journal of Water Resources Planning and Management, ASCE*, 115(2), 212-229

Boulos, P.F. and Wood, D. J. (1990). Explicit network calculation of pipe network parameters. *Journal of Hydraulic Engineering, ASCE*, 116(11), 1329-1344

Bouchart, F. and Goulter, I. (1991). Reliability improvements in design of water distribution networks recognizing valve location. *Water Resources Research*, 27(12), 3029-3040.

Huang, G.H., Baetz, B.W. and Patry, G.G. (1994). Grey fuzzy dynamic programminng: application to municipal solid waste managemet planning problems. *Civil Engineering Systems*, 11(1), 43-74.

Kaufmann, A. and Gupta, M.M. (1991). *Introduction to Fuzzy Arithmetic*. Van Nostrand Reinhold, New York.

Lansey, K. E., Duan, N., Mays, K.W., and Tung, Y-K. (1989). Water distribution design under uncertainty. *Journal of Water Resources Planning and Management, ASCE*, 115(5), 630-645.

Lansey, K. and Basnet, C. (1991). Parameter estimation for water distribution networks. *Journal of Water Resources Planning and Management, ASCE*, 117(1), 126-144.

Ligget, J.A. and Chen, L-C. (1994). Inverse transient analysis in pipe networks. *Journal of Hydraulic Engineering, ASCE*, 120(8), 934-955.

Yen, B.C., Chen, S.T. and Melching, C.S. (1986). First-order reliability analysis. In *Stochastic and Risk Analysis in Hydraulic Engineering* Ed. by B.C. Yen, Water Resources Publications, Littleton, Colo., 1-36

Zadeh, L.A. (1965). Fuzzy sets. *Information and Control*, 8, 338-353.

Stochastic Hydraulics'96, Tickle, Goulter, Xu, Wasimi & Bouchart (eds) © 1996 Balkema, Rotterdam. ISBN 90 5410 817 7

An inverse analysis of model parameters for heterogeneous aquifer based on Genetic Algorithm

Takao Suzuki
Gifu National College of Technology, Japan

Morihiro Harada
Meijo University, Nagoya, Japan

Amin Hammad & Yoshito Itoh
Nagoya University, Japan

ABSTRACT: Aquifer parameters such as transmissivity T and storativity (or effective porosity) S are generally identified on the basis of a criterion that minimizes the difference between observed heads and computed heads at some locations. In heterogeneous aquifer, however, the observed head includes random components reflected by locality of the aquifer properties. Therefore, by the conventional method it is not always possible to identify the suitable values of parameters. In this research, in order to overcome the difficulties found in the conventional method, a new identification procedure based on a Genetic Algorithm (GA) is proposed and its applicability is considered.

1 INTRODUCTION

In the regional groundwater analyses for quantitative management of aquifer basin, two dimensional horizontal flow model is usually adopted since thickness of the aquifer is much smaller than the other two dimensions of the flow region. The model parameters of two dimensional analysis are transmissivity T and storativity (or effective porosity in unconfined case) S of each aquifer. Usually, these parameters are identified by trial and error so that results of analysis fit with observed head data. Inverse analysis methods, which estimate the optimum values of parameters from the observed data, have been developed for more rational identification of parameters (Yeh 1986). Even in these methods, the parameters are identified on the basis of criterion that minimizes difference between the observed heads and the calculated heads. For the identification of aquifer parameters, it is necessary to correct observational information as much as possible. In the usual observation of groundwater, however, the head data are measured only at a few observation wells. Moreover, since an actual aquifer is a heterogeneous flow region in which hydraulic properties are distributed irregularly, behavior of the head data is often affected by the locality of the aquifer properties.

In addition, the calculated heads are numerical solutions of the governing equation including macroscopic parameters under simplified boundary conditions. That is, representative scales in space between these heads are so different from each other. Therefore, it may be not always possible to identify exact values of parameters by the conventional method that compares the heads of different spatial scales.

Genetic algorithms (GAs) are optimization programs based on the mechanism of natural selection and natural genetics. In GAs, natural selection is implemented through selection and recombination operators. A population of candidate solutions, usually coded as bit strings, is modified from one generation to the next by the probabilistic application of the genetic operators. GAs have been used in many domains of civil engineering such as road maintenance planning (Chan 1994) as well as other fields. GAs are used in this research for the parameters' identification for the following reasons (Goldberg 1989):

1) GAs do not work with the optimization parameters themselves, but with a discrete coding of the parameter set in the form of finite length strings that represent the artificial chromosomes. GAs process populations of these strings in successive generations. They only use the objective function information without

617

the information of the derivatives or other auxiliary knowledge.

2) In the case of the parameters' identification
problem, although it is possible to obtain the
derivatives of the objective function, the assumption of the initial values to obtain the true
solution is not always easy. GAs search from a
number of points at a time, in contrast to the
single point approach of the traditional optimization methods. This means that GAs can
process a large number of assumptions at the
same time. Furthermore, GAs use probabilistic transition rules, not deterministic rules to
guide the search towards regions of the search
space with expected improvement. The operators including reproduction, crossover, and mutation improve the search process in an adaptive manner.

2 PHENOMENA CONSIDERED IN THE INVERSE ANALYSIS

2.1 Generation of observed "actual phenomena"

In two dimensional unconfined aquifer as shown
in Fig.1, where groundwater heads fluctuate due
to rainfall infiltration, an inverse problem for estimating the aquifer parameters is considered based
on the observed heads at some locations. When a
recharge by rainfall infiltration exists, it is possible
to identify the parameters T and S independently.
However, in the case that the intensity of recharge
is too small, or in the case that aquifer properties
are not uniform, it may become difficult to identify precise values of these two parameters. In this
research, the conventional method and a new procedure based on GA are compared and discussed
for their applicability for the parameter identification under such situation.

The actual phenomena is simulated in the computer through the linearized fundamental equation
given by Eq.(1)

$$S(x,y)\frac{\partial h(x,y,t)}{\partial t} = \frac{\partial}{\partial x}[T(x,y)\frac{\partial h(x,y,t)}{\partial x}] +$$

$$\frac{\partial}{\partial y}[T(x,y)\frac{\partial h(x,y,t)}{\partial y}] + Re(t) \quad (1)$$

where x, y are the coordinates of heads as shown in
Fig. 1(a), t: time, $h(x,y,t)$: unconfined head,

$T(x,y)$: transmissivity, $S(x,y)$: effective porosity,
and $Re(t)$: recharge intensity due to rainfall of 1.0
mm/day. For the sake of simplicity, two cases are
considered here for the aquifers' parameters frequency distributions. In the first case, the region
is ideally homogeneous with identical values of the
aquifer parameters. In the second case, the parameters T and S follow the log-normal and the normal
distributions, respectively. The values of transmissivity $T(x,y)$ and effective porosity $S(x,y)$ are
generated from the following distributions for each
finite element of the analysis domain. (I) Homogeneous region: $T(x,y) = 100m^2/day$ and $S(x,y) =$
0.10. (II) Heterogeneous region: $T(x,y)$ follows
log-normal probability distribution function with
$E[log_{10}T] = 2.0$ and $Var[log_{10}T] = 0.50$, $S(x,y)$
follows the normal probability distribution function with $E[S(x,y)] = 0.10$ and $Var[S(x,y)] =$
0.001. The initial and boundary conditions are
given as follows:
Initial condition: $h(x,y,0) = h_0 = 10m$
Boundary conditions: $\partial h/\partial n = 0 : \Omega \in \Omega_1$,
and $h(x,y,t) = h_b = 10m : \Omega \in \Omega_2$
where n is the normal direction to Ω_1. Ω_1 and Ω_2
are the boundaries of the region shown in Fig.1(a).
Under these conditions, the virtual observed head
data are generated by sampling from the numerical
results of FEM analysis based on the fundamental
equation Eq.(1).

2.2 Identified equivalent aquifer model

As mentioned above, the aquifers considered in
this analysis are a homogeneous aquifer (I) with
the constant parameters, and a heterogeneous aquifer (II) with the parameters fluctuating a little in
the neighborhood of their averages. In the analysis
of these aquifers, it is adequate from the engineering point of view to treat the aquifer as an equivalent homogeneous model. The simulated head values at several fixed points in the region are considered as the virtual observed heads. Then, based
on those observed head data, the equivalent parameters of the model which represent the whole
region are identified. The fundamental equation of
the equivalent homogeneous model is:

$$S\frac{\partial h^*(x,y,t)}{\partial t} = T[\frac{\partial^2 h^*(x,y,t)}{\partial x^2} +$$

$$\frac{\partial^2 h^*(x,y,t)}{\partial y^2}] + Re(t) \quad (2)$$

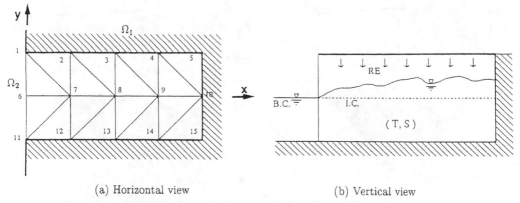

(a) Horizontal view (b) Vertical view

Fig. 1. Virtual unconfined aquifer considered in this research

where $h^*(x, y, t)$ is the head in the equivalent aquifer, T and S are the equivalent parameters identified, and $Re(t)$ is the recharge intensity. Because the target of the inverse problem is to estimate the parameters T and S, it is assumed that initial and boundary conditions and $Re(t)$ values are known and same as the "actual" aquifers (I) and (II).

3 PARAMETER IDENTIFICATION BY CONVENTIONAL PROCEDURE

3.1 Formulation of inverse problem

The conventional criterion for the parameter identification may be stated as to minimize the following objective function,

$$f(P_1, P_2, ..., P_l) = \sqrt{\Sigma_{i=1}^m (h_i^{obs} - h_i^{cal})^2} \to min. \tag{3}$$

In this equation, h_i^{obs} describes the observed head, h_i^{cal}: the calculated head, m: the number of observed values, P: the model parameter and l: the number of parameters. In the case of Eq.(2), $l = 2$, $P_1 = T$, and $P_2 = S$. To overcome the difficulty of the nonlinearity of the optimization problem as expressed by Eq.(3), the Gauss-Newton method is usually adopted. That is, after linearizing the function $h(P_j)$ in the neighborhood of parameter P_j, the following normal equation is derived by the condition of $[\partial f/\partial P_j] = 0$.

$$\Sigma_{k=1}^l \; \Sigma_{i=1}^m Q_{ij} Q_{ik} \cdot \Delta P_k = \Sigma_{i=1}^m (h_i^{obs} - h_i^{cal}) \cdot Q_{ij},$$
$$j = 1, 2, ..., l \tag{4}$$

in which ΔP is the correction value of P, and Q is the Jacobian matrix of head for parameters given by:

$$Q_{ij} = \frac{\partial h_i^{cal}}{\partial P_j} \tag{5}$$

The Jacobian Q may be calculated by the following forward difference approximation.

$$Q_{ij} = \{h_i^{cal}(P_j + \delta P_j) - h_i^{cal}(P_j)\}/\delta P_j \tag{6}$$

where $\delta P_j = \delta_j \cdot P_j$, and δ_j is a constant. This inverse analysis is carried out by the following procedures. At first, after giving the initial values (first approximation) of P_j and δ_j, the Jacobian matrix is calculated by the numerical analysis of fundamental equation Eq.(2) for the equivalent model. Then, the correction values of parameters are estimated by the normal equation Eq.(4). The initial values are improved by using the correction values, and the process is iterated to get the correction values which satisfy the following condition,

$$\Delta P_j < \varepsilon_j \tag{7}$$

where ε_j is a constant index of convergence ($\varepsilon_j = P_j/100$). The values of δ_j are here assumed to be $\delta_1 = 0.001$ and $\delta_2 = 0.1$.

3.2 Behavior of objective function

Let us consider how the objective functions of the homogeneous and heterogeneous aquifer behave for values of parameters T and S in the equivalent model. Because it is practically impossible to establish many wells for the head observation in the whole region, the observation points are assumed

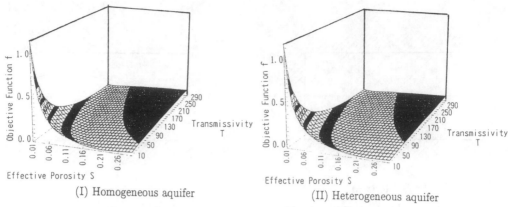

(I) Homogeneous aquifer   (II) Heterogeneous aquifer

Fig. 2. Behavior of objective function f (bird's eye view)

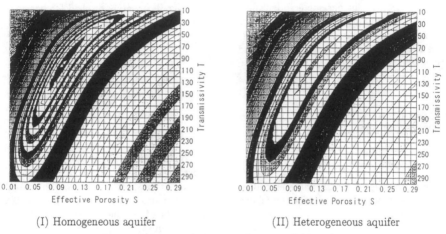

(I) Homogeneous aquifer   (II) Heterogeneous aquifer

Fig. 3. Behavior of objective function f (detail)

here to be only three nodes with numbers 2, 7 and 12 as shown in Fig.1(a). The observations are carried out 5 times at each 2 days from $t = 0$ to $t = 10$ days. Fig.2 shows the behavior of the objective functions based on the observed head data. Fig.2(I) and Fig.2(II) are for homogeneous or heterogeneous cases, respectively. In these figures, the root values of function f given by Eq.(3) are calculated for the sets of T and S in ranges : $T = 0 \sim 300 m^2/day$ and $S = 0 \sim 0.30$.

Fig.3 shows the plan graphs drawn in detail using finer contour lines than Fig.2. In Fig.3(I), values of f approach to zero at $T = 100 m^2/day$ and $S = 0.10$. Because the aquifer (I) is homogeneous, T and S in the equivalent model are equal to the

estimated T and S in the actual aquifer. Therefore, if the parameters are identified correctly, the computed heads agree perfectly with the observed heads, and then the objective function becomes equal to zero. In this case, the objective function behaves under a curved surface with a single peak.

In Fig.3(II), however, a curved surface of f does not have clearly a single peak point such as in Fig.3(I). Because of the aquifer heterogeneity, the equivalent model can not reproduce the fluctuation of observed heads which reflects the locality of hydraulic properties. Therefore, the observed heads fluctuate differently from the homogeneous case, and the objective function becomes a curved surface with a flat peak.

3.3 *Inverse analyses by non-linear least square procedure*

In the conventional procedure, the parameters T and S of the equivalent model are identified using Gauss-Newton method so that the objective function f is minimized. As stated before, the convergence in the iteration process is judged under the condition given by Eq.(7). As the initial values of parameters, eight sets of T_0 and S_0 are selected in this research as shown in Table 1.

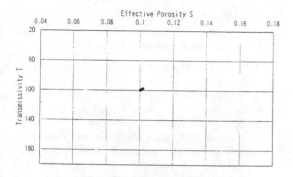

(I) Homogeneous aquifer

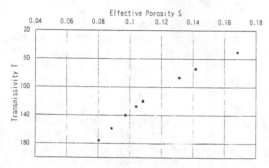

(II) Heterogeneous aquifer

Fig. 4. Distributions of parameter values identified by conventional method

Results of the identification for homogeneous and heterogeneous aquifers are indicated in Table 1 and Fig.4. In the homogeneous case (case I), the parameters converge to the point of the true values: $T = 100m^2/day$ and $S = 0.10$. Even if the convergence criteria is severe as in Eq.(7), the identification succeeded within only a few iterations.

In the case of heterogeneous aquifer, however, the identification process becomes entirely differ-

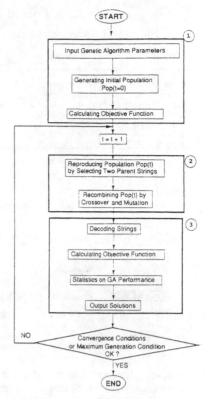

Fig. 5. Flowchart of the GA

ent. As shown in Table 1, the number of iterations reaches several hundreds because of the ill condition of convergence. The values of T and S are very different according to their initial values T_0 and S_0, and scattered in a wide region as shown in Fig.4 (II). This distribution of the results corresponds to the flat region of the objective function in Fig.3 (II).

Results identified for the heterogeneous aquifer may become different according to the initial values of parameters T_0 and S_0, as in this example. In other words, the identification in heterogeneous case may be unstable according to the initial values. In practical work of identification, only one set of initial values are assumed considering the physical properties of sediments in the aquifer and the local permeability by pumping tests etc. Therefore, it may be said that the results discussed above are unreliable in practical use. In addition, the behavior of the objective function is unknown

Table 1. Initial values and results of identification

(I) Homogeneous aquifer

initial value		identified value		number of
S0	T0	S	T	iteration
0.05	25	0.10166	97.4845	12
0.25	25	0.10077	98.32655	13
0.05	50	0.10039	99.07894	5
0.25	50	0.10011	99.41087	5
0.05	150	0.10012	99.8443	4
0.25	150	0.10015	99.91972	7
0.05	250	0.10004	99.93468	4
0.25	250	0.10012	99.89167	6

(II) Heterogeneous aquifer

initial value		identificated value		number of
S0	T0	S	T	iteration
0.05	25	0.10388	127.5241	428
0.25	25	0.0802	175.674	476
0.05	50	0.08823	158.7075	656
0.25	50	0.09713	140.0282	396
0.05	150	0.14169	74.18479	18
0.25	150	0.16802	50.51004	114
0.05	250	0.10818	120.1743	23
0.25	250	0.13115	86.47234	84

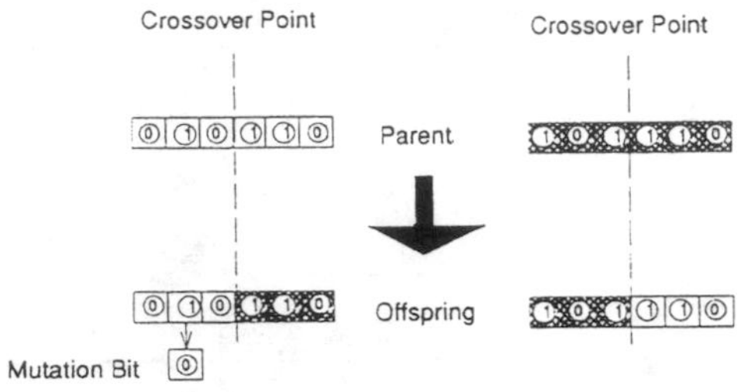

Fig. 6. Crossover and mutation operators

in the inverse analysis and consequently, it is necessary to evaluate the distribution of the results identified without any information of the objective function.

4 NEW IDENTIFICATION PROCEDURE BASED ON GA

4.1 *Outline of the GA*

The flowchart illustrating the GA implemented in the present study is shown in Fig. 5. The three modules numbered in this figure are the main processes of the GA, that are the production of the initial generation, the reproduction by selection, crossover, and mutation operators, and the decoding and evaluation. These modules are explained in the following sections. The GA starts from the parameter definition, and the coding structure is determined at the same time. For each generation, the GA first generates strings by selecting two parents from the previous generation and reproducing them by crossover and mutation until the population of this generation is made. Then, the GA will decode and evaluate the strings of this generation.

The objective function is calculated by decoding every string. This process will repeat many times before the near-optimum solution is found.

1) Code design: An essential characteristic of GAs is the coding of the variables that describe the problem. The most common coding method is to transform the variables into a binary string of a specific length. In this research, each of the parameters T and S is represented by nine bits' string. This representation allows to discretize the value of each of the parameters into 512 values (from 2 to 1024 m^2/day with a step of 2 for T, and from 0.001 to 0.512 with a step of 0.001 for S).

2) Crossover and mutation operators: In the reproduction procedure of new generations, the operators of *crossover* and *mutation* are performed. In the present research, at each new generation, the best 20% of the population of the previous generation will survive. The remaining 80% will be selected from the strings whose objective function values are less than the average objective function value of the population of the previous generation. The crossover and mutation operators are shown in Fig. 6. Once the crossover point is randomly selected, the parts to the left of the crossover point will be retained and the parts to the right of the crossover point will be exchanged. Similarly, with a probability of mutation, one bit is altered into another within a string after the mutation position is randomly selected.

3) Decoding and evaluation: The decoding process of a string is inverse to the generation process. After the creation of every generation, by decoding the strings of a population, the values of T and S are obtained. After many generations, the best member of the population with

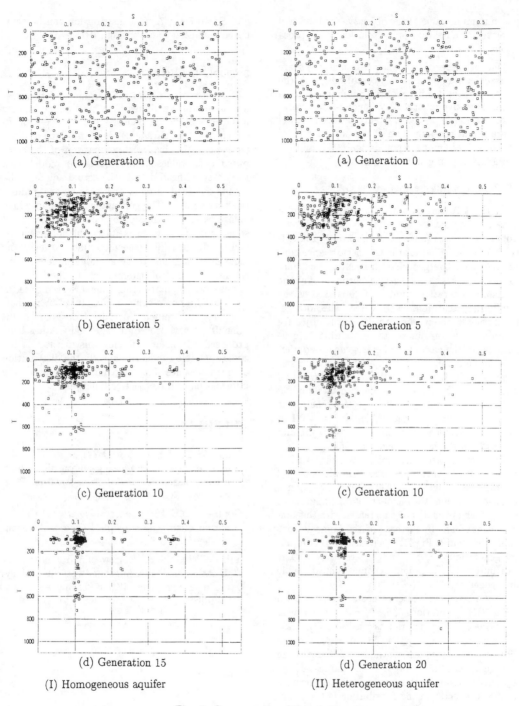

(a) Generation 0   (a) Generation 0

(b) Generation 5   (b) Generation 5

(c) Generation 10   (c) Generation 10

(d) Generation 15   (d) Generation 20

(I) Homogeneous aquifer   (II) Heterogeneous aquifer

Fig. 7. Convergence of T and S

a minimum value of the objective function will represent a near-optimum solution of the problem.

One or more of the following conditions are used as the terminating conditions: (1) the convergence condition of the minimum value of the objective function (1% among 3 successive generations), (2) the difference between the average and the minimum of the objective function values among a population (1%), and (3) the maximum generation number.

4.2 Results of inverse analysis by GA

In the GA calculation, it is assumed that a moderate population size (400), a high crossover probability (80%), and a low mutation probability (0.05%) are good for the GA performance considering the convergence requirement and calculation time. For the comparison purpose, only the maximum generation number of 20 is used as the terminating condition. In the case of the homogeneous aquifer, the minimum value of the objective function converges quickly to zero within 10 generations. In the case of the heterogeneous aquifer, the minimum value of the objective function converges quickly to a value of about 0.01973 within 8 generations. The values of T and S corresponding to the minimum values of the objective function at generation 20 are $T = 100m^2/day$ and $S = 0.1$ in the homogeneous aquifer, and $T = 96m^2/day$ and $S = 0.124$ in the heterogeneous aquifer.

In order to check the convergence of T and S, the scattograms of their values are drawn at several generations in Fig.7. It is clear from these figures that the number of the members of the populations having near-optimal values increases with the generation number. In the case of the homogeneous aquifer, shown in Fig.7(I), generations 0 (the initial generation), 5, 10, and 15 are drawn. The distribution of the points becomes almost stable after generation 15.

In the case of the heterogeneous aquifer, shown in Fig.7(II), generations 0, 5, 10, and 20 are drawn. The same initial generation is used as in the case of the homogeneous aquifer. In this case, the distribution of the points continues to change even after generation 15. Moreover, the flatness of the objective function in the case of the heterogeneous aquifer is reflected in the distribution of the population of generation 10 where many points are concentrated in the ranges of $T = 70 \sim 200m^2/day$

and $S = 0.7 \sim 1.7$. This distribution can be compared with that found in Fig. 4(II).

5 CONCLUSIONS

The inverse problem for the parameters of equivalent model in two-dimensional unconfined aquifer has been considered. The criterion for the identification of model parameters is to minimize the objective function by direct comparison of the computed head values in the equivalent model with the observed head data in the actual aquifer. The results obtained through this research are summarized as follows: The conventional method which identifies the parameters by the non-linear least square method is much useful for the homogeneous aquifer. In the heterogeneous aquifer, however, the identification is apt to be unstable depending on the initial values of parameters and the locations of observation point because the observed data include local random-components caused by the heterogeneity of aquifer. In order to overcome this difficulty, a new procedure based on GA has been proposed. By means of using this method, it becomes possible to identify quickly the distribution of optimal values of model parameters without any information of the objective function.

REFERENCES

Chan, W.T., T.F.Fwa, & C.Y.Tan 1994. Road-maintenance planning using genetic algorithms. I: Formulation. *Journal of Transportation Engineering*, ASCE, 120(5): 693-709.

Goldberg, D.E. 1989. *Genetic algorithms in search, optimization, and machine learning.* Addison-Wesley Inc., U.S.A.

Yeh, W.W-G 1986. Review of parameter identification procedure in groundwater hydrology: the inverse problem. *Water Resour. Res.,* 22(2): 95-108.

9 Sediment transportation

Stochastic Hydraulics'96, Tickle, Goulter, Xu, Wasimi & Bouchart (eds) © 1996 Balkema, Rotterdam. ISBN 90 5410 817 7

Sediment gradation analysis for aggragation/degradation process

Shou-young Chang

National Taipei Institute of Technology, Taiwan, China

ABSTRACT : In an alternating aggadation-degradation experiment, the sediment mixture were sampled at various locations, including the sand wave crest and trough. Two specific layers, called *transport material layer* and *coarser surface layer*, are proposed in this study through the analyses of sampling data. The physical meaning of these two specific layers can be reasonably explained and the thickness of each layer for various stages can be well estimated based on sampling results. Dimensional analyses were applied to establish the relationship between the size distribution of these two layers and the relevant dimensionless variables. Based on this established relationship, regression equations were developed and they agree reasonably with the experimental data.

1. INTRODUCTION

In alluvial river, if any of the factors, such as sediment discharge, Channel geometry, water and sediment properties, is changed, the state of equilibrium will be consequently disturbed and a process of reaching another state of equilibrium will begin. This process is inevitably accompanied by aggradation and/or degradation along the riverbed. In fact, aggradation and degradation are primarily two of most important nonequilibrium phenomena of alluvial rivers. Furthermore, due to nonuniformity of bed material, hydraulic sorting and armouring will take place in these process, and the whole process becomes very complicated.

Recent research in mechanics of sediment transport has led to the development of mathematical and numerical models for simulation of river bed degradation (Litter & Meyer, 1972, Karim, 1982, Misri, 1984) and aggradation(Soni., 1980, Proffitt, 1983, Yen et al. 1987, Chang, 1990, Park, 1990). However, the techniques have not proven practical. The primary difficulty is that the fundamental mechanism and the process of overloading or underloading are seldom investigated. Consequently, the sorting phenomena are not well understood basically. Yen et al. (1985, 1987, 1990)performed a series of aggradation and degradation experiments with both uniform and graded coarse sand.

They found that the cycle of aggradation followed by degradation with nonuniform sediment is irreversible (Yen et al. ,1989 1992); besides, a kind of sediment wave named as sorting wave, had also been observed(Yen 1987, 1990). The sorting wave was considered to be due to a longitudinal sorting of the nonuniform sediment and was different from the ordinary typical sand wave. Numerous sampling work and sieve analyses were performed during Yen's experiments. Those valuable data are to explore the fundamental phenomena of sorting process; however, only a few preliminary investigations were examined.

The present study aims to illuminate some fundamental phenomena and behaviors of channel-bed evolution from Yen's experimental data (Yen, 1988) mainly under the condition of overloading followed by underloading with various nonuniform sediments.

2. DIMENSIONAL ANALYSIS

The physical parameters affecting the size distribution of sediment mixture Ω under a nonequilibrium process, such as aggradation and degradation, are fluid properties (density ρ ,fluid kinematic viscosity ν), sediment properties (including density ρ_s, initial median grain size D_i, and the initial grain size gradation σ_i, defined as the geometric standard deviation), flow conditions (including mean flow velocity U, shear velocity U_*), time t and gravitational acceleration g. Through dimensional analysis performed on the 10 variables mentioned, the following functional relation for dimensionless parameters is obtained (Chang, 1993):

$$\Omega = f_1(\frac{\rho\, U_*^2}{(\rho_s - \rho)g\, D_i},\ \frac{U_*D_i}{\nu},\ \frac{U}{U_*},\ \frac{t\,\nu}{D_i^2},\ \frac{\rho_s}{\rho},\ \sigma_i\ )\quad (1)$$

The former two terms in the brackets are the dimensionless shear stress F_* and boundary Reynolds number R_* for the size D_i respectively. These are exactly the most important parameters in the Shields' diagram. The F_*, ratio of threshold force and resistant force of the particle, is an index of the strength of bed load movement. The R_* can be treated as the turbulent intensity in a movable bed. The third term in Eq.(1), U/U_*, indicates the roughness n of the river bed since the coefficients of roughness , such as Manning's n ,Chezy's c and Darcy-Weisbach f can all be derived into a function of U/U_* .

The fourth term, $t\,\nu\,/D_i^2$, is a dimensionless time parameter T concerning the process from an equilibrium state to another state. The size distributions of sediment mixture Ω (t) may be assembled into median diameter D(t) and geometric standard deviation σ (t) corresponding to a specific time t . In this study, ρ and ρ_s were kept constant. Therefore, Eq.(1) is simply reduced to

$$D(t)/D_i = f_2(F_*, R_*, n, T, \sigma_i) \qquad (2)$$
$$\sigma(t) = f_3(F_*, R_*, n, T, \sigma_i) \qquad (3)$$

Eqs.(2) and (3) are used to construct empirically the regression relationship regarding the sorting behavior by applying the sampling data from Yen's experiments (Yen et al. 1988) in this study.

3. EXPERIMENTS

The flume used for Yen's experiments is 72 m long and 1 m wide. The water discharge was maintained at a constant rate of 0.12 m³/s for all experiments. Bed and water surface elevations were measured at six different locations respectively during the experiments. A sediment supplier was located at the upstream end to feed the sediment and a weight-measuring device was installed downstream of the test section to measure the

cumulative weight of sediment discharge. A sluice gate at the downstream end of the flume was to maintain a constant tailwater level. The layout of the flume is shown elsewhere(Yen et al. 1988, 1992).

All graded sediments used in the experiments had the same initial median diameter D_i of 1.8 mm and three different gradations, with the initial geometric standard deviation σ_i being 2.0, 2.6 and 3.2 respectively. Size fraction by percent for each Gradation is shown in Table 1.

Table 1. Size fraction for each Gradation

Sieve no	3/8"	4	6	10	16	20
Size(mm)	9.52	4.76	3.36	2.00	1.19	0.84
σ_i=2.0	--	8.0	11.0	25.0	28.0	14.0
σ_i=2.6	3.8	11.2	11.0	20.0	20.0	13.0
σ_i=3.2	5.0	8.0	--	27.0	--	30.0

Sieve no	30	40	60	100	bott.	Total
Size(mm)	0.59	0.42	0.25	0.15		(%)
σ_i=2.0	8.0	4.3	--	--	1.7	100
σ_i=2.6	--	15.0	--	--	6.0	100
σ_i=3.2	--	17.0	8.0	3.0	2.0	100

The initial bed slope was set at 0.0035 At the beginning of an experiment, a sediment supply rate of 3.3 kg/min. was continuously released from the upstream end until the channel bed reached a state of equilibrium(referred to here as EQU test). The sediment supply rate was then increased to 9.9 kg/min until a new equilibrium was reached(hereafter called AGG test). The rate of sediment supply was thereafter reduced back to and kept at 3.3 kg/min until another new equilibrium was fully developed(referred to here as DEG test). Finally, the sediment supply was cut off, and only clear water was released from the upstream end until the channel bed was fully armored(hereafter called CLE test). The same procedure was applied to experimental runs having different sediment gradations. Identification of the experimental runs is given in Table 2. The accumulated duration time of each test is also provided at the table.

Table 2. Identification of Experiments

σ_i	EQU Q_s=3.3	AGG Q_s=9.9	DEG Q_s=3.3	CLE Q_s=0.0
2.0	B2(28.0)	B3(62.6)	B4(94.1)	B5(133.2)
2.6	C2(32.9)	C3(65.8)	C4(102.9)	C5(134.2)
3.2	A2(19.8)	A3(43.1)	A4(80.4)	A5(85.0[*])

Q_s:Sediment supply rate (kg/min)
* :Accumulated duration time(hr)

628

The flow was carefully stopped, and the flume was slowly drained in order to take bed material sample at some chosen times during an experiment. Three different sampling areas were chosen in each stop. At each chosen area, samples were taken from two locations; one at the trough and the other at the crest of sorting wave(Yen et al. 1988). The sediment sampler was pushed into the channel bed, and the top six layers of the bed material, each being 1 cm in thickness, were scooped out by a spoon. Sieve analysis was then performed to determine the vertical variation of the median particle size D and its geometric standard deviation σ . Having been analyzed, the samples were then placed back into the sampling holes from where they were taken, and the experiment was continued. Details of the experimental procedure are given elsewhere(Yen et al. 1988,1992).

4. RESULTS

4-1. Hydraulic Characteristics

During the experiments, the hydraulic characteristics vary significantly. The range of flow parameters for each run from equilibrium test to clear water test are summarized as shown in Table 3. The range of energy slope S_f is between 0.00295 and 0.00469; the Froude number N_f defined by mean velocity and mean depth ranges from 0.60 to 0.82.

Table 3. Range of Parameters in Experiments

Run	Q_s kg/m.	$S_f \times$ 1000	Velo m/s	N_f	U_* cm/s	$n \times 10^2$
B Max.	10.88	4.497	0.893	0.778	7.743	2.189
B Min.	0.11	2.949	0.753	0.602	6.742	1.832
C Max.	10.29	4.690	0.898	0.785	8.250	2.445
C Min.	0.13	3.464	0.788	0.644	7.033	1.919
A Max.	10.56	4.599	0.926	0.821	7.700	2.101
A Min.	1.46	3.871	0.836	0.704	7.244	1.872

Sieve analyses were performed not only from the sediment sampler but also from the sediment supplier during the experiment. The standard deviation of σ_i and D_i of graded sediment used in the experiment are listed in Table 4. From the table, one can see both the standard deviations of σ_i and those of D_i increase as σ_i increase. All the values in the table are small indicate that the sediment

used in the experiment were mixed thoroughly. Additionally, the deviation of σ_i and D_i can be referred to examine the accuracy or/and the sensitivity of sieve analyses while sorting process are to be investigated.

Table 4 Standard deviation of the sediment used in the experiment

Runs	σ_i	D_i (mm)
B	0.096	0.062
C	0.130	0.121
A	0.280	0.140

The bed elevation change for each test, as shown in Table 5, is calculated from the mean bed surface elevations of six bed-level gages. Though each EQU test as mentioned previously, has same initial bed slope, flow discharge and sediment supply rate, the differences of bed elevation between the initial stage and the final stage of EQU test are highly discrepant. Test B2 scoured 2.9 cm deep while Test C2 and A2 deposit 0.4 cm and 0.8 cm high respectively; that is, if an experiment begins with all identical conditions, including initial bed slope, sediment supply rate, flow discharge, and fixed tailwater etc., only except for sediment gradation σ_i, the aggradation tends to substitute degradation while the σ_i employed in the experiment become larger.

As previously reported(Yen et al. 1988, 1992), that the equilibrium channel bed in an aggradation-degradation cycle with nonuniform sediment is not reversible can also be found at Table 5. When an aggradation-degradation cycle is completed, the bed elevation deposit about 0.6, 0.5 and 2.1 cm for Run B, C and A respectively.

Table 5. Extent of bed elevation change

σ_i(Run)	EQU	AGG	DEG	CLE
2.0(B)	−2.9cm	+4.0cm	−3.4cm	−0.8cm
2.6(C)	+0.4cm	+2.4cm	−1.9cm	−1.3cm
3.2(A)	+0.8cm	+4.4cm	−2.3cm	-----

−:scour +:deposition

It is worth also noting that as each test is completed, the variation of bed elevation should also result in the gradation change of bed materials, e.g. there exists a so-called deposition layers around 4.0, 2.4 and 4.4 cm thickness for Run B, C and A after the aggradation test is completed. In other words, the material scoured resulting from the degradation test are mainly from the deposition layers.

4-2. Variation of size distribution

As described above, bed materials were sampled at some specific time during experiment by virtue of stopping the flow. At each stop, six layers of bed material at six various locations, each being 1 cm in thickness, were scooped out. Neglecting the spatial variations, the average sieve data was taken for trough and crest respectively in this study; consequently, only the temporal changes of gradation at the nonequilibrium process were taken into account. Detailed size distributions for each layer and location have been described in the report by author (Chang, 1993). Fig. 1 and Fig. 2 show the temporal variations of size distribution at the first layer for trough and crest respectively. From those figures,

one can see each size fraction changes along the nonequilibrium circle. The time periods for each test, such as EQU AGG DEG and CLE test, are given in Table 1.

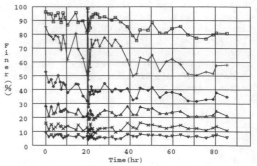

Fig. 1-3. Temporal variations of grain size for the first layer (Run A, wave trough)

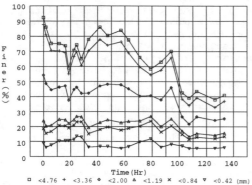

Fig. 1-1. Temporal variations of grain size for the first layer (Run B, wave trough)

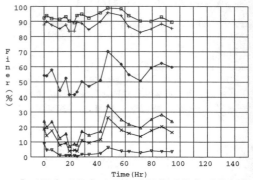

Fig. 2-1. Temporal variations of grain size for the first layer (Run B, wave crest)

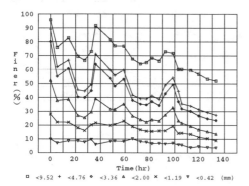

Fig. 1-2. Temporal variations of grain size for the first layer (Run C, wave trough)

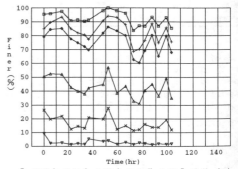

Fig. 2-2. Temporal variations of grain size for the first layer (Run C, wave crest)

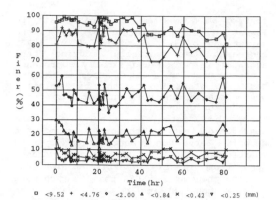

Fig. 2-3. Temporal variations of grain size for the first layer (Run A, wave crest)

Via the equilibrium test, the bed materials are no longer homogeneous vertically at the beginning of the aggradation test; furthermore, the bed material of AGG and DEG test become more complex since they compose both scouring and deposition phenomena alternately. Therefore the analyses for this study was much more difficult than other similar studies.

To simplify the complicated measurement,

median diameter D and geometric standard deviation σ are employed to interpret the size distribution curve for each layer. D can be treated as concentrated tendency of sediment size and σ indicates the deviation of size distribution. Detailed gradation hydrograph of D and σ are also given elsewhere (Chang, 1993). Some of those results are shown in Fig. 3 to illustrate the variations of D and σ for Run A respectively. In general, all of the minimum σ for each run appear at the first layer of wave crest, while, all of the maximum D show up at the first layer of wave trough.

During equilibrium test, diameter D from the first to the second layers for both wave trough and crest are increasing. On the other hand, the σ increases at the wave trough but decreases at the wave crest. Beneath those layers ,from the 3rd. to 6th. layer, their diameter D and σ remains almost unchanged. This may suggest ,especially from Fig. 3-1 and Fig. 3-3, the bed material influenced by fluid flow is about two layers in depth.

As for the AGG test, the σ of wave crest decreases at the beginning of aggradation; thereafter, an increasing σ was followed. The σ of the surface layer is generally smaller than that of its lower layer, which results in a vertical

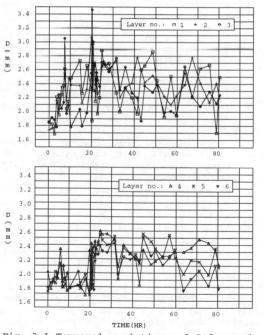

Fig. 3-1 Temporal variations of D for each layer(Run A, wave crest)

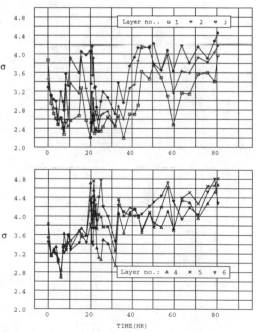

Fig. 3-2 Temporal variations of σ for each layer(Run A, wave crest)

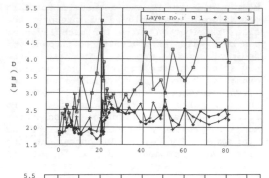

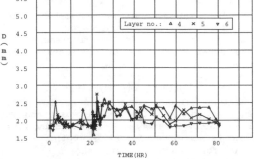

Fig. 3-3 Temporal variations of *D* for each layer(Run A, wave trough)

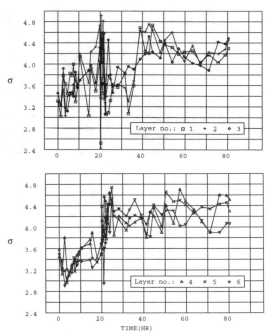

Fig. 3-4 Temporal variations of *σ* for each layer(Run A, wave trough)

sorting with maximum *σ* located at the lowest layer. Diameter *D*, like the variations of bed elevation (Yen, 1992) increases exponentially during the aggradation period and finally approaches a steady state at a later stage of the AGG test. In other words, the bed material become coarser and more uniform during the aggradation process.

The bed materials scoured in the degradation test was the previous sediment deposit at the aggradation period Therefore, due to the nonhomogeneous deposition layers in this study, the sorting phenomena at DEG test is much more complex than that at AGG test. That may be investigated in the future studies.

The bed armor is produced rapidly while the Clear water test begins. The sorting wave then disappears accordingly.

4-3. *transport material layer* and *coarser surface layer*

As illustrated in Fig. 3, the size distributions at the top layers of wave crest, such as *σ* are quite different from those of wave trough. Moreover, those size distribution of sediment located at bed surface, including both crest and trough, also deviate significantly with their subsurface layer material. Therefore, from the point of mechanism view as illustrated as Fig. 3, the material at the upper layers of wave crest should be the transport material, and those layers may be defined as "*transport material layer*". In contrast, the coarser material at the top layers of wave trough is analogous to the coarse *coarser surface layer* existed at movable bed(Diplas, 1987, 1988), and may be named as "*coarser surface layer*".

From the preliminary estimation(Yen, 1988, Chang, 1993), the average thickness of *transport material layer* of Test A3 and A4 are around 2.37 cm and 0.80 cm respectively, meanwhile, they are 3.1 cm and 1.02 cm deep for Test B3 and B4. Because the wave crest indicate a high amplitude of sediment discharge(Yen, 1992, Chang 1993), the thickness of transport material at the wave crest should be more or less greater than the average thickness calculated by Yen and Chang. Therefore, the *transport material layer*, in this study, are supposed to comprise three or four top layers of wave crest in AGG test and one or two top layers in DEG test.

The *coarser surface layer* play as a skin resistance role within a certain range of sediment transport rate. While *coarser surface layer* hold the most part of bed surface area, they dominate the friction roughness, such as in DEG test. Therefore, from the observation in the experiment and analyses from Fig. 3, one or two top layers are taken as the *coarser surface layer*

4-4 Regression Analysis

In order to explore qualitatively the primary factor relevant to the size distributions of *transport material layer* and *coarser surface layer*, a multivariable regression equation of the following form can be chosen for Eqs.(2) and (3):

$$\frac{D(t)}{D_i} = a_1 \times F_*^{p_1} \times R_*^{\varsigma_1} \times n^{d_1} \times T^{e_1} \times \sigma_i^{f_1} \tag{4}$$

$$\sigma(t) = a_2 \times F_*^{p_2} \times R_*^{\varsigma_2} \times n^{d_2} \times T^{e_2} \times \sigma_i^{f_2} \tag{5}$$

By fitting Eqs.(4) and (5) with the experimental data, the exponents of each dimensionless parameter can be calculated using multiple regression analyses. For example ,if the *transport material layer* had a thickness of 3 cm(three layers), the regression equation for the median diameter, $D(t)$, during the aggradation period can be obtained as

$$\frac{D(t)}{D_i} = a_1' \times F_*^{0.689} R_*^{-0.482} n^{-0.375} T^{-0.043} \sigma_i^{0.543}$$

$$N = 23, \ r = 0.907 \tag{6}$$

if the *transport material layer* had a thickness of 4 cm, the regression equation for the median diameter, $D(t)$, during the aggradation period becomes

$$\frac{D(t)}{D_i} = a_1'' \times F_*^{0.509} R_*^{-0.406} n^{-0.203} T^{-0.028} \sigma_i^{0.500}$$

$$N = 23, \ r = 0.900 \tag{7}$$

in which, N is the number of data points; r is the correlation coefficient for the regression analysis and can be computed by :

$$r^2 = 1 - \sum (D(t) - \hat{D}(t))^2 / \sum D(t)^2 \tag{8}$$

where $\hat{D}(t)$ is the predicted value based on the regression equation and estimated for the particular sample at hand.

The exponents in Eqs. (4) and (5) for various tests are listed in Table 6. From the positive or negative sign of the exponents, the correlation between size distribution and those parameters can be obtained qualitatively.

One can see from Table 6-1, the diameter D increase with increasing F_* and increasing σ_i. This may attributable to the fact that when F_* increase, the ability to overthrow bed material will increase accordingly; in addition, with large σ_i, the probability to move coarser

sediment will arise; therefore, the tendency of σ_i and F_* as shown in Table 6 is reasonable. The exponent of parameter T with a tiny value exhibit the T influence the sediment gradation insignificantly. Parameter n has an opposite response between AGG test and DEG test. This may be concerned with sediment transport rate. As mentioned previously, the *coarser surface layer* produce dominant resistance in less sediment discharge. If sediment supply increase, *transport material layer* will strengthen their influence as they occupy most of bed surface area. Thus, the influence of parameter n in the DEG test will be greater than that in the AGG test.

The regressive analyses for σ of *transport material layer* are listed in Table 6-2. From the table, one can see that parameter R_* has a different feeble trend at AGG and DEG test, and also can see the σ increase with decreasing F_*. In light of larger σ_i with larger deviation, as mentioned previously, the σ for *transport material layer* increase as σ_i increase.

Table 6-3 and Table 6-4 exhibit the regression analyses for *coarser surface layer*. It is worth to note that the trends of most parameters for DEG test are quite different from those of CLE test even though the CLE test can be generally treated as a special case of the DEG test. This may be explained by the fact that the existence of sediment supply will produce great discrepancy on both hydraulic characteristics and bed form. DEG test will induce hydraulic parameters change periodically and accompany with a sorting wave. Meanwhile, in CLE test, the hydraulic parameters vary smoothly and the bed form changes instead of armor coat.

The number of data points used in regression analyses and correlation coefficient r for regression equation are shown in Table 7. In general, the correlation between the size distribution

Table 6-1 The exponents in Eq.(4) for *transport material layer*

Test	t_L	F_*	R_*	n	T	σ_i
EQU	1	-0.026	0.499	0.328	0.095	0.481
	2	0.048	0.465	0.170	0.101	0.462
AGG	3	0.689	-0.482	-0.375	-0.043	0.543
	4	0.509	-0.406	-0.203	-0.028	0.500
DEG	1	1.372	-0.763	0.723	-0.013	0.377
	2	1.708	-0.742	1.141	0.014	0.111

t_L:thickness of *transport material layer* or *coarser surface layer*(unit:cm or layers)

Table 6-2 The exponents in Eq.(5) for *transport material layer*

Test	t_L	F_*	R_*	n	T	σ_i
EQU	1	1.956	-0.498	2.503	-0.019	0.936
	2	2.202	-0.417	2.830	0.023	1.033
AGG	3	-0.340	-0.021	-1.145	0.031	0.992
	4	-0.381	-0.099	-1.873	0.035	1.136
DEG	1	-0.844	0.142	0.450	-0.007	1.418
	2	-1.061	0.037	-0.349	-0.025	1.442

Table 6-3 The exponents in Eq.(4) for *coarser surface layer*

Test	t_L	F_*	R_*	n	T	σ_i
EQU	1	4.099	-0.772	4.427	0.239	0.685
	2	2.758	-0.477	2.803	0.152	0.568
AGG	1	2.334	-2.510	-1.254	0.116	0.528
	2	2.058	-1.814	-0.886	0.076	0.308
DEG	1	-2.159	-0.514	-9.200	0.205	1.796
	2	-1.603	-0.305	-6.138	0.145	1.352
CLE	1	0.555	1.403	0.976	0.131	0.683
	2	0.378	1.103	0.949	0.093	0.716

Table 6-4 The exponents in Eq.(5) for *coarser surface layer*

Test	t_L	F_*	R_*	n	T	σ_i
EQU	1	-1.369	-0.236	-2.271	0.051	0.165
	2	-0.036	-0.250	-0.701	0.088	0.536
AGG	1	-1.397	0.517	-0.121	-0.063	0.973
	2	0.268	-0.314	-0.774	-0.024	0.605
DEG	1	-0.715	-0.069	-1.453	0.095	0.950
	2	-1.028	-0.159	-3.049	0.113	1.187
CLE	1	1.871	-0.807	2.004	-0.088	-0.918
	2	1.322	-0.166	1.522	-0.037	-0.228

Table 7 Correlation Coeff. for regression analyses

Test	N	t_L	Table 6-1	Table 6-2	t_L	Table 6-3	Table 6-4
EQU	27	1	0.776	0.957	1	0.771	0.895
	27	2	0.874	0.933	2	0.797	0.943
AGG	23	3	0.907	0.785	1	0.724	0.956
	23	4	0.900	0.829	2	0.770	0.976
DEG	28	1	0.847	0.904	1	0.611	0.699
	28	2	0.829	0.934	2	0.755	0.721
CLE	11	1			1	0.863	0.914
	11	2			2	0.866	0.985

and five dimensionless parameters mentioned for *transport material layer* is better than that for *coarser surface layer*. From the table, one can see that most of correlation coefficients r are greater than 0.8. This can be inferred that the parameters chosen in Eqs.(4) and (5) has a reasonably good agreement with the experimental data. Therefore, Eqs.(4) and (5) are suitable to correlate with the size distribution.

From experimental observation, the particles sampling from the first layer of the wave crest are particular ones, e.g. at the early stage of AGG test, the *coarser surface layer* is the rare part at bed surface. Thus, From Table 7, the correlation for two-layers sampling are better than those for one-layer sampling.

5. CONCLUSIONS

From the results presented, the following conclusions can be drawn:

1. In the EQU test with the same experimental set-up and initial conditions except for the specified σ_i, it is found that at certain σ_i, bed degradation occurs, and the sediment regime switches to bed aggradation as σ_i increase.

2. Several upper layers at the wave crest and trough may be defined as *transport material layer* and *coarser surface layer* respectively. Not only the physical meaning can be easily appreciated, but also the thickness of each layer for various stage can be well estimated based on sampling results.

3. By utilizing dimensional analysis and regression analyses applied to the experimental data, regression relations between size distribution and the relevant independent variables for *coarser surface layer* and *transport material layer* has been established, (see Table 6). These regression relations agree reasonably with the experimental data.

4. Because the existence of sediment supply will produce a big discrepancy on both hydraulic characteristics and bed form, though the CLE test can be treated as a special case of the DEG test, the responses of most parameters in the DEG test are quite different from those of the CLE test.

ACKNOWLEDGMENTS

This research was primarily sponsored by the National Science Council(NSC) of the Republic of China(under Grant No. NSC 81-410-E-027-511). The author would like to express his sincere appreciation to NSC. Valuable comments by Prof. C.L. Yen of the National Taiwan University and Prof. H.W. Shen of U.C. Berkeley during preparation of this manuscript are also highly appreciated.

REFERENCE

Chang, S.Y. (1990). "A Study on the Characteristic of Aggradation Process." A thesis for the degree of Doctor of Philosophy, Dept. of Civil Engrg. ,National Taiwan University.

Chang, S.Y. (1993). "Nonequilibrium Bedload Transport of Sediment Mixture." Report No. NSC-81-410-E-027-511, National Science Council, R.O.C. (in Chinese)

Diplas, P. and Sutherland, A.J. (1988). "Sampling Techniques for Gravel Sized Sediments." ASCE Jour. of Hyd. Div. Vol. 114(5) 484-501.

Little, W.C. and Mayer, P.G. (1972). "The Role of Sediment Gradation on Channel Armoring." School of Civil Engineering in Cooperation with Environment Resources Center, Georgia Institute of Technology, Atlanta, Georgia.

Misri, R.L., Grade, R.J. and Ruju, R.K.G. (1984). "Bed-Load Transport of Coarse Nonuniform Sediment." ASCE Jour. of Hyd. Div. Vol. 110, 312-329.

Park,.G. (1990). "Surface-Based Bedload Transport Relation for Gravel Rivers." Jour. of Hyd. Res., Vol. 28(4) , 417-436.

Proffitt, G.T. and Sutherland, A.T. (1983). "Transport on Nonuniform Sediments," Jour. IAHR, Vol. 21(1) , 33-43.

Soni, J.P., Garde, R.J. and Ranga, R.K.G. (1980). "Aggradation in streams due to overloading." J. Hydra. Div., ASCE, 106(1), 117-132.

Yen, C.L., Lee, H.Y., and Chang, S.Y. (1986). "Mechanics and Numerical Simulation of Aggrading/Degrading Streams with Application to the Choshi River (I)." Report No. 74-65, Disaster Prevention Research program, National Science Council, R.O.C. (in Chinese)

Yen, C.L., Lee, H.Y., and Chang, S.Y. (1987). "Some Aspects of Channel Aggradation Due to Overloading of Coarse Material." Proc. of ROC-JAPAN Joint Seminar on Water Resources Engineering.

Yen, C.L., Wang, R.Y., and Chang, S.Y. (1988). "A Study on Aggradation of Coarse Sediment in Open Channel." 6th Congress of IAHR-APD.

Yen, C.L., Lee, H.Y., and Chang, S.Y. (1988). "Mechanics and Numerical Simulation of Aggrading/Degrading Streams with Application to the Choshi River (II)." Report No. 76-52, Disaster Prevention Research program, National Science Council, R.O.C. (in Chinese)

Yen, C.L., Lee, H.Y., and Chang, S.Y. (1989). "Recovery of Channel Bed in Aggradation/Degradation Process," XXIII Congress of IAHR, Ottawa, Canada.

Yen, C.L., Lee, H.Y., and Chang, S.Y. (1990). "Mechanics and Numerical Simulation of Aggrading/Degrading Streams with Application to the Choshi River (III)." Report No. 78-64, Disaster Prevention Research program, National Science Council, R.O.C. (in Chinese)

Yen,C.L., Chang,S.Y., and Lee,H.Y. (1992). "Aggradation Degradation Process in Alluvial Channels." J. Hydra. Engrg. ASCE, 118(12), 1651-1669.

Stochastic Hydraulics'96, Tickle, Goulter, Xu, Wasimi & Bouchart (eds) © 1996 Balkema, Rotterdam. ISBN 90 5410 817 7

Uncertainty analysis of sediment transport formulas

Keh-Chia Yeh & Sen-Long Deng
Department of Civil Engineering, National Chiao Tung University, Hsinchu, Taiwan, China

Chian-Min Wu
Water Resources Planning Commission, Ministry of Economic Affairs, Taipei, Taiwan, China

ABSTRACT: Three uncertainty analysis methods were applied to sediment transport models to analyze the uncertainty of the calculated sediment transport rates estimated by the Einstein bedload function and Yang's formula. All the stochastic input parameters considered were assumed to be independently and uniformly distributed. Analysis results of both sediment formulas revealed that the flow velocity was the most important parameter that contributed to the uncertainty of the sediment transport rate.

1 INTRODUCTION

Accurate determination of the sediment transport rate for a sediment-laden flow over a bed composed of nonuniform sediments still remains a difficult task because of lack of sound governing equations that describe the movements of sediments. From the physical viewpoint, the sediment transport rate is directly related to the flow intensity and bed composition. Hence, all of the existing sediment transport formulas can be obtained empirically by including the flow and sediment properties based on physical laws and available observed data. If one considers a sediment transport formula as a model for estimating the sediment transport rate, then he would face the problem of uncertainty inherently resided in the model. As was pointed out by Yen et al. (1986), the possible sources of uncertainty in a model output can be the model itself and its parameters. Model uncertainty reflects the inability of the model to represent the true physical behavior of the system considered. On the other hand, model-parameter uncertainties result from the inability to quantify accurately the model parameters. To simplify the analysis in this paper, model uncertainty and the model coefficients are assumed to be deterministic. In other words, only the input parameters, such as flow intensity and sediment composition, are considered stochastic with uncertainty.

Two sediment transport formulas are used and compared in this paper. One is the Einstein bedload function (Einstein 1950), and the other is

Yang's formula (Yang 1973). The former (called Einstein's formula hereinafter) was derived by considering the hydraulics and the stochastic characteristics of the sediment movement, a more theoretically based approach. The latter was obtained by using the idea of unit stream power as well as the multiple regression technique. For sand bed cases, 436 sets of laboratory sand data were used in the development of Yang's formula.

The object of uncertainty analysis in this paper is to estimate the uncertainty of the model output (sediment transport rate, Q_s) subject to random or stochastic input parameters. Such information provides the basis for the further reliability analysis of the model performance. Applications of uncertainty analysis to the alluvial sediment transport problems are few: Yeh et al. (1992) and Yeh and Tung (1993) studied the uncertainty of a pit-migration model, Chang et al. (1993) analyzed the uncertainty of the HEC2-SR model, in addition, Tung and Yeh (1993) further evaluated the reliability of hydraulic structures affected by migration pits, and Bechteler and Maurer (1992) used the first order reliability method to evaluate the reliability of sediment transport formulas.

In this paper, three uncertainty-analysis methods are applied to analyze and compare the uncertainty features between Einstein's and Yang's formulas. Two reaches of different bed slopes, one in the Missouri River in the U.S.A. and the other in the Cho-Shui River in Taiwan, were used as the testing examples. Moreover, to check the uncertainty associated with the variations of water discharge, a low and a high discharge for each reach were considered in the analysis.

2 SEDIMENT TRANSPORT FORMULAS

2.1 Einstein's Formula

Einstein's formula has the capability of computing the nonuniform bed-material load by separately considering the bed-load and suspended-load modes. The details of the physical consideration and probability concept for the movement of the sediment can be referred to Einstein's (1950) original work. The bed-material load per unit width, q_T, based on Einstein's formula is as follows:

$$q = \sum_{i=1}^{ns} (i_B q_{Bi} + i_s q_{si})$$

$$= \sum_{i=1}^{ns} i_B q_{Bi} [\ 1 + \frac{0.216 A_i^{zi-1}}{(1 - A_i)^{zi}} (\ln\frac{30.2 \ r_b}{\Delta} J_1 + J_2)]$$

(1)

where ns = total number of sediment size classes; $i_s q_{si}$ = suspended load per unit width for sediment size d_{si}; $i_B q_{Bi}$ = bedload per unit width for sediment size d_{si}, which is determined by using several figures prepared by Einstein (1950); $A_i = 2d_{si}/r_b$, in which r_b = hydraulic radius associated with the channel bed; $zi = w_i/\kappa u_*'$, in which w_i = fall velocity of d_{si}, $\kappa = 0.4$, and u_*' = shear velocity based on the shear stress due to grain roughness; $\Delta = k_s/x$, in which $k_s = d_{65}$, the representative bed roughness and x = a dimensionless quantity in the logarithmic velocity distribution law; and J_1 and J_2 are two integrals of the forms

$$J_1 = \int_{A_i}^{1} (\frac{1-\eta}{\eta})^{zi} d\eta$$

(2)

and

$$J_2 = \int_{A_i}^{1} (\frac{1-\eta}{\eta})^{zi} \ln\eta \ d\eta$$

(3)

where $\eta = y/r_b$, in which y = water depth. As was pointed out by Einstein and Chien (1953), the original hiding-factor curve that accounted for the sheltering of small particles by the larger particles gives too large a hiding effect for small d_{si}/X values in which X is the characteristic grain size of the mixture. Hence, the modified hiding-factor curve proposed by Odgaard and Lee (1986) was adopted in the computation for q_T.

2.2 Yang's Formula

Yang's (1973) formula for the bed-material load can be expressed as

$$\log C = 5.435 - 0.286 \log\frac{wd_s}{v} - 0.457 \log\frac{u_*}{w} +$$

$$(1.799 - 0.409 \log\frac{wd_s}{v} - 0.314 \log\frac{u_*}{w}) \cdot$$

$$\log(\frac{VS}{w} - \frac{V_{cr}S}{w})$$

(4)

where C = concentration (ppm by weight) of the bed-material load; v = kinematic viscosity; u_* = shear velocity; $d_s = d_{50}$; VS = unit stream power, in which V = average velocity and S = water surface or energy slope; and $V_{cr}S$ = critical unit stream power required at the incipient motion. The V_{cr}/w value can be determined by

$$\frac{V_{cr}}{w} = \frac{2.5}{\log(\frac{u_* d_s}{v}) - 0.06} + 0.66; \quad 1.2 < \frac{u_* d_s}{v} < 70$$

(5)

and

$$\frac{V_{cr}}{w} = 2.05; \quad 70 \leq \frac{u_* d_s}{v}$$

(6)

Note that when the bed material is nonuniform, Eq. (4) is still valid for each particle size d_{si}.

3 METHODS OF UNCERTAINTY ANALYSIS

3.1 The First-Order Variance Estimation (FOVE) Method

This method estimates uncertainty of the model output as a function of the variances of stochastic input parameters. Assume that the output Y (Q_s) of a model (Einstein's or Yang's formula) can be expressed as a function of stochastic input parameters Xs as

$$Y = f(X^t) = f(X_1, X_2, ..., X_m)$$

(7)

where X = an m-dimensional column vector of stochastic input parameters; the superscript t = the matrix or vector transpose; and f( ) denotes a functional relationship. The FOVE method considers the first-order Taylor series expansion term of Eq. (7)

$$Y \cong f(\mathbf{x_0}^t) + \mathbf{s_0}^t (\mathbf{X} - \mathbf{x_0}) \qquad (8)$$

where $\mathbf{s_0}$ = m-dimensional column vector of local sensitivity coefficients with elements $s_{io} = (\partial f/\partial X_i)_{\mathbf{x_0}}$ being the local sensitivity coefficient of Y with respect to the i-th input parameter X_i at the expansion point $\mathbf{x_0}$. Applying the expectation and variance operators to Eq. (8) with $\mathbf{x_0} = \mu$, the mean and variance of Y can be estimated as

$$E(Y) = \mu_Y \cong f(\mu^t) \qquad (9)$$
and
$$Var(Y) = \sigma_Y^2 \cong \mathbf{s}^t \Omega \mathbf{s} \qquad (10)$$

where μ_Y and σ_Y = the mean and standard deviation of the model output, respectively; μ and Ω = the vector of the mean and the covariance matrix of stochastic input parameters, respectively; and $\mathbf{s}$ = the sensitivity coefficient vector evaluated at $\mathbf{x_0} = \mu$. If all stochastic input parameters are independent, then

$$Var(Y) \cong \sum_{i=1}^{m} s_i^2 \, \sigma_i^2 \qquad (11)$$

According to Eq. (11), the contribution of each stochastic input parameter, C_i, to the overall uncertainty of the model output can be computed as

$$C_i = \frac{s_i^2 \sigma_i^2}{\sigma_Y^2}; \qquad i = 1 \text{ to } m \qquad (12)$$

3.2 POINT ESTIMATION (PE) METHOD WITH REGRESSION

The PE method was originally proposed by Rosenblueth (1975, 1981) to deal with symmetric or asymmetric stochastic input parameters. The idea was to approximate the original probability density function (PDF) of a random variable by discrete probability masses located at two points, such that the first three statistical moments of the original PDF were preserved. For a model with m stochastic input parameters, 2^m model evaluations were required to estimate the statistical moments of the model output. Harr (1989) proposed a modification that reduced the required model evaluations from 2^m to 2m, and greatly enhanced the practical applicability of the PE method.

By Harr's modification, the correlation matrix C of m stochastic input parameters can be decomposed as

$$C = \mathbf{VLV}^t \qquad (13)$$

where $\mathbf{V}$ = eigenvector matrix with m eigenvectors; and $\mathbf{L}$ = a diagonal matrix with λ_1, λ_2, ..., λ_m being the corresponding eigenvalues. The correlation matrix can be represented geometrically by a hypersphere of radius $\sqrt{m}$ centered at μ. Each eigenvector passing through the origin of the hypersphere intersects the sphere surface at two points. Thus, these 2m intersection points are used to estimate the statistical moments of the model output. For the detailed procedure of Harr's PE method refer to Yeh and Tung (1993).

To examine the relative importance of stochastic input parameters (X_is), one can use the multiple linear regression based on the X_i values located at the 2m intersection points and their respective model output values:

$$y_k = a_0 + \sum_{i=1}^{m} a_i x_{ki} + e_k; \quad k = 1 \text{ to } 2m \qquad (14)$$

where a_0 = intercept; a_i, in which i = 1 to m, are the regression coefficients which represent the global sensitivity coefficients; and e_k = model error term. If the coefficient of determination, R^2, of Eq. (14) does not approach one, it may imply that the linear function is insufficient to describe the relationship between X_i and Y. In such a case, other regression models are considered. In the linear regression analysis, R^2 represents the variation explained by the model. The larger the value of R^2 is, the better the regression model fits the observed data. The value of R^2 can be expressed as

$$R^2 = \frac{SSR}{SST} = 1 - \frac{SSE}{SST} \qquad (15)$$

where SSR = the residual sum of squares; SST = the total sum of squares; and SSE = the error sum of squares. The purpose of the type I sum of squares (SS) analysis is to determine the sequential sum of squares by adding one stochastic input parameter at a time in sequence to the regression model and computing the incremental SSR value associated with the parameter. When the value of R^2 nears one, the contribution of each stochastic input parameter, C_i, to the overall uncertainty of the model output can be approximated by

$$C_i = \frac{SSR_i}{SSR} R^2; \quad i = 1 \text{ to } m \qquad (16)$$

where SSR is the summation of SSR_i. If all stochastic input parameters are uncorrelated, then the SSR_i value associated with the i-th parameter is not affected by its order in the sequence, and can fully reflect its contribution to the overall uncertainty in the model output.

3.3 LATIN HYPERCUBIC SAMPLING (LHS) TECHNIQUE WITH REGRESSION

The basic concept of the LHS technique is to select random samples for each stochastic input parameter over its range in a stratified manner such that the overall uncertainty of a model output can be reasonably described by finite samples. Its procedure is summarized in Yeh and Tung (1993).

The description of the LHS technique assumes that the m stochastic input parameters are uncorrelated. If they are correlated, the joint probability density functions of the variables are required. According to Mckay (1988), a sample size of $K \cong 2m$ would be sufficient in the LHS procedure for the purpose of uncertainty analysis.

Similarly, the contribution of each stochastic input parameter, in a global sense, to the overall uncertainty of the model output, based on the K sets of stochastic inputs and model outputs, can be obtained by using Eq. (16).

4 UNCERTAINTIES IN SEDIMENT FORMULAS AND EXAMPLES

In Einstein's formula, $(5 + ns)$ stochastic input parameters were selected: flow velocity (V), water depth (D), bed slope (S_o), representative bed roughness (d_{65}), kinematic viscosity (ν), and ns sediment sizes. The reason for selecting ν as a stochastic input is that its variation will affect the fall velocity and concentration of the sediment, and the thickness of the laminar sublayer. In Yang's formula, $(4 + ns)$ of the above parameters (except for d_{si}) were selected as the stochastic input parameters. Note that the effect of the water depth appears indirectly in the calculation of sediment concentration. In this paper, all the selected stochastic input parameters were assumed to be independently and uniformly distributed on the interval $[0.9\mu_i, 1.1\mu_i]$, in which μ_i is the mean value of the i-th stochastic input parameter.

To demonstrate the uncertainty of the model output (Q_s) subject to the stochastic input parameters, we selected two representative cross-sections as the testing examples: one from the

Missouri River in the U.S.A. and the other from the Cho-Shui River in Taiwan.

The reach of the Missouri River between the Gavins Point Dam and the Iowa-Missouri border was used. The width of this channelized reach was about 700 ft with the average bed slope being 0.00015 with $d_{50} = 0.41$ mm. The nonuniform bed mixture was discretized into seven (ns = 7) discrete sediment size classes (Table 1). Low and high water discharges were assumed to be 5.5×10^4 cfs and 2.75×10^5 cfs, respectively. The estimated flow velocities and water depths corresponding to these two discharges are listed in Table 1.

The cross-section near the Tzu-Chiang Bridge represented the characteristics of the interested reach of the Cho-Shui River. The bed slope of this reach was about 0.00104 with $d_{50} = 0.31$ mm. The bed mixture was discretized into six discrete sediment size classes and is shown in Table 2. The river widths corresponding to the low discharge $(4.7 \times 10^4$ cfs) and the high discharge $(3.1 \times 10^5$ cfs) were 2,070 ft and 5,020 ft, respectively. The estimated velocities and water depths for these two discharges are also listed in Table 2.

Because the number of the discrete sediment size classes influences the estimation of Q_s, we further considered the case of ns = 1, i.e., using d_{50} as the representative size of the nonuniform sediment mixture. There were 2m sets of stochastic parameters required to estimate the uncertainty of Q_s by using Harr's PE method. For the convenience of comparison, we generated 30 sets of stochastic parameters, i.e., $K > 2m$, in the LHS procedure.

5 RESULTS OF UNCERTAINTY ANALYSIS

5.1 Uncertainty of Q_s

The expected value $(\overline{Q_s})$, standard deviation (σ_{Q_s}), and coefficient of variation (CV_{Q_s}) of Q_s computed by Einstein's and Yang's formulas can be determined based on the stochastic parameter values listed in Tables 1 and 2. The value of σ_{Q_s} or CV_{Q_s} provides an indication about the degree of uncertainty in the model output.

1. FOVE Method

When the relationship between the stochastic input parameters and the model output is complicated, the local sensitivity coefficients in Eq. (8) is approximated by the central finite difference. Using Eqs. (9) and (10), the values of $\overline{Q_s}$, σ_{Q_s}, and CV_{Q_s} were computed and are summarized in Table 3. The conclusions are as follows:

Table 1. Flow and bed-material characteristics in the Missouri River

Parameters		Mean	Min.	Max.
Low Q	Vel. (ft/s)	4.59	4.13	5.04
	Dep. (ft)	17.14	15.42	18.85
High Q	Vel.(ft/s)	9.01	8.11	9.91
	Dep. (ft)	43.58	39.22	47.94
Slope		0.00015	0.000135	0.000165
v (ft^2/s)		1.06×10^{-5}	9.5×10^{-6}	1.17×10^{-5}
d_{65} (mm)		0.831	0.748	0.915
d_{50} (mm)		0.412	0.371	0.453
d_{s1} (mm) (4%)		0.149	0.134	0.164
d_{s2} (mm) (34%)		0.247	0.222	0.272
d_{s3} (mm) (41%)		0.590	0.531	0.649
d_{s4} (mm) (11%)		1.190	1.071	1.309
d_{s5} (mm) (5%)		2.40	2.16	2.64
d_{s6} (mm) (3%)		4.80	4.32	5.28
d_{s7} (mm) (2%)		9.52	8.57	10.47

Table 2. Flow and bed-material characteristics in the Cho-Shui River

Parameters		Mean	Min.	Max.
Low Q	Vel. (ft/s)	5.93	5.34	6.53
	Dep. (ft)	4.24	3.82	4.67
High Q	Vel.(ft/s)	9.02	8.12	9.93
	Dep. (ft)	6.78	6.10	7.46
Slope		0.00104	0.00093	0.00114
v (ft^2/s)		1.06×10^{-5}	9.5×10^{-6}	1.17×10^{-5}
d_{65} (mm)		0.48	0.432	0.528
d_{50} (mm)		0.31	0.279	0.341
d_{s1} (mm) (19%)		0.088	0.080	0.097
d_{s2} (mm) (27%)		0.193	0.173	0.212
d_{s3} (mm) (21%)		0.419	0.377	0.460
d_{s4} (mm) (16%)		0.838	0.754	0.922
d_{s5} (mm) (12%)		1.676	1.508	1.843
d_{s6} (mm) (5%)		3.516	3.164	3.868

(a) Estimation of Q_s using Einstein's formula is very sensitive to the number of sediment size classes (ns). However, it is not the case when Yang's formula is used.

(b) According to the CV_{Q_s} value, the uncertainty of Q_s based on Einstein's formula is larger than that based on Yang's formula.

(c) The uncertainty of Q_s based on Einstein's formula has the tendency to increase as the bed slope increases.

2. Harr's PE Method

The $\overline{Q}_s$, σ_{Q_s}, and CV_{Q_s} based on the 2m sets of stochastic input parameters and the model outputs are summarized in Table 4. Comparing Table 3 to Table 4, the three statistical values using Harr's PE

method are a little larger than those obtained by the FOVE method. Furthermore, the conclusions mentioned in the FOVE method are valid here.

3. LHS technique

The $\overline{Q}_s$, σ_{Q_s}, ,and CV_{Q_s} based on the 30 sets of stochastic input parameters generated by the LHS technique and the associated model outputs are summarized in Table 5. The conclusions are the same as the above-mentioned.

It can be seen from the above analysis that essential differences do not occur between the local uncertainty results (by the FOVE method) and the global uncertainty results (by Harr's PE method and the LHS technique).

5.2 Contribution of Stochastic Parameters to the Uncertainty of Q_s

To compare the relative importance of each stochastic input parameter for their contribution to the uncertainty of the model output, the value of the i-th stochastic parameter was standardized as

$$x_{ji}' = \frac{x_{ji} - \overline{x}_i}{\sigma_i}; \quad j = 1 \text{ to } 2m \text{ or } K \tag{17}$$

where $\overline{x}_i$ and σ_i represent the mean value and the standard deviation of the i-th stochastic input parameter, respectively. Also, the model output y_j (Q_s) was centralized as

$$y_j' = y_j - \overline{y}; \quad j = 1 \text{ to } 2m \text{ or } K \tag{18}$$

where $\overline{y}$ is the mean value of the model output.

1. Einstein's Formula

In the local uncertainty analysis, the contribution of each stochastic input parameter (C_i) to the overall uncertainty of Q_s using the FOVE method was evaluated by Eq. (12) for various cases. Results show that the C_i value associated with the flow velocity was the largest one. The C_i values corresponding to the two smallest particle sizes were large, too. Fig. 1, as a demonstration, plots the result for the case in the Cho-Shui River, given a high discharge.

In the global uncertainty analysis, Eq. (16) was used for the comparison of C_i for each parameter. By Harr's PE method, the C_i value associated with the flow velocity, again, was the largest one, and the contribution from the kinematic viscosity was visible, in addition to those from the two smallest particle sizes. Fig. 2 illustrates this behavior. On the basis of the LHS technique, the relative magnitudes of the C_i values were similar to those by Harr's PE method, except for the water depth being another important parameter. Fig. 3 shows one of the results.

Table 3. The values of $\overline{Q}_s$, σ_{Qs}, and CV_{Qs} by the FOVE method

Cases	$\overline{Q}_s$ (lb/s-ft)		σ_{Qs} (lb/s-ft)		CV_{Qs}	
	Einstein	Yang	Einstein	Yang	Einstein	Yang
MO-LN*	0.23	0.64	0.041	0.107	0.18	0.17
MO-LU	0.24	0.59	0.041	0.102	0.17	0.17
MO-HN	2.38	8.00	0.696	1.238	0.29	0.15
MO-HU	1.43	7.73	0.283	1.231	0.20	0.16
CS-LN	10.40	3.29	3.429	0.548	0.33	0.17
CS-LU	1.44	2.98	0.313	0.485	0.22	0.16
CS-HN	65.55	13.38	19.27	2.153	0.29	0.16
CS-HU	7.46	12.07	2.370	1.911	0.32	0.16

*MO - Missouri River; CS - Cho-Shui River;

LN - low discharge, ns ≠ 1; LU - low discharge, ns = 1;

HN - high discharge, ns ≠ 1; HU - high discharge, ns = 1.

Table 4. The values of $\overline{Q}_s$, σ_{Qs}, and CV_{Qs} by Harr's PE method

Cases	$\overline{Q}_s$ (lb/s-ft)		σ_{Qs} (lb/s-ft)		CV_{Qs}	
	Einstein	Yang	Einstein	Yang	Einstein	Yang
MO-LN	0.23	0.64	0.044	0.110	0.19	0.17
MO-LU	0.24	0.60	0.040	0.107	0.16	0.18
MO-HN	2.51	8.01	0.790	1.274	0.31	0.16
MO-HU	1.46	7.73	0.309	1.299	0.21	0.17
CS-LN	10.83	3.34	3.681	0.569	0.34	0.17
CS-LU	1.48	3.00	0.351	0.513	0.23	0.17
CS-HN	67.82	13.56	20.46	2.266	0.30	0.17
CS-HU	7.90	12.16	2.729	2.020	0.35	0.17

Table 5. The values of $\overline{Q}_s$, σ_{Qs}, and CV_{Qs} by the LHS technique

Cases	$\overline{Q}_s$ (lb/s-ft)		σ_{Qs} (lb/s-ft)		CV_{Qs}	
	Einstein	Yang	Einstein	Yang	Einstein	Yang
MO-LN	0.23	0.64	0.043	0.112	0.19	0.18
MO-LU	0.24	0.59	0.038	0.098	0.16	0.16
MO-HN	2.50	8.03	0.745	1.314	0.30	0.16
MO-HU	1.46	7.74	0.282	1.179	0.19	0.15
CS-LN	10.53	3.31	3.162	0.574	0.30	0.17
CS-LU	1.47	2.99	0.311	0.397	0.21	0.13
CS-HN	66.78	13.44	17.79	2.274	0.27	0.17
CS-HU	7.71	8.27	2.295	1.569	0.30	0.13

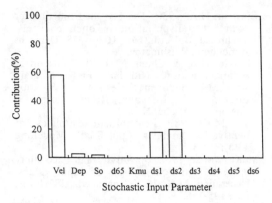

Fig. 1   The Ci values for the CS-HN case
(Einstein's formula, FOVE method)

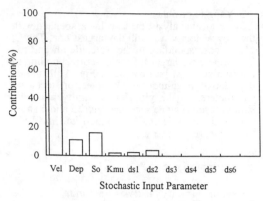

Fig. 4   The Ci values for the CS-HN case
(Yang's formula, FOVE method)

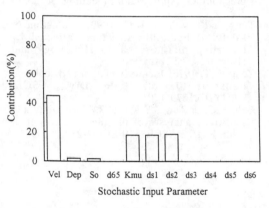

Fig. 2   The Ci values for the CS-HN case
(Einstein's formula, Harr's PE method)

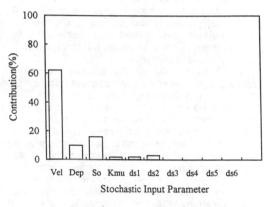

Fig. 5   The Ci values for the CS-HN case
(Yang's formula, Harr's PE method)

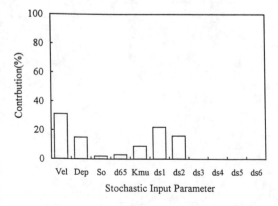

Fig. 3   The Ci values for the CS-HN case
(Einstein's formula, LHS technique)

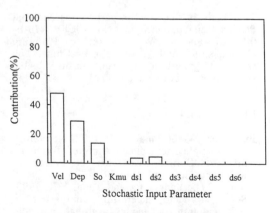

Fig. 6   The Ci values for the CS-HN case
(Yang's formula, LHS technique)

2. Yang's Formula

In the local analysis, the C_i value associated with the flow velocity was still the largest one. The C_i values corresponding to the water depth and the bed slope were larger than those corresponding to the two smallest particle sizes. Fig. 4 illustrates the relative magnitudes of C_i values for stochastic parameters. In the global uncertainty analysis, similar results can be seen, and Figs. 5 and 6 show two of them by Harr's PE method and the LHS technique, respectively.

6 CONCLUSIONS

Three uncertainty analysis methods, i.e., the FOVE method, Harr's PE method, and the LHS technique, were applied to analyze the uncertainty associated with Einstein's and Yang's formulas for estimating the sediment transport rate subject to stochastic input parameters. All the stochastic input parameters in both formulas were assumed to be independently and uniformly distributed. The analysis was applied to two rivers with different bed slopes, sediment mixture, and low and high water discharges.

Results showed that the flow velocity is the most important parameter for both sediment formulas to the uncertainty of the sediment transport rate. The sediment transport rate computed by Einstein's formula was very sensitive to the number of sediment size classes for nonuniform sediment mixture. This results in the important role played by the small particle sizes in the uncertainty of the sediment transport rate. Yang's formula, however, does not have such a phenomenon, and thus the water depth and the bed slope were the next two important parameters to the uncertainty of the calculated sediment transport rate.

ACKNOWLEDGMENTS

This study is supported in part by the National Science Council, under grant No. NSC81-0410-E-009-602, and the Water Resources Planning Commission, MOEA, Republic of China. The writers are grateful to Y.K. Tung for his valuable help during the course of this study.

REFERENCES

Bechteler, W. & Maurer M. 1992. Reliability theory applied to sediment transport formulae. 5th Int. Symp. on River Sedimentation, Karlsruhe, Germany, 311-317.

Chang, C.H., Yang, J.C., and Tung, Y.K. 1993. Sensitivity and uncertainty analysis of a sediment transport model: a global approach. Stochastic Hydrology and Hydraulics, 7:299-314.

Einstein, H.A. 1950. The bed load function for sediment transportation in open channels. Technical Bulletin No. 1026, US Dept. of Agriculture, Washington.

Einstein, H.A. & Chien, N. 1953. Transport of sediment mixtures with large ranges of grain sizes. MRD Sediment Series No. 2, U.S. Army Engrg. Div., Missouri River, Corps of Engineers, Omaha, Neb.

Harr, M.E. 1989. Probabilistic estimates for multivariate analysis. Appl. Math. Modelling 13(5):313-318.

McKay, M.D. 1988. Sensitivity and uncertainty using a statistical sample of input values. Uncertainty Analysis, Y. Ronen, ed., CRC Press, Fla., 145-186.

Rosenblueth, E. 1975. Point estimates for probability moments. Proc. Nat. Academy of Science, 72(10):3812-3814.

Rosenblueth, E. 1981. Two-point estimates in probabilities. Appl. Math. Modelling, 5(5):329-335.

Tung, Y.K. & Yeh, K.C. 1993. Evaluation of safety of hydraulic structures affected by migrating pits. Stochastic Hydrology and Hydraulics, 7:131-145.

Yang, C.T. 1973. Incipient motion and sediment transport. J. of Hydr. Div., ASCE, 99(HY10):1679-1704.

Yeh, K.C. & Tung, Y.K. 1993. Uncertainty and sensitivity analyses of pit-migration model. J. Hydr. Engrg., ASCE, 119(2):262-283.

Stochastic Hydraulics'96, Tickle, Goulter, Xu, Wasimi & Bouchart (eds) © 1996 Balkema, Rotterdam. ISBN 90 5410 817 7

Power spectrum of flows of hyperconcentrated clay suspensions

Zhaoyin Wang
International Research and Trainaing Center on Erosion and Sedimentation, Beijing, China

Erich J. Plate
Institut fur Hydrologie und Wasserwirschaft, Universität Karlsruhe, Germany

ABSTRACT: Turbulence structure of flows of hyperconcentrated clay suspensions were experimentally studied in a smooth-boundary open channel. Power spectra of the flows show that following increase in concentration of clay, the frequency of energy-carrying eddies is lower and lower. If the fluid exhibits large yield stress, only low-frequency turbulence exists. The chain of the turbulent energy transition is shorter and the lifetime of eddies is much less than the flow of water.

1 INTRODUCTION

Flows of hyperconcentrated clay suspensions often occurred in the middle reaches of the Yellow Rivers. Lahars, or hyperconcentrated flows composed of fine volcanic deposits, took place in Philippines and the North Tortul River, USA (Pierson and Janda, 1992). These flows were turbulent if the velocity was large enough. Turbulence structure of the flows was different from the turbulent water flow. For flow of hyperconcentrated clay suspension, Pierce (1917) stated: "At the peak of the flood and for almost an hour afterward, the river ran with a smooth, oily movement and presented the peculiar appearance of a stream of molten red metal instead of its usual rough, choppy surface." Lane (1940) quoted a statement by Eliassen, " When a river gets more than five percent of silt by weight, all eddy currents become damped, and water flows in a straight line motion one never sees when the water is clear." Vanoni (1946) suggested that turbulence is damped in the presence of suspended sediment, yet the nature of such a damping is obscure. Hino (1963), in contrast, predicted from his theory "... the decrease in the turbulence intensity of the damping effect due to suspension of sediment particles will be very small, contrary to the commonly accepted view, whereas a rather rapid decrease in lifetime of eddies will result." Some literatures on mud flows seem to indicate that turbulent eddies were suppressed. Some other lituratures, however, stated that if short term, small scale turbulence is not present long term, large scale turbulence is surely present. It was found from previous studies that turbulence may develop if the velocity is high

enough, and the higher the clay concentration, the higher the velocity is needed for development of turbulence.

Hyperconcentrated turbulent flow has extremely high competence of carrying sediment. Significant erosion and siltation is often associated with such flow. In some cases a whole river stopped flowing for several hours, in another case the whole river bed was eroded 9 meters in a few days (Wan and Wang, 1994). To investigate the laws of erosion and sedimentation of coarse sediment in hyperconcentrated flow and lahar, detailed information about the turbulence structure of these flows is needed.

2 EXPERIMENTS

Experiments were conducted in a smooth-boundary tilting flume 24 m-long, 60 cm-wide and 65 cm-high. Clay suspensions were used in the experiments. The clay particles were very fine, and the rheologic properties of the suspensions were studied with a rotating viscometer (Wang, Larsen and Xiang, 1994). The suspensions were non-Newtonian fluid and roughly followed the Bingham equation

$$\tau = \tau_B + \eta \varepsilon \tag{1}$$

in which τ_B is yield stress, η rigidity coefficient, ε shear rate which equals velocity gradient.

Fluctuating velocity of the flows was measured with a total pressure velocimeter (Wang, Ren and Wang, 1995). In the experiments measurement was done at the central profile of the channel. The aspect ratio of the flows (width/depth) was 4 to 10. The measured signals were transmitted to a computer through an analog/digital converter. With a sampling frequency of 256 s⁻¹, 8192 velocity values were obtained in 32 seconds in each sample. At each point two samples were taken initially and compared. If the values of the mean velocity and the root mean square of fluctuating velocity (rms-u') of those were the same, no third sample was taken; otherwise a third sample was taken and the computed values were averaged for the three samples. Only on a few occasions was the third sample required.

A Reynolds number for the non-Newtonian fluid flows is defined as follows:

$$Re_m = \frac{4\rho UH}{\eta[1 + \dfrac{1}{2}\dfrac{\tau_B H}{\eta U}]} \qquad (2)$$

in which ρ is density of the fluid, U and H are average velocity and depth of the flow, respectively. It is found that a flow of clay suspension was fully developed turbulent if the Reynolds number was over 10,000, and the flow was in a transitional region between laminar and turbulent if the Reynolds number was between 2,000 and 10,000. In the transitional region the upper flow was laminar while the lower flow exhibited high turbulence intensity and is intermittently turbulent. The whole flow was laminar if the Reynolds number was smaller than 2,000. It seems that turbulence of the flows are mainly affected by the yield stress of the suspensions. Fig.1 shows the fluctuating velocity curves of flows of clear water and clay suspensions measured at the same elevation, in which u' is the fluctuating velocity in the flow direction and U. is shear velocity. The most striking difference of clay suspension flows from clear water flow is much lower turbulence frequency and much smoother curves. The curves of the clay suspensions exhibit

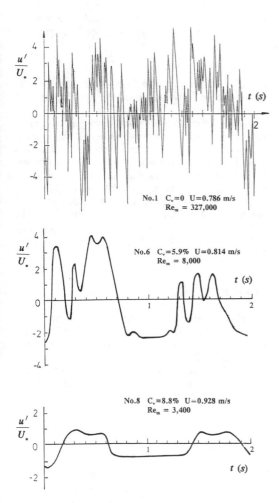

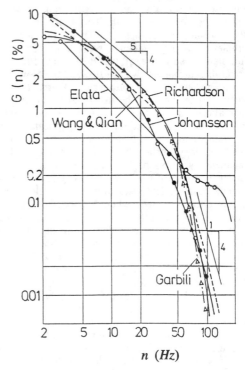

Fig. 1 Fluctuating velocity of flows of clear Water and clay suspensions

Fig.2 Distributions of spectral density for flows of water

646

intermittence of turbulence. At high clay concentration the yield stress was so high that turbulence was suppressed and the turbulence intensity was much lower than that of clear water flow.

3 POWER SPECTRUM

By using Fast Fourier Transformation, turbulent energy of eddies of various frequencies, $E(n)$, are calculated. Normalized by the total turbulent energy, a spectral density of turbulent energy, $G(n)$, is defined as

$$G(n) = \frac{E(n)}{u'^2} \times 100\% \tag{3}$$

Fig.2 shows the turbulent energy spectra in clear water flows measured by different researchers

(Wang and Qian, 1989; Elata et al., 1961; Garbili et al., 1982; Johansson et al., 1982; Richardson et al., 1968). Most of the data in the figure follow the trends:

In the low-frequency domain, $n < 2$ Hz:
$$G(n) = 10\%$$
In the main frequency domain, $5 < n < 20$ Hz:
$$G(n) = c_1 \, n^{-0.8}$$
In the high frequency domain, $n > 50$ Hz:
$$G(n) = c_2 \, n^{-4}$$

Fig.3 presents the distributions of the spectral density for flows of low clay concentrations in fully developed turbulent region, in which different symbols represent the measured values at different elevations. The distribution of $G(n)$ for flow of water is similar to those in Fig.2. Generally speaking, the distributions at different elevations differ not much. Following increase in clay concentration, the turbulent energy in low-frequency domain ($n < 2$ Hz) increases and in high-frequency

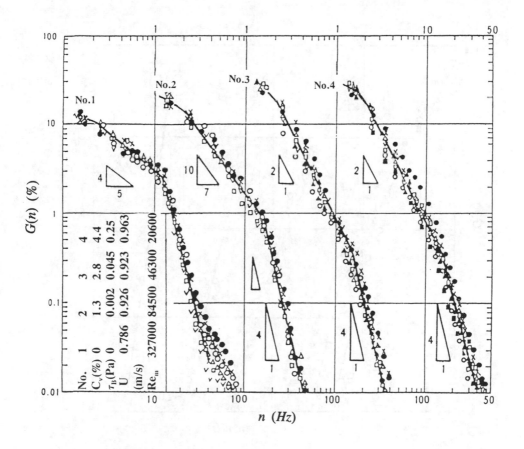

Fig.3 Distributions of spectral density for flows of clay suspension in the turbulent region

647

domain (n>20 Hz) decreases. It is summarized as follows:

Clear water:
  low-frequency domain, n<2 Hz, G(n)=10-13%;
  main-frequency domain, 2<n<20 Hz,
  $G(n)=c_1 n^{-0.8}$;
  high-frequency domain, n>20 Hz, $G(n)=c_2 n^{-4}$;
1.3% clay suspension:
  low-frequency domain, n<2 Hz, G(n)=20%;
  main-frequency domain, 2<n<20 Hz,
  $G(n)=c_1 n^{-1.4}$;
  high-frequency domain, n>20 Hz, $G(n)=c_2 n^{-4}$;
2.8% clay suspension:
  low-frequency domain, n<2 Hz, G(n)=30%;
  main-frequency domain, 2<n<20 Hz,
  $G(n)=c_1 n^{-2.0}$;
  high-frequency domain, n>20 Hz, $G(n)=c_2 n^{-4}$;
4.4% clay suspension:
  low-frequency domain, n<2 Hz, G(n)=30%;
  main-frequency domain, 2<n<20 Hz,
  $G(n)=c_1 n^{-2}$

high-frequency domain, n>20 Hz, $G(n)=c_2 n^{-4}$;

in which c_1 and c_2 are constants. They have different values for different liquids and can be determined from the power spectral density curves.

The distributions of G(n) in the high-frequency domain for different liquids are the same, i.e. G(n) is proportional to the -4 power of frequency. The percentage of the total turbulent energy for frequency higher than 20 Hz can be calculated by:

$$P(n>20Hz) = \int_{20}^{\infty} c_2 n^{-4} dn = \frac{20}{3} G(20) \qquad (4)$$

where $c_2 = 20^4 G(20)$. P(n>20 Hz) decreases with increasing clay concentration. It is 3% in clear water flow, 2% in the flow of 1.3% clay suspension, 0.6% in the flow of 4.4% clay suspension, and only 0.3% in the flow of 5.9% clay suspension.

Fig.4 shows distributions of spectral density of turbulent energy in the transitional region. Because the clay concentration is higher, the percentage of

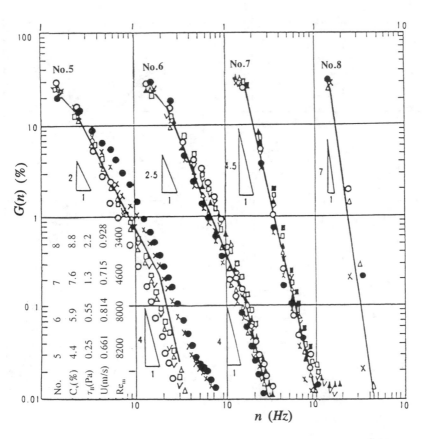

Fig.4 Distributions of spectral density for flows of clay suspension in the transitional region

turbulent energy in the low-frequency domain is still greater and the slope of the G(n) curve in the main-frequency domain is still steeper. The main results are:

4.4% clay suspension:
  low-frequency domain, $G(n)=30\%$;
  main-frequency domain, $G(n)=c_1 \, n^{-2}$;
5.9% clay suspension:
  low-frequency domain, $G(n)=30\%$;
  main-frequency domain, $G(n)=c_1 \, n^{-2.5}$;
7.6% clay suspension:
  low-frequency domain, $G(n)=30\%$;
  main-frequency domain, $G(n)=c_1 n^{-4.5}$;
8.8% clay suspension:
  low-frequency domain, $G(n) = 30\%$;
  main-frequency domain, $G(n) = c_1 n^{-7}$;

There is no or almost no high frequency turbulence in these cases. The experiments No.4 and No.5 were with the same clay concentration but different Reynolds numbers. The distributions of G(n) in the two experiments were roughly the same, which indicates that the distribution of the spectral density of turbulence energy relies mainly on the rheological properties of the fluid but not much on the flow velocity.

Because the flows in the transitional region were intermittently turbulent, the power spectra were affected by the intermittency. Employing a window function, Plate separated an intermittent record into two processes. The window function is generated from the record and represents the intermittency. Power spectrum associated with the intermittency were given by (Plate, 1977)

$$G_w(n) = \frac{G_{wo}}{1 + [\phi(1-\phi)\frac{2\pi n}{f}]^2} \qquad (5)$$

where G_{wo} is a constant, f is the frequency of the intermittency, and φ is the ratio of accumulated time of turbulent flow to the total time for statistics and is called as intermittency factor. φ equals 0 for laminar flow and equals 1 for fully developed turbulent flow. The equation indicates that the intermittency has the greatest influence on the spectrum if the intermittency factor equals 0.5. Because the intermittency frequency f is usually less than 2 Hz, intermittency mainly enhances turbulent energy of low frequency turbulence.

4 MECHANISM OF SPECTRAL STRUCTURE

In turbulent flow the energy is transferred from the energy-carrying eddies to the energy-dissipating eddies by a cascade of large eddies breaking up to form eddies of smaller size that become unstable in their turn and break up into still smaller ones. Kolmogorov (1962) made explicit the consequences of the cascade hypothesis for the structure of the small eddies that contain only a small part of the whole turbulent energy. If the Reynolds number of the flow is sufficiently large, he showed that the small eddies must be in a state of absolute equilibrium in which the rate of receiving energy from large eddies is nearly equal to the sum of the rates of loss to even smaller eddies by breaking up and by working against viscous stresses.

An equation in terms of spectral functions, with wave number k instead of frequency n, was obtained (Xia, 1982):

$$\frac{\partial}{\partial t} \int_0^k E(k,t)dk = \int_0^k e(k,t)dk -$$

$$\frac{2\mu}{\rho} \int_0^k k^2 E(k,t)dk - \int_0^k F(k,t)dk \qquad (6)$$

in which μ is the viscosity of the fluid; $E(k,t)$ represents the turbulent energy for eddies with wave number k at time t; $e(k,t)$ represents the energy transferred from mean flow to wave number k; the third term on the right hand side represents the energy transferred from wave numbers less than k to wave numbers larger than k. The wave number k is related to the frequency n approximately by

$$k = \frac{2\pi n}{\bar{u}} \qquad (7)$$

where $\bar{u}$ is the mean velocity. The second term on the right hand side is a negative term involving the viscosity and is thus the energy dissipation.

In a steady turbulent flow the rate of change of turbulent energy is zero, and Eq.(6) can be rewritten as

$$e(k) - 2\frac{\mu}{\rho}k^2 E(k) - F(k) = 0 \qquad (8)$$

F(k) represents the energy transferred from turbulence of wave number k to those of high wave numbers. According to the relationship between the wave number and turbulence frequency, the wave spectrum function E(k) is proportional to k^{-m} for the frequency spectrum function E(n) is proportional to n^{-m}. Therefore, $E(k) \sim k^{-0.8}$ in the main frequency domain in flow of clear water. The second term in Eq.(8), or the energy dissipation term, is proportional to $k^{1.2}$, which means that the higher the frequency, the higher the rate of energy dissipation. For low frequency turbulence, the energy dissipation term is very small and

$$e(k) \approx F(k) \tag{9}$$

The formula indicates that almost all the energy from the mean flow is transferred to high frequency. For high frequency turbulence, however, very little energy is directly transferred from the mean flow and

$$F(k) \approx 2\frac{\mu}{\rho}k^2 E(k) \tag{10}$$

Almost all energy dissipated by high frequency turbulence comes from low frequency turbulence.

In high frequency domain, $E(k) \sim k^{-4}$, the energy dissipation term in Eq.(8) is proportional to k^{-2}, which suggests that the higher the frequency, the less the rate of energy dissipation. Accordingly, the eddies of between the high frequency domain and the main frequency domain $(20 \sim 50 \text{ Hz})$ has the maximum rate of dissipation of turbulence energy for flow of water. For flows of low clay concentrations, $E(k) \sim k^{-(0.8 \sim 2)}$ in the main frequency domain, the above laws are still held.

In the flows of high clay concentrations the viscous shear stress is $\tau_B + \eta \dot{\epsilon}$ and it is quite large even at very small shear rate for the existence of the yield stress. Therefore, not only high frequency eddies, that involves high shear rate, but also low frequency eddies, cause energy dissipation of turbulence. By using the effective viscosity

$$\eta_e = \eta + \frac{\tau_B}{\epsilon} \tag{11}$$

to substitute the viscosity μ, the equation (8) can be rewritten as

$$e(k) - \frac{2}{\rho}(\eta + \frac{\tau_B}{\epsilon})k^2 E(k) - F(k) = 0 \tag{12}$$

The existence of τ_B and the large viscosity changed the cascade energy transition. For clay concentration higher than 5%, $E(k) \sim k^{-(2.5 \sim 7)}$ in the main frequency domain, the energy dissipation term in Eq.(12) is proportional to $k^{-0.5}$ to k^{-5}. Following increase in frequency, the rate of energy dissipation decreases, and the maximum rate of turbulent energy dissipation occurs between the low frequency domain and main frequency domain. Most of turbulent energy is converted into heat by the eddies in the low and main frequency domains, and high frequency eddies receives very little energy. For instance the eddies of frequency higher than 10 Hz in 7.6% clay suspension has only 0.01% of the turbulent energy. The higher the clay

concentration, the more turbulent energy consumed in low frequency turbulence. These results prove that for turbulent flow of clay suspensions the chain of the turbulence energy transition is shorter and the lifetime of eddies is much less than the flow of water.

5 CONCLUSIONS

Turbulence may develop in the flows of clay suspensions if the velocity is high enough. The power spectra shows that following increase in viscosity and yield stress of the fluid, the frequency of energy-carrying eddies is lower and lower. If the fluid exhibited large yield stress, only low-frequency turbulence exists. Turbulent energy is mostly dissipated by eddies of frequency about $20 \sim 50$ Hz for flow of water but by eddies about $2 \sim 5$ Hz for flows of high clay concentrations. For turbulent flow of clay suspensions the chain of the turbulence energy transition is shorter and the lifetime of eddies is much less than the flow of water.

ACKNOWLEDGEMENT

This research project is supported by the National Natural Science Foundation of China (No.59425005) and the Alexander von Humboldt Foundation of Germany.

REFERENCES

Elata, C. and Ippen, A.T., 1961, The dynamics of open channel flow with suspensions of neutrally buoyant particles, Tech. Rept, No.45, Hydrodynamics Lab., M.I.T..

Garbili, J.L., Forster, F.K. and Jorgensen, J.F., 1982, Measurement of fluid turbulence based on pulsed ultrasound techniques, J. Fluid Mcch., Vol. 118, 471-505.

Hino, M., 1963, Turbulent flow with suspended particles, J. of Hydraulic Division, ASCE, Vol. 89, No,4, Proc. Paper, pp.161-185.

Johansson, A.V. and Alfredsson, P.H., 1982, On the structure of turbulent channel flow, J. Fluid Mech., Vol. 122, 295-314.

Kolmogorov, A.N., 1962, A refinement of previous hypotheses concerning local structure in a viscous incompressible fluid at high Reynolds numbers, J. Fluid Mech., Vol.13, 82-5.

Lane, E.W., 1940, Notes on limit of sediment concentration, J. of Sedim. Petrology, Vol.10, No.2, pp.95-96.

Pierce, R.C., 1917, The measurements of silt-laden streams, Water-Supply Paper 400, Deol. Survey,

U.S. Dept. of Interior, Washington, D.C., pp.39-59.

Pierson, T.C. and Janda, R.J., 1992, Immediate and long-term hazards from lahars and excess sedimentation in Rivers draining Mt.Pinatubo, Philippines, Water Resources Investigations Report 92-4039, U.S. Geological Survey.

Plate, E.J., 1977, Intermittent processes and conditional sampling, in "Stochastic Processes in Water Resources Engineering, Water Resources Publications, Fort Collins, Colorado 80522, pp.1-1 - 1-27.

Richardson, E.V. and McQuivey, R.S., 1968, Measurements in turbulence in water, Proc. ASCE, J. Hydraulic Div., Vol.94, Hy2, 411-430.

Vanoni, V.A., 1946, Transportation of suspended sediment by water, Trans. of the American Society of Civil Engineers, 111, 67-133.

Wang, XingKui and Qian, N, 1989, Turbulence characteristics of sediment-laden flow, J. Hydraulic Engineering, Vol. 115, No.6, 781-800.

Wang, ZhaoYin, Larsen, P. and Xiang, W., 1994, Rheological properties of sediment suspensions and their implications, Journal of Hydraulic Research, IAHR, No,5.

Wang, ZHaoYin, Ren, Y.M. and Wang, X., 1995, Total pressure probe for measurements of turbulence in sediment-laden flows, Proceedings of 2nd International Conference on Hydro-Science and Engineering, Beijing, China.

Wan, ZhaoHui & Wang, ZhaoYin, 1994, Hyperconcentrated Flow, IAHR Monograph Series, Balkema Publishers.

Xia, ZhenHuan, 1982, Introduction to Turbulent Flow Dynamics, Tsinghua Press, Chapter 12 (in Chinese).

Stochastic Hydraulics'96, Tickle, Goulter, Xu, Wasimi & Bouchart (eds) © 1996 Balkema, Rotterdam. ISBN 90 5410 817 7

Stochastic analysis of pressure fluctuations for sediment laden flow

Wang Mu-Lan
Hohai University, Nanjing, China

Wang De-Chang & Zhu Jie
Institute of Hydraulic Research of Yellow River Conservancy Commission, China

ABSTRACT: In this paper, the laboratory experimental study for the wall pressure fluctuations of the generalized model for the suspended sediment laden flow is presented. The data processing FFT have been applied for experimental samples. The pressure fluctuation equations for the sediment laden flow were deduced. At last the wall pressure fluctuation characteristics for the different sediment content laden flow were obtained.

1 INTRODUCTION

It is necessary to know the variation in pressure fluctuations of the sediment laden flow for estimation of hydraulic load, for solution of the cavitation problems as well as for studying the mechanism of the inter action between the flow and sediment in design of hydraulic projects on sediment laden rivers. Therefore study of hydraulic characteristics including parameters of wall pressure fluctuations of sediment laden flow is very important both in theory and in practice. In this paper the laboratory experimental study for the wall pressure fluctuations of the generalized model for the suspended sediment laden flow is presented. Twelve signals of the wall pressure fluctuations for different sediment content were recorded by means of tape memory. The data processing FFT have been applied in stochastic analysis of the experimental samples. The pressure fluctuation equations for the sediment laden flow were deduced based on the precursory works. At last the wall pressure fluctuation characteristics for the different sediment content laden flow were compared with the wall pressure fluctuation characteristics for the clear flow.

2 EXPERIMENTATION AND DATA PROCESSING

2.1 Generalized model (cheng 1992, wang 1992)

Experiments were conducted in a pressure conduit with 8cm wide, 10cm deep and 640cm long. A baffle plate with 2.5cm high was mounted on the top of the conduit to form a suddenly expansion flow as shown in Fig. 1 (a). The measuring points for velocities, time-average pressures and fluctuating pressures were located as shown in Fig. 1 (b). The measurements were made for discharges, $Q = 17.5$, 14.5, 12 l/s, as well as sediment concentrations, $S = 0$, 85, 117, 138, 170, 180, 242, 269, 331 kg/m^3 repectively.

According to the precursory works it is known that the flow belongs to the low sediment laden flow, as well as the variations of shear stresses in this flow obey the Newton's Law when $s < 170$kg/m^3, but the flow get into the high sediment laden flow, as well as the variations of shear stresses in this flow obey the Bingham Law when s has reached $170 \sim 200$kg/m^3. The pressure fluctuations of sediment laden flow were measured by means of CYG06 type fluctuating pressure tranducer with special device to prevent plugging of the probe. The output signals were recorded on a tape recorder.

2.2 Pattern, velocity and time-average pressure of clear flow

The flow which rushed out from the baffle and formed submerged jet with high velocity contracts continuosly till to the minimum section 1-1 behind the baffle, then begins to expand and finally, spread

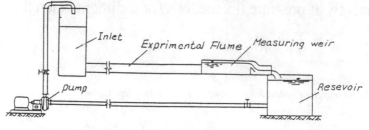

a) Circulating System

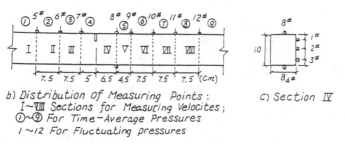

b) Distribution Of Measuring Points:
I~Ⅷ Sections for Measuring Velocites;
①~⑨ For Time-Average Pressures
1~12 For Fluctuating pressures

c) Section Ⅳ

Fig.1 Schematic Diagram of Equipment

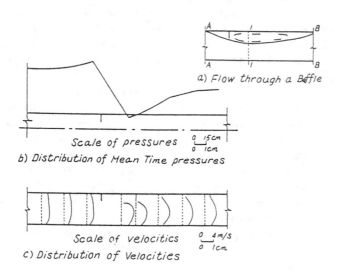

a) Flow through a Baffle

Scale of pressures   0 15cm / 0 1cm
b) Distribution of Mean Time pressures

Scale of velocities   0 4m/s / 0 1cm
c) Distribution of Velocities

Fig. 2 Test for clear water, Q =17.5=ℓ/s

654

to whole space of the tunnel at the section B-B. Additionaly, a large amount of intensly mixed eddies with large size is generated at the interface of main flow and recirculating flow where there are great velocity gradients along the flow path, these eddies accumulate, grow and then gradually attenuate. The jet, especially the tail of recirculation region consequently oscillate, as shown in Fig2 (a).

The variations of time-average pressure and velocities are shown in Fig. 2 (b), (c) and Table 1.

Table 1. The variation of time-average velocities and head losses of baffle

Discharge L/S	Cross-section average velocity m/s						Head loss of baffle ΔH (m)	coefficient of head loss of baffle K
	I	III	IV	V	VII	V_0 (for baffled section)		
17.5	2.14	2.10	2.52	2.37	2.24	2.92	0.22	0.51
14.5	1.76	1.79	2.02	1.97	1.86	2.42	0.16	0.54
12.0	1.47	1.50	1.73	1.64	1.52	2.00	0.11	0.54

Note: $K = \Delta H / \dfrac{v_0^2}{2g}$,

v_0—— average velocity of baffled section for clear flow.

2.3 Data processing and main characteristics of pressure fluctuations of sediment laden flow. (wang 1989)

The process of pressure fluctuation was assumed to be stochastic and ergodic for steady flow. The main parameters of data processing were selected as follows: time step $\Delta t = 0.005 \sim 0.01$s, sample size $N = 1024$. The data processing were performed on Real Time-signal Analyser 5451c.

1. Feature of pressure fluctuations in amplitude domain

The intensity and the average amplitude of pressure fluctuations are defined as $D_{p'} = \overline{p'^2}$ and $\sigma_{p'} = \sqrt{D_{p'}}$, respectively.

A large amount of third and fourth monents of the measured samples of pressure fluctuations, some of them were given in table 2, show the probability density function of amplitude fluctuation can be approximately described by a normoa distribution function.

Table 2. The coefficient of third and fourth monents of pressure fluctuations

Discharge Q·L/S	Sediment concentration S kg/m³	3#		5#		10#	
		C_s	C_e	C_s	C_e	C_s	C_e
14.5	0	0.1758	3.1846	0.3459	2.5941	3.3754	3.7263
	117	0.1059	3.0476	0.2591	3.6424	0.2048	3.1721
	170	0.3630	2.5849	0.1036	2.9387	0.1481	3.0964
	242	0.2534	2.7539	0.2438	2.8479	0.2861	3.4367
	331	0.2260	3.2789	0.2142	2.8723	0.2238	2.6400

Table 2 shows that the normal distributions are independent of sediment concentrations. The intensity coefficient β' of pressure fluctuations without taking into account the influence of specific gravity of sediment laden flow can be written as

$$\beta' = \sqrt{D_{p'}} / \frac{v_0^2}{2g} = \sigma_{p'} / \frac{v_0^2}{2g} \qquad (1)$$

Table 3. Q = 14.5 L/S The intensity and intensity coefficient of pressure fluctuations

Sediment concentration S kg/m³	3#		5#		10#	
	$\sigma_{p'}$·cm	β' %	$\sigma_{p'}$·cm	β' %	$\sigma_{p'}$·cm	β' %
0	0.75	2.52	1.31	4.40	1.36	4.55
117	0.80	2.68	1.36	4.56	2.04	4.70
170	0.82	2.75	1.32	4.43	1.52	5.10
242	0.86	2.89	1.49	5.01	1.56	5.23
331	0.91	3.05	1.63	5.47	1.65	5.54

Table 3 shows some $\sigma_{p'}$ and β' for different sediment concentration. It is found in table 3 that the $\sigma_{p'}$ and β' increase with the growth of sediment concentrations S.

The coefficient of pressure fluctuation taking into account the influence of specific gravity of sediment laden flow can be expressed as

$$\beta = \sigma_{p'} / (\frac{r_m}{r} \cdot \frac{v_0^2}{2g}) \qquad (2)$$

where r_m——specific gravity of sediment laden flow.

Fig. 3 (a), (b) show the variation of β' and β with the sediment concentration S. It is known from the table that β' increases with the growth of sediment concentration S, while the tendency increase with sediment concentration S of β disappeared, and the points of β for the clear flow and sediment laden flow

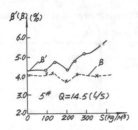

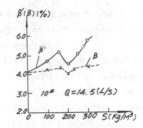

Fig. 3 Variations of β', β with S.

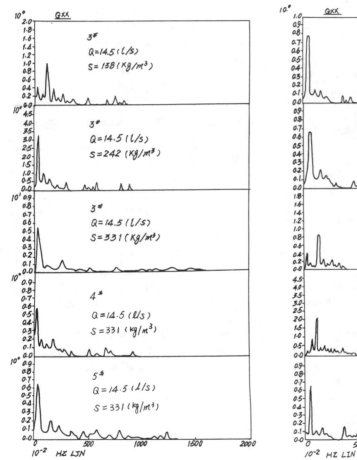

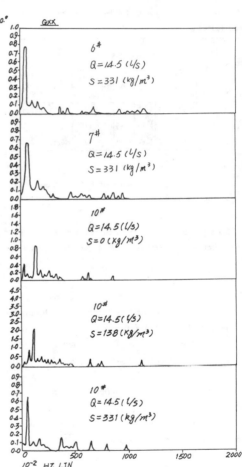

Fig. 4 Power Spectrum

gathered round the line paralleling to x axis. That means, generally, the parameter β is independent of the sediment concentration S. Moreover, test gives the gradually increasing tendency of extremal difference between the maximum and minimum values of pressure fluctuation, $B = p'_{max} - p'_{min}$ with the increase of sediment concentrations S. The variation of the parameter B is found within the range of $B / \frac{v_0^2}{2g} = 0.1 \sim 0.3$.

2. Feature of pressure fluctuations in frequency domain

The signal of pressure fluctuation may be considered as a composition of harmonic components with different amplitudes, frequencies and powers. The power spectrum is one of important characteristics of pressure fluctuations. Spectral analysis can provide the predominant frequency and variation of spectral densities with the frequencies, they are relevant to some important engineering problems, such as dynamic loads and vibration induced by flow etc. In this paper the FFT is applied to the spectral analysis of pressure fluctuations. Fig 4 shows the spectrum of points $3^\#$, $4^\#$, $5^\#$, $6^\#$, $7^\#$ and $10^\#$, respectively.

From the Fig 4 it is found that the fluctuating pressures of sediment laden flow belong to the fluctuation realm of low frequences. The predominant frequency obviously turned to the lower frequencies with the increment of sediment concentration. Generally the predominant frequency varies in the range from 0.5 to 3.0 Hz.

3. Feature of pressure fluctuations in time domain.

The auto-correlation function R (t) expresses the correlative degree of pressure fluctuation at time t and $t+\tau$ in a point, it can be defined as

$$R(\tau) = \frac{\overline{p'_i(t) \cdot p'_i(t+\tau)}}{\sqrt{\overline{p'^2_i}} \cdot \sqrt{\overline{p'_i(t+\tau)}}}, 0 < R(\tau) < 1 \quad (3)$$

R (τ) will attenuate with the increment of τ. The time lag τ_0 corresponds to the fact that the value of R (τ) becomes zero for the first time at the moment τ_0. The larger the time lag τ_0 the greater the sizes of large eddies motivating the fluctuations of pressures.

Table 4 gives the variations of τ_0 at some measuring points for different sediment concentration. From the table 4 it is found that τ_0 is gradually increased with the increment of sediment concentrations. This tendency is in agreament with the variations of ampli-

tudes and predominant frequencies against the sediment concentrations.

Table 4. The time lag τ_0 of pressure fluctuations

τ_0 s point	Sediment contentra- -tions $_3$ kg/m^3 0	85	117	170	202	242
$3^\#$	0.015	0.011	0.048	0.078	0.131	0.120
$10^\#$	0.025	0.032	0.013	0.048	0.053	0.048

3 BASIC EQUATIONS OF PRESSURE FLUCTU- ATIONS

3.1 Basic Equations of pressure fluctuations for clear flow

According to the N-S equation for non-compressing fluid, a certain mathematical translations have been conducted. The basic equation of pressure fluctuation for steady flow can be obtained as

$$\nabla^2_{p'} = - \rho [2 \frac{\partial \bar{u}_i}{\partial x_j} \frac{\partial u'_j}{\partial x_i} + (- \frac{\partial^2 u'_i u'_j}{\partial x_i \partial x_j} + \frac{\partial^2 \overline{u'_i u'_j}}{\partial x_i \partial x_j})]$$

(4)

From the above equation it can be found that the pressure fluctuation composed of two parts: the first part is the first term in the right side of the equality depended on the fluctuating velocities and average shear-stresses, i. e. so called "fluctuation shcar" term, the second part is the term in the second brackets of the right side depended on the fluctuating velocities in the two directions, i. e. so-called "fluctuation-fluctuations" term.

3.2 Basic Equation of pressure fluctuations for high sediment laden flow

The small particles of fine sand united together can form some wadding groups by electronic actions between the small particles and media. With the increment of sediment concentration the combined wadding groups possess certain shear strengths, the water does not flow when the shear stresses in the fluid are less than the shear-strengths, the corresponding critical shear-strength is defined as bingham's extreme shear strengths.

Suppose the sign τ_B expresses the Bingham's shear-strengths, as well as the sign η-coefficient of stiffness which is much greater than the viscocity coefficient of fluid. The continuity equation and movement equations of high sediment laden flow can be written as.

$$\frac{\partial \rho u_i}{\partial x_i} = 0 \tag{5}$$

$$\frac{\partial}{\partial x_j}(\rho u_i u_j) = \rho g_i - \frac{\partial p}{\partial x_i} + \frac{\partial \tau_B}{\partial x_j} + \frac{\partial}{\partial x_j}(\eta \frac{\partial u_i}{\partial x_j}) \tag{6}$$

Consedering the τ_B and η as parameters without fluctuations, and neglecting the interchange of momentum induced by the density fluctuations of high sediment concentration flow, taking the Reynolds average the above equation may be expressed as

$$\frac{\partial \overline{\rho u_i}}{\partial x_i} = 0 \tag{7}$$

$$\frac{\partial}{\partial x_j}(\overline{\rho u_i}\,\overline{u_j} + \overline{\rho' u'_i u'_j}) = \rho g_i \frac{-\partial \overline{p}}{\partial x_i} + \frac{\partial \tau_B}{\partial x_j} + \frac{\partial}{\partial x_j}(\eta \frac{\partial \overline{u_i}}{\partial x_j}) \tag{8}$$

Substituting $u_i = \overline{u}_i + u'_i, u_j = \overline{u}_j + u'_j, p = \overline{p} + p'$ into Eq. (5), (6) and reducing Eq. (7), (8) respectively give

$$\frac{\partial \overline{\rho u'_i}}{\partial x_i} = 0 \tag{9}$$

$$\frac{\partial}{\partial x_j}(\rho \overline{u}_i u'_j + \rho \overline{u}_j u'_i + \rho u'_i u'_j - \overline{\rho u'_i u'_j})$$

$$= -\frac{\partial p'}{\partial x_i} + \frac{\partial}{\partial x_j}(\eta \frac{\partial u'_i}{\partial x_j}) + \rho' g_i \tag{10}$$

From Eq (10) and noticing Eq (9), with several mathematical deductions it follows then

$$\nabla^2 p' = -\left[2\frac{\partial \rho \overline{u}_i}{\partial x_j}\cdot\frac{\partial u'_j}{\partial x_i} + (-\frac{\partial^2 \overline{\rho u'_i u'_j}}{\partial x_i \partial x_j} + \frac{\partial^2 \rho u'_i u'_j}{\partial x_i \partial x_j})\right]$$

$$+ \frac{\partial^2}{\partial x_i \partial x_j}(\eta \frac{\partial u'_i}{\partial x_j}) + \frac{\partial}{\partial x_i}(\rho' g_i) \tag{11}$$

Especially, when the fluid density ρ may be considered as a constant, Eq(11) can be translated into

$$\nabla^2_{p'} = -\rho\left[2\frac{\partial \overline{u}_i}{\partial x_j}\frac{\partial u'_j}{\partial x_i} + (-\frac{\partial^2 \overline{u'_i u'_j}}{\partial x_i \partial x_j} + \frac{\partial^2 u'_i u'_j}{\partial x_i \partial x_j})\right] \tag{12}$$

It appears that Eq. (12) and Eq. (4) for clear flow are formal resemblance.

3.3 Basic equation of pressure fluctuations for low sediment laden flow

Suppose

$$\rho = \rho_t(1 - s) + \rho_s s \tag{13}$$

$$u_i = \frac{1}{\rho}[\rho_s u_{s,i} s + \rho_f u_{f,i}(1 - s)] \tag{14}$$

$$\tau_{ij} = (1 - s)\tau_{f,ij} + s\tau_{s,ij} \tag{15}$$

in which the subscript f corresponds to the param-

eters of the clear fluid as well as the subscript s corresponds to laden sediment parameters.

Applying the similar supposition for high sediment laden flow, i. e. neglecting the density fluctuation induced by the interchanges of momentums, but retaining the density fluctuation induced by the gravity, the basic equation of pressure fluctuations for low sediment laden flow can be obtained

$$\nabla^2 p' = -\left[2\frac{\partial \rho \overline{u}_i}{\partial x_j}\cdot\frac{\partial u'_j}{\partial x_i} + (-\frac{\partial \overline{u'_i u_j}}{\partial x_i \partial x_j} + \frac{\partial^2 u'_i u'_j}{\partial x_i \partial x_j})\right]$$

$$+ \frac{\partial^2 \tau'_{ij}}{\partial x_i \partial x_j} + \frac{\partial \rho' g_i}{\partial x_i} \tag{16}$$

4 QUALITATIVE ANALYSIS OF MEASURED PRESSURE FLUCTUATIONS

Under the guidance of basic equations of pressure fluctuations for clear flow and for sediment laden flow the qualitative analysis of the measured pressure fluctuations in this paper can be taken as follows.

Analysis of the equntion (4) for clear flow and equation (11) for high sediment laden flow indicated that the terms "fluctuation-shears" and "fluctuations-fluctuations" in the right side of equations (4) and (11) are formal resemblance, the third term and fourth term in the right side of equation (11) are related to the "viscosity" and the variation of gravity induced by the density fluctuations in high sediment laden flow respectively. From the comparison between the equation (4) and equation (11) it is obvious that Eq(11) for high sediment concentration flow has third and fourth terms, whil Eq. (4) for clear flow did not content them. Similarly, it is known, the Eq. (16) for low sediment concentration has third and fourth terms flow whil the Eq(4) for clear flow did not content them. similarly, it is known that Eq (16) for low sediment concentration has third and fourth terms whil the Eq(4) for clear flow did not content them. Suppose the influence of sediment laden on the velocities fluctuations can be neglected when the sediment concentration is not very high, then it is known from Eq. (11) and Eq. (16) the amplidutes of pressure fluctuations for sediment laden flow increase with the increment of sediment concentrations. This analytical results can be considered as the theoretical discription of the variations in intensity coefficient of pressure fluctuations for sediment laden flow, β', obtained by experiments in this paper.

5 CONCLUSIONS

The following main results can be provided:

1. The probability density function of amplitudes of pressure fluctuations can be approximately described by a normal distribution function. The intensity coefficient of pressure fluctuations β' increases with the increment of sediment concentration, s, but the intensity coefficient of pressure fluctuations, taking into account the influence of specific gravity of sediment laden flow, β, probably is a constant, being unvarying with the sediment concentration.

2. The pressure fluctuations of sediment laden flow belongs to the fluctuation realm of low frequncy. The predominant frequency obviously turns to the lower frequency with the increment of sediment concentration.

3. The time lag τ_0 expressing the average size of large eddies motivating the pressure fluctuations gradually increases with the increment of sediment concentration. This tendency is in agreement with the variations of amplitudes and predominant frequencies against the sediment concentrations.

4. The basic equations of pressure fluctuations for high sediment laden flow and for low sediment laden flow have been deduced in this paper. These basic equations can guide the investigators analysis and study of pressure fluctuations of sediment laden flow as well as check obtained exprimental results in the future.

Nevertheless, when the sediment concentrations go beyond the experimental rang applied in this paper or reach very hight, since the materials avariable are not yet sufficient to proof the vilidity of above applied supposition and maintioned variations in intensity coefficient, β', an investigation inclyding more experiments with wider range of sediment concentrations and more complete theoretical analysis remian to be carried on.

References

Cheng Jie-ren 1992. Turbulent Transportation of High Sediment Laden Flow. PHD Dissertation of Hohai University.

Wang De-Chang et. al. 1992 Experimental Study of Pressure Fluctuations of Sediment Laden Flow. Report of Hydraulic Research Institute of YRC.

Wang Mu-Lan 1989. Development of Study of Pressure Fluctuations in Fluid. "Discharge Project and High Speed Flow".

Author index